Neotropical Owls

Paula L. Enríquez
Editor

Neotropical Owls

Diversity and Conservation

Springer

Editor
Paula L. Enríquez
El Colegio de la Frontera Sur
San Cristóbal de Las Casas, Chiapas, Mexico

Translation from the Spanish language edition: Los Búhos Neotropicales: Diversidad y Conservación by Paula L. Enríquez, © ECOSUR 2015. All rights reserved

ISBN 978-3-319-57107-2 ISBN 978-3-319-57108-9 (eBook)
DOI 10.1007/978-3-319-57108-9

Library of Congress Control Number: 2017943672

Printed on acid-free paper

This Springer imprint is published by Springer Nature
The registered company is Springer International Publishing AG
The registered company address is: Gewerbestrasse 11, 6330 Cham, Switzerland

To my parents
To Dr. Leon H. Kelso and Dr. Joe T. Marshall Jr.,
pioneers in the study of the Neotropical owls

Chapter Reviewers

José Luis Alcántara Carvajal
Ganadería-Área de Fauna Silvestre, Colegio de Postgraduados, Texcoco, Estado de México, México

Alberto Anzures Dadda†
Departamento Conservación de la Biodiversidad, El Colegio de la Frontera Sur, San Cristóbal de Las Casas, Chiapas, México

María del Coro Arizmendi Arriaga
UBIPRO-FES Iztacala, Universidad Nacional Autónoma de México, Tlalnepantla, Estado de México, México

Richard Cannings
Naramata, BC, Canada

Gilberto Chávez León
Instituto Nacional de Investigaciones Forestales, Agrícolas y Pecuarias, Uruapan, Michoacán, México

Richard J. Clark
York College of Pennsylvania, York, PA, USA

Patricia Escalante Pliego
Departamento de Zoología Instituto de Biología, Universidad Nacional Autónoma de México, Coyoacán, México

Inigo Fajardo
Umbrete, Sevilla, Spain

Claus König
Ludwigsburg, Germany

Huw Lloyd
Department of Environmental and Geographical Sciences, Manchester Metropolitan University, Manchester, UK

Bruce Marcot
USDA Forest Service. Pacific Northwest Research Station, Portland, OR, USA

Miguel Ángel Martínez Morales[1]
Departamento Conservación de la Biodiversidad, El Colegio de la Frontera Sur, Barrio María Auxiliadora, San Cristóbal de Las Casas, Chiapas, México

Alejandro Meléndez Herrada
Laboratorio de Ecología y Conservación de Fauna Silvestre, Departamento El Hombre y su Ambiente, Universidad Autónoma Metropolitana Unidad Xochimilco, Coyoacán, México

Borja Milá
Museo Nacional de Ciencias Naturales Consejo Superior de Investigaciones Científicas (CSIC), Madrid, Spain

Patricia Ramírez Bastida
Facultad de Estudios Superiores Iztacala (FES-Iztacala), Universidad Nacional Autónoma de México, Tlalnepantla, Estado de México, México

José Luis Rangel-Salazar[1]
Departamento Conservación de la Biodiversidad, El Colegio de la Frontera Sur, San Cristóbal de Las Casas, Chiapas, México

Jaime R. Rau Acuña
Universidad de los Lagos-Campus Osorno, Osorno, Chile

Luis Fabio Silveira
Departamento de Zoología, Universidad de Sao Paulo, São Paulo, SP, Brazil

Jean-Marc Thiollay
CERSP, CRBPO Museum Nacional d'Historie Naturelle, Paris, France

Fernando Urbina Torres
Laboratorio de Ornitología, Centro de Investigaciones Biológicas, Universidad Autónoma del Estado de Morelos, Cuernavaca, Morelos, México

José Raúl Vázquez Pérez
Departamento Conservación de la Biodiversidad, El Colegio de la Frontera Sur, San Cristóbal de Las Casas, Chiapas, México

James W. Wiley
Patuxent Wildlife Research Center, U. S. Fish and Wildlife Service, Laurel, MD, USA

Inigo Zuberogoitia
Departamento de Medio Natural de Estudios Medioambientales, ICARUS, S. L. Leioa, Bizkaia, Spain

[1] Reviewed more than one chapter

Preface

Many years ago, I had the pleasure of working with the National Wildlife Federation on the design and distribution of a bilingual poster about birds of prey. The poster was to dispel some of the common myths about raptors. For distribution throughout the Americas, most of the myths were familiar; eagles do not kill lambs (especially those in the tropics), vultures do not spread disease, and all hawks do not eat chickens (in fact many feed on snakes). But when it came to the owls, not being familiar with any myths, I had to do some research, and I was astounded with the results.

According to the Kwakiutl people in the Canadian Northwest, if you hear an owl call your name, your death is imminent. In the American Southwest in the culture of the Uto-Aztecan tribe, the Hopi, taboos about owls are associated with sorcery and other evils. Further south, the Aztecs and Mayan, along with other natives of Mesoamerica, considered the owl a symbol of death and destruction. In fact, the Aztec god of death, Mictlantecuhtli, was often depicted with owls. There is an old saying in Mexico that is still in use: *Cuando el tecolote canta, el indio muere* ("When the little owl sings, the Indian dies"). The Popol Vuh, a Mayan religious text, describes owls as messengers of Xibalba (the Mayan "place of fright"). In modern day some people believe owls will tear out your eyes or heart. The mother of a friend living in Merida, Yucatan, Mexico, simply thought owls were bad luck. So it would appear that there is no shortage of myths about owls. Most people fear anything that moves about in the night. Modern-day reports of nesting territorial owls attacking people likely just reinforce the myths. Perhaps such attacks in earlier times resulted in the creation of the various myths in the first place. As with most myths, they reflect a lack of accurate knowledge about the animal. Not only do most laypeople lack even the basic knowledge about owls, but many species have also been little studied by scientist.

There is certainly no shortage of books on owls. A few of the most recent include two titled *Owls of the World* (Burton, John A. 1992. Eurobook Ltd; Third Edition 208 pp. and Claus König, Friedhelm Weick, and Jan-Hendrik Becking. 2009 Second Edition, Yale University Press 528 pp.). Heimo Mikkola's 2012 book titled *Owls of the World: A Photographic Guide* published by Firefly Books is likely the best book available for identification using photographs. For a more general overview of all the owls, one should consult Josep Del Hoyo, Andrew Elliot, and Jordi Sargatal's *Handbook of the Birds of the World Vol. 5: Barn Owls to Hummingbirds*. Until now no singular source existed for detailed scientifically accurate information on the owls throughout Mexico and Central and South America. The scientific literature contains some information as does a recently published book by the Peregrine Fund (David F. Whitacre and J. Peter Jenny 2012. *Neotropical Birds of Prey: Biology and Ecology of a Forest Raptor Community*. Cornell University Press, Ithaca, NY. pp. 428). While the vast majority of the Peregrine Fund's book is on diurnal birds of prey, several owls are included in the treatment. Considering our lack of knowledge about this interesting family of nocturnal birds, *Neotropical Owls: Their Diversity and Conservation* will certainly fill in some of the major gaps.

Books of this nature are only as valuable as the knowledge of their contributors. Clearly not simply another book written by a single author who reviewed the literature, *Neotropical Owls: Their Diversity and Conservation*, contains contributions from the region's authorities on owls. Rather than presented in a species format, the book contains country profiles, which include tables, lists, and even some species accounts of the owls that have been recorded within the various countries. Available in two versions, its [1]downloadable bilingual (Spanish-English) version allows individuals conversant in either language to extract important information contained within the accounts. If the current data contained within the English hardcopy book was not enough to justify adding it to your ornithological libraries, the artwork certainly should provide such justification. Throughout the book, illustrations by renowned artists Lynn Delvin and Rina Pellizzari Raddatz are liberally distributed. A graduate of the arts program at Western Michigan University, Mr. Delvin developed an interest in owls at an early age. His work is included in the *Michigan Breeding Birds Atlas* as well as in various magazines and can be viewed at numerous private shows in Southern Michigan. Born in Santiago de Chile, Ms. Pellizzari studied arts at the University of Chile and design at the Metropolitan Technological University of Santiago. Some of her projects include scientific illustration and editorial design for scientific publications and scientific outreach, in Mexico and also in Central America and Chile.

Without a doubt, *Neotropical Owls Diversity and Conservation* will make a lasting contribution to our knowledge of this little known group. Kudos to Paula Enríquez and her staff for producing such a stellar product which will likely become a classic reference on these nocturnal birds in the Neotropics.

Director, Center for the Study of Tropical Birds, Inc. Jack Clinton Eitniear
San Antonio, TX, USA

[1] http://bibliotecasibe.ecosur.mx/sibe/book/000012610

Acknowledgments

This book is the culmination of many years of effort by numerous individuals. Its realization would not be possible without the enthusiasm of numerous authors and colleagues that collaborated on various chapters. To all of them I appreciate their patience and support in this long process. But also, to everyone for their interest in this project resulting in a significant increase in the knowledge of this fascinating group, especially given that they are little studied in Neotropical areas.

I would like to thank to all the reviewers; their comments and suggestions increased the quality of the manuscripts. I especially would like to thank Jack C. Eitniear for writing the preface and all his help in different phases of this project. To Jose Raúl Vázquez Pérez for his help in the editorial process. To Brock Huffman for his unconditional help in this project. To Mateo Ruiz Taylor for his comments and revisions to all the English abstracts. To Lynn Delvin and Rina Pellizzari Raddatz for their exceptional illustrations.

Thanks to all the photographers who allowed us to use their images. To Departamento de Difusión de El Colegio de la Frontera Sur, especially to Laura López Argoytia for her interest in this project. To Springer Nature, but especially Dr. Joao Pildervasser, associate editor, for believing in this book, Raagai Priya Chandra Sekaran and Sujitha Logaganesan for their assistance. To all my family, especially to my daughters Natalia and Paula Rangel, for always being present and their love. Finally, thanks to everyone involved in making this dream possible.

Contents

Chapter 1
Introduction: The Birds in the Neotropical Region

José Luis Rangel-Salazar and Paula L. Enríquez

Abstract The class Aves is very diverse being distributed in almost all environments. This is particularly true in the Neotropical region as it is characterized by wide extensions of humid tropical forests, high climatic heterogeneity, and an important geological history. This history has produced a high diversity of habitats resulting in great avian diversity. Approximately 3500–4037 species of birds have been reported in this region, of which 80 are owls. Nocturnal raptors have evolved to adapt and survive displaying nocturnal activity and mainly inhabiting forested environments. These adaptions have resulted in the evolution of unique morphological characteristics (e.g., forward-directed eyes and asymmetric ears). However, despite this diversity, little biological information has been published on these birds. More research is needed on the ecological requirements of this group allowing us to better establish strategies to conserve them.

Keywords Bird conservation • Nocturnal raptors • Biological diversity • Evolutionary characteristics

J.L. Rangel-Salazar (✉) • P.L. Enríquez
Departamento Conservación de la Biodiversidad, El Colegio de la Frontera Sur, San Cristóbal de Las Casas, P.O. Box 63, San Cristóbal de Las Casas 29290 Chiapas, Mexico
e-mail: jlrangel@ecosur.mx

P.L. Enríquez (ed.), *Neotropical Owls*, DOI 10.1007/978-3-319-57108-9_1

Striped Owl (*Asio clamator*)

Birds have fascinated mankind since both have coexisted on the planet. There are at least three simple reasons: birds are conspicuous and diverse, they strike both visual and auditory senses, and they are present in virtually all environments. Birds are a representation and indicators of the variety of life on Earth. The fascination that humans have had with birds during the last two centuries has enabled and encouraged the definition and classification of most species in the world. This clarity in identifying bird species has allowed us to describe and evaluate patterns of variation in the number and types of birds at different spatial and temporal scales. For instance, it has allowed us to understand the variation in species composition between major biogeographic regions or, at a finer spatial scale, between ecosystems that make up a landscape or different landscapes that make up a region. We partially have managed to understand variation over time, through the history of the Earth and through annual cycles such as during periods of reproduction and migration (Attenborough 1998; but see Claramunt and Cracraft 2015).

Another fascinating aspect of birds is that they have enabled us to understand how they are affected by contemporary ecological factors such as variation in the distribu-

Fig. 1.1 Map of the Neotropical region

tion of habitat, resources, and climate and the migration patterns of bird species across large biogeographic regions (Newton 2003). But it is just as important to know and understand which human activities are resulting in direct and indirect changes in the distribution and abundance of species. This is especially true because, while few species of birds are favored by human activities, most of them are negatively affected.

The Neotropical biogeographic region, or the Neotropics, is defined as the tropical region of the American continent, ranging from 20° latitude north to near 57° south latitude and from 112° to 35° longitude west, an area representing approximately 18.2 million km^2 (Newton 2003, Fig. 1.1). This region includes the tropical parts of Mexico, Central America, and the Caribbean and South America down to Tierra del Fuego. The Neotropics is characterized as having high biological diversity often including lineages of species that are shared with other continents; this is because at some point in the past, all continents were joined. The region has mainly tropical forests but also includes temperate forests, mountain environments (e.g., paramo, Yungas), and even deserts and plains. These ecosystems contain, and maintain, the greatest diversity of the planet's terrestrial vertebrates (Jenkins et al. 2013). This high biological diversity or biodiversity represents a variation of the different life forms, from genes to individuals, species, populations, communities, and ecosystems. It also represents the variation in the mechanisms and dimensions that limit them, such as spatial biological units and weather. However, biodiversity is not only what it is now but what it was and what it will be over time. The concept of biodiversity still appears to be controversial (e.g., Martínez-Meyer et al. 2014), and it has even been suggested that biological lineages are biodiversity (e.g., Navarro-Sigüenza et al. 2014), even though biological lineages are units and are therefore far from being biodiversity. Perhaps this is because our understanding of the patterns of biodiversity has been a challenge since the eighteenth and nineteenth centuries. A variety of mechanisms and factors have been proposed to explain the variation (Ricklefs 2004) that has generated several convergent and divergent points of view.

Several authors have recognized birds as one of the indicators of biodiversity (e.g., del Hoyo et al. 1999; Newton 2003). In the Neotropical region, these authors have registered around 3500 bird species (Stotz et al. 1996). Yet the diversity of birds varies within the region, where the greatest diversity is around the equator in countries like Colombia and Peru, with 1879 and 1800 species, respectively (Schulenberg et al. 2010; McMullan et al. 2014).Understanding this bird diversity and distribution has been a difficult task (Newton 2003; Ribas et al. 2012), involving history and geography. This wealth of bird species varies both latitudinally and longitudinally in the region, as well as temporally due to latitudinal and altitudinal annual migratory movements. Approximately 340 species of birds belonging to 38 families travel from their Neotropical environments to breed in temperate environments of North America. Apparently, climate change promoted the displacement of environmental blocks that were followed by organisms, often migratory species, then forced to move by the increase in breeding density in tropical habitats. The hypothesis of reproductive density dependence (Newton 1998) appears to be complementary to the hypothesis of reproductive background (Greenberg and Marra2005). This hypothesis is often considered an explanation for the migration of birds that breed in temperate environments and winter in tropical environs.

Owls belong to the order Strigiformes and are a good example of the process of diversification of bird species. Currently, there are a total of 250 species of owls worldwide (*secundum* Konig et al. 2008) or more depending the authority followed (see Chap. 2). They are distributed in virtually all environments except Antarctica and some oceanic islands. Owls can be found in a variety of ecosystems, and although many species are arboreal, some are also fossorial. Owls are mainly nocturnal, though a few species are diurnal, and their size varies greatly, from the smallest species measuring 14 cm or less (elf owl, *Micrathene whitneyi*) to the largest species measuring 80 cm (eagle owl, *Bubo bubo*). Despite being widely distributed throughout the world, most owls live in tropical areas with approximately 35% of the total species inhabiting the Neotropical region. There are three genera that are well represented in this region (i.e., *Megascops*, *Glaucidium*, and *Strix*) and six are endemic (i.e. *Megascops*, *Gymnoglaux*, *Lophostrix*, *Pseudoscops*, *Pulsatrix*, and *Xenoglaux*).

The fossils found indicate that owls date back to the Cenozoic Eocene, 65 million years ago. At that time, there was possibly a greater diversification of birds, as most modern orders were present during that period (Welty 1982). During this time, the Earth underwent important changes in both climate, with a long-term cooling, and geology, where continents moved to their current positions and formed mountain chains. There were many small mammals during the first half of the Tertiary period which possibly contributed to the diversification and adaptive radiation of owls (Burton 1973).

Fossils that have been found exhibit characteristics of the families Tytonidae and Strigidae being both from extinct species and recent extant genera. A characteristic of these fossils is that they do not represent true owls or are similar enough to identify a common ancestor. But it is recognized that owls were diverse and have their own evolutionary characteristics. For example, they are the only avian species with eyes directed forward; the vision is stereoscopic and is highly developed. Despite limited eye mobility, a wide field of vision is achieved by moving the head up to 270°, and the eye structure allows sight at low light intensities. The tarsi are *semi-zygodactylous*, with the outer toe having mobility forward or backward, and the plumage is extremely soft and very dense. Another important feature is that some species evolved asymmetrical ears, possibly to improve their strategies to locate prey at night and in dense vegetation (Norberg 1987). There are six genera that have this characteristic [i.e., *Tyto*, *Phodilus*, *Strix*, *Asio* (*Rhinoptyx*), *Pseudoscops*, and *Aegolius*]. However, the degree of asymmetry varies among species of *Strix* and the structures involved. Although in the other species, the degree of asymmetry does not vary among species; it surprisingly varies in structure and form. Thus, it is considered that this feature developed independently in each group or genus (Norberg1977, 2002).

There is still much to learn about this diverse group of birds in the Neotropical region. Clearly studying them is a challenge because they are uncommon, have nocturnal habits, usually go unnoticed, and live in places difficult to access. Like other groups of birds with nocturnal habits (e.g., Caprimulgiformes), owls should be included in future ecological and evolutionary studies. Only then will we improve our understanding about these birds and to be able to better conserve them.

References

Attenborough D (1998) The life of birds. Dominio Books Ltd, London

Burton JA (1973) Owls of the world: their evolution, structure and ecology. P. Lowe, Holland

Claramunt S, Cracraft J (2015) A new time tree reveals Earth historys imprint on the evolution of modern birds. Sci Adv 1(11). doi:10.1126/sciadv.1501005

Greenberg R, Marra PP (eds) (2005) Birds of two worlds: the ecology and evolution of migration. The Johns Hopkins University Press, Baltimore

del Hoyo J, Elliott A, Sargatal J et al (eds) (1999) Handbook of the birds of the world. Lynx Editions, Barcelona. 1992–2013
Jenkins CN, Pimm SL, Joppa LN (2013) Global patterns of terrestrial vertebrate diversity and conservation. Proc Natl Acad Sci U. S. A 110:E2602–E261O
König C, Weick F, Becking J-H (2008) Owls of the world, 2nd edn. Christopher Helm Publisher, London
Martínez-Meyer E, Sosa-Escalante E, Álvarez F (2014) El estudio de la biodiversidad en México: ¿Una ruta con dirección? Rev Mex Biodivers, Supl 85:S1–S9
McMullan M, Donegan TM, Quevedo A (2014) Guía de las aves de Colombia. Fundación Proaves, Bogotá
Navarro-Sigüenza AG, Rebón-Gallardo MF, Gordillo-Martínez A et al (2014) Biodiversidad de aves en México. Rev Mex Biodivers, Supl 8S:S476–8495
Newton I (1998) Population limitation in birds. Academic Press Limited, London
Newton I (2003) The speciation and biogeography of birds. Academic Press Limited, London
Norberg RA (1977) Occurrence and independent evolution of bilateral ear asymmetry in owls and implications on owl taxonomy. Philos Trans R Soc Lond B 280:375–408
Norberg RA (1987). Evolution, structure, and ecology of northern forest owls. In:Biology and Conservation of Northern Forest Owls. USDA, Fort Collins.
Norberg RA (2002) Independent evolution of outer ear asymmetry among five owl lineages; morphology, function and selection. In: Newton I, Kavanagh R, Olsen J et al (eds) Ecology and conservation of owls. CSIRO Publishing, Collingwood, pp 329–342
Ribas CC, Maldonado-Coelho M, Smith BT et al (2012) Towards an integrated historical biogeography of the Neotropical lowland avifauna: combining diversification analysis and landscape evolution. Ornitol Neotrop 23:189–208
Ricklefs RE (2004) Community diversity: relative roles of local and regional processes. Science 24:167–171
Schulenberg TS, Stotz DF, Lane DF et al (2010) Aves de Perú. Series Biodiversidad Corbidí 01. Perú, Lima
Stotz DF, Fitzpatrick JW, Parker TA III et al (1996) Neotropical birds: ecology and conservation. Conservation international and the field Museum of natural history. The University of Chicago Press, Chicago
Welty J (1982) The life birds. Sounders College Publishing, Philadelphia

Chapter 2
A Review of the Systematics of Neotropical Owls (Strigiformes)

Paula L. Enríquez, Knut Eisermann, Heimo Mikkola, and José Carlos Motta-Junior

Abstract Although birds belong to the best known animal groups, their systematics has not been fully resolved yet. Among the approximately 80 Neotropical owl species, there are monotypic genera such as *Lophostrix*, complex and diverse genera such as *Megascops* or *Glaucidium*, as well as widespread and variable taxa such as *Bubo virginianus* and *Tyto furcata*. Based on a literature review, we provide here an overview of the current taxonomy and nomenclature of Neotropical owls, and indicate knowledge gaps as focus points for future research.

Keywords Owl evolution • Strigiformes classification • Molecular studies • Phylogeny

P.L. Enríquez (✉)
Departamento Conservación de la Biodiversidad, El Colegio de la Frontera Sur, P. O. Box 63, San Cristóbal de Las Casas 29290, Chiapas, México
e-mail: penrique@ecosur.mx

K. Eisermann
PROEVAL RAXMU Bird Monitoring Program, Cobán, Alta Verapaz, Guatemala

H. Mikkola
Department of Biology, University of Eastern Finland, FIN -70211, Kuopio, Finland

J.C. Motta-Junior
Laboratorio de Ecología de Aves – LABECOAVES Departamento de Ecología, Instituto de Biociências da Universidad de São Paulo, 05508-090 São Paulo, SP, Brazil

P.L. Enríquez (ed.), *Neotropical Owls*, DOI 10.1007/978-3-319-57108-9_2

Birds (Aves) belong to the best studied animal groups, but much remains to be learned about the number of species, taxonomy and systematic position (Barrowclough et al. 2016). Systematics should reflect evolutionary history, but the use of different taxonomic methods has resulted in contradictory phylogenetic positioning (Navarro 1988). Birds in the order Strigiformes (owls) present specific characteristics (i.e. soft plumage, predatory adaptations, binocular vision and nocturnal habits) defining them as a particular group. The phylogenetic position of owls in relation to other birds is controversial (Cracraft 1981; Sibley and Ahlquist 1990; Ericson et al. 2006; Livezey and Zusi 2007; Prum et al. 2015). This chapter provides an overview of the current knowledge on the taxonomy and systematics of Neotropical owls.

The oldest known bird fossil dates back some 225 million years (Pacheco et al. 2011), but the oldest owl-like fossils are from the Paleocene (57–65 million years ago), and more fossil records are known from the Eocene, 34–57 million years ago

(Brodkorb 1971; Mayr 2010). Eocene was obviously the time of the appearance and diversification of many modern groups of organisms including mammals and birds, although the real strigid owls were recorded only 23–24 million years ago (Mikkola 2014). In this time, 80% of the modern bird orders developed and diversified (Welty 1982). None of the real owl fossils recorded so far has provided details on which other bird groups could share the common ancestor with the owls (Grossman and Hamlet 1988). Most of the owl fossils have been found in North America and Europe, and very few in the Neotropical region (Mayr 2010). There are fossil records of pygmy owls in the Americas from Pleistocene deposits (between 2 million and 13,000 years ago) in Mexico and Brazil (Mikkola 2014).

The order Strigiformes (owls) is subdivided into two families: barn owls (Tytonidae) and typical owls (Strigidae). There are several morphological differences separating these families, including the structure of sternum and shape of ears, relative length of toes and serrated or smooth cutting edges to the claw of the middle toe (Ridgway 1914; Sibley and Ahlquist 1972; König et al. 2008). Both families are traditionally divided into two subfamilies (Peters 1940): the Tytonidae is subdivided into the Tytoninae and the Pholinae which together have approximately 20 species (Table 2.1). The genera *Tyto* and *Pholidus* are supported by modern molecular sequence data, and they have diverged from a common ancestor more than 10 million years ago (Wink et al. 2008). The family Strigidae has been divided into the subfamilies Buboninae and Striginae with together more than 200 species (Table 2.1). Some northern species of the latter subfamily have developed bilateral asymmetry of the external ears which help these owls to catch prey hiding underneath of the snow (Norberg 1987). Norberg (1978) indicated that the outer ears are symmetric in the majority of Strigidae genera and that an asymmetrical arrangement is known to involve parts of the skull in four species: Ural owl, *Strix uralensis*; great grey owl, *Strix nebulosa*; Boreal owl, *Aegolius funereus*; and northern saw-whet owl, *Aegolius acadicus*. Research on the skulls of all owl species could prove asymmetry in other species as well, although the majority of owls may have no asymmetry in the hard parts of the head. Barn owls (Tytonidae) of the genera *Tyto* and *Pholidus* show a bilateral asymmetry of the external ears, thus making the skull parts unreliable factors in separating the Strigidae and the Tytonidae (Mikkola 1983). Based on other osteological characters, it has been proposed to divide *Strigidae* family into three subfamilies: Surniinae, Striginae and Asioninae (Ford 1967; Marks et al. 1999). Based on molecular analyses, Wink et al. (2008) recommended a new classification of subfamilies: Surniinae, Striginae and Ninoxinae, in which the subfamily Asioninae should be seen as part of the Striginae to avoid a paraphyletic assemblage.

Sibley and Ahlquist (1972) established a historic revision of owl classification marking the most important similarities and differences between the two families. They also mentioned family similarities between Strigiformes and Caprimulgiformes (nightjars and relatives, according to Fürbringer 1888, Shufeldt 1904), Falconiformes (hawks and eagles, according to Seebohm 1890, Cracraft 1981) and also Psittaciformes (parrots and parakeets, according to Gadow 1892). Ericson et al. (2006) used a large dataset of five nuclear genes showing that owls are members of

Table 2.1 Number of species of barn owls (Tytonidae) and typical owls (Strigidae) in the world according to different authors

Author(s) Family	Peters (1940)	Burton (1973)	Sibley and Monroe (1990)	Marks et al. (1999)	König et al. (1999)	König et al. (2008)	Dickinson and Remsen (2013)	Mikkola (2014)	Clements et al. (2016)	Gill and Donsker (2016)
Tytonidae	11	10	17	16	11	26	19	27	18	19
Strigidae	133	120	159	189	201	224	191	241	207	223
Total	144	130	176	205	212	250	210	268	225	242

the Coronaves in which owls are in a same clade with diurnal raptors (Accipitridae), vultures (Cathartidae), trogons (Trogonidae) and others, but excluding falcons (Falconidae), which cluster as a sister group to parrots and songbirds. Although there are noticeable similarities between owls and nightjars, and morphological and anatomical similarities between owls and hawks (Livezey and Zusi 2007), other authors have stated that these close family relationships are rather based on convergence since they are not supported by sequence data (Gibb et al. 2007; Wink et al. 2008; Pratt et al. 2009). More recently, Pacheco et al. (2011) have found closer relationship between owls and Psittaciformes, but Prum et al. (2015) established Strigiformes as sister group to the coraciimorph clade including Coliiformes, Leptosomiformes, Trogoniformes, Bucerotiformes, Coraciiformes and Piciformes. Our understanding of the phylogeny of birds keeps developing with the improvement of methodologies of molecular analyses.

The classification of Strigiformes has changed considerably in the last decades, especially in complex groups with highly variable plumages and vocalizations, such as the *Otus* complex (Marshall 1967), including the American species recently separated as genus *Megascops* (Banks et al. 2003). Similar changes have occurred in the genus *Glaucidium* (Howell and Robbins 1995; Robbins and Stiles 1999). Our current knowledge of both genera is based on taxonomic revisions in the last 70 years (Moore and Peters 1939; Buchanan 1964; Howell and Robbins 1995; Wink et al. 2008; Eisermann and Howell 2011). Molecular studies have increased the number of owl species, but the number of accepted species differs considerably between authors (Table 2.1). Gill and Donsker (2016) listed 53 subspecies in the Tytonidae and 432 subspecies in the Strigidae.

In a recent compilation, Mikkola (2014) listed 268 owl species including newly proposed species. Since that time, some new owl species have been proposed (e.g. Kirwan et al. 2015), so the world total could now be over 270 different living owl species.

The scientific names (biological nomenclature) of Neotropical Strigiformes have not been standardized. Remsen et al. (2016) listed 44 owl species in South America, and American Ornithologists' Union (AOU 1998 and supplements) listed 44 owl species for Mexico and Central America. In this chapter, we use the nomenclature according to König et al. (2008) but also include in the discussion the American Ornithologists' Union (AOU 1998) and supplements (most recent supplement Chesser et al. 2016). It follows an account of taxonomic changes proposed for the Neotropics in the recent literature:

American Barn Owl (*Tyto furcata*): Previously this owl species was considered a subspecies of Common Barn Owl (*Tyto alba*) of the Old World. Recent molecular studies support the separation of the populations in the Americas, for a high degree of genetic variation which Wink et al. (2008) recognize four subspecies and Mikkola (2014) six. This large number of subspecies is an indication that several of them could be distinct species. AOU (1998 and supplements) and Remsen et al. (2016) have not accepted the separation of *Tyto furcata* from *Tyto alba* and do not recognize Curaçao barn owl (*Tyto bargei*), Lesser Antilles barn owl (*Tyto insularis*) and Galápagos barn owl (*Tyto punctatissima*). Wink et al. (2008) anticipated also the split of *Tyto tuidara* and *Tyto pratincola* from *Tyto furcata*, and Mikkola (2014) mentioned that *Tyto contempta* could also be a new species. However, recent DNA molecular studies suggest three main clades from *Tyto alba* complex: *Tyto alba* (Africa, Europe), *Tyto furcata* (New World; including *bargei*) and *Tyto javanica* (Australasia; including *delicatula* and *stertens*) (Aliabadian et al. 2016).

Flammulated Owl (*Psiloscops flammeolus*): This species was formerly classified as *Otus flammeolus* but differs according to Wink and Heidrich (1999), Penhallurick (2002), König et al. (2008) and Wink et al. (2008) in vocalization and genetics so much that it is now separated into a monotypic genus *Psiloscops* (as originally classified by Coues 1899). Nucleotide sequences of the mitochondrial cytochrome b gene study showed that flammulated owl is more directly related to *Megascops* species in the New World rather than being a sister group to *Otus* owls in the Old World (Proudfoot et al. 2007). The closest living relative to the flammulated owl among the Neotropical screech owls appears to be the Puerto Rican endemic *M. nudipes* (Dantas et al. 2016).

The New World screech owl genus *Megascops*, recently split from *Otus* based on vocal and molecular evidence, currently includes 22 species (Dantas et al. 2016) divided into some 63 taxa according to Marks et al. (1999).

Complex of Pacific Screech Owl (*Megascops cooperi*): König et al. (2008) recognize two species: Oaxaca screech owl (*Megascops lambi*) endemic to Oaxaca, Mexico and Pacific screech owl (*M. cooperi*) on the Pacific slope from southern Mexico to Costa Rica. It is not known if this species hybridizes with Oaxaca screech owl in areas where the two overlap (Mikkola 2014). Unfortunately, a genetic study only included samples of the Pacific screech owl (Dantas et al. 2016). AOU (1998 and supplements) does not include *M. lambi*, and Dickinson and Remsen (2013) consider it as subspecies of *M. cooperi*.

Complex of Northern *Megascops watsonii* and Southern Tawny-Bellied Screech Owl (*M. w. usta*): König et al. (2008) separated two species: northern Tawny-bellied screech owl (*Megascops watsonii*) in northern parts of South America as well as in northern Amazonian part of Brazil and southern Tawny-bellied screech owl (*M. usta*) in Amazonian Colombia, Ecuador, Peru and Brazil south to lowland forests in northern Bolivia and Brazilian Mato Grosso. The Brazilian Committee for the Ornithological Records (Piacentini et al. 2015) accepted the status of these two species, but Remsen et al. (2016) recognize only one species as did a recent molecular study, probably due to the lack of broader geographic and population level sampling (Dantas et al. 2016). These last authors considered both *M. watsonii* and *M. usta* as paraphyletic, highlighting the urgent need to a taxonomic review of the *M. watsonii-usta* complex.

Complex of Guatemalan Screech Owl (*Megascops guatemalae*) and Vermiculated Screech Owl (*M. vermiculatus*): König et al. (2008) recognized five species based on morphological and vocal differences:

- Guatemalan screech owl (*Megascops guatemalae*) from Mexico to northern Costa Rica
- Vermiculated screech owl (*M. vermiculatus*) from Costa Rica to north-western Colombia and northern Venezuela
- Roraima screech owl (*M. roraimae*) from northern Colombia and Venezuela to Roraima and Duida mountains (northern Brazil and Guyana)

- Rio Napo screech owl (*M. napensis*) from the eastern slopes of the Andes in eastern Colombia to northern Bolivia
- Tumbes screech owl (*M. pacificus*) from the lowland areas in south-western Ecuador and north-western Peru.

Marks et al. (1999), but not Banks et al. (2003), accepted the split between *M. guatemalae* and *M. vermiculatus*, and Hilty (2003) mentioned *M. roraimae* as a separated species. Piacentini et al. (2015) listed only *M. guatemalae* until more complete analysis are provided. Only one species, *M. guatemalae*, was recognized for South America by Remsen et al. (2012) and for Mexico and Central America by AOU (1998 and supplements). The recent molecular data support not only the split between *M. guatemalae* and *M. vermiculatus* but also indicate that three other splits (*M. roraimae*, *napensis* and *pacificus*) are probably best treated as distinct species (Dantas et al. 2016). The southern edge of distribution of *M. guatemalae* is controversial. König et al. (2008) and Mikkola (2014) indicate that this owl ranges into northern Costa Rica, which may be erroneous. According to Marks et al. (1999), *M. guatemalae* and *M. vermiculatus* are allopatric, and the limit between the ranges of both species is located in Nicaragua.

Complex of Great Horned Owl (*Bubo virginianus*): König et al. (2008) separated Magellanic horned owl (*Bubo magellanicus*) in the Andes of Peru, Bolivia, Chile and Argentina and the southern cone of South America, but Remsen et al. (2016) considered these populations part of *B. virginianus*. DNA sequence difference between *B. magellanicus* and *B. virginianus* is 1.6%, maybe justifying their separation as two distinct species, as they also differ clearly in size and colour (König et al. 1996).

Genus *Pulsatrix*: König et al. (2008) list four *Pulsatrix* species: Spectacled owl (*P. perspicillata*), Short-browed owl (*P. pulsatrix*), Band-bellied owl (*P. melanota*), and Tawny-browed owl (*P. koeniswaldiana*). Bencke (2001), Ramírez-Llorens and Bellocq (2007) and Remsen et al. (2016) consider *P. pulsatrix* in eastern Brazil from Bahia south to northeast Argentina too premature to separate as full species as it is rather seen only as the subspecies of the spectacled owl. A study on genetic and voice differences would be needed to confirm the species status of the short-browed owl.

Genus *Ciccaba*: Norberg (1977), Sibley and Ahlquist (1990), Sibley and Monroe (1990), Howell and Webb (1995), Wink and Heidrich (1999), Norberg (2002), Weick (2006), Wink et al. (2008) and König et al. (2008) have incorporated all ex-*Ciccaba* species into the genus *Strix* based on external ear asymmetry and on molecular analysis. Some authors, however, maintain the genus *Ciccaba* (Dickinson and Remsen 2013; Clements et al. 2016; Remsen et al. 2016), and the taxonomic committee of the American Ornithologists' Union has been considering the change but did not approve it (Banks et al. 2003).

Complex of Mottled Owl (*Strix virgata*): König et al. (2008) recognized two species, Mexican wood owl (*Strix squamulata*) from México, south to north-western Colombia and western Ecuador, and mottled owl (*S. virgata*) in most parts of northern and central South America east of the Andes. Only one species, *Ciccaba virgata*, has been accepted by Remsen et al. (2016) for South America and by AOU (1998 and supplements) for Mexico and Central America.

Fulvous Owl (*Strix fulvescens*): This owl is morphologically similar to the barred owl (*Strix varia*) of North America and Northern Mexico and was until recently recognized only as subspecies of barred owl (J.T. Marshall, pers.com.; Enríquez et al. 1993). Both species form possibly a superspecies (AOU 1998). The voice of fulvous owl resembles that of spotted owl (*Strix occidentalis*); it has been speculated that fulvous owl may form a superspecies together with spotted and barred owls which are known to hybridize in North America (Hamer et al. 1994). Recent records from Mexico confirm the distribution of fulvous owl west of the Isthmus of Tehuantepec in Oaxaca suggesting sympatric distribution with the barred owl in that area (Gómez de Silva 2010; Ramírez-Julián et al. 2011). Superspecies speculation requires comparative studies, molecular and biological (Mikkola 2014).

Complex of Mountain Pygmy Owl (*Glaucidium gnoma*): König et al. (2008) separated three different Neotropical species out of this complex, Baja pygmy owl (*Glaucidium hoskinsii*; in Baja California peninsula), Mountain pygmy owl (*G. gnoma*; in Mexico west of the Isthmus of Tehuantepec) and Guatemalan pygmy owl (*G. cobanense*; in México east of the Isthmus of Tehuantepec, Guatemala and Honduras). Other authors have classified all or some of the four taxa as subspecies of *G. gnoma* (Weick 2006; Dickinson and Remsen 2013; Clements et al. 2016). A recent comparative study on vocalizations of *G. cobanense* and *G. gnoma* (Eisermann and Howell 2011; Howell and Eisermann 2011) found differences supporting the species status of *G. cobanense*, as originally proposed (Sharpe 1875; Griscom 1931). Molecular studies are required in order to confirm the new taxonomic status of pygmy owls in northern Central America (Heidrich et al. 1995). AOU (1998 and supplements) recognizes only one species *Glaucidium gnoma*.

Complex of Least Pygmy Owl (*Glaucidium minutissimum*): König et al. (2008) recognized five species:

- Tamaulipas pygmy owl (*Glaucidium sanchezi*) in the mountains of north-eastern Mexico
- Colima pygmy owl (*G. palmarum*) along the Pacific coast of Mexico
- Central American pygmy owl (*G. griseiceps*) from south-eastern Mexico to northern and western Colombia and north-western Ecuador
- Sick's pygmy owl (*G. sicki*) in eastern Brazil south to eastern Paraguay and eastern Peru, possibly extending to north-eastern Argentina
- Pernambuco pygmy owl (*G. minutissimum*) from the state of Pernambuco in north-eastern Brazil

Earlier the least pygmy owl complex was considered to be polymorphic with eight different subspecies (Peters 1940). Buchanan (1964) classified five subspecies in Mexico based on morphological differences. Later Howell and Robbins (1995) proposed distribution limits of four species: *G. palmarum*, which included three subspecies *palmarum*, *oberholseri* and *griscomi*, occurring in western Mexico; *G. sanchezi*, with distribution in north-eastern Mexico; *G. griseiceps*, occurring from south-eastern Mexico through Central America to the Pacific coast of South America (including three subspecies: *griseiceps*, *rarum* and *occultum*); and *G. minutissimum*, with distribution in south-eastern Brazil and Paraguay. Piacentini et al. (2015)

and Remsen et al. (2016) consider *G. sicki* a synonym of *G. minutissimum*. Populations, which König et al. (2008) named as Pernambuco pygmy owl (*Glaucidium minutissimum*), were referred to as *G. mooreorum* by Remsen et al. (2016). Recently, Grantsau (2010) proposed *Glaucidium pumila* as a pygmy owl species living in south-eastern and south-western Brazil, but other authors have not recognized this species (Sigrist 2006; König et al. 2008; Mikkola 2014; Piacentini et al. 2015; Remsen et al. 2016).

Complex of Ferruginous Pygmy Owl (*Glaucidium brasilianum*): König et al. (2008) separated three species, Ridgway's pygmy owl (*Glaucidium ridgwayi*) from southern Arizona and Texas, USA, throughout Mexico and Central America and south- to north-western Colombia; ferruginous pygmy owl (*G. brasilianum*) in South America east of the Andes; and Chaco pygmy owl (*G. tucumanum*) from Bolivia, Paraguay and northern Argentina, possibly to south-western Brazil. Other authors accept also *G. ridgwayi* (Heidrich et al. 1995; Wink and Heidrich 1999; Weick 2006; Proudfoot et al. 2006; Wink et al. 2008). American Ornithologists' Union (AOU 1998 and supplements) and Remsen et al. (2012) recognize only *G. brasilianum*.

Burrowing Owl (*Athene cunicularia*): This owl was once separated from *Athene* into its own monotypic genus *Speotyto* (Clark 1997), but based on recent anatomic, molecular, behavioural, vocal and osteological data, it was reclassified back into the genus *Athene* (AOU 1998; König et al. 1999, 2008).

Unspotted Saw-whet Owl (*Aegolius ridgwayi*): Briggs (1954) and Mayr and Short (1970) consider this as a subspecies of northern saw-whet owl (*A. acadicus*). AOU (1998) mentioned that both taxa form a superspecies. Ecology and biology of unspotted saw-whet owl are little known explaining the lack of feather or blood samples, so the molecular status of *A. ridgwayi* remains unknown (Wink and Heidrich 1999; Wink et al. 2008) although a detailed study of its vocalizations is now available (Eisermann 2013).

Striped Owl (*Asio clamator*): This species has been placed previously in the genera *Rhinoptynx* and *Pseudoscops* (Olson 1995). It has been placed in the genus *Asio* together with long-eared owl (*Asio otus*), marsh owl (*A. capensis*) and short-eared owl (*A. flammeus*) based on molecular studies (Wink et al. 2008; König et al. 2008). AOU (1998 and supplements) kept this species in the genus *Pseudoscops*, and Remsen et al. (2016) listed it in the genus *Asio*.

Owls are much more difficult to find and study than many other birds, especially diurnal birds, explaining why they are relatively little known, with even new species still discovered. During the last 40 years, several new owl species have been found in the Neotropical region including a new genus *Xenoglaux* (O'Neill and Graves 1977). A list of these newly found or reclassified (based on molecular and/or vocal differences) Neotropical owl species in chronological order starting from 1977 follows:

- Long-whiskered owl (*Xenoglaux loweryi*) described by O'Neill and Graves (1977)
- Cloud-forest screech owl (*Megascops marshalli*) by Weske and Terborgh (1981)
- Tumbes screech owl (*Megascops pacificus*) and Koepcke's screech owl (*M. koepckeae*) by Hekstra (1982)

- Cinnamon screech owl (*Megascops petersoni*) by Fitzpatrick and O'Neill (1986)
- Montane forest screech owl (*Megascops hoyi*) by König and Stranek (1989)
- Amazonian pygmy owl (*Glaucidium hardyi*) by Vielliard (1989)
- Yungas pygmy owl (*Glaucidium bolivianum*), Peruvian pygmy owl (*G. peruanum*) and Chaco pygmy owl (*G. tucumanum*) by König (1991)
- Subtropical pygmy owl (*G.`parkeri*) by Robbins and Howell (1995)
- Cloud-forest pygmy owl (*G. nubicola*) by Robbins and Stiles (1999)
- Pernambuco pygmy owl (*Glaucidium mooreorum*) by Silva et al. (2002)
- Sick's pygmy owl (*Glaucidium sicki*) by König and Weick (2005)

New description of the Koepcke's screech owl (*Megascops koepckeae*) and its subspecies *M. k. hockingi* was made by Fjeldså et al. (2012). Recent molecular studies have now reconfirmed the independent species status of *M. koepckeae* (Dantas et al. 2016). In 2007, a new species of *Megascops* was located in Minca Village, Sierra Nevada of Santa Marta mountain range in northern Colombia, but this species is still waiting to be officially described (König et al. 2008; Chaparro-Herrera et al. 2015). *Megascops vermiculatus pallidus* from northern Venezuela and the Sierra Perija of northern Colombia may deserve species status based on distinct vocalizations (N. K. Krabbe, unpub. data, see Dantas et al. 2016).

Systematics and taxonomy of the Neotropical owl species are still developing. New samples of feathers, tissue and blood, voice recordings and new photographs are extending our knowledge and understanding on the evolution, taxonomy and molecular phylogeny of these birds. "The last word on owl taxonomy is yet to be spoken!" as so well concluded König et al. (2008).

References

Aliabadian M, Alaei-Kakhki N, Mirshamsi O et al (2016) Phylogeny, biogeography, and diversification of barn owls (Aves:Strigiformes). Biol J Linn Soc 119:904-918

AOU (American Ornithologists' Union) (1998) The American Ornithologists' Union check-list of North American birds, 7th edn. American Ornithologists' Union, Washington, DC

Banks RC, Cicero C, Dunn JL et al (2003) Forty-fourth supplement to the American Ornithologists' Union check-list of North American birds. Auk 120:923–931

Barrowclough GF, Cracraft J, Klicka J et al (2016) How many kinds of birds are there and why does it matter? PLoS One. doi:10.1371/journal.pone.0166307

Beneke GA (2001) Lista de referência das aves do Rio Grande do Sul. Fundação Zoobotânica do Rio Grande do Sul, Porto Alegre

Briggs MA (1954) Apparent neoteny in the saw-whet owls of Mexico and Central America. Proc Biol Soc Wash 67:179–182

Brodkorb P (1971) Catalogue of fossil birds: part 4 (Columbiformes through Piciformes). Bull Fla State Mus Biol Sci 15:163–266

Buchanan OM (1964) The Mexican races of the Least Pygmy Owl. Condor 66:103–112

Burton JA (ed) (1973) Owls of the world: their evolution, structure, and ecology. Peter Lowe/Eurobook Ltd, Wallingford

Chaparro-Herrera S, Córdoba-Córdoba S, López-Ordóñez JP et al (2015) Los búhos de Colombia. In: Enríquez PL (ed) Los búhos neotropicales: diversidad y conservación. ECOSUR, San Cristóbal de Las Casas, pp 277–329

Chesser RT, Burns KJ, Cicero C et al (2016) Fifty-seventh supplement to the American Ornithologists' Union check-list of North American birds. Auk 133:544–560
Clark RJ (1997) A review of the taxonomy and distribution of the burrowing owl (*Speotyto cunicularia*). J Raptor Res 9:14–23
Clements JF, Schulenberg TS, Iliff MJ et al (2016) The eBird/Clements checklist of birds of the world: v2016. http://www.birds.cornell.edu/clementschecklist/download/. Accessed 26 Nov 2016
Coues E (1899) On certain generic and subgeneric names in the A.O. U. Check-list Osprey 3:144
Cracraft J (1981) Toward a phylogenetic classification of the recent birds of the word (class Aves). Auk 98:681–714
Dantas SM, Weckstein JD, Bates JM et al (2016) Molecular systematics of the new world screech-owls (*Megascops*: Aves, Strigidae): biogeographic and taxonomic implications. Mol Phylogenet Evol 94:626–634
Dickinson EC, Remsen Jr JV (2013) The Howard and Moore complete checklist of the birds of the world, 4th edn, vol. 1, non-passerines. Aves Press, Eastbourne
Eisermann K (2013) Vocal field marks of unspotted saw-whet owl and Guatemalan pygmy owl. Neotropical Birding 13:7–13
Eisermann K, Howell SNG (2011) Vocalizations of the Guatemalan pygmy-owl (*Glaucidium cobanense*). J Raptor Res 45:304–314
Enríquez PL, Rangel-Salazar JL, Holt DW (1993) Presence and distribution of Mexican owls: a review. J Raptor Res 27:154–160
Ericson PGP, Anderson CL, Britton T et al (2006) Diversification of Neoaves: integration of molecular sequence data and fossils. Biol Lett 2:543–547
Fitzpatrick JW, O'Neill JP (1986) *Otus petersoni*, a new screech-owl from the eastern Andes, with systematic notes on *O. colombianus* and *O. ingens*. Wilson Bull 98:1–14
Fjeldså J, Baiker J, Engblom G et al (2012) Reappraisal of Koepcke's screech owl *Megascops koepckeae* and description of a new subspecies. Bull Brit Ornithologists Club 132:180–193
Ford NL (1967) A systematic study of the owls based on comparative osteology. Ph.D. thesis, University of Michigan, Ann Arbor
Fürbringer M (1888) Untersuchungen zur Morphologie und Systematik der Vögel, zugleich ein Beitrag zur Anatomie der Stütz- und Bewegungsorgane, vol 1–2. Van Holkema, Amsterdam
Gadow H (1892) On the classification of birds. Proc Zool Soc London 1892:229–256
Gibb GC, Kardailsky O, Kimball RT et al (2007) Mitochondrial genomes and avian phylogeny: complex characters and resolvability without explosive radiations. Mol Biol Evol 24:269–280
Gill F, Donsker D (eds) (2016) IOC World Bird list (v 6.4). doi: 10.14344/IOC.ML.6.4
Gómez de Silva H (2010) Spring migration, March through May 2010: Mexico. N Am Birds 64:503–507
Grantsau R (2010) Guia completo para identificação das aves do Brasil, vol 1. Vento Verde, São Carlos
Griscom L (1931) Notes on rare and little-known Neotropical pygmy-owls. Proc New England Zool Club 12:37–43
Grossman M, Hamlet J (1988) Birds of the prey of the world. Bonanza Books, New York
Hamer TE, Forsman ED, Fuchs AD et al (1994) Hybridization between barred and spotted owls. Auk 111:487–492
Heidrich P, König C, Wink M (1995) Bioakustik, Taxonomie und molekulare Systematik amerikanischer Sperlingskäuze (Strigidae: *Glaucidium* spp.) Stuttg Beitr Naturk Ser A 534:1–47
Hekstra GP (1982) Description of twenty-four new subspecies of American *Otus* (Aves, Strigidae). Bull Zool Mus Amsterdam 9:49–63
Hilty SL (2003) Birds of Venezuela. Princeton University Press, Princeton
Howell SNG, Eisermann K (2011) Guatemalan pygmy-owl *Glaucidium cobanense* is a good species. Neotropical Birding 9:74–76
Howell SNG, Robbins MB (1995) Species limits in the least pygmy owl (*Glaucidium minutissimum*) complex. Wilson Bull 107:7–25

Howell SNG, Webb S (1995) A guide to the birds of Mexico and northern Central America. Oxford University Press, New York

Kirwan GM, Schweizer M, Copete JL (2015) Multiple lines of evidence confirm that Hume's owl *Strix butleri* (A. O. Hume, 1878) is two species, with description of an unnamed species (Aves: Non-Passeriformes: Strigidae). Zootaxa 3904:28–50

König C (1991) Zur Taxonomie und Ökologie der Sperlingskäuze (*Glaucidium* spp.) des Andenraumes. Ökologie der Vögel 13:15–76

König C, Straneck R (1989) Eine neue Eule (Aves: Strigidae) aus Nordargentinien. Stuttg Beitr Naturk Ser A 428:1–20

König C, Weick F (2005) Ein neuer Sperlingskauz (Aves: Strigidae) aus Südostbrasilien. Stuttg Beitr Naturk Ser A 688:1–12

König C, Heidrich P, Wink M (1996) Zur Taxonomie der Uhus (Strigidae: *Bubo* spp.) im südlichen Südamerika. Stuttg Beitr Naturk Ser A 540:1–9

König C, Weick F, Becking J-H (1999) Owls: a guide to the owls of the world. Yale University Press, New Haven

König C, Weick F, Becking J-H (2008) Owls of the world, 2nd edn. Christopher Helm, London

Livezey BC, Zusi RL (2007) Higher-order phylogeny of modern birds (Theropoda, Aves: Neornithes) based on comparative anatomy. II. Analysis and discussion. Zool J Linnean Soc 149:1–95

Marks JS, Cannings RJ, Mikkola H (1999) Family Strigidae: typical owls. In: del Hoyo J, Elliott A, Sargatal J (eds) Handbook of the birds of the world, Barn-owls to hummingbirds, vol 5. Lynx Edicions, Barcelona, pp 76–242

Marshall JT Jr (1967) Parallel variation in North and Middle American screech-owls, Monograph no. 1. Western Foundation Vertebrate Zoology, Los Angeles

Mayr G (2010) Paleogene fossil birds. Springer, Heidelberg/Berlin

Mayr E, Short LL (1970) Species taxa of North American birds. Pub Nutall Ornithol Club, no.9, Cambridge, MA

Mikkola H (1983) Owls of Europe. T. & A.D Poyser, Calton

Mikkola H (2014) Owls of the world: a photographic guide, 2nd edn. Firefly Books, Richmond Hill

Moore RT, Peters JL (1939) The genus *Otus* of Mexico and Central America. Auk 56:38–56

Navarro A (1988) Filogenia y clasificación de aves. Ciencia 12:16–29

Norberg RÅ (1977) Occurrence and independent evolution of bilateral ear asymmetry in owls and implications on owl taxonomy. Phil Trans R Soc B 280:375–408

Norberg RÅ (1978) Skull asymmetry, ear structure and function, and auditory localization in Tengmalm's owl, *Aegolius funereus* (Linné). Phil Trans R Soc B 282:325–410

Norberg RÅ (1987) Evolution, structure, and ecology of northern forest owls. In: Nero E, Clark R, Knapton R, Hamre R (eds) Biology and conservation of northern forest owls. U.S. Dept. of Agriculture Forest Service, Rocky Mountain Forest and Range Experiment Station, Fort Collins, pp 3–34

Norberg RÅ (2002) Independent evolution of outer ear asymetry among five owl lineages, morphology, function, and selection. In: Newton I, Kavanagh R, Olsen J, Taylor I (eds) Ecology and conservation of owls. CSIRO, Collingwood, pp 329–342

O'Neill J, Graves G (1977) A new genus and species of owl (Aves: Strigidae) from Peru. Auk 94:409–416

Olson SL (1995) The genera of owls in the Asioninae. Bull Brit Ornithol Club 115:35–39

Pacheco MA, Battistuzzi FU, Lentino M et al (2011) Evolution of modern birds revealed by mitogenomics: timing the radiation and origin of major orders. Mol Biol Evol 28:1927–1942

Penhallurick JM (2002) The taxonomy and conservation status of the owls of the world: a review. In: Newton I, Kavanagh R, Olsen J, Taylor I (eds) Ecology and conservation of owls. CSIRO, Collingwood, pp 343–354

Peters JL (1940) Check-list of birds of the world, vol 4. Harvard University Press, Cambridge, MA

Piacentini VQ, Aleixo A, Agne CA et al (2015) Annotated checklist of the birds of Brazil by the Brazilian Ornithological Records Committee/Lista comentada das aves do Brasil pelo Comitê Brasileiro de Registros Ornitológicos. Rev Bras Ornit 23(2):91–298

Pratt RC, Gibb GC, Morgan-Richards M et al (2009) Toward resolving deep neoaves phylogeny: data, signal enhancement, and priors. Mol Biol Evol 26:313–326

Proudfoot GA, Honeycutt RL, Slack RD (2006) Mitochondrial DNA variation and phylogeography of the Ferruginous Pygmy-Owl (*Glaucidium brasilianum*). Conserv Genet 7:1–12

Proudfoot GA, Gehlbach FR, Honeycutt RL (2007) Mitochondrial DNA variation and phylogeography of the eastern and western screech-owls. Condor 109:617–628

Prum RO, Berv JS, Dornburg A et al (2015) A comprehensive phylogeny of birds (Aves) using targeted next-generation DNA sequencing. Nature 526:569–573

Ramírez-Julián R, González-García F, Reyes-Macedo G (2011) Registro del búho leonado *Strix fulvescens* en el estado de Oaxaca, México. Rev Mex Biodiv 82:727–730

Ramírez-Llorens P, Bellocq MI (2007) New records clarify the southern distribution of the spectacled owl (*Pulsatrix perspicillata*). J Raptor Res 41:268–276

Remsen, JV Jr, Areta JI, Cadena CD et al (2016) A classification of the bird species of South America. American Ornithologists' Union. http://www.museum.lsu.edu/~Remsen/SACCBaseline.htm. Accessed 26 Nov 2016

Ridgway R (1914) The birds of North and Middle America. Part 6. Bull US Nat Mus 50:594–825

Robbins MB, Howell SNG (1995) A new species of pygmy-owl (Strigidae: *Glaucidium*) from the eastern Andes. Wilson Bull 107:1–6

Robbins MB, Stiles FG (1999) A new species of pygmy-owl (Strigidae: *Glaucidium*) from the Pacific slope of the northern Andes. Auk 116:305–315

Seebohm H (1890) Classification of birds: an attempt to diagnose the subclasses, orders, suborders, and some of the families of existing birds. R. H Porter, London

Sharpe RB (1875) Contributions to a history of the Accipitres. Notes on birds of prey in the museum at the Jardin des Plantes and in the collection of Mons A Bouvier. Ibis 17:253–261

Shufeldt RW (1904) An arrangement of the families and the higher group of birds. Amer Nat 38:833–857

Sibley CG, Ahlquist JE (1972) A comparative study of the egg white proteins of non-passerine birds. Bull Peabody Mus Nat Hist 39:593–624

Sibley CG, Ahlquist JE (1990) Phylogeny and classification of birds. Yale University Press, New Haven

Sibley CG, Monroe BL Jr (1990) Distribution and taxonomy of birds of the world and its 1993 supplement. http://www.ornitaxa.com/SM/SMOrg/sibley4.html. Accessed 5 Mar 2012

Sigrist T (2006) Aves do Brasil: uma visão artística. Editora Avis Brasilis, São Paulo

Silva JM, Coelho G, Gonzaga LP (2002) Discovered on the brink of extinction: a new species of pygmy-owl (Strigidae: *Glaucidium*) from Atlantic forest of northeastern Brazil. Ararajuba 10:123–130

Vielliard J (1989) Uma nova especie de *Glaucidium* (Aves: Strigidae) da Amazonia. Rev Brasileira Zool 6:685–693

Weick F (2006) Owls (Strigiformes): annotated and illustrated checklist. Springer, Berlin

Welty J (1982) The life of birds. Saunders College Publishing, Philadelphia

Weske JS, Terborgh JW (1981) *Otus marshalli*, a new species of screech-owl from Peru. Auk 98:1–7

Wink M, Heidrich P (1999) Molecular evolution and systematics of owls (Strigiformes). In: König C, Weick F, Becking J-H (eds) Owls: a guide to the the owls of the world. Pica Press, Sussex, pp 39–57

Wink M, Heidrich P, Sauer-Gürth H et al (2008) Molecular phylogeny and systematics of owls (Strigiformes). In: König C, Weick F, Becking J-H (eds) Owls of the world, 2nd edn. Christopher Helm, London, pp 42–63

Chapter 3
The Owls of Argentina

Ana Trejo and María Susana Bó

Abstract Between 18 and 24 Strigiformes species have been registered in Argentina, with some variations due to the different taxonomic criteria and to take as valid or not some uncertain records. Strigiformes are distributed in all the territory (including islands), although the largest number of species is found in the subtropical forest regions: Paraná rainforest (15 species), Chaco (11 species), and Yungas (9 species). The species are associated with three types of habitat: subtropical forest, temperate forests, and grasslands. There are no endemic species recorded in Argentina. The conservation status has been reconsidered recently—five species have been classified as threatened and four as vulnerable. Alteration and destruction of habitats (especially deforestation and agricultural expansion on natural grasslands) and toxicology are considered the main threats. We reviewed 456 publications, 44% of them are about distribution status, 32% about foraging, 12% about breeding ecology, and 5% about taxonomy and nomenclature. The distribution of those species for which there is better information suggests that studies have focused on the Pampa and northern Patagonia Regions (mainly Patagonian). There is at least minimum knowledge of different biological aspects for all the species, and they are categorized as nonthreatened. Of the nine endangered species, six exclusively inhabit the Paraná rainforest, one species (*Strix chacoensis*) is in the Chaco subtropical dry forest, one (*S. rufipes*) is in the forested Patagonian Andes, and one species is found in the Paraná rainforest, Chaco, and Yungas (*Asio stygius*).

Keywords Distribution • Owl taxonomy • Conservation • Biology

A. Trejo (✉)
Centro Regional Bariloche, Universidad Nacional del Comahue, 8400 Bariloche, Argentina
e-mail: ana.r.trejo@gmail.com

M.S. Bó
Departamento de Biología, Facultad de Ciencias Exactas y Naturales, Universidad Nacional de Mar del Plata, Funes 3250, 7600 Mar del Plata, Argentina
e-mail: msbo@mdp.edu.ar

P.L. Enríquez (ed.), *Neotropical Owls*, DOI 10.1007/978-3-319-57108-9_3

Burrowing Owl (*Athene cunicularia*)

3.1 Introduction

In this paper, a review is made of the taxonomic diversity, distribution, conservation, and status of knowledge on biological and ecological aspects of the Argentinian Strigiformes. The data presented in this review is fully based on bibliography collected by the authors, consisting of books, national and international publications, and our own field trips and observations, as well as from the so-called "gray" literature, that is, scattered literature that have little or very limited distribution.

3.2 Taxonomic Diversity

In Argentina, there are between 18 and 24 Strigiformes species that have been registered, depending on the source consulted. An analysis of the most widely used lists of species, field guides, and books (e.g., de la Peña and Rumboll 1998; König et al. 1999; Marks et al. 1999; Mazar Barnett and Pearman 2001; Clements 2007; Narosky and Yzurieta 2010, Remsen et al. 2017) shows that the number of species varies according to whether a dubious record is included or not (JC Chébez, pers. com.), such as the case of *Glaucidium minutissimum* in the NE part of the country (Misiones Province), but overall the number of species varies when applying different taxonomic criteria. In brief, the differences will depend if the following taxa are considered or not as valid species: *Glaucidium bolivianum* (separate from *G. jardinii*), *G. tucumanum* (separate from *G. brasilianum*), *Strix chacoensis* (separate from *S. rufipes*), *Megascops hoyi* and *M. sanctaecatarinae* (separate from *M. atricapilla*), and *Bubo magellanicus* (separate from *B. virginianus*).

Glaucidium bolivianum was recently described (König 1991a). This taxon is traditionally included in *G. jardinii* (e.g., Meyer de Schauensee 1970). König (1991b) and Heidrich et al. (1995b) provide bioacoustic and molecular evidence to consider it as a valid species.

G. brasilianum tucumanum subspecies has been described by Chapman (1922) from specimens of the Salta Province in the northwest of Argentina. The author based his work on a morphological description of the specimens, and although he recognizes them as a separate taxon, he appears doubtful as to whether the taxon qualifies as a species or subspecies. This taxon is considered as a different species of from *G. brasilianum* by Heidrich et al. (1995b) and Wink and Heidrich (1999) based on genetic data and vocal differences. Marks et al. (1999) and König et al. (1999) follow this criterion, also noting differences in habitat and plumage.

Strix rufipes was described from a specimen of the southern part of Chile. Later on Cherrie and Reichenberger (1921) studied specimens from the Chaco region in Paraguay and classified them as *S. rufipes*. However, these authors recognized differences in color compared with the type. This difference was ratified by Dabbene (1926) who studied specimens from the north and south. Straneck and Vidoz (1995) separated the northern taxon (*S. chacoensis*) from the southern one (*S. rufipes*) based not only on the differences in color but also in size and vocalizations. Although *S. chacoensis* has been considered a subspecies of *S. rufipes*, it is possible that it is more related to *S. hylophila* due to vocalization similarities (Straneck and Vidoz 1995; König et al. 1999; Marks et al. 1999).

Previously, the *sanctaecatarinae* taxon was included within the *M. atricapilla* species (e.g., Meyer de Schauensee 1970; Olrog 1979). This criterion is followed by some more recent authors (Sibley and Monroe 1990), which also include the recently described *M. hoyi* (König and Straneck 1989, a description based on individuals from Salta Province in the northwest of Argentina) within *M. atricapilla*. The criteria for this classification are based on an alleged wide intra-specific variation of sizes and in considering the vocalizations as identical. However, König (1994), Heidrich et al. (1995a), and Marks et al. (1999) consider *M. sanctaecatarinae* and *M. hoyi* as valid species, based on molecular, morphological, and acoustic differences.

Traylor (1958) recognizes two subspecies of the *Bubo* genus in Argentina: *B. virginianus nacurutu* (NW and center) and *B. v. magellanicus* (mountain region). This author identifies significant morphological differences between the individuals of both subspecies, being *B. v. magellanicus* much smaller in size. Later, König et al. (1996) propose to consider the *magellanicus* taxon as a valid species, separating it from *B. v. nacurutu* specifically based on genetic and vocalization differences.

Even though the species *Bubo magellanicus* has been widely accepted in many general books (Marks et al. 1999; König et al. 1999), the American Ornithologists' Union has not included it in their classification of birds from South America (Remsen et al. 2017) due to the valid consideration that critical intermediate populations have not been studied.

In this chapter, we used the recognized taxa for Argentina by Mazar Barnett and Pearman (2001) (see Appendix 3.1). With respect to the nomenclature of the species, we follow Remsen et al. (2017).

3.3 Distribution and Habitat Associations

Argentina is constituted by the Great Plains, delimited at the west by the Andes Mountain Range, with a variation in altitude that goes from 7000 m down to sea level. The country has a great variety of climates and environments, such as subtropical forests, xerophytic forests and shrubland, temperate forests, desert areas, grasslands, and wetlands. The total area of the country is 3.7 million km^2 including Tierra del Fuego, Antarctica, and the South Atlantic Islands. The continental areas cover a total of 2.7 million km^2 (Brown et al. 2006). Within this diversity of environments and climates, some ornithogeographical regions have been defined (according to Mazar Barnett and Pearman 2001 based on Nores 1987, Fig. 3.1) that attempt to establish consistent bird distribution areas and characterized by the presence of local endemisms (i.e., species that are only found in the considered region). These ornithogeographical regions vary according to the Strigiformes species richness and the number of local endemic species (Table 3.1). In a global sense, Argentina does not have endemic Strigiformes species. *Asio flammeus* has some resident populations in some islands (Couvé and Vidal 2003), and the resident population of *Athene cunicularia* in Tierra del Fuego is currently extinct (Clark 1986). The island region has an endemic subspecies, *A. flammeus sanfordi*, which nests in the Malvinas Islands (Couvé and Vidal 2003). Furthermore, there are occasional records of the continental Patagonian species in Tierra del Fuego and the South Atlantic Islands.

The Strigiformes species of Argentina have different levels of association with plant communities. One group of species is associated mainly with subtropical or tropical rainforest like *Megascops hoyi*, *M. atricapilla*, *M. sanctaecatarinae*, *Pulsatrix koeniswaldiana*, *Strix hylophila*, *Ciccaba virgata*, *C. huhula*, *Glaucidium bolivianum*, and *Aegolius harrisii*. Other species are associated with temperate or xerophytic forests and shrubland, such as *Megascops choliba*, *Pulsatrix perspicillata*, *Glaucidium nana*, *G. brasilianum*, *S. rufipes*, *G. bolivianum*, and *Asio stygius*. A group of species like *Tyto alba*, *Bubo virginianus*, *B. magellanicus*, *Strix chacoensis*, and *Asio clamator* are less specific and are associated with forested and open environments. Finally, the species that frequent mainly open environments are *Athene cunicularia* and *Asio flammeus*.

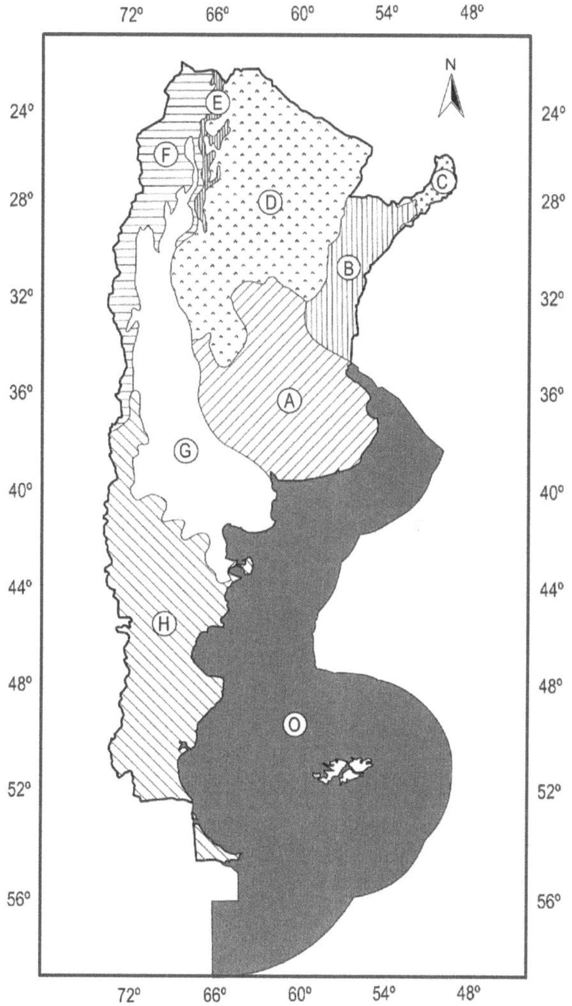

Fig 3.1 Ornithogeographical regions of Argentina. In parenthesis, the main plant formations. (**a**) Pampas (wet grasslands); (**b**) Mesopotamian savanna (xerophilus and gallery forests, grasslands, and wetlands); (**c**) Paraná forest (subtropical forest); (**d**) Chaco (xerophilus forests, grasslands); (**e**) Yungas (subtropical forest); (**f**): Prepuna, Puna, and Highlands (deserts and high grasslands); (**g**) Monte Desert (shrubland); (**h**) Patagonia (temperate Andean forest and steppe); (**o**) Extra-continental Argentina (Reproduced from Mazar Barnett and Pearman 2001)

3.4 Conservation Status at the National Level and Conservation Strategies

All the Strigiformes species found in Argentina are included in Apendix II of the CITES (2016), that is, species that are not necessarily endangered, but their status might change if the illegal trade is not controlled. At the world level, only one species of Strigiformes distributed in Argentina (*Strix hylophila*) is found to be in any category of risk (Near Threatened) according to the criteria of the IUCN (International Union for the Conservation of Nature 2016). At the national level, bird species have been recently recategorized (López-Lanús et al. 2008, Res 348/10 Secretaría de

Table 3.1 Distribution of Strigiformes species in different ornithogeographical continental regions of Argentina according to Mazar Barnett and Pearman (2001)

Species	Pampas	Mesopotamian savanna	Paraná forest	Chaco	Yungas	Prepuna, Puna, and Highlands	Monte desert	Patagonia
Tyto alba	X	X	X	X	X	X	X	X
Megascops choliba[a]	X	X	X	X	X		X	
Megascops atricapilla			X					
Megascops hoyi					X			
Megascops sanctaecatarinae			X					
Pulsatrix perspicillata[b]			X	X	X			
Pulsatrix koeniswaldiana			X					
Bubo virginianus	X	X		X				
Bubo magellanicus				X		X	X	X
Strix hylophila			X					
Strix rufipes								X
Strix chacoensis				X			X	
Ciccaba virgata			X					
Ciccaba huhula			X		X			
Glaucidium bolivianum					X			
Glaucidium brasilianum	X	X	X	X	X		X	

Species	Pampas	Mesopotamian savanna	Paraná forest	Chaco	Yungas	Prepuna, Puna, and Highlands	Monte desert	Patagonia
Glaucidium nana								X
Athene cunicularia	X	X	X	X		X	X	X
Aegolius harrisii			X		X			
Asio clamator	X	X	X	X				
Asio stygius			X	X	X			
Asio flammeus[c]	X	X	X	X		X	X	X
Species richness	7	7	15	11	9	4	7	6
Local endemisms	–	–	5	–	2	–	–	2

[a]Occasional presence in Monte Desert (Casares 1944; Mazar Barnett and Pearman 2001)
[b]Occasional presence in the Paraná forest (see references in Ramírez Llorens and Bellocq 2007)
[c]Occasional presence in the Paraná forest (Chébez 1996)

Ambiente y Desarrollo Sustentable-Fauna Silvestre), based on biological and ecological information, following the criteria explained by Reca et al. (1994). Of the Strigiformes species considered for this recategorization, five were categorized as Threatened and four as Vulnerable (Appendix 3.1).

Of the nine species under conservation risk categories, six (*M. atricapilla, M. sanctaecatarinae, P. koeniswaldiana, S. hylophila, C. virgata, and C. huhula*) exclusively inhabit the Paraná forest, one (*Strix chacoensis*) in the Chaco dry forests, one (*S. rufipes*) in the forested mountain area of Patagonia, and one (*Asio stygius*) is distributed among the Paraná forest, Chaco, and Yungas regions. In general, these species are locally endemic, with distribution that goes from restricted to very restricted (such is the case of *Megascops sanctaecatarinae*), and low abundance and exclusively inhabit one type of environment (mainly forests or rainforests).

A fundamental strategy for the national conservation of biodiversity in Argentina is the Protected Areas Federal System, which contains 300 areas under private or governmental protection (covering approximately 7% of the national territory; Burkart 2006). There are 33 national parks, 6 natural monuments, 6 natural reserves, and 4 protected marine areas under the National Parks Administration (Administración de Parques Nacionales, APN) and an even greater number of areas that are managed by national, provincial, and municipal government agencies, non-governmental organizations (NGO), or private organizations. However, the relative effort to preserve the different regions has been unequally balanced (Brown and Pacheco 2006), and the effective level of protection that the different areas receive is deficient, due to the absence of control on the field, lack of budget, equipment, or human resources, among other reasons (Burkart 2006). However, all the Strigiformes species in a risk category (Appendix 3.1) are under some protection level (Chébez et al. 1998; López-Lanús et al. 2008).

3.5 Threats

Firstly, it is convenient to differentiate between the threats that Strigiformes suffer as individuals, as opposed to those that affect them as a population. Among the former, it has been mentioned hunting or intentional killing, electrocution, and being run over on the road (Newton et al. 1997). Other factors, such as persecution due to fear, revulsion, or superstitious or mistaken beliefs, need to be investigated. Although studies made in Argentina are scarce, it would seem that one of the most important dangers at an individual level is being run over by vehicles on the roads.

A study carried out in the northern Patagonia forest region (Trejo and Seijas 2003) shows that of the birds found dead on the road over a 3-year period, 42% corresponded to Strigiformes (35% to *Strix rufipes* and 7% to *Tyto alba*). It has been mentioned that among the elements that predispose these birds to have said accidents are the use of snares placed near the roads, their movement at ground level when chasing their prey, and "tunnel vision," with frontal eyes and almost no peripheral vision, which prevents them from perceiving approaching vehicles when they cross the road.

At the population level, the main impacts that affect or could affect Strigiformes are the alteration and destruction of their habitat and contamination by toxic substances. The transformation of the habitat and its fragmentation, especially on forest habitats, has a negative impact on the abundance, the richness, and the diversity of the birds of prey (Carrete et al. 2009). The destruction of the habitat increases the risk of mortality, changes the trophic resources, and reduces the nesting sites with a drastic effect on productivity. The fragmentation of the habitat interferes with the natural dispersion of the species as well as that of their prey (Marks et al. 1999).

Deforestation, along with desertification and fragmentation, is an important problem in Argentina. In the past 75 years, the reduction of the natural forested surface reached 66% of the original surface (Pérez Pardo 2006). Deforestation currently reaches a rate of roughly 2500–2000 km^2/year, mainly in the Chaco, Paraná forest, and Yungas regions (Gasparri and Grau 2006). At present, there are few spaces in the forestry systems that are not impacted by human activities. The main causes of forest coverage loss are livestock and agricultural production, overgrazing, fire, and forestry exploitation for wood or energetic purposes. These activities occupy more than 80% of the Argentinean territory and have a substantial effect on the natural resources (Pérez Pardo 2006).

The Patagonian forests are the exception since they have a high percentage of protected areas in a good state of conservation. However, the risks of fire hazards are high, especially during the dry season (Prémoli et al. 2006). These fires are very frequently deliberate and, along with other impacts of anthropic origin (logging and replacement with exotic species, pollution, ranching) produced by the expansion of urban centers and touristic developments, are a threat particularly for *Strix rufipes*, the only species of the Patagonian owl that is strictly confined to the temperate austral forest (both in Chile and in Argentina). In Argentina, recent studies carried out in the Patagonia mountain range show healthy populations (Trejo et al., unpublished data). However, because of their requirements of a restricted habitat, the conservation situation for this bird is uncertain, especially in Chile where its population seems to be in decline (Jaksic et al. 2001), probably due to a reduced availability of the habitat because of deforestation (Omland et al. 2001).

In the last few years, natural grasslands exclusively used for livestock grazing have been replaced by extensive agricultural systems (Viglizzo et al. 2006). These land-use changes along with the agricultural expansion have an impact on various birds, including *Asio flammeus*, a species that is closely dependent on natural or semi-natural grasslands for feeding as well as for nesting. Some studies indicate that there is a noticeable retraction of this species in the Buenos Aires Province (Bilenca et al. 2009). Nevertheless, some of the species can be positively affected (Filloy and Bellocq 2007). The transformation of natural environments into agroecosystems and urban areas can result to be beneficial for generalist species such as *Tyto alba* and *Athene cunicularia*, most likely because the number of prey increases. Recent studies demonstrate that there is a positive association between *Athene cunicularia* and croplands and pastures (Filloy and Bellocq 2007; Pedrana et al. 2008), although agricultural machinery can also have a negative impact by destroying nests (Bellocq 1993). The presence of this owl was also documented in urban zones of the Buenos Aires Province (Pedrana et al. 2008; Sánchez et al. 2008; Baladrón 2010).

Regarding the subject of contamination by chemical substances (both from direct exposure as well as secondary when feeding on prey poisoned with pesticide or rat poison), it has not been studied on Strigiformes in Argentina. Some pesticides, whose toxicity on birds of prey has been demonstrated (Iolster and Krapovickas 1999), are prohibited in the country (organochlorides, DDT, organophosphorus such as Monocrotofós, Res. SENASA 256/03). However the new second-generation anticoagulant rat poisons such as Brodifacoum are widely used both as a poison for rodents and toxic bait for hares and rabbits (Bonino 2004). There is currently evidence at the world level that at least *Tyto alba* can die when eating rodents poisoned with Brodifacoum (Eason and Spurr 1995).

3.6 The Biological and Ecological State of Knowledge

Strigiformes are a difficult group to study given their habits and features (nocturnal species, many times cryptic, frequently inhabiting forests, and jungles that are not easily accessible) which, along with the lack of specific research, turns them into one of the least known groups of the Argentinian avifauna. Since none of these species are endemic to Argentina, in some cases there is information on different aspects of their biology that is related to other parts of their geographic distribution (especially in Brazil and Chile). However, in this chapter we refer exclusively to the existing knowledge of the populations that live in Argentina. The revision was made of 456 publications (until 2010) that cover different aspects of the distribution and biology of Strigiformes in Argentina. Most publications are about distribution (44%), aspects related to feeding (32%), behavior—mostly timely observations—(12%), and lastly on reproductive biology (7%). The remainder 5% corresponds to taxonomic and nomenclature discussions. Among the most recent behavioral work is a study about the connection between the selection of a nesting habitat by *A. cunicularia* and the differences in individual tolerance to human presence (Carrete and Tella 2010).

Trophic ecology is the most studied aspect in Argentina, as well as in other Neotropical countries. These studies go from simple prey enumeration (specifically with the purpose of increase the knowledge on the distribution of small mammals) to a more complex diet analysis that consists of trophic parameters estimations and other aspects of feeding ecology such as hunting strategies (collected in Pardiñas and Cirignoli 2002; Bó et al. 2007). In the past years, research has been done on interactions and sympatric species (Leveau et al. 2004; Trejo et al. 2005b; Trejo 2006; Donadio et al. 2009; Baladrón 2010), prey selection aspects (Bellocq 1987; Trejo and Guthmann 2003; Trejo et al. 2005a), and diet studies taking into account spatial and temporal variations (Travaini et al. 1997; Leveau et al. 2006; Trejo and Lambertucci 2007, J.H. Sarasola and M.A. Santillán, personal communication). A study has also been made on the functional response of *Bubo magellanicus* in relation to density variations of an introduced species, the European hare (*Lepus europaeus*) in northern Patagonia (Monserrat et al. 2005), a species that has become an important prey for diurnal birds of prey, and carrion for vultures and condors. These kinds of studies are of great importance for the conservation of this group of birds of prey.

The most well-known species with respect to feeding habits is *Tyto alba* (Bó et al. 2007). Its diet has been studied in almost all the regions of the country. This is probably because it is a very common species, which inhabits all environments, even suburban and urban, and also because usually an accumulation of pellets is found on their nests or perch. Other species that have been studied less thoroughly are *Bubo magellanicus* (references in Pardiñas and Cirignoli 2002; Nabte et al. 2006), *Asio clamator* (references in Pardiñas and Cirignoli 2002; Pautasso 2006), *Athene cunicularia* (references in Pardiñas and Cirignoli 2002; Andrade et al. 2004, 2010; Nabte et al. 2008; Sánchez et al. 2008; Cavalli 2011; Cavalli et al. 2013), *Glaucidium brasilianum* (Carrera et al. 2008), and *G. nana* (Santillán et al. 2010). In some cases, such as *Strix rufipes*, the feeding habits have been studied from a low number of pellets and with a focus based especially on small mammals (Udrizar Sauthier et al. 2005). Studies regarding the feeding ecology of this species in the northeast of Patagonia are currently underway (Beaudoin and Trejo, unpublished data).

The only species that have information about their reproductive biology (nests, eggs, clutch size and brood size, phenology, and description of the chicks) are *T. alba*, *M. choliba*, *B. virginianus*, *G. brasilianum*, *A. cunicularia*, *A. clamator*, and *A. flammeus* (citation in Trejo 2007; Carrera et al. 2008; Salvador 2012). Regarding *A. cunicularia*, some studies have analyzed the relationship between habitat modification and the breeding behavior of the species (Baladrón, Cavalli, Bó and Isacch, unpublished data). Recently, observations on the breeding biology of *Glaucidium nanum* (Santillán et al. 2010) have been carried out; this last work is of special interest since it extends the distribution of breeding populations of the species in Eastern Argentinian Patagonia. Although most of the data has not been published yet, currently studies are being carried out in northwestern Patagonia on the nesting and breeding biology of *S. rufipes* (Beaudoin and Ojeda 2011). All the other species are practically unknown.

Habitat selection in Strigiformes species is a more recent studied topic. Pioneer study is done by the CECARA (Center for the Study and Conservation of Birds of Prey in Argentina) which, in cooperation with Spanish researchers, has studied the selection of the breeding habitat of *G. brasilianum* at different spatial scales (Campioni et al. 2012). With respect to the study of the distribution of the different species, diverse methodological techniques have begun to be developed, such as the response to playback methods (Trejo et al. 2011) and the use of nest boxes (Liébana et al. 2013).

Observation of the distribution of the species, for which there is a higher degree of knowledge, shows that research has been centered in the Pampa and north Patagonia regions (mainly the Patagonian Steppe). Also, it is worth noting that all the species for which there is at least a minimum degree of knowledge of different elements of their biology are in the Not Threatened conservation category (Appendix 3.1).

3.7 Final Comments

Strigiformes are a little known group, given the difficulties inherent to their life mode and the environments in which they inhabit. These features, along with the shortage of researchers in Argentina specialized on this group of birds make it so that there is a minimal knowledge for most of the species. However, this is a group with great ecological importance, given its key role in the trophic networks as top predators and as main predators of rodents. This last feature makes them of particular importance for control of vector species for zoonosis such as hemorrhagic fever and hantavirus pulmonary syndrome (HPS). It is imperative that measures be taken toward conservation of Strigiformes, accompanied by the preservation of the ecosystems they inhabit. In order to do this, it is important to know not only what they eat but also many other aspects of their basic biology, such as breeding biology and preferences of their habitat.

Population studies and the response of these predators to the fluctuations in their prey are also of paramount importance since this information is essential for any management plan. Although in the last few decades progress has been made in this direction, there is still a lot to do in the future if we want to preserve these birds. As happens with all birds and fauna in general, threats like the destruction of their natural habitats threaten their persistence. However, and even more important, one of the main threats Strigiformes have in Argentina is the lack of knowledge about their status and ecology, which are fundamental features for the development of strategies for the conservation of the species.

Burrowing Owl (*Athene cunicularia*)

Acknowledgments The authors wish to thank JC Chébez who willingly shared his extensive knowledge about the distribution and conservation state of the Argentinean fauna and sadly is no longer with us.

Appendix 3.1

Listing of the Strigiformes species found in Argentina, scientific and common names, and conservation category

Species	English common name	Category
Tyto alba	Barn owl	NT
Megascops choliba	Tropical screech owl	NT
Megascops hoyi	Montane forest screech owl	NT
Megascops atricapilla	Black-capped screech owl	VU
Megascops sanctaecatarinae	Long-tufted screech owl	T
Bubo virginianus	Great horned owl	NT
Bubo magellanicus[a]	Lesser horned owl	–
Pulsatrix perspicillata	Spectacled owl	NT
Pulsatrix koeniswaldiana	Tawny-browed owl	VU
Strix hylophila	Rusty-barred owl	VU
Strix rufipes	Rufous-legged owl	VU
Strix chacoensis	Chaco owl	T
Ciccaba virgata	Mottled owl	T
Ciccaba huhula	Black-banded owl	T
Glaucidium bolivianum	Yungas pygmy owl	NT
Glaucidium brasilianum	Ferruginous pygmy owl	NT
Glaucidium nana	Austral pygmy owl	NT
Athene cunicularia	Burrowing owl	NT
Aegolius harrisii	Buff-fronted owl	NT
Asio clamator	Striped owl	NT
Asio stygius	Stygian owl	T
Asio flammeus	Short-eared owl	NT

For the listing of the species, we have followed Mazar Barnett and Pearman (2001). Nomenclature follows Remsen et al. (2017). Conservation categories were taken from López-Lanús et al. (2008)
NT not threatened, *T* threatened, *VU* vulnerable
[a]Not categorized because it was included in *Bubo virginianus*

References

Andrade A, Udrizar Sauthier DE, Pardiñas UFJ (2004) Vertebrados depredados por la lechucita vizcachera (*Athene cunicularia*) en la Meseta de Somuncurá (Río Negro, Argentina). Hornero 19:91–93

Andrade A, Nabte MJ, Kun ME (2010) Diet of the Burrowing Owl (*Athene cunicularia*) and its seasonal variation in Patagonian steppes: implications for biodiversity assessments in the Somuncurá Plateau Protected Area, Argentina. Stud Neotropical Fauna Environ 45:101–110

APN (Administración de Parques Nacionales). http://www.parquesnacionales.gob.ar/_aapp_.htm. Accessed 25 April 2017

Baladrón AV (2010) Impacto de la depredación de aves rapaces sobre micromamíferos en el sudeste bonaerense. Tesis Doctoral, Universidad Nacional de Mar del Plata

Beaudoin F, Ojeda V (2011) Nesting of Rufous-legged Owls in evergreen Nothofagus forests. J Raptor Res 45:272–274

Bellocq MI (1987) Selección de hábitat de caza y depredación diferencial de *Athene cunicularia* sobre roedores en ecosistemas agrarios. Rev Chil Hist Nat 60:81–86

Bellocq MI (1993) Reproducción, crecimiento y mortalidad de la lechucita vizcachera (*Speotyto cunicularia*) en agrosistemas pampeanos. Hornero 13:272–312

Bilenca D, Codesido M, González Fischer C (2009) Cambios en la fauna pampeana. Ciencia Hoy 18:8–17

Bó MS, Baladrón AV, Biondi LM (2007) Ecología trófica de Falconiformes y Estrigiformes: Tiempo de síntesis. Hornero 22(Número Especial Aves Rapaces):97–115

Bonino N (2004) Métodos utilizados para el control de liebres y conejos introducidos en la Patagonia. http://www.inta.gov.ar/bariloche/info/documentos/natural/fauna/liebres.htm. Accessed 9 Feb 2011

Brown A, Pacheco S (2006) Propuesta de actualización del Mapa Ecorregional de la Argentina. In: Brown A, Martínez Ortiz U, Acerbi M et al (eds) La Situación Ambiental Argentina 2005. Fundación Vida Silvestre Argentina, Buenos Aires, pp 28–31

Brown A, Martínez Ortiz U, Acerbi M et al (2006) Un enfoque ecorregional de la problemática ambiental en la Argentina. In: Brown A, Martínez Ortiz U, Acerbi M (eds) La Situación Ambiental Argentina 2005. Fundación Vida Silvestre Argentina, Buenos Aires, pp 25–27

Burkart R (2006) Las áreas protegidas de la Argentina. In: Brown A, Martínez Ortiz U, Acerbi M et al (eds) La Situación Ambiental Argentina 2005. Fundación Vida Silvestre Argentina, Buenos Aires, pp 399–409

Campioni L, Sarasola JH, Santillan M et al (2012) Breeding season habitat selection by Ferruginous Pygmy Owls *Glaucidium brasilianum* in central Argentina. Bird Study. doi:10.1080/00063657.2012.733336

Carrera JD, Fernández FJ, Kacoliris FP et al (2008) Field notes on the breeding biology and diet of Ferruginous Pygmy-owl (*Glaucidium brasilianum*) in the dry chaco of Argentina. Ornitol Neotrop 19:315–319

Carrete M, Tella JL (2010) Individual consistency in flight initiation distances in burrowing owls: a new hypothesis on disturbance-induced habitat selection. Biol Lett 6:167–170

Carrete M, Tella JL, Blanco G et al (2009) Effects of habitat degradation on the abundance, richness and diversity of raptors across Neotropical biomes. Biol Conserv 142:2002–2011

Casares J (1944) Aves de Estanzuela, San Luis. Hornero 8:379–463

Cavalli M (2011) Influencia de la modificación del ambiente sobre la conducta de caza de la Lechucita de las Vizcacheras (*Athene cunicularia*) en la Region Pampeana. Tesis de Grado Universidad Nacional de Mar del Plata

Cavalli M, Baladron AV, Isacch JP et al. (2013) Prey selection and food habits of breeding Burrowing Owls (Athene cunicularia) in natural and modified habitats of Argentina pampas. Emu. http://dx.doi.org/10.1071/MU13040

Chapman FM (1922) Descriptions of apparently new birds from Colombia, Ecuador and Argentina. Am Mus Novitates 31:1–8

Chébez JC (1996) Fauna misionera. L.O.L.A., Buenos Aires

Chébez JC, Rey NR, Barbaskas M et al (1998) Las aves de los parques nacionales de la Argentina. L.O.L.A, Buenos Aires

Cherrie GJ, Reichenberger EMB (1921) Description of proposed new birds from Brazil, Paraguay and Argentina. Am Mus Novitates 27:1–6

CITES (Convention on International Trade in Endangered Species of Wild fauna and Flora) (2016) http://www.cites.org/. Accessed 25 April 2017

Clark R (1986) Aves de Tierra del Fuego y Cabo de Hornos. Guía de campo. L.O.L.A, Buenos Aires

Clements JF (2007) The Clements checklist of birds of the world, 6th edn. Cornell University Press, Ithaca

Couvé E, Vidal C (2003) Aves de Patagonia, Tierra del Fuego y Península Antártica, Islas Malvinas y Georgias del Sur. Fantástico Sur, Punta Arenas

Dabbene R (1926) Sobre la distribución de la lechuza *Strix rufipes*. Hornero 3:405–407

De la Peña MR, Rumboll M (1998) Birds of southern South America and Antartica. Harper Collins, London

Donadío E, Merino ML, Bolgeri MJ (2009) Diet of two coexisting owls in the high Andes of northwestern Argentina. Ornitol Neotrop 20:137–141

Eason CT, Spurr EB (1995) The toxicity and sub-lethal effects of Brodifacoum in birds and bats. A literature review. Sci Conserv 6:1–16

Filloy J, Bellocq MI (2007) Respuesta de las aves rapaces al uso de la tierra: un enfoque regional. Hornero 22:131–140

Gasparri I, Grau R (2006) Patrones regionales de deforestación en el subtrópico argentino y su contexto ecológico y socio-económico. In: Brown A, Martínez Ortiz U, Acerbi M et al (eds) La Situación Ambiental Argentina 2005. Fundación Vida Silvestre Argentina, Buenos Aires, pp 442–446

Heidrich P, König C, Wink M (1995a) Molecular phylogeny of South American Screech Owls of the *Otus atricapillus* complex (Aves: Strigidae) inferred from nucleotide sequences of the mitochondrial cytochrome *b* gene. Z Naturforsch 50:294–302

Heidrich P, König C, Wink M (1995b) Bioakustik Taxonomie und molekulare Systematik amerikanischer Sperlingskäuze (Strigidae: *Glaucidium* spp.) Stutt Beitr Naturkd Serie A (Biologie) 534:1–47

Iolster P, Krapovickas S (1999) Los plaguicidas en uso en la Argentina: riesgos para las aves silvestres. Temas de naturaleza y Conservación 2. Aves Argentinas (AOP), Buenos Aires

IUCN (International Union for the Conservation of Nature) (2016) IUCN Red List of Threatened Species. http://www.iucn.redlist.org/. Accessed 25 April 2017

Jaksic FM, Pavez E, Jiménez JE et al (2001) The conservation status of raptors in the metropolitan región, Chile. J Raptor Res 35:151–158

König C (1991a) Zur Taxonomie und Ökologie der Sperlingskäuze (*Glaucidium* spp.) des Andenraumes. Oekol. Vögel 13:25–76

König C (1991b) Biological patterns in owl taxonomy, with emphasis on bioacoustical studies on Neotropical Pygmy (*Glaucidium*) and Screech Owls (*Otus*). In: Meyburg BU, Chancellor RD (eds) Raptor conservation today. WWGBP/The Pica Press, Sussex, pp 1–19

König C (1994) Lautäusserungen als interspezifische Isolationsmechanismen bei Eulen der Gattung *Otus* (Aves: Strigidae) aus dem südlichen Südamerika. Stutt Beitr Naturkd Serie A (Biologie) 511:1–35

König C, Straneck R (1989) Eine neue Eule (Aves: Strigidae) aus Nordargentinien. Stutt Beitr Naturkd. Serie A (Biologie) 428:1–20

König C, Heidrich P, Wink M (1996) Zur Taxonomie der Uhus (Strigidae: *Bubo* spp.) im südlichen Südamerika. Stutt Beitr Naturkd Serie A (Biologie) 540:1–9

König C, Weick F, Becking JH (1999) Owls. A guide to the owls of the world. Yale University Press, New Haven/London

Leveau LM, Leveau CM, Pardiñas UFJ (2004) Trophic relationships between White-tailes kites (*Elanus leucurus*) and barn owls (*Tyto alba*) in southern Buenos Aires Province, Argentina. J Raptor Res 38:178–181

Leveau LM, Teta P, Bogdaschewski R et al (2006) Feeding habits of the Barn Owl (*Tyto alba*) along a longitudinal-latitudinal gradient in central Argentina. Ornitol Neotrop 17:353–362

Liébana MS, Sarasola JH, Santillan MA (2013) Nest- box occupancy by neotropical raptors in a native forest of central Argentina. J Raptor Res 47:208–213

López-Lanús B, Grilli P, Coconier E et al (eds) (2008) Categorización de las aves de la Argentina según su estado de conservación. Informe de Aves Argentinas/AOP y Secretaría de Ambiente y Desarrollo Sustentable, Buenos Aires

Marks JS, Cannings RJ, Mikkola H (1999) Family Strigidae (typical owls). In: del Hoyo J, Elliott A, Sargatal J (eds) Handbook of the birds of the world, Barn-owls to hummingbirds, vol 5. Lynx Edicions, Barcelona, pp 76–243

Mazar Barnett J, Pearman M (2001) Lista comentada de las aves argentinas. Lynx Edicions, Barcelona

Meyer de Schauensee R (1970) A guide to the birds of South America. Livingston Publishing Company, Pennsylvania
Monserrat AL, Funes MC, Novaro AJ (2005) Respuesta dietaria de tres rapaces frente a una presa introducida en Patagonia. Rev Chilena Hist Nat 78:425–439
Nabte MJ, Saba SL, Pardiñas UFJ (2006) Dieta del búho magallánico (Bubo magellanicus) en el Desierto del Monte y la Patagonia argentina. Ornitol Neotrop 17:27–38
Nabte MJ, Pardiñas UFJ, Saba SL (2008) The diet of the burrowing owl, *Athene cunicularia*, in the arid lands of northeastern Patagonia, Argentina. J Arid Environ 72:1526–1530
Narosky T, Yzurieta D (2010) Guía para la identificación de las aves de Argentina y Uruguay. Vázquez Mazzini, Buenos Aires
Newton I, Wyllie I, Dale L (1997) Mortality causes in British Barn owls (*Tyto alba*), based on 1,101 carcasses examined during 1963–1996. In: Duncan JR, Johnson DH, Nicholls TH (eds) Biology and conservation of owls of the northern hemisphere. 2nd international symposium, GTR NC-190, US Department of Agriculture, Forest Service, Saint Paul, pp 299–307
Nores M (1987) Zonas ornitogeográficas de Argentina. In: Narosky T, Yzurieta D (eds) Guía para la identificación de las aves de Argentina y Uruguay. Vázquez Mazzini, Buenos Aires, pp 295–303
Olrog CC (1979) Nueva lista de la avifauna argentina. Opera Lilloana 27:1–323
Omland K, Martínez D, Gaillard D et al (2001) Conservación de depredadores: áreas extensas y bosques antiguos. In: Primack R, Rozzi R, Feinsinger P et al (eds) Fundamentos de conservación biológica. Perspectivas latinoamericanas. Fondo de Cultura Económica, México, pp 459–461
Pardiñas UFJ, Cirignoli S (2002) Bibliografía comentada sobre los análisis de egagrópilas de aves rapaces en Argentina. Ornitol Neotrop 13:31–59
Pautasso AA (2006) Dieta del lechuzón orejudo (*Asio clamator*) en el centro y este de la provincia de Santa Fe, Argentina. Ornitol Neotrop 17:289–293
Pedrana J, Isacch JP, Bó MS (2008) Habitat relationships of diurnal raptors at local and landscape scales in southern temperate grasslands of Argentina. Emu 108:301–310
Pérez Pardo O (2006) La desertificación en la República Argentina. In: Brown A, Martínez Ortiz U, Acerbi M et al (eds) La Situación Ambiental Argentina 2005. Fundación Vida Silvestre Argentina, Buenos Aires, pp 433–435
Prémoli AC, Aizen MA, Kitzberger T et al (2006) Situación Ambiental de los Bosques Patagónicos. In: Brown A, Martínez Ortiz U, Acerbi M et al (eds) La Situación Ambiental Argentina 2005. Fundación Vida Silvestre Argentina, Buenos Aires, pp 281–291
Ramírez-Llorens P, Bellocq MI (2007) New records clarify the southern distribution of the Spectacled Owl (*Pulsatrix perspicillata*). J Raptor Res 41:268–276
Reca A, Úbeda C, Grigera D (1994) Conservación de la fauna de tetrápodos. I. Un índice para su evaluación. Mastozool Neotrop 1:17–28
Remsen JV, Areta JI, Cadena CD et al. (2017) A classification of the bird species of South America. American American Ornithologists' Union. http://www.museumlsu.edu/~Remsen/SACCBaseline.html/. Accessed 25 April 2017
Salvador SA (2012) Reproducción del lechuzón de campo (*Asio flammeus suinda*) en Cordoba, Argentina. Biologia 15:79–83
Sánchez KB, Malizia AI, Bó MS (2008) Trophic ecology of the burrowing owl (*Athene cunicularia*) in urban environments of Mar Chiquita Biosphere Reserve (Buenos Aires Province, Argentina). Ornitol Neotrop 19:71–80
Santillán M, Travaini A, Procopio DE et al (2010) New records indicate that Austral Pygmy Owl *Glaucidium nanum* breeds in eastern Patagonia. Cotinga 32:1–4
Sibley CG, Monroe BL Jr (1990) Distribution and taxonomy of birds of the world. Yale University Press, London
Straneck R, Vidoz F (1995) Sobre el estado taxonómico de *Strix rufipes* (King) y de *Strix chacoensis* (Cherrie y Reichenberger) (Aves: Strigidae). Nótulas Fauníst 74:1–5
Travaini A, Donázar JA, Ceballos O et al (1997) Food habits of common barn owls along an elevational gradient in Andean Argentine Patagonia. J Raptor Res 31:59–64

Traylor MA (1958) Variation in South American great horned owls. Auk 75:143–149
Trejo A (2006) Segregation by size at the individual prey level between barn owls and Magellanic horned owls in Argentina. J Raptor Res 40:168–172
Trejo A (2007) Identificación de especies y áreas prioritarias para el estudio de la reproducción de aves rapaces de Argentina. Hornero 22(Número Especial Aves Rapaces):85–96
Trejo A, Guthmann N (2003) Selectivity of owls on size and sex classes of rodents: activity and microhabitat use of prey. J Mammal 84:652–658
Trejo A, Lambertucci S (2007) Feeding habits of Barn Owls along a vegetative gradient in northern Patagonia. J Raptor Res 41:277–287
Trejo A, Seijas S (2003) Una estimación de aves muertas en ruta en el Parque Nacional Nahuel Huapi, noroeste de la Patagonia argentina. Hornero 18:97–101
Trejo A, Guthmann N, Lozada M (2005a) Seasonal selectivity of Magellanic Horned Owl (*Bubo magellanicus*) on rodents. Eur J Wildlife Res 51:185–190
Trejo A, Kun M, Sahores M et al (2005b) Diet overlap and prey size of two owls in the forest-steppe ecotone of southern Argentina. Ornitol Neotrop 16:539–546
Trejo A, Beaudoin F, Ojeda V (2011) Response of Rufous-legged Owls to broadcast of conspecific calls in southern temperate forest of Argentina. J Raptor Res 45:267–271
Udrizar Sauthier DE, Andrade A, Pardiñas UFJ (2005) Predation of small mammals by Rufous-legged Owl, Barn Owl, and Magellanic Horned Owl in Argentinean Patagonia forests. J Raptor Res 39:163–166
Viglizzo EF, Frank FC, Carreño L (2006) Situación ambiental en las ecorregiones Pampa y Campos y Malezales. In: Brown A, Martínez Ortiz U, Acerbi M et al (eds) La Situación Ambiental Argentina 2005. Fundación Vida Silvestre Argentina, Buenos Aires, pp 263–269
Wink M, Heidrich P (1999) Molecular evolution and systematic of the owls (Strigiformes). In: König C, Weick F, Owls BJH (eds) A guide to the owls of the world. Yale University Press, New Haven/London, pp 39–57

Chapter 4
The Owls of Belize

H. Lee Jones and Jan C. Meerman

Abstract Belize is one of the smallest countries in Central America and has one of the lowest human population densities of any country in the world. More than 40% of its land is in protected open space. Of its eleven resident owl species, *Ciccaba virgata*, a woodland species, is the most common and widespread, followed roughly in order by *Megascops guatemalae* of woodland habitats; *Tyto alba*, an inhabitant of open areas; and the more locally distributed *Glaucidium brasilianum* of open woodlands and edges. *Pulsatrix perspicillata*, *C. nigrolineata*, *G. griseiceps*, and *Lophostrix cristata* are confined largely to mature broadleaf forests in the interior. *Asio stygius* is confined to pine woodlands; *Bubo virginianus* to a variety of habitats on the Ambergris peninsula, and along the northern coastal strip where it is rare; and *Pseudoscops clamator* to open meadows and savannas on the coastal plain south of Belize City. *Athene cunicularia* and *Asio flammeus* have been recorded in Belize as vagrants. The percentage of each owl species' distributional range that lies within designated protected areas is examined, and the extent to which these lands are managed for protection of their natural resources is explored. The most serious current anthropogenic threats to owls and their habitats are discussed, as are potential impacts on owls from global climate change. In light of these documented and perceived threats, *A. stygius* and *B. virginianus mayensis* are the most vulnerable, the former from potential forest fires, bark beetle infestations, timber extraction, and land clearing for milpas, and the latter because of its very limited distribution in Belize, its small world population, and uncertainties about the viability of its source population in the Yucatan Peninsula. Protection of owl habitats through improved management and patrolling of the country's vast network of protected areas, better enforcement of its environmental laws and regulations, and expanded environmental education programs appears at this time to be the most effective conservation strategies for owls in Belize.

Keywords Owl • Belize • Conservation • Deforestation

H.L. Jones (✉)
P.O. Box 158, Punta Gorda, Belize
e-mail: bzbirdman@gmail.com

J.C. Meerman
P.O. Box 208, Belmopan, Belize
e-mail: meerman@biological-diversity.info

P.L. Enríquez (ed.), *Neotropical Owls*, DOI 10.1007/978-3-319-57108-9_4

Black and White Owl (*Ciccaba nigrolineata*)

4.1 Introduction

At 22,960 km^2, Belize is the second smallest country in mainland Latin America. With only 12 people per km^2, its population density is the lowest in Central America and among the lowest in the world (United Nations 2008). However, its annual population growth rate, at 2.7%, is the highest in Central America and one of the highest in the Western Hemisphere. In light of this dynamic, while Belize has 78 terrestrial protected areas (including archeological reserves, extractive reserves, and private reserves), giving 42% of its land some form of protected status (Meerman and Wilson 2005), its rapidly expanding population will be placing greater and greater demands on these presently secured natural resources in the coming decades.

Of the seven countries that comprise Central America, Belize is the only one without a Pacific coastline and one of only two that does not border on both the Pacific and Caribbean oceans. It is the only country in Central America without a true cloud forest. A few montane and semi-montane bird species, however, such as *Chlorospingus ophthalmicus* (common chlorospingus) and *Xiphorhynchus erythropygius* (spotted woodcreeper), occur on Belize's highest peaks, such as Doyle's Delight (16° 29′ 40″ N, 89° 02′ 43″ W) and Victoria Peak (16° 48′ 50″ N, 88° 36′ 27″ W), at elevations of 1124 and 1120 meters above mean sea level (amsl).

4.2 Taxonomy Diversity

The nomenclature in this chapter follows that of the American Ornithologists' Union (1998) and its annual supplements through 2016. At the end of 2009, Belize's avifaunal list stood at approximately 584 species (Jones 2003a, b, 2004a, b, Jones and Komar 2005, 2007, 2008, 2010), 13 of which are owls. Eleven owls are resident and two have been recorded as vagrants (Jones 2003a).

The first comprehensive treatment of the avifauna of Belize (Russell 1964) listed ten species of owls. Twenty-two years later, Wood, Leberman, and Weyer (1986) also listed ten species; however, they included *Pseudoscops clamator* (striped owl), which was not known to occur in Belize prior to the 1980s (and not adequately documented until after Wood, Leberman, and Weyer's 1986 list was published), and did not include *Athene cunicularia* (burrowing owl), which Russell had accepted based on two apparent specimens that have never been located. Howell and Webb (1995) accepted 11 species, adding *Lophostrix cristata* (crested owl), a species that was not documented in Belize until 1990, and retaining *P. clamator*, but rejecting *A. cunicularia*. Miller and Miller (1998, 2000), on the other hand, included *A. cunicularia* (presumably based on a convincing sight record in 1998), as well as *L. cristata* and *P. clamator*, bringing the total to 12 species. With the 1999 salvaging of a partial specimen of *Asio flammeus* (short-eared owl), Jones and Vallely (2001) and Jones (2003a) recognized 13 species, the currently accepted number.

Three species, *Tyto alba* (barn owl), *Megascops guatemalae* (vermiculated screech owl), and *Ciccaba virgata* (mottled owl), are widespread, with the first found primarily in open areas and the latter two restricted to wooded areas. Three species, *L. cristata*, *Pulsatrix perspicillata* (spectacled owl), and *P. clamator*, are restricted to roughly the southern half of the country. *Glaucidium griseiceps* (central American pygmy owl) is confined to the western half of the country, *Bubo virginianus* (great horned owl) to the northeast, and *Asio stygius* (stygian owl) to the central and coastal pine woodlands. *Glaucidium brasilianum* (ferruginous pygmy owl) and *Ciccaba nigrolineata* (black-and-white owl) have more complex distributions, and much remains to be learned about their distributional limits in the country. No owls have been recorded on the true cayes, although two species, *B. virginianus* and *C. virgata*, are found on Ambergris Caye, a long peninsula separated from the mainland only by a narrow canal that was constructed across its base by the Maya about 1500 years ago. Two species have been recorded as vagrants, *A. cunicularia* on three occasions and *A. flammeus* once (Jones et al. 2000; Appendix 4.1).

4.3 Distribution and Habitat Association

In most of Belize, *C. virgata* is the most frequently encountered owl. It is found in virtually all wooded habitats and elevations on the mainland throughout the country. *Megascops guatemalae* is equally widely distributed and found in most wooded habitats, but it is distinctly less common. Within its somewhat limited range in Belize, *G. brasilianum* is both conspicuous and, for an owl, rather common in many areas (e.g., Corozal District and portions of Orange Walk and Cayo districts). Its apparent numbers, though, are likely skewed upward in relation to the first two because it is highly vocal, it is largely diurnal, and it is found in open, as well as wooded, habitats (Jones 2003a).

Tyto alba is conspicuous but not common in human-inhabited areas and is less conspicuous in sparsely uninhabited areas such as abandoned quarry sites and savannas. *P. clamator* is locally relatively common in savannas and clearings in the coastal plain south of Belize City, but it is often overlooked, perhaps because of its simple, relatively nondescript vocalizations. The same may be true of *A. stygius* to some extent; however, it appears to be genuinely scarce, or even absent, in many areas of suitable habitat.

Bubo virginianus is one of Belize's most conspicuous owls but also the rarest of the 11 resident species. It is highly vocal and to some extent crepuscular; however, it is confined to Ambergris peninsula and a narrow strip of coastline in the northern half of the country where it is very local and seldom reported.

The remaining four resident species are forest dwellers. They are generally absent from disturbed or highly fragmented forests and regenerating forests with extensive second growth. *Ciccaba nigrolineata* may be found in open woodland, even partially cleared forests, as long as an ample supply of large, mature trees are present (Howell and Webb 1995, Jones unpublished data). *Pulsatrix perspicillata* is generally associated with the gallery forest and other woodland with large trees (König et al. 1999, Jones unpublished data). *Glaucidium griseiceps* and *L. cristata* are confined to the forest interior, and the latter is further confined to the Maya Mountains and associated foothills, much of which are accessible only on foot.

4.4 Vegetation Types

Based on a combination of remote sensing analysis and extensive fieldwork, 85 distinct vegetation types have been recognized in Belize (Meerman and Sabido 2001), but many of the finer divisions are not likely distinguished by most bird species and thus not relevant for the purposes of this review. Thus, attempting to mimic habitat classifications typically used in various Belize natural history literatures, we have merged many of these vegetation types into eight broad terrestrial ecosystems, or habitats, used by owls (Fig. 4.1):

4.4.1 Lowland Broadleaf Forest (<500 m amsl; 1,044,000 ha)

It includes a variety of broadleaf forest types including, in addition to high tropical moist forest, swamp forests and low, scrubby forests locally called bajos. Only one species, *C. nigrolineata*, is primarily confined to mature stands of this forest type.

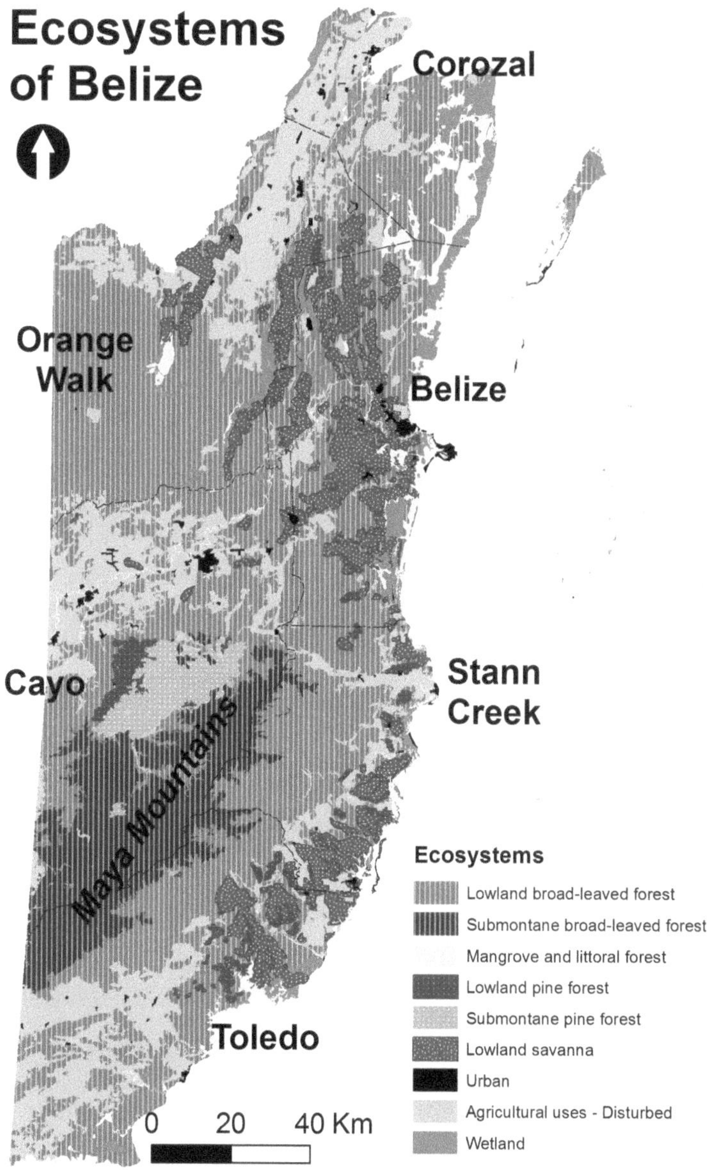

Fig. 4.1 Terrestrial ecosystems of Belize

Other species found here are *M. guatemalae*, *P. perspicillata*, *G. brasilianum*, *C. virgata*, and, locally, *G. griseiceps* and *L. cristata*.

4.4.2 Submontane Broadleaf Forest (>500 m amsl; 223,000 ha)

It includes all higher elevation forests up to Belize's maximum elevation of 1124 m amsl. No owl species are confined to this forest type, but *L. cristata* and *G. griseiceps* are found predominantly in the submontane broadleaf forest. Other species found in this vegetation zone are *M. guatemalae*, *P. perspicillata*, *G. brasilianum*, and *C. virgata*.

4.4.3 Lowland Savanna (194,000 ha)

It comprises a mosaic of open grassy terrain with herbaceous swamps and hammocks of broadleaf trees and pines. The distinction between this and the following is not always straightforward. Savannas with scattered trees are, to a limited extent, the habitat of *A. stygius* and *P. clamator*. *Ciccaba virgata* occurs along the edges of savannas where they merge with broadleaf forest.

4.4.4 Lowland Pine Forest (<500 m amsl; 28,000 ha)

It is essentially a savanna area with a more or less closed stand of *Pinus caribaea* (Caribbean pine). The pine forest, both lowland and submontane, is the home of the much sought-after *A. stygius* in Belize. *C. virgata* in the north and *G. brasilianum* occur along the edges of pine forests where they merge with the broadleaf forest.

4.4.5 Submontane Pine Forest (>500 m amsl; 47,000 ha)

It comprises a mixture of upland savanna-like habitats and relatively closed stands of *P. caribaea* and *P. patula* var. *tecunumanii* that define the region known as the Mountain Pine Ridge. It includes extensive silvicultural stands of these species, but also scattered natural stands. *A. stygius* is most frequently encountered in the submontane pine forest. Although not confined to pines, *G. brasilianum* is the most conspicuous owl in the Mountain Pine Ridge. *M. guatemalae* is found in the pine forest, at least in the Mountain Pine Ridge, where it is relatively scarce and may be confined to its fringes near broadleaf forest.

4.4.6 Mangrove and Littoral Forests (80,000 ha)

They include mangrove swamps as well as broadleaf forests on young marine deposits (littoral forest). Although quite different, these forest types usually occur in close proximity. Littoral forests, especially, are under heavy pressure, as they are the prime real estate for tourism-related development. In Belize, *B. virginianus* is largely confined to coastal areas with the littoral forest and mixed mangrove stands, primarily on the Ambergris peninsula, but it is also found in the bajo-type lowland broadleaf forest and in suburban areas, especially around San Pedro.

4.4.7 Disturbed Rural Landscapes (456,000 ha)

They include sparsely developed areas with scattered trees, various stages and forms of second-growth scrub and early-stage regenerating forests, and agricultural areas (cropland, pastureland, and orchards). Although not man-altered, hurricane-devastated forests are also included in this category. *P. clamator* is confined to open areas with scattered trees. *T. alba* and *G. brasilianum* are also found in many open areas with relatively sparsely distributed trees, and the former, along with *P. clamator*, often forages in treeless areas. Also *M guatemalae*, *B. virginianus*, and *C. virgata* occur in man-altered habitat types that are recovering or transitional to an actual forest.

4.4.8 Urban and Suburban Landscapes (21,000 ha)

Tyto alba is an inhabitant of urban and suburban areas throughout the country. In the northern half of the country, especially in Corozal District and along the Western Highway corridor in Cayo District, *G. brasilianum* is frequently encountered, and in the northeast (and formerly in Belize City) *B. virginianus* can be found in urban settings.

4.5 Owl Conservation

While no conservation measures in Belize are directed specifically toward owls, the de facto conservation of their natural habitats is of direct benefit. As it is, most Belizean owls live in habitats and ecosystems that are at least for now sufficiently protected from the most serious anthropogenic disturbances. For example, of the range-restricted *L. cristata*, no less than 91% of its Belize population lies within currently protected areas (PAs) (Table 4.1). The species that has the smallest

Table 4.1 Representation of owl species within protected areas (PAs) in Belize

Species	Estimated distributional range in the country (ha)	Within PAs (ha)	Proportion within PAs (%)
Tyto alba	799,200	48,600	6
Megascops guatemalae	2,184,200	791,500	36
Lophostrix cristata	469,600	425,500	91
Bubo virginianus	81,000	8900	11
Pulsatrix perspicillata	966,800	620,700	64
Glaucidium brasilianum	1,336,000	500,000	37
Glaucidium griseiceps	1,093,100	644,900	59
Ciccaba virgata	2,193,500	796,800	36
Ciccaba nigrolineata	1,234,400	687,900	56
Asio stygius	388,700	126,300	32
Pseudoscops clamator	534,400	135,200	25

proportion of its range within PAs is *T. alba* with only about 6% of its assumed range included. But this is a species that is largely associated with human-modified habitats. More worrisome is *B. virginianus*, which has only about 11% of its very small range within existing PAs. Widespread species *M. guatemalae* and *C. virgata* have nearly two-thirds of their populations outside the PAs and thus remain vulnerable to substantial population declines, even if illegal forest destruction can be curtailed in PAs like the Columbia River Forest Reserve (see below).

BirdLife International (www.birdlife.org) recognizes six important bird areas (IBAs) in Belize, which collectively cover most of the country (Table 4.2). The IBA with the most owl species is the Maya Mountains and Southern Reserves with ten. The only IBA lacking owls is the "offshore and barrier islands" IBA; otherwise, all owl species are represented within multiple IBAs.

Similarly, Conservation International(www.conservation.org) recognizes a number of key biodiversity areas (KBAs). These are a subset of PAs that have the highest concentrations of IUCN-listed species. Eight KBAs have been identified for Belize (Meerman 2007, Fig. 4.2), each of which includes owls (however, no owl species in Belize are IUCN listed). With the notable exception of *B. virginianus*, which is not represented, all owl species are represented in multiple KBAs. Maya Mountains, north, and Peccary Hills showed high owl diversity with nine species each (Table 4.3).

4.6 Threats

Because much of the country's land is protected in forest reserves, archeological reserves, national parks, wildlife sanctuaries, and the like, most owl species and their habitats are relatively free of major anthropocentric threats such as deforestation and poaching, at least in theory. Probably the most important direct threat is

Table 4.2 Important bird areas (IBAs) in Belize with owl species represented

Species	Coastal and inland islands	Crooked Tree and associated wetlands	Maya Mountains and Southern Reserves	Northeastern Belize	Rio Bravo and Gallon Jug	Offshore and barrier islands
Tyto alba	x	x	x	x	x	
Megascops guatemalae	x	x	x	x	x	
Lophostrix cristata			x			
Pulsatrix perspicillata		x	x		x	
Bubo virginianus	x			x		
Glaucidium griseiceps		x	x		x	
Glaucidium brasilianum		x	x	x	x	
Ciccaba virgata	x	x	x	x	x	
Ciccaba nigrolineata	x	x	x		x	
Asio stygius	x	x	x		x	
Pseudoscops clamator	x		x			

persecution out of fear or because of the belief that owls will prey on chickens. Such persecution will typically take place in and around human settlements. Protected areas that in Belize are free of human settlements, thus, provide a direct protection to owls. Outside of PAs, and even within some PAs such as forest reserves, owls are vulnerable to habitat loss from logging and other forms of deforestation (Table 4.4), often the predecessor of human settlements and agriculture. With the former comes increasing hunting pressure, and with the latter, pesticide use. Levels of deforestation in Belize are very low in comparison with neighboring Mesoamerican countries (Meerman et al. 2010).

Ironically, barn owls which thrive in human-modified habitats have also the highest level of negative human conflicts. Persistently, barn owls are considered to be evil and harbingers of misery (Matola 2012) and consequently prosecuted.

Many PAs, especially those in more remote areas, as along the western border, lack sufficient manpower for their management and protection. Laws protecting them are all but unenforceable. Illegal encroachment into Belize's vast forest reserves and national parks in western Toledo and Cayo districts remains virtually unchecked; based on 2007 data (Meerman unpublished data), a total of 85,600 ha of virgin forest in PAs along the Guatemalan border has been cut to make way for agricultural expansion. Illegal extraction of forest resources such as mahogany (*Swietenia macrophylla*), the ornamental Xaté palm (*Chamaedorea ernesti-augusti*), and game meat penetrates even farther into Belize's PAs. The intruders are often

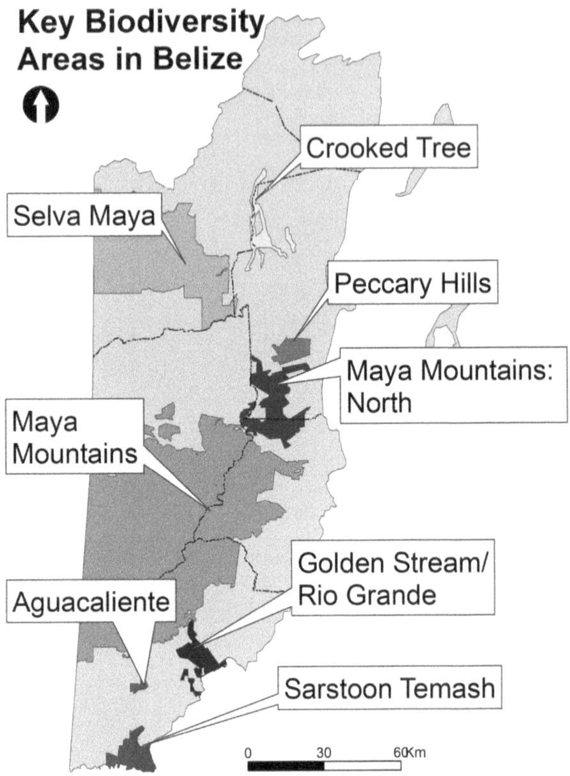

Fig. 4.2 Key biodiversity areas in Belize

armed and have fired upon unarmed NGO personnel charged with managing and protecting these areas. The situation is further complicated by the difficulty in accessing these areas from the Belizean side of the border, which has few roads and rugged mountain terrain.

As is becoming increasingly apparent, global climate change is posed to have enormous impacts on habitats and species (e.g., Karl and Trenberth 2003; Intergovernmental Panel on Climate Change 2007a, b), and owls will be no exception (Table 4.4). Habitat alteration over time is likely to be the most important primary result of climate change. In the case of Belize, predicted rise in temperature, more erratic rainfall regimes, and increasing frequency and severity of hurricanes all have the potential to radically alter forest ecosystems (Anderson et al. 2008). Secondary effects could include disrupted plant reproductive cycles and the increased threat of forest fires. Frequent forest fires will ultimately lead to a replacement of forests with savanna ecosystems. Consequently, forest owls (*M. guatemalae, L. cristata, P. perspicillata, G. griseiceps, C. virgata, A. stygius*, and *C. nigrolineata*) can be expected to decline, while species adapted to open areas (*T. alba, P. clamator*) can be expected to benefit in some way (Table 4.4).

Table 4.3 Key biodiversity areas (KBAs) in Belize with owl species represented

	Agua caliente	Crooked Tree, south	Golden Steam/ Rio Grande	Maya Mountains	Maya Mountains, north	Peccary Hills	Sarstoon/Temash	Selva Maya
Tyto alba	x	x	x			x		x
Megascops guatemalae	x	x	x	x	x	x	x	x
Lophostrix cristata				x	x			
Pulsatrix perspicillata				x	x	x		x
Bubo virginianus								
Glaucidium griseiceps				x	x	x		x
Glaucidium brasilianum				x	x	x		x
Ciccaba virgata	x	x	x	x	x	x	x	x
Ciccaba nigrolineata	x		x	x	x	x	x	x
Asio stygius		x		x	x	x		x
Pseudoscops clamator	x		x		x	x	x	

Table 4.4 Perceived adverse and beneficial impacts on owls from anthropogenic causes

	T. alba	*M. guatemalae*	*L. cristata*	*P. perspicillata*	*B. virginianus*	*G. griseiceps*	*G. brasilianum*	*C. virgata*	*C. nigrolineata*	*A. stygius*	*P. clamator*
I. Deforestation											
(a) Clear-cut logging		–	–	–		–		–	–	–	
(b) Settlements	+	–	–	–		–		–	–	–	+
(c) Agriculture, ranching	+	–	–	–		–	–	–	–	–	+
II. Insufficient management											
(a) Poaching											
(b) Persecution	–	–	–	–	–	–	–	–	–	–	–
(c) Illegal resource extraction			–	–		–			–		
III. Pests and Pesticides											
(a) Bark beetle infestations										–	
(b) Pesticide use	–	?			?		–				–
IV. Climate change											
(a) Hurricanes		–	–	–	–		–	–	–	–	
(b) Disrupted reproduction	–	–	–	–	–	–	–	–	–	–	–
(c) Forest fires	+	–	–	–		–		–	–	–	+
(d) Floods, erosion			–	–							

Code: + beneficial impact; – negative impact; ? unknown effect

Neither Miller and Miller (1997) nor Jones and Vallely (2001) listed any owl species of concern in their respective reports on birds of conservation concern. While there is no official red data list for Belize, in a draft list of "Critical Species" prepared in 2005, *Asio stygius* was listed as vulnerable (Meerman 2005). The Belize population occupies lowland and foothill areas and is essentially isolated from other populations in Central America, which are confined to higher elevations (König et al. 1999); therefore, there may be no effective source population to sustain the local populations in case of a population crash due to causes such as the bark beetle infestation and forest fires or deforestation for timber, agriculture, and settlements.

The subspecies *B. virginianus mayensis* is confined to the Yucatan Peninsula and reaches its southernmost distributional limit in northeastern Belize (Bangs and Peck 1908, Webster and Orr 1958). The reasons for its apparent scarcity in the country away from the Ambergris peninsula are unknown. Although anecdotal evidence indicates a decline in this species on mainland Belize in the past half century (see Species Accounts), the incomplete historical record makes it almost impossible to document a steady, or even significant, decline in numbers historically. Without sufficient data on population numbers to the north in Quintana Roo and Campeche in Mexico, it is also impossible to know if there is an adequate source population to sustain the Belize population.

Lophostrix cristata, *P. perspicillata*, *G. griseiceps*, and *C. nigrolineata* are all relatively uncommon and more or less restricted to primary forest, with population centers in Belize within the Maya Mountains and Vaca Plateau, both currently under threat from illegal encroachment by Guatemalan loggers and ranchers.

Two other species, *T. alba* and *M. guatemalae*, are found throughout the country in appropriate habitat, and although relatively uncommon, their populations appear to be stable based on empirical evidence. *P. clamator* is found in open and disturbed habitats but is restricted to the southern third of the country. Nevertheless, threats to this species are also seen as minimal. *G. brasilianum* and *C. virgata* are widespread and, for owls, common and thus are of least concern, at least in the short term.

4.7 Conservation Strategies

Since most owls are not targeted by poachers and the superstitious in Belize as they are in some neighboring countries, the most effective conservation strategy in Belize would appear to be the protection of their habitats through better management and enforcement within the network of PAs that collectively cover 42% of the country's terrestrial territory. Approximately 78% of Belize is still covered with relatively undisturbed or infrequently disturbed ecosystems. Belize has an adequate legal framework for protection of owls and other birds but lacks sufficient manpower and funding for enforcement of existing laws. These PAs vary from forest reserves, which are actually extractive reserves maintained and managed for the extraction of timber resources, to nature reserves, which provide the strictest level of protection. In addition, there are a number of private PAs, the largest of which is the 105,000-ha

Rio Bravo Conservation and Management Area. Although all non-private terrestrial PAs are technically managed by the Forest Department within the Ministry of Natural Resources, a number of comanagement agencies conduct the actual on-the-ground management.

Best known among these is the Belize Audubon Society, which manages ten PAs. Like most comanagement agencies, the Audubon Society is also responsible for raising the funds required for management. Another important in-country funding source is the Protected Areas Conservation Trust (PACT), which charges a small conservation fee to every tourist leaving the country.

Despite the legal framework established to protect its natural resources, with its small population (= tax base) and limited sources of revenue, the government lacks adequate funds to sufficiently manage and patrol its vast system of PAs. Thus, many PAs are effectively unmanaged or suffer from severe understaffing. It also does not have the military capacity to adequately patrol and secure its western border or the law enforcement capacity to effectively enforce its environmental laws and regulations designed to protect its natural resources from exploitation, even by its own government. The government has repeatedly circumvented its own environmental review process and ignored court-ordered mandates to comply with well-established environmental law in pushing through its most lucrative projects, some of which have been deemed highly environmentally destructive (see Barcott 2008).

But, while the rapidly growing population is putting the PAs under increasing pressure and the government is sometimes less than cooperative, the present situation in Belize is still favorable compared with its Central American neighbors where PA resources are under much greater threat of exploitation. Actually, as discussed above, one of the largest and most severe pressures on Belize's PA system comes from across the border in Guatemala, from where rapid population expansion and unplanned development are spilling over into Belize.

In addition to protection of forestlands, environmental education plays an important role in the conservation of both habitats and species. Examples of organizations that are particularly active in this field are the Belize Audubon Society, Belize Zoo and Tropical Education Center, Birds Without Borders/Aves Sin Fronteras, and Belize Foundation for Research and Environmental Education (BFREE). All these have programs designed to educate landowners, school children, or the public at large in the protection and conservation of Belize's diverse natural resources. Specific to owls, the Belize Zoo has published a series of children's books on conservation and protection of Belize's natural heritage in which an owl (*Pulsatrix perspicillata*) is the central character (Matola 1988, 1993, 2000). The Belize Zoo also uses an orphaned barn owl as an "ambassador" for its species, visiting schools and demonstrating it in media outreach presentations (Matola 2012).

The Belize Audubon Society (www.belizeaudubon.org), founded in 1969, has been instrumental in protecting the country's natural resources while educating the public about their value and sustainable use. Its environmental education activities are closely tied to the ten PAs that it manages.

The Belize Zoo's (www.belizezoo.org) education department was established in 1986 with a commitment to heighten the awareness and increase the participation of all Belize citizens in the natural world, to provide them with a greater understanding of the country's unique ecosystems and diverse wildlife, and to instill in them the positive attitudes and valuable skills that will ultimately aid in the preservation of Belize's natural resources. More than 15,000 schoolchildren visit the Belize Zoo each year, all of whom receive environmental education on-site.

Birds Without Borders/Aves Sin Fronteras (www.zoosociety.org/Conservation/BWB-ASF), in addition to its conservation research programs, recently published two booklets for landowners with recommendations on how to manage their land to help birds, one edition for Belizeans (Piaskowksi et al. 2006) and another for landowners in the USA (Piaskowski et al. 2008). It also regularly posts educational material on its website.

BFREE (www.bfreebz.org), located in a forest setting in northern Toledo District, offers watershed ecology programs of various lengths and levels of comprehension that are designed to provide participants with an enjoyable, hands-on educational experience. Many more NGOs are active in environmental education on a local level.

4.8 Species Accounts

Tyto alba pratincola **(Bonaparte 1838) Barn Owl**

Tyto alba is an uncommon to locally fairly common resident on the mainland (Jones 2003a). It is one of the most widespread owl species away from forested land, being most prevalent in urban, suburban, and agricultural areas, meadows, marshland, savannas, and around quarry sites.

Deforestation may have benefited this open country species. But while cleared forested areas for agriculture have provided additional foraging habitat for the species, pesticide use associated with most agriculture may have negated any benefits accrued from habitat expansion. Urbanization has also benefited the species through the provision of artificial nest sites and high rodent prey populations. The most serious threat to this species appears to be superstition among the local population which considers barn owls harbingers of evil and misery. As a result, urban barn owls are routinely killed. The Belize Zoo has embarked on an education campaign to change this superstition (Fig. 4.3).

Megascops guatemalae **(Sharpe 1875a) Vermiculated Screech Owl**

Two color morphs have been documented in Belize, the common gray morph and a rufous morph that has been observed on at least two occasions in the Mountain Pine Ridge (J. Hortsmann 2004 photograph, *fide* L. Jones; R. Martinez 2009 pers. comm.).

M. guatemalae is found throughout mainland Belize in forested areas and forest edges, including recovering second growth with young trees. While found primarily in broadleaf forest, it also has been found in pine woodlands, albeit less commonly (Jones unpublished data; R. Phillips pers. comm.). It occurs to at least 1100 m elevation, as at Little Quartz Ridge in western Toledo District (Jones and Gardner 1997).

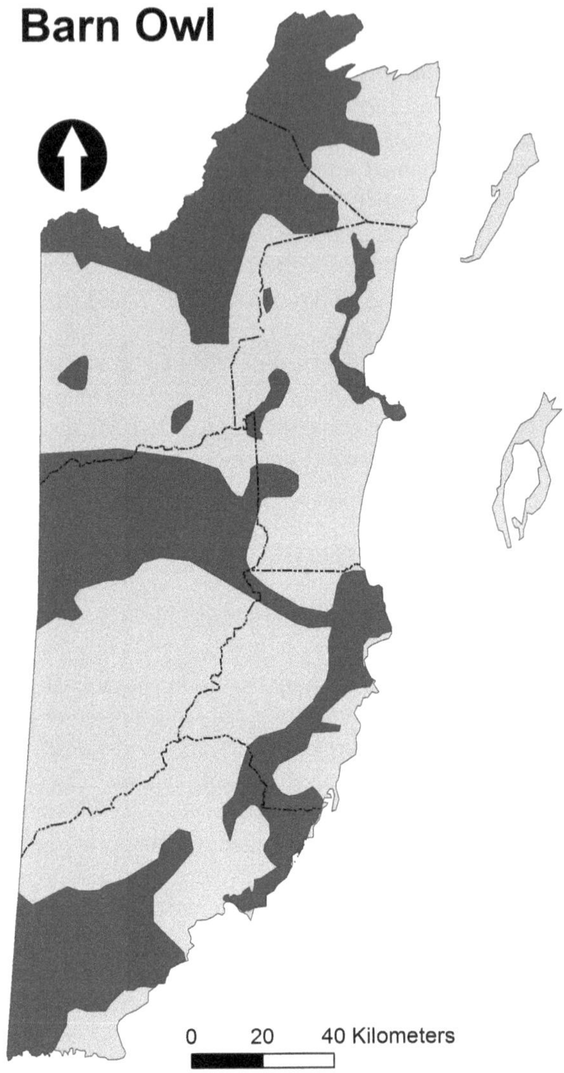

Fig. 4.3 Distribution of Barn owl

In many areas, it is the second most common species. It is absent from the cayes and Ambergris peninsula.

Forest destruction is perhaps the greatest threat to this species in Belize; however, with 36% of its assumed range within PAs, this is not seen as a serious threat at this time (Fig. 4.4).

Lophostrix cristata stricklandi **(Sclater Salvin 1859) Crested Owl**
Lophostrix cristata was not discovered in Belize until 1990 (Miller and Miller 1992). Since then, it has been found on several occasions in the submontane and

Fig. 4.4 Distribution of Vermiculated screech owl

lowland broadleaf forest in the southern part of the country (Parker et al. 1993; Jones and Gardner 1997). Because it is restricted to mature forests far from major roads, it has been largely overlooked and remains Belize's least familiar owl species. Until recently, it was thought to be confined to mid-elevations in the Maya Mountains (to at least 800 m), but a recent report from Five Blues Lake, southwestern Belize District, on 23 February and 4 March 2007 (P. and I. Jones pers. comm.), and a report from the Billy Barquedier National Park, northern Stann Creek District, on 21 May 2009 (B. Miller pers. comm.), extend its known range in Belize about 30 km to the northeast and its lower elevational limit to 81 amsl.

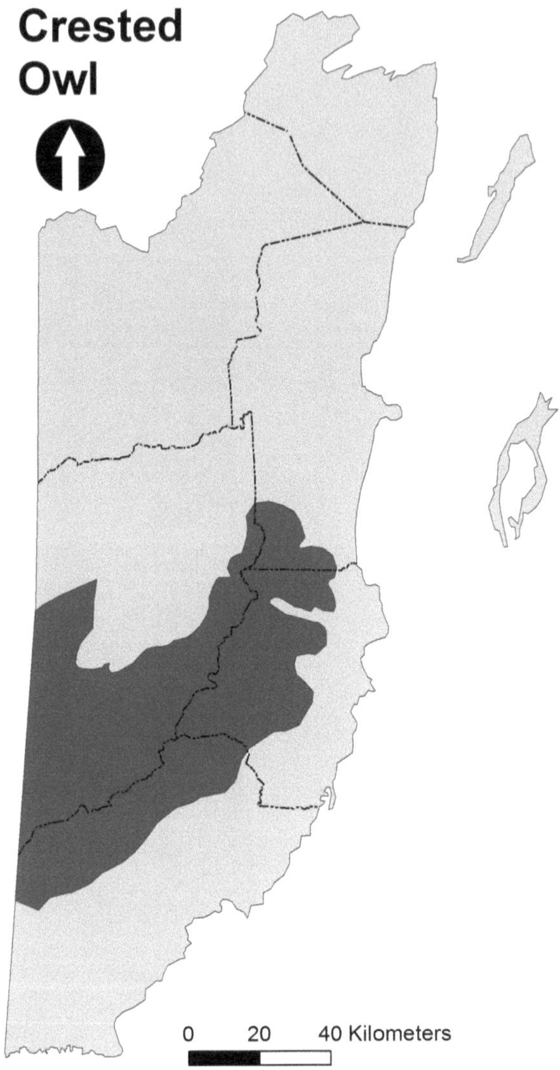

Fig. 4.5 Distribution of Crested owl

Logging in selected portions of the Maya Mountains poses a local, but expanding, threat to *L. cristata*; however, as long as the bulk of the forests in this largely inaccessible region remain intact, the threats to the species should remain minimal. The largest single threat at the moment may be the extensive encroachment into its habitat by loggers, ranchers, and settlers from across the western border in Guatemala (Fig. 4.5).

Pulsatrix perspicillata saturata **(Ridgway 1914) Spectacled Owl**

Pulsatrix perspicillata is an uncommon and local resident in the interior north to the Rio Bravo area of western Orange Walk District and along the Sibun River, east to near the Cayo-Belize District border. Like *L. cristata*, it is found in the dense

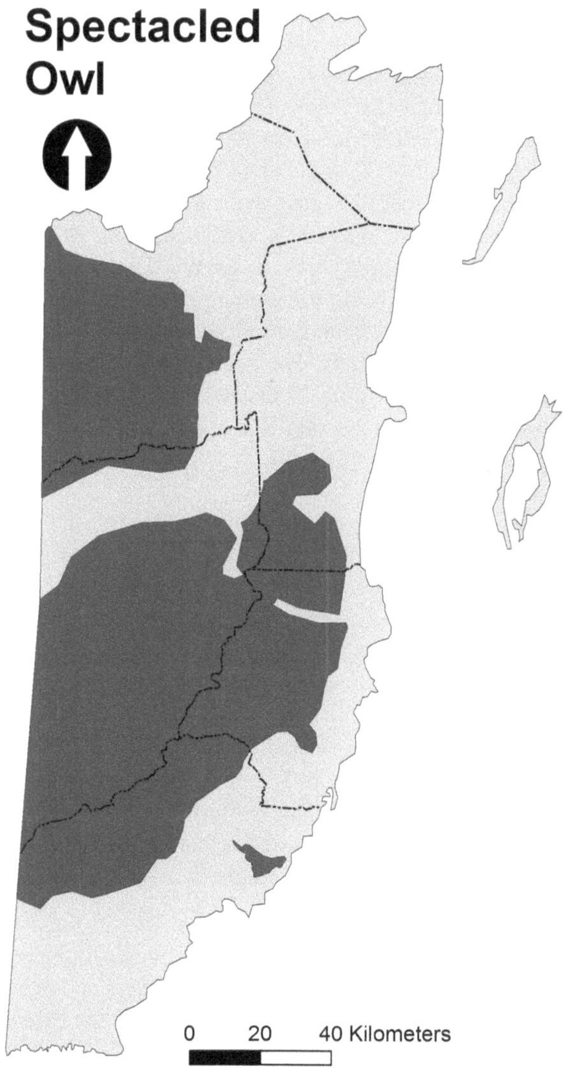

Fig. 4.6 Distribution of Spectacled owl

interior of mature forests but also frequents clearings with a partial cover of large trees, such as shady plantations, and in gallery forests to near sea level along rivers (Howell and Webb 1995; König et al. 1999).

As with other forest owls in Belize, the major threat to *P. perspicillata* is deforestation. It does not appear to be adversely impacted by selective clearing of land for resorts and ranches as long as a partial cover of large trees remains. However, selective harvesting of large trees such as mahogany may have a negative impact on this species, although the severity of this impact is not known (Fig. 4.6).

Bubo virginianus mayensis **(Gmelin 1788) Great Horned Owl**
Bubo virginianus is an uncommon to fairly common resident on the Ambergris peninsula; however, few records exist from elsewhere, and these are confined to the coastal strip from Sarteneja, Corozal District, to the vicinity of Manatee Lagoon in southern Belize District (Russell 1964), plus an undated sight record from the rice mill at Big Falls in Toledo District (M. Meadows pers. comm.), possibly of a stray. The historical record of this species in Belize away from the Ambergris peninsula is sparse. It was apparently fairly common in Belize City until perhaps three decades ago, but there are no recent records. According to R. Burgos (pers. comm.): "I do remember in the past seeing them often enough here in Belize City in the coconut and, in particular, an old date tree near where I presently live. I have not seen [it] in many years." The most recent record from the mainland was of two at the Sarteneja airstrip on 27 October 2014 (R. Martinez, pers. comm.). The record from near Manatee Lagoon was of a bird collected 20 May 1906 that still retained some of its downy plumage and thus was thought to have been reared locally (Russell 1964).

B. virginianus occupies both open and closed broadleaf forest, the fringes of towns, and for foraging, agricultural fields, pastures, orchards, and wetlands. There are no perceived threats to the species in Belize other than, perhaps, persecution by humans and pesticide use. The cause of its decline on the mainland is not known; however, populations of many species at the fringes of their range are often unstable and subject to periodic expansions and contractions, even local extinction and recolonization (Fig. 4.7).

B. v. mayensis is restricted to the Yucatan Peninsula where it is considered to be rare (Enriquez Rocha and Rangel-Salazar 1996; Semarnat 2008). As the range of *mayensis* extends only marginally into Belize, serious population declines in the Yucatan may have a profound effect on its status in Belize, especially if source populations immediately to the north in southern Quintana Roo are not sustained.

The Genus *Glaucidium*
In the past there has been some confusion as to the relative distribution and habitat preferences of the two pygmy owls in Belize and even as to which species occur in the country. This is in no small part due to the extensive plumage variation in *G. brasilianum*. For example, Walters (1993) claimed two records of *Glaucidium gnoma* banded and photographed in the hand on the edge of the Mountain Pine Ridge. Howell (1995), however, refuted Walters' claims based primarily on the improbability of their occurrence in Belize, and the senior author, having examined one of the photographs, concurs with Howell that the photographed bird was a brown morph *G. brasilianum*. Also, despite claims to the contrary, *G. griseiceps* is unrecorded from the coastal plain. Although Wood and Leberman (1987) salvaged a carcass of an immature *Glaucidium* in Belmopan on 23 March 1984 that they identified as *griseiceps* and mist-netted and photographed another *Glaucidium* identified by them as *griseiceps* "southeast of Mile 35 on the Western Highway," both of these were later determined to be *brasilianum* by Parkes (1995).

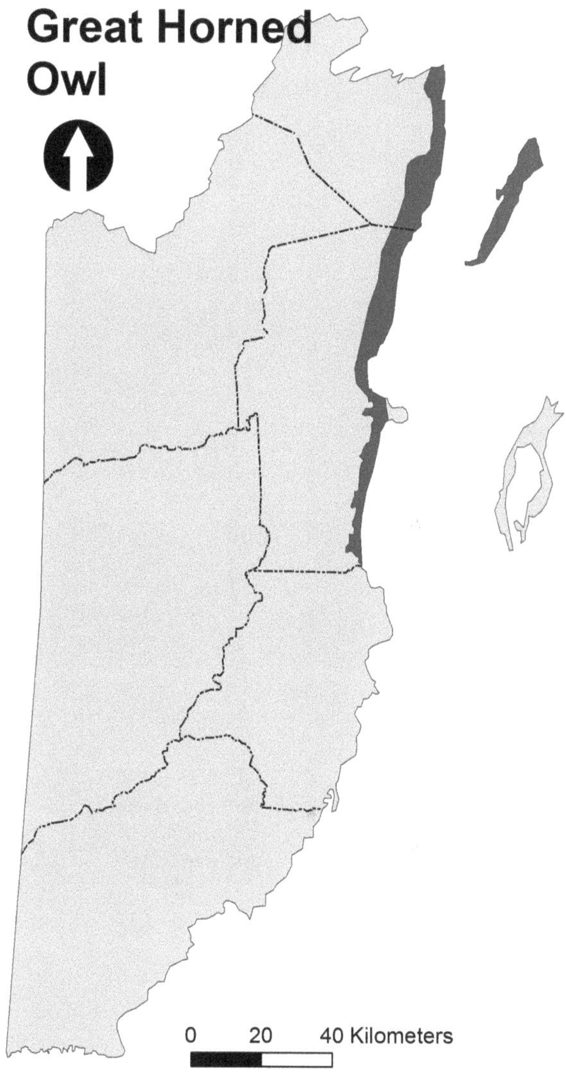

Fig. 4.7 Distribution of Great horned owl

Glaucidium griseiceps **(Sharpe 1875b) Central American Pygmy Owl**
Glaucidium griseiceps is an uncommon to locally fairly common resident of the mainland interior north to northern Cayo District and southwestern Belize District. It is also found in western Orange Walk District and locally in eastern Orange Walk District, as at Hill Bank (Vallely and Whitman 1997) and the Lamanai archeological site on the New River (England 2000).

G. griseiceps is found in the submontane broadleaf forest and locally in the lowland broadleaf forest, generally within the forest interior but also near clearings and the forest edge. Like other forest-dwelling owls in Belize, it is threatened by logging

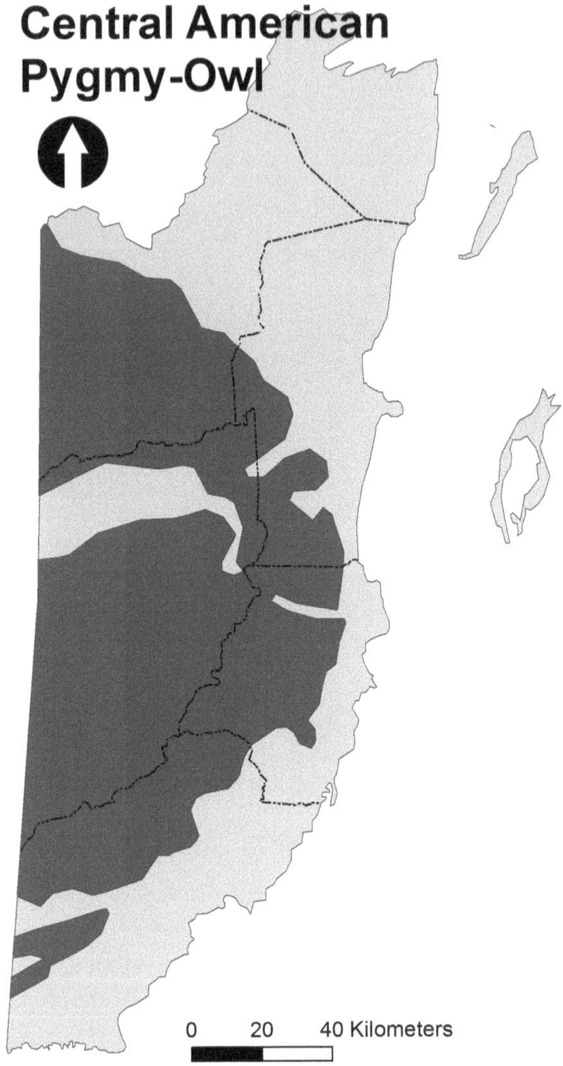

Fig. 4.8 Distribution of Central American pygmy owl

and encroaching civilization, but in most areas this threat has been minimal. As with *L. cristata*, the illegal encroachment of Guatemalan loggers, ranchers, and settlers into western Belize where some of Belize's finest examples of primary forest are found is having a local, but profound, impact on this species' habitat (Fig. 4.8).

Glaucidium brasilianum ridgwayi **(Sharpe 1875b) Ferruginous Pygmy Owl**
The distribution of *G. brasilianum* in Belize is complex. It is a relatively common resident in Corozal District south to northeastern and western Orange Walk District and most of Cayo District, especially the Mountain Pine Ridge east to west-central Belize District. It is less common in the Maya Mountains and foothills of western

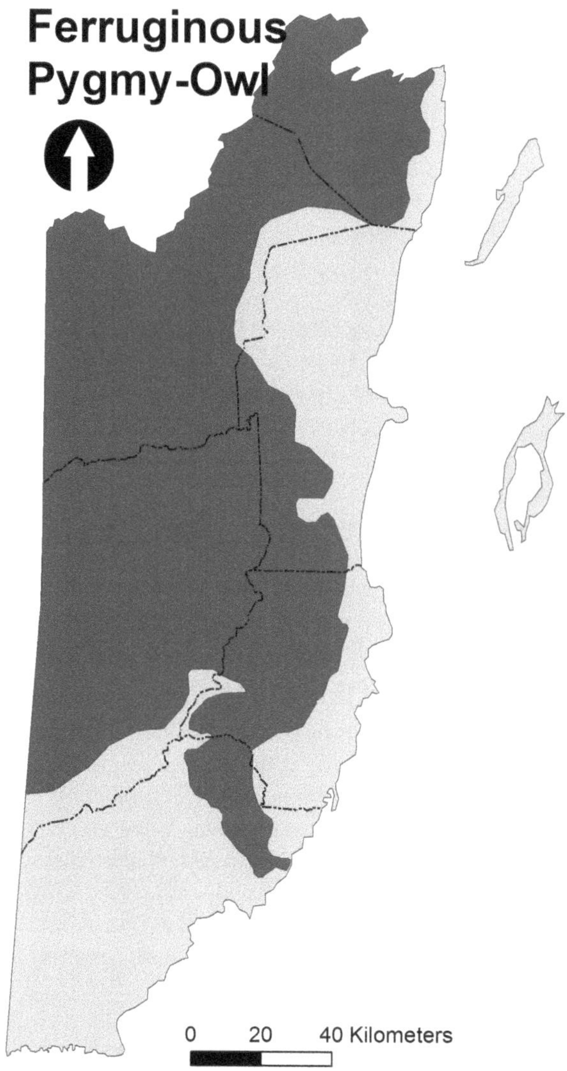

Fig. 4.9 Distribution of Ferruginous pygmy owl

Stann Creek District and northern Toledo District. A female collected on 9 March 1956 had enlarged gonads (Russell 1964).

G. brasilianum has a very broad ecological niche, occupying dense forest (typically forest edges) and thickets to open, semi-urban and urban settings with scattered trees. It is especially common in Corozal District and in the Mountain Pine Ridge where it frequents transitional areas between open pinelands and rainforests (Russell 1964). In the Maya Mountains and western Orange Walk District, its range and habitat overlap broadly with that of *G. griseiceps*. Because of its broad habitat preferences, including urban areas and its relative abundance in portions of Belize, threats to this species from habitat loss are believed to be minimal at this time (Fig. 4.9).

Athene cunicularia hypugaea **(Bonaparte 1825) Burrowing Owl**
Athene cunicularia has been recorded in Belize on three occasions. Peck collected two individuals on the beach at the mouth of the Manatee River, Belize District, in January 1901 (Russell 1964), but the disposition of the specimens is unknown. On 8 and 9 April 1998, Meerman observed and carefully documented a single *A. cunicularia* at a quarry site next to the main highway just outside Indian Creek Village, Toledo District (Jones et al. 2000). And on 27 March 2016 R. Martinez photographed one at Crooked Tree Wildlife Sanctuary (from eBird).

A. cunicularia is highly migratory in parts of its range and winters regularly south to southern Mexico (Howell and Webb 1995) and formerly to Guatemala (Eisermann and Avendaño 2015), mainly in the interior. It has been recorded as a vagrant in the Yucatán Peninsula (Paynter 1955; Storer 1961; Lopez Ornat et al. 1989; MacKinnon 1992), Honduras (Monroe 1968; Bonta and Anderson 2002; *contra* Howell and Webb 1995), Costa Rica, and Panama. Specimens taken in Costa Rica on 20 December 1900 (Slud 1964; Stiles and Skutch 1989) and Panama on 13 December 1900 (Wetmore 1968; Ridgely and Gwynne 1989), along with the two taken in Belize in January 1901, suggest a significant influx of migrants into Central America in the winter of 1900–1901.

The subspecies recorded in Belize is almost certainly the widespread migratory *A. c. hypugaea* from western North America and Mexico, which has been taken in Central America as far south as Honduras (Monroe 1968), Costa Rica (Slud 1964), and Panama (Wetmore 1968).

Ciccaba virgata centralis **(Griscom 1929) Mottled Owl**
Ciccaba virgata is a fairly common resident on the mainland, including the northern half of the Ambergris peninsula (Meadows 1994). It is the only owl other than *B. virginianus* that has been recorded on Ambergris Peninsula. In most areas, this is the most common owl in Belize, equaled or surpassed in numbers only by *G. brasilianum* in Corozal District and the Mountain Pine Ridge, and perhaps locally in northern Cayo Districts. Russell (1964) cited several instances of breeding or suspected breeding, including a nest with two eggs observed by Peck near Ycacos Lagoon, Toledo District, on 16 March 1907 (Fig. 4.10).

C. virgata is restricted to forested areas, although it is relatively common along forest edges and in most second growth with a moderate amount of young trees (Russell 1964, Jones 2003a). Like the other forest-dwelling owls in Belize, the mottled owl may be locally threatened by forest destruction, but because of its relative abundance and wide distribution in the country, such threats are seen as minimal at this time.

Ciccaba nigrolineata **(Sclater 1859) Black-and-White Owl**
Ciccaba nigrolineata is an uncommon resident in the foothills and adjacent lowlands of the Maya Mountains, including the Vaca Plateau, north to western, and locally, eastern Orange Walk District. It occurs south locally in the coastal plain to Toledo District. It has not been documented in Corozal or Belize Districts where suitable habitat is sparse.

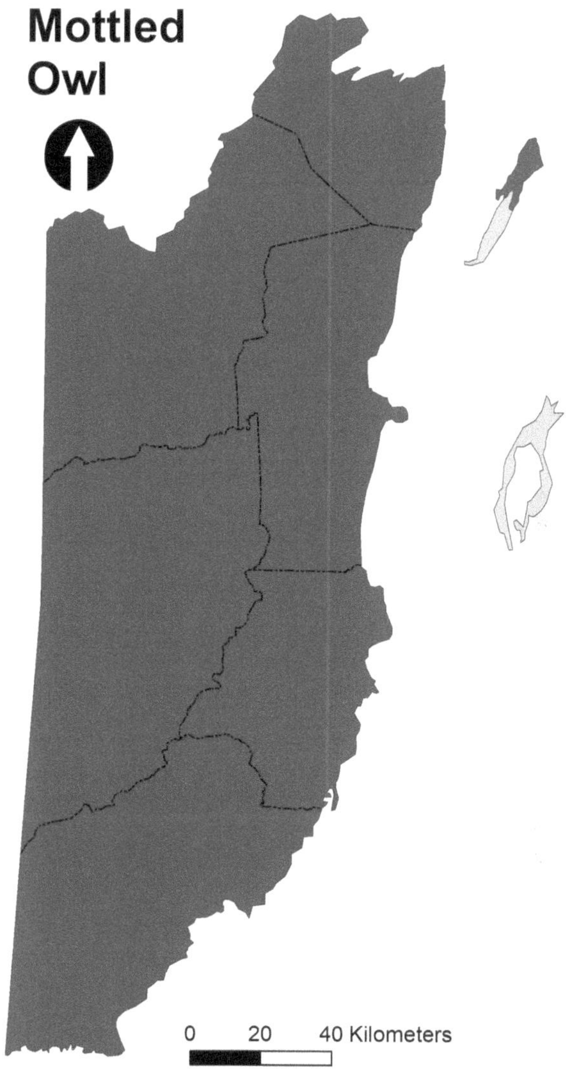

Fig. 4.10 Distribution of Mottled owl

This species is found in the mature lowland broadleaf forest, including the gallery forest, relatively open forested areas with scattered large trees, and forest clearings that retain a modest number of large trees such as *Ficus spp.* and *Ceiba pentandra. C. nigrolineata* has been observed hunting bats under artificial lights in a forested area near San Ignacio, Cayo District (Meerman pers. obs.), and on the outskirts of Orange Walk Town, Orange Walk District (J. Urbina pers. comm.).

Like other forest-dwelling owls in Belize, it is threatened locally by forest destruction. Impacts on this species are likely similar to those affecting *P. perspicillata*, as the two species have similar habitat preferences (Fig. 4.11).

Fig. 4.11 Distribution of Black-and-white owl

Asio stygius robustus **(Kelso 1934) Stygian Owl**
In Belize, *A. stygius* is strictly confined to pine woodlands. It is probably resident throughout the pine belt, but actual localities of record are relatively few outside the Mountain Pine Ridge where it is eagerly sought by tourists. In the lowland pine tracts of the coastal plain, it has been recorded near Hill Bank, Orange Walk District (Meadows pers. comm.), in central and western Belize District (Wood and Leberman 1987), and in northeastern Toledo District (Jones unpublished

notes). Because it is strictly nocturnal, and not especially vocal, it is easily overlooked, and this has undoubtedly been the case in relatively poorly birded Stann Creek District, which is bisected from north to south by a patchy network of pineland habitat. Although it is restricted to montane areas above 700 m over much of its range, its occurrence near sea level in Belize is seen as an anomaly (König et al. 1999).

Because *A. stygius* in Belize is a habitat specialist with a restricted distribution, it is vulnerable to habitat destruction, whether it is logging, land clearing, fire, disease, or a combination of these events. For example, from 1999 through 2001, pine stands throughout Belize, and especially in the Mountain Pine Ridge, experienced a severe bark beetle (*Dendroctonus spp.*, Coleoptera: Scolytidae) infestation that destroyed approximately 80% of the pine woodland occupied by the stygian owl in the Pine Ridge. The severity of the pine bark beetle outbreak has been attributed to an abundance of dense, susceptible pine stands, a period of unusual drought, and failure to recognize and respond to the beetle outbreak in its early stages (Billings et al. 2004). A consequence of this devastation was that the remaining stygian owls were concentrated in the few areas where clusters of healthy pines remained and were thus relatively easy to find (Fig. 4.12).

Recent research by the Belize Raptor Research Institute (Phillips 2011) suggested that the habitat preferences of the stygian owl may be more complicated than expected as a radio-tagged male routinely departed from the pine forest to hunt in the adjacent broadleaf forest nearly 15 km from the nesting site.

Asio flammeus **(Pontoppidan 1763) Short-eared Owl**

The only record of *A. flammeus* in Belize is of a mummified and half-buried carcass found on a levy at Aqua Mar Shrimp Farm, Toledo District, on 4 March 1999 (Jones et al. 2000). It is impossible to know how long before the carcass was found that the bird had died. This specimen is in all likelihood the nearly cosmopolitan *A. flammeus*.

Pseudoscops clamator forbesii **(Lowery and Dalquest 1951) Striped Owl**

Pseudoscops clamator is an uncommon resident east of the Maya Mountains in Toledo and Stann Creek districts, becoming less common and local north to the Sibun River drainage in southwestern Belize District (Piaskowski et al. 2003). Although not rare, *P. clamator* was not discovered in Belize until 1988. Its preferred habitats are forest edges abutting marshes, seasonally wet meadows, early-stage second growth with a scattering of trees, and open pine savannas (Fig. 4.13).

Locally, it may be expanding its range (and increasing its numbers) following the clearing of forest land for agriculture and urban development (König et al. 1999), which may at least partially account for the lack of records in Belize prior to 1988. Use of pesticides in agricultural areas, however, may be curtailing its expansion to some extent.

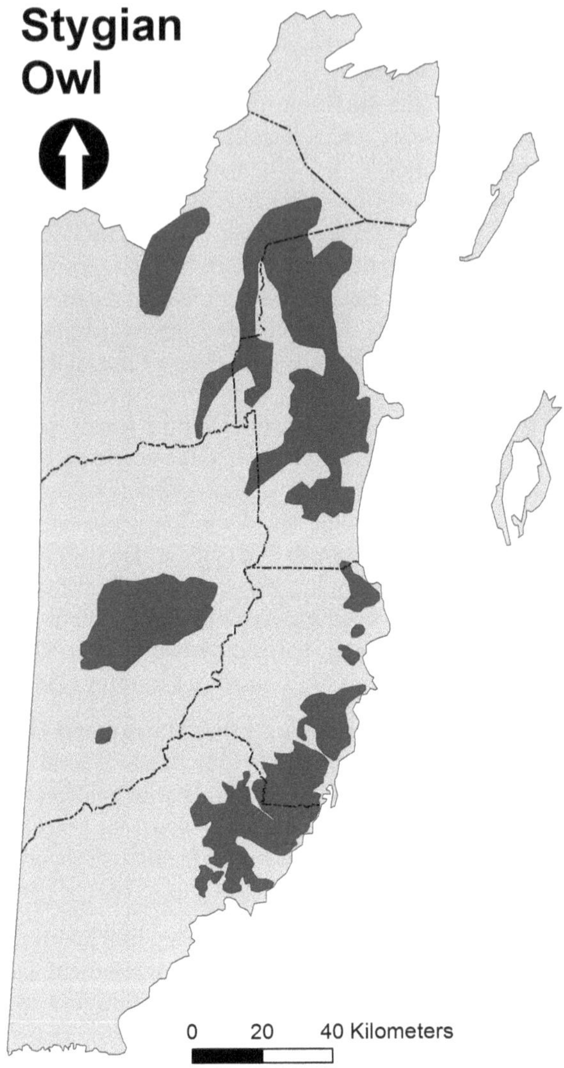

Fig. 4.12 Distribution of Stygian owl

4.9 Conclusions

Belize has one of the lowest population densities in the region and thus a larger percentage of forested land, much of which has been conserved as either extractive forest reserves or fully protected nature reserves. Additionally, most owls are not shot as a result of superstitious beliefs, as in neighboring Guatemala, nor are they captured for the pet trade. Ecotourism provides one of the largest sources of revenue, especially foreign exchange, which is in short supply, thereby providing a

Fig. 4.13 Distribution of Striped owl

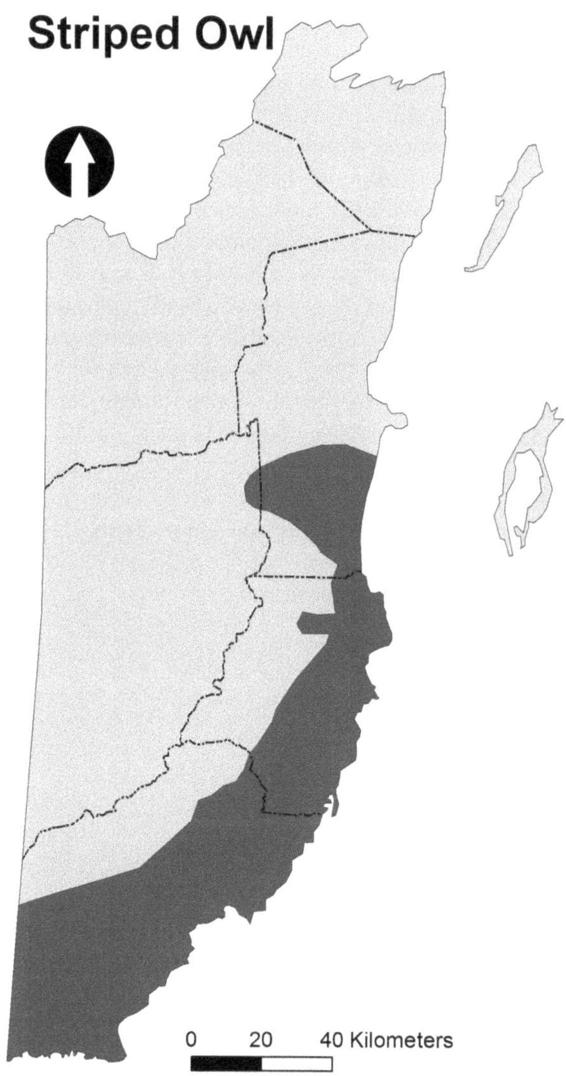

strong incentive for the country to continue protecting its natural resources. Funds from tourism, managed by PACT, have been used primarily to support baseline and other environmental studies within Belize's PA system. Thus, compared with neighboring countries with much higher population densities (United Nations 2008) and significantly more deforestation, it would appear that fewer of Belize's owls are in jeopardy from anthropogenic causes. However, the country still lacks sufficient manpower or funds to adequately staff, manage, and protect its forest resources, the government has shown a tendency to enforce its environmental laws selectively, and law enforcement and the Belize Defense Force have been ineffective in their attempts to curb incursions of Guatemalan settlers and poachers into forest reserves along its western border. In fact, the largest single threat to forest-dwelling owls in Belize at this time appears to be habitat destruction from illegal timber harvesting, ranching, farming, and settlements along and near the country's western border.

The two owls at greatest risk, though, are not those inhabiting forestlands in the west but *Asio stygius* and the Yucatan endemic *Bubo virginianus mayensis*, the former from threats to its limited pineland habitat, and the latter from a small, and perhaps declining, source population to the north, as well as unknown factors that have apparently caused its decline in Belize over the past half century or so. *A. stygius* was decimated by a pine bark beetle infestation recently and is further threatened by potentially non-sustainable timber harvesting in the Mountain Pine Ridge and clearing of pine woodlands and savannas for settlements and milpas in the lowlands. *B. v. mayensis* has always had a restricted range in Belize, and its future security in the country is likely dependent on the stability of the population to the north in Quintana Roo and Campeche, Mexico. Belize's current efforts to protect its owls and their habitats, however, could be for naught if the long-term effects of climate change are not addressed and ultimately reversed and if the country's population continues to expand at its present rate, thus placing greater and greater demand on its natural resources.

Ferruginous Pygmy Owl (*Glaucidium brasilianum*)

Crested Owl (*Lophostrix cristata*)

Spectacled Owl (*Pulsatrix perspicillata*)

Stygian Owl (*Asio stygius*)

Guatemalan Screech Owl (*Megascops guatemalae*)

Spectacled Owl (*Pulsatrix perspicillata*)

Stygian Owl (*Asio stygius*)

Appendix 4.1

English and local names for owls in Belize

Species	English name	Creole[a]	Q'eqchi[a, b]	Mopan[a]	Yucatec
	"Owl"	Monkey bird	Quarom	Bouh	Ak'ah ch'ich
Tyto alba	Barn owl				
Megascops guatemalae	Vermiculated screech owl				
Lophostrix cristata	Crested owl				
Pulsatrix perspicillata	Spectacled owl		Bubu te'		
Bubo virginianus	Great horned owl				Tunkuruchu
Glaucidium griseiceps	Central American pygmy owl	Screech owl		Ton ton	Ton kaj xnuc
Glaucidium brasilianum	Ferruginous pygmy owl	Screech owl		Ton ton	Ton kaj xnuc
Athene cunicularia	Burrowing owl				
Ciccaba virgata	Mottled owl	Screech owl			

Species	English name	Creole[a]	Q'eqchi[a, b]	Mopan[a]	Yucatec
Ciccaba nigrolineata	Black-and-white owl				
Asio stygius	Stygian owl				
Asio flammeus	Short-eared owl				
Pseudoscops clamator	Striped owl				

[a]From Jones (2003a)
[b]Also spelled K'ekchi

References

American Ornithologists' Union (1998 and supplements published by the North American Checklist Committee in 2000 and annually from 2002-2016) The American Ornithologists' Union check-list of North American birds, 7th edn. American Ornithologists' Union, Washington, DC

Anderson ER, Cherrington EA, Flores AI et al (2008) Potential impacts of climate change on biodiversity in Central America, Mexico and the Dominican Republic. CATHALAC/USAID, Panama City

Bangs O, Peck ME (1908) On some rare and new birds from British Honduras. Proc Biol Soc Wash 21:43–46

Barcott B (2008) The last flight of the scarlet macaw: one woman's fight to save the world's most beautiful bird. Random House, New York

Billings RF, Clarke SR, Espino Mendoza V et al (2004) Bark beetle outbreaks and fire: a devastating combination for Central America's pine forests. Unasylva 217:15–18. 55 (2004/2)

Bonaparte CL (1825) American ornithology or the natural history of birds inhabiting the United States. I. Mitchell, Philadelphia

Bonaparte CL (1838) A geographical and comparative list of the birds of Europe and North America. J. van Voorst, London

Bonta M, Anderson DL (2002) Birding Honduras: a checklist and guide. EcoArte, Tegucigalpa

Eisermann K, Avendaño C (2015) Los Búhos de Guatemala. In: Enríquez P (ed) Los Búhos Neotropicales, diversidad y conservación. ECOSUR, San Cristóbal de las Casas, pp 381–434

England MC (2000) The landbird monitoring programme at Lamanai, Belize: a preliminary accessment. Cotinga 13:32–43

Enríquez Rocha PL, Rangel-Salazar JL (1996) Nest site records of the Yucatan great horned owl in Sian Ka'an, Mexico. Ornitol Neotrop 7:149–151

Gmelin JF (1788) System a Naturae, vol 1. G.E. Beer, Leipzig

Griscom L (1929) A collection of birds from Cana. Darien Bull Mus Comp Zool 69:149–190

Howell SNG (1995) A critique of Walters' (1993) new bird records from Belize. Bull Br Ornithol Club 115(3):177–180

Howell SNG, Webb S (1995) A guide to the birds of Mexico and northern Central America. Oxford University Press, New York

Intergovernmental Panel on Climate Change (2007a) In: Solomon S, Qin D, Manning M et al (eds) Climate change 2007: the physical science basis. Contribution of Working Group I to the Fourth Assessment Report of the Intergovernmental Panel on Climate Change. Cambridge Univ. Press, New York

Intergovernmental Panel on Climate Change (2007b) In: Parry ML, Canziani OF, Palutikof JP et al (eds) Climate change 2007: impacts, adaptation and vulnerability. Contribution of Working Group II to the Fourth Assessment Report of the Intergovernmental Panel on Climate Change. Cambridge University Press, New York

Jones HL (2003a) Birds of Belize. University of Texas Press, Austin
Jones HL (2003b) The nesting season, June through July 2003: Central America. North Am Birds 57(4):554–555
Jones HL (2004a) Fall migration, August through November 2003: Central America. North Am Birds 58(1):155–157
Jones HL (2004b) The nesting season, June through July 2004: Central America. North Am Birds 58(4):611–612
Jones HL, Gardner D (1997) Birds recorded in the vicinity of Little Quartz Ridge, 10–24 February 1997. In: Meerman J, Matola S (eds) The Columbia river forest reserve Little Quartz Ridge expedition: a biological assessment. Columbia University Press, New York
Jones HL, Komar O (2005) Spring migration, March through May 2005: Central America. North Am Birds 59(3):506–510
Jones HL, Komar O (2007) Spring migration, March through May 2007: Central America. North Am Birds 61(3):521–525
Jones HL, Komar O (2008) Fall migration, August through November 2007: Central America. North American Birds 62(1):163–170
Jones HL, Komar O (2010) Fall migration, August through November 2009: Central America. North American Birds 64(1):164–168
Jones HL, Vallely AC (2001) Annotated checklist of the birds of Belize. Lynx Edicions, Barcelona
Jones HL, McRae E, Meadows M et al (2000) Status updates for selected species in Belize, including several species previously undocumented from the country. Cotinga 13:17–31
Karl TR, Trenberth KE (2003) Modern global climate change. Science 302:1719–1723
Kelso L (1934) A new stygian owl. Auk 51:522–523
König C, Weick F, Becking JH (1999) Owls: a guide to owls of the world. Yale University Press, New Haven
Lopez Ornat A, Lynch JF, MacKinnon de Montes B (1989) New and noteworthy records of birds from the eastern Yucatan Peninsula. Wilson Bull 101:390–409
Lowery GH, Jr, Dalquest WW (1951) Birds from the state of Veracruz, Mexico. University of Kansas Publications, Mus Nat Hist 3. Lawrence, Kansas pp 531–649
MacKinnon B (1992) Check-list of the birds of the Yucatan Peninsula. Amigos de Sian Ka'an, Mérida
Matola S (1988) The story of Hoodwink the owl. Macmillan Publishing, New York
Matola S (1993) The further adventures of Hoodwink the owl. Conservation International, Arlington
Matola S (2000) Hoodwink the owl meets Mac the Macaw. Impresiones Profesionales del Sureste, Mérida
Matola S (2012) Words of wisdom from happy the owl. BRC Pinting, Kansas 42 pp
Meadows M (1994) Bacalar Chico bird report, Ambergris Caye, Belize. Sponsored by the International Tropical Conservation Foundation; administered by Chief Fisheries Administrator, Fisheries Department, Government of Belize
Meerman JC (2005) National list of critical species: taskforce on Belize's protected areas. Ministry of Natural Resources and the Environment, Belmopan
Meerman JC (2007) Key biodiversity area (KBA) analysis for Belize (draft). Report to Conservation International, Arlington
Meerman JC, Sabido W (2001) Central American ecosystems mapping project: Belize, vol 2. Programme for Belize, Belize City
Meerman JC, Wilson JR (2005) The Belize national protected areas system plan. Taskforce on Belize's protected areas. Ministry of Natural Resources and the Environment, Belmopan
Meerman JC, Epting J, Steiniger M et al (2010) Forest cover and change in Belize Circa 1990–2000–2005. Conservation International, Arlington
Miller BW (1997) Avian risk assessment: bird species of conservation concern. Wildlife Conservation Society, New York
Miller BW, Miller CM (1992) Distributional notes and new species records for birds in Belize. Occas Pap Belize Nat Hist Soc 1:6–25

Miller BW, Miller CM (1998) Birds of Belize: a checklist. Wildlife Conservation Society, Gallon Jug
Miller BW, Miller CM (2000) Birds of Belize: a checklist, 2nd rev. Wildlife Conservation Society, Gallon Jug
Monroe BL Jr (1968) A distributional survey of the birds of Honduras. Ornithological Monographs No.7, American Ornithologists' Union. Allen Press, Lawrence
Parker III TA, Holst BK, Emmons LH et al (1993) A biological assessment of the Columbia River forest Reserve, Toledo District, Belize. Rapid Ecological Assessment (RAP) Working Papers 3, Conservation International, Arlington
Parks KC (1995) Correction. Ann Carnegie Mus 64(3):235
Paynter RA (1955) The Ornithogeography of the Yucatan Peninsula, Peabody Mus. Nat. Hist., Yale University 9, Kansas, pp 1–347
Phillips R (2011) Studying the stygian owl *Asio stygius robustus* in Mountain Pine Ridge, Belize. Spizaetus 12:2–6
Piaskowski VD, Teul M, Cal RN et al (2003) The birds of central Belize. In: Day MJ, Barker D (eds) Caribbean geography. Special issue: Belize and the wider Caribbean, part 1: selection of papers from 4th International Belize Conference, San Ignacio, 2002
Piaskowski VD, Teul M, Cal RN et al (2006) The birds without Borders – Aves Sin Fronteras® recommendations for landowners: how to manage your land to help birds (Belize and Mesoamerica edition). Foundation for Wildlife Conservation, Inc., and Zoological Society of Milwaukee, Milwaukee
Piaskowski VD, Williams KM, Boese GK et al (2008) The birds without Borders – Aves Sin Fronteras® recommendations for landowners: how to manage your land to help birds (Wisconsin, Midwest, and eastern United States edition). Foundation for Wildlife Conservation, Inc., and Zoological Society of Milwaukee, Milwaukee
Pontoppidan E (1763) Den danske atlas eller konge-riget Dannemark, med dets naturlige egenskaber, elementer, indbyggere, vaexter, dyr og andre affodninger, dets gamle tildrageiser of naervaerene omstaendigheder i alle provintzer, staeder, kirker, slotte of herregaarde. Kopenhagen
Ridgely RS, Gywnne JA Jr (1989) A guide to the birds of Panama, with Costa Rica, Nicaragua, and Honduras. Princeton University Press, Princeton
Ridgway R (1914) The birds of North and Middle America. Part 6. Bulletin of the United States National Museum 50, part IV. Smithsonian Institution, Washington, DC
Russell SM (1964) A distributional study of the birds of British Honduras. Ornithological Monographs No.1, American Ornithologists' Union, Allen Press, Lawrence
Sclater PL (1859) Description of a new species of owl of the genus *Ciccaba*. Proc Zool Soc Lond 27:131
Sclater PL, Salvin O (1859) On the ornithology of Central America. Part III. Ibis 1:213–234
Semarnat (2008) Norma Oficial Mexicana NOM-059-SEMARNAT-2001, Protección ambiental-especies nativas de México de flora y fauna silvestres-Categorías de riesgo y especificaciones para su inclusión, exclusión o cambio-Lista de especies en riesgo. Diario Oficial de la Federación, México
Sharpe RB (1875a) Catalogue of birds in the British Museum, vol. 2 Striges. Trustees of the British Museum (Natural History), London
Sharpe RB (1875b) Contributions to the history of the accipiters. The genus *Glaucidium*. Ibis 5(ser 3):35–59
Slud P (1964) The birds of Costa Rica: distribution and ecology. Bull Am Mus Nat Hist 128, 430 pp.
Stiles FG, Skutch AF (1989) A guide to the birds of Costa Rica. Cornell University Press, Ithaca
Storer RW (1961) Two collections of birds from Campeche, Mexico. Occas. pap. Mus. Zool. Univ. Mich. 621. Mich Museum of Zoology, University of Michigan 1961, Ann Arbor, pp 1–20
United Nations (2008) Population division of the Department of Economic and Social Affairs of the United Nations Secretariat, World Population Prospects: the 2008 revision, http://esa.un.org/unpp

Vallely AC, Whitman AA (1997) The birds of Hill Bank, northern Belize. Cotinga 8:39–49
Walters R (1993) Some records of birds from Belize, Central America, including three first records. Bull Br Orn Club 113:145–147
Webster JD, Orr RT (1958) Variation in the great horned owls of middle America. Auk 75:134–142
Wetmore A (1968) The birds of the Republic of Panama, part 2: Columbidae (pigeons) to Picidae (woodpeckers). Smithsonian Institution Press, Washington, DC
Wood DS, Leberman RC (1987) Results of the Carnegie Museum of Natural History expedition to Belize. III. Distributional notes on the birds of Belize. Carnegie Museum of Natural History, Pittsburgh
Wood DS, Leberman RC, Weyer D (1986) Checklist of the birds of Belize, Special Publication No. 12. Carnegie Museum of Natural History, Pittsburgh

Chapter 5
The Owls of Bolivia

Diego R. Méndez

Abstract In Bolivia there are 25 owl species within 12 genera, representing 33% of the neotropical owl species. In each of the 12 ecoregions of the country, there are at least three owl species, the Yungas being the most diverse ecoregion, with 15 species. Most of the owls in Bolivia (15 species) have a distribution that ranges from moderately restricted to widely distributed, while ten species have a restricted range. Information on the biology and ecology of owls from Bolivia is scarce or nonexistent. One species, *Megascops marshalli*, is currently classified as vulnerable at a national level. The main threats to these owls are habitat loss and human persecution. Although all species are found in at least one protected area, there are no specific conservation measures for them. Systematic studies on all aspects of owl biology and ecology are needed, in order to fill information gaps and promote their conservation.

Keywords Strigiformes • Conservation • Folkloric use • Research priorities

D.R. Méndez (✉)
Programa de Investigación de Aves Rapaces en Bolivia,
Calle El Villar 369, Barrio San Cristóbal, Sucre, Bolivia
e-mail: diemndez@gmail.com

P.L. Enríquez (ed.), *Neotropical Owls*, DOI 10.1007/978-3-319-57108-9_5

Rufous banded Owl (*Ciccaba albitarsis*)

5.1 Diversity

Due to its geographical location and extent, Bolivia is a mega-diverse and multicultural country. Its territory covers many of the habitats that exist in South America and with 1414 bird species recorded; it is the richest country in bird species among those without access to the sea (Maillard et al. 2009). There are 25 species of Strigiformes in Bolivia (Table 5.1), representing 2% of the birds in the country, which account for 33% and 10% of Neotropical and global owl species, respectively (König et al. 1999), showing a remarkable diversity.

The taxonomy of this order is particularly complex as most species have barely discernible differences in their cryptic appearance (Wink et al. 2009). This chapter follows the taxonomy and phylogeny proposed by Wink et al. (2009). The two families of the order, Tytonidae and Strigidae, are present in Bolivia. The Tytonidae family consists of a subfamily, Tytoninae, and one genus, *Tyto*. The Strigidae family consists of the Striginae subfamily, which includes five tribes (Bubonini, Strigini, Pulsatrigini, Megascopini, Asionini) and eight genera (*Megascops*, *Lophostrix*, *Pulsatrix*, *Bubo*, *Strix*, *Ciccaba*, *Pseudoscops*, *Asio*), and Surniinae subfamily that

Table 5.1 Distribution of owl species in Bolivia

Name	Common or indigenous names[a]	Species	Subspecies[a]
Barn Owl	Suindá (Gni)	*Tyto alba*	tuidara
Tropical Screech Owl	Alilikuku, pire-kúi (Gni); Sumburukuku (Gni, Gyo)	*Megascops choliba*	suturutus, wetmorei?, choliba?
Rufescent Screech Owl	Tiula (Aym)	*Megascops ingens*	ingens
Cloud-forest Screech Owl		*Megascops marshalli*	monotypic
Tawny-bellied Screech Owl		*Megascops watsonii*	usta
Montane Forest Screech Owl		*Megascops hoyi*	monotypic
Vermiculated Screech Owl		*Megascops guatemalae*	napensis
White-throated Screech Owl		*Megascops albogularis*	remotus
Crested Owl		*Lophostrix cristata*	cristata
Spectacled Owl	Urukereía-guasu, murucututu (Gni)	*Pulsatrix perspicillata*	perspicillata, boliviana
Band-bellied Owl		*Pulsatrix melanota*	philoscia, melanota
Great Horned Owl	Chuseka (Aym); nakurutu (Gni); jucu (Que)	*Bubo virginianus*	nakurutu, magellanicus
Chaco Owl		*Strix chacoensis*	monotypic
Mottled Owl		*Ciccaba virgata*	superciliaris?
Black-banded Owl	Ñacurutú-hû (Gni)	*Ciccaba huhula*	huhula
Rufous-banded Owl		*Ciccaba albitarsis*	monotypic
Yungas Pygmy Owl		*Glaucidium bolivianum*	monotypic
Subtropical Pygmy Owl		*Glaucidium parkeri*	monotypic
Amazonian Pygmy Owl	Kavureĩi (Gni)	*Glaucidium hardyi*	monotypic
Ferruginous Pygmy Owl	Caburé-í, kaure (Gni); caute (Gyo); caboré (Tup)	*Glaucidium brasilianum*	pallens, ucayalae
Burrowing Owl	Pejpera, tiptiri (Aym); urukureía (Gni); pesperi (Tup)	*Athene cunicularia*	cunicularia, juninensis, boliviana
Buff-fronted Owl	Kavure, kavure-pyta (Gni)	*Aegolius harrisii*	iheringi, dabbenei?
Striped Owl	Ñakurutuĩi (Gni)	*Pseudoscops clamator*	midas
Stygian Owl		*Asio stygius*	stygius
Short-eared Owl	Ch'iuseka (Aym); suinda (Gni)	*Asio flammeus*	suinda

[a]Taken from Hennessey et al. (2003). *?:* not confirmed subspecies, *Aym:* Aymara, *Gni: Guaraní*, *Gyo:* Guarayo, *Que:* Quechua, *Tup:* Tupi-Guaraní

includes three tribes (Surnini, Athenini, Aegolini) and three genera (*Glaucidium*, *Athene*, *Aegolius*).

In accordance with the pattern of the Neotropics, *Megascops* (seven species), *Glaucidium* (four species), and *Strix* (four species) are the genera with more species in Bolivia, whereas of the 13 Neotropical genera, the only absence is *Xenoglaux* (endemic to Peru). The genera *Pulsatrix* and *Ciccaba* are represented by three and two species, respectively, while *Tyto*, *Lophostrix*, *Bubo*, *Strix*, *Athene*, *Aegolius*, and *Pseudoscops* by one species each.

Bolivia is divided into nine administrative and political regions known as *departamentos*, and in all of there are at least three species of owls. La Paz, Santa Cruz, and Cochabamba with 22, 21, and 20 species, respectively, are the *departamentos* with the largest number of owl species. In contrast, in Oruro and Potosí, there are only three species recorded. In the remaining four *departamentos* (i.e., Pando, Beni, Chuquisaca, and Tarija), there are between eight and 15 species (Hennessey et al. 2003).

Bolivia features 12 ecoregions (see Fig. 5.1 and Appendix 5.1 for a description of each ecoregion). With 15 species, the *Yungas* are the most diverse ecoregion in terms of owl species, followed by the *Sabanas inundables*, the *Sudoeste de la Amazonía*, and *Cerrado* with 11 species each. In the remaining eight ecoregions (*Chiquitanía*, *Gran Chaco*, *Boliviano-Tucumana*, *Chaco Serrano*, *Bosques Secos Interandinos*, *Prepuna*, *Puna Norteña*, and *Puna Sureña*), there are between three and ten species. The *Prepuna*, with three species, and both *Puna* ecoregions, with four species each, are the least diverse ecoregions (Hennessey et al. 2003).

Most species (19) occur in three or more *departamentos*. Of the five species found in only two of them, three are only found in the *Yungas*: *Megascops marshalli* and *M. albogularis* in Cochabamba and La Paz, and *Glaucidium parkeri* in Beni and La Paz; *Strix chacoensis* is present only in the *Gran Chaco* ecoregion of Santa Cruz and Tarija, and *Asio stygius* has only been recorded in Santa Cruz. In Oruro and Potosí, where *Puna* ecoregions are predominant and few forested areas exist, there is the smallest number of owl species, with four in each *departamento* (Hennessey et al. 2003).

There are 29 types of Neotropical habitats in Bolivia (Stotz et al. 1996; Appendix 5.2), in 17 of which owls have been recorded. Most habitats where there are no records are bodies of water or are situated nearby (e.g., *bofedales*, riparian weeds, beaches, and others). Twenty-three owl species, or 92% of the species in the country, are forest dwellers; and of the 11 types of forest found in Bolivia, there are no owl records in white sand forests and palm groves, both of which are Amazonian forests. Montane evergreen forests are home to the greatest diversity of owls in the country, a total of 15 species. The tropical lowland evergreen forests, which are found in all the Amazonian ecoregions, represent the habitat with the second highest diversity of owls, with 11 species. Tropical deciduous forests are present mainly in the *Gran Chaco*, *Chiquitanía*, and *Cerrado* ecoregions, and 11 owl species are found in this type of forest. The remaining forest habitats (i.e., evergreen, riparian, dwarf, *Polylepis*, gallery, and secondary forests) feature between two and six owl species. Finally, in the non-forest habitats (i.e., bushes, closed, field, grassland, highlands, agricultural lands, and second grade weeds), between one and four species are recorded (Hennessey et al. 2003).

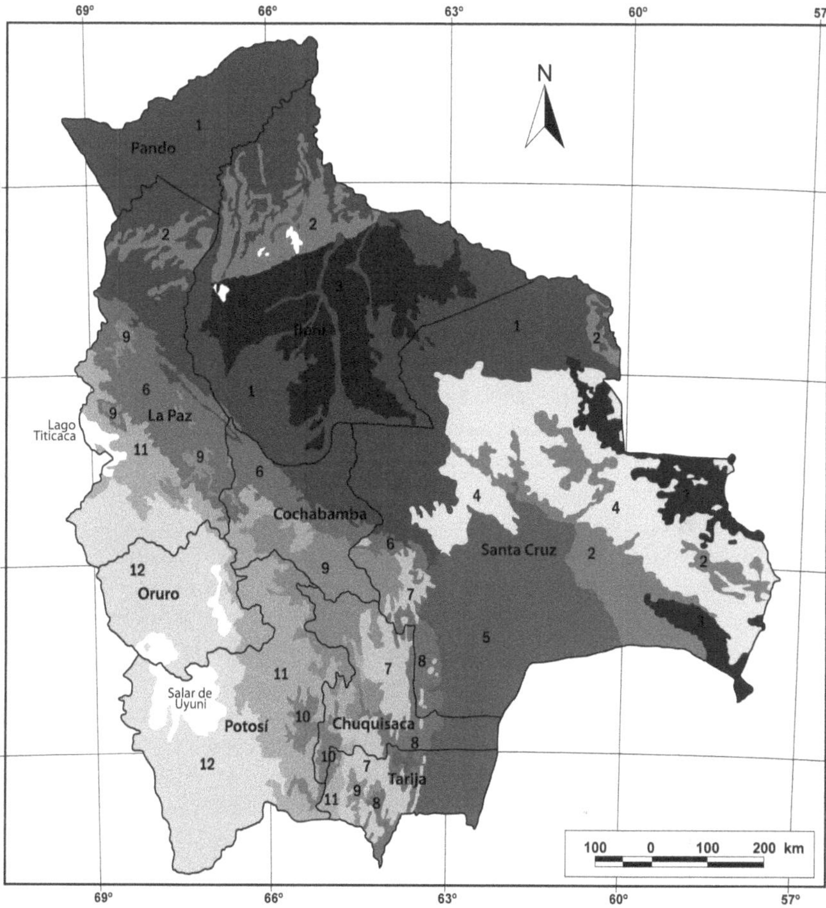

Fig. 5.1 Ecoregions of Bolivia, based on Ibisch et al. (2003): (**1**) Sudoeste de la Amazonía, (**2**) Cerrado, (**3**) Sabanas inundables, (**4**) Chiquitanía, (**5**) Gran Chaco, (**6**) Yungas, (**7**) Boliviano-Tucumana, (**8**) Chaco Serrano, (**9**) Valles Secos Interandinos, (**10**) Prepuna, (**11**) Puna Norteña, (**12**) Puna Sureña (see Appendix 5.1 for the description of each Eco region)

5.2 Distribution

The distribution range of Bolivian owls can be described according to the distribution criterion developed for the vertebrates of Bolivia (Aguirre et al. 2009a) in combination with the records compiled in Hennessey et al. (2003) and the range of the country's ecoregions (Ibisch et al. 2003). Most of the species have a wide range of distribution (the range includes more than 40% of the country) and are present in three to seven ecoregions and two to nine habitat types (see Appendices 5.1 and 5.2): *Tyto alba*, *Megascops choliba*, *M. watsonii*, *Pulsatrix perspicillata*, *Bubo virginianus*, *Ciccaba virgata*, *C. huhula*, *Glaucidium brasilianum*, *Athene cunicularia*,

Pseudoscops clamator, and *Lophostrix cristata*. The species that have a moderately restricted distribution range (the range includes 15–40% of Bolivia) are: *Ciccaba albitarsis*, *Glaucidium bolivianum*, *G. hardyi*, and *Asio flammeus*. Finally, there exist ten species that present a restricted distribution range (it includes between 5% and 15% of Bolivia): *Megascops ingens*, *M. marshalli*, and *G. parkeri* (exclusive species of the evergreen mountainous forests of the *Yungas*), *M. hoyi* (registered only in the *Boliviano-Tucumana* ecoregion), *M. guatemalae*, *M. albogularis*, and *Pulsatrix melanota* (registered in 2–4 types of *Yungas*' habitats), *Strix chacoensis* (registered only in tropical deciduous forests of the Gran Chaco), *Aegolius harrisii* (registered only in the *Boliviano-Tucumana* and *Yungas* ecoregions), and *Asio stygius*, registered only in the central eastern extreme of Santa Cruz.

Bolivia does not have any endemic species of owl; it shares 23 species with Peru, 17 with Brazil, 15 with Argentina, and 13 with Paraguay (BirdLife International 2012a). In relation to a shared endemism, *Megascops marshalli* occurs only in Bolivia and Peru. Previously considered endemic to Peru, the records of *M. marshalli* in Bolivia were confirmed by comparing vocalizations and collected specimens (Herzog et al. 2009).

Compared to other Neotropical countries (e.g., Trejo 2007), there are very few published owl records in Bolivia, particularly those that complement the distribution information contained in Hennessey et al. (2003) (e.g., Miserendino 2007); thus it is necessary to update the information on the distribution of owls in Bolivia. In this respect it is particularly necessary to carry out intensive searches of the least common species and/or those that have a restricted range, as well as to increase the frequency of ornithological explorations in regions little known such as the *Yungas*, the *Puna,* or in the *departamentos* of southern Bolivia (i.e., Chuquisaca, Potosi, and Tarija).

5.3 Conservation Status

One owl species was categorized as Vulnerable in the most recent assessment of the conservation status of birds in Bolivia (Balderrama 2009). The species is *Megascops marshalli* and it is categorized as Vulnerable at the national level because its populations are declining as a consequence of habitat degradation and because of its patchy distribution inside a narrow geographical range (Herzog and Balderrama 2009). This species is one of the six species that only live in the *Yungas*, where it occupies cloud forests with abundant epiphytes (Herzog and Balderrama 2009). Though worldwide *M. marshalli* is considered to be Near Threatened (Bird Life International 2012b), a detailed evaluation of the threats it faces might contribute to its categorization as Vulnerable (Herzog et al. 2009). The remaining 24 species of owls of Bolivia are globally categorized as Least Concern (Bird Life International 2012a), and all of them are listed in the Appendix II of CITES (CITES 2012).

According to the model Conservation Status of Bolivia (WCS 2008), the most disturbed and/or threatened habitats in the country – which are particularly diverse in terms of owls – are:

1. The tropical lowland evergreen forests in central eastern Cochabamba (the southernmost part of the Amazon) and east and west of Pando, in the area of influence of the cities of Cobija and Riberalta/Guayaramerín, respectively
2. The tropical deciduous forests, in the center of Santa Cruz (including the ecotones between the *Sudoeste de la Amazonía*, *Chiquitanía*, and *Gran Chaco* ecoregions)
3. The evergreen montane forests in the eastern Andean slope

The lack of biological studies in these important areas for owl conservation, together with the lack of information on the population parameters and distribution of all owl species in the country, impedes a precise evaluation of the conservation status of these species. For example, the Method for Assessing the Degree of Threat to a Species (MEGA, in Spanish) in Bolivia considers five criteria: (1) distribution, (2) habitat conservation status, (3) population status, (4) intrinsic biological vulnerability of the species, and (5) major threats (Aguirre et al. 2009b); and in general, such information is nonexistent or limited for Strigiformes (this chapter).

5.4 Major Threats

The principal threats to the birds in Bolivia are habitat loss and the extraction of individuals from the wild; the first one related particularly to agricultural processes, logging, and introduction of exotic vegetation, whereas the second one is related to the pet trade, hunting, and cultural uses (Balderrama 2009). Particularly for owls, there are certain factors that influence the threats of greatest importance, which are deforestation related to logging and to the expansion of the agricultural border (92% of the owl species in Bolivia live in some type of forest), the use of individuals to make folkloric costumes (masks elaborated with complete individuals of *Bubo virginianus* and *Tyto alba*, pers. obs. Figs. 5.2, 5.3, and 5.4), and direct persecution since owls may be considered sign of bad luck.

The rate of deforestation in Bolivia is one of the highest worldwide with about 300,000 hectares deforested annually (Urioste 2010). The loss of forested areas occurs markedly in Santa Cruz (76% of the total deforested area in the country in 2005), followed by Pando, Beni, Cochabamba, Tarija, and La Paz (Muñoz 2006). These *departamentos* include the most diverse ecoregions in terms of owls (e.g., *Yungas*, *Sabanas inundables*), thus demanding special attention in controlling the deforestation in these regions, as well as the need to elaborate effective strategies for the exploitation of forest resources so that impact on the biodiversity is the least possible, especially focusing on the ecological requirements and natural history of owls.

The use of owls to make folkloric costumes seems to be a common practice that is probably increasing. Though these birds have likely been part of rituals and

Fig. 5.2 Girls dancing the *Tobas*, wearing headdresses that include complete individuals of Barn Owl (*Tyto alba*) (picture taken at the 2012 *Carnaval de Oruro festivity*)

beliefs for a long time, certainly they were never used in the numbers that are used today (e.g., >50 birds, including barn owl (*T. alba*) and great horned owl (*B. virginianus*) in the headdresses of a single folkloric group of dancers). Since there are no studies that quantify the impact of these practices, nor studies on the mortality of owls that are eliminated because they are considered evil omen birds, it is possible that the impact that these threats have on owl populations is being underestimated.

5.5 Conservation Strategies

All owl species in Bolivia occur in at least one protected area (Hennessey et al. 2003); besides this, there is any specific conservation measure focused on owls. The lack of basic information about the biology and ecology of owls is a factor that prevents the development of effective strategies for their conservation.

5.6 Status of Biological and Ecological Knowledge

Raptors are some of the least studied birds in Bolivia, particularly the nocturnal raptors (Strigiformes). To evaluate the status of biological and ecological knowledge on these birds, detect information gaps and set investigation and conservation

Fig. 5.3 Two dancers of the *Tobas* wearing masks made of complete individuals of Great Horned owls (*Bubo virginianus*) (picture taken at the 2012 *Carnaval de Oruro* festivity)

Fig. 5.4 Close-up of a *Tobas* headdress made with an individual of a Great Horned Owl (*Bubo virginianus*) (picture taken at the 2012 *Carnaval de Oruro* festivity)

priorities; a bibliographical review was conducted of the publications on the owls of Bolivia. The review considered all the publications referring to any aspect of owl biology and ecology (i.e., distribution, habitat preferences, movements, abundance, diet, reproduction, behavior, conservation status, others). Databases (e.g., Global Raptors Information Network, SORA, SciELO, Google Scholar), local and international journals, books, theses, newspapers, and non-conventional literature were reviewed. Pertinent publications were classified according to relevancy (i.e., major relevancy if an owl topic was part of the publication objectives, minor relevancy if owls were only mentioned in the work).

There are only seven publications that document studies on owls in Bolivia and that include these as a central part of the investigations. The publications are isolated temporally and spatially, and continuity does not exist in any of the studies. The species on which more investigations have been conducted is *Tyto alba* (three publications), although only referring to its diet in two localities of the Salar de Uyuni (Reboledo and Lartigau 1998), the Estación Biológica del Beni (Vargas et al. 2002), and in a disturbed valley in La Paz (Aliaga-Rossel and Tarifa 2005). There are two publications about *Athene cunicularia*, one on the species' diet in Ulla-Ulla Reserve (currently renamed as Área Natural de Manejo Integrado Nacional Apolobamba), La Paz (Serrano and Anderson 1986), and the first record of the species in Pando (Miserendino 2007). There is one publication on the diet of *Bubo virginianus* in Chuquisaca (Pokines 2007) and another on the distribution of *Megascops marshalli*, which includes considerations on its ecology, vocalizations, and conservation status (Herzog et al. 2009).

Additionally, two newspaper notes contribute information, one on captive caring of *B. virginianus* in the zoo of Oruro, a city located in the *Altiplano* of Bolivia (Ecológico Kiswara 2011), and another that documents a notable record of *Asio flammeus* in the Laguna de Albarrancho, in the southwest of the city Cochabamba, now a new locality for the species (Redacción Central 2014).

5.7 Conclusions

Knowledge on the biology and ecology of the owls of Bolivia is very limited; just a few aspects have been studied, without continuity, in spite of the fact that the potential for investigating this group of raptors in the country is great. Systematic studies on the owls of Bolivia are needed, since it is important to cover the significant information gaps in order to develop effective conservation, monitoring, and management plans for the species and their habitats.

Acknowledgments To José Balderrama, for his encouragement on conducting this work and to Paula Enríquez for her ongoing assistance. Víctor Pérez collaborated in the edition of the map of Bolivian ecoregions. Many thanks to Nicole Sault and Nathaly Camargo, who assisted me in the translation of the chapter from Spanish to English.

Striped Owl (*Asio clamator*)

Appendix 5.1

Description of the ecoregions of Bolivia according to Ibisch et al. (2003)

				Owl species		
Eco region	Area [Km^2]	Altitudinal range	Description	N°	Species	*Departamentos*
1. *Sudoeste de la Amazonía*	276,547	100–1000 m	Terrain: plains and slightly wavy hills. Next to the eastern Andean slope. Vegetation: humid forests, mostly evergreen, tall (30–45 m). Seasonally flooded forests (*Várzea* and *Igapó*). Palm forests, bamboo forests. Transitional forests to the *Chiquitanía*	11	*Tyto alba* *Megascops choliba* *M. watsonii* *Lophostrix cristata Pulsatrix perspicillata* *Ciccaba virgata* *C. huhula* *Glaucidium hardyi* *G. brasilianum* *Athene cunicularia* *Pseudoscops clamator*	Beni, Cochabamba, La Paz, Pando y Santa Cruz
2. *Cerrado*	84,967	100–1100 m [1000–2000 m, "mountain savannahs"]	Terrain: plains (some floodable), hills, low mountains, inselbergs. Vegetation: savannahs (sartenejales), forest islands, wooded plains, *cerradao*, campo grasslands	11	*Tyto alba* *Megascops choliba* *M. watsonii* *Pulsatrix perspicillata* *Bubo virginianus* *Ciccaba huhula* *Glaucidium hardyi* *G. brasilianum* *Athene cunicularia* *Pseudoscops clamator* *Asio stygius*[a]	Beni, La Paz, Santa Cruz

3. *Sabanas inundables*	127,988	100–800 m	Terrain: floodplains, large seasonal lagoons. Vegetation: marshy savannahs (*yomomos*, *curichis*), forest islands, open forest (*tajibales*, palm forests), *tusecales*, gallery forests	11	*Tyto alba* *Megascops choliba* *M. watsonii* *Lophostrix cristata* *Pulsatrix perspicillata* *Bubo virginianus* *Ciccaba virgata* *C. huhula* *Glaucidium brasilianum* *Athene cunicularia* *Pseudoscops clamator*	Beni, Santa Cruz
4. *Chiquitanía*	101,769	100–1400 m	Terrain: plains, mountains, inselbergs (Precambrian shield). Vegetation: semi-deciduous forest, moderately tall (15–25 m). Flood savannahs. Palm forests, evergreen forests, and bamboo forests	8	*Tyto alba, Megascops choliba, M. watsonii, Pulsatrix perspicillata, Bubo virginianus, Glaucidium brasilianum, Athene cunicularia, Pseudoscops clamator*	Santa Cruz
5. *Gran Chaco*	105,006	200–600 m	Terrain: plains with few hills and small mountains. Vegetation: dry deciduous forest, short (5–15 m) with columnar cactus. *Algarrobales*. Palm forests, seasonally flooded	9	*Tyto alba* *Megascops choliba* *M. watsonii* *Pulsatrix perspicillata* *Bubo virginianus* *Strix chacoensis*[a] *Glaucidium brasilianum* *Athene cunicularia* *Pseudoscops clamator*	Chuquisaca, Santa Cruz, Tarija

(continued)

Eco region	Area [Km²]	Altitudinal range	Description	Owl species		*Departamentos*
				N°	Species	
6. *Yungas*	55,556	1000–4200 m (including the *Páramo Yungueño*)	Relief: craggy hillsides, steep slopes. Dissected valleys. Vegetation: Different types of humid evergreen forests, short to medium high (5–30 m)	15	*Megascops choliba* *M. ingens*[a] *M. marshalli*[a] *M. watsonii* *M. guatemalae*[a] *M. albogularis*[a] *Lophostrix cristata* *Pulsatrix perspicillata* *P. melanota*[a] *Ciccaba huhula* *C. albitarsis* *Glaucidium bolivianum* *G. parkeri*[a] *Aegolius harrisii* *Pseudoscops clamator*	
7. *Boliviano-Tucumana*	29,386	800–3900 m	Terrain: steep slopes, valleys, mountains. Vegetation: semi-humid forests, semi-deciduous, deciduous, and evergreen; Puna-like shrubbery and grasslands	8	*Megascops choliba* *M. hoyi*[a] *Pulsatrix perspicillata* *Ciccaba albitarsis* *Glaucidium bolivianum* *Glaucidium brasilianum* *Athene cunicularia* *Aegolius harrisii*	Chuquisaca, Santa Cruz, Tarija
8. *Chaco Serrano*	23,176	700–2000 m	Low mountainous ranges, dry deciduous forest (~25 m)	5	*Megascops choliba* *Pulsatrix perspicillata* *Glaucidium bolivianum* *G. brasilianum* *Athene cunicularia*	Chuquisaca, Santa Cruz, Tarija

9. *Bosques secos interandinos*	44,805	500–3300 m	Terrain: rather dissected valleys. Small plains. Vegetation: dry deciduous forest (10–20 m), with tall columnar cactus	7	*Tyto alba* *Megascops choliba Pulsatrix perspicillata* *Bubo virginianus* *Glaucidium bolivianum* *G. brasilianum* *Athene cunicularia*	Chuquisaca, Cochabamba, La Paz, Potosi, Santa Cruz, Tarija
10. *Prepuna*	8516	2300–3400 m	Terrain: rather wide valleys, dissected valleys. Small plains. Vegetation: deciduous thorny chaparral with columnar cacti	3	*Tyto alba* *Bubo virginianus* *Athene cunicularia*	Chuquisaca, Potosi, Tarija
11. *Puna Norteña*	84,606	3200–5100 m	Terrain: hilly plains, high plateaus, valleys. Mountainous topography. Vegetation: grasslands with shrubs. Thickets. High Andean evergreen forests. Bogs (*Bofedales*).	4	*Tyto alba, Bubo virginianus* *Athene cunicularia* *Asio flammeus*	Chuquisaca, Cochabamba, La Paz, Oruro, Potosi, Tarija
12. *Puna Sureña*	136,177	Approx. 3500–4100 m 3800 to almost 7000 m	Mountainous areas, high plateaus (*Altiplano*). Very wide valleys. Volcanos. Dunes and sandbanks. Salt flats. Vegetation: grasslands with shrubs and/or thickets (*tolares*), halophyte thickets. Open Andean forests. Bogs (*Bofedales*)	4	*Tyto alba* *Bubo virginianus* *Athene cunicularia* *Asio flammeus*	La Paz, Oruro, Potosí

[a]This species have only been recorded in one ecoregion.

Appendix 5.2

Description of the ecosystems of Bolivia according to Stotz et al. (1996) and Navarro and Maldonado (2002). Synthesizes the characteristics of the series of vegetation that are representative of each type of habitat and mentions the owl species present.

Habitat	Description	Ecoregions	Owl species	
			N°	Species
Tropical lowland evergreen forests	Tall and dense canopy (25–35 m), scattered emerging trees (30–45 m). More or less dense arboreal understory (10–25 m). Shrubby stratus of variable density (3–6 m) and herbaceous stratus up to 2 m. Palms. Moderately abundant lianas and relatively few epiphytes	*Sudoeste de la Amazonía, Sabanas inundables, Cerrado*	10	*Tyto alba* *Megascops choliba* *M. watsonii* *Lophostrix cristata* *Pulsatrix perspicillata* *Bubo virginianus* *Ciccaba virgata* *C. huhula* *Glaucidium hardyi*[a] *G. brasilianum*
Flooded tropical evergreen forest	Moderately open to almost continuous – medium to high (15–35 m) – canopy, with emerging trees of 40 m. Arboreal understory of one or two strata (10–20 m). Shrubby understory and small trees (5–8 m). Herbaceous understory 0.5–3.5 m). Palms. Moderately abundant lianas and epiphytes	*Sudoeste de la Amazonía, Sabanas inundables*	3	*Megascops watsonii* *Pulsatrix perspicillata* *Ciccaba huhula*
River-edge forests	Variable canopy (10–20 m, 20–30 m; open to dense). Almost nonexistent arboreal understory. There may be open to moderately dense shrubby understory. Generally dense herbaceous understory (1–3 m)	*Sudoeste de la Amazonía, Cerrado, Sabanas inundables, Chiquitanía, Yungas, Boliviano-Tucumana, Chaco Serrano*	2	*Megascops choliba* *Bubo virginianus*

(continued)

Habitat	Description	Ecoregions	Owl species	
			N°	Species
Montane evergreen forests	Variable canopy (10 m, 10–15 m, 15–30 m), open, semi-open moderately closed, or dense. Arboreal understory mostly nonexistent. Variable shrubby and herbaceous understories, open to more or less dense. Important presence of epiphytes, moderate to abundant. Variable presence of lianas, fairly nonexistent to abundant. Tree ferns and palms in several regions	*Yungas, Boliviano-Tucumana, Chaco Serrano, Bosques secos interandinos, Prepuna*	15	*Megascops ingens*[a] *M. marshalli*[a] *M. hoyi* *M. guatemalae* *M. albogularis* *Lophostrix cristata* *Pulsatrix perspicillata* *P. melanota* *Bubo virginianus* *Ciccaba virgata* *C. huhula* *C. albitarsis*[a] *Glaucidium bolivianum* *G. parkeri*[a] *Aegolius harrisii*
Elfin forest	Evergreen. Short (<12 m), semi-open canopy. Open shrubby understory and short herbaceous understory	*Yungas, Puna Norteña*	3	*Megascops hoyi* *Bubo virginianus* *Glaucidium bolivianum*
Polylepis forests	Short to medium canopy (5–12 m), generally semi-open. Variable open shrubby understory (<3 m). The presence of epiphytes is very important in some places	*Yungas, Prepuna, Puna Norteña, Puna Sureña*	3	*Megascops hoyi* *Bubo virginianus* *Glaucidium bolivianum* *Asio stygius*
Tropical deciduous forests	Semi-open to almost dense canopy, moderately high (15–25 m). Arboreal understory (7–10 m). Moderately dense shrubby and herbaceous understory. Presence of lianas and palm trees	*Cerrado, Chiquitanía, Gran Chaco, Boliviano-Tucumana, Chaco Serrano, Bosques secos interandinos*	11	*Tyto alba* *Megascops choliba* *M. watsonii* *M. hoyi* *M. guatemalae* *Pulsatrix perspicillata* *Bubo virginianus* *Strix chacoensis*[a] *Ciccaba virgata* *Glaucidium brasilianum* *Aegolius harrisii*

(continued)

Habitat	Description	Ecoregions	Owl species	
			N°	Species
Gallery forests	Medium canopy (20–25 m) with some emerging trees (30–35 m). Arboreal understory that has an upper and lower strata	*Sudoeste de la Amazonía, Cerrado, Sabanas inundables, Chiquitanía*	4	*Megascops choliba* *Pulsatrix perspicillata* *Bubo virginianus* *Ciccaba virgata*
White sand forests	Reduced diameter trees, semi-open canopy. Short and irregularly open arboreal and shrubby understories. Herbaceous understory with ferns	*Sudoeste de la Amazonía*	–	–
Palm forests	Open canopy (20–25 m, some > 30 m). Mostly open arboreal and shrubby understories (<5 m). Variable herbaceous understory (1–3 m)	*Sabanas inundables*	–	–
Second-growth forests	It represents any type of successional forest that originally developed on a location that has been seized and/or destroyed by human activities; with variable level of recovery	All ecoregions	5	*Megascops choliba* *M. guatemalae* *Pulsatrix perspicillata* *P. melanota* *Glaucidium brasilianum*
Arid lowland scrubs	Thorny bushes. Cactus. Scattered arboreal cactus.	*Chaco, Bosques secos interandinos, Sabanas inundables, Prepuna*	4	*Tyto alba* *Bubo virginianus* *Glaucidium brasilianum* *Athene cunicularia*
Arid montane scrubs	Microforests and/or xeromorphic bushes. Open canopy (2–5 m). Open understory dominated by cactus, bromeliads, and woody plants. Arboreal cactus	*Bosques secos interandinos, Sabanas inundables, Prepuna*	4	*Tyto alba* *Bubo virginianus* *Glaucidium brasilianum* *Athene cunicularia*
Semi-humid/ humid montane scrubs	Semi-open scrublands and grasslands (0.7–1 m), with two strata	*Puna Norteña*	–	–
Cerrado	Medium canopy (± 20 m). Mostly dense understory. Several rocky outcrops	*Cerrado*	1	*Asio stygius*

(continued)

Habitat	Description	Ecoregions	Owl species	
			N°	Species
Campo grasslands	Forbs are dominant (1.5–2.5 m). Forest islands (medium height, closed canopy) and wooded plains (scattered medium-sized trees)	*Sabanas inundables*	2	*Tyto alba* *Athene cunicularia*
Low, seasonally wet grasslands	Prairies of ± 1 m high grasses, sometimes adjacent to water bodies, forests, and/or palm forests	*Sabanas inundables*	2	*Bubo virginianus* *Pseudoscops clamator*
Puna	High Andean short vegetation <0.5 m. Few scattered woody shrubs (sometimes absent). Grasses and tussocks	*Puna Norteña, Puna Sureña*	3	*Tyto alba* *Athene cunicularia* *Asio flammeus*
Riparian thickets	Forbs, usually graminoids. Variable size (1–5 m) depending on the location	All ecoregions	–	–
River island scrub	Islets formed by large grasses and sedges with intertwined stems and roots, sometimes with small trees and palm trees	*Sabanas inundables*	–	–
Pastures/ agricultural lands	Cleared land sometimes with some trees left intentionally, where grasses dominate. Mosaics of crops and remaining climax and/or successional vegetation	All ecoregions	2	*Athene cunicularia* *Asio flammeus*
Second growth brushwood	Successional thickets, shrubs, and grasslands.	All ecoregions	3	*Tyto alba* *Glaucidium brasilianum* *Pseudoscops clamator*
Freshwater marshes	Generally flat areas, permanently or seasonally flooded by shallow water. Sparse aquatic vegetation	All ecoregions	–	–
Riverine sand beaches	Jetties and floodplains	*Sudoeste de la Amazonía*	–	–

(continued)

Habitat	Description	Ecoregions	Owl species	
			N°	Species
Freshwater lakes and ponds	Lentic water bodies that vary in size and depth	*Sudoeste de la Amazonía, Sabanas inundables, Puna Norteña, Puna Sureña*	–	–
Alkaline lakes	Lentic water bodies with large amount of dissolved salts	*Puna Sureña*	–	–
Rivers	Lotic water bodies within a wide range of length, depth, etc.	All ecoregions	–	–
Streams	Lotic water bodies, reduced flow, generally fast	All ecoregions	–	–
Bogs (*Bofedales*)	Semiaquatic vegetation growing at ground level, usually they form of peat bogs	*Puna Norteña, Puna Sureña*	–	–

[a]These species have been recorded in only one habitat type.

References

Aguirre LF, Aguayo R, Balderrama J et al (2009a) El método de evaluación del grado de amenaza de especies (MEGA). In: Ministerio de Medio Ambiente y Agua 2009. Libro rojo de la fauna silvestre de vertebrados de Bolivia. La Paz, pp 7–18

Aguirre LF, Peñaranda DA, Rocha O (2009b) Evaluación del estado de conservación de los vertebrados de Bolivia. In: Ministerio de Medio Ambiente y Agua 2009. Libro rojo de la fauna silvestre de vertebrados de Bolivia. La Paz, pp 19–22

Aliaga-Rossel E, Tarifa T (2005) *Cavia sp.* Como principal presa de la lechuza de campanario (*Tyto alba*) al final de la estación seca en una zona intervenida al norte del departamento de La Paz Bolivia. Ecol Bolivia 40(1):35–42

Balderrama JA (2009) Aves. In: Ministerio de Medio Ambiente y Agua 2009. Libro rojo de la fauna silvestre de vertebrados de Bolivia. La Paz, pp 305–314

BirdLife International (2012a) IUCN Red list for birds. http://www.birdlife.org. Accessed 01 May 2012

BirdLife International (2012b) Species factsheet: *Megascops marshalli.*http://www.birdlife.org. Accessed 01 May 2012

CITES (2012) Apéndices I, II y III. En vigor a partir del 2 de abril de 2012. Convención sobre el Comercio Internacional de Especies Amenazadas de Fauna y Flora. http://www.cites.org/esp/app/appendices.php. Accessed 01 May 2012

Ecológico Kiswara (2011) Búhos tienen renovado nido y se adaptaron sin problemas. La Patria. http://lapatriaenlinea.com/index.php?t=buhostienen-renovado-nido-y-se-adaptaron-sin-problemas&nota=92208. Accessed 13 Mar 2014

Hennesey, AB, Herzog SK, Sagot F (2003) Lista Anotada de las Aves de Bolivia, Quinta Edición. Asociación Armonía/BirdLife International, Santa Cruz de la Sierra

Herzog SK, Balderrama JA (2009) *Megascops marshalli*. In: Ministerio de Medio Ambiente y Agua 2009. Libro rojo de la fauna silvestre de vertebrados de Bolivia. La Paz, pp 399–400
Herzog SK, Ewing SR, Evans KL et al (2009) Vocalizations, distribution, and ecology of the Cloud-Forest Screech Owl (*Megascops marshalli*). Wilson J Ornithol 121(2):240–252
Ibisch PL, Beck SG, Gerkmann B et al (2003) Ecoregiones y ecosistemas. In: Ibisch PL, Mérida G (eds) Biodiversidad: La riqueza de Bolivia. Estado de conocimiento y conservación. Ministerio de Desarrollo Sostenible. Editorial FAN, Santa Cruz de la Sierra, pp 47–88
König C, Weick F, Becking JH (1999) Owls: a guide to owls of the world, 2nd edn. Helm, London
Maillard Z, Davis O, Hennessey AB (2009) Bolivia. In: Devenish C, Díaz Fernández DF, Clay RP et al (eds) Important Bird Areas Americas – Priority sites for biodiversity conservation, BirdLife conservation series no. 16. BirdLife International, Quito, pp 91–98
Miserendino RS (2007) Registro del Chiñi (*Athene cunicularia*) para la Amazonía boliviana. Kempffiana 3(2):23–24
Muñoz A (2006) Apuntes de la Audiencia Técnica – Avances de la Deforestación en Bolivia. Proyecto BOLFOR II, Santa Cruz de la Sierra, p2
Navarro G, Maldonado M (2002) Geografía ecológica de Bolivia: Vegetación y Ambientes Acuáticos. Edit.: Centro de Ecología Simón I. Patiño-Departamento de difusión, Cochabamba
Pokines JT (2007) Prey remains form a great Horned Owl (*Bubo virginianus*) roost in the Icla valley, Bolivia. J Raptor Res 41:174–175
Reboledo CA, Lartigau B (1998) Análisis de regurgitados de *Tyto alba* (Gray, 1829) en el Salar de Uyuni, Departamento de Potosí, República de Bolivia. Biol Neotrop 1(2):19–22
Redacción Central (2014) Cazadores invaden Albarrancho. Los Tiempos. http://www.lostiempos.com/actualidad/local/20140312/cazadores-invaden-albarrancho?page=2. Accessed 13 Mar 2014
Serrano P, Anderson S (1986) Datos sobre la alimentación de *Athene cunicularia* en la Reserva de Fauna Altoandina de Ulla-Ulla, Bolivia. Acta Vertebrata 13:180–182
Stotz DF, Fitzpatrick JW, Parker TA et al (1996) Neotropical birds: ecology and conservation. University of Chicago Press, Chicago
Trejo A (2007) Bibliografía comentada sobre aves rapaces de Argentina. Hornero 22(2):185–217
Urioste A (2010) Deforestación en Bolivia: Una amenaza mayor al cambio climático. Documento de Trabajo. Fundación Friedrich Ebert. Foro de Desarrollo y Democracia, 29 p
Vargas JA, Landaeta AC, Simonetti JA (2002) Bats as prey of barn owls (*Tyto alba*) in a tropical savanna in Bolivia. J Raptor Res 36:146–147
Wildlife Conservation Society (2008) Mapa del Estado de Conservación de Bolivia. Versión 2. Elab. Siles TM, Domic E, Painter L, Wallace RB. Wildlife Conservation Society, La Paz
Wink MA, El-Sayed E, Sauer-Gürth H et al (2009) Molecular phylogeny of owls (Strigiformes) inferred from DNA sequences of the mitochondrial cytochrome b and the nuclear RAG-1gene. Ardea 97(4):581–591

Chapter 6
The Owls of Brazil

Jose Carlos Motta-Junior, Ana Claudia Rocha Braga, and Marco Antonio Monteiro Granzinolli

Abstract In this chapter, we have compiled the literature for the 23 Brazilian owl species, which represent almost one-third of all Neotropical owls. Our main objective here was to review the taxonomy, distribution, ecology, and conservation of each species, both revealing gaps in our knowledge and supplying directions to promote more research on Brazilian owls. Even though in the last 15 years there was an impressive increasing number of specific publications on owls in Brazil, particularly on food habits, we have detected that other essential data about distribution, taxonomy, population density, habitat requirements, reproductive biology, and anthropogenic effect on owl populations are needed for most owl species. Probably the lack of basic biological information on most species is responsible for the relative absence of owls in the Brazilian official lists of threatened fauna in regional and national scales.

Keywords Brazilian owls • Literature review • Ecology • Conservation

J.C. Motta-Junior (✉) • A.C.R. Braga • M.A.M. Granzinolli
Laboratorio de Ecología de Aves – LABECOAVES, Departamento de Ecologia, Instituto de Biociências da Universidade de São Paulo, 05508-090 São Paulo, SP, Brazil
e-mail: labecoaves@yahoo.com

P.L. Enríquez (ed.), *Neotropical Owls*, DOI 10.1007/978-3-319-57108-9_6

Ferruginous Pygmy Owl (*Glaucidium brasilianum*)

6.1 Introduction

In this chapter, we examine the scientific literature for the 23 owl species currently recorded in Brazil, providing a basis for further studies. We have compiled Brazilian owl biological literature from studies originally made within Brazil or from birds collected/trapped in Brazil. Some major works such as textbooks, guides, or revisions including Brazilian species were also considered (e.g., Clark et al. 1978; Sick 1993, 1997; Holt et al. 1999; König et al. 1999, 2008; Mikkola 2012). Data on distribution of owls in Brazil were acquired from field guides or catalogs (e.g., Meyer de Schauensee 1966, 1982; Pinto 1978; Sick 1993, 1997; Holt et al. 1999; König et al. 2008; Grantsau 2010; Minns et al. 2010), from regional listings (e.g., Belton 1994; Scherer-Neto and Straube 1995; Rosário 1996; Bencke 2001; Pacheco 2003; Tubelis and Thomas 2003; Willis and Oniki 2003; Naka et al. 2006), and from Internet databases (http://www.wikiaves.com.br/, http://www.hbw.com/ibc/, and http://www.xeno-canto.org/). Our main goal here was to review the taxonomy, distribution, ecology, and conservation of Brazilian owls, revealing gaps in our knowledge and supplying some analysis and directions to promote more research on owls in Brazil.

6.2 Taxonomic Diversity

Brazilian owls include ten (37%) of all 27 recognized genera of Strigiformes in the world (König et al. 2008). The genera with the most species are *Megascops* (6) and *Glaucidium* (4), though other two genera (*Asio and Strix*) have three species each (Table 6.1, Appendix 6.1). There are currently 23 recognized species of owls in Brazil (Sigrist 2006; Piacentini et al. (2015), representing 30.7% of all 75 Neotropical owl species (Enríquez et al. 2006) and 9.2% of the world's 250 species (König et al. 2008; Mikkola 2012, Appendix 6.1).

Some genera have had taxonomic revisions in the last decade. The burrowing owl (*Athene cunicularia*) was included within *Speotyto* (Clark 1997), but studies on several features including DNA evidence proposed reclassification to *Athene* (Penhallurick 2002; Wink et al. 2004; König et al. 2008). Norberg (2002) and König et al. (2008), based, respectively, on morphology and DNA evidence, placed *Ciccaba* species within genus *Strix*. The striped owl (*Asio clamator*) has been placed in the genus *Rhinoptynx* (Burton 1984; Remsen et al. 2016) and in the genus *Pseudoscops* (Olson 1995; Sibley 1996). On the other hand, Penhallurick (2002) and König et al. (2008) recommend its return to the genus *Asio*, based on DNA evidence. Heidrich et al. (1995a) confirmed by DNA analysis the specific status of four Brazilian *Megascops*: *M. choliba*, *M. atricapilla*, *M. usta*, and *M. sanctaecatarinae*. More recently, a study using three mitochondrial and three nuclear genes analyses in the genus *Megascops* has proposed *M. usta* to be a subspecies of *M. watsonii* and *M. roraimae* to be a subspecies of *M. vermiculatus* (Dantas et al. 2016). König et al. (2008), based on recent DNA evidence, suggested a new species: the American barn owl *Tyto furcata*, split from the common barn owl (*T. alba*).

In a similar way as stated by Enríquez et al. (2006) for Neotropical owls, our understanding of Brazilian owl taxonomy also continues to change rapidly (e.g., König et al. 1999, 2008; Mikkola 2012; Remsen et al. 2016). However, we need additional data on field recordings of vocalizations, DNA studies, distribution, and detailed revisions on specimens at museums to elucidate some pending questions on Brazilian owl taxonomy (e.g., *Megascops*, *Pulsatrix*, and *Glaucidium*). As an example, of the 20 Brazilian owl species with recorded songs and calls in Hardy et al. (1999), we found only seven with recordings made in Brazilian territory. More recently Boesman (2006) have included ten species with recordings made inside Brazil. These informations are vital for a more solid taxonomic foundation of the group and for their conservation (Enríquez et al. 2006).

6.3 Distribution and Biome Association

Brazil is the largest country (8,514,877 km^2) in the Neotropics possessing currently 26 states (Fig. 6.1). According to Stotz et al. (1996), in Brazil, there are five major zoogeographic regions (Amazonia North, Amazonia South, Central South America, Atlantic Forest, and Pampas) and, within these, 13 subregions in Brazil.

Table 6.1 Owl species distribution per state in Brazil

Scientific name	Common name	Amazonas	Acre	Rondonia	Roraima	Amapá	Pará	Maranhão	Piauí	Rio Grande do Norte	Pernambuco	Sergipe	Paraíba	Alagoas	Ceará	Bahia	Tocantins	Mato Grosso	Mato Grosso do Sul	Goiás	Minas Gerais	Espírito Santo	Rio de Janeiro	São Paulo	Paraná	Santa Catarina	Rio Grande do Sul
Tyto furcata	American barn owl	X	X	X	X	X	X	X	X	X	X	X	X	X	X	X	X	X	X	X	X	X	X	X	X	X	X
Megascops choliba	Tropical screech owl	X	X	X	X	X	X	X	X	X	X	X	X	X	X	X	X	X	X	X	X	X	X	X	X	X	X
Megascops watsonii	Northern tawny-bellied screech owl	X			X	X	X											X									
Megascops usta	Southern tawny-bellied screech owl	X	X	X			X	X	X								X	X									
Megascops atricapilla	Black-capped screech owl										X			X		X				X	X	X	X	X	X	X	
Megascops sanctaecatarinae	Long-tufted (or Santa Catarina) screech owl																								X	X	X
Megascops roraimae[a]	Roraima (or foothill) screech owl				X		X																				
Lophostrix cristata	Crested owl	X	X	X	X	X	X	X									X	X									
Pulsatrix perspicillata[b]	Spectacled owl	X	X	X	X	X	X	X	X	X	X	X	X	X	X	[b]	X	X	X	[b]	[b]	[b]	[b]	[b]	[b]	[b]	[b]
Pulsatrix koeniswaldiana	White-chinned (or tawny-browed) owl															X					X	X	X	X	X	X	X
Strix virgata	Mottled owl	X	X	X	X	X	X	X	X		X			X		X		X	X	X	X	X	X	X	X	X	X
Strix hylophila	Rusty-barred owl																				X	X	X	X	X	X	X
Strix huhula	Black-banded owl	X	X	X	X	X	X	X	X							X	X	X	X	X	X	X	X	X	X	X	

Bubo virginianus	Great horned owl	X		X	X	X	X	X	X	X	X			X	X	X	X	X	X	X	X	X	X	X	X	X	X
Glaucidium minutissimum[c]	Least pygmy owl						X									X		X		X	X	X	X	X	X	X	
Glaucidium hardyi	Amazonian pygmy owl	X	X	X	X	X	X	X										X									
Glaucidium brasilianum	Ferruginous pygmy owl	X	X	X	X	X	X	X	X	X	X	X	X	X	X	X	X	X	X	X	X	X	X	X	X	X	X
Glaucidium mooreorum[d]	Pernambuco pygmy owl										X																
Athene cunicularia	Burrowing owl	X	X	X	X	X	X	X	X	X	X	X	X	X	X	X	X	X	X	X	X	X	X	X	X	X	X
Aegolius harrisii	Buff-fronted owl	X									X		X	X	X	X	X	X		X	X			X	X	X	X
Asio stygius	Stygian owl	X	X	X	X	X	X									X		X	X	X	X	X	X	X	X	X	X
Asio clamator	Striped owl	X	X	X	X	X	X	X	X	X	X	X	X	X	X	X	X	X	X	X	X	X	X	X	X	X	X
Asio flammeus	Short-eared owl															X		X		X	X			X	X	X	X
Number of species		15	12	13	14	13	16	12	10	7	11	6	7	10	8	15	11	17	10	14	16	14	14	16	17	17	14

Compilation from Meyer de Schauensee (1966, 1982), Pinto (1978), Dubs (1992), Sick (1993, 1997), Scherer-Neto and Straube (1995), Bencke (2001), Naka et al. (2006, 2007), Rosário (1996), Cândido-Junior (1999), Holt et al. (1999), Pacheco (2003), Silva et al. (2003), Tubelis and Tomas (2003), Willis and Oniki (2003), Pacheco and Olmos (2006), König et al. (2008), Legal et al. (2009), Pinheiro and Dornas (2009), Grantsau (2010), Minns et al. (2010), and Guilherme (2012) consults to the sites http://www.wikiaves.com.br/, http://www.hbw.com/ibc/, and http://www.xeno-canto.org/ and several more specific distribution references in Appendix 6.2. The Distrito Federal was grouped with the state of Goiás, since its tiny territory is located within this state. Nomenclature according to König et al. (2008), except when indicated

[a]Vermiculated screech owl *M. guatemalae* after Piacentini et al. (2015)

[b]Includes short-browed owl *P. pulsatrix* after König et al. (2008) and state's columns indicate occurrence of *P.p. pulsatrix*

[c]Sick's pygmy owl *G. sicki* according to König et al. (2008)

[d]*G. minutissimum* after König et al. (2008)

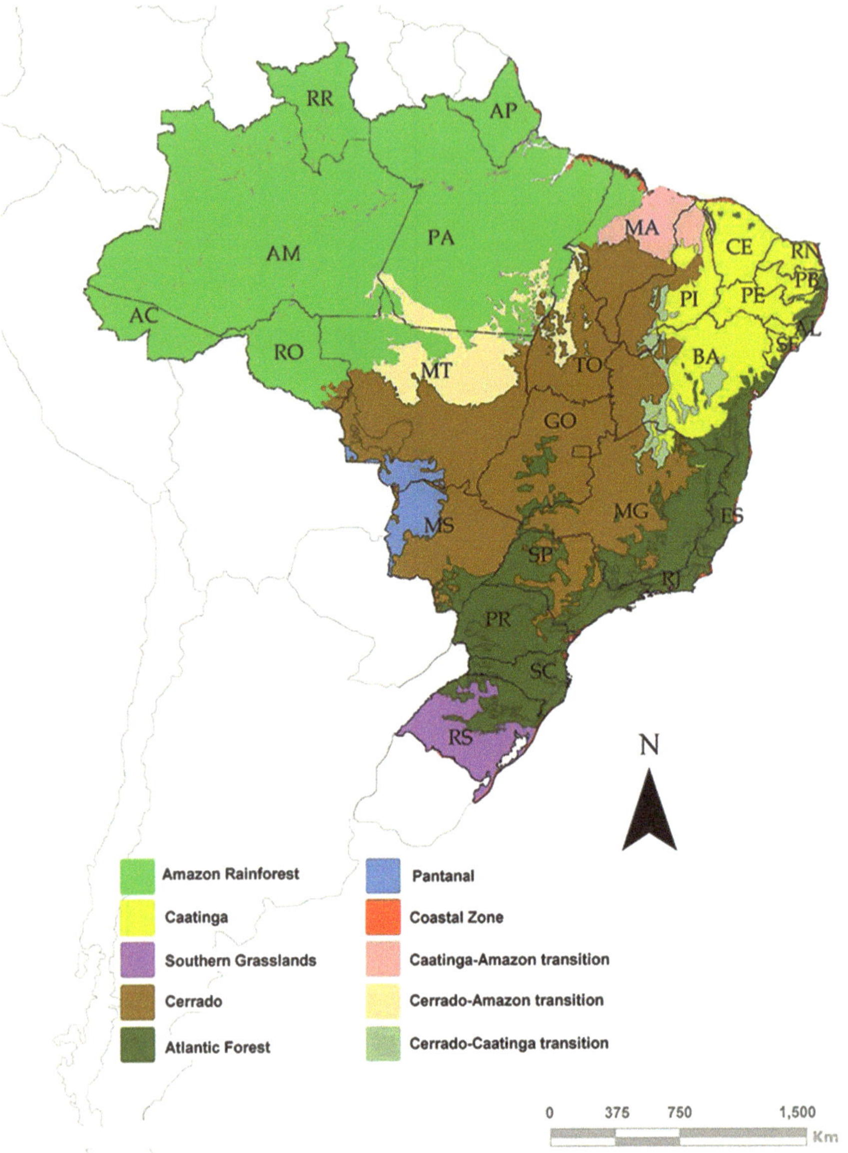

Fig. 6.1 Brazilian six major biomes and 26 states (Adapted from WWF (http://www.wwf.org.br/))

State abbreviations: RR, Roraima; AM, Amazônia; AC, Acre; AP, Amapá; RO, Rondônia; PA, Pará; MA, Maranhão; MT, Mato Grosso; PI, Piauí; CE, Ceará; RN, Rio Grande do Norte; PB, Paraíba; PE, Pernambuco; AL, Alagoas; SE, Sergipe; BA, Bahia; TO, Tocantins; GO, Goiás; MG, Minas Gerais; ES, Espírito Santo; RJ, Rio de Janeiro; MS, Mato Grosso do Sul; SP, São Paulo; PR, Paraná; SC, Santa Catarina; RS, Rio Grande do Sul.

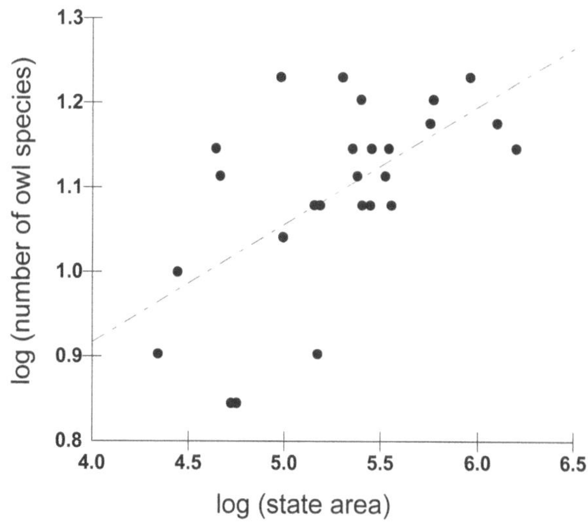

Fig. 6.2 The (log) of total number of owl species as a function of the (log) area in km^2 of the 26 Brazilian states according to a simple linear regression model: Log(number of owl species) = 0.260 + 0.156(Log (Area km^2)), R^2 = 0.348, F = 12.809, n = 26, p = 0.002) (Data of territorial area of states are from IBGE (2007))

Six major biomes are recognized for Brazil (Overbeck et al. 2007): the Amazon Rainforest, Atlantic Forest, Cerrado, Caatinga, Pantanal, and Southern Grasslands (or Pampas) (Fig. 6.1). Detailed descriptions of zoogeographic regions, biomes, and bird habitat diversity in Brazil can be found in Sick (1993, 1997) and Stotz et al. (1996).

Brazil has 23 recognized species of owls (Sigrist 2006; Piacentini et al. (2015), with one endemic (Pernambuco pygmy owl*Glaucidium mooreorum*) and five "quasi-endemic" (species having distributions restricted to two or three countries, sensu Escalante et al. 1993 and Enríquez et al. 2006): black-capped screech owl (*Megascops atricapilla*), long-tufted (or Santa Catarina) screech owl (*M. sanctae-catarinae*), white-chinned owl (*Pulsatrix koeniswaldiana*), rusty-barred owl (*Strix hylophila*), and least pygmy owl (*Glaucidium minutissimum*). It is noteworthy that these species have more than 80% of their distribution inside Brazil. Other six species also have most of their distribution (up to 70%) within Brazil: Southern tawny-bellied screech owl (*M. usta*), crested owl (*Lophostrix cristata*), spectacled owl (*P. perspicillata*, including ssp. *pulsatrix*), Amazonian pygmy owl (*G. hardyi*), black-banded owl (*S. huhula*), and ferruginous pygmy owl (*G. brasilianum* accepting its separation from *G. ridgwayi*, *G. peruanum*, and *G. tucumanum*; see Heidrich et al. 1995b, König et al. 1999, and Proudfoot et al. 2006).

Although state territories are an artificial or political construct, the number of owl species in Brazil increases with state territorial area (Fig. 6.2). However, as stated by Enríquez et al. (2006), in a similar analysis for all Neotropical countries, other factors such as latitude, habitat variability or heterogeneity, and climate (Diniz-Filho et al. 2004) influence the number of species in states. For Brazil, we also suppose that the states with larger population and presumably larger number of research centers and ornithologists (see Motta-Junior and Braga 2012), as those from Minas Gerais south to Rio Grande do Sul, have an advantage on owl species detection (Table 6.1). In fact, though statistically significant (Fig. 6.2), the regression for species-area relationship for Brazilian states has a lower coefficient of determination ($R^2 = 0.348$) when compared to Enríquez et al.'s (2006) findings for Neotropical countries ($R^2 = 0.45$).

According to the analysis of Enríquez et al. (2006) for owl distribution in Neotropical region, Mexico and Peru have the majority of species (32 each), followed by Ecuador (29) and Colombia (27). Although these four countries are highly variable in topography and habitat types, which can explain its high taxonomic diversity (Enríquez et al. 2006), why does Brazil, by far the largest country in Neotropics and also presenting great variability in topography and habitats (Stotz et al. 1996), have only 23 owl species? We suppose the relatively low number of owl species found in Brazil in comparison with those countries (each with less than one-fourth of Brazil's territorial area) can be perhaps partially explained by possible gaps in collections and field research, as at the northwestern Amazonian region (see Fig. 2 in Diniz-Filho et al. 2004). We expect that, with additional biological research on owls in Brazil, more species should be reported to the Brazilian list or even described. For example, according to text and maps in König et al. (2008), there is an indication that band-bellied owl (*Pulsatrix melanota*) and Chaco pygmy owl (*Glaucidium tucumanum*) perhaps occur in extreme western Brazil. Just new collections or field observations with vocalization recordings in this region can confirm these suspicions.

6.4 Endangered Species

Thiollay (1989, 1994) has reported that raptors, including Strigiformes, are among the most endangered and neglected groups of tropical birds. Surely this overlooking is reflected on the Brazilian official lists of threatened fauna in national and even regional scales: just three owl species are listed for Brazil (national list), two as "vulnerable" (the spectacled owl and the black-banded owl) and one as "extinct" (the Pernambuco pygmy owl) (ICMBio 2014). Additionally only six of eight official state lists of threatened fauna have owls included (Table 6.2). In three of these states (São Paulo, Paraná, and Rio Grande do Sul), six species are listed as "data deficient" (Table 6.2), but this category is not considered legally for protection in Brazil (Machado et al. 2005). We are wondering why the rusty-barred owl is listed globally as "near threatened" by IUCN since 2004 (Birdlife International 2016a, IUCN 2016), but in Brazil (approximately 90% of species distribution) this owl continues to be ignored.

Judging from the comments on status and conservation of Brazilian owls in both Holt et al. (1999), König et al. (1999, 2008), one can imagine if there are at least some threatened species not listed in the Brazilian list of endangered fauna. As suggested by Penhallurick (2002) in relation to the need of revision for owls in the IUCN Red List, we think the Brazilian list is also in need of revision concerning owls. Fortunately, some initiatives have been proposed, as that of the Brazilian

National Plan to the conservation of birds of prey (Soares et al. 2008), which have recommended the inclusion of Pernambuco pygmy owl as "critically endangered," and new evaluations of buff-fronted owl (*Aegolius harrisii*), short-eared owl (*Asio flammeus*), and rusty-barred owl as possible candidates to be included in the Brazilian Red List of endangered fauna. But, as perceived by the last national list (ICMBio 2014), the National Plan's suggestion was not considered.

Regional lists in Brazil also possibly suffer some inconsistencies. As an example, the state of Minas Gerais' list of threatened fauna (COPAM 2010, Table 6.2) has no owl species included, even though both species nationally listed as "vulnerable" (the spectacled owl and the black-banded owl) have few records in this state (Wikiaves 2016). We suggest that these species should be included at least as "data deficient" because of the paucity of records at the state, while we do not confirm this paucity to be a result of true rarity or simply a difficult detection of species.

Penhallurick (2002) made a plea for more attention to subspecies as an important unity for conservation strategies. The Brazilian examples of the current taxonomic dispute on short-browed owl (*Pulsatrix perspicillata pulsatrix*) and Roraima screech owl (*Megascops guatemalae roraimae*/*M. roraimae*) status (CBRO 2001a; Ramirez-Llorens and Bellocq 2007; König et al. 2008; Remsen et al. 2016) truly support Penhallurick's concerns (Penhallurick 2002). In this case, the Brazilian previous Red List (Machado et al. 2005) has pioneered in considering subspecies of birds when listing threatened taxa, not the case for the most recent listing (ICMBio 2014).

6.5 Threats

6.5.1 Habitat Destruction

The most serious risk for many species of raptors is habitat loss (Thiollay 1984, 1989, 1994; Bierregaard 1998; König et al. 2008). König et al. (1999, 2008) report habitat destruction (mostly forests) as one of the main concerns for at least 18 of the 23 Brazilian species.

The Atlantic Forest presents the highest number of endemic species including six owls: black-capped screech owl, long-tufted screech owl, rusty-barred owl, white-chinned owl, least pygmy owl, and Pernambuco pygmy owl (Bencke et al. 2006, Table 6.3). One important center of endemism of the Atlantic Forest is located in northeastern Brazil, often called Center Pernambuco (Silva and Castelleti 2003). The North Atlantic Forest currently has less than 5% of its original vegetation (Ribeiro et al. 2009), which severely reduced the habitat of Permanbuco pygmy owl, recently considered extinct (ICMBio 2014; Pereira et al. 2014, Table 6.2). Most of the studies on owls in Brazil concentrate on this biome (Motta-Junior and Braga 2012).

Table 6.2 Status of Brazilian owls in global, zoogeographic, national, and regional (state) scales

Scientific name	Common name	IUCN 2016	CITES 2016	Sensibility in Neotropics	Conservation Priority in Neotropics	Brazil 2014	State of Pará 2008	State of São Paulo 2014	State of Espírito Santo 2007	State of Paraná 2004	State of Santa Catarina 2011	State of Rio Grande do Sul 2014	State of Rio de Janeiro 2000	State of Minas Gerais 2010
Tyto furcata	American barn owl		II	L	4									
Megascops choliba	Tropical screech Owl		II	L	4									
Megascops watsonii	Northern tawny-bellied screech Owl		II	H	4									
Megascops usta	Southern tawny-bellied screech owl		II											
Megascops atricapilla	Black-capped screech owl		II	L	4									
Megascops sanctaecaterinae	Long-tufted screech owl		II											
Megascops roraimae[a]	Roraima screech owl		II	M	4									
Lophostrix cristata	Crested owl		II	H	4									
Pulsatrix perspicillata[b]	Spectacled owl		II	M	4	VU		DD		DD		EN		
Pulsatrix koeniswaldiana	White-chinned owl		II	H	3									
Strix virgata	Mottled owl		II	M	4					DD		DD		
Strix hylophila	Rusty-barred owl	NT	II	H	3									

Strix huhula	Black-banded owl		II	M	4	VU		DD	VU	DD	EN			
Bubo virginianus	Great horned owl		II	L	4			TH						
Glaucidium minutissimum[c]	Least pygmy owl		II	M	4				EN					VU
Glaucidium hardyi	Amazonian pygmy owl		II	H	4									
Glaucidium mooreorum[d]	Pernambuco pygmy owl	CR	II			EXT								
Glaucidium brasilianum	Ferruginous pygmy owl		II	L	4									
Athene cunicularia	Burrowing owl		II	M	4									
Aegolius harrisii	Buff-fronted owl		II	H	4					DD		DD		
Asio stygius	Stygian owl		II	M	3					DD				
Asio clamator	Striped owl		II	L	4									
Asio flammeus	Short-eared owl		II	L	3			NT		DD	VU	NT		

Categories used after IUCN: *CR* critically endangered, *EN* endangered, *VU* vulnerable, *NT* near threatened, *DD* data deficient, *TH* "threatened," without categorization (São Paulo 2014). Sensibility in the Neotropics (Stotz et al. 1996): *L* low, *M* medium, *H* high. Conservation Priority in the Neotropics (Stotz et al. 1996): *1* urgent, *2* high, *3* medium, *4* low. Scientific and common names after König et al. (2008), except when indicated

Bibliographic sources for lists: IUCN (2016), CITES (UNEP-WCMC 2016), Brazil - 2014 (ICMBio, 2014), Pará (Pará 2008), Sao Paulo (São Paulo 2014), Espírito Santo (Simon et al. 2007), Paraná (Straube et al. 2004), Santa Catarina (CONSEMA 2011), Rio de Janeiro (Alves et al. 2000), Rio Grande do Sul (Rio Grande do Sul 2014), Minas Gerais (COPAM 2010)

Subspecies: ([1]) *B. v. deserti;* ([2]) *P. p. pulsatrix;* ([3]) *S. v. borelliana;* ([4]) *S. h. albomarginata*

[a]Vermiculated screech owl (*M. guatemalae*) after Piacentini et al. (2015)

[b]Includes short-browed owl (*P. pulsatrix*) after König et al. (2008)

[c]*G. sicki* after König et al. (2008)

[d]*G.minutissimum*after König et al. (2008)

Especially after the 1960s the Cerrado has been rapidly converted to agriculture and livestock, which was also promoted by the government (Silva 2000; Müller 2003; Azevedo and Monteiro 2006). The growing demand for food and agricultural commodities, including the recent development of new techniques for producing biofuel, has added additional challenges to biodiversity conservation in Brazil. Today, the Cerrado, which formerly covered 21% of the country, has less than 45% of its original vegetation remaining (Klink and Machado 2005) and has been rapidly converted to vast agricultural fields (Machado et al. 2004; Klink and Machado 2005; Macedo and Tetti 2007). The natural vegetation has been converted in a rate of more than 1% per year (Machado et al. 2004), and besides landcover conversion, soil degradation, the intense use of pesticides, the intensification of the fire regimes, and the invasion of exotic plant species composed the severe threat that the Cerrado is facing (Klink and Machado 2005). The Cerrado region is the second in number of references on owls in Brazil (Motta-Junior and Braga 2012).

The Amazon Rainforest is also suffering a growing pressure from agriculture and livestock expansion (Laurance et al. 2004), mostly in its southern border, called arc of deforestation (Fearnside 2005). According to Fearnside (2005), the Amazonian deforestation rates are associated with the global macro-economy, which reveals the importance of large landowners, especially cattle ranchers, in the deforestation rates. In general, Amazonian owl species are not well known, and few studies have focused on the species from the region, such as Barros (2003) and Borges et al. (2004). In the Amazon Rainforest, there are five owl species, which occur exclusively in this biome in Brazil: Northern tawny-bellied screech owl (*Megascops watsonii*), Southern (or Austral) tawny-bellied screech owl, Roraima (or foothill) screech owl (*M. roraimae*), Amazonian pygmy owl (*Glaucidium hardyi*), and crested owl (Table 6.3).

The Caatinga is the only exclusive Brazilian biome (Fig. 6.2), and the estimates of its converted area range from 27.5% to 51.7% (Leal et al. 2005). The main threats to this biome are the intensification of agriculture and livestock and the increase in the use of wood for fuel. The environmental degradation has changed the rain levels, and together with the bad soil conservation, techniques employed in agriculture have promoted around 15% of the area to desertification (Leal et al. 2005). The representativeness of bird studies in the Caatinga is very poor (Nascimento 2000; Pacheco 2003; Silva et al. 2003) which makes it even more difficult for the establishment of conservation plans for the group in the region.

The Pantanal is the largest wetland in the planet (Brandon et al. 2005) and was defined as a biosphere reserve by UNESCO in 2000. Besides the notable species abundance, the small number of endemic species is of a great contrast (Swartz 2000). The European colonization of this region started around two centuries ago with cattle ranching in very large properties, which led to a low human density that preserved relatively well the natural environments (Alho and Lacher Jr. 1991). However, in the last years, with the development of new techniques of cattle ranching and division of the former large properties, as well as the expansion of high-technology agriculture, Pantanal wetlands have been under increasing threat (Harris et al. 2005).

Table 6.3 Distribution of owls in Brazilian biomes

Scientific name	Common name	Amazon Rainforest	Cerrado	Pantanal	Caatinga	Atlantic forest	Southern Grasslands
Tyto furcata	American barn owl	O	X	X	X	O	X
Megascops choliba	Tropical screech owl	X	X	X	X	X	W
Megascops watsonii	Northern tawny-bellied screech owl	X	F				
Megascops usta	Southern tawny-bellied screech owl	X	F				
Megascops atricapilla [1]	Black-capped screech-owl					X	
Megascops sanctaecaterinae [1]	Long-tufted screech owl					X	
Megascops roraimae[a]	Roraima screech owl	X					
Lophostrix cristata	Crested owl	X					
Pulsatrix perspicillata[b]	Spectacled owl	X	F	F	F	X	
Pulsatrix koeniswaldiana [1]	White-chinned owl					X	
Strix virgata	Mottled owl	X			F	X	
Strix hylophila [1]	Rusty-barred owl					X	
Strix huhula	Black-banded owl	X	F	F		X	
Bubo virginianus	Great horned owl	X	X	X	X	X	W
Glaucidium minutissimum[c 1]	Least pygmy owl					X	
Glaucidium hardyi	Amazonian pygmy owl	X		F			
Glaucidium mooreorum[d 2]	Pernambuco pygmy owl					X	
Glaucidium brasilianum	Ferruginous pygmy owl	X	X	X	X	X	W
Athene cunicularia	Burrowing owl	O	X	X	X	O	X
Aegolius harrisii	Buff-fronted owl		X	X	X	X	

(continued)

Table 6.3 (continued)

Scientific name	Common name	Amazon Rainforest	Cerrado	Pantanal	Caatinga	Atlantic forest	Southern Grasslands
Asio stygius	Stygian owl	X	X	X	X	X	W
Asio clamator	Striped owl	O	X	X	X	O	X
Asio flammeus	Short-eared owl		X	X		O	X

X regular occurrence, *O* needs enclaves of open to semi-open habitats or colonizing deforested areas, *F* only occurs in enclaves of forested habitats, *W* needs also some patches of wooded and semi-open habitats. Data compilation from our field experience and from Sick (1993,1997), Stotz et al. (1996), Holt et al. (1999), Pacheco (2003), Silva et al. (2003), Tubelis and Tomas (2003), König et al. (2008), and Menq (2016b). Scientific and common names according to König et al. (2008), except when indicated

Biome endemism: [1] – Endemic of South Atlantic Forest sub-region (sensu Stotz et al. 1996) (Bencke et al. 2006); [2] – Endemic of North Atlantic Forest sub-region (sensu Stotz et al. 1996) (Bencke et al. 2006)

[a]*M. guatemalae* after Piacentini et al. (2015)

[b]Includes *P. pulsatrix* after König et al. (2008)

[c]*G. sicki* according to König et al. (2008)

[d]*G. minutissimum* after König et al. (2008)

The Southern Grasslands (Pampas) are restricted to the south of Brazil and the neighbor countries (MMA - Ministério Do Meio Ambiente 2000; Overbeck et al. 2007). This biome is composed of a mosaic of grasslands, shrublands, and, in a lesser extent, different forest types (Overbeck et al. 2007). The traditional human activity in the region is cattle ranching, using the natural pastures for this purpose in large properties, which guaranteed the sustainability of the region for a long period of time (Pillar and Quadros 1997; Pillar et al. 2006). However, in the last two decades, the cattle stock started to increase considerably, and concerns of overgrazing started to rise (Overbeck et al. 2007). In the last three decades, the expansion of agriculture converted approximately 25% of the natural grasslands into agricultural fields, as rice, corn, and more recently soybean plantations are expanding rapidly in the region (Overbeck et al. 2007). Another great threat to this region is the demand for exotic tree plantations, such as *Eucalyptus* spp., *Pinus* spp., and *Acacia* spp., which are also expanding in the region (Pillar et al. 2006). More than converting natural areas in pine plantations, the pine species are invading other natural areas, mostly grasslands (as found by Bustamante and Simonetti 2005 in Chile), which are also threatened by the invasion of exotic grasses (Overbeck et al. 2007). Biodiversity information of this biome is also lacking (Overbeck et al. 2007). Thinking about owls, the converting process of open natural areas to nonnatural forest areas can be a negative factor for owl populations that inhabit this biome, which is naturally a grassland area.

As exposed, the main threats to the Brazilian biodiversity are the conversion of the natural habitats to the production of agricultural and cattle goods and the consequent infrastructure necessary to this production, such as roads and energy (Macedo and Tetti 2007). Definitely, the ultimate challenge for the next years to Brazilian biodiversity conservation in general and owls in particular will be to find a balance between internal and external demand for increasing agricultural production and the requirements of conservation of globally important habitats. Additionally, it is important to note the comparatively low number of studies about Strigiformes detected in four Brazilian biomes: Southern Grasslands, Pantanal, Amazon Rainforest, and Caatinga (Motta-Junior and Braga 2012).

6.5.2 Illegal Animal Trade

Although owl species apparently are not an important target of the poachers for illegal market in Brazil (RENCTAS 2001; Ferreira and Glock 2004), we have found some records of illegally commercialized Strigiformes. Common species such as the American barn owl and the tropical screech owl were found in street markets in the state of Pernambuco (Pereira and Brito 2005). Additionally, the American barn owl and burrowing owl were found in street markets in Rio Grande do Sul (Ferreira and Glock 2004). The most common use of these owls is as pet animals.

The main routes of the illegal animal trade starts in the north, northeast, and center-west of the country and goes to the south and southeast, mainly São Paulo and Rio de Janeiro, where they are usually shipped to Europe, Asia, or North America (RENCTAS 2001). The difficulty of gathering data about the illegal animal market makes extremely difficult the understanding of the actual extent of this threat to owl species.

6.5.3 Road Mortality and Other Accidents

According to Bencke and Bencke (2000), roadkills may be an important cause of mortality among large forest owls on southern Brazil. For example, road-killed individuals of white-chinned owl (Bencke and Bencke 1999, 2000), black-banded owl (Gonzaga and Gastiglione 2004), mottled owl (*Strix virgata*) (Lau 2008), and rusty-barred owl (Joppert 2007) have been reported. Small and/or open country owls are also vulnerable to roadkills as recorded by Coelho et al. (2008) for tropical screech owl, burrowing owl, American barn owl, and striped owl in south Brazil; ferruginous pygmy owl (Melo and Santos-Filho 2007) in the state of Mato Grosso; burrowing owl, great horned owl, least pygmy owl, ferruginous pygmy owl, striped owl, white-chinned owl, and American barn owl in Cerrado and Pantanal areas (Fisher et al. 2003); and American barn owl and short-eared owl in southeastern Brazil (A Martensen and ACR Braga unpubl. Data). In a general statement, König et al. (2008) described tropical screech owl's hunting habits along roadsides as the cause of roadkills. One of the most complete studies to date is that of Coelho et al. (2008) in the state of Rio Grande do Sul, where four owl species were 6% of all 67 bird species and 14 owl individuals were 8.3% of all 169 bird individuals road-killed, whereas the 14 owl species recorded for this state represent only 2.2% of all 624 bird species of the state (Bencke 2001).

Roadkills are not restricted to one region of Brazil, and in spite of the relatively small number of records in this country, it may be an important cause of mortality for owls (Bencke and Bencke 1999, 2000; Bencke et al. 2003; Coelho et al. 2008; Favreto and Santos 2009, Rebelato et al. 2011). A 3-year study on 180 km of a highway between Rio de Janeiro (RJ) and Juiz de Fora (MG) revealed that, for the 92 birds recorded, 37% were owls from seven species, mainly the American barn and striped owls (Freitas et al. 2013). Owls are probably attracted to the roads because of prey conspicuity/density (Duncan 2003) or because of forest edges, a preferred foraging habitat for forest species (Bencke and Bencke 1999). Some species are locally rare, such as the mottled owl in the state of Rio Grande do Sul (Lau 2008), which suggests roadkills can have a strong effect in owl populations (Bencke and Bencke 1999, 2000). Hence, the enhancement of road density could increase the pressure over owl populations.

The different centers of wildlife rehabilitation receive owls from the illegal market, but mainly from the requirements of the local community to remove individuals (Borges et al. 2006), mostly due to noise or repulsion. A large fraction of these individuals are probably from falls from the nest (Borges et al. 2006; Joppert 2007; Souza et al. 2014, and ACR Braga unpubl. data). Among the most common accidents with owls, Joppert (2007) includes collision with energy transmission lines (one tropical screech owl, two striped owls), collision with buildings (two striped owls), collision with barbed wire fences (one striped owl), and collision with automobile or find on roads (three tropical screech owls and one rusty-barred owl). In addition, we report at least an observation of an American barn owl impaled in barbed wire at Minas Gerais (see photos). These injuries or deaths are mostly related to urban sites (Joppert 2007). Collision with buildings can be related to constructions near natural areas, as Bornschein and Reinert (2000) report for a black-banded owl in the state of Paraná.

6.5.4 Myths and Superstitions

There is all over Brazil, especially in countryside areas, the misperception of owls as bad omens (Ihering 1940; Costa-Neto 1999; Petroff 2002) or predatory pests, which could bring threats to owls, as unfortunately corroborated by some reports of human persecution, such as the mottled owl (Reinhardt 1870), tropical screech owl (Santos 1979), and striped owl (Joppert 2007). The vernacular name in Portuguese, "mocho-do-diabo" (literally "devil's owl"), for the stygian owl (*Asio stygius*) is suggestive of the misperception of part of the population.

Santos (1979) reports some indigenous myths such as the ferruginous pygmy owl's hypnotic ability with respect to other birds, becoming virtually a "tyrant." One of the common names of the American barn owl ("suindara" in indigenous language, not Portuguese) literally means "one who does not eat" (Santos 1979). Another curious but also negative indigenous myth about the great horned owl states that contact with this owl spreads laziness to humans (Santos 1979). In Amazon, Costa-Neto (1999) reports on the use of ferruginous pygmy owl feathers by indigenous people as amulets of good fortune in health and love. Some scientific divulgation articles and sites have been published to better inform people about Brazilian owls (e.g., Motta-Junior and Alho 1998; Costa- Neto 1999; Petroff 1999, 2002; Motta-Junior et al. 2004a, b; Sibinelli 2010; Menq 2016a).

On the other hand, popular knowledge on fauna can be very useful to the careful ethnobiologist reasearcher. As an example, Loss et al. (2014), in their ethnotaxonomic study, described a new locality record for mottled owl in central-west state of Bahia.

6.5.5 Pesticides

With the modernization of the agricultural production in the 1960s, the development and use of chemical agents have grown substantially in Brazil (Silva 2000). Nowadays, the growth of agricultural production and the expansion of cultivated areas are bringing more quantities and types of chemicals such as pesticides, insecticides, and herbicides to the environment. Owls occupy upper trophic levels in various food chains (Blus 1996); therefore, they are susceptible to the effect of these substances. The effects of pesticides on diurnal birds of prey are well known (Thiollay and Meyburg 1988; Goldstein et al. 1999) and can be a consistent threat for owls (Burton 1984; Blus 1996; Peakall 2002). According to König et al. (1999) and König et al. (2008), at least six Brazilian species can be affected by the use of pesticides: American barn owl, tropical screech owl, great horned owl, burrowing owl, striped owl, and short-eared owl. However, unfortunately, no specific studies on the effect of pesticides or toxicology in Brazilian owls have been published (Motta-Junior and Braga 2012), and these studies are urgently needed.

6.5.6 Species Benefited by Habitat Disturbance

Although natural habitat destruction surely affects negatively most owl species, some can be more common or are expanding its distribution after forest and savanna deforestation. The tropical screech owl (*Megascops choliba*) isn't rare in suburban or even urban areas where some woodlots occur (Goeldi 1894; Sick 1993; Sick 1997; Petroff et al. 2000; Sigrist 2006). Firebreaks can affect positively the occurrence of burrowing owls by clearing the vegetation (Tubelis and Delitti 2010; Reis 2015). Sick (1993, 1997) and Aleixo and Poletto (2007) suggested that the burrowing owl could be expanding its distribution in areas formerly occupied by forest in the Amazon and Atlantic Forests. Sigrist (2006) made a similar suggestion for the striped owl. This expansion to deforested areas is true also for the American barn owl (Sick 1993, 1997), and at least locally, the short-eared owl is very common in soybean plantations in the state of Goiás (Luis F. Silveira, pers. comm.), probably because of the high abundance of rodents in these plantations (Engel and Mello 1993). However, the same species is rare or occasional in the states of Rio Grande do Sul and São Paulo (Belton 1994, Motta-Junior 2009b).

6.6 Conservation Strategies

Brazil has suffered a rapid progress in conservation actions and in conservation policy, stimulated by the onslaught of Amazon and other unexploited regions (Mittermeier et al. 2005). Since the early 1970s, Brazil has seen the proliferation of national parks and reserves, which is a clear evidence of the rapid growth in conservation awareness. An example was the great commitment to parks and other protected areas at national and regional scales, between 1976 and 1990 (Mittermeier et al. 2005). However, it was only in 1988 that there began a major review of the Brazilian protected area system, and after more than 10 years, in 2000, it became law, the "Sistema Nacional de Unidades de Conservacão da Natureza" (SNUC) or National Protected Areas System (Silva 2005).

The SNUC defines and regulates protected area categories at federal, state, and municipal levels. The protected areas are divided by SNUC into two types: strictly protected areas and sustainable use areas (Silva 2005; Rylands and Brandon 2005). The former have the biodiversity conservation as the main objective, including national parks, biological reserves, ecological stations, natural monuments, and wildlife refuges. These categories have different levels of public activities allowed inside the areas, as some are closed to the public (e.g., biological reserves, ecological stations). National parks, which are the largest strictly protected areas, allow educational and recreational activities, besides scientific research.

The sustainable use areas allow for a variety of forms of use or extraction, with biodiversity conservation as a secondary objective (Silva 2005, Rylands and Brandon 2005). Like the strictly protected areas, it has different categories of uses: national forests (FLONA), mainly used for silviculture, sustainable logging, research, and recreation; environmental protection area (APA), which is more like a mechanism of land use than an actual protected area; extractivity reserve, mostly common in Amazon, which is used to promote the sustainable use of natural resources by local communities; and fauna reserve, sustainable development reserve, and RPPN, which are private natural reserves (Rylands and Brandon 2005).

Across Brazil, the proportion of strictly protected areas and sustainable use areas is relatively balanced, but its distribution in the Brazilian main biomes is not (Rylands and Brandon 2005). As indicated in Table 6.4, the sustainable use areas are more concentrated on the Amazon Rainforest, with 7.7% of its natural area on this type of conservation unit, while on Pantanal there are no protected areas of sustainable use.

In spite of the fact that Brazilian Amazon has 13.4% of its total area preserved in protected areas (Table 6.4), all other biomes have the protection area below the 10% minimum cover for effective conservation of biodiversity, recommended as a global strategy by the Secretariat of the Convention on Biological Diversity (2002). The biomes with less proportion of protected area are the Pantanal, the Caatinga, and the Southern Grasslands (Table 6.4).

Table 6.4 Major types of protected areas in Brazilian biomes – strictly protected areas and protected areas of sustainable use

Biome	Area (km^2)	Strictly protected areas[a](%)	Sustainable use areas[a](%)
Amazon Rainforest	4,239,000	5.7	7.7
Cerrado	2,116,000	2.2	1.9
Atlantic Forest	1,076,000	1.9	0.11
Pantanal	142,500	1.1	0
Caatinga	736,800	0.8	0.11
Southern Grasslands	171,300	<0.5[b]	0
All biomes	8,534,000	3.5	3.4

Adapted from Klink and Machado (2005)

[a]Values are given as percentages of the original extent of the biome and with the combination of federal and state protected areas

[b]Data from Overbeck et al. (2007)

Protected areas, actually, are the major tool available for conserving natural areas in Brazil. However, the public protected area system in Brazil has been suffering deficiencies in personal, infrastructure, and management procedures. One example is that, of the 53 national parks in the country, only 20 are open to the public. Another major problem is that around 50% of the areas, designated as federal protected areas, require some type of land tenure regularization (Silva 2005). The situation is not different for other governmental units.

The problems and challenges are even greater for the sustainable areas. Some of these areas, like APAs, function more like a mechanism for land-use management, including zones with some strictly protected areas. However, little is done for preserving natural areas. Difficult issues with the local human community are what and how much can be exploited in these sustainable use areas (Rylands and Brandon 2005). Some authors indicate that the hunting pressure inside the sustainable use areas or around protected areas is too much for conserving the biota (Peres and Zimmerman 2001).

An increasing category of sustainable use area in Brazil is the private reserve (RPPN), which currently amounts to around 450 reserves in a relatively small total area (500,000 lan2; Rylands and Brandon 2005). However the RRPNs are protecting key habitats of a variety of threatened species in the Atlantic Forest, Cerrado, and Pantanal, with participation of private landowners that are exempted of land taxes if they transform part of their land into a permanent conservation area (Silva 2005). This is a great combination of interests and brings the landowners to the conservation side.

Brazil has experienced major growth in nongovernmental conservation capacities and has developed a strong community of world-class conservation scientists and practitioners (Mittermeier et al. 2005). A great movement of conservationist NGOs appeared on the last years. These organizations are getting a fundamental position in the conservation and social movements in Brazil, as they are able to link and catalyze the relationship between academic studies on biological conservation and its practical application, including the linkage with the governmental sphere (Mittermeier et al. 2005).

Recently, many initiatives to improve connectivity between protected areas and manage land uses over large areas have been discussed. Some of the large landscape-scale initiatives include biosphere reserves and conservation or ecological corridors (Rylands and Brandon 2005), and also some priority areas are proposed (such as MMA - Ministério do Meio Ambiente 2000). These discussions have been conducted in collaboration with numerous nongovernmental organizations, universities, and research institutions, as well as different levels of the government.

In addition, there are specific conservation actions, like the selection of Important Bird Areas (IBAs) on Brazilian biomes, which is part of the BirdLife International Program (beginning in Brazil in 2000). The government has initiatives as well, such as the elaboration of the National Plan to the Conservation of Birds of Prey (Soares et al. 2008), which main goal is to plan actions with different time and regional scales, for effective conservation of red-listed species of raptors. This document had a special concern with the recently discovered Pernambuco pygmy owl (Silva et al. 2002), recommending its inclusion in the Brazilian Red List as critically endangered (CR), because of its small distribution area (< 100 km^2). However, recent field evaluations revealed a sad truth, since this species appears to be already extinct (Pereira et al. 2014; ICMBio 2014). Soares et al. (2008) also recommended both rusty-barred owl and buff-fronted owl as "near threatened" for Brazil, but as seen they were not considered by the last National Red List (ICMBio 2014).

Conservation strategies used by the different governmental agencies include lists of threatened species, which have lawful value to make illegal the destruction of natural environments inhabited by these species. Some states, recognizing the usefulness of such lists, have adopted their own assessments and published Red Lists for conservation planning and priorities (Alves et al. 2000; Bencke et al. 2003; Straube et al. 2004; Simon et al. 2007; Para 2008; Bressan et al. 2009; COPAM 2010; CONSEMA 2011; Rio Grande do Sul 2014; São Paulo 2014).

6.7 Status of Biological and Ecological Knowledge

The first naturalistic studies including Brazilian owls (e.g., Marcgrave 1648; Descourtilz 1852; Euler 1869, 1900; Reinhardt 1870; Goeldi 1889, 1894; Inhering 1900; Snethlage 1928) had descriptions and anecdotal observations on habitat, general behavior, breeding, and food habits. However, most references on owls from the mid-nineteenth to mid-twentieth centuries were on taxonomy and distribution (e.g., Berlepsch 1902; Kelso 1934a, 1934b, 1940). Even today a brief analysis of references in Appendix 6.2 reveals that an important part of literature refers to new records or geographic extensions of species (e.g., Gonzaga and Castiglioni 2004; Bernardi et al. 2008; Lau 2008; Vasconcelos and Diniz 2008; Kaminski 2009; Rebelato et al. 2011).

Thiollay (1994) stated that little information about tropical diurnal raptors had been published, but a poorer situation exists for Neotropical owls (Bierregaard 1998; Enríquez et al. 2006). Ecological data on Neotropical owls, such as population densities and breeding, are rudimentary or even nonexistent, mostly for forest species (Bierregaard 1998; Enríquez et al. 2006; Motta-Junior and Braga 2012, Appendix 6.2). Even adequate quantitative information on food habits, one of the relatively most studied aspects in Brazilian owl biology, is absent for virtually all species, except perhaps for the American barn owl and the burrowing owl (see review on Motta-Junior and Braga 2012).

Some owl species with broad distributions in both Neotropical and Nearctic regions are inadequately known in their tropical ranges (Enríquez et al. 2006). While significant ecological data were accumulated in the Nearctic region for the American barn owl, great horned owl, short-eared owl, and even the burrowing owl (Clark et al. 1978; Clark 1997), our understanding of the biology and ecology of these species is comparatively poor in the Neotropics (Enríquez et al. 2006), including in Brazil. At least partially, this can be explained because studying owls in tropical areas is not an easy task: several habitats are physically demanding, and most owls have nocturnal activities (Enríquez et al. 2006).

6.7.1 Food Habits and Trophic Ecology

The diet of some species has been studied by the analysis of stomach contents, particularly until the mid-twentieth century (e.g., Moojen et al. 1941; Hempel 1949; Schubart et al. 1965). However, though useful, these studies were generally based on few stomachs with very limited information. As an example, Schubart et al. (1965), based only on one stomach of the stygian owl, were led to state that this species in Brazil is virtually insectivorous, when in fact this is a highly ornithophagous owl (Motta-Junior 1996, 2006).

Field studies of Neotropical owls have raised information mainly on food habits, because of the relatively easy collection and analysis of regurgitated pellets. Most information has been published for the American barn owl (e.g., Lange 1981; Scheibler and Christoff 2004; Motta-Junior 2006; Roda 2006; Magrini and Facure 2008) and for the burrowing owl (e.g., Silva-Porto and Cerqueira 1990; Motta-Junior and Bueno 2004; Motta-Junior 2006; Vieira and Teixeira 2008). However, many of these studies neither identify prey to species/genus level nor estimate biomass consumption (e.g., Lange 1981; Motta-Junior 1988; Soares et al. 1992b; Jordao et al. 1997; Teixeira and Melo 2000; Menezes and Meira 2012). In addition, many of the published information on diet are from anecdotal observations on one event of predation or one stomach content (e.g., Granzinolli and Motta-Junior 2003; Martins et al. 2003; Dornas and Pinheiro 2007; Garcia et al. 2008; Ramos et al. 2011; Carvalho et al. 2011; Silva et al. 2013; Anza and Zilio 2015). Only recently have been published the first quantitative studies in Brazil on the diet of tropical screech owl (Motta-Junior 2002), great horned owl (Tomazzoni et al. 2004), striped owl (Motta-Junior et al. 2004a, b), and stygian owl (Motta-Junior 2006). However, most of these studies are restricted to one locality.

Some studies in the Cerrado biome on American barn owl, striped owl, and burrowing owl have detected that insects are mainly preyed on rainy and warm months, while rodents are mainly captured in the dry and cold season (Motta-Junior 1988; Motta-Junior and Alho 1998, 2000; Motta-Junior and Bueno 2004; Motta-Junior et al. 2004a, b). This pattern fits with the natural cycles of abundance of the respective prey in the Cerrado, suggesting an opportunistic temporal feeding behavior by these owl species.

Studies on the trophic ecology of sympatric owls and other raptors are scarce in Brazil and only recently have been published (Motta-Junior and Alho 2000; Zilio 2005, 2006; Motta-Junior 2006; Silva 2006; Scheibler 2007). In general, these studies reveal that there are mechanisms of ecological isolation operating in the relationships among owl species, especially in trophic niches and choice of foraging habitats. The differences do not occur only on proportions of food types in the diet but also in activity sites of preferential prey (Motta-Junior 1996).

Similarly scarce are studies on prey selection and prey vulnerability to owl predation on small mammals (Motta-Junior 1996; Bueno 2003; Bueno and Motta-Junior 2008, 2015; Magrini 2006; Magrini and Facure 2008) and mostly on insects (Siervi 2014). In general, these researches have showed that juvenile or subadult rodents are selected by both the American barn owl and the burrowing owl due to their greater physical and behavioral vulnerability to predation when compared to adults. Additionally, in general, rodents of the genus *Calomys* are preferred in relation to the rodent *Necromys* (= *Bolomys*) *lasiurus* (Motta-Junior 1996; Magrini and Facure 2008), though no selection or contrary results were found too (Bueno and Motta-Junior 2015).

Trophic ecology studies focusing on a species' geographical variation in diet were the purpose of only two studies: burrowing owl in southeastern Brazil (Motta-Junior and Bueno 2004) and American barn owl in Cerrado region (Motta-Junior 2004). The first species showed a highly variable diet, but though numerically based on insects, in terms of biomass consumption, vertebrates were important, particularly rodents. The American barn owl showed a similar pattern but with a higher proportion of rodent biomass than the burrowing owl.

Some researchers, mainly mastozoologists, have reported pellets of American barn owl as a useful additional tool for small mammal inventories (e.g., Bonvicino and Bezerra 2003; Escarlate-Tavares and Pessoa 2005; Scheibler and Christoff 2007; Souza et al. 2010; Rocha et al. 2011). As a revealing example, some Brazilian small mammal species were firstly described by the collection and analysis of American barn owl pellets in caves (Lund 1950; Voss and Myers 1991). Even for recording birds, owl pellets can be useful; as an example, a gallinule was firstly recorded for Distrito Federal by stygian owl pellet remains (Lopes et al. 2012).

6.7.2 Reproductive Biology

One of the major gaps in the natural history of Brazilian owls is breeding information. From the 23 owl species that occur in Brazil, only the American barn owl in the state of Paraná (Lange 1981; Silva 2006), the striped owl in the state of Minas Gerais (Pereira and Oliveira 2010), the ferruginous pygmy owl in the state of Bahia (Lima and Lima-Neto 2008, 2009a), and the tropical screech owl in central Brazil (Dias and Lima 2015) have adequate data including all breeding cycles wherein there are more than one nest accompanied. Seven species are studied with partial breeding cycle information or based on observations of one nest only: American barn owl (Ihering, 1900), burrowing owl (Ihering 1900; Silva 2002; Jacobucci 2007; Lima 2007; Martinelli 2010), ferruginous pygmy owl (Euler 1900; Dubs 1992; Castro et al. 2010), stygian owl (Oliveira 1981; Scherer-Neto 1985; Lopes et al. 2004; Motta-Junior et al. 2010), striped owl (Aguiar and Naiff 2009), buff-fronted owl (Studer and Teixeira 1994; Girao and Albano 2010), great horned owl (Lisboa et al. 2005), and tropical screech owl (Euler 1900; Oliveira 1984; Marini et al. 2007; Lima and Lima-Neto 2009b; Motta-Junior et al. 2010). In addition, some general ornithology textbooks and a few studies with focus on a single breeding characteristic (e.g., egg size, egg number or nest substrate) report information for the American barn owl (Sick 1997), burrowing owl (Snethlage 1928; Andrade 1992;

Belton 1994; Sick 1997; Aguiar and Perrini 2004), ferruginous pygmy owl (Snethlage 1928; Sick 1997), great horned owl (Snethlage 1928; Sick 1997; Antas 2009), stygian owl (Albuquerque 1983), short-eared owl (Sick 1997), buff-fronted owl (Antas 2009), tropical screech owl (Andrade 1992, Sick 1997), and spectacled owl (Euler 1869; Snethlage 1928), as well as general comments about the genera *Otus* (=*Megascops*), *Rhinoptynx* (=*Asio*), and *Ciccaba* (=*Strix*) (Sick 1993, 1997).

Nest site selection of Brazilian owls is virtually absent in the literature, except for a recent study by Tubelis and Delitti (2010), which suggests a preference of grasslands managed by fire (firebreaks) over unmanaged grasslands for active burrows of burrowing owls in central Brazil. These managed grasslands are frequently burned, contributing to the maintenance of a short herbaceous stratum, facilitating movements and visibility for the burrowing owls. A case of interspecific competition for tree cavity resulted in the killing of an American barn owl by a pair of hyacinth macaws (*Anodorhynchus hyacinthinus*) (Tortato and Bonanomi 2012). Another possible case of competition for nest involved tropical screech owl and guira cuckoos (*Guira guira*) (Claudino et al. 2012).

Some studies conducted in Brazil of species with limited distribution, or with scarce breeding information throughout the entire range, are summarized here. For example, there is scattered information about the breeding of the stygian owl in the literature. The four studies made in Brazil by Oliveira (1981), Albuquerque (1983), Scherer-Neto (1985), and Lopes et al. (2004) revealed that stygian owl nests may be located in the ground, as is common in other regions, or in the branches of the Paraná pine (*Araucaria angustifolia*). In the case of the ferruginous pygmy owl, Lima and Lima-Neto (2008) reported artificial nest utilization and daily weight gain of the nestlings and observed that only the female provides food to the nestlings. These authors also demonstrated that the nestlings' diet in the first five days is composed exclusively of small lizards and later complemented by birds, rodents, and insects. Observations of a single nest of the buff-fronted owl in northeastern Brazil showed that this owl nests in a dead palm tree in March, probably in an abandoned parrot's nest. The hole (60 cm deep x 15 cm diameter x 10 cm entrance) was 6 m above the ground, and the clutch consisted of three eggs (see Studer and Teixeira 1994).

It is worst to notice that consistent data about owl breeding biology is lacking for entire regions of Brazil, with virtually no breeding studies from Amazon Rainforest and Caatinga, for example. The few available studies are limited to the eastern, central, and southern regions. Even in these regions, the breeding biology of abundant species with wide distribution ranges (e.g., burrowing owl, American barn owl, and tropical screech owl) is poorly understood. Considering that variation in latitude can directly affect the breeding biology of birds of prey (e.g., period, number of eggs), and that Brazil has the largest variation in latitude in the southern hemisphere (5° N to 33° S), this gap is inexcusable.

The lack of knowledge about the breeding biology of most Brazilian owls is a serious gap, mainly in a country where natural areas are suffering drastic and rapid alterations. The information about breeding biology is one of the basic steps in understanding the ecology of any species. Without this information, the implantation of conservation plans for owls becomes difficult, with high probability of failure.

6.7.3 Behavioral Studies

While brief observations on general behavior of Brazilian owls can be found scattered in the literature (e.g., Euler 1869; Snethlage 1928; Ihering 1940; Santos 1979; Sick 1993, 1997; Sigrist 2006), some researches have specifically treated behavior as a main focus. The few examples include studies on burrowing owl's general behavior (Soares et al. 1992a; Perillo et al. 2011), hunting tactics (Lourenço 1980; Martins and Egler 1990; Corbo et al. 2013; Specht et al. 2013), and defensive behavior (Silva 2002; Aguilar and Perini 2004; Jacobucci 2007). Sazima (2015a) reported the chasing behavior of ferruginous pygmy owl on foraging hummingbirds. Other studies on general behavior included juvenile tropical screech owl (Oniki 1984) and striped owl (Oliveira 1980). Petroff (2001), Azevedo (2004), and Specht (2007) suggested methods and behavioral studies with owls in captivity for better reintroduction to the wild. Lange (1981) focused on the American barn owl's diet, but his data included some general observations on the behavior of wild and captive individuals. Coelho (1979) reported an observation of autophagy by a captive tropical screech owl possibly due to stress.

Home range and territory estimates for owls are virtually absent in Brazil. Only recently the first telemetry study on home range for the tropical screech owl was completed in a Brazilian savanna (Barros 2011; Barros and Motta-Junior 2014). The mean (±SD) home range size was 51.2 (±26.9) ha using the 95% minimum convex polygon method and 80.8 (±40.2) ha using the 95% fixed kernel method.

Prey reactions against owls, such as the widespread mobbing behavior, were until recently only the object of general or anecdotal observations (Descourtilz 1852; Euler 1869; Ruschi 1960; Sick 1993, 1997; Sigrist 2006; Motta-Junior 2007; Specht et al. 2008; Cunha et al. 2009, 2013; Cunha and Fontenelle 2014). Cunha and Vasconcelos (2009) have compiled an extensive list of 127 bird species attracted by vocalizations of the ferruginous pygmy owl in Brazil. Only recently have been published the first experimental studies in Brazil on mobbing behavior against owls (Cunha 2012; Motta-Junior and Santos-Filho 2012; Meireles et al. 2015). Another recent field experiment showed that frugivorous bats significantly decreased the foraging frequency when a visual cue of the American barn owl was present. No reaction however occurred when burrowing owl cues were present (Breviglieri et al. 2013).

Vocal activity of owls in Brazil has been rarely studied. Castiglioni (2014) conducted one of the first specific studies in Brazil about the temporal distribution of vocal activity of four forest owl species. The vocal responsiveness to playback of the burrowing owl and tropical screech owl according to atmospheric conditions and lunar phases was a subject of study in a savanna (Cerrado) in southeastern Brazil. The response rate of the former species increased under higher temperature and humidity, whereas the response rate of the latter species was higher in full moon nights.

6.7.4 *Morphology, Anatomy, Histology, Genetics, and Physiology*

Relatively few studies have been published on body and tissue structure, as osteology (Mahecha and Oliveira 1998; Hofting and Alvarenga 2001) and hematology (Sanches et al. 2004; Goulart 2015). A histological study of the stomach of burrowing owls collected in Brazil was published by Rocha and Lima (1998). Iris color variation in tropical screech owl was discussed in Oliveira (1984). Kelso (1940) pioneered the first comprehensive comparative analysis of external ear variation in several species of Brazilian owls, such as tropical screech owl, crested owl, spectacled owl, mottled owl, and rusty-barred owl. A recent study described the eye morphology of striped owl (Rodarte-Almeida et al. 2013). Information on body mass, an important ecological variable, is relatively scattered in a few studies (e.g., Marini et al. 1997; Willis and Oniki 2002; Motta-Junior 2006). Squarzoni et al. (2010) performed a biometric analysis of striped owl eyes.

Leucism in birds is described as a complete or partial lack of pigmentation in feathers but with normal coloration of the legs, feet, bill, and eyes (Grouw 2006). This disorder is rare in owls (Alaja and Mikkola 1997). The first (Motta-Junior et al. 2010) and second (Nogueira and Alves 2011) reported cases of leucism for South American owls were found in the burrowing owl. Apparently for this species, there is only one case reported for the United States (Alaja and Mikkola 1997). A recent case of chromatic aberration of plumage was reported for the same species in central Brazil (Carvalho et al. 2015).

Cytogenetic information for five Brazilian Strigiformes was compiled by Santos and Gunski (2006), though mostly from data outside Brazil, while Oliveira et al. (2008) have described the chromosome painting of spectacled owl. Apart from general DNA studies including Brazilian species but rarely individuals (e.g., Heidricht et al. 1995a, 1995b; Wink 2000; Wink et al. 2004; König et al. 2008), in general Brazilian studies on owl genetics are very rare and need attention, mostly because of its importance for owl taxonomy. Noteworthy is the recent DNA evaluation of new world screech owls (*Megascops*) including five Brazilian species (Dantas et al. 2016).

Surprisingly we have not detected any true Brazilian physiological study on owls. For example, metabolic and bioenergetic studies would be important to any further ecological research about an owl's impact on its prey and its exact role in food webs.

6.7.5 *Parasitology*

Parasitological studies on Brazilian owls are generally scattered and focusing on parasite descriptions, occurrence, and/or listings (Freitas and Lent 1937; Strachan 1957; Amaral 1962; Pinto 1968; Valim et al. 2005; Mascarenhas et al. 2009; Andery et al. 2013; Vitaliano 2012; Gallas and Silveira 2013; Pedroso 2015; Sacchi 2015). On the other hand, some studies have focused on owls, particularly assessing causes of mortality (Joppert 2007; Ecco et al. 2012; Echenique et al. 2016), or reviewed owl parasites in a regional scale (Graciolli and Bispo 2005; Silva et al. 2009; Silva et al. 2014).

6.7.6 Macroecology

The new approach of macroecology to South American owls has been proposed considering mostly body lengths and distribution patterns (Sant'ana and Diniz-Filho 1997, 1999; Sant'ana 1998; Diniz-Filho and Sant'ana 1998, 2000; and Diniz-Filho et al. 2004). The control of the phylogenetic effects in these cross-species data normally has been done using phylogenetic autocorrelation analysis (Sant'ana and Diniz-Filho 1997; Diniz-Filho and Sant'ana 1998; Diniz-Filho and Sant'ana 2000). Sant'ana and Diniz-Filho (1999) revealed a positive correlation between body length and geographic range size for 29 South American owl species, including several Brazilian species. Diniz-Filho et al. (2004) concluded that both climate and environmental heterogeneity should be retained as possible explanations for the owl diversity gradient in South America.

6.7.7 Population and Habitat use

Owl density studies in Brazil are rare. Some inventories on raptors are only qualitative (e.g., Petroff 2003) or semiquantitative (e.g., Azevedo et al. 2003; Zorzin et al. 2008; Fink et al. 2012), and many records of owls have been published by researchers that were not studying specifically owls, like ornithologists interested mostly in all bird communities (e.g., Bornschein and Reinert 2000; Antunes 2005). In most cases, these researchers stayed on the field on the first hours of day and/or last hours of afternoon. Unfortunately, in most of these inventories, the taxonomic diversity of owls is lower than expected (e.g., Olmos et al. 2005), because the main interest is diurnal birds or most of the study is conducted during daylight hours. Consequently, there are few studies specifically on census of owls in Brazil, four in the Amazon Rainforest (Barros 2003; Borges et al. 2004; Barros and Cintra 2009; Esclarski and Cintra 2014), three in the Atlantic Forest (Amaral 2007; Claudino 2013; Menq and Anjos 2015), and two in the Cerrado of the state of São Paulo (Braga 2006; Motta-Junior 2006).

Barros (2003) and Barros and Cintra (2009) studies were conducted on Ducke Reserve, near the city of Manaus, which has a goal of assessing the spatial distribution and effects of the forest structural components on the habitat use and structure of an owl assemblage of six species. The most abundant species (75% of observations) were the crested owl and the Amazonian pygmy owl. Areas with high abundance of dead tree trunks (still erect) and live trees were significantly higher in density of crested owl and Northern tawny-bellied screech owl, whereas the Amazonian pygmy owl was more often found in sites with larger canopy openness (Barros and Cintra 2009). Further studies on habitat use by forest owls in the same reserve have been conducted by Esclarski (2014) and Esclarski and Cintra (2014).

Borges et al. (2004) also studied owls of Amazon region on Jaú National Park. In an assemblage of six forest owls, the most abundant species were the Northern tawny-bellied screech owl and the spectacled owl (*P. perspicillata*). Some species showed clear tendencies to use different habitats, as the Northern tawny-bellied owl was more abundant in "terra-firme" forest, while others were more common (tropical screech owl) or exclusively recorded in "igapó" forest. A more recent study conducted also in Ducke Reserve showed some different findings. Esclarski and Cintra (2014) found crested owl to be uniformly distributed within the study area, as tawny-bellied screech owl. Their analysis also revealed relationships between four owl species and components of forest structure associated with food availability: Amazonian pygmy owl and distance to nearest stream, Northern tawny-bellied screech owl and leaf-litter depth, crested owl and dead fallen trunks on forest floor, and spectacled owl and dead fallen trunks on forest floor.

Still in Amazon, Sberze et al. (2010) made a comparison of secondary forests and old growth forest site occupancy by nocturnal birds 70lan north of the city of Manaus. They tested species-specific occupancy predictions for Caprimulgiformes and Strigiformes, including crested owl, Amazonian pygmy owl, and to a lesser extent other five owl species. Results showed that owls were relatively indifferent to forest type.

From six owl species detected in an Atlantic Forest area in the state of Minas Gerais (Amaral 2007), only the least pygmy owl occurred significantly more in primary forest than in secondary forest. This species and the mottled owl appeared to prefer sites with higher canopy heights. The white-chinned owl occurred in areas with higher tree abundance, while the ferruginous pygmy owl was the most generalist species. More recently, a study in a seasonal semi-deciduous forest in southern Brazil (Menq and Anjos 2015) revealed that the canopy height and the presence of hollow trees, fallen trees, and glades are the most important structural components influencing habitat use of six owl species.

Motta-Junior's (2006) study within the Cerrado region in southeastern Brazil had the main goal of evaluating trophic ecology of five sympatric owls, secondarily reporting habitat use and density by these species. The burrowing owl and the tropical screech owl were the most abundant species, and while the first used disturbed and natural grasslands, the latter occupied mostly areas with some trees. In a more specific study, Braga (2006) analyzed the habitat selection of the same species in Cerrado reserve, reporting a preference of wooded savannas for the tropical screech owl and more open Cerrado physiognomies for the burrowing owl. Besides the preference of burrowing owl to less wooded physiognomies, this species seems to choose areas with the presence of sentinel perches (Braga 2006; Tubelis and Delitti 2010). Similar habitat selection results for the tropical screech owl were found by Barros (2011) and Barros and Motta-Junior (2014) using telemetry methods.

Most of these studies used special methods for detecting the presence of owls including waiting for spontaneous owl's vocalizations (Barros 2003; Amaral 2007; Barros and Cintra 2009; Castiglioni 2014) or eliciting owl's vocalizations using playback (Borges et al. 2004; Motta-Junior 2006; Braga 2006; Sberze et al. 2010). The use of a specific method is extremely necessary to study or census owls. Standardization of techniques for owl census is vital for reliable data, but unfortunately, there are only two recent studies in Brazil that stress this issue (Braga and Motta-Junior 2009; Granzinolli and Motta-Junior 2010). A rare study in the Pantanal region on habitat selection by the black-banded owl revealed that this species needs forest fragments larger than 2000 ha, taller trees with >50 cm trunk circumference, and a forest density above 250 trees/ha (Tomas et al. 2013).

The better understanding of owl distribution and owl relationships with habitats is very important to make reliable conservation plans. Thus, the critical lack of census and habitat use studies of Brazilian owls must be addressed by Brazilian ornithologists. More studies are needed in all biomes but particularly in the Caatinga, Pantanal, Amazon Rainforest, and Southern Grasslands (Motta-Junior and Braga 2012).

6.8 Conclusions

Although in the last 15 years there was an impressive increasing number of publications about Brazilian owls (see Albuquerque and Motta-Junior 2006; Motta-Junior and Braga 2012), we have still important gaps in our knowledge of this raptor group. Crucial information is desirable for most owl species such as distribution, population density, habitat and home range requirements, reproductive biology, and anthropogenic effect on owl populations, including studies on the effect of pesticides. Even food habit studies are needed, since most published papers lack quantitative analysis and have low level of taxonomic identification of prey. New studies on food habits must invest in more refined prey identification (Granzinolli and Motta-Junior 2010). Better understanding of owl taxonomy is also very important, since conservation laws have focused almost exclusively at the species level (Enríquez et al. 2006). Hence, ornithologists should promote more collections of tissues, feathers, or blood and vocalization recordings to clarify current taxonomic uncertainties.

As a provisory measure to alleviate the critical lack of census and distribution information, we suggest that future bird community/inventory studies should include just 1 or 2 hours after sunset (or before sunrise) to evaluate owl occurrence. Moreover, more specific owl censuses should be undertaken using standard techniques (Braga and Motta-Junior 2009; Granzinolli and Motta-Junior 2010).

Thirteen of the 23 Brazilian owl species are cited as "poorly known" or "need further study" in König et al. (1999, 2008), most of them forest living species. A rapid observation of Appendix 6.2 reveals to us the critical lack of information within Brazil for ten owl species: Northern tawny-bellied screech owl, Southern tawny-bellied screech owl, black-capped screech owl, long-tufted screech owl,

Roraima screech owl, crested owl, Amazonian pygmy owl, Pernambuco pygmy owl, buff-fronted owl (see review in Girao and Albano 2010), and short-eared owl. As showed in this chapter and by the quantitative review of Motta-Junior and Braga (2012), even the remaining 13 species, including the common ones, have gaps in many aspects of biological knowledge.

The lack of basic biological information on most species perhaps might be responsible for the relative absence of owls in the Brazilian official lists of threatened fauna in regional and national scales. Thus, owl conservation and management practices only will be improved if we expand biological and ecological data on these species. This biological and ecological information associated with habitat protection through governmental and private reserves (Enríquez et al. 2006) will continue to be the key conservation strategy for Brazilian owls.

Acknowledgments We are grateful to FAPESP (Fundação de Amparo à Pesquisa do Estado de São Paulo), CNPq (Conselho Nacional de Desenvolvimento Científico e Tecnológico), PPGE-USP (Programa de Pós-Graduacão em Ecologia da Universidade de São Paulo), and WWF-Brazil for supporting our projects about owls during the last 20 years. We appreciated the critical readings of C. König, L.F. Silveira, and P.L. Enríquez and the English revision provided by A.A. Bueno. We are indebted to several people that sent to us missing references: P.T.Z. Antas, S.H. Borges, R. Cintra, P.L. Enríquez, L. Fukui, L.A.P. Gonzaga, M. Guentert, A.M. Joppert, P.*C. lima*, G.A.B. Mahecha, C. Marti, J.F. Pacheco, I. Sazima, L.F. Silveira, G.V.A Specht, F.C. Straube, and M.F. Vasconcelos. This chapter is dedicated to the memory of Carl D. Marti, an excellent owl researcher as a professional biologist and a gentleman as a person.

Barn Owl (*Tyto alba*)

Barn Owl (*Tyto alba*)

Burrowing Owl (*Athene cunicularia*)

Burrowing Owl (*Athene cunicularia*)

Ferruginous Pygmy Owl *(Glaucidium brasilianum)* x *Tyrannus savana*

Rusty-barred Owl (*Strix hylophila*)

Striped Owl (*Asio clamator*)

Striped Owl (*Asio clamator*)

Tropical Screech Owl (*Megascops choliba*)

Appendix 6.1

Brazilian common names of the 23 owl species currently recognized in Brazil according to Willis and Oniki (1991), Sick (1997), and Piacentini et al. (2015)

Scientific name	Common names in Brazil
Tyto furcata (Temminck, 1827)	Suindara, coruja-da-igreja, coruja-branca, rasga-mortalha
Megascops choliba (Vieillot, 1817)	Corujinha-do-mato, corujinha-orelhuda, corujinha-de-orelha
Megascops watsonii (Cassin, 1849)	Corujinha-orelhuda, corujinha-amazônica
Megascops usta (Sclater, 1858)	Corujinha-relógio
Megascops atricapilla (Temminck, 1822)	Corujinha-sapo
Megascops sanctaecatarinae (Salvin, 1897)	Corujinha-do-sul
Megascops roraimae (Salvin, 1897)[a]	Corujinha-de-roraima, corujinha-do-norte
Lophostrix cristata (Daudin, 1800)	Coruja-de-crista, coruja-de-carapuça
Pulsatrix perspicillata (Latham, 1790)[b]	Murucututu, corujão, coruja-de-garganta-preta, mocho-mateiro
Pulsatrix koeniswaldiana (Bertoni and Bertoni, 1901)	Murucututu-de-barriga-amarela, corujão-de-garganta-branca
Strix hylophila (Temminck, 1825)	Coruja-listrada, coruja-pintada
Strix virgata (Cassin, 1849)	Coruja-do-mato, coruja-de-bigodes, mocho-carijó
Strix huhula (Daudin, 1800)	Coruja-preta, mocho-negro
Bubo virginianus (Gmelin, 1788)	Corujão-orelhudo, jacurutu, jucurutu, mocho-orelhudo
Glaucidium hardyi (Vielliard, 1990)	Caburé-da-amazônia
Glaucidium mooreorum (Silva, Coelho, and Gonzaga, 2002)[c]	Caburé-de-pernambuco
Glaucidium minutissimum (Wied, 1830)[d]	Caburé-miudinho, caburezinho
Glaucidium brasilianum (Gmelin, 1788)	Caburé, caburé-do-sol, caburé-ferrugem
Athene cunicularia (Molina, 1782)	Coruja-buraqueira, coruja-do-campo
Aegolius harrisii (Cassin, 1849)	Caburé-acanelado, caburé-canela
Asio clamator (Vieillot, 1808)	Coruja-orelhuda, coruja-gato, mocho-orelhudo
Asio stygius (Wagler, 1832)	Mocho-diabo, coruja-diabo
Asio flammeus (Pontoppidan, 1763)	Mocho-dos-banhados, coruja-dos-banhados, coruja-dos-campos

Scientific names according to König et al. (2008), except when indicated
[a]*M. guatemalae* after Piacentini et al. (2015)
[b]Includes *P. pulsatrix*
[c]*G. sicki* after König et al. (2008)
[d]*G. minutissimum* after König et al. (2008)

Appendix 6.2

Literature found for the 23 species of Brazilian owls

American barn owl (*Tyto furcata*)	Descourtilz (1852), Euler (1869), Reinhardt (1870), Goeldi (1894), Ihering (1900), Lutz et al. (1915), Gliesch (1933), Freitas and Lent (1937), Ihering (1940), Moojen et al. (1941), Lund (1950), Ruschi (1953), Schubart et al. (1965), Valle and Carnevalli (1973), Pinto (1978), Santos (1979), Valle (1980), Lange (1981), Sazima and Caramaschi (1986), Motta-Junior (1996), (1988), (2004) (2006), Antas and Cavalcanti (1988), Novelli et al. (1988), Voss and Myers (1991), Sick (1993), Belton (1994), Talamoni (1996), Motta-Junior and Talamoni (1996), Motta-Junior et al. (1996), Bergamini (1997), Jordão et al. (1997), Marini et al. (1997), Sick (1997), Mahecha and Oliveira (1998), Motta-Junior and Alho (1998), (2000), Costa-Neto (1999), González et al. (1999), Höfling and Camargo (1999), König et al. (1999), Bergamini (2002), Willis and Oniki (2002), Almeida et al. (2003), Bueno (2003), Bonvicino and Bezerra (2003), Fischer et al. (2003), Petroff (2003), Azevedo (2004), Ferreira and Glock (2004), Prada (2004), Scheibler and Christoff (2004), Bueno and Motta-Junior (2005), Escarlate-Tavares and Pessoa (2005), Graciolli and Bispo (2005), Pereira and Brito (2005), Magrini (2006), Roda (2006), Santos and Gunski (2006), Silva (2006), Scheibler and Christoff (2007), Krabbe (2007)[e], Scheibler (2007), Joppert (2007), Pinto et al. (2007), Bueno and Motta-Junior (2008), Coelho et al. (2008), König et al. (2008), Magrini and Facure (2008), Zorzin et al. (2008), Antas (2009), Sigrist (2009), Silva et al. (2009), Wink et al. (2009), Catroxo et al. (2010), Martinelli and Prado (2010), Sibinelli (2010), Brasil et al. (2010), Grantsau (2010), Gwynne et al. (2010), Souza et al. (2010), Rocha et al. (2011), Andery (2011), Ramos et al. (2011), Vargas et al. (2011), Fink et al. (2012), Tortato and Bonanomi (2012), Andery et al. (2013), Breviglieriet al. (2013), Costa Silva and Henderson (2013), Bueno and Motta-Junior (2015), Goulart (2015), Pedroso (2015), Sacchi (2015)

(continued)

Tropical screech owl (*Megascops choliba*)	Euler (1869), Reinhardt (1870), Goeldi (1894), Euler (1900), Berlepsch (1902), Snethlage (1928), Hellmayr (1929), Kelso (1940), Moojen et al. (1941), Bequaert (1955), Strachan (1957), Schubart et al. (1965), Pinto (1978), Coelho (1979), Santos (1979), Hekstra, (1982), Oliveira (1984), Oniki (1984), Andrade (1992), Sick (1993), Belton (1994), König (1994a,b), Heidrich et al. (1995a), Motta-Junior (1996), Marini et al. (1997), Sick (1997), Mahecha and Oliveira (1998), Motta-Junior and Alho (1998), Höfling and Camargo (1999), Holt et al. (1999), König et al. (1999), Petroff et al. (2000), Petroff (2001), Motta-Junior (2002), Almeida et al. (2003), Azevedo et al. (2003), Graciolli and Carvalho (2003), Petroff (2003), Willis and Oniki (2002), Willis (2003), Borges et al. (2004), Prada (2004), Sanches et al. (2004), Graciolli and Bispo (2005), Pereira and Brito (2005), Borges et al. (2006), Braga (2006), Motta-Junior (2006), Roda and Pereira (2006), Sigrist (2006), Amaral (2007), Krabbe (2007), Marini et al. (2007), Joppert (2007), Specht (2007), Accordi and Barcellos (2008), Coelho et al. (2008), König et al. (2008), Zorzin et al. (2008), Antas (2009), Braga and Motta-Junior (2009), Martinelli (2009), Martinelli and Volpi (2009), Mascarenhas et al. (2009), Lima and Lima-Neto (2009b), Sigrist (2009), Wink et al. (2009), Grantsau (2010), Gwynne et al. (2010), Motta-Junior et al. (2010), Sberze et al. (2010), Andery (2011), Barros (2011), Claudino et al. (2012), Fink et al. (2012), Andery et al. (2013), Claudino (2013), Barros and Motta-Junior (2014), Dias and Lima (2015), Hernandes and O'Connor (2015), Menq and Anjos (2015), Pedroso (2015), Sacchi (2015), Vieria et al. (2015), Dantas et al. (2016)
Northern tawny-bellied screech owl (*Megascops watsonii*)	Schubart et al. (1965), Pinto (1978), Sick (1993), Heidrich et al. (1995b), Sick (1997), Holt et al. (1999), König et al. (1999), (2008), CBRO - Comite Brasileiro de Registros Ornitológicos (2001b), Barros (2003), Borges et al. (2004), Barros and Cintra (2009), Grantsau (2010), Sberze et al. (2010), Esclarski (2014), Esclarski and Cintra (2014), Dantas et al. (2016)
Southern tawny-bellied screech owl (*Megascops usta*)	Pinto (1978), Sick (1993), Heidrich et al. (1995a), Sick (1997), König et al. (1999), (2008), , Wink and Heidrich (2000), CBRO (2001b), Grantsau (2010), Gwynne et al. (2010), Barbosa et al. (2015)
Black-capped screech owl (*Megascops atricapilla*)	Euler (1869), Reinhardt (1870), Pinto (1978), Hekstra (1982), König (1991), (1994a, b), Sick (1993), Heidrich et al. (1995a), Sick (1997), Holt et al. (1999), König et al. (1999), (2008), Wink and Heidrich (2000), Willis and Oniki (2002), Graciolli and Bispo (2005), Roda and Pereira (2006), Sigrist (2006), Krabbe (2007), Grantsau (2010), Castiglioni (2014), Gonzaga and Castiglioni (2015), Menq and Anjos (2015), Dantas et al. (2016)
Long-tufted screech owl (*Megascops sanctaecaterinae*)	Pinto (1978), Hekstra (1982), König (1991), (1994a, b), Sick (1993), Belton (1994), Heidrich et al. (1995a), Sick (1997), Holt et al. (1999), König et al. (1999), (2008), Wink and Heidrich (2000), Graciolli and Bispo (2005), Sigrist (2006), Accordi and Barcellos (2008), Legal et al. (2009), Fink et al. (2012), Dantas et al. (2016)
Roraima screech owl (*Megascops roraimae*)[a]	Pinto (1978), Sick (1993), Heidrich et al. (1995b), Sick (1997), Holt et al. (1999), König et al. (1999), (2008), CBRO (2001a), Robbins (2006), Soares et al. (2008), Grantsau (2010), Aleixo et al. (2011), Dantas et al. (2016), Remsen et al. (2016)

(continued)

Crested owl (*Lophostrix cristata*)	Pinto (1978), Sick (1993,1997), Holt et al. (1999), Barros (2003), König et al. (2008), Barros and Cintra (2009), Grantsau (2010), Gwynne et al. (2010), Sberze et al. (2010), Rocha and López-Baucells (2014), Esclarski (2014), Esclarski and Cintra (2014)
Spectacled owl (*Pulsatrix perspicillata*)[b]	Euler (1869), Goeldi (1894), Berlepsch (1902), Snethlage (1928), Kelso (1934a,1940), Schubart et al. (1965), Pinto (1978), Santos (1979), Sick (1993), Belton (1994), Sick (1997), Holt et al. (1999), König et al. (1999), Bencke (2001), Barros (2003), Fischer et al. (2003), Borges et al. (2004), Valim et al. (2005), Roda and Pereira (2006), Santos and Gunski (2006), Ramirez-Llorens and Bellocq (2007), Sigrist (2006), König et al. (2008), Oliveira et al. (2008), Soares et al. (2008), Antas (2009), Wink et al. (2009), Grantsau (2010), Gwynne et al. (2010), Sberze et al. (2010), Carvalho et al. (2011), Esclarski (2014), Esclarski and Cintra (2014), Barbosa et al. (2015)
White-chinned owl (*Pulsatrix koeniswaldiana*)	Goeldi (1894), Kelso (1934a,1940), Hempel (1949), Ruschi (1953), Schubart et al. (1965), Pinto (1978), Sick (1993,1997), Bencke and Bencke (1999), Holt et al. (1999), König et al. (1999), (2008), Bornschein and Reinert (2000), Bencke (2001), Willis and Oniki (2002), Ribeiro and Vasconcelos (2003), Almeida et al. (2003), Borges et al. (2006), Piacentini et al. (2006), Amaral (2007), Zorzin et al. (2008), Accordi and Barcellos (2008), Legal et al. (2009), Sigrist (2009), Wink et al. (2009), Grantsau (2010), Fink et al. (2012), Kanegae et al. (2012), Claudino (2013), Castiglioni (2014), Garcia et al. (2015), Menq and Anjos (2015), Pedroso (2015), Vasconcellos et al. (2016)
Mottled owl (*Strix virgata*)	Reinhardt (1870), Voous (1964), Schubart et al. (1965), Pinto (1978), Sick (1993,1997), Belton (1994), Holt et al. (1999), König et al. (1999), (2008), Bencke and Bencke (2000), Bencke (2001), Hofting and Alvarenga (2001), Willis and Oniki (2002), Barros (2003), Graciolli and Carvalho (2003), Antunes (2005), Graciolli and Bispo (2005), Voous (1964), Silveira and Belmonte (2005), Roda and Pereira (2006), Amaral (2007), Krabbe (2007), Zorzin et al. (2008), Lau (2008), Bernardi et al. (2008), Soares et al. (2008), Legal et al. (2009), Grantsau (2010), Gwynne et al. (2010), Sberze et al. (2010), Andery (2011), Andery et al. (2013), Claudino (2013), Castiglioni (2014), Loss et al. (2014), Rocha and López-Baucells (2014)[f], Menq and Anjos (2015)
Rusty-barred owl (*Strix hylophila*)	Goeldi (1894), Kelso (1940), Bequaert (1955), Voous (1964), Schubart et al. (1965), Pinto (1978), Sick (1993), Belton (1994), Sick (1997), Holt et al. (1999), König et al. (1999), (2008), Willis (2003), Azevedo et al. (2003), Graciolli and Carvalho (2003), Petroff (2003), Antunes et al. (2006), Piacentini et al. (2006), Santos and Gunski (2006), Joppert (2007), Accordi and Barcellos (2008), Soares et al. (2008), Legal et al. (2009), Favretto and Santos (2009), Grantsau (2010), Ramos et al. (2011), Claudino (2013), Mazzoni et al. (2016)
Black-banded owl (*Strix huhula*)	Reinhardt (1870), Goeldi (1894), Hellmayr (1929), Schubart et al. (1965), Pinto (1978), Santos (1979), Sick (1993,1997), Holt et al. (1999), König et al. (1999), (2008), Bornschein and Reinert (2000), Willis and Oniki (2002), Barros (2003), Borges et al. (2004), Gonzaga and Castiglioni (2004), Krabbe (2007), Soares et al. (2008), Vasconcelos and Diniz (2008), Vasconcelos et al. (2008), Marques (2009), Grantsau (2010), Gwynne et al. (2010), Sberze et al. (2010), Andery (2011), Andery et al. (2013), Tomas et al. (2013), Castiglioni (2014)

(continued)

Great horned owl (*Bubo virginianus*)	Marcgrave (1648), Euler (1869), Goeldi (1894), Euler (1900), Ihering (1900), Snethlage (1928), Traylor (1958), Schubart et al. (1965), Belton (1978), Pinto (1978), Santos (1979), Sick (1993), Belton (1994), König et al. (1996), (1999), (2008), Sick (1997), Mahecha and Oliveira (1998), Holt et al. (1999), Wink and Heidrich (2000), Almeida et al. (2003), Azevedo et al. (2003), Fischer et al. (2003), Petroff (2003), Tomazzoni et al. (2004), Lisboa et al. (2005), Crozariol and Almeida (2006), Piacentini et al. (2006), Roda and Pereira (2006), Sigrist (2006), Dornas and Pinheiro (2007), Soares et al. (2008), Zorzin et al. (2008), Antas (2009), Motta-Junior (2009a), Peters et al. (2009), Silva et al. (2009), Sigrist (2009), Grantsau (2010), Gwynne et al. (2010), Andery (2011), Andery et al. (2013), Gallas and Silveira (2013), Silva et al. (2013), Anza and Zilio (2015), Echenique et al. (2016)
Least pygmy owl (*Glaucidium minutissimum*)[c]	Goeldi (1894), Schubart et al. (1965), Pinto (1978), Santos (1979), Hekstra, (1982), Vielliard (1989), Sick (1993), König (1994a), Heidrich et al. (1995b), Howell and Robbins (1995), Sick (1997), Holt et al. (1999), König et al. (1999), (2008), Silva et al. (2002), Willis and Oniki (2002), Fischer et al. (2003), Petroff (2003), König and Weick (2005), Robbins (2006), Amaral (2007), Zorzin et al. (2008), Legal et al. (2009), Grantsau (2010), Fink et al. (2012)
Amazonian pygmy owl (*Glaucidium hardyi*)	Hekstra (1982), Vielliard (1989), Sick (1997), König (1994a), Heidrich et al. (1995b), Howell and Robbins (1995), Holt et al. (1999), König et al. (1999), (2008), Silva et al. (2002), Barros (2003), Barros and Cintra (2009), Grantsau (2010), Gwynne et al. (2010), Sberze et al. (2010), Esclarski (2014), Esclarski and Cintra (2014)
Ferruginous pygmy owl (*Glaucidium brasilianum*)	Descourtilz (1852), Euler (1869,1900), Reinhardt (1870), Goeldi (1894), Ihering (1900), Lutz et al. (1915), Snethlage (1928), Hellmayr (1929), Moojen et al. (1941), Mitchell (1957), Ruschi (1960), Schubart et al. (1965), Pinto (1978), Santos (1979), Hekstra (1982), Dubs (1992), Sick (1993), Belton (1994), König (1994a), Heidrich et al. (1995b), Sick (1997), Mahecha and Oliveira (1998), Costa-Neto (1999), Holt et al. (1999), König et al. (1999), (2008), Silva et al. (2002), Willis and Oniki (2002), Fischer et al. (2003), Petroff (2003), Borges et al. (2004), Graciolli and Bispo (2005), Proudfoot et al. (2006), Roda and Pereira (2006), Sigrist (2006), Amaral (2007), Krabbe (2007), Motta-Junior (2007), Lima and Lima-Neto (2008), Specht et al. (2008), Zorzin et al. (2008), Antas (2009), Legal et al. (2009), Lima and Lima-Neto (2009a), Cunha and Vasconcelos (2009), Cunha et al. (2009), Sigrist (2009), Wink et al. (2009), Castro et al. (2010), Grantsau (2010), Gwynne et al. (2010), Andery (2011), Cunha (2012), Andery et al. (2013), Cunha et al. (2013), Cunha and Fontenelle (2014), Menq and Anjos (2015), Sazima (2015a)
Pernambuco pygmy owl (*Glaucidium mooreorum*)[d]	Silva et al. (2002), König and Weick (2005), Robbins (2006), Roda and Pereira (2006), König et al. (2008), Soares et al. (2008), Grantsau (2010), Pereira et al. (2014)

(continued)

Burrowing owl (*Athene cunicularia*)	Euler (1869), Reinhardt (1870), Goeldi (1894), Euler (1900), Ihering (1900), Snethlage (1928), Hellmayr (1929), Ihering (1940), Moojen et al. (1941), Hempel (1949), Bequaert (1955), Mitchell (1957), Amaral (1962), Schubart et al. (1965), Pinto (1968), Lourenço et al. (1975), Lourenço and Dekeyser (1976), Lourenço (1977), Pinto (1978), Lourenço and Bastos (1979), Santos (1979), Lourenço (1980), Antas and Cavalcanti (1988), Novelli et al. (1988), Silva-Porto and Cerqueira (1990), Martins and Egler (1990), Andrade (1992), Soares et al. (1992a,1992b), Rocha (1993), Santos (1993), Sick (1993), Belton (1994), Stafford and Ferreira (1995), Motta-Junior (1996), Motta-Junior et al. (1996), Clark (1997), Marini et al. (1997), Sick (1997), Mahecha and Oliveira (1998), Motta-Junior and Alho (1998), Rocha and Lima (1998), Costa-Neto (1999), Höfling and Camargo (1999), Holt et al. (1999), König et al. (1999), Valdujo and Nogueira (2000), Teixeira and Melo (2000), Motta-Junior and Alho (2000), Serpa and Monteiro (2001), Silva (2002), Willis and Oniki (2002), Almeida et al. (2003), Azevedo et al. (2003), Bueno (2003), Fischer et al. (2003), Graciolli and Carvalho (2003), Granzinolli and Motta-Junior (2003), Martins et al. (2003), Petroff (2003), Sawaya et al. (2003), Aguilar and Perini (2004), Ferreira and Glock (2004), França et al. (2004), Moraes et al. (2004), Oliveira et al. (2004), Motta-Junior and Bueno (2004), Kokubum and Zacca (2004), Prada (2004), Bueno and Motta-Junior (2005), Tozetti et al. (2005), Zilio (2005,2006), Borges et al. (2006), Braga (2006), Motta-Junior (2006), Sigrist (2006), Silva (2006), Santos and Gunski (2006), Aleixo and Poletto (2007), Jacobucci (2007), Joppert (2007), Krabbe (2007), Lima (2007), Bueno and Motta-Junior (2008), Bastian et al. (2008), Coelho et al. (2008), Garcia et al. (2008), König et al. (2008), Vieira and Teixeira (2008), Zorzin et al. (2008), Antas (2009), Braga and Motta-Junior (2009), Silva et al. (2009), Turci and Bernardes (2009), Sigrist (2009), Wink et al. (2009), Branco et al. (2010), Motta-Junior et al. (2010), Tubelis and Delitti (2010), Martinelli (2010), Grantsau (2010), Gwynne et al. (2010), Andery (2011), Nogueira and Alves (2011), Perillo et al. (2011), Ramos et al. (2011), Cunha (2012), Menezes and Meira (2012), Vitaliano (2012), Andery et al. (2013), Breviglieri et al. (2013), Cunha et al. (2013), Menezes and Ludwig (2013), Siervi (2014), Corbo et al. (2013), Specht et al. (2013), Silva et al. (2014), Bueno and Motta-Junior (2015), Carvalho et al. (2015), Pedroso (2015), Reis (2015), Sacchi (2015), Sazima (2015b)
Buff-fronted owl (*Aegolius harrisii*)	Pinto (1978), Sick (1993), Belton (1994), König (1994b), Studer and Teixeira (1994), Sick (1997), Holt et al. (1999), König et al. (1999), Ribas and Santos (2007), König et al. (2008), Lima and Salles (2008), Antas (2009), Kaminski (2009), Santos (2009), Wink et al. (2009), Girão and Albano (2010), Grantsau (2010), Gwynne et al. (2010), Rebelato et al. (2011), Pereira et al. (2012), Ubaid et al. (2012), Claudino (2013)

(continued)

Stygian owl (*Asio stygius*)	Reinhardt (1870), Kelso (1934b), Schubart et al. (1965), Pinto (1978), Santos (1979), Oliveira (1981), Albuquerque (1983), Scherer-Neto (1985), Motta-Junior and Taddei (1992), Sick (1993), Belton (1994), Melo-Junior et al. (1996), Motta-Junior (1996), Motta-Junior et al. (1996), Marini et al. (1997), Sick (1997), Carrano (1998), Mahecha and Oliveira (1998), Motta-Junior and Alho (1998), Holt et al. (1999), König et al. (1999), Silveira et al. (2001), Almeida et al. (2003), Azevedo et al. (2003), Borges et al. (2004), Lopes et al. (2004), Motta-Junior (2006), Sigrist (2006), Krabbe (2007), Naka et al. (2007), König et al. (2008), Soares et al. (2008), Legal et al. (2009), Crozariol(2010), Motta-Junior et al. (2010), Grantsau (2010), Gwynne et al. (2010), Andery (2011), Fink et al. (2012), Lopes et al. (2012), Meireles et al. (2015), Menq and Anjos (2015), Sacchi (2015)
Striped owl (*Asio clamator*)	Descourtilz (1852), Goeldi (1889,1894), Lutz et al. (1915), Snethlage (1928), Schubart et al. (1965), Pinto (1978), Santos (1979), Oliveira (1980), Sick (1993), Belton (1994), Motta-Junior (1996), Motta-Junior et al. (1996), Sick (1997), Motta-Junior and Alho (1998), Mahecha and Oliveira (1998), Höfting and Camargo (1999), Holt et al. (1999), König et al. (1999), Bencke (2001), Petroff (2003), Willis (2003), Fischer et al. (2003), Motta-Junior et al. 2004a, b, Almeida et al. (2003), Azevedo et al. (2003), Prada (2004), Sanches et al. (2004), Valim et al. (2005), Borges et al. (2006), Motta-Junior (2006), Amaral (2007), Sigrist (2006), Joppert (2007), Coelho et al. (2008), König et al. (2008), Lau (2008), Zorzin et al. (2008), Aguiar and Naiff (2009), Braga and Motta-Junior (2009), Sigrist (2009), Silva et al. (2009), Wink et al. (2009), Catroxo et al. (2010), Grantsau (2010), Gwynne et al. (2010), Pereira and Oliveira (2010), Squarzoni et al. (2010), Andery (2011), Ramos et al. (2011), Ecco et al. (2012), Andery et al. (2013), Claudino (2013), Cunha et al. (2013), Rodarte-Almeida et al. (2013), Goulart (2015), Sacchi (2015)
Short-eared owl (*Asio flammeus*)	Pinto (1978), Sick (1993,1997), Belton (1994), Bagno and Rodrigues (1998), Holt et al. (1999), König et al. (1999), Sigrist (2006), Faria (2007), Braz (2008), König et al. (2008), Motta-Junior et al. (2008), Soares et al. (2008), Motta-Junior (2009b), Wink et al. (2009), Grantsau (2010), Gwynne et al. (2010), Cunha et al. (2013), Meireles et al. (2015)

Here we Present a list of the Available Literature on the 23 Brazilian owl Species. While our list may not be Exhaustive, we have Attempted to Provide the main References for each Species. We Include References for Studies Actually Conducted within Brazil or Using Owls Captured/Collected in Brazil. We have Included some References that Only Marginally were About Owls (e.g., Bird Community Studies) but at Least Contained Brief Specific Texts on one or more owl Species. Some Major Revisions or Books Considering owl Species Occurring in Brazil were also Included, Mainly when Specific Studies on Certain Species are Lacking. Common and Scientific names According to König et al. (2008), except when Indicated

[a]*M. guatemalae* after Piacentini et al. (2015)

[b]Includes *P. pulsatrix*

[c]*G. sicki* after König et al. (2008)

[d]*G. minutissimum* after König et al. (2008)

[e]This author cites *A. clamator* in the text, but the correct species according to original sources is *T. furcata*

[f]The authors erroneously identified a *Strix virgata* as an immature *Lophostrix cristata* (see Aleixo 2014)

References

Accordi IA, Barcellos A (2008) Novas ocorrências e registros notáveis sobre distribuicão de aves em Santa Catarina, sul do Brasil. Biotemas 21(1):85–93

Aguiar KMO, Naiff RH (2009) Aspectos reprodutivos e dieta alimentar dos ninhegos de *Rhinoptynx clamator* (Aves: Strigidae) no campus Marco Zero da Universidade Federal do Amapá, Macapá-AP. Acta Amaz 39(1):221–224

Aguilar TM, Perini F (2004) Observacões sobre o comportamento da coruja-buraqueira, *Athene cunicularia* (Strigiformes/Strigidae) em ambiente perturbado no campus da UFMG. Atual Ornitol 118:6–7

Alaja P, Mikkola H (1997) Albinism in the great Gray owl (*Strix nebulosa*) and other owls. In: Duncan JR, Johnson DH, Nicholls TH (eds) Biology and conservation of owls of the northern hemisphere, General technical report NC-190. USDA Forest Service, North Central Research Station, St. Paul, pp 33–37

Albuquerque EP (1983) Notas sobre a ocorrência de *Asio stygius* (Wagler, 1832) – Aves, Strigidae) e de *Macropsalis creaga* (Bonaparte, 1849) – Aves, Caprimulgidae no Estado do Rio Grande do Sul, Brasil. Roessleria 5:165–166

Albuquerque JLB, Motta-Junior JC (2006) Editorial. Rev Bras Ornit 14(4):315

Aleixo A (2014) Erratum. Rev Bras Ornit 22(2):244

Aleixo A, Poletto F (2007) Birds of an open vegetation enclave in southern Brazilian Amazonia. Wilson J Ornithol 119(4):610–630

Aleixo A, Poletto F, Lima MFC et al (2011) Notes on the vertebrates of northern Pará, Brazil: a forgotten part of the Guianan region, II. Avifauna. Bol Mus Para Emilio Goeldi Cienc Nat 6(1):11–65

Alho CJR, Lacher TE Jr (1991) Mammalian conservation in the Pantanal of Brazil. In: Mares MA, Schmidly DJ (eds) Latin American mammalogy: history, biodiversity and conservation. University of Oklahoma Press, Norman, pp 280–294

Almeida AF, Carvalho MAS, Summa MEL (2003) Levantamento da avifauna Metropolitana de São Paulo, atendida pela Divisão Técnica de Medicina Veterinaria e Manejo da Fauna Silvestre/DEPAVE/PMSP. Boletim CEO 15:16–26

Alves MAS, Pacheco JF, Gonzaga LAP et al (2000) Aves. In: Bergallo HG, Rocha CF, Alves MAS et al (orgs) A fauna ameacada de extincão do Estado do Rio de Janeiro. Editora UERJ, Rio de Janeiro, pp 113–124

Amaral KF (2007) Composicão e abundãncia de corujas em Floresta Atlântica e sua relacão com variáveis de habitat. Dissertation, Universidade Federal do Rio Grande do Sul, Porto Alegre

Amaral V (1962) *Stemostoma augei* n. sp. (Acarina: Rhinonyssidae Vitz, 1935) e identificacão das especies *Rhinoecius bisetosus* Strandtmann, 1952 e *Rhinoecius grandis*. Arq Inst Biol 29:69–81

Andery DA (2011) Perfil sanitário de rapinantes de cativeiro e recolhimento em um centro de triagem de animais silvestres, Belo Horizonte/MG. Dissertation, Universidade Federal de Minas Gerais, Belo Horizonte

Andery DA, Ferreira FC Jr, Araújo AV et al (2013) Health assessment of raptors in triage in Belo Horizonte, MG, Brazil. Braz J Poultry Sci 15(3):169–286

Andrade MA (1992) Aves Silvestres: Minas Gerais. CIPA, Belo Horizonte

Antas PTZ (2009) Pantanal, guia de aves, 2nd edn. Editora SESC, Depto. Nacional, Rio de Janeiro

Antas PTZ, Cavalcanti RB (1988) Aves comuns do Planalto Central. Editora UnB, Brasilia

Antunes AZ (2005) Alteracões na composicão da comunidade de aves ao longo do tempo em um fragmento florestal no sudeste do Brasil. Ararajuba 13(1):47–61

Antunes AZ, Eston MR, Santos ASR et al (2006) Presenca da coruja-listrada *Strix hylophila* Temminck, 1825 (Aves, Strigidae) no Parque Estadual Carlos Botelho, São Miguel Arcanjo, Estado de São Paulo. Rev Inst Flor 18:167–171

Anza J, Zilio F (2015) Molina's hog-nosed skunk as prey of the great horned owl: predation or opportunist scavenging? Rev Bras Ornit 23(4):377–379

Azevedo MAG, Machado DA, Albuquerque JLB (2003) Aves de rapina na ilha de Santa Catarina, SC: composicão, freqüência de ocorrência, uso de habitat e conservacão. Ararajuba 11(1):75–81

Azevedo AA, Monteiro JLG (2006). Análise dos Impactos Ambientais da Atividade Agropecuária no Cerrado e suas inter-relacões com os Recursos Hídricos na Região do Pantanal. WWF – World Wildlife Fund. http://www.wwf.org.br/informacoes/bliblioteca/index.cfm?uNewsID=3000. Accessed 18 Nov 2010

Azevedo VC (2004) Criacão, condicionamento e soltura de ninhegos de *Tyto alba* (Strigiformes: Tytonidae) resgatados em área urbana na cidade de Salvador, Bahia. Boletim ABFPAR 7(2):79–85

Bagno MA, Rodrigues FHG (1998) Novos registros de espécies de aves para o estado de Goiás, Brasil. Ararajuba 6(1):64–65

Barbosa MO, Dornas T, Dionísio C et al (2015) Novos registros ornitológicos para a região de Palmas, Tocantins. Atual Ornitol 188:59–65

Barros FM (2011) Área de vida, uso e selecão de habitat pela corujinha-do-mato *Megascops choliba* (Strigiformes: Strigidae) em uma área de cerrado na região central do Estado de São Paulo. Dissertation, Universidade de São Paulo, São Paulo

Barros FM, Motta-Junior JC (2014) Home range and habitat selection by the tropical screech-owl in a Brazilian savanna. J Raptor Res 48(2):142–150

Barros OG (2003) O uso de habitat pela comunidade de corujas (Strigidae) em uma floresta de terra firme na Amazonia central, Brasil. Dissertation, INPA/UFAM, Manaus

Barros OG, Cintra R (2009) The effects of forest structure on occurrence and abundance of three owl species (Aves: Strigidae) in the Central Amazon forest. Fortschr Zool 26(1):85–96

Bastian AMS, Fraga ED, Mäder A et al (2008) Análise de egagropilas decoruja-buraqueira, *Athene cunicularia* (Molina, 1782) no campus da UNISINOS, São Leopoldo - RS (Strigiformes: Strigidae). Biodivers Pampeana 6(2):70–73

Belton W (1978) Supplementary list of new birds for Rio Grande do Sul, Brazil. Auk 95:413–415

Belton W (1994) Aves do Rio Grande do Sul. Editora UNISINOS, São Leopoldo

Bencke GA (2001) Lista de referencia das aves do Rio Grande do Sul. Fundacão Zoobotânica do Rio Grande do Sul, Porto Alegre

Bencke GA, Bencke SC (1999) The potential importance of road deaths as a cause of mortality for large forest owls in southern Brazil. Cotinga 11:17–18

Bencke GA, Bencke SC (2000) More road-killed owls and a new record for Santa Catarina, Brazil. Cotinga 13:69

Bencke GA, Fontana CS, Dias RA et al (2003) Aves. In: Fontana CS, Bencke GA, Reis RE (orgs) Livro Vermelho da Fauna Ameacada de Extincão no Rio Grande do Sul. EDIPUCRS, Porto Alegre, pp 189–479

Bencke GA, Mauricio GN, Develey PF et al (2006) Areas Importantes para a Conservação das Aves no Brasil: Parte I - estados do Domínio da Mata Atlântica. SAVE Brasil, São Paulo

Bequaert J (1955) The Hippoboscidae or louse-flies (Diptera) of mammals and birds part II. Taxonomy, evolution and revision of American genera and species. Entomol Am 35:233–416

Bergamini FS (1997) Hábitos alimentares de *Tyto alba tuidara* (Gray, 1829) na região de Indaiatuba – São Paulo, Brasil. Atual Ornitol 80:10–11

Bergamini FS (2002) Hábitos alimentares de *Tyto alba tuidara* (Gray, 1829) (Strigiformes, Tytonidae) na região de Indaiatuba - São Paulo, Brasil. Bioikos 16:45–51

Bernardi IP, Miranda JMD, Abreu KC et al (2008) Novo registro de *Strix virgata* (Cassin, 1849) para o estado do Rio Grande do Sul, Brasil (Strigiformes: Strigidae). Bioci, Porto Alegre 16(2):138–141

Berlepsch HV (1902) Exhibition of south American owls. Bull Br Orn Club 12:4–10

Bierregaard ROJ (1998) Conservation status of birds of prey in south American tropics. J Raptor Res 32:19–27

Birdlife International 2016a. Species factsheet: *Strix hylophila.*http://www.birdlife.org. Accessed 23 Nov 2016

Birdlife International 2016b. Species factsheet: *Glaucidium mooreorum*.http://www.birdlife.org. Accessed 23 Nov 2016
Blus LJ (1996) Effects of pesticides on owls in North America. J Raptor Res 30(4):198–206
Boesman P (2006) Birds of Brazil: MP3 sound collection (version 1.0). CD-ROM. Bird-Jesounds, Winsum
Bonvicino CR, Bezerra AMR (2003) Use of regurgitated pellets of Barn owl (*Tyto alba*) for inventorying small mammals in the Cerrado of Central Brazil. Stud Neotrop Fauna E 38:1–5
Borges SH, Henriques LM, Carvalhaes A (2004) Density and habitat use by owls in two Amazonian forest types. J Field Ornithol 75:176–182
Borges RC, Oliveira AO, Bernardo N et al (2006) Diagnostico da fauna silvestre apreendida e recolhida pela Polícia Militar de Meio Ambiente de Juiz de Fora, MG (1998 e 1999). Rev Bras Zoocienc 8(1):23–33
Bornschein MR, Reinert BL (2000) Aves de três remanescentes florestais do norte do estado do Paraná, sul do Brasil, com sugestões para conservacão e manejo. Rev Bras Zool 17(3):615–636
Braga ACR (2006) Uso do hábitat da corujinha-do-mato (*Otus choliba*) e da coruja-buraqueira (*Athene cunicularia*) em remanescentes de cerrado da região central do Estado de São Paulo. Dissertation, Universidade de São Paulo, São Paulo
Braga ACR, Motta-Junior JC (2009) Weather conditions and moon phase influence on tropical screech owl and Burrowing owl detection by playback. Ardea 97(4):395–401
Branco JO, Hillesheim JC, Fracasso HAA, Christoffersen ML, Evangelista CL (2010) Bioecology of the ghost crab. *Ocypode quadrata* (Fabricius, 1787) (Crustacea: Brachyura) compared with other intertidal Crabs in the southwestern Atlantic. J Shellfish Res 29(2):503–512
Brandon K, Fonseca GAB, Rylands AB et al (2005) Introduction: Brazilian conservation: challenges and opportunities. Conserv Biol 19(3):595–600
Brasil AO, Ribeiro S, Niederauer GI et al (2010) Quirópteros encontrados na dieta de *Tyto alba* (Strigiformes, Tytonidae) no Centro de Porto Alegre, RS, Brasil. Chirop Neotrop, Supl 16(1):35–36
Braz VS (2008) Ecologia e conservacão das aves campestres do bioma Cerrado. Thesis, Universidade de Brasília, Brasília, Brazil
Bressan PM, Kierulff MCM, Sugieda AM (eds) (2009) Fauna ameaçada de extinção no estado de São Paulo - vertebrados. Fundação Parque Zoológico de São Paulo. Secretaria do Meio Ambiente, São Paulo
Breviglieri CPB, Piccoli GCO, Uieda W et al (2013) Predation-risk effects of predator identity on the foraging behaviors of frugivorous bats. Oecologia 173:905–912
Bueno AA (2003) Vulnerabilidade de pequenos mamíferos a vertebrados predadores em campos cerrados da estação Ecológica de Itirapina, SP. Dissertation, Universidade de São Paulo, São Paulo
Bueno AA, Motta-Junior JC (2005) A lua e os pequenos mamíferos. Ciência Hoje 37(219):64–66
Bueno AA, Motta-Junior JC (2008) Small mammal prey selection by two owl species in southeastern Brazil. J Raptor Res 42(4):248–255
Bueno AA, Motta-Junior JC (2015) Behavioural and morphological strategies by small savannah rodents to avoid predation. Mamm Biol 80:401–408
Burton JA (1984) Owls of the world, revised edn. Tanager Books, Dover
Bustamante RO, Simonetti JA (2005) Is *Pinus radiata* invading the native vegetation in central Chile? Demographic responses in a fragmented forest. Biol Invas 7:243–249
Cândido-Junior JF (1999) Efeito de alterações ambientais sobre a avifauna do estado de Rondônia, Brasil. Thesis, Universidade Estadual Paulista, Rio Claro
Carrano E (1998) Registros do mocho-diabo *Asio stygius* (Wagler, 1832) no estado do Paraná. Atual Ornitol 85:1
Carvalho CB, Freitas EB, Araújo R et al (2015) Chromatic aberration in the plumage of *Athene cunicularia* (Strigiformes: Strigidae) for the Cerrado Biome, Central Brazil. Atual Ornitol 187:21

Carvalho LFAC, Cunha NL, Fischer E et al (2011) Predation on broad-eared bat *Nyctinomops laticaudatus* by the spectacled owl *Pulsatrix perspicillata* in southwestern Brazil. Rev Bras Ornit 19(3):417–418

Castiglioni, GAD (2014) Uso de vocalização de corujas em estudos científicos, distribuição temporal da atividade vocal de quatro espécies de corujas florestais e individualidade vocal de *Megascops atricapillus* (Strigiformes: Strigidae) na Reserva Biológica União, Rio de Janeiro. Thesis, Universidade Federal do Rio de Janeiro, Rio de Janeiro

Castro WS, Franchin AG, Marçal O Jr (2010) Reprodução de *Glaucidium brasilianum* (Gmelin, 1788) em área urbana de Uberlandia. Minas Gerais Rev Bras Ornit 18(1):55–58

Catroxo MHB, Taniguchi DL, Melo NA et al (2010) Viral research in Brazilian owls (*Tyto alba* and *Rhinoptynx clamator*) by transmission electron microscopy. Int J Morphol 28(2):627–636

CBRO - Comitê Brasileiro de Registros Ornitológicos (2001a) Resolução N° 44 - Incluir *Otus guatemalae* (Sharpe, 1875) na lista principal de aves brasileiras, inserindo-o imediatamente após *Otus atricapillus* (Temminck, 1822). Nattereria 2:52

CBRO - Comite Brasileiro de Registros Ornitológicos (2001b) Resolução N° 45 - Incluir *Otus usta* (Sclater, 1859) na lista principal de aves brasileiras, inserindo-o imediatamente após *Otus watsonii* (Cassin, 1848). Nattereria 2:52–53

Clark RJ (1997) A review of the taxonomy and distribution of the Burrowing owl (*Speotyto cunicularia*). J Raptor Res Reports 9:14–23

Clark RJ, Smith DG, Kelso LH (1978) Working bibliography of owls of the world. NWF Scient./Tech.Ser. No.1, Raptor Information Center, National Wildlife Federation, Washington

Claudino RM (2013) Como a fragmentação da Mata Atlântica no sudoeste de Minas Gerais afeta aspectos da comunidade de população e assembleia de corujas (Aves: Strigiformes). Dissertation, Universidade Federal de Ouro Preto, Ouro Preto

Claudino RM, Rodrigues RC, Silva MRB (2012) New record of nesting site of tropical screech owl (*Megascops choliba*) from Brazil. Ornitol Neotrop 23(1):137–141

CONSEMA (2011) Resolução CONSEMA N° 002, de 06 de dezembro de 2011. Lista das espécies da fauna ameaçadas de extinção em Santa Catarina. Available at: http://www.fatma.sc.gov.br/upload/Fauna/resolucao_fauna__002_11_fauna.pdf. Accessed 30 Nov 2016

Corbo M, Macarrão A, D'Angelo GB et al (2013) Aves do campus da Unicamp e arredores. Avis Brasilis, Valinhos

Crozariol MA (2010) Ocorrência do mocho-diabo, *Asio stygius* (Aves: Strigiformes), no município de Taubaté, Vale da Paraíba Paulista, com alguns dados sobre a sua alimentação. Atual Ornitol 156:4–5

Crozariol MA, Almeida FC (2006) Primeiro registro de *Bubo virginianus* (Gmelin, 1788) (Aves: Strigidae) no litoral norte do estado de São Paulo, sendo o terceiro registro documentado da espécie para o estado. Atual Ornitol 132:20–21

Coelho AGM (1979) Autophagism in a captive *Otus choliba* Decussatus (Aves: Strigidae). Notulae Biol 1:1

Coelho IP, Kindel A, Coelho AVP (2008) Roadkills of vertebrate species on two highways through the Atlantic Forest biosphere reserve, southern Brazil. Eur J Wildl Res 54:689–699

COPAM (2010) Lista das Especies da Fauna Ameacadas de Extinção no Estado de Minas Gerais. Deliberação Normativa COPAM N° 147, de 30 de abril de 2010. Diario do Executivo – "Minas Gerais" - 04/05/2010, Belo Horizonte, MG. http://www.siam.mg.gov.br/sla/download.pdf?idNorma=13192. Accessed 02 May 2011

Costa-Neto EM (1999) As corujas e o homem. Ciência Hoje 26(156):74–76

Costa Silva P, Henderson RW (2013) Barn owl (*Tyto alba*) predation on *Corallus hortulanus* (Squamata, Boidae). Herpetol Notes 6:35

Cunha FCR 2012. Tumulto de aves contra *Glaucidiu brasilianum* (caburé) e *Athene cunicularia* (coruja-buraqueira). Dissertation, Universidade Federal de Ouro Preto, Ouro Preto

Cunha FCR, Vasconcelos MF (2009) Aves que são atraídas pela vocalização do cabure *Glaucidium brasilianum* (Strigidae). Rev Bras Ornit 17(2):144–149

Cunha FCR, Fontenelle JCR (2014) Registros de tumulto em aves no Brasil: uma revisão usando a plataforma WikiAves. Atual Ornitol 177:46–53
Cunha FCR, Vasconcelos MF, Specht GVA (2009) Alerta vermelho! Cabure na area! Ciência Hoje 43(257):26–29
Cunha FCR, Specht GVA, Brites FR (2013) Owls and mobbing behavior: anecdotal observations. J Life Sci 7(11):1204–1208
Dantas SM, Weckstein JD, Bates JM et al (2016) Molecular systematics of the new world screech-owls (*Megascops*: Aves, Strigidae): biogeographic and taxonomic implications. Mol Phylogenet Evol 94:626–634
Descourtilz JT (1852) Ornithologie Brésilienne, ou Histoire des Oiseaux du Brésil remarquables par leur plumage, leur chant ou leurs habitudes. Thomas Reeves, Rio de Janeiro
Dias RI, Lima MR (2015) Breeding biology and nest survival in tropical screech-owls (*Megascops choliba*) in the Brazilian Cerrado. J Wilson Ornithol 127(3):432–440
Diniz-Filho JAF, Sant'ana CER (1998) Macroecological analysis of south American owls (Strigiformes) using phylogenetic independent contrasts. J Comp Biol 3:27–32
Diniz-Filho JAF, Sant'ana CER (2000) Phylogenetic correlograms and the evolution of body size in south American owls (Strigiformes). Genet Mol Biol 23:285–292
Diniz-Filho JAF, Rangel TFL, Hawkins BA (2004) A test of multiple hypotheses for the species richness gradient of south American owls. Oecologia 140:633–638
Dornas T, Pinheiro RT (2007) Predação de *Opisthocomus hoazin* por *Spizaetus ornatus* e de *Bubulcus ibis* por *Bubo virginianus* em Tocantins, Brasil. Rev Bras Ornit 15(4):601–604
Dubs B (1992) Birds of southwestern Brazil. Catalogue and guide to the birds of the Pantanal of Mato Grosso and its border areas. Betrona-Verlag, Küsnacht
Duncan JR (2003) Owls of the world: their lives, behavior and survival. Firefly Books, Buffalo
Ecco R, Preisa IS, Vilela DAR et al (2012) Molecular confirmation of *Trichomonas gallinae* and other parabasalids from Brazil using the 5.8S and ITS-1 rRNA regions. Vet Parasitol 190:36–42
Echenique JVZ, Bandarra PM, Brauner RK et al (2016) Infecção por pox vírus e *Aspergillus fumigatus* em *Bubo virginianus* (Coruja jacurutu). Pesquisa Vet Brasil 36(7):630–633
Engel L, Mello DA (1993) Rodents in agroecosystems in the Cerrado Province of the Federal District (Brasilia/DF, Brazil). Ciênc Cult 45:128–133
Enríquez PL, Johnson DH, Rangel-Salazar JL (2006) Taxonomy, distribution and conservation of owls in the neotropics: a review. In: Rodríguez-Estrella R (ed) Current raptor studies in México. Centro de Investigaciones Biológicas del Noroeste y CONABIO, La Paz, pp 254–307
Escalante P, Navarro AG, Peterson AT (1993) A geographic, historical and ecological analysis of land bird diversity in Mexico. In: Ramamoorthy TP, Bye R, Lot A et al (eds) Biological diversity in Mexico, origins and distribution. Oxford University Press, New York, pp 281–307
Escarlate-Tavares F, Pessoa LM (2005) Bats (Chiroptera, Mammalia) in Barn owl (*Tyto alba*) pellets in northern Pantanal, Mato Grosso, Brazil. Mastozool Neotrop 12(1):61–67
Esclarski P (2014) Efeitos da estrutura da floresta de terra-firme no uso de habitat em corujas (Aves, Strigiformes) na Amazônia central. Dissertation, Instituto Nacional de Pesquisas da Amazônia, Manaus
Esclarski P, Cintra R (2014) Effects of Terra Firme-forest structure on habitat use by owls (Aves: Strigiformes) in central Brazilian Amazonia. Ornitol Neotrop 25(4):433–458
Euler C (1869) Beiträge zur Naturgeschichte tier Vögel Brasiliens. J Ornithol 17:241–255
Euler C (1900) Descrição de ninhos e ovos das aves do Brasil. Rev Mus Paul 4:9–148
Faria LP (2007) Registros de aves globalmente ameaçadas, raras e endêmicas para a região de Vicente Pires, Distrito Federal, Brasil. Rev Bras Ornit 15(1):117–122
Favretto MA, Santos EB (2009) Coleção de aves do Laboratorio de Zoologia da UNOESC, Santa Catarina, Brasil: observações sobre o inicio da construção do acervo. Atual Ornitol 152:33–35
Fearnside PM (2005) Deforestation in Brazilian Amazonia: history, rates and consequences. Cons Biol 19(3):680–688
Ferreira CM, GLOCK L (2004) Diagnostico preliminar sobre a avifauna traficada no Rio Grande do Sul, Brasil. Bioci Porto Alegre 12(1):21–30

Fink D, Brandt CS, Rupp AE et al (2012) Ocorrência de corujas (Aves: Strigiformes) na RPPN Bugerkopf, Blumenau, Santa Catarina. Biotemas 25(2):75–80

Fischer WA, Ramos-Neto MB, Silveira L et al (2003) Human transportation network as ecological barrier for wildlife on Brazilian Pantanal-Cerrado corridors. In: Irwin CL, Garrett P, McDermott KP (eds) Proceedings of the international conference on ecology and transportation. Center for Transportation and the Environment, North Carolina State University, Raleigh, pp 182–194

França FGR, Braz VS, Balbino SF (2004) *Briba brasiliana* (Lagartixa). Predation Herptol Rev 35(4):386–387

Freitas JFT, Lent H (1937) Novo trematodeo parasito parasito de *Strix flammea perlata* (Licht.) Mem Inst Oswaldo Cruz 32:535–541

Freitas SR, Sousa COM, Bueno C (2013) Effects of landscape characteristics on roadkill of mammals, birds and reptiles in a highway crossing the Atlantic Forest in southeastern Brazil. Proceedings of the 2013 International Conference on Ecology and Transportation, Scottsdale, Arizona, USA, 2013. http://www.icoet.net/ICOET_2013/documents/papers/ICOET2013_Paper201D_Freitas_at_al.pdf. Accessed 30 Nov 2016

Gallas M, Silveira EF (2013) *Neodiplostomum reflexum* Chandler & Rausch, 1947 (Digenea, Diplostomidae) em *Bubo virginianus* Gmelin, 1788 (Aves, Strigidae): primeiro registro para o Estado do Rio Grande do Sul, Brasil. Cad Pesqui Sér Biol 25(1):50–58

Garcia SA, Cristofoli SL, Bastian AMS et al (2008) Sobre mortandade de anfibios em área de reprodução de Coruja-buraqueira, *Athene cunicularia* (Molina, 1782) em São Leopoldo - RS. Biodivers Pampeana 6(2):11–13

Garcia SA, Santos CR, Corrêa LLC et al (2015) Registros de *Pulsatrix koeniswaldiana* (Strigiformes: Strigidae) em fragmentos de mata-atlântica no Vale do Rio dos Sinos e primeiro espécime em coleção científica no estado do Rio Grande do Sul. Cad Pesqui Sér Biol 27(3):31–36

Girao W, Albano C (2010) Sinopse da história, taxonomia, distribuição e biologia do caboré *Aegolius harrisii* (Cassin, 1849). Rev Bras Ornit 18(2):102–109

Gliesch R (1933) Animais úteis aos homens. Egatea 18(3):101–126

Goeldi EA (1889) Die amerikanische Ohreule (*Otus americanus* Gmélin). Schwei Blatt Ornithol, Zürich 13:331–332,346–347,363–364

Goeldi EA (1894) As aves do Brasil. Livraria Clássica de Alves e Cia, Rio de Janeiro

Goldstein ML, Lacher TE Jr, Woodbridge B et al (1999) Monocrotophos-induced mass mortality of Swainson's hawks in Argentina, 1995-96. Ecotoxicology 8(3):201–214

Gonzaga LP, Castiglioni GDA (2004) Registros recentes de *Strix huhula* no Estado do Rio de Janeiro (Strigiformes: Strigidae). Ararajuba 12:141–144

Gonzaga LP, Castiglioni GDA (2015) Four hundred and fifty years in the dark: Black-capped screech-owl *Megascops atricapilla* (Temminck, 1822) recorded for the first time in the city of Rio de Janeiro (Strigiformes: Strigidae). Atual Ornitol 185:7–9

Gonzalez EM, Claramunt SJ, Saralegui AM (1999) Mamíferos hallagos en egagrópilas de *Tyto alba* (Aves, Strigiformes, Tytonidae) en Bagé, Rio Grande do Sul, Brasil. Iheringia Sér Zool 86:117–120

Goulart MA (2015) Parâmetros hematológicos, bioquímicos e pesquisa de cepas produtoras de ESBL e carbapenemases em aves de rapina em cativeiro no Brasil. Dissertation, Universidade Federal do Paraná, Curitiba

Graciolli G, Carvalho CJB (2003) Hippoboscidae (Diptera, Hippoboscoidea) no Estado do Paraná, Brasil: chaves de identificação, hospedeiros e distribuição geográfica. Rev Bras Zool 20(4):667–674

Graciolli G, Bispo AA (2005) Ectoparasitic Hippoboscidae (Diptera) on six species of owls (Strigiformes) in the state of Paraná, southern Brazil. Ararajuba 13(2):81–182

Grantsau R (2010) Guia completo para identificação das aves do Brasil, vol 1. Vento Verde, São Carlos

Granzinolli MAM, Motta-Junior JC (2003) *Tantilla melanochepala* (Black-headed snake) predation. Herptol Rev 34(2):156–157

Granzinolli, MAM, Motta-Junior JC (2010) Aves de rapina: levantamento, seleção de habitat e dieta. Matter S von, Straube F, Accordi I et al (orgs) Ornitologia e Conservação: Ciência Aplicada, Técnicas de Pesquisa e Levantamento. Editora Technical Books, Rio de Janeiro, pp 169–187

Grouw H v (2006) Not every white bird is an albino: sense and nonsense about colour aberrations in birds. Dutch Birding 28:79–89

Guilherme E (2012) Birds of the Brazilian state of Acre: diversity, zoogeography, and conservation. Rev Bras Ornit 20(4):393–442

Gwynne JA, Ridgely RS, Tudor G et al (2010) Aves do Brasil- Pantanal & Cerrado. Editora Horizonte, São Paulo

Hardy JW, Coffey BB, Reynard GB (1999) Voices of the new world owls (Strigiformes: Tytonidae, Strigidae), revised edn. ARA Records, Gainesville

Harris MB, Tomas W, Mourão G et al (2005) Safeguarding the Pantanal wetlands: threats and conservation initiatives. Cons Biol 19(3):714–720

Heidrich P, König C, Wink M (1995a) Molecular phylogeny of South America screech owls of the *Otus atricapillus* Complex (Aves: Strigidae) inferred from nucleotide sequences of the mitochondrial cytochrome b Gene. Z Naturforsch 50:294–302

Heidrich P, König C, Wink M (1995b) Bioakustik, taxoniomie und molekulare systematic amerikanischer sperlingskäuze (Strigidae: *Glaucidium* spp.). Stutt Beitr Naturkd. Ser A (Biologie) 534:1–47

Hekstra GP (1982) Description of twenty four new subspecies of American *Otus* (Aves, Strigidae). Bull Zool Mus Univ Amsterdam 9(7):49–63

Hellmayr CE (1929) A contribution to the ornithology of north-eastern Brazil. Field Mus Nat Hist Publ 255. Zool Ser 12:235–501

Hempel A (1949) Estudo da alimentação natural de aves silvestres do Brasil. Arq Inst Biol 19:237–268

Hernandes FA, O'Connor BM (2015) *Cystoidosoma hermaphroditus* sp. n. The first representative of the quill mite family Ascouracaridae (Acari: Astigmata: Pterolichoidea) from an owl (Aves: Strigiformes). Folia Parasitol 62:037

Höfting E, Camargo HFA (1999) Aves no campus, 3rd edn. EDUSP /IBUSP, São Paulo

Hofting E, Alvarenga HMF (2001) Osteology of the shoulder girdle in the piciformes, passeriformes and related groups of birds. Zool Anz 240:196–208

Holt DW, Berkley R, Deppe C et al (1999) Strigidae species accounts. In: del Hoyo J, Elliott A, Sargatal J (eds) Handbook of the birds of the world, Barn-owls to hummingbirds, vol 5. Lynx Edicions, Barcelona, pp 152–242

Howell SNG, Robbins MB (1995) Species limits of the least pygmy owl *Glaucidium minutissimum* Complex. Wilson Bull 107:7–25

IBGE (2007) Censo Demográfico de 2007. Instituto Brasileiro de Geografia e Estatística, Rio de Janeiro. http://www.ibge.gov.br/home/estatistica/populacao/contagem2007/default.shtm. Accessed 01 Dec 2012

ICMBio (2014) Lista das espécies terrestres e mamíferos aquáticos ameaçados de extinção do Brasil. Portaria MMA n° 444, de 17 de dezembro de 2014. http://www.icmbio.gov.br/portal/images/stories/biodiversidade/fauna-brasileira/avaliacao-do-risco/PORTARIA_N%C2%BA_444_DE_17_DE_DEZEMBRO_DE_2014.pdf. Accessed 30 Oct 2016

Ihering H v (1900) Catálogo crítico-comparativo dos ninhos e ovos das aves do Brasil. Rev Mus Paul 4:191–300

Ihering R v (1940) Dicionário dos animais do Brasil. Secretaria da Agricultura, Industria e Comercio, São Paulo

IUCN 2016. IUCN Red List of Threatened Species. http://www.iucnredlist.org . Accessed 25 Oct 2016

Jacobucci GB (2007) Comportamento de alarme em corujas buraqueiras (*Athene cunicularia*) durante o período reprodutivo no sudeste do Brasil. Rev Bras Zoociências 9(2):145–150

Joppert AM (2007) Estudo prospectivo das causas de morte de Falconiformes e Strigiformes de vida livre no município de São Paulo. Thesis, Universidade de São Paulo, São Paulo

Jordão FS, Reis ML, Louzada DS (1997) Análise do conteúdo de pelotas de *Tyto alba* (Strigiformes, Tytonidae) na Estação Ecológica de aguas Emendadas (Brasilia, DF). In: Leite LL, Saito CH (eds) Contribuição ao conhecimento ecológico do cerrado. Universidade de Brasília Brasília, Brasília, pp 207–209

Kaminski N (2009) Primeiro registro documentado de *Aegolius harrisii* para o estado de Santa Catarina, Brasil. Cotinga 31:79

Kanegae MF, Camacho I, Fernandes BC et al (2012) Ocorrência da murucututu-de-barriga-amarela (*Pulsatrix koeniswaldiana*) em fragmentos florestais no estado do Rio de Janeiro. Ornitol Neotrop 23:499–505

Kelso LH (1934a) A key to the owls of the genus Pulsatrix Kaup. Auk 51:234–236

Kelso LH (1934b) A new stygian owl. Auk 51:522–523

Kelso LH (1940) Variation of the external ear-opening in the Strigidae. Wilson Bull 52:24–29

Klink CA, Machado RB (2005) Conservation of the Brazilian Cerrado. Cons Biol 19(3):707–713

Kokubum MNC, Zacca W (2004) *Tropidurus torquatus* (Calango). Predation. Herptol Rev 35(3):270

König C (1991) Taxonomische und ökologische Untersuchungen an Kreischeulen (*Otus* spp.) des südlichen Südamerika. J Ornith 132:209–214

König C (1994a) Biological patterns in owl taxonomy, with emphasis on bioacoustical studies on Neotropical pygmy (*Glaucidium*) and screech owls (*Otus*). In: Meyburg BU, Chancellor RD (eds) Raptor conservation today. World Working Group on Birds of Prey and Owls and The Pica Press, Berlin, pp 1–19

König C (1994b) Lautäusserungen als interspezifische Isolationsmechanismen bei Eulen der Gattung *Otus* (Aves: Strigidae) aus dem südlichen Südamerika. Stuttgarter Beitrr Naturk, Ser A 511:1–35

König C, Weick F (2005) Ein neuer Sperlingskauz (Aves: Strigidae) aus Südostbrasilien. Stuttgarter Beitrr Naturk, Ser A 688:1–12

König C, Heidrich P, Wink M (1996) Zur Taxonomie der Uhus (Strigidae: *Bubo* spp.) im südlichen Südamerika. Stuttgarter Beitrr Naturk, Ser A 540:1–9

König C, Weick F, Becking JH (1999) Owls: a guide to the owls of the world. Yale University Press, New Haven

König C, Weick F, Becking JH (2008) Owls of the world. A guide to the owls of the world, 2nd edn. Christopher Helm, London

Krabbe N (2007) Birds collected by P.W. Lund and J.T. Reinhardt in south-eastern Brazil between 1825 and 1855, with notes on P.W. Lund's travels in Rio de Janeiro. Rev Bras Ornit 15:331–357

Lange RB (1981) Contribuiçao ao conhecimento da bionomia de aves II. Observações sobre o comportamento de *Tyto alba tuidara* (J.E.Gray). Est Biol. Univ Católica Paraná 7:1–27

Lau R (2008) Redescoberta de coruja-do-mato (*Strix virgata,* Strigidae) na região centro-leste do estado do Rio Grande do Sul, Brasil. Atual Ornitol 144:19

Laurance WF, Albernaz AKM, Fearnside PM et al (2004) Deforestation in Amazônia. Science 304:1109

Leal IR, Silva JMC, Tabarelli M et al (2005) Changing the course of biodiversity conservation in the Caatinga of northeastern Brazil. Cons Biol 19(3):701–706

Legal E, Cadorin TJ, Kohler GU (2009) Strigiformes e Caprimulgiformes em Santa Catarina, sul do Brasil: Registros relevantes e novas localidades. Biotemas 22(4):125–132

Lima B, Salles O (2008) Novo registro documentado do caburé-acanelado (*Aegolius harrisii*) para o Estado de São Paulo. Atual Ornitol 143:23

Lima PC (2007) Comportamento reprodutivo da coruja-buraqueira *Athene cunicularia grallaria* (Temminck, 1822) em um encrave de cerrado no litoral norte da Bahia. Atual Ornitol 135:12–13

Lima PC, Lima-Neto TNC (2008) O comportamento reprodutivo do caburé *Glaucidium brasilianum* (Gmelin, 1788) no litoral norte da Bahia: um ensaio fotográfico. Atual Ornitol 141:65–85

Lima PC, Lima-Neto TNC (2009a) Comprovação da existência de cainismo no comportamento reprodutivo do caburé *Glaucidium brasilianum* (Strigiformes: Strigidae) e novos itens alimentares de sua dieta. Atual Ornitol 151:15

Lima PC, Lima-Neto TNC (2009b) O comportamento reprodutivo da corujinha-do-mato *Otus choliba choliba* (Strigiformes: Strigidae) no litoral norte da Bahia: um ensaio fotográfico. Atual Ornitol 152:41–65

Lisboa JS, Pallinger F, Leão C (2005) Nidificação de *Bubo virginianus* em MG. Boletim ABFPAR 8(2):83–85

Lopes LE, Goes R, Souza S (2004) Observations on a nest of the stygian owl (*Asio stygius*) in the central Brazilian Cerrado. Ornitol Neotrop 15:423–427

Lopes LE, Pinho JB, Gaiotti MG et al (2012) Range and natural history of seven poorly known Neotropical rails. Waterbirds 35(3):470–478

Loss ATG, Costa-Neto EM, Machado CG et al (2014) Ethnotaxonomy of birds by the inhabitants of Pedra Branca Village, Santa Teresinha municipality, Bahia state, Brazil. J Ethnobiol Ethnomed 10:55. Available at https://ethnobiomed.biomedcentral.com/articles/10.1186/1746-4269-10-55. Accessed 30 Nov 2016

Lourenço WR (1977) Notes complémentaires sur la prédation de scorpións par la chouette *Speotyto cunicularia grallaria* (Temminck, 1822) (Aves, Strigidae). Oiseau Rev Fr Ornith 47:211–213

Lourenço WR (1980) Sur le comportement alimentaire de *Speotyto cunicularia grallaria* en particulier sur sa condition de prédateur des scorpions. Rev Nordest Biol 3(1):115–119

Lourenço WR, Dekeyser PL (1976) Deux oiseaux predateurs de scorpions. Oiseau Rev Fr Ornith 46:167–172

Lourenço WR, Dekeyser PL, Bastos EK (1975) Notas sobre a biologia de *Speotyto cunicularia grallaria* (Temminck), 1822 (Aves, Strigidae). Cerrado 7(30):22–24

Lourenço WR, Bastos EK (1979) Sur la predation specifique du scorpion *Bothriurus araguayae* par la chouette *Speotyto cunicularia grallaria*. Oiseau Rev Fr Ornith 49(3):231–233

Lund PW (1950) Memórias sobre a paleontologia brasileira, revistas e comentadas por Carlos de Paula Couto. Instituto Nacional do Livro, Rio de Janeiro

Lutz A, Neiva A, Lima AC (1915) Sobre "Pupipara" ou "Hippoboscidae" de aves brasileiras. Mem Inst Oswaldo Cruz 7:1–29

Macedo I, Tetti L (2007) A energia da cana-de-açúcar. Workshop - A Expansão da Agro-energia e seus impactos sobre os ecossistemas brasileiros. Conservation International and Fundaçao Brasileira do Desenvolvimento Sustentável, Rio de Janeiro. http://www.conservation.org.br/publicacoes/workshop_agronegocios.php. Accessed 20 Nov 2008

Machado ABM, Martins CS, Drumont GM (2005) Lista da Fauna Brasileira Ameaçada de Extinção: incluindo as listas das espécies quase ameaçadas e deficientes dados. Biodiversitas, Belo Horizonte

Machado RB, Ramos-Neto MB, Pereira PGP et al (2004) Estimativas de perda da área do Cerrado brasileiro. Relatório técnico não publicado. Conservação Internacional, Brasília. http://www.conservation.org.br/arquivos/RelatDesmatamCerrado.pdf. Accessed 28 Feb 2012

Magrini L (2006) Predação de pequenos mamíferos por suindara (*Tyto alba*) e seu papel no controle de reservatórios naturais de hantavirus em uma área periurbana do municipio de Uberlandia, Minas Gerais, Brasil. Dissertation, Universidade Federal de Uberlândia, Uberlândia, MG, Brazil

Magrini L, Facure KG (2008) Barn owl (*Tyto alba*) predation on small mammals and its role in the control of hantavirus natural reservoirs in a periurban area in southeastern Brazil. Braz J Biol 68(4):733–740

Mahecha GAB, Oliveira CA (1998) An additional bone in the sclera of the eyes of owls and the common Potoo (*Nictibius griseus*) and its role in the contraction of the nictitating membrane. Acta Anat 163:201–211

Marcgrave G (1648) Liber Quintus, Qui agit de Avibus. In: Piso W, Marcgrave G (eds) Historia Naturalis Brasiliae. Fr. Hackium et Lud. Elzevirium, Leiden/Amsterdam, pp 190–220

Marini MA, Motta-Junior JC, Vasconcellos LA et al (1997) Avian body masses from the cerrado region of central Brazil. Ornitol Neotrop 8(1):93–99

Marini MA, Aguilar TM, Andrade RD et al (2007) Biologia da nidificação de aves do sudeste de Minas Gerais, Brasil. Rev Bras Ornit 15(3):367–376

Marques AB (2009) Novos registros de *Strix huhula* (Strigiformes: Strigidae) no Estado do Rio de Janeiro. Atual Ornitol 151:33–34
Martinelli AG (2010) Observaciones sobre selección y reutilizacion de sítios de nidificación de La lechucita Vizcachera *Athene cunicularia* (Strigiformes:Strigidae) en el Parque Municipal do Sabiá, Uberlandia, Estado de Minas Gerais, Brasil. Notulas Faunísticas 50:1–6
Martinelli AG, Volpi TA (2009) Conteúdos estomacais de *Megacosps choliba,* atropeladas ao longo de uma rodovia no Sul do estado do ES. An Congr Ecol Brasil 9:1–2. http:/ /www.seb-ecologia.org.br/2009/resumos_ixceb/887.pdf . Accessed 10 Oct 2011
Martinelli AG, Prado GM (2010) *Leptodactylus ocellatus* (butter frog). Barn owl predation. Herptol Rev 41(1):64–65
Martinelli MM (2009) *Tropidophis paucisquamis* (Brazilian Dwarf Boa). Diet and caudal luring. Herpetol Rev 40(1):104–105
Martins M, Egler SG (1990) Comportamento de caça em urn casal de corujas-buraqueiras (*Athene cunicularia*) na região de Campinas, São Paulo, Brasil. Rev Bras Biol 50:579–584
Martins M, Spina F, Monteiro C et al (2003) *Bothrops altematus* (Urutu). Predation. Herptol Rev 34(2):147–148
Mascarenhas CS, Coimbra MAA, Müller G et al (2009) Ocorrência de *Ornithonyssus bursa* (Berlese, 1888) (Acari: Macronyssidae) em filhotes de *Megascops choliba* (corujinha-do-mato) e *Pitangus sulphuratus* (bem-te-vi), no Rio Grande do Sul, Brasil. Rev Bras Parasitol Vet 18(4):69–70
Mazzoni LG, Vasconcelos MF, Perillo A et al (2016) Filling gaps in the distribution of Atlantic Forest birds in Minas Gerais, southeastern Brazil. Atual Ornitol 190:33–47
Meireles RC, Silva GGR, Cunha FCR (2015) Aves são capazes de diferenciar predadores simpátricos de predadores alopátricos? Atual Ornitol 184:55–58
Melo ES, Santos-Filho M (2007) Efeitos da BR-070 na Província Serrana de Cáceres, Mato Grosso, sobre a comunidade de vertebrados silvestres. Rev Bras Zoocienc 9:185–192
Melo-Junior TA, Pacheco JF, Diniz MG (1996) Ocorrência de *Asio stygius* (Strigiformes: Strigidae) na região metropolitana de Belo Horizonte e em outras localidades do estado de Minas Gerais. Ararajuba 4:34–38
Menezes LN, Meira N (2012) Análise da ecologia alimentar da *Athene cunicularia* (Aves, Strigidae) numa área sob influência antrópica no município de Assis – SP. Arq Ciênc Vet Zool UNIPAR 15(1):37–41
Menezes LN, Ludwig PR (2013) Diversidade alimentar da coruja-buraqueira (*Athene cunicularia*) em ambiente antropomorfizado no município de Maracaí/SP. J Health Sci Inst 31(4):347–350
Menq W (2016a) Portal Aves de Rapina Brasil. http://www.avesderapinabrasil.com. Accessed 20 Nov 2016
Menq W (2016b). Aves de rapina da Mata Atlântica. http://www.avesderapinabrasil.com/arquivo/artigos/avesderapina_mataatlantica.pdf. Accessed 23 Nov 2016
Menq W, Anjos L (2015) Habitat selection by owls in a seasonal semi-deciduous forest in southern Brazil. Braz J Biol 75(4):143–149
Meyer de Schauensee R (1966) The species of birds of South America and their distribution. Livingston Publ. Co, Narberth/Livingston
Meyer de Schauensee R (1982) A guide to the birds of South America, revised edn. Academy of Natural Sciences, Philadelphia/Intercollegiate Press, Philadelphia
MMA - Ministério Do Meio Ambiente (2000) Avaliação e Ações prioritárias para conservação da biodiversidade da Mata Atlântica e Campos Sulinos. Relatório técnico, Brasília
Mikkola H (2012) Owls of the world. A photographic guide. Christopher Helm, London
Minns J, Buzzetti D, Albano C et al. (2010). Aves do Brasil. Vozes e Fotografias. Vol. 1, Floresta Atlântica, Cerrado, Caatinga, Pantanal, Campos Sulinos e Costa. DVD ROM. Avis Brasilis, São Paulo
Mitchell MH (1957) Observations on birds of southeastern Brazil. University of Toronto Press, Toronto
Mittermeier RA, Fonseca GAB, Rylands AB et al (2005) A brief history of biodiversity conservation in Brazil. Cons Biol 19(3):601–607

Moojen J, Carvalho JC, Lopes HS (1941) Observações sobre o conteúdo gástrico das aves brasileiras. Mem Inst Oswaldo Cruz 36(3):405–444

Moraes VS, Pedroso-Junior NN, Bandeira DLC (2004) Aspectos ecológicos da coruja-buraqueira (*Speotyto cunicularia*) agregados a uma análise sócio-econômica visando a conservação de dunas costeiras em pontal do Sul, PR. Bioikos (Campinas) 18(2):11–19

Motta-Junior JC (1988) Alimentação diferencial da suindara (*Tyto alba*) (Aves, Strigiformes) em duas estações do ano em São Carlos, estado de São Paulo. An Sem Reg Ecol 5:357–364

Motta-Junior JC (1996) Ecologia alimentar de corujas (Aves, Strigiformes) na região central do estado de São Paulo: biomassa, sazonalidade e seletividade de suas presas. Thesis, Universidade Federal de São Carlos, São Carlos

Motta-Junior JC (2002) Diet of breeding screech-owls (*Otus choliba*) in southeastern Brazil. J Raptor Res 36(4):332–334

Motta-Junior JC (2004) Ecologia alimentar da suindara (*Tyto alba*) (Aves: Tytonidae) em áreas do Cerrado Brasileiro. In: Coutinho LM, WBC D (eds) Cerrado. CD-ROM. Departamento de Ecologia, Instituto de Biociências da Universidade de São Paulo, São Paulo, SP, Brazil, pp 1–10

Motta-Junior JC (2006) Relações troficas entre cinco Strigiformes simpátricas na região central do Estado de São Paulo, Brasil. Rev Bras Ornit 14(4):359–377

Motta-Junior JC (2007) Ferruginous pygmy-owl (*Glaucidium brasilianum*) predation on a mobbing fork-tailed flycatcher (*Tyrannus savana*) in south-east Brazil. Biota Neotropica 7(2):321–324. http://www.biotaneotropica.org.br/v7n2/en/fullpaper?bn04407022007+en. Accessed 30 Nov 2011

Motta-Junior JC (2009a) *Bubo virginianus* (Gmelin, 1788). Strigiformes, Strigidae. In: Bressan, PM, Kierulff MCM, Sugieda AM (eds.) Fauna ameaçada de extinção no estado de São Paulo - vertebrados. Fundação Parque Zoológico de São Paulo. Secretaria do Meio Ambiente, São Paulo, p 176

Motta-Junior JC (2009b) *Asio flammeus* (Pontoppidan, 1763). Strigiformes, Strigidae. In: Bressan, PM, Kierulff MCM, Sugieda AM (eds.) Fauna ameaçada de extinção no estado de São Paulo - vertebrados. Fundação Parque Zoológico de São Paulo. Secretaria do Meio Ambiente, São Paulo, p 177

Motta-Junior JC, Taddei VA (1992) Bats as prey of stygian owls in southeastern Brazil. J Raptor Res 26:259–260

Motta-Junior JC, Talamoni SA (1996) Biomassa de presas consumidas por *Tyto alba* (Strigiformes, Tytonidae) durante a estação reprodutiva no Distrito Federal. Ararajuba 4(1):38–41

Motta-Junior JC, Alho CJR (1998) Corujas: o que elas comem? Ciência Hoje 23(136):60–62

Motta-Junior JC, Alho CJR (2000) Ecologia alimentar de *Athene cunicularia* e *Tyto alba* (Aves: Strigiformes) nas Estações Ecológica e Experimental de Luiz Antonio, SP. In: Santos JE, Pires JSR (eds) Estação Ecológica de Jataí, vol 1. RIMA Editora, São Carlos, pp 303–315

Motta-Junior JC, Bueno AA (2004) Trophic ecology of the burrowing owl in Southeast Brazil. In: Chancellor RD, Meyburg BU (eds) Raptors worldwide. World Working Group on Birds of Prey and Owls and MME BirdLife, Berlin/Budapest, pp 763–775

Motta-Junior JC, Braga ACR (2012) Estado del conocimiento sobre la ecología y biología de búhos en Brasil. Ornitol Neotrop 23:233–240

Motta-Junior JC, Santos-Filho PS (2012) Mobbing on the striped owl (*Asio clamator*) and Barn owl (*Tyto alba*) by birds in southeast Brazil: do diet influence mobbing? Ornitol Neotrop 23:159–168

Motta-Junior JC, Bueno AA, Braga ACR (2004a) Corujas brasileiras. http:// http://www.ib.usp.br/labecoaves/PDFs/pdf30CorujasIBC.pdf. Accessed 20 Nov 2008

Motta-Junior JC, Alho CJR, Belentani SCS (2004b) Food habits of the striped owl *Asio clamator* in Southeast Brazil. In: Chancellor RD, Meyburg BU (eds) Raptors worldwide. World Working Group on Birds of Prey and Owls and MME BirdLife, Berlin/Budapest, pp 777–784

Motta-Junior JC, Granzinolli MAM, Develey PF (2008) Aves da Estação Ecológica de Itirapina, Estado de São Paulo, Brasil. Biota Neotrop 8(3):207–227

Motta-Junior JC, Granzinolli MAM, Monteiro AR (2010) Miscellaneous ecological notes on Brazilian birds of prey and owls. Biota Neotrop 10(4):355–360

Motta-Junior JC, Talamoni SA, Vasconcellos LAS (1996) Levantamento dos mamíferos do campus da Universidade Federal de São Carlos, estado de São Paulo. An Sem Reg Ecol 7:173–182

Müller C (2003) Expansion and modernization of agriculture in the Cerrado - the case of soybeans in Brazil's center-west. Department of Economics Working Paper 306, Universidade de Brasília, Brasília

Nascimento JLX (2000) Estudo comparativo da avifauna em duas estações ecológicas da Caatinga: Aiuaba e Seridó. Melopsittacus 3(1):12–35

Naka LM, Cohn-Haft M, Mallet-Rodrigues F et al (2006) The avifauna of the Brazilian state of Roraima: bird distribution and biogeography in the Rio Branco Basin. Rev Bras Ornit 14(3):197–238

Naka LM, Cohn-Haft M, Whittaker A et al (2007) Avian biogeography of Amazonian flooded forests in the Rio Branco Basin, Brazil. Wilson J Ornith 119:439–449

Nogueira DM, Alves MAS (2011) A case of leucism in the Burrowing owl *Athene cunicularia* (Aves: Strigiformes) with confirmation of species identity using cytogenetic analysis. Fortschr Zool 28:53–57

Norberg RA (2002) Independent evolution of outer ear asymmetry among five owls lineages, morphology, function and selection. In: Newton I, Kavanagh R, Olsen J et al (eds) Ecology and conservation of owls. CSIRO Publishing Australia, Collingwood, pp 329–342

Novelli R, Takase E, Castro V (1988) Estudo das aves mortas por atropelamento em um trecho da rodovia BR-471, entre os distritos de Quinta e Taim, RS, Brasil. Rev Bras Zool 5(3):441–454

Oliveira EH, Moura SP, Anjos LJ et al (2008) Comparative chromosome painting between chicken and spectacled owl (*Pulsatrix perspicillata*): implications for chromosomal evolution in the Strigidae (Aves, Strigiformes). Cytogenet Genome Res 122(2):157–162

Oliveira RB, Pontes GMF, Solé M et al (2004) *Lystrophis dorbignyi* (Nariguda) and *Helicops infrataeniatus* (cobra-d'água). Predation. Herptol Rev 35(1):70–71

Oliveira RG (1980) Observações sobre a coruja-orelhuda (*Rhinoptynx clamator*), no Rio Grande do Sul (Aves,Strigidae). Rev Bras Biol 40:599–604

Oliveira RG (1981) A ocorrência do "Mocho-diabo" *Asio stygius,* no Rio Grande do Sul. An Soc Sul Riograndense Ornitol 2:9–12

Oliveira RG (1984) Diferença na cor da íris em do is "cabores-de-orelhas" (*Otus choliba* Viellot, 1817), irmãos de ninho. An Soc Sul Riograndense Ornitol 5:15–19

Olmos F, Silva WAG, Albano CG (2005) Aves em oito áreas de Caatinga no sul do Ceará e oeste de Pernambuco, nordeste do Brasil: composição, riqueza e similaridade. Pap Avulsos Zool 45(14):179–199

Olson SL (1995) The genera of owls in the Asioninae. Bull Br Orn Club 115:35–39

Oniki Y (1984) Behaviour and moult of a young captive tropical screech owl *Otus choliba*. Avic Mag 90(2):104–108

Overbeck GE, Müller SC, Fidelis A et al (2007) Brazil's neglected biome: the south Brazilian Campos. Perspect Plant Ecol Evol Syst 9:101–116

Pacheco JF (2003) As aves da Caatinga - Apreciação historica do processo de conhecimento. In: JMC S, Tabarelli M, Fonseca MT et al (eds) Biodiversidade da Caatinga: áreas e ações prioritárias para a conservação. Ministério do Meio Ambiente, Universidade Federal de Pernambuco, Fundação de Apoio ao Desenvolvimento da UFPE, Conservation International do Brasil, Fundação Biodiversitas, EMBRAPA Semi-Árido, Brasilia, pp 189–250

Pacheco JF, Olmos F (2006) As aves do Tocantins 1: região sudeste. Rev Bras Ornitol 14(2):85–100

Pará (2008) Programa Estadual de Especies Ameaçadas de Extinção - Programa Extinção Zero. Decreto no 802 de 20/02/2008. http://www.cemave.net/listavermelha/index.php?option=comjdownloads&Itemid=33&view=finish&cid=20&catid=5. Accessed 02 May 2011

Peakall DB (2002) Poisoning in wild (free-living) raptor. In: Cooper JE (ed) Birds of prey: health & disease, 3rd edn. Blackwell, Oxford, pp 163–170

Pedroso LGA (2015) Ácaros de penas (Acari: Astigmata) em aves não Passeriformes do Brasil. Dissertation, Universidade Estadual Paulista, Rio Claro

Penhallurick JM (2002) The taxonomy and conservation status of the owls of the world: a review. In: Newton I, Kavanagh R, Olsen J et al (eds) Ecology and conservation of owls. CSIRO Publishing Australia, Collingwood, pp 343–354

Pereira GA, Brito MT (2005) Diversidade de aves silvestres comercializadas nas feiras livres da região metropolitana de Recife, Pernambuco. Atual Ornitol 126:14

Pereira MF, Oliveira LR (2010) Aspectos reprodutivos da coruja orelhuda Rhinoptynx *clamator* (Aves: Strigidae) em urn fragmento de cerrado na Fazenda São Josě, Patrocínio, MG. Rev UNICERP, Saúde e Meio Ambiente 1:107–115

Pereira GA, Lobo-Araújo LW, Leal S et al (2012) Important bird records from Alagoas, Pernambuco and Paraíba, north-east Brazil. Cotinga 34:17–22

Pereira GA, Dantas SDM, Silveira LF et al (2014) Status of the globally threatened forest birds of northeast Brazil. Pap Avulsos Zool 54(14):177–194

Peres C, Zimmerman B (2001) Perils in parks or parks in peril? Reconciling conservation in Amazonian reserves with and without use. Cons Biol 15(3):793–797

Perillo A, Queiroz MB, Mazzoni LG et al (2011) Padrões de atividade da coruja-buraqueira, *Athene cunicularia* (Strigiformes: Strigidae), no campus da Pontifícia Universidade Católica de Minas Gerais, Belo Horizonte, e comentários sobre um peculiar comportamento de estocagem de alimento. Atual Ornitol 160:55–58

Peters FB, Roth PRO, Jeronimo AA et al (2009) Predação de Lepus europaeus (Lagomorpha: Leporidae) por Bubo virginianus (Strigiformes: Strigidae) no sul do Brasil. Biodivers Pampeana 7(1):31–34

Petroff MADS (1999) Corujas, rapinantes noturnas! Boletim ABFPAR 2(2):25–30

Petroff MADS (2001) Processo de aprendizagem incluído na reabilitação de corujas insetívoras. Boletim ABFPAR 4(1):21–23

Petroff MADS (2002) Introdução a filosofia, paleornitologia e sistemática das corujas brasileiras. Boletim ABFPAR 5(1):9–17

Petroff MADS (2003) Levantamento dos rapinantes do Parque Estadual de Itaúnas e as interferencias em seus hábitos. Dissertation, Universidade de São Paulo, São Paulo

Petroff MADS, Souza JM, Oyamagushi HM (2000) Aumento na incidência da corujinha-do-mato (*Otus choliba*) na cidade de Itaunas. Boletim ABFPAR 3(1):6–8

Piacentini VQ, Ghizoni-Junior IR, Azevedo MAG et al (2006) Sobre a distribuição de aves em Santa Catarina, Brasil, parte I: registros relevantes para o estado ou inéditos para a Ilha de Santa Catarina. Cotinga 26:25–31

Piacentini VQ, Aleixo A, Agne CE et al (2015) Annotated checklist of the birds of Brazil by the Brazilian ornithological records committee / Lista comentada das aves do Brasil pelo Comitê Brasileiro de Registros Ornitológicos. Rev Bras Ornit 23(2):91–298

Pillar VD, Quadros FLF (1997) Grassland-forest boundaries in southern Brazil. Coenoses 12:119–126

Pillar VD, Boldrini I, Hasenack H et al. (2006) Workshop estado atual e desafios para a conservação dos campos. UFRGS, Porto Alegre. http:/ /ecoqua.ecologia.ufrgs.br. Accessed 20 Nov 2008

Pinheiro RT, Dornas T (2009) Distribuição e conservação das aves na região do Cantão, Tocantins: ecótono Amazônia/Cerrado. Biota Neotrop 9(1):187–205

Pinto FAS, Luz HR, Fonseca CF et al (2007) Hábitos alimentares da coruja-das-torres, *Tyto alba* (Aves, Strigiformes, Tytonidae) no municipio de Seropédica, RJ. An Congr Ecol Brasil 8:1–2. http://www.sebecologia.org.br/viiiceb/pdf/41O.pdf. Accessed 10 Oct 2011

Pinto OMO (1978) Novo catálogo das aves do Brasil. Primeira parte Empresa Gráfica da Revista dos Tribunais, São Paulo

Pinto RM (1968) Sobre urn novo nematódeo paras ito de caboré do campo (Nematoda, Subuluroidea). Atas Soc Biol Rio de Janeiro 11:189–190

Prada CS (2004) Atropelamento de vertebrados silvestres em uma região fragmentada no nordeste no estado de São Paulo: Quantificação do impacto e análise de fatores envolvidos. Dissertation, Universidade Federal de São Carlos, São Carlos

Proudfoot GA, Honeycutt RL, Slack D (2006) Mitochondrial DNA variation and phylogeography of the ferruginous pygmy-owl (*Glaucidium brasilianum*). Conserv Genet 7:1–12
Ramirez-Llorens P, Bellocq MI (2007) New records clarify the southern distribution of the spectacled owl (*Pulsatrix perspicillata*). J Raptor Res 41(4):268–276
Ramos CCO, Benedito E, Zawadzki CH (2011) Dieta e conteúdo calórico de aves atropeladas na região central do estado do Paraná, Brasil. Biotemas 24(4):153–170
Rebelato MM, Cunha GG, Machado RF et al (2011) Novo registro do caburé-acanelado (*Aegolius harrisii*) no Bioma Pampa, sul do Brasil. Biotemas 24:105–107
Reinhardt J (1870) Bidrag til Kundskab om Fuglefaunaen i Brasiliens Campos. Videnskabelige Meddelelser fra den Naturhistoriske Forening i København. Copenhagen, Denmark
Reis MG (2015) Efeitos do fogo sobre assembleias de aves de Cerrado. Thesis, Universidade Federal de São Carlos, São Carlos
Remsen, JV Jr, Areta JI, Cadena CD et al (2016) A classification of the bird species of South America. American Ornithologists' Union. http://www.museum.lsu.edu/~Remsen/SACCBaseline.htm. Accessed 26 Nov2016
RENCTAS - Rede Nacional de Combate ao Tráfico de Animais Silvestres (2001) 1º Relatório Nacional sobre o Tráfico de Fauna Silvestre. Renctas, Brasilia. http://www.renctas.org.br/files/REL_RENCTAS_pt_final.pdf. Accessed 20 Nov 2008
Ribas CR, Santos REF (2007) Novo registro documentado do caburé-acanelado *Aegolius harrisii* (Cassin 1849) para o estado do Paraná. Atual Ornitol 140:4–5
Ribeiro MC, Metzger JP, Martensen AC et al (2009) The Brazilian Atlantic Forest: how much is left, and how is the remaining forest distributed? Implications for conservation. Biol Conserv 142(6):1141–1153
Ribeiro RCC, Vasconcelos MF (2003) Ocorrência de *Pulsatrix koeniswaldiana* e de *Nyctibius grandis* em área urbana no leste de Minas Gerais, Brasil. Ararajuba 11(2):233–234
Rio Grande do Sul (2014) Táxons da fauna silvestre do Estado Rio Grande do Sul ameaçadas de extinção. Decreto N° 51797, de 8 de setembro de 2014. http://www.al.rs.gov.br/filerepository/repLegis/arquivos/DEC%2051.797.pdf. Accessed 30 Oct 2016
Robbins MB (2006) Proposal (#243) to South American Classification Committee. Change the name of Least Pygmy-Owl to *Glaucidium sicki*.http://www.museum.lsu.edu/ -remsen/SACCprop243.html. Accessed 30 Nov 2008
Rocha CFD (1993) The set of defence mechanisms in a tropical sand lizard (*Liolaemus lutzae*) of southeastern Brazil. Ciên Cult 45(2):116–122
Rocha RG, Ferreira E, Leite YLR et al (2011) Small mammals in the diet of Barn owls, *Tyto alba* (Aves: Strigiformes) along the mid-Araguaia River in central Brazil. Fortschr Zool 28(6):709–716
Rocha R, López-Baucells A (2014) Opportunistic predation by Crested Owl *Lophostrix cristata* upon Seba's Short-tailed Bat *Carollia perspicillata*. Rev Bras Ornit 22(1):35–37 (Aleixo A (2014) Erratum. Rev Bras Ornit 22(2):244)
Rocha SO, Lima MAI (1998) Histological aspects of the stomach of Burrowing owl (*Speotyto cunicularia,* Molina, 1782). Rev Chil Anat 16(2):191–197
Roda SA (2006) Dieta de *Tyto alba* na Estação Ecológica do Tapacura, Pernambuco, Brasil. Rev Bras Ornit 14(4):449–452
Roda SA, Pereira GA (2006) Distribuição recente e conservação das aves de rapina florestais do Centro Pernambuco. Rev Bras Ornit 14(4):331–344
Rodarte-Almeida ACV, Machado M, Baldotto SB et al (2013) O olho da coruja-orelhuda: observações morfológicas, biométricas e valores de referência para testes de diagnóstico oftálmico. Pesq Vet Bras 33(10):1275–1289
Rosário LA (1996) As aves em Santa Catarina: distribuição geográfica e meio ambiente. FATMA, Florianópolis
Ruschi A (1953) Dois casos de sanguivorismo de *Desmodus rotundus* (E. Geoffroy) e *Dyphilla ecaudata spix*, no homem e outras observações sobre quirópteros hematófagos e acidentalmente hematofagos. Bol Mus Biol Mello Leitão 13:1–8

Ruschi A (1960) Métodos de capturar e transportar pássaros vivos a longas e curtas distâncias e alguns tipos de gaiolas e viveiros para mantê-los em condições de reprodução em cativeiro. Bol Mus Biol Mello Leitão (Sér Divulgação) 2:1–32

Rylands AB, Brandon K (2005) Brazilian protected areas. Cons Biol 19(3):612–618

Sacchi ABV (2015) Diagnóstico sorológico e molecular de agentes transmitidos por artrópodes em aves carnívoras. Thesis, Universidade Estadual Paulista, Jaboticabal

Sanches TC, Joppert AM, Gattamorta MA et al (2004) Preliminary study about hematologic profile of some wildlife Brazilian birds of prey in São Paulo city. In: Proceedings American Association of zoo Veterinarians, American Association of Wildlife Veterinarians, wildlife disease Association: health and conservation of captive and free-ranging wildlife: joint conference, San Diego, August 28–September 3, 2004, pp 621–622

Sant'ana CER (1998) Macroecologia e evolução do tamanho do corpo em corujas (Aves: Strigiformes) da America do Sul. Dissertatio, Universidade Federal de Goiás, Goiânia

Sant'ana CER, Diniz-Filho JAF (1997) Autocorrelação filogenética para o tamanho do corpo em corujas (Strigiformes) da America do Sul. Ararajuba 5(1):39–43

Sant'ana CER, Diniz-Filho JAF (1999) Macroecologia de corujas (Aves: Strigiformes) da America do Sul. Ararajuba 7(1):3–11

Santos E (1979) Da ema ao beija-flor. Editora Itatiaia, Belo Horizonte

Santos CM (1993) Ecologia trófica e alguns aspectos do comportamento alimentar de *Athene cunicularia* (Molina, 1782) (Strigidae, Aves), na região de São Carlos, SP. Dissertation, Universidade Federal de São Carlos, São Carlos

Santos LP, Gunski RJ (2006) Revisão de dados citogenéticos sobre a avifauna brasileira. Rev Bras Ornit 14(1):35–45

Santos REF (2009) Ampliação da distribuição de *Aegolius harrisii* a partir de coleta por atropelamento. Atual Ornitol 147:46–47

São Paulo (2014) Espécies da fauna silvestre ameaçadas de extinção no Estado de São Paulo. Decreto N° 60.133, de 7 de fevereiro de 2014. Available at: http://www.al.sp.gov.br/repositorio/legislacao/decreto/2014/decreto-60133-07.02.2014.html. Accessed 28 Nov 2016

Sawaya RJ, Ariedi-Junior VR, Monteiro C, Spina F (2003) *Oxyrhopus rhombifer* (false coral snake). Predation. Herpetol Rev 34(2):152–153

Sazima I (2015a) Lightning predator: the ferruginous pygmy owl snatches flower-visiting hummingbirds in South-western Brazil. Rev Bras Ornit 23(1):12–14

Sazima I (2015b) Bee in the nose: raptors let or let not stingless bees enter their nostrils. Rev Bras Ornit 23(3):323–326

Sazima I, Caramaschi U (1986) Descrição de *Physalaemus deimaticus, sp. n.*, e observações sobre comportamento deimatico em *P. nattereri* (Steindn.) - Anura, Leptodactylidae. Rev Biol 13:91–101

Sberze M, Cohn-Haft M, Ferraz G (2010) Old growth and secondary forest site occupancy by nocturnal birds in a Neotropicallandscape. Anim Conserv 13:3–11

Scheibler DR (2007) Food partitioning between breeding white-tailed kites (*Elanus leucurus;* Aves; Accipitridae) and Barn owls (*Tyto alba;* Aves; Tytonidae) in southern Brazil. Braz J Biol 67(1):65–71

Scheibler DR, Christoff AU (2004) Small mammals in the diet of Barn owls (*Tyto alba*) in agroecosystems of southern Brazil. Ornitol Neotrop 15:65–70

Scheibler DR, Christoff AU (2007) Habitat associations of small mammals in southern Brazil and use of regurgitated pellets of birds of prey for inventorying a local fauna. Braz J Biol 67(4):619–625

Scherer-Neto P (1985) Notas bionômicas sobre o "mocho-diabo" (*Asio stygius* Wagler, 1832), no Paraná. An Soc Sul Riograndense Ornitol 6:15–18

Scherer-Neto P, Straube FC (1995) Aves do Paraná: historia, lista anotada, e bibliografia. Editora dos Autores, Curitiba

Schubart O, Aguirre AC, Sick H (1965) Contribuição para o conhecimento da alimentação das aves brasileiras. Arq Zool S Paulo 12:95–249

Secretariat of the Convention on Biological Diversity (2002) Global strategy for plan to conservation. Secretariat of the convention on biological diversity. CBD, UMEP, Botanical Gardens Conservation International, Montreal

Serpa LE, Monteiro AR (2001) Composição da dieta alimentar da coruja-buraqueira *Athene cunicularia* (Temmick, 1822) em Juiz de Fora - MG: variação no uso das presas. Revista UNIVAP 8(12):46–50

Sibinelli V (2010) A coruja do coração. Terra da Gente 77:24–31

Sibley CG (1996) Birds of the world. Thayer Birding Software. Version 2.0, Cincinnati

Sick H (1993) Birds in Brazil: a natural history. Princeton University Press, Princeton

Sick H (1997) Ornitologia brasileira, uma introdução. Nova Fronteira, Rio de Janeiro

Siervi TC (2014) Dieta e seleção de coleópteros copro-necrófagos (Scarabaeidae: Scarabaeinae) pela coruja-buraqueira (*Athene cunicularia*, Strigiformes: Strigidae) em campos da Estação Ecológica de Itirapina, Estado de São Paulo, Brasil. Dissertation, Universidade de São Paulo, São Paulo

Sigrist T (2006) Aves do Brasil: uma visão artística. Editora Avis Brasilis, São Paulo

Sigrist T (2009) Iconografia das aves do Brasil: Bioma Cerrado. Avis Brasilis Editora, Vinhedo

Silva FCA (2006) Ecologia alimentar de *Athene cunicularia* e *Tyto alba* (Aves, Strigiformes) na cidade de Curitiba e Região Metropolitana, Estado do Paraná. Crescimento dos filhotes da coruja das torres *Tyto alba* (Aves, Strigiformes) nos municípios de Curitiba e Colombo, Estado do Paraná. Dissertation, Universidade Federal do Paraná, Curitiba

Silva LL (2000) O papel do Estado no processo de ocupação das areas de Cerrado entre as decadas de 60 e 80. Caminhos de Geografia 1(2):24–36

Silva M (2005) The Brazilian protected areas program. Cons Biol 19(3):608–611

Silva RVL (2002) Influência da ação humana no comportamento agonístico da coruja-buraqueira. Ararajuba 10(2):237–240

Silva DE, Corrêa LLC, Pazinato DMM et al (2013) Observação do comportamento predatório de *Bubo virginianus* (Gmelin, 1788) (Strigidae), em *Falco sparverius* (Linnaeus, 1758) (Falconidae), no sul do Brasil. Cad Pesqui, Sér Biol 24(3):39–43

Silva JMC, Coelho G, Gonzaga LP (2002) Discovered on the brink of extinction: a new species of pygmy-owl (Strigidae: *Glaucidium*) from Atlantic Forest of northeastern Brazil. Ararajuba 10(2):123–130

Silva JMC, Castelletti CHM (2003) Status of the biodiversity of the Atlantic Forest of Brazil. In: Galindo-Leal C, Câmara IG (eds) The Atlantic Forest of South America: biodiversity status, trends, and outlook. Center for Applied Biodiversity Science and Island Press, Washington, DC, pp 43–59

Silva JMC, Souza MA, Bieber AGD et al (2003) Aves da Caatinga: status, uso do hábitat e sensitividade. In: Leal IR, Tabarelli M, Silva JMC (eds) Ecologia e Conservação da Caatinga. Editora Universitaria da UFPE, Recife, pp 237–274

Silva AS, Zanette RA, Lara VM et al (2009) Gastrointestinal parasites of owls (Strigiformes) kept in captivity in the southern region of Brazil. Parasitol Res 104:485–487

Silva TM, Okamoto AS, Silva LAF et al (2014) New record of *Pelecitus* sp. (Nematoda, Onchocercidae) as a parasite of *Athene cunicularia* (Strigiformes, Strigidae) in southeastern Brazil. Brazil J Vet Parasitol 23(2):274–275

Silva-Porto F, Cerqueira R (1990) Seasonal variation in the diet of the Burrowing owl *Athene cunicularia* in a restinga of Rio de Janeiro state. Ciênc Cult 42(12):1182–1186

Silveira LF, Belmonte FJ (2005) Comportamento reprodutivo e hábitos da ararajuba, *Guarouba guarouba*, no município de Tailândia, Pará. Rev Bras Ornit 13(1):89–93

Silveira LF, Calonge-Mendez A, Brito GRR (2001) Range extensions and new records for birds in Piauí state, Brazil. Internat J Ornithol 4(2):219–224

Simon JE, Antas PTZ, Pacheco JF et al (2007) As Aves Ameaçadas de Extinçâo no Estado do Espírito Santo. In: Mendes SL, Passamani M (orgs) Livro Vermelho das Espécies da Fauna Ameaçada de Extinção no Estado do Espírito Santo. IPEMA, Vitória, pp 47–68

Snethlage H (1928) Meine reise durch Nordostbrasilien. III. Bausteine zur biologie der angetroffenen arten. J Ornithol 76:668–738

Soares ES, Amaral FSR, Carvalho-Filho EPM, Granzinolli MAM et al (2008) Plano de ação nacional para a conservação de aves de rapina. ICMBio, Brasilia

Soares M, Schiefler AF, Ximenez A (1992a) Aspectos do comportamento de *Athene cunicularia* (Molina, 1782) (Aves: Strigidae), na restinga da Praia da Joaquina, Ilha de Santa Catarina, SC. Biotemas 5:71–74

Soares M, Schiefler AF, Ximenez A (1992b) Hábitos alimentares de *Athene cunicularia* (Molina, 1782) (Aves: Strigidae) na restinga da Praia da Joaquina, Ilha de Santa Catarina, SC. Biotemas 5:85–89

Souza DP, Asfora PH, Lira TC et al (2010) Small mammals in Barn owl (*Tyto alba* - Aves, Strigiformes) pellets from northeastern Brazil, with new records of *Gracilinanus* and *Cryptonanus* (Didelphimorphia, Didelphidae). Mamm Biol 75:370–374

Souza TO, Vilela DAR, Câmara BGO (2014) Pressões sobre a avifauna brasileira: Aves recebidas pelo CETAS/IBAMA, Belo Horizonte, Minas Gerais. Ornithologia 7(1):1–11

Specht GVA (2007) Treinamento antipredação em corujinha-do-mato (*Megascops [Otus] choliba*) - Strigiformes - Aves para program a de translocação. Dissertation, Pontificia Universidade Católica de Minas Gerais, Belo Horizonte

Specht GVA, Cunha FCR, Akaki GS (2008) Observações sobre o comportamento reprodutivo do pavó, *Pyroderus scutatus* (Shaw, 1792) (Aves: Cotingidae), em Antônio Dias, Minas Gerais. Atual Ornitol 142:10–11

Specht GVA, Gonçalves GL, Young RJ (2013) Comportamento de caça da coruja buraqueira, *Athene cunicularia* (Molina, 1782) (Aves: Strigiformes) em ambiente urbano em Belo Horizonte, Minas Gerais, Brasil. Lundiana 11(1–2):17–20

Squarzoni R, Perlmann E, Antunes A et al (2010) Ultrasonographic aspects and biometry of striped owl's eyes (*Rhinoptynx clamator*). Vet Ophthalmol 13:86–90

Stafford BJ, Ferreira FM (1995) Predation attempts on callitrichids in the Atlantic coastal rain forest of Brazil. Folia Primatol 65:229–233

Stotz DF, Fitzpatrick JW, Parker TA III et al (1996) Neotropical birds ecology and conservation. The University of Chicago Press, Chicago

Strachan AA (1957) Eye worms of the family Thelaziidae from Brazilian birds. Can J Zool 35:179–187

Straube FC, Urben-Filho A, Kajiwara D (2004) Aves. In: Mikich SB, Bérnils RS (orgs) Livro vermelho da fauna ameaçãda no Estado do Paraná. Instituto Ambiental do Paraná, Curitiba, pp 135–150

Studer A, Teixeira DM (1994) Notes on the buff-fronted owl *Aegolius harrisii* in Brazil. Bull Br Orn Club 114(1):62–63

Swartz FA (2000) The Pantanal in the 21st century - for the Planet's largest wetland, an uncertain future. In: Swartz FA (ed) The Pantanal of Brazil, Paraguay and Bolivia. Hudson MacArthur Plubishers, Gouldsboro, pp 1–24

Talamoni SA (1996) Ecologia de uma comunidade de pequenos mamíferos da Estação Ecológica de Jataí, Municipio de Luiz Antônio, SP. Thesis, Universidade Federal de São Carlos, São Carlos

Teixeira FM, Melo C (2000) Dieta de *Speotyto cunicularia* Molina, 1782 (Strigiformes) na região de Uberlândia, Minas Gerais. Ararajuba 8(2):127–131

Thiollay JM (1984) Raptor community structure of a primary rain forest in French Guiana and effect of human pressure. J Raptor Res 18(4):117–122

Thiollay JM (1989) Area requirements for the conservation of rain forest raptors and game birds in French Guiana. Cons Biol 3(2):128–137

Thiollay JM (1994) A world review of tropical forest raptors current trends, research objectives and conservation strategy. In: Meyburg BU, Chancellor RD (eds) Raptor conservation today. World Working Group on Birds of Prey and Owls. World Working Group on Birds of Prey and Owls and The Pica Press, Berlin, pp 231–239

Thiollay JM, Meyburg BU (1988) Forest fragmentation and the conservation of raptors: a survey on the island of java. Biol Conserv 44:229–250

Tomas WM, Freitas GO, Pereira GMF (2013) Efeito de área e densidade de árvores sobre a probabilidade de ocupação de manchas florestais no Pantanal por corujas pretas (*Strix huhula*). Simpósio sobre Recursos Naturais e Socioeconômicos do Pantanal 6:124–129. Available at: http://www.cpap.embrapa.br/agencia/simpan6/Resumos_6Simpan_1Evinci_2013.pdf. Accessed 20 Nov 2016

Tomazzoni AC, Pedo E, Hartz SM (2004) Food habits of great horned owls (*Bubo virginianus*) in the breeding season in Lami Biological reserve, southern Brazil. Ornitol Neotrop 15(2):279–282

Tortato FR, Bonanomi J (2012) Disputa por cavidade entre *Anodorhynchus hyacinthinus* (Latham, 1790) (Psittacidae) e *Tyto alba* (Scopoli, 1769) (Tytonidae) na região do Pantanal de Paiaguás, Corumbá, Mato Grosso do Sul, Brasil. Rev Bras Ornit 20(1):22–25

Tozetti AM, Hulle N, Vetorazzo V (2005) *Ameiva ameiva* (Giant Ameiva). Predation. Herptol Rev 36(4):443–444

Traylor MA (1958) Variation in south American great horned owls. Auk 75(2):143–149

Tubelis DP, Thomas WM (2003) Bird species of the wetland, Brazil. Ararajuba 11(1):5–37

Tubelis DP, Delitti WBC (2010) Fire management and the nesting of *Athene cunicularia* (Aves, Strigidae) in grasslands in central Cerrado, Brazil. Biota Neotrop 10(2):93–101

Turci LCB, Bernarde PS (2009) Vertebrados atropelados na Rodovia Estadual 383 em Rondonia, Brasil. Biotemas 22(1):121–127

Ubaid FK, Maffei F, Moya GM et al (2012) Range extension for Buff-fronted owl *Aegolius harrisii* in south-east Brazil. Bull Br Orn Club 132(3):175–179

UNEP-WCMC (2016) CITES-Listed Species 2016. .UNEP-WCMC Species Database. http://www.cites.org. Accessed 25 Oct 2016

Valdujo PH, Nogueira C (2000) *Bothrops neuwiedipauloensis* (jararaca rabo-de-osso). Predation. Herptol Rev 31(1):45

Valim MP, Teixeira RHF, Amorim M et al (2005) Malófagos (Phthiraptera) recolhidos de aves silvestres no Zoológico de São Paulo, SP, Brasil. Rev Bras Entomol 49(4):584–558

Valle C (1980) História Natural dos "Glires" da Bacia do Rio das Velhas, segundo Peter Lund (1ª Parte: Rodentia Cricetidae). Lundiana 1:7–21

Valle C, Camevalli N (1973) O papel das corujas na interpretação de Lund. Revista S O M 5:8–9

Vargas GD, Albano AP, Fischer G et al (2011) Avian pox virus infection in a common Barn owl (*Tyto alba*) in southern Brazil. Pesq Vet Brasil 31(7):620–622

Vasconcelos MF, Diniz MG (2008) 170 years after Lund: rediscovery of the Black-banded owl *Strix huhula* in the metropolitan region of Belo Horizonte, Minas Gerais, Brazil (Strigiformes: Strigidae). Rev Bras Ornit 16(3):277–280

Vasconcelos MF, Lopes LE, Hoffmann D et al (2008) Noteworthy records of birds from the Pantanal, Chiquitano dry forest and Cerrado of south-western Brazil. Bull Br Orn Club 128:57–67

Vasconcelos MF, Guimarães PRL, Marques LO et al (2016) Recomendações para aproveitamento científico de aves encontradas mortas em campo. Atual Ornitol 189:57–65

Vieira LA, Teixeira RL (2008) Diet of *Athene cunicularia* (Molina, 1782) from a sandy coastal plain in southeast Brazil. Bol Mus Biol Mello Leitão N Ser 23:5–14

Vieira W, Bezerra DMM, Vieira KS et al (2015) *Megascops choliba* (Strigiformes: Strigidae) predation on *Scinax x-signatus* (Anura: Hylidae) in the semiarid, northeastern. Brazil. Herpetol Notes 8:275–276

Vielliard J (1989) Uma nova espécie de *Glaucidium* (Aves, Strigidae) da Amazônia. Rev Bras Zool 6(4):685–693

Vitaliano SN (2012) Isolamento e caracterização biológica e genotípica de *Toxoplasma gondii* em animais selvagens do Brasil. Thesis, Universidade de São Paulo, São Paulo

Voss RS, Myers P (1991) *Pseudoryzomys simplex* (Rodentia: Muridae) and the significance of Lund's collections from the caves of Lagoa Santa, Brazil. Bull Am Mus Nat Hist 206:414–432

Voous KH (1964) Wood owls of the genera *Strix* and *Ciccaba*. Zool Meded Rijksmus Nat Hist, Leiden 39:471–478

Wikiaves (2016) Portal Wikiaves – a enciclopedia das aves do Brasil. http://www.wikiaves.com.br/. Accessed 20 Nov 2016

Willis EO (2003) Birds of a eucalyptus woodlot in interior São Paulo. Braz J Biol 63(1):141–158
Willis EO, Oniki Y (1991) Nomes gerais para as aves brasileiras. Gráfica da Região, Américo Brasileinse
Willis EO, Oniki Y (2002) Birds of Santa Teresa, Espírito Santo, Brazil: do humans subtract species? Pap Avulsos Zool 42(9):193–264
Willis EO, Oniki Y (2003) Aves do Estado de São Paulo. Divisa, Rio Claro
Wink M (2000) Advances in DNA studies of diurnal and nocturnal raptors. In: Chancellor RD, Meyburg BU (eds) Raptors at risk. WWGBP/ Hancock House, Surrey, pp 831–844
Wink M, Heidrich P (2000) Molecular systematics of owls (Strigiformes) based on DNA sequences of the mitochondrial cytochrome b gene. In: Chancellor RD, Meyburg BU (eds) Raptors at risk. WWGBP/Hancock House, Surrey, pp 819–828
Wink M, El-Sayed A, Sauer-Gürth H et al (2009) Molecular phylogeny of owls (Strigiformes) inferred from DNA sequences of the mitochondrial cytochrome b and the nuclear RAG-1 gene. Ardea 97:582–591
Wink M, Sauer-Gurth H, Fuchs M (2004) Phylogenetic relationships in owls based on nucleotide sequences of mitochondrial and nuclear marker genes. In: Chancellor RD, Meyburg BU (eds) Raptors worldwide. World Working Group on Birds of Prey and Owls and MME BirdLife, Berlin/Budapest, pp 517–526
Zilio F (2005) Estudo do nicho ecológico de duas aves de rapina (*Falco sparverius* e *Athene cunicularia*) em uma região de dunas do Rio Grande do Sul, Brasil. Dissertation, Universidade Estadual Paulista, Rio Claro
Zilio F (2006) Dieta de *Falco sparverius* (Aves: Falconidae) e *Athene cumcularia* (Aves: Strigidae) em uma região de dunas no sul do Brasil. Rev Bras Ornit 14(4):379–392
Zorzin G, Canuto M, Carvalho-Filho EM et al (2008) Aves de rapina noturnas do Parque Estadual do Rio Doce, Minas Gerais. MG Biota 1:44–57

Chapter 7
The Owls of Chile

Ricardo A. Figueroa R., Sergio Alvarado O., E. Soraya Corales S., Daniel González-Acuña, Roberto Schlatter V., and David R. Martínez P.

Abstract To achieve a better understanding of the life histories of Chilean owls and to generate a useful source of information for future studies, we undertook an exhaustive review of all available information relating to their taxonomy, natural history, ecology, genetic and conservation biology. Studying these topics we gathered information on morphology, morphometrics, distribution, abundance, habitat, reproduction, longevity, behaviour, diet, feeding ecology, population ecology, community ecology, functional ecology, ecophysiology, endemism, conservation status, threats, human perception, legislation, education and outreach, physical rehabilitation, and habitat management. During our review, we rediscovered and retrieved naturalist's observations that had remained totally unknown to contemporary ornithologists. Some of the challenges that must be addressed to achieve a clearer understanding of the biology of Chilean owls and strengthen conservation strategies are (i) to definitively determine the validity of *Bubo magellanicus* as a separate species and establish the true extent of its distributional hiatus; (ii) to determine the geographic boundaries for *Glaucidium nana* and *G. peruanum*; (iii) to detect the ocurrence of population size variations and to identify possible causal factors; (iv) to identify variables that promote the use of, and preference for, habitats; (v) to ascertain, in detail, their reproductive characteristics, home ranges, and dynamic movements; (vi) to assess their diets in poorly studied ecosystems; (vii) to deter-

R.A. Figueroa R. (✉) • E.S. Corales S.
Escuela de Graduados, Facultad de Ciencias Forestales y Recursos Naturales, Universidad Austral, Valdivia, Chile
e-mail: ra_figueroa_rojas@yahoo.com

S. Alvarado O.
División de Epidemiología, Escuela de Salud Pública, Facultad de Medicina, Universidad de Chile, Santiago, Chile

D. González-Acuña
Facultad de Ciencias Veterinarias, Universidad de Concepción, Chillán, Chile

R. Schlatter V.
Instituto de Ciencias Marinas y Limnológicas, Facultad de Ciencias, Universidad Austral, Valdivia, Chile

D.R. Martínez P.
Departamento de Ciencias Biológicas y Biodiversidad, Universidad de los Lagos, Osorno, Chile

P.L. Enríquez (ed.), *Neotropical Owls*, DOI 10.1007/978-3-319-57108-9_7

mine their relevance in local food webs and roles in ecosystems; (viii) to evaluate the genetic structure of owls in highly fragmented landscape; and (ix) to promote much more education about their natural history. We think that our review will be useful both in guiding new conservation efforts and opening new research perspectives that will help fill information gaps.

Keywords Natural history • Owl ecology • Owl conservation • Owl taxonomy

Austral Pygmy Owl (*Glaucidium nana*)

7.1 Introduction

Owls (Strigiformes) represent 3.3% of terrestrial bird species present in Chile (7 out of 213 species; Araya and Bernal 1995; Araya et al. 1995; Vilina and Cofré 2008). Despite being a very small set of species, owls inhabiting our country (hereafter Chilean owls) have received considerable attention from naturalists and biologists because of their direct relationship to agriculture, forest management, and their utility as models to test ecological theories (Reed 1905; Housse 1945; Muñoz and Murúa 1990; Schlatter and Murúa 1992a; Jaksic 1997). However, most of their breeding, behavioural and ecological characteristics remain unknown. In addition, the knowledge corpus on Chilean owls has developed asymmetrically with much of the research work concentrating on areas of central Chile and mainly focused on food habits (Jaksic 1997; Muñoz-Pedreros et al. 2004; Raimilla et al. 2012). Moreover, relevant information provided by natural historians since the mid-nineteenth century have been cast into oblivion or remain unknown to contemporary ornithologists. It is partially explained by the difficulty of retrieving original documents from public record systems either because of the time passed, discontinuity and material loss, or even, lack of specialized personnel.

During recent decades, Chilean citizens have been increasingly aware of the importance of owls for public health, role in ecosystems, educational value, and the necessity of protect them (Jaksic and Jiménez 1986; Figueroa et al. 2001a; Tala and Iriarte 2004; Rivas and Figueroa 2009). Taking into account that biological conservation of organisms is, in part, based on how much we know about them, our principal aim is to report the knowledge status about diverse biological aspects of Chilean owls. In addition, we have synthesized various perspectives on conservation and management of owls in Chile. Because our review is aimed at a wide audience, we have decided to avoid a purely analytical or academic focus. With respect to the latter, Muñoz-Pedreros and Norambuena (2011) and Raimilla et al. (2012) provide brief analytical reviews on the knowledge status of Chilean raptors. We hope our review will contribute to a better understanding of the life history, ecology and the overall importance of owls for humans. Through our review we also hope to stimulate new research avenues and further conservation efforts for Chilean owls.

7.2 Methods

7.2.1 Topics

The present review constitutes an expanded and enriched version of previous syntheses of the natural history, ecology and conservation of Chilean owls (Figueroa and Alvarado 2012; Figueroa et al. 2015). Here, we provide an updated review on the following topics: (1) taxonomy, (2) natural history, (3) ecology, (4) genetics, and (5) biological conservation. The first topic consists of an analysis of the taxonomic status of each owl species plus a review of their morphology and morphometry. In the natural history topic, we include information on distribution, residence status, abundance, habitats, reproduction, longevity, behaviour, and diet. In the ecology topic, we summarize information on trophic ecology, population ecology, community ecology, functional ecology, spatial ecology and ecophysiology. In the genetic topic we summarize results of recent studies on the genetic divergence among some owl species. In biological conservation, we summarize information on population status, threats, conservation priority, legislation and protection, rehabilitation, human perception, ecosystemic services, bio-indication, education and outreach, and habitat management and restoration. Complementarily, we also compiled information on parasites and taphonomy. Finally, an overview of some methods used in the study of Chilean owls is provided as complementary material (Appendix 7.2).

7.2.2 Review of Literature

We thoroughly reviewed the available literature on Chilean owls by covering a period of 231 years (1782–2017). Our search process benefited largely from previous compilations about Chilean bird's literature (e.g. Paynter 1988; Silva-Aránguiz 2012). We retrieved bibliographical material from public, private, and electronic systems. The oldest literature and most difficult to access in our country was retrieved from electronic open access libraries such as Biodiversity Heritage Library (www.biodiversitylibrary.org), Internet Archive Digital Library (www.archive.org), Searchable Ornithological Research Archive (elibrary.unm.edu/sora) and Google Books (books.google.cl). When documents were unavailable on previous file systems, we directly requested it from colleagues.

The revised material included books, book chapters, research papers, outreach papers, technical reports, congress proceedings, ornithological webpages, and audio-visual material. Research papers were considered at any stage of advance: published, in press or in revision. We only considered technical reports when we could corroborate their origin and reliability. Information derived from general reviews of owls was considered with caution because it was mostly based on secondary sources and, in many cases, authors provided no direct original references in the text (e.g. Fjeldsa and Krabbe 1990; Marks et al. 1999). However, these reviews were valuable in two ways: (1) as a source of bibliographic references and (2) as a source of additional information (e.g. König et al. 1999). In some cases, we included information collected in boundary areas within neighbouring countries, particularly when there were biomes shared with Chile (e.g. Patagonian steppe, southern temperate forest; Scott and Sharpe 1912). We complemented information from all of these sources with unpublished data or personal observations of colleagues and own.

7.2.3 Special Treatments

Due to the considerable number of dietary studies, we considered it necessary to synthesize as much as possible the available information. We separated information into two categories of analysis: overall and seasonal. Overall analysis included combined results obtained over one or more years. Seasonal analysis included results separately obtained for different climatic seasons or for breeding vs nonbreeding seasons. Based on this information, we characterized owl diets as follows: (i) proportion in consumption of vertebrate vs invertebrate prey, (ii) seasonal composition of prey taxa, (iii) number frequency of prey taxa, and (iv) mean body size of vertebrate prey. For all these analyses, we only took into account studies based on pellets because prey quantification resulted more reliable. Exceptionally, we included information based on prey remains.

The biomass contribution of prey was estimated by multiplying the number of individuals of each prey type by its respective mean body mass and dividing the products by the combined biomass of all prey types. Although these calculations were made at the prey species level, for economy, we provided only results at the level of taxonomic classes. Mean body masses of prey were taken from literature or from our unpublished information. Temporal variations in diet were evaluated based on two extensive climate seasons: spring-summer vs fall-winter season. Because some owl species can breed in both spring-summer and fall-winter, we do not use the terms "breeding season" and "nonbreeding season" to refer to seasonality.

For its behavioural meaning, we also reviewed information about vocalizations and circadian activity. The voices of Chilean owls are perceived differently by different authors, so there are a great variety of phonetic interpretations on the literature. In order to detect vocal patterns and identify typical voices, we searched for syllabic similarities among all reported phonetic interpretations. Indeed, audios of Chilean owls are available in compact disc format (Voces de las Aves de Chile, Egli 2002) or can be retrieved from www.avesdechile.cl. Because in our country there is a seasonal time change (daylight vs standard time) resulting in 1 h difference between spring-summer and autumn-winter, we present separately the information on circadian activity for both seasonal periods.

7.2.4 Bioclimatic Zones

Following Jaksic and Jiménez (1986), we divided information according to four extensive bioclimatic zones: (1) northern Chile (Arica-Copiapó, 18–28°S) which includes hyperarid desert areas with oases and puna, (2) central Chile (Copiapó-Chillán, 28–36°S) which is characterized by a Mediterranean climate and includes semiarid scrub, and sclerophyllous shrublands and forests, (3) southern Chile (Chillán-Palena, 36–43°S) which is characterized by a humid climate and comprises mainly the southern temperate forest ecoregion, and (4) southernmost Chile (Palena-Cabo de Hornos, 43–56°S) which comprises mainly cold *Nothofagus* forests interspersed with Patagonian steppe areas.

7.2.5 Presentation of Information

The information is summarized primarily on Tables. At the foot of each we provide all respective references listed in chronological order. When references are too numerous, we present them as a simple list under the table or as additional material. The above had two aims: to indicate the precise origin of the information and to minimize the number of citations on the text.

7.2.6 *Conceptual Clarifications*

7.2.6.1 Natural History and Ecology

The separation between natural history and ecology was based on the descriptive mode of the former versus the use of hypothetical-deductive method of the latter (Beehler 2010). However, because many aspects of natural history are inevitably intertwined with ecological aspects, it was difficult to separate categorically the information into one of the topics. To avoid arbitrary divisions, we included together those aspects of the natural history and ecology that were strongly linked (e.g. diet and trophic ecology).

7.2.6.2 Chromatic Variations

Many species of raptors show variations in the plumage colouration both at individual and population level as a consequence of genetic expression or environmental adaptations (Ferguson-Lees and Christie 2001; Roulin 2003). Since such variations can lead to species misidentification, we considered it important to describe the owls' chromatic diversity.

7.2.6.3 Habitat and Habitat Types

Here we use the term "habitat type" to simply refer to a clearly distinct vegetation type (e.g. forest, steppe, pastureland) or forming part of a vegetational gradient (e.g. old-growth forest, secondary forest; Morrison et al. 2006). We also included as habitat types human-origin gradients such as crops, towns, and cities. We are aware that this cannot fully meet the modern concept of "habitat" which refers to a combination of abiotic (e.g. water) and biotic (e.g. food) features and environmental conditions (e.g. temperature, precipitation) promoting the occupation of a given physical space by a particular species (Block and Brennan 1993; Morrison et al. 2006). However, individuals of a species occupy a particular type of vegetation because they find many conditions that ensure their survival in that area.

7.2.6.4 Behaviour

Although we have treated behavioural aspects as a separate topic, much of the information is inevitably linked to other biological aspects such as feeding ecology and reproduction. Thus, additional information on behaviour is included in several other topics throughout this chapter. In the case of reproductive behaviour, we have decided to include it specifically in the topic of reproduction to better gather the information.

7.2.6.5 Hunting and Flight Modes

Hunting modes of raptors are dichotomized often as active (aerial-hunting) or passive ("sit-and-wait") searching (Jaksic 1985; Jaksic and Carothers 1985; Dellacasa et al. 2011). Because many raptors have specialized for active searching, they have diversified their flight modes, hunting with several techniques that vary both in rate of energy expenditure and return (Videler 2005). Several of these flight techniques can be clearly recognizable in the field when raptors are searching for prey. Here, we describe at least six flight techniques we have observed in the field and that several other authors have described in the raptor literature: (1) gliding, a flight at variable speed with extended wings and their tips relatively tucked, without wing flapping and making use of horizontal wind and thermal updrafts; (2) cruising, a horizontal high-speed flapping flight; (3) quartering, a low-speed, to-and-for flight; (4) windhovering, a stationary flight wherein owls face into wind updrafts and control their position with wing beats and tail movements; (5) soaring, a vertical flight in circles on thermal or wind updrafts; and (6) diving, a straight high-speed flight with closed wings.

7.2.6.6 Owl Personality

Overall, the behaviour pattern of a species is determined by the similarities and differences of personality of all member individuals (e.g. Groothuis and Carere 2005). Years ago, Housse (1945) characterized Chilean owls taking into account the personality of some individuals he observed both in the wild and in captivity. Here, we describe qualitatively Chilean owls' personality based on observations from different natural historians.

7.3 Results

7.3.1 Taxonomy

The taxonomic history and diversity of Chilean owls is summarized in Table 7.1. During the last two decades, the taxonomic status of some species of Neotropical owls, including some inhabiting Chile, has undergone changes that are not yet widely accepted. Below, we briefly discuss the taxonomy of Chilean owls in the light of these changes.

König and Weick (2008), based on phylogenetic analyses of Wink et al. (2008), determined that the American population of *Tyto alba*, excepting *Tyto bargei*, corresponds to a distinct species: *Tyto furcata*. New molecular evidence supports König and Weick (2008). Studies based on distinct molecular marker show a consistent genetic separation between the New World and Old World populations of *Tyto alba*

Table 7.1 Taxonomic history of Chilean owls

Current denomination	Original denomination	Holotype	Type locality	Author	Synonymy
Tyto furcata tuidara American barn owl [Lechuza blanca, chiwëd]	*Strix tuidara*	?	Southern Brazil	J.E. Gray	*S. perlata*, *S. flammea*, *T. alba zottae*, *T. a. hauchecorni*, *T. a. tuidara*
Bubo magellanicus Magellanic owl [Tucúquere, toutou]	*Strix magellanicus*	?	Tierra del Fuego, Magallanes	R.R. Lesson	*B. virginianus*, *B. v. nacurutu*, *B. v. andicolus*, *B. v. magellanicus*
Athene cunicularia cunicularia Burrowing owl [Pequén, peken]	*Strix cunicularia*	?	Coquimbo	G.I. Molina	*Strix coquimbana*, *Noctua cunicularia*, *Pholeoptynx cunicularia*, *Speotyto cunicularia*
Athene c. nanodes [Pequén]	*Speotyto c. nanodes*	FMNH	Lima, Perú	H. von Berlepsch, J. Stolzmann	*S. c. intermedia*
Glaucidium nanum[a] Austral pygmy owl [Chuncho, kilkil]	*Strix nana*	NMNH	Pto. del Hambre, Magallanes	P.P. King	*Noctua pumilia*, *N. nana*, *G. n. vafrum*, *G. (brasilianum)? nanum*
Glaucidium peruanum Peruvian pygmy owl [Chuncho del norte]		BMNH	Apurimae, Perú	C. König	*Strix brasiliana*, *Glaucidium brasilianum*
Asio flammeus suinda Short-eared owl [Nuco, nuku]	*Strix suinda*	MNBHU	Paraguay y Río de la Plata	L.J.P. Vieillot	*Otus brachyotus*, *Ulula otus*, *Asio brachyotus*, *Otus breviauris*, *A. f. breviauris*
Strix rufipes rufipes Rufous-legged owl [Concón, konkon]	*S. rufipes*	BMNH	Pto. del Hambre, Magallanes	P.P. King	*Ulula rufipes*, *U. fasciata*, *Syrnium rufipes*

(continued)

Table 7.1 (continued)

Current denomination	Original denomination	Holotype	Type locality	Author	Synonymy
Strix r. sanborni [Concón de Chiloé]	*S. rufipes*	FMNH	Quellón, Isla de Chiloé	L. Wheeler	–

Current scientific names are followed by English conventional common names. Spanish common names frequently used and names in Mapudungún are in brackets. Holotypes: FMNH, Field Museum of Natural History, Chicago; BMNH, British Museum of Natural History, London; NMNH, National Museum of Natural History, Washington, DC; MNBHU, Museum für Naturkunde, Berlin (ZMB, Zoologisches Museum Berlin)

[a]Although the denomination *nanum* has been extensively used throughout the literature, Remsen et al. (2016) clarify that the correct denomination is *nana*. Henceforth we use the latter denomination. References: *Tyto furcata* = 4, 5, 10, 12, 15, 16, 18, 23, 25, 33, 34, 36, 39, 44, 49, 54, 55, 58, 59; *Bubo magellanicus* = 2, 3, 7, 18, 27, 29, 33, 38, 40, 45, 46, 49, 53, 55–58; *Athene cunicularia cunicularia* = 1, 8–10, 13, 16, 17, 20, 21, 24, 25, 33, 55, 56, 58; *Athene c. nanodes* = 22, 28, 29, 32, 41, 43, 47, 55, 56, 58; *Glaucidium nana* = 6, 10, 12, 13, 16, 19, 31, 33, 37, 39, 41, 42, 44, 45, 48, 51, 55, 60; *G. peruanum* = 2, 26, 45, 47, 50–52, 56–58; *Asio flammeus* = 3, 10, 11, 14, 28, 30, 33, 37, 39, 58; *Strix rufipes rufipes* = 6, 10, 18, 33, 37, 39, 51, 55, 56, 58; *Strix r. sanborni* = 35, 37, 39, 41, 47, 55, 56, 58. [1]Molina 1782, [2]Gmelin 1788, [3]Vieillot 1817, [4]Lichtenstein 1823, [5]Gray 1828, [6]King 1828, [7]Lesson 1828, [8]Poeppig 1829, [9]Lafresnaye and D'Orbigny 1837, [10]Des Murs 1847, [11]d'Orbigny 1847, [12]Cassin 1855, [13]Kaup 1862, [14]Schlegel 1862, [15]Sclater 1867, [16]Philippi 1868, [17]Hudson 1874, [18]Sharpe 1875a, [19]Ridgway 1876, [20]Sharpe 1881, [21]Sclater 1891, [22]von Berlepsch and Stolzmann 1892, [23]Lataste 1895a, [24]Reed 1896, [25]Schalow 1898, [26]Chubb 1910, [27]Ridgway 1914, [28]Cory 1915, [29]1918, [30]Bangs 1919, [31]Wetmore 1922, [32]Zimmer 1930, [33]Hellmayr 1932, [34]Kelso 1938, [35]Wheeler 1938, [36]Kleinschmidt 1940, [37]Peters 1940, [38]Kelso 1941, [39]Goodall et al. 1951, [40]Traylor 1958, [41]Johnson 1965, [42]Meyer de Schauensee 1966, [43]Johnson 1967, [44]Burton 1973, [45]Clark et al. 1978, [46]Vuilleumier 1985, [47]Araya and Millie 1986, [48]Marín et al. 1989, [49]Fjeldsa and Krabbe 1990, [50]König 1991, [51]Araya et al. 1995, [52]Heidrich et al. 1995, [53]König et al. 1996, [54]Bruce 1999, [55]König et al. 1999, [56]Marks et al. 1999, [57]Torres-Mura 2004, [58]Weick 2006, [59]König and Weick 2008, [60]Remsen et al. 2016

(Alaie and Aliabadian 2012; Nijman and Aliabadian 2013; Colihueque et al. 2015; Aliabadian et al. 2016). In addition to the genetic divergence, *T. furcata* also distinguishes specimens of *T. alba* by its head and body larger and stouter and more powerful talons (König and Weick 2008). Despite this evidence, *T. furcata* has not still been recognized as a separate species by the South American Classification Committee (Remsen et al. 2016) or by the American Ornithologists' Union (americanornithology.org). However, recently many owl specialists in the Neotropics have given validity to *T. furcata* (see Enríquez et al. 2015). We here also follow König and Weick (2008).

Long before in Chile, *T. furcata* (≈ *T. alba*) was also subject to changes regarding its taxonomic classification. Because of nomenclatural confusions, in the middle of the nineteenth century, this species was named *Strix flammea* and *S. perlata* (e.g. Des Murs 1847; Philippi 1868; Lataste 1895a). Even though the nomination *S. flammea* corresponded to the European form, Des Murs (1847) assumed that it was also present in Chile. Although *S. perlata* was considered a distinct species because of their longer tarsi (Lichtenstein 1823), Des Murs (1847) believed that *perlata* was simply a variety of *S. flammea*. However, Lataste (1895a) noted that specimens collected in Chile had tarsi consistently longer than those of the European form. In addition, Raspail (1895) stated that eggs of Chilean specimens were different from those of the European form (see sect. 7.3.9.3). Thus, the nomination *S. flammea* was discarded for Chilean morphs (Lataste 1895a). Subsequently, *S. perlata* also became an invalid nomination because it had already been proposed for other owl species (Kaup 1862). Finally, Mathews (1916) and Chubb (1916) resolved that the correct nomination for the southern South American type should be *Tyto tuidara*. At present, *tuidara* is only considered a subspecies of *T. furcata*. However, König and Weick (2008) suggest that *T. furcata*, given its paraphyletic origin, could be split into distinct species. One of such new species may be *T. tuidara*. In the past, some authors suggested the existence of other subspecies in Chile (Table 7.1).

Bubo magellanicus was described as a valid species as early as the end of the nineteenth century (Sharpe 1875a; Crawshay 1907), but some authors had already considered it only a subspecies of *B. virginianus* (Sclater and Salvin 1873; Table 7.1). König et al. (1996) proposed to return it to the category of full species, arguing that *magellanicus* has evident morphometric, morphological, vocal and genetic differences from *nacurutu* and *nigrescens*, subspecies geographically closer to *virginianus* (Weick 2006). *B. magellanicus* tends to be smaller than *B. virginianus* (total length 45 vs >45 cm, wing chord 218–358 mm vs 330–376 mm; Traylor 1958) and has weaker talons and shorter ear tufts. The rim around the facial disc is more pronounced, and underparts have a finer dark barring, denser and more regular than in *virginianus*. Sharpe (1875a) affirmed that in *magellanicus* the final joint of the toe is always bare (sometimes the entire toe), whereas in *virginianus* the toes are more thickly feathered. The males of *magellanicus* utter a "wubúh-worrrr" double, with the tremolo on "or" being more prolonged in females. In contrast, males of *nacurutu* utter a "wu-bubú buh buh" and females a "wu-bububú" followed by a long "buh". The nucleotide sequences of *virginianus* and *magellanicus* differ by 1.6%, a difference that König et al. (1996) thought was sufficient to consider them separate species. Despite this evidence, König et al. (1996) were questioned by several authors who consider that the evidence provided is insufficient (Robbins 2011), leading even to reject the proposal of the South American Classification Committee to recognize *magellanicus* as a valid species by its North American counterpart (Remsen et al. 2016). However, a more recent genetic analysis seems to confirm that *Bubo virginianus* and *B. magellanicus* are separate species (Wink et al. 2008). In support of the validity of *magellanicus*, A. Jaramillo (Remsen et al. 2016) states that the species shows more noticeable vocal and morphological differences from *virginianus* than some sister species of *Glaucidium*. Because the joint evidence suggests that *B. magellanicus* is indeed a separate species, here we follow the classification of König et al. (1996).

Athene cunicularia is represented in Chile by the subspecies *nanodes* and *cunicularia* (Table 7.1). While *nanodes* (von Berlepsch and Stolzmann 1892) has a marginal distribution in northern Chile (see sect. 7.3.5.1), *cunicularia* (Molina 1782) is widely distributed in our country. According to some authors, there is no noticeable

Table 7.2 Range of body measurements of subspecies belonging to *Strix rufipes* and *Athene cunicularia* in Chile

	Length (mm)				
Subspecies	Total	Wing chord	Tail	Culmen	Tarsi
Strix rufipes rufipes	330–429	250–290	144–184	16–20	38–51
S. r. sanborni[a]	360	241	141	16.5	44
Athene cunicularia cunicularia	18–30	135–200	78–114	15–20	46–60
A. c. nanodes	20–24	164–173	77–83	21–23	40–41

[a]Measurements correspond to a juvenile (see Wheeler 1938). References: *Strix* rufipes = 1, 4–6, 10–13, 15–19; *Athene* cunicularia = 2, 3, 5–9, 14, 15, 17. [1]Scott and Sharpe 1912, [2]Zimmer 1930, [3]Hellmayr 1932, [4]Wheeler 1938, [5]Goodall et al. 1951, [6]Burton 1973, [7]Jaksic et al. 1977, [8]Araya and Millie 1986, [9]Fjeldsa and Krabbe 1990, [10]Martínez 1995, [11]Straneck and Vidoz 1995, [12]Marks et al. 1999, [13]Figueroa et al. 2001a, [14]Jaramillo 2003, [15]Pavez 2004a, [16]Brito 2005, [17]Weick 2006, [18]Figueroa and Alvarado 2007, [19]authors, unpublished data

distinction in plumage colouration between these two subspecies (Hellmayr 1932; Johnson 1967). However, Zimmer (1930) noted that individuals of *nanodes* are somewhat darker. Moreover, *nanodes* tends to be smaller (Table 7.2). Another subspecies that might be present in Chile is *juninensis*, but up to now its presence has not been corroborated by ornithologists; this subspecies is distributed from Argentine northwest to Peru, and it could enter Chilean territory through the Andean high plateau (Fjeldsa and Krabbe 1990). Recent records of *A. cunicularia* on the Andean high plateau of Arica and Iquique could correspond to the subspecies *juninensis* (Martínez and González 2005).

Glaucidium nanum is currently recognized as a valid species, but its validity has not been free from question. Until the mid-twentieth century, ornithologists considered *G. nanum* a separate species from *G. brasilianum* (Table 7.1), but some authors thought that the former was only a variation of the latter (Hellmayr 1932; Peters 1940). After Peters (1940), most ornithologists treated *nanum* as a subspecies (e.g. Olrog 1963; Johnson 1967; Clark et al. 1978; Sibley and Monroe 1990) or a geographical variation of *brasilianum* (Burton 1973; Marín et al. 1989). After Meyer of Schauensee (1966, 1970), *nanum* was again treated as a separate species (e.g. Araya and Millie 1986). In the midst of this dichotomy, and following Clark et al. (1978), Jaksic and Jiménez (1986) treated the entire Chilean population of *Glaucidium* as *G. brasilianum*. Historically, ornithologists have differentiated the Chilean specimens of *nanum* from those of *brasilianum* of southern Peru by variations in plumage colouration, now valid to distinguish between *nanum* and *peruanum* (see sect. 7.3.4). Recent genetic analyses have confirmed that *nanum* is a separate species from *brasilianum* (Wink et al. 2008, 2009). Although ornithologists consider *nanum* monotypic (Marks et al. 1999), there is considerable variability in the width of the tail bars (Marín et al. 1989). This also led Wetmore (1922) to believe that there were at least two "races" within the Chilean population of *G. nanum*: *n. vafrum* (central Chile) and *n. nanum* (southern and southernmost Chile). The first one is with the tail's dark bars twice as wide as the pale bars and the latter with the tail's bars of similar width. Hellmayr (1932) rejected this proposal, stating that he was unable to separate *G. nanum* into two "races", given the many exceptions to the rule. Due to the great variability in the colouration of *nanum*, Des Murs (1847) also believed that individuals present in southernmost Chile (*Noctua nana = Strix nana*) corresponded to a different species (*Noctua pumila*). Although the nomination *nanum* has been widely used in the owl literature Remsen et al. (2016) state that the correct nomination is *nana*. From here on, we follow this corrected nomination.

Glaucidium peruanum represents the *Glaucidium* population distributed on the western slope and part of the eastern slope of the Andes between Ecuador and northern Chile. *G. peruanum* is distinguished from *G. nana* and *G. brasilianum* by vocal and genetic differences (König 1991; Heidrich et al. 1995). However, its current status seems not to be definitive. Morphological and vocal differences suggest that *G. peruanum* could include two species, one on the Pacific slope (including Chile) and another on the Amazon slope, being the nomination *peruanum* applicable to the Amazon population (König 1991; Marks et al. 1999; Torres-Mura 2004).

In Chile, *Asio flammeus* is represented by the subspecies *suinda* (Table 7.1). The nomination *suinda* is taxonomically stable, but it was not free of nomenclatural confusion (Table 7.1). Hellmayr (1932), arguing geographical reasons, provisionally adopted the name subspecific *breviauris* of Schlegel (1862) for the form of southern South America. Kelso (1934) proposed to apply *suinda* to those birds known as *breviauris*. After Peters (1940), *breviauris* was replaced by *suinda*. The race *suinda* differs from *bogotensis* (Colombia-Peru) for its lighter plumage and stronger legs and bill and from the nominal race *flammeus* for its darker plumage (König et al. 1999; Weick 2006). Individuals present in the Juan Fernández Archipelago also correspond to the subspecies *suinda* (Goodall et al. 1951; Araya and Millie 1986; Hahn et al. 2006). Recently, Colihueque et al. (2015) found a strong genetic divergence (3.1%) among individuals of *A. flammeus* from southern South America (Chile and Argentina) and those from North America, Europe, and Asia. As these authors suggest, it is possible that the current taxonomic status of South American populations of *A. flammeus* requires further analysis.

Wheeler (1938) divided the Chilean population of *Strix rufipes* into two subspecies: *rufipes* and *sanborni* (Wheeler 1938). The first one is widely distributed and corresponds to the type described by King (1828, Table 7.1). The second, described by Wheeler (1938), would be restricted to Chiloe Island. According to Wheeler (1938), *sanborni* differs from *rufipes* for its smaller body size (Table 7.2), darker plumage and upperparts much less barred. Because Wheeler (1938) described *sanborni* based only on an immature specimen, the validity of this subspecies has been questioned (König et al. 1999; Marks et al. 1999). However, Wheeler (1938) argued that all specimens of *rufipes* he had examined consistently differed in colour from the type of Chiloé Island, adding that a direct comparison with immature specimens of *rufipes* allowed him to confirm such differences. König et al. (1999) suggested that *sanborni* could represent rather a dark morph. The confinement of *sanborni* to Chiloe Island is also questionable because the species inhabits several others very nearby islands. A recent analysis based on ecological niche suggest that *S. rufipes* is a sister taxa of *Strix hilophylla*, an Amazonian forest-specialist owl (Girini et al. 2017).

7.3.2 *Morphology and Morphometrics*

The morphology and morphometrics of Chilean owls have been described with variable level of detail (e.g. Des Murs 1847; Goodall et al. 1951). Scott and Sharpe (1912) provided a masterly description of Patagonian owls. During recent decades, Chilean ornithologists have published a number of field guides and books on birds which include descriptions with diagnostic characters and high-quality images to facilitate identification in the field (e.g. Egli and Aguirre 2000; Couve and Vidal 2000, 2003; Martínez and González 2005; Rivas and Figueroa 2009; Celis-Diez et al. 2011). Some authors have also provided identification keys for Chilean raptors, including owls (González 1980; Núñez and Meriggio 2004; Sanhueza and Muñoz-Pedreros 2007). Good descriptions and pictures are available on some webpages such as www.avesdechile.cl, www.avesdevaldivia.cl, www.flickr.com/photos/tomas_rivas/, or http://www.flickr.com/photos/dias_de_vias. We summarize diagnostic characters and plumage colouration of Chilean owls in Table 7.3. For much more details, we recommend reviewing references on the foot of same Table. Figures 7.1, 7.2, 7.3, 7.4, 7.5, 7.6 and 7.7 complement descriptions given in Table 7.3.

7.3.2.1 Body Size and Geographical Variations

Among Chilean owls, *Tyto furcata* is a medium-sized species (Tables 7.4 and 7.5). Its wings are long relative to its body, and they extend something beyond the tail when at rest (Jaramillo 2003; Rivas and Figueroa 2009). Tarsi are proportionally long (Fig. 7.1). *T. furcata* tends to be lighter than other owl species with similar body size (Tables 7.4 and 7.5). Patagonian specimens appear to be smaller (Scott and Sharpe 1912; Cory 1918; Kelso 1938).

Bubo magellanicus is the largest owl species in Chile (Tables 7.4 and 7.5, Fig. 7.2). At rest, its wings do not align with the tip of the tail, and tarsi are proportionally shorter (Couve and Vidal 2000; Jaramillo 2003). There is no evidence of geographical variations in body size.

Athene cunicularia is one of the smallest species among Chileans owls (Tables 7.4 and 7.5). At rest, its wings almost align to the tip of the tail (Jaramillo 2003; wPavez 2004a). Legs are evidently long (Fig. 7.3). Its body size tends to increase slightly towards south (Hellmayr 1932; see sect. 7.3.3). According to Wetmore (1926), Chilean specimens appear to be smaller than those from Argentina.

Glaucidium nana and *G. peruanum* (Figs. 7.4 and 7.5) are the smallest owl species in Chile. The total length of *G. nana* is almost 20–40% shorter than that of *A. cunicularia*, and its wings are proportionately short (Tables 7.4 and 7.5). In Chile, *G. nana* tends to be smaller southwards. Based on literature, we provided the following measurements: northern Chile, wing chord = 103.3 ± 4.5 mm and tail length = 78.7 ± 4.0 mm (N = 4); central Chile, wing chord = 108 ± 6.1 mm (N = 15) and tail length = 76.6 ± 4.2 (N = 14); southern Chile, wing chord = 99.8 ± 6.8 mm and tail length = 70.2 ± 5.3 mm (N = 23); and southernmost Chile, wing chord = 97.8 ±

Table 7.3 Basic morphology of Chilean owls

Species	Aspect	Head	Upperparts	Underparts	Tail (rectrices)	Legs	Juvenile
Tyto furcata	Medium-sized, slender aspect, legs and wings proportionately long (Fig. 7.1). Eyes dark. It reflects bright-white when it is illuminated at night	Eyes small and opaque blue. Heart-shaped facial disk dirty-white with rufous and blackish below (Fig. 7.1). Grayish-white beak	Grayish-brown with yellowish spots. Back, neck, and wing coverts with dark-brown fringes covered by fine white speckling. Wings normally with a rather large, pale area on secondaries Secondaries and tail feathers often unmarked whitish. F, darker	Chest, belly, and wing coverts creamy-white with fine black speckles (Fig. 7.1). Primaries with 4–5 incomplete black bars. F, yellowish belly, pale ocher chest. M, paler and finer speckles	Dark-brown above and white below, with 4–5 thin gray-brown bars (Fig. 7.1)	Thighs and tarsi feathered white. Yellow toes (Fig. 7.1)	Like adult, or more mottled
Bubo magellanicus	Large-sized and bulky body. Broad and long wings. Notorious triangular ear tufts. Black stripes between eyes and ear tufts give it an "aggressive" appearance (Fig. 7.2)	Big eyes, yellow iris. Facial disk pale-brown with black rim. Dark-brown beak. Thin and whitish eyebrows (Fig. 7.2)	Grayish-brown spotted with blackish-gray	Creamy-white with fine and narrow blackish-gray barred. Throat whitish. Chest with broad black spots (Fig. 7.2)	Grayish-brown above, with 7–8 broad dark-brown bars (≈1 cm width); below yellowish-brown with brown fine bars	Thighs, tarsi, and toes feathered yellowish-brown (Fig. 7.2). Dark fine bars on thighs	Like adult but pale-brown shaded; without ear tufts

Athene cunicularia	Medium-sized, long-legged owl. Upright when is on a perch. Prominent eyebrows. Long and rounded wings (Fig. 7.3)	Earthy-brown with whitish flecks. Small brown facial disk. Dark-brown around eyes. Yellow iris. Pale-yellow beak (Fig. 7.3)	All earthy-brown with abundant and noticeable white spots. M, more shaded of grayish-brown in summer?	Chest, belly, and flanks creamy-white with broad brown bars (Fig. 7.3). Throat and neck whitish separated by a dark-brown necklace. F, more barred than males	Rectrices with 4–5 brown broad bars (Fig. 7.3)	Evidently long (Fig. 7.3). Thighs feathered pale-brown. Tarsi feathered creamy-white (Fig. 7.3)	Upperpart grayish-brown. Crown unflecked. Underparts unbarred. Head, chest, and throat, dark-brown. Dirty-yellow belly. White chin
Glaucidium nana	Small-sized and stocky with head visibly large. Short legs and large feet in relation to its body size. Eyebrows tilted inward give it an "angry" face (Fig. 7.4)	Brown crown with abundant fine creamy-white streaks (Fig. 7.4). Hindneck with two black patches bordered white above ("false eyes"). White eyebrows. Beak and iris yellow	Cinnamon brown or grayish-brown with elongated and abundant whitish spots, particularly over wings (Fig. 7.4)	All whitish with brown lanceolate streaks covered with fine white spots particularly over chest	Both sides with 7–10 cinnamon thin bars which tend to meet at feather shaft (Fig. 7.4)	Thighs and tarsi feathered white. Bare and yellow toes	Like adult but with crown, nape, mantle, and chest uniform-brown. Underparts with broader speckles

(continued)

Table 7.3 (continued)

Species	Aspect	Head	Upperparts	Underparts	Tail (rectrices)	Legs	Juvenile
Glaucidium peruanum	Very similar to *G. nana* (Fig. 7.5)	Crown with abundant white fine speckles (Fig. 7.5). "False eyes" like *G. nana*, but with an ochre band immediately below. White eyebrows. Yellow iris	All grayish-brown with whitish spots (Fig. 7.5)	All off-white with brown streaks toward the flanks	Both sides with <6–7 white broad bars that do not meet at feather shaft (Fig. 7.5)	Thighs feathered dirty-white. Tarsi and toes yellow	Like adult but crown unspotted
Asio flammeus	Medium-sized. Belly "flamed." Long wings. Blackish ring around the eyes give it a "surprised" face (Fig. 7.6)	Pale-yellow facial disk without black rim (Fig. 7.4). Eyebrows and bristles white (Fig. 7.6). Yellow iris. Small ear tufts sometimes visible at sunset	Yellowish-brown mottled blackish-brown (Fig. 7.6). Pale-yellow or dirty-white speckles on wings (Fig. 7.6). F, darker	Creamy yellowish-brown streaked dark-brown; broad streaks on throat and chest, and thin streaks on belly. Dark carpal "comma" on the leading edge under wings (Fig. 7.6) F, darker. Streaks on chest like "Y"	Creamy-yellow with 4–5 notorious chestnut-brown bars (Fig. 7.6)	Thighs, tarsi, and toes feathered creamy-yellow (Fig. 7.6)	Crown and rump dark-brown. Dark-brown facial disk. Mantle barred dirty-yellow. Yellow-dirty belly

Strix rufipes	Large-sized and bulky body. Plumage and dark eyes (Fig. 7.7). Rounded wings. Short broad tail. Gentle appearance (Fig. 7.7)	Dark-brown. Grayish facial disk faintly delineated. Dark-brown eyes, each with a cinnamon patch on external edge, and a whitish patch like "half-moon" on inner edge (Fig. 7.7). Eyebrows and bristles white. Grayish peak (Fig. 7.7)	Chocolate-brown with fine whitish and yellow bars. Wings dark-brown with brown and white fine bars (Fig. 7.7). M, darker	Pale-brown with profuse barred white. Whitish throat with cinnamon collar. Chest with irregular bars alternately pale-brown, brown, and yellowish (Fig. 7.7). M, darker	Above cinnamon brown and whitish below, with 5 dark-brown bars on both sides	Thighs, tarsi and toes feathered rufous (Fig. 7.7)	Grayish with abundant black bars and rufous-shaded spots (Fig. 7.7). Dirty-white facial disk. Rufous patches around eyes may be already visible

F = female, M = male

References: Molina 1782, Des Murs 1847, Cassin 1855, Sharpe 1875b, Ridgway 1876, Crawshay 1907, Scott and Sharpe 1912, Chapman 1922, Wetmore 1922, 1926, Hellmayr 1932, Wheeler 1938, Goodall et al. 1951, 1957, Traylor 1958, Johnson 1965, 1967, Burton 1973, Venegas and Jory 1979, González 1980, Meyer de Schauensee 1982, Araya and Millie 1986, Marín et al. 1989, Fjeldsa and Krabbe 1990, Martínez 1995, Straneck and Vidoz 1995, de la Peña and Rumboll 1998, Couve and Vidal 1999, 2000, 2003, König 1991, Heidrich et al. 1995, König et al. 1996, 1999, Marks et al. 1999, Egli and Aguirre 2000, Figueroa et al. 2001a, Jaramillo 2003, Pavez 2004a, Martínez and González 2005, Weick 2006, Figueroa and Alvarado 2007, Rivas and Figueroa 2009, Celis-Diez et al. 2011, Altamirano et al. 2012, Alvarado et al. 2015

Fig. 7.1 Adult *Tyto furcata* held for physical recovery at the Center for Rehabilitation of Wildlife of the Faculty of Veterinary Science at the University of Concepcion in Chillán, southern Chile (7 July 2015) (Photograph Daniel González-Acuña)

Fig. 7.2 Adult *Bubo magellanicus* registered in Waldorf, Nevados de Chillán, south-central Chile (9 January 2007) (Photograph Daniel González-Acuña)

Fig. 7.3 Adult *Athene cunicularia* observed near Fray Jorge National Park, north-central Chile (10 December 2010) (Photograph Daniel González-Acuña)

Fig. 7.4 Adult *Glaucidium nana* (brown morph) registered in Chiloé National Park, southern Chile (9 April 2006) (Photograph Daniel González-Acuña)

Fig. 7.5 Adult *Glaucidium peruanum* (brown morph) registered at km 17 of the Azapa Valley, northern Chile (12 August 2012) (Photograph Daniel González-Acuña)

5.8 mm and tail length = 65.7 ± 4.1 mm (N = 8) (Sharpe 1875b; Scott and Sharpe 1912; Wetmore 1922; Peters 1923; Hellmayr 1932; Marín et al. 1989). Jiménez and Jaksic (1989) documented a similar trend (see sect. 7.3.3). These results confirm perception of Scott and Sharpe (1912) and Peters (1923) that Patagonian individuals tend to be smaller. Apparently, *G. peruanum* is smaller and lighter than *G. nana* (Table 7.4).

The body size of *Asio flammeus* partially overlaps with that of *T. furcata* and *Strix rufipes* (Table 7.4 and 7.5). Tail and wings are proportionately long (Fig. 7.6);

Fig. 7.6 Adult *Asio flammeus* flying over an abandoned pasture and perching on a fence post in an area near city of Talcahuano, southern Chile (10 December 2012) (Photographs Daniel González-Acuña)

Fig. 7.7 Adult (**a**) and juvenile (**b**) *Strix rufipes* held for physical recovery at the Center for Rehabilitation of Wildlife of the Faculty of Veterinary Science at the University of Concepcion in Chillán, southern Chile. The juvenile individual is approximately 1 month old (Photographs Daniel González-Acuña)

at rest, wings are beyond the tip of the tail (Jaramillo 2003; Pavez 2004a). Tarsi are relatively short relative to its body size. It is unknown whether there are geographical variations in body size. The only available information corresponds to a female collected in central Chile which had a wing chord greater than a female collected in southern Chile (324 mm and 317 mm, respectively; D. González-Acuña, unpublished data).

Table 7.4 Range of body measurements and mass of Chilean owls

	Length (mm)						
Species	Total	Wing chord	Tail	Bill	Tarsi	Wingspan	Mass (g)
Tyto furcata	322–485	253–338	113–175	21–30	50–85	680–1080	207–590
Bubo magellanicus	450–550	305–368	162–216	27–42	50–52	1100–1180	517–1500
Athene cunicularia	180–300	135–200	78–114	15–20	46–60	500–682	150–370
Glaucidium nana	150–220	90–117	50–86	10–16	21–30	250–400	49.8–100
Glaucidium peruanum	150–210	95–109	67.5–79.4	11	–	4000	53–76.3
Asio flammeus	330–430	250–330	141–180	15–25	5–5.9	900–1100	240–450
Strix rufipes	330–429	240–290	120–191	16–30	38–51	900–980	300–620

References: *Tyto furcata* = 1, 5, 9, 12, 15, 19–28, 31, 34, 37, 39, 40, 44–52, 54–58, 59; *Bubo magellanicus* = 1, 5, 12, 14, 15, 19, 20, 22–28, 31, 34, 37, 38, 40–44, 47–52, 54, 55, 58, 59; *Athene cunicularia* = 1, 5, 7, 8, 12, 15, 18, 19, 23–27, 31, 37, 39, 42, 43, 48, 50, 51, 52, 54, 55, 58, 59; *Glaucidium nana* = 1–6, 8, 11–13, 15, 17, 20, 25, 27, 29–31, 37, 39, 40, 42, 43, 45, 47–52, 54, 55, 58, 59; *Glaucidium peruanum* = 12, 13, 29, 30, 32, 42, 43, 51, 52, 55; *Asio flammeus* = 1, 5, 12, 15, 17, 18, 20, 25, 26, 31, 33, 34, 37, 39, 42, 43, 48–52, 54, 55, 57–59; *Strix rufipes* = 1, 5, 10, 12, 13, 15, 16, 18, 20, 25–27, 31, 33–37, 39, 42, 43, 47, 48, 50–55, 57–59. [1]Gay 1847, [2]Sharpe 1875b, [3]Ridgway 1876, [4]Crawshay 1907, [5]Scott and Sharpe 1912, [6]Wetmore 1922, [7]Zimmer 1930, [8]Hellmayr 1932, [9]Griscom and Greenway 1937, [10]Wheeler 1938, [11]Olrog 1948, [12]Goodall et al. 1951, [13]1957, [14]Traylor 1958, [15]Johnson 1965, [16]1967, [17]Humphrey et al. 1970, [18]Burton 1973, [19]Jaksic et al. 1977, [20]Venegas and Jory 1979, [21]Herrera and Jaksic 1980, [22]Jaksic and Yáñez 1980a, [23]Jaksic et al. 1981, [24]Jaksic 1983, [25]Araya and Millie 1986, [26]Clark 1986, [27]Morgado et al. 1987, [28]Jaksic 1988, [29]Jiménez and Jaksic 1989, [30]Marín et al. 1989, [31]Fjeldsa and Krabbe 1990, [32]König 1991, [33]Ortiz et al. 1994, [34]Venegas 1994, [35]Martínez 1995, [36]Straneck and Vidoz 1995, [37]Chester 1995, [38]König et al. 1996, [39]de la Peña and Rumboll 1998, [40]Couve and Vidal 1999, [41]Donázar et al. 1997, [42]König et al. 1999, [43]Marks et al. 1999, [44]Santibáñez and Jaksic 1999, [45]Egli and Aguirre 2000, [46]Hoffmann and Lazo 2000, [47]Couve and Vidal 2000, [48]Figueroa et al. 2001a, [49]Jaksic et al. 2002, [50]Couve and Vidal 2003, [51]Jaramillo 2003, [52]Pavez 2004a, [53]Brito 2005, [54]Martínez and González 2005, [55]Weick 2006, [56]Sanhueza 2007, [57]Figueroa and Alvarado 2007, [58]Rivas and Figueroa 2009, [59]authors, unpublished data

Strix rufipes is a medium-sized owl (Tables 7.4 and 7.5). At rest, its wings align with the tip of the tail (Jaramillo 2003), and tarsus are proportionately short (Fig. 7.7). Due to the small sample size, it is not possible to know whether there are geographical variations in body size, but there appears to be an increase in body size southwards (see below).

7.3.3 Sexual Dimorphism

In *Tyto furcata tuidara*, females tend to have upperparts darker than males, yellowish belly, and chest with an ochre tint. The ventral part of males is almost all white with small tiny dark spots (Scott and Sharpe 1912; Jaramillo 2003; Pavez 2004a).

Table 7.5 Body measurements and mass (mean ± SD) of Chilean owls

	Length (mm)						
	Total	Wing chord	Tail	Bill	Tarsi	Wingspan	Mass (g)
Tyto furcata	360–380[a]	307.5 ± 4.1(8)	131.9 ± 3.88(8)	29 ± 0.1(8)	–	–	–
	466 ± 9(5)	–	–	–	–	–	–
	388 ± 40(8)	302.3 ± 12.4(7)	138.4 ± 18.2(7)	25 ± 3.3(9)	69.1 ± 13.5(7)	916 ± 120(9)	340 ± 87.6(18)
Bubo magellanicus	480–500[a]	339.8 ± 5.0(10)	182.9 ± 4.0(10)	25.6 ± 0.35(10)	–	–	–
	538 ± 7(7)	–	–	–	–	–	–
	461 ± 0.0(1)	325 ± 0.0(1)	191 ± 0.0(1)	42 ± 0.0(1)	52 ± 0.0(1)	1115 ± 21(2)	886.1 ± 296(6)
Athene cunicularia	260[a]	184.7 ± 1.54(13)	91.6 ± 2.6(13)	16.9 ± 0.39(13)	–	–	–
	289.4 ± 6.2(9)	–	–	–	–	–	–
	–	183.3 ± 5.2(19)	–	–	–	–	–
	280 ± 0.0(1)	164 ± 22.4(3)	96 ± 0.0(1)	16 ± 0.0(1)	60 ± 0.0(1)	591 ± 129(2)	291 ± 111(2)
Glaucidium nana	200–210[a]	103.6 ± 1.0(27)	69.3 ± 0.7(27)	13 ± 0.2(27)	–	–	–
	–	101.9 ± 7.7(29)	70.8 ± 5.6(29)	–	–	–	–
	–	100 ± 8(5)	74.9 ± 10.8(5)	11.5 ± 0.7(5)	–	–	61 ± 1(3)
	168.8 ± 15.6(13)	–	–	–	–	–	
	–	–	–	–	–	–	73.5 ± 10.6(10)
	–	–	–	–	–	–	85 ± 0.0 (1)
Glaucidium peruanum	190–200[a]	95 ± 0.0(1)	68 ± 0.0(1)	11 ± 0.0(1)	–	–	–
	–	99.9 ± 2.8(4)[b]	74.7 ± 0.5(4)[b]	10.6 ± 0.4(3)[b]	–	–	65 ± 5.6(3)[b]
Asio flammeus	420[a]	319.5 ± 1.7(15)	154.2 ± 2.1(15)	31.5 ± 0.4(15)	–	–	–
	380 ± 28(2)	285 ± 49.5(2)	173.5 ± 9.2(2)	25.5 ± 0.7(2)	54.5 ± 6.4(2)	925 ± 21(2)	379.5 ± 95(4)
Strix rufipes	380[a]	271.2 ± 3.1 (12)	157.3 ± 1.7 (12)	29.9 ± 0.4(12)	–	–	–
	381 ± 40(5)	248 ± 26(9)	149.1 ± 22.4(9)	21.6 ± 3.4(5)	41 ± 4.2(2)	785 ± 163(2)	434 ± 181.6(3)

The number of measured specimens is indicated in parentheses

[a]Information provided by the original source. [b]Includes only individuals analyzed by Marín et al. (1989) that showed typical morphological characteristics. References: *Tyto furcata* = 5, 7, 8, 11, 12, 16, 18; *Bubo magellanicus* = 5, 7, 10, 11, 16, 18; *Athene cunicularia* = 2, 3, 5, 10, 11, 18; *Glaucidium nana* = 1, 3, 5, 6, 13, 14, 16, 18; *Glaucidium peruanum* = 5, 14; *Asio flammeus* = 5, 6, 16, 18; *Strix rufipes* = 2, 4, 5, 11, 15, 17, 18. [1]Crawshay 1907, [2]Scott and Sharpe 1912, [3]Hellmayr 1932, [4]Olrog 1948, [5]Goodall et al. 1951, [6]Humphrey et al. 1970, [7]Jaksic et al. 1977, [8]Herrera and Jaksic 1980, [9]Jaksic and Marti 1981, [10]1984, [11]Morgado et al. 1987, [12]Jaksic 1988, [13]Jiménez and Jaksic 1989, [14]Marín et al. 1989, [15]Martínez 1995, [16]Jaksic et al. 2002, [17]Brito 2005, [18]authors, unpublished data

Table 7.6 Range of body measurements and mass of Chilean owls by sex

	Total length (mm)		Wing length (mm)		Tail length (mm)		Mass (g)	
Species	Female	Male	Female	Male	Female	Male	Female	Male
Tyto furcata	361–380	349–360	271–393	253–295	133	124	207–490	259–450
Bubo magellanicus	461–508	457	330–368	305–356	216	203	975–1333	825–1074
Athene cunicularia	254	226	173–198	172–190	81–99	81–90	120–250	130–185
Glaucidium nana	165–210	150–195	100–116	90–112	60–91.5	50–80	50–100	55–85.5
Glaucidium peruanum	–	–	101–109	97–102.5	74.5–81.5	67–80	64.5–76.3	53–75
Asio flammeus	380–430	330–380	318	305–306	165	152	280–500	200–450
Strix rufipes	429	330–394	250–275	233–264	155–191	141–163	425	257

References: *Tyto furcata* = 1, 5, 18, 19, 22, 23; *Bubo magellanicus* = 1, 8, 18, 19, 20, 23; *Athene cunicularia* = 1, 3, 4, 17, 23; *Glaucidium nana* = 1, 2, 4, 7, 9, 11, 12, 15–18, 20, 23; *Glaucidium peruanum* = 12, 16; *Asio flammeus* = 1, 9, 10, 18, 19, 23; *Strix rufipes* = 1, 6, 13, 14, 19, 20, 23. [1]Scott and Sharpe 1912, [2]Wetmore 1922, [3]Zimmer 1930, [4]Hellmayr 1932, [5]Griscom and Greenway 1937, [6]Wheeler 1938, [7]Olrog 1948, [8]Traylor 1958, [9]Humphrey et al. 1970, [10]Burton 1973, [11]Jiménez and Jaksic 1989, [12]Marín et al. 1989, [13]Martínez 1995, [14]Straneck and Vidoz 1995, [15]de la Peña and Rumboll 1998, [16]König et al. 1999, [17]Marks et al. 1999, [18]Jaksic et al. 2002, [19]Pavez 2004a, [20]Weick 2006, [21]Figueroa and Alvarado 2007, [22]Sanhueza 2007, [23]authors, unpublished data

Females also are distinguished by having darker and larger spots on the wings and tail and a greater number of fringes on the back (Bruce 1999). Adult females tend to be larger than adult males, but there is ample overlap in body size (Table 7.6). Unexpectedly, we found that the mean body mass of adult females tends to be smaller than that of adult males. In southernmost Chile, females and males had a mean body mass of 303.3 g (range 295–310 g, N = 3) and 310 g (259–383, N = 2), respectively (Jaksic et al. 2002). In Chillán, central Chile, the mass of two females ranged 207–300 g and that of a male was 300 g (D. González-Acuña, unpublished data). This suggests that body mass would not be a good indicator of the sex in subspecies *tuidara*, but our results could also be an artefact of small sample sizes or body condition of individuals sampled.

Individuals of *Bubo magellanicus* exhibit no obvious colour differences between the sexes (e.g. Scott and Sharpe 1912). Possibly, as in *virginianus*, adult males have ear tufts perpendicular to head, and adult females inclined laterally (Fjeldsa and Krabbe 1990). Although there is not a clear segregation, females tend to be larger and heavier than males (Table 7.6).

Athene cunicularia exhibits a subtle sexual dimorphism. Adult females appear to have the ventral part more barred than males and with a more marked black tint (Scott and Sharpe 1912; Marks et al. 1999). According to Marks et al. (1999), the adult males show a colouration more coffee-greyish on upperparts during summer.

Adult females tend to be slightly larger and heavier than adult males (Table 7.6). A reanalysis of wing chord measurements provided by Hellmayr (1932) resulted in the following figure: northern Chile, males = 177.3 ± 2.2 mm (range = 175–180 mm, N = 4) and females = 180 ± 2.8 mm (range = 178–182 mm, N = 2); central Chile, male = 183.3 ± 2.9 (range = 180–185, N = 3) and females = 184.3 ± 1.2 mm (range = 183–185 mm, N = 3); and southern Chile, males = 185.5 ± 3.6 mm (range = 180–190 mm, N = 6) and females = 197 mm (N = 1). The wing chord of an adult male and female collected in southern Patagonia was 172 mm and 185 mm, respectively (Scott and Sharpe 1912). Moreover, one female and two males collected in northern Chile showed no obvious differences; the length of the wing chord and tail in the female was 188 mm and 81 mm, respectively, whereas in males was 176–198 mm and 81–93 mm, respectively (D. González-Acuña, unpublished data).

Glaucidium nana and *G. peruanum* show no obvious sexual dimorphism in plumage colour. However, Sharpe (1875b) mentions that plumage of adult females of *G. nana* tends to be browner. Adult females of both species tend to be larger than adult males (Tables 7.6 and 7.7). A female and three male adults of *G. peruanum* collected in northern Chile had the following measurements: female, wing chord = 104 mm, tail length = 74.5 mm, and body mass = 69 g, and males, wing chord = 98.6 ± 1.1 mm, tail length = 74.8 ± 0.6 mm, and body mass = 63.7 g (Marín et al. 1989).

Asio flammeus and *Strix rufipes* are sexually dimorphic, both in colour and body size. Adult females of *A. flammeus* tend to be darker and larger than adult males (Scott and Sharpe 1912; Martínez and González 2005; Table 7.6). The adult males of *S. rufipes* tend to be darker than the females, particularly on the back and chest (Des Murs 1847; Jaramillo 2003); however, females tend to be larger than males (Table 7.6), independent of geographical distribution (Table 7.8).

7.3.4 Chromatic Variations

The colour variation within the Chilean population of *Tyto furcata* is unclear. The dark morphs mentioned by some authors (e.g. Goodall et al. 1951; Couve and Vidal 2003) may instead correspond to adult females, and the markedly white colouring of some individuals could reflect the existence of pale morphs (Kleinschmidt 1940).

Bubo magellanicus show a plumage chromatically diverse presenting either pale, dark or intermediate morphs (Scott and Sharpe 1912; Goodall et al. 1951; Traylor 1958; Araya and Millie 1986; Martínez and González 2005). According to Scott and Sharpe (1912), variation in plumage colour could be related to habitat characteristics; i.e. to the extent that vegetation is more dense and complex, individuals tend to be darker. These authors found (i) that individuals inhabiting open forests of southern Patagonia are more greyish and silvered than those inhabiting dense forests of northern Patagonia, which tend to be more brown-greyish, particularly on the back and (ii) that individuals from central Patagonia have a complete deficiency of grey-brown tint. Traylor (1958) described similar colour morphs, but he did not detect

Table 7.7 Body lengths and mass (mean ± SD) of females and males of *Glaucidium nana* in Chile

	Females			Males			
Zones	Wing chord (mm)	Tail (mm)	Mass (g)	Wing chord (mm)	Tail (mm)	Mass (g)	Source
Northern	103.8 ± 3.8(4)	79.4 ± 3.5(4)	76.3 ± 6(3)	97.5(1)	75.5(1)	62(1)	3[a]
Central	112 ± 2(2)	74.6 (1)	–	105 ± 1.3(2)	73.1 ± 4.4(2)	–	1[b]
	112.2 ± 6.2(4)	79 ± 2.9(4)	–	105 ± 6.2(3)	73 ± 3(3)	–	2[b]
	110.7 ± 4.5(22)	81.2 ± 6.8(19)	75 ± 0.0(1)	102.6 ± 3.9(31)	77.4 ± 6.5(25)	74 ± 0.0(1)	3[a]
	112(1)	91.5(1)	–	–	–	–	4[b]
Southern	105.4 ± 4.8 (8)	73.4 ± 4.1(8)	–	96.1 ± 4.7(9)	67.2 ± 3.6(9)	–	2[b]
	102.9 ± 4.7(52)	68.9 ± 5.9(49)	95.5 ± 58.7(2)	96.3 ± 3.9(31)	63.7 ± 4.8(22)	66.5 ± 6.4(3)	3[a]
	–	–	–	94.5 ± 1.7(3)	67.8 ± 3.2(3)	61 ± 1(3)	4[b]
Southernmost	102 ± 0(2)	68.5 ± 0.7(2)	–	92.7 ± 1.5(3)	63.3 ± 1.1(3)	–	2[b]
	101.7 ± 4.8(10)	68.6 ± 4.5(10)	72.8 ± 3.2(2)	95.2 ± 2.5(13)	61.9 ± 5.5(13)	59 ± 3.6(3)	3[a]

The number of measured specimens is indicated in parentheses

[a]Values directly taken from original source. [b]Values estimated with raw data. References: [1]Wetmore 1922, [2]Hellmayr 1932, [3]Jiménez and Jaksic 1989, [4]Marín et al. 1989

Table 7.8 Body lengths (mean ± SD) of females and males of *Strix rufipes* in Chile

	Females			Males			
Zones	Total (cm)	Wing chord (mm)	Tail (mm)	Total (cm)	Wing chord (mm)	Tail (mm)	Source
Central	41.2(1)	–	172(1)	–	255.5 ± 32(2)	153 ± 14(2)	5
Southern	–	254.5 ± 6.4(2)	141.5 ± 20.5(2)	33(1)	241.5 ± 0.7(2)	133.5 ± 11(2)	2, 3, 5
	–	263.3(2)	151.6(2)	–	–	–	4[a]
Southernmost	42.9(1)	290(1)	191(1)	39.4(1)	245(1)	152.4(1)	1[b]

The number of measured specimens is indicated in parentheses

[a]Authors did not report the standard deviation. [b]Specimens were collected in the Argentina-Chile boundary. References: [1]Scott and Sharpe 1912, [2]Wheeler 1938, [3]Olrog 1948, [4]Straneck and Vidoz 1995, [5]authors, unpublished data

geographical variations. Sharpe (1881) noticed that a female collected on an oceanic island of southern Chile was markedly dark and he suggested the occurrence of melanism. Recently, an entirely albine individual was registered in Tierra del Fuego (Kusch and Donoso 2017).

Athene cunicularia can present the two chromatic extremes. Hellmayr (1932) found an individual with upperparts almost black and a very yellowish belly. Recently, Fuentes and González-Acuña (2011) recorded an individual with plumage plentifully spotted with white. Possibly, such records represent cases of melanism and leucism, respectively. Wetmore (1926) stated that the population of *A. cunicularia* of southern South America is much more variable in plumage colour compared to the North American population.

Glaucidium nana shows a great variation in plumage colour. At least three basic morphs are known: (i) a predominant brown morph, (ii) a less common grey morph, and (iii) a rare rufous morph (Sharpe 1875b; Ridgway 1876; Scott and Sharpe 1912,;Wetmore 1922; Goodall et al. 1951; Marín et al. 1989; Jaramillo 2003). In all these morphs, the bill is yellowish-green, and the "light" bars of the tail (>6) tend to be narrower than the "dark" bars (Jaramillo 2003). We summarize the distinctive features of each morph as follows: Brown morph: predominantly brown with reddish tinge, but the "light" tail bars are rufous, and the "dark" tail bars are dark-brown; number of "light" bars tends to be higher than in the other morphs (Fig. 7.4; see Couve and Vidal 2003; Jaramillo 2003; Martínez and González 2005). Grey morph: overall colour grey-brown, but the "light" tail bars tend to be whitish and less numerous than in the brown morph (Figueroa et al. 2001a; Jaramillo 2003). Rufous morph: plumage richer in reddish tinge than in the brown morph; colour pattern of the tail bars is similar to the brown morph, but sometimes the bars are not so distinguishable appearing in the tail almost entirely reddish (Jaramillo 2003; Martínez and González 2005; Rivas and Figueroa 2009). The origin of this polymorphism is intriguing because it is unclear whether it is due to individual, geographical, or age variations. Rufous individuals have been observed in different parts of the distribution range of *G. nana* (Scott and Sharpe 1912; Wetmore 1922). Sharpe (1875b) mentions that juvenile individuals are characterized by greyish-brown heads and suggests that the rufous-brown plumage may correspond to an intermediate stage towards maturity.

Glaucidium peruanum also exhibits three basic morphs: grey (holotype), brown, and rufous (Heidrich et al. 1995; König et al. 1999; Marks et al. 1999; Jaramillo 2003; Martínez and González 2005). Here, we summarize the distinctive features of these morphs based on König et al. (1999): Grey morph: dark parts of plumage are brown grey; upperparts with whitish spots and underparts prominently mottled. Crown can have whitish specks in individuals from higher altitudes and streaks or lanceolate spots in individuals from lower altitudes (Fig. 7.5). "Light" tail bars are whitish (Jaramillo 2003; Martínez and González 2005). Brown morph: similar to grey morph but richer in brown tint. Red morph: dark parts of plumage rich in reddish tint; upperparts with fringes and ochre or whitish spots and belly with orange-brown vertical streaks; crown with pale-ochre lanceolate streaks; the "light" tail bars are rufous and slightly wider than bars "dark", the latter being dark brown

(Martínez and González 2005). The grey morph would be predominant in the Chilean population of *G. peruanum* (Marín et al. 1989; Jaramillo 2003).

Asio flammeussuinda, regardless of sex, presents a dark morph which is a result of increased expression of reddish tinge (Scott and Sharpe 1912). This morph would be common in northernmost Chile (Jaramillo 2003). In southern Patagonia, the rufous plumage seems to predominate (Scott and Sharpe 1912).

Strix rufipes apparently presents chromatic variations associated with climatic conditions and habitat. Individuals inhabiting the southern temperate rainforest are usually darker than those inhabiting sclerophyllous or *Nothofagus* forests of central Chile (Jaramillo 2003; Figueroa and Alvarado 2007). According to Burton (1973), individuals in southern Chile have dark-orange eye discs, and those of central Chile, heavily barred dark-brown and white. Possibly, the subspecies *sanborni* corresponds to a dark morph.

7.3.5 Distribution and Residence

The quantity of references published from the late nineteenth century to the present has allowed a better understanding of the geographical distribution of Chilean owls. The following provides valuable contributions because they include distribution maps: Traylor (1958), Fjeldsa and Krabbe (1990), de la Peña and Rumboll (1998), Bruce (1999), König et al. (1999), Marks et al. (1999), Couve and Vidal (2003), Jaramillo (2003), Martínez and González (2005), and Bonacic and Ibarra (2010). Some maps even include boundaries of dispersal and temporal use areas (Fjeldsa and Krabbe 1990; Jaramillo 2003; Martínez and González 2005).

7.3.5.1 Mainland Distribution

Tyto furcata is found in almost all mainland Chile (Table 7.9), but with some altitudinal restrictions. According to Fjeldsa and Krabbe (1990), its reproductive distribution range is restricted to a western fringe between 21 and 40°S, with an altitudinal limit <2000 m. Jaramillo (2003) suggests a continuous breeding distribution throughout Chile, except in the Andean high plateau, Atacama Desert, icefields, and Andean peaks. This same author indicates an altitudinal limit of 1500 m, although its distributional map covers areas between 1500 and 2500 m possibly, the latter is due to the reduced scale of the maps. There are some specific records confirming an altitudinal distribution >1500 m (Torres-Mura and Contreras 1989: 1700 m a.s.l., Jaksic et al. 1999: 2500 m a.s.l., authors: 1700–2000 m a.s.l.).

Bubo magellanicus is widespread in Chile with an altitudinal distribution range >4000 m (Table 7.9). However, its distributional limits are unclear. On its distribution map, Traylor (1958) leaves out *B. magellanicus* on much of the western margin of northern Chile. Fjeldsa and Krabbe (1990) delimitate a breeding range between Copiapó and Cape Horn (27–56°S), and they outline a dispersion range on the

Table 7.9 Biogeographic distribution of Chilean owls

Species	Bioclimatic zones	Administrative regions	Latitudinal range (°S)	Altitude (m)
Tyto furcata	N, C, S, SM	Par-Mag	17°30′–56°00′	0–2500
Bubo magellanicus	N, C, S[a], SM	Par-Mag	17°30′–56°00′	0–4500
Athene cunicularia	N, C, S, [SM]	Par-Lag, [Ays]	18°00′–41°30′ [45°30′S]	0–3000, 4000?[c]
Glaucidium nana	N, C, S, SM	Tar?, Ant-Mag[b]	21°00′?/27°00′–56°00′[b]	0–2000
Glaucidium peruanum	N	Par-Ant	17°30′–21°30′	0–3600
Asio flammeus	N, C, S, SM	[Par], Ata-Mag	[18°00′–18°30′] 28°00′–55°00′	0–700[d]
Strix rufipes	C, S, SM	Coq-Mag	32°00′–56°00′	0–1200[e]

Bioclimatic zones: N = northern Chile, C = central Chile, S = southern Chile, SM = southernmost Chile. Administrative regions from north to south: Par = Parinacota, Tar = Tarapacá, Ant = Antofagasta, Ata = Atacama, Coq = Coquimbo, Lag = Los lagos, Ays = Aysén, Mag = Magallanes. Brackets indicate distributional discontinuity

[a]It has a distributional gap between 38 and 44°S but maintains distributional continuity along Andean areas. [b]According to Araya and Millie (1986), the northern distribution of *Glaucidium nana* reaches up Tarapaca; the latitudinal range given in the table is an arbitrary estimate. However, several authors consistently indicate that northern limit of this species reaches almost 27° south latitude. [c]Although there is no concrete evidence, some authors indicate that the species can be registered up to nearly 4000 m altitude in the Andes mountain range (Fjeldsa and Krabbe 1990; Marks et al. 1999). [d]Housse (1945) affirms that *Asio flammeus* may be found up to 1200 m altitude, but it seems to preferentially occupy areas <700 m.a.s.l. [e]Although some authors suggest that the altitudinal limit of *Strix rufipes* reaches up 2000 m, we think that the most probable limit is 1000–1200 m.a.s.l. References: *Tyto furcata* = 1, 2, 4, 5, 7, 9, 11–16, 18, 19, 22–24, 27–32, 34–41, 44, 46, 49, 51; *Bubo magellanicus* = 1–5, 7, 9, 10, 12–19, 22–24, 26–28, 30, 32–35, 38–41, 43, 46, 47, 49, 51, 52, 53; *Athene cunicularia* = 1–5, 7, 9, 11–16, 18, 19, 22–24, 27, 30–33, 35, 36, 38–41, 44, 46, 49, 51; *Glaucidium nana* = 1–5, 7–9, 12, 14–24, 27, 28, 30, 32–36, 38–41, 44, 46, 49, 51; *Glaucidium peruanum* = 5, 9, 12, 18, 19, 22, 32, 33, 41, 44, 46, 49; *Asio flammeus* = 1, 2, 4, 5, 7, 9, 12, 14–16, 18, 19, 22–24, 27, 28, 31–35, 38–42, 44, 46, 49–52; *Strix rufipes* = 1, 2, 4–7, 9, 12–19, 22–25, 27, 28, 30, 32–35, 38–41, 44–46, 48–51. [1]Gay 1847, [2]Philippi 1868, [3]Lane and Sclater 1897, [4]Scott and Sharpe 1912, [5]Hellmayr 1932, [6]Wheeler 1938, [7]Housse 1945, [8]Barros 1950, [9]Goodall et al. 1951, [10]Traylor 1958, [11]Barros 1963, [12]Johnson 1965, [13]1967, [14]Humphrey et al. 1970, [15]Meyer de Schauensee 1970, [16]Venegas and Jory 1979, [17]Vuilleumier 1985, [18]Araya and Millie 1986, [19]Jaksic and Jiménez 1986, [20]1989, [21]Marín et al. 1989, [22]Fjeldsa and Krabbe 1990, [23]Venegas 1994, [24]Chester 1995, [25]Straneck and Vidoz 1995, [26]König et al. 1996, [27]de la Peña and Rumboll 1998, [28]Venegas and Sielfeld 1998, [29]Bruce 1999, [30]Couve and Vidal 1999, [31]Jaksic et al. 1999, [32]König et al. 1999, [33]Marks et al. 1999, [34]Mella 1999, [35]Couve and Vidal 2000, [36]Egli and Aguirre 2000, [37]Hoffmann and Lazo 2000, [38]Figueroa et al. 2001a, [39]Jaksic et al. 2002, [40]Couve and Vidal 2003, [41]Jaramillo 2003, [42]Escobar and Vukasovic 2004, [43]Lazo 2004, [44]Pavez 2004a, [45]Martínez 2005a, [46]Martínez and González 2005, [47]Mella 2005, [48]Trejo et al. 2006, [49]Weick 2006, [50]Figueroa and Alvarado 2007, [51]Rivas and Figueroa 2009, [52]Bonacic and Ibarra 2010, [53]Mella et al. 2016

Andean mountain range from 27°S northwards. Jaramillo (2003) gives *B. magellanicus* a continuous breeding range, but discards its presence along the Atacama Desert. Martínez and González (2005) suggest the existence of a western dispersion area between Antofagasta and the Chile-Peru boundary (17–24°S). The map in Bonacic and Ibarra (2010) excludes *B. magellanicus* along the Andean frontier between 17 and 27°S. Unlike other authors, Couve and Vidal (2003) affirm that there is a distributional gap between 38 and 44°S. Although we have confirmed the absence of *B. magellanicus* in much of this gap, there are at minimum two recent records of nesting pairs in Temuco, southern Chile (38°44′S–72°35′W; A. García, personal communication). Moreover, several records suggest a distributional continuity for *B. magellanicus* along the Andean frontier of the hiatus area (Elgueta et al. 2006; R.A. Figueroa, personal observation). Note that the gap extension indicated by Couve and Vidal (2003) coincides with the original distribution of the Valdivian forest. It is possible that the very dense vegetation of this forest type had imposed on the past a barrier to movements and flight manoeuvrability of *B. magellanicus*. However, the current landscape conditions such as the decreased forest cover and high abundance of European hare (*Lepus europaeus*) may be attracting the *B. magellanicus* towards the hiatus area.

Athene cunicularia presents three distinguishable distributional cores throughout mainland Chile: (1) a small population in northern Chile, (2) an extensive population between northern and southern Chile, and (3) a small marginal population in southernmost Chile (Table 7.9). The disjunction between the first two populations seems due more to segregation in altitude than in latitude (Jaramillo 2003). The first population, represented by the subspecies *nanodes*, extends throughout lower areas between Iquique and the Chile-Peru boundary (18–20°S; Jaramillo 2003; Martínez and González 2005). The second population extends from southern Arica or Pica (19–20°30′S) to near Puerto Montt (38–42°S; Couve and Vidal 2003; R.A. Figueroa, personal observation). There are at least three confirmed records in northern Chile: Pampa del Tamarugal (20°24′S–69°44′W), Playa Los Verdes (20°49′S–70°09′W), and Ojo Opache, Calama (22°29′S–68°55′W; Carevic 2005, 2011, S. Alvarado and J. Cabot, personal observation). The altitudinal distribution of this population seems to depend on the orographic conditions. Barros (1963) documents a maximum distribution altitude of 1750 m. However, Lane and Sclater (1897) recorded an individual in the Andes of Tarapacá at nearly 3000 m a.s.l. Coincidentally, Martínez and González (2005) recorded *A. cunicularia* in San Pedro de Atacama, at an altitude of about 2700 m. These authors suggested that the species could also be present on the Altiplano of Arica and Iquique (18°S; >4000 m a.s.l.). Both the Andean high plateau and the Andean mountain range up to Puerto Montt (41°S) are believed to constitute dispersal areas (Martínez and González 2005). The population of southernmost Chile would represent the distributional continuity of the Argentine population. The steppe plains would facilitate the entrance of individuals of *A. cunicularia* from Argentina to Chile. The existence of this population is supported by Olrog (1948), who recorded a female nesting in grassland in Coyhaique Alto, Chile (45°30′S–71°53′W).

Glaucidium nana is widespread in mainland Chile, but its distribution range in northernmost Chile is not so clear. Whereas some authors claim that *G. nana* distributes continuously throughout Chile (e.g. Araya and Millie 1986), others indicate that it extends from southern Atacama Desert up to Cape Horn (Cabo de Hornos) (26–56°S, Table 7.9). The presence of this species in northernmost Chile is supported by the specimens collected in Quebrada de Parca (20°01′S–69°12′W, 2700 m a.s.l.) and Lluta Valley, Tarapacá (18°25′S–70°06′W, 940 m a.s.l.; Marín et al. 1989). However, these individuals had a plumage colouration intermediate between *peruanum* and *nana*, but closer to the latter (Marín et al. 1989: Table 7.1). This ambiguity in colour patterns makes the presence of *G. nana* in northern Chile doubtful. After the work of Marín et al. (1989), several surveys have been unsuccessful in verifying presence of *G. nana* in northernmost Chile (e.g. Estades 1995; Sielfeld et al. 1996; Peredo and Miranda 2001). For further uncertainty, Chapman (1922) collected an individual of *Glaucidium* in southernmost Peru (Moquegua, 17°10′S–70°55′W) which he identified as *G. nana*, but it had rather intermediate plumage colouration. Thus, the geographical disjunction between population of *G. nana* and *G. peruanum* (Johnson 1967; Short 1975; Vuilleumier 1985) is not still resolved.

In Chile, *Glaucidium peruanum* is distributed between the Chile-Peru boundary and the northern Atacama region (18–22°S) and from the Pacific coast to nearly 3600 m a.s.l. (Jaramillo 2003; Martínez and González 2005; Table 7.9). The specific locations where the species has been recorded are Putre (18°11′S–69°33′W, 3500 m a.s.l.), Valle de Lluta (Chapisca 18°22′S–69°54′W, 1100 m a.s.l.; Tocontasi 18°27′S–70°04′W, 1050 m a.s.l.; an unidentified locality 18°25′S–70°06′W, 940 m a.s.l.), Caleta Buena (19°53′S–70°07′W, 30 m a.s.l.), Quebrada de Parca (20°01′S–69°12′W, 2700 m a.s.l), Iquique (20°14′S–70°07′W, 35 m a.s.l.), and Pica (20°28′S–69°22′W, 1200 m a.s.l.; Hellmayr 1932; Goodall et al. 1951; Marín et al. 1989; Martínez and González 2005; Barros and Díaz 2008). In addition, Marín et al. (1989) described specimens with plumage colour between *nana* and *peruanum*, but closer to the latter, in Valle del Lluta, Quebrada de Camarones (28°40′ W–70°39′W, 500–900 m a.s.l.), and Punitaqui (30°50′S–71°15′W, 200 m a.s.l.). If these records actually correspond to *G. peruanum*, it could represent a considerable expansion of the southern distribution range of this species.

Asio flammeus is widespread in mainland Chile (Table 7.9). So far, we have identified two main distributional cores: estuary of Lluta River (18°24′S) and area between Copiapó and Magallanes (26–54°S). Several ornithologists have observed individuals of *A. flammeus* all year round for at least a decade in Lluta suggesting that the species has a permanent residence in the area (Peredo and Miranda 2001; Jaramillo 2003; Martínez and González 2005; Peredo et al. 2007). These records would represent the southern limit of the distribution range of the Peruvian population. Regarding of population that extends between Copiapó and Magallanes, there are different approaches. Whereas some authors indicate a continuous latitudinal distribution of resident individuals (de la Peña and Rumboll 1998; Martínez and González 2005; Bonacic and Ibarra 2010), others suggest the existence of breeding areas interspersed between dispersion areas (Fjeldsa and Krabbe 1990). In addition, the residence status of the southernmost fraction of this population is unclear. Although local ornithologists mention that *A. flammeus* is resident in Magallanes (Venegas and Jory 1979; Venegas 1994; Venegas and Sielfeld 1998), other authors suggest that it is only a summer visitor (Couve and Vidal 2003; Jaramillo 2003). Possibly, some individuals remain throughout the year in favourable sites while others partially migrate.

The distribution of *Strix rufipes* is restricted to the current distribution of native forest (Table 7.9). The southern limit of its distribution range is well known, but its northern limit is uncertain. According to literature, the northern latitudinal limit of *S. rufipes* is Los Vilos (31°55′S; Johnson 1965, 1967; Martínez 2005a), but its presence has not been confirmed and it could be locally extinct. It is possible that some individuals remain in nearby mountain ravines. A study conducted in the Santa Inés relict forest (32°10′S–71°30′W, 68.2 ha), 28 km south of Los Vilos, was unsuccessful in detecting the presence of *S. rufipes* (Reid et al. 2002). Martínez and González (2005) indicate that northern latitudinal limit could be near Cachagua (32°34′S–71°27′W), 75 km south of Los Vilos. One of us (S. Alvarado, personal observation) recorded the species in the Quebrada del Tigre (32°31′S, 71°24′W), 6-km northeast of Cachagua. Pavez (2004a) mentions that the species is regular from Quillota (32°50′S–71°14′W) southwards. Several other records near Quillota support Pavez (2004a): Lago Peñuelas (32°09′S–71°31′W; Brito 2005), La Campana (32°55′S–71°04′W; Elortegui and Torres-Mura 2002), and Oasis La Campana (32°54′S–71°04′W; Alvarado et al. 2007). These records suggest a stable population core that would keep the northern limit of *S. rufipes* up to at least 33°S. Jaramillo (2003) defines two main areas of resident individuals: (1) a coastal fringe between Los Vilos and San Antonio (32–34°S) and (2) an area between Metropolitan region and Cape Horn (33–56°S). Although the distribution map in Jaramillo (2003) suggests the absence of *S. rufipes* on the coast of the Maule region (34°43′–36°00′S), some investigators have documented the presence of breeding pairs in that area (35°26′S–72°17′W; Estades et al. 1998; Vukasovic et al. 2006).

7.3.5.2 Insular Distribution

Up to now *Tyto furcata* has been registered in an oceanic island in northern Chile and at least in four oceanic islands in southern Chile (Table 7.10). It is absent in the Juan Fernández Archipelago and Easter Island (Isla de Pascua), although in the latter there is a prehistoric record (Steadman 1995). *Bubo magellanicus* is present in several oceanic islands of southern and southernmost Chile (Table 7.10), but its presence in the Chiloé Archipelago is uncertain. Housse (1933) mentions that *B. magellanicus* is present in this archipelago, but he does not clarify whether its records are from islands or the mainland. In addition, the Chiloé Archipelago is part of the distributional gap indicated by Couve and Vidal (2003). *Athene cunicularia* has a resident population in the islands belonging to the Pingüino de Humboldt National Reserve, near the coastal edge of the Coquimbo region, at almost 29°S (Cruz-Jofré and Vilina 2014; Table 7.10). Although there was a resident population on the Isla Grande of Tierra del Fuego (Crawshay 1907), this seems to have become extinct during the second decade of the twentieth century (see sect. 7.3.20.1). Housse (1945) mentioned that *A. cunicularia* is present in Santa Maria Island, Gulf of Arauco, but ornithologists have not reported new records. *Glaucidium nana* is present in several oceanic islands of southern and southernmost Chile (Table 7.10). *Asio flammeus* inhabits the Juan Fernández Archipelago, Chiloé Archipelago, and

Table 7.10 Records of owls on Chilean oceanic islands

Island	Loc	Lat	Alt (m.a.s.l.)	Area (km^2)	Isol (km)	Tfur	Bmag	Acun	Gnan	Afla	Sruf	Source
Chañaral	PHR	29°01′–71°34′	0–130	5.2	6.3	–	–	R, I	–	–	–	19, 33, 34
Damas	PHR	29°13′–71°31′	0–15	0.6	5.2	–	–	R, I	–	–	–	19, 33, 34
Gaviota	PHR[a]	29°15′–71°26′	0–30	1.8	0.4	–	–	R	–	–	–	19, 33, 34
Choros	PHR	29°15′–71°32′	0–100	3	6	–	–	R	–	–	–	19, 21, 22, 33, 34
Chungungo	NLS	29°24′–71°21′	0–30	0.15	1.5	NR	–	–	–	–	–	28
Tilgo	NLS	29°32′–71°20′	0–35	0.45	0.2	–	I	–	–	–	–	33
Robinson Crusoe	JFA	33°38′–78°51′	0–915	96	667	–	–	–	–	R	–	15, 16, 25, 29, 36
Santa Clara	JFA	33°42′–78°56′	0–375	5	614	–	–	–	–	+	–	16, 25, 29, 37
Eastern (Isla de Pascua)	RNP	27°07′–109°21′	0–525	164	3526	†	–	–	–	–	–	18, 30
Santa María	ARG	37°01′–73°31′	0–70	35	24	–	–	?	–	–	–	6
Mocha	TIR	38°23′–73°52′	0–370	48	34	R	–	–	–	NR[b]	–	3, 4
Chiloé	CHA	42°28′–73°46′	0–780	9000	79	R	?	–	R	?	R	5, 6, 8, 24, 26, 27
Guafo	CHA	43°35′–74°42′	0–300	214	120	–	–	–	–	+	–	32
Las Huichas	LCA	45°09′–73°71′	0–90	3.6	4	–	–	–	–	–	+	36
Serrano	BOP	48°24′–74°41′	0–900	1060	20	–	–	–	+	–	–	36
Wellington	BOP	71°18′–74°36′	0–1100	5600	2	–	–	–	+	–	–	36
Evans	ALR	51°19′–74°01′	0–520	165	16	–	+	–	–	–	–	1
Piazzi	ALR	51°42′–74°05′	0–600	340	11	–	+	–	–	–	–	35

(continued)

Table 7.10 (continued)

Island	Loc	Lat	Alt (m.a.s.l.)	Area (km^2)	Isol (km)	Tfur	Bmag	Acun	Gnan	Afla	Sruf	Source
Newton	ALR	51°52′–73°44′	0–500	44	5	–	–	–	+	–	–	1
Riesco	ALR	53°05′–72°38′	0–1500	5039	0.5	–	+	–	–	–	–	11
Tierra del Fuego	TFA	53°50′–68°37′	0–2500	48,000	3.5	R	R	E	R	+	R	2, 7–10, 15, 17, 20, 23
Hoste	TFA	55°11′–69°00′	0–600	3500	2*	–	–	–	–	–	R	13, 14, 20, 26
Navarino	TFA	55°05′–67°36′	0–1200	2470	3*	R	R	–	R	–	R	14, 17, 26, 31
Nueva	TFA	55°13′–66°32′	0–450	111	11*	–	–	–	–	+	–	12, 14, 23, 26

Loc = locality, Lat = latitude, Alt = altitude (m.a.s.l.), Isol = degree of isolation (distance to the mainland in km); the asterisk indicates distance to the Isla to the Isla Grande de Tierra del Fuego. Tfur = *Tyto furcata*, Bmag = *Bubo magellanicus*, Acun = *Athene cunicularia*, Gnan = *Glaucidium nana*, Sruf = *Strix rufipes*. Locality: PHR = Pingüino de Humboldt National Reserve, NLS = 57 km north La Serena, JFA = Juan Fernández Archipelago, RNP = Rapa Nui National Park, ARG = Arauco Gulf, TIR = off coast of Tirúa, CHA = Chiloé Archipelago, LCA = Los Chonos Archipelago, BOP = Bernardo O'Higgins National Park, ALR = Alacalufes National Reserve, TFA = Tierra del Fuego Archipelago. Residence status: R = resident (i.e., it reproduces on the island), NR = nonresident (i.e., it is present, but is not reproduced on the island), I = winter record, + = registered, but residence status unknown,? = presence unconfirmed, E = extinct, † = prehistorical record

[a]Island located outside the park boundary. [b]Corresponds to a single record by Bullock (1935), but currently there is no evidence of their presence; possibly it is an accidental visitor. References: [1]Sharpe 1881, [2]Crawshay 1907, [3] Housse 1925, [4]Bullock 1935, [5]Wheeler 1938, [6]Housse 1945, [7]Olrog 1948, [8]Goodall et al. 1951, [9]Humphrey et al. 1970, [10]Keith 1970, [11]Texera 1973, [12]Barros 1976, [13]Sielfeld 1977, [14]Venegas and Jory 1979, [15]Araya and Millie 1986, [16]Fuentes et al. 1993, [17]Venegas 1994, [18]Steadman 1995, [19]Vilina et al. 1995, [20]Venegas and Sielfeld 1998, [21]Contreras et al. 1999, [22]Zunino and Jofré 1999, [23]Couve and Vidal 2000, [24]Jiménez 2000, [25]Hahn and Römer 2002, [26]Couve and Vidal 2003, [27]Jaramillo 2003, [28]Hertel et al. 2005, [29]Hahn et al. 2006, [30]Jaramillo et al. 2008, [31]Ippi et al. 2009, [32]Reyes et al. 2009, [33]Luna-Jorquera et al. 2012, [34]Cruz-Jofré and Vilina 2014, [35]E.S. Corales, personal observation, [36]R.A. Figueroa, personal observation, [37]D. González-Acuña, personal observation

the Isla Grande of Tierra del Fuego and adjacent islands (Table 7.10). The existence of a resident population in Juan Fernández Archipelago is well known (e.g. Hahn et al. 2006). The presence of *A. flammeus* in the Chiloé Archipelago is unclear; while some authors suggest it is present in the archipelago (Jaramillo 2003; Martínez and González 2005), others believe it is actually absent (Couve and Vidal 2003). However, a recent record in Guafo Island (Reyes et al. 2009) partly confirms its presence in this archipelago. Individuals observed on the Chilean territory of the Isla Grande of Tierra del Fuego would correspond to summer visitors (Humphrey et al. 1970; Venegas and Jory 1979; Couve and Vidal 2000; Jaramillo 2003). Barros (1976) reported an accidental record of the species in Isla Nueva. *Strix rufipes* is present on several oceanic islands in southern and southernmost Chile, being resident in many of them (Table 7.10).

7.3.5.3 Bioclimatic Distribution

We summarize the bioclimatic distribution of Chilean owls in Table 7.11. *Tyto furcata*, *Bubo magellanicus* and *Glaucidium nana* are present in all bioclimatic ecoregions of Chile. Despite its geographic discontinuity, *Athene cunicularia* is present in most of the bioclimatic ecoregions (12 out of 16). *G. peruanum* occupies all bioclimatic ecoregions of northern Chile. In northern Chile, *Asio flammeus* is present only in the coastal desert; however, in its southern distribution range, it occupies most of the bioclimatic ecoregions (11 out of 16). *Strix rufipes* occupies all bioclimatic ecoregions within its distribution range. Excepting areas with extreme weather conditions (e.g. absolute desert, icefields), bioclimatic variations do not appear to impose greater restrictions on the particular distribution of each owl species.In the particular case of *S. rufipes*, however, precipitation level (400-1500 mm) appear to be a determinant factor in the continental distribution of this species (Girini et al. 2017).

7.3.6 Geographical Diversity

Jaksic et al. (2002) analysed the geographical diversity of Chilean raptors along a latitudinal gradient including four locations: Fray Jorge (30°38′S), Aucó (31°30′S), Apoquindo (33°21′S), Rio Clarillo (33°46′S), and Torres del Paine (51°S). Although the authors found that the total species richness increased steadily towards the south, the owl species richness remained almost static (three species at 30°38′–33°46′S and four species at 51°S). Subsequently, Rau and Jaksic (2004a) standardized the raptor species diversity in function of the area of each study site and found that the richest and poorest site was Apoquindo and Fray Jorge, respectively. This trend was consistent with owl species richness. Although this analysis is helpful, the latitudinal discontinuity in sampling (no locality was studied between 34 and 50°S) and the very low detectability of *A. flammeus* because of its secretive habits (see sect. 7.3.7.1 or 7.3.11) make results inconclusive.

Table 7.11 Distribution of owls across bioclimatic ecoregions of Chile

Zone	Ecoregion	Latitude (°S)	Altitude (m.a.s.l.)	Tfur	Bmag	Acun	Gnan	Gper	Afla	Sruf
Northern	Desert									
	Coastal	18°00′–25°00′	0–1100	+	+	+	+[a]	+	+	–
	Interior	18°00′–28°00′	1100–3000	+	+	+	+[a]	+	–	–
	Tropical									
	Marginal	20°00′–25°30′	3000–4000	+	+	+	+[a]	+	–	–
	Highland	17°30′–28°00′	4000–4600	+	+	+	+[a]	+[a]	–	–
Central	Continental									
	Andean	27°30′–34°30′	1500–4000	+	+	–	+	–	–	–
	Mediterranean									
	Per-arid	25°00′–30°00′	0–4000	+	+	+	+	–	+	–
	Arid	29°00′–33°00′	0–4000	+	+	+	+	–	+	–
	Semiarid	32°30′–34°00′	0–2000	+	+	+	+	–	+	+
	Subhumid	33°30′–37°00′	0–2500	+	+	+	+	–	+	+
	North-humid	35°30′–36°45′	700–1200	+	+	+	+	–	+	+
South	South-humid	36°45′–39°00′	0–1800	+	+[b]	+	+	–	+	+
	Per-humid	37°00′–39°00′	0–1400	+	+[b]	+	+	–	+	+
	Oceanic									
	Mediterranean	37°00′–43°30′	0–1800	+	+[b]	+	+	–	+	+

Table 7.11 (continued)

Zone	Ecoregion	Latitude (°S)	Altitude (m.a.s.l.)	Tfur	Bmag	Acun	Gnan	Gper	Afla	Sruf
Southernmost	Cold temperate	43°30′–51°00′	0–1500	+	+[b]	–	+	–	+	+
	Sub-antarctic	47°30′–56°00′	0–700	+	+	–	+	–	–	+
	Tras-Andean	43°30′–54°00′	0–1000	+	+	–	+	–	+	+

Lat = ecoregion latitudinal range, Alt = ecoregion altitudinal range. Tfur = *Tyto furcata*, Bmag = *Bubo magellanicus*, Acun = *Athene cunicularia*, Gnan = *Glaucidium nana*, Gper = *Glaucidium peruanum*, Afla = *Asio flammeus*, Sruf = *Strix rufipes*. Sign "plus" indicates it is present, and sign "minus" indicates it has not been registered

Ecoregions: Desert, coastal = semidesert with chamaephytes, cacti, and annuals; interior = chamaephytes, open savannas, absolute desert. Tropical, marginal = jaral, hygrophilous shrubs; andean = in increasing altitude: tolar, scrubland, and llaretal, vegas, or marshes. Continental, andean = steppe and tundra grasses. Mediterranean, per-arid = jaral, thorny shrub; arid = coastal sclerophyllous forest and interior thorny scrubland; semiarid = shrub steppes, thorny savannah, coastal higrophylous forest; subhumid = sclerophyllous forest; north-humid = thorny savannah; south-humid = Valdivian-type woodland; per-humid = Valdivian rainforest. Oceanic, mediterranean = mainly Valdivian forest; cold temperate = Valdivian forest with conifers; sub-antarctic = marshy tundra (peatland), Magellanic forest; tras-Andean = transitional Valdivia-type forest, Magellanic forest, and Patagonian steppe. Notes: [a]Reported by some authors, but it requires verification; [b]ecoregions within distributional gap of the species. References: Sharpe 1875a, 1881, Ridgway 1876, Lane and Sclater 1897, Crawshay 1907, Scott and Sharpe 1912, Wetmore 1922, Bullock 1929, Hellmayr 1932, Wheeler 1938, Housse 1945, Olrog 1948, Goodall et al. 1951, Philippi et al. 1954, Bernath 1965, Cody 1970, Humphrey et al. 1970, Meyer de Schauensee 1970, Texera 1973, Venegas and Jory 1979, Vuilleumier 1985, Araya and Millie 1986, Jaksic and Jiménez 1986, Jiménez and Jaksic 1989, Marín et al. 1989, Fjeldsa and Krabbe 1990, Estades 1992, 1997, Martínez and Jaksic 1996, Estades et al. 1998, de la Peña and Rumboll 1998, Venegas and Sielfeld 1998, Brito 1999, Couve and Vidal 1999, 2000, 2003, König et al. 1999, Jaksic et al. 1999, 2001, 2002, 2004, López and Figueroa 1999, López and Domínguez 2002, Mella 1999, Egli and Aguirre 2000, Figueroa et al. 2001a, Díaz et al. 2002, Mella 2002, Jaramillo 2003, Pavez 2004a, Schlatter 2004, Martínez and González 2005, Brito 2005, Elgueta et al. 2006, Trejo et al. 2006, Figueroa and Alvarado 2007, Rivas and Figueroa 2009, Luna-Jorquera et al. 2012, Carevic et al. 2013, Cruz-Jofré and Vilina 2014, Mella et al. 2016

7.3.7 *Abundance*

7.3.7.1 Perceived Abundance

According to their own perception, several authors have established some abundance category for Chilean owls. Historically, *Tyto furcata*, *Bubo magellanicus*, *Athene cunicularia*, and *Glaucidium nana* have been considered as common, frequent, or abundant throughout Chile (Table 7.12). In general, these species are easy to detect both visually and aurally. Some authors mention that *B. magellanicus* tends to be more numerous towards southern latitudes (e.g Hellmayr 1932; Housse 1945; Goodall et al. 1951), but Jaksic and Jiménez (1986) listed this species as "common" between central and southern Chile. Ornithologists have always perceived *A. cunicularia* as a "common" or "abundant" owl in Chile (Table 7.12). Johnson (1967) reported that in the early twentieth century, it was possible to see hundreds of individuals perched on fence posts along the railway line that crossed the prairies of central and southern Chile. Although currently is not possible to see so large a number of individuals, *A. cunicularia* is still perceived as a common species (Egli and Aguirre 2000; Couve and Vidal 2003). However, it appears to be "scarce" in northern and southern Chile (Jaksic and Jiménez 1986). The scarce abundance of this owl species in southernmost Chile seems to be due to drastic changes in land use (see sect. 7.3.20.1).

The perception of abundance of *Asio flammeus* and *Strix rufipes* is more variable. Before 1980, *A. flammeus* was considered "moderately abundant" in central and southern Chile (Table 7.12). Conversely, Jaksic and Jiménez (1986) stated that this species is rather "scarce" in those zones. This is interpretable as a historical change in abundance, but it could also be an artefact of the species' low detectability due to their secretive habits (see sect. 7.3.11). Observations made during 3 years in agroecosystems of southern Chile made it possible to register at least 1–2 individuals daily (R.A. Figueroa, unpublished data). Thus, *A. flammeus* could be categorized rather as a "common" owl (sensu Jaksic and Jiménez 1986) in southern Chile. In addition, in the same zone it is usual to observe individuals of *A. flammeus* crossing roads or perching on fence posts along roadsides (R. A. Figueroa, personal observation). According to Schlatter (2004), *A. flammeus* seems to be "common" in peat bogs of southernmost Chile. *Strix rufipes* is perceived as "scarce" in central Chile, but "common" or "abundant" in southern Chile (Table 7.12). In southernmost Chile, the perception of its abundance tends to be inconsistent (Table 7.12). According to our observations, *S. rufipes* is similarly detectable in southern and southernmost Chile, we being able to detect aurally at least one individual over a period of 1–5 days.

7.3.7.2 Temporal Fluctuations

Long-term studies have shown that abundance of local owls in the semiarid scrub of central Chile fluctuates temporally (Jaksic et al. 1992, 1993, 1997; Arim and Jaksic 2005). Because of the difficulty of observing owls directly during the

Table 7.12 Relative abundance of Chilean owls according to the criterion of different authors

		Source																
Species	Zone	1	2	3	4	5	6	7	8	9	10	11	12[b]	13	14	15	16	17
Tyto furcata	Northern	–	–	–	–	–	–	–	–	–	A	–	E	–	–	–	–	–
	Central	–	I	–	–	–	C	–	–	–	A	–	F	–	–	–	C	–
	South	–	–	–	–	C	C	–	–	–	A	–	F	–	–	–	–	LC
	Southernmost	–	–	–	–	–	–	–	–	–	A	S/LC	S	–	NLC	LC/R	–	LC
Bubo magellanicus	Northern	–	–	–	NI[a]	–	A	–	–	–	C	–	F	–	–	–	–	–
	Central	–	LC	–	–	–	A	–	–	–	C	–	C	–	–	–	–	–
	South	–	–	–	–	–	–	–	–	–	–	–	F	–	–	–	–	S
	Southernmost	–	–	VC	–	–	A	N	–	N	F	C	C	C	C	C	–	C
Athene cunicularia	Northern	–	–	C	U	–	–	–	–	–	–	–	S	–	–	–	–	–
	Central	C	VC	C	A	–	C	–	–	–	–	–	C	–	F	–	–	–
	South	–	–	C	A	C	C	–	–	–	–	–	F	–	–	–	–	S/C
	Southernmost	–	–	C	–	–	–	–	–	–	–	–	S	–	–	–	–	C
Glaucidium nana	Northern	–	–	VC	–	–	–	–	–	S	S	–	S?	–	–	–	–	–
	Central	–	F	VC	C	–	C	–	–	C	A	–	C	–	–	–	–	–
	South	–	–	VC	A	–	C	–	C	C	A	–	F	–	–	–	–	C/A
	Southernmost	–	–	VC	–	–	C	–	C	C	A	C	C	–	–	C	–	C/A
Asio flammeus	Northern	–	–	–	–	–	–	–	–	M	–	–	–	–	–	–	–	–
	Central	–	R	NR[a]	–	–	NR[a]	F	–	M	M	–	S	–	–	–	–	–
	South	–	–	–	–	C	–	C	–	M	M	–	S	–	–	–	–	S/C
	Southernmost	–	–	–	–		NLC[a]	–		M	M	–	F	–	F	C	–	S/C

(continued)

Table 7.12 (continued)

		Source																
Species	Zone	1	2	3	4	5	6	7	8	9	10	11	12[b]	13	14	15	16	17
Strix rufipes	Northern	–	–	–	–	–	–	–	–	–	–	–	–	–	–	–	–	–
	Central	–	–	–	–	–	S	S	–	VS	R	–	R	S	–	–	–	–
	South	–	–	–	–	C	NLC[a]	A	R	A	A	–	C	S	–	–	–	R
	Southernmost	–	–	–	–	–	–	S	–	A	A	LC	C	S	R	R/S	–	R

1 = Gay 1847, 2 = Cassin 1855, 3 = Philippi 1868, 4 = Lane and Sclater 1897, 5 = Bullock 1929, 6 = Hellmayr 1932, 7 = Housse 1945, 8 = Olrog 1948, 9 = Goodall et al. 1951, 10 = Johnson 1965, 1967, 11 = Venegas and Jory 1979, 12 = Jaksic and Jiménez 1986, 13 = Fjeldsa and Krabbe 1990, 14 = Venegas 1994, 15 = Venegas and Sielfeld 1998, 16 = Egli and Aguirre 2000, 17 = Couve and Vidal 2003. Abundance categories: I = inusual, C = común, VC = very common, F = frequent, R = rare, NI = not unfrequent, U = uncommon, A = abundant, NR = not rare, NLC = not little common, S = scarce, N = numerous, M = moderately abundant or moderate number, VS = very scarce, LC = little common. All above categories were given by the authors themselves

[a]These categories should perhaps be reinterpreted as "frequent" or "common". [b]C = 1–5 individuals can be detected daily, F = 1 individual can be detected weekly, E = 1 individual can be detected monthly, R = <5 individuals can be detected annually

Table 7.13 Temporal variation in the relative abundance of owls in the semiarid scrub of central Chile estimated from the range of variation in the number of pellets collected per weather season (PPWS), per breeding season (PPBS), and per year (PPY)

	Aucó			Fray Jorge
Species	PPWS (1987–1990)[1]	PPBS (1987–2001)[2]	PPY (1987–2001)[3]	PPBS (1989–1994)[4]
Tyto furcata	0–157	0–181	1–247	12–633
Bubo magellanicus	0–245	10–433	28–833	34–271
Athene cunicularia	0–156	25–217	16–1415	32–207
Glaucidium nana	0–100	23–129	1–247	0–35

[1]Jaksic et al. 1992, [2]1993, [3]Arim and Jaksic 2005, [4]Jaksic et al. 1997

night, Chilean ecologists have estimated owl abundance based on standardized indexes of relative abundance, e.g. number of pellets collected per unit time. In two study areas, Aucó and Fray Jorge, the collected number of pellets of *Tyto furcata*, *Bubo magellanicus*, *Athene cunicularia,* and *Glaucidium nana* varied considerably among seasons, breeding periods, and years (Table 7.13). In the case of *A. cunicularia*, such variations were concurrent with temporal variations in population density (2–7 individuals/15 ha between 1990 and 1993, Silva et al. 1995) and the mean number of sighted individuals (2–6.8 between 1989 and 1994, Jaksic et al. 1997). We discussed the causes and implications of these changes in the sect. 7.3.13.

7.3.7.3 Local Abundance Relative to Habitat

Several studies have estimated the abundance of local owls in particular habitat types. In a coastal forest-scrubland mosaic of central Chile, Muñoz-Pedreros et al. (2010) estimated a natural density for *Tyto furcata* of 0.02 individuals/km^2. In rural areas of Aysén, southern Chile, the most frequently sighted species during nocturnal road surveys from a motorized vehicle were *T. furcata* and *Bubo magellanicus* (66.6% and 31% of all records [N = 48], Cruces and Cerda 1999). In grasslands of the Juan Fernández Archipelago, Hahn et al. (2006) reported that the density of *Asio flammeus* reaches 0.003 individuals/ha (8 individuals/2723 ha) and that its population size is <50 individuals. Martínez and Jaksic (1996) estimated that the mean abundance of *Strix rufipes* in secondary and old-growth Valdivian forest remnants is 0.13 and 0.22 pairs/km linear, respectively. Consistently, Ibarra et al. (2012) found that *S. rufipes*, along with *Glaucidium nana*, tend to be more abundant in old-growth forest remnants. Although these studies are spatially restricted, their results constitute benchmarks that might allow future comparisons.

7.3.8 Habitat

7.3.8.1 Habitat Types

Tyto furcata uses different habitat types including from old-growth forests to urban centres (Table 7.14). In central and southern Chile, this species is frequently observed in agricultural areas with abandoned pastures, farmland, forest remnants, scattered trees, human residences and fence lines (e.g. Simeone 1995; Figueroa et al. 2009). This species is also often sighted in suburban areas with grasslands, trees, fruit plants, water courses, and buildings (e.g. Housse 1945; Jaksic et al. 2001; González-Acuña et al. 2004).

Bubo magellanicus inhabits mountainous areas with forest/shrub patches (Table 7.14). In northern Chile, this species tends to be restricted to Andean ravines (Hellmayr 1932; Goodall et al. 1951; Traylor 1958), but it can occupy wooded plains (e.g. Pampa del Tamarugal; Torres-Mura et al. 1997). In Malleco province, southern Chile, one of us registered *B. magellanicus* in pine plantations interspersed with native forest (R.A. Figueroa, personal observation). This species has occasionally been recorded in urban parks inside Santiago city (Jaksic et al. 2001; S. Alvarado, personal observation). In southernmost continental Chile (Punta Dungeness), a pair was located inhabiting coastal stabilized dunes where the cover were calafate (*Berberis buxifolia*) and mataverde *(Lepidophyllum cupressiforme)* shrubs. At Tierra del Fuego island (Cullen area), and on a steppe environment, a pair also used foothills covered with sparse calafate shrubs, adjacent to a tuco-tuco colony (*Ctenomys magellanicus*). In both sites shrubs never grew higher than 1.5 m (D. R. Martínez, personal observation).

Athene cunicularia typically inhabits sites with low vegetation (Table 7.14). In central Chile, this species commonly inhabits sandy beaches and dunes along the coastline and open shrublands in Andean foothills (Table 7.14). In agricultural areas of southern Chile, *A. cunicularia* occupies pasturelands with low grass and a scarce presence of cattle. In urban areas of southern Chile (e.g. Temuco, Valdivia), pairs of *A. cunicularia* may occupy pastures or marshes that have been isolated by urban sprawl (Fernández et al. 1980; R.A. Figueroa, personal observation). The terrains occupied by this owl can be either flat, slightly sloping, or sinuous (Table 7.14).

Glaucidium nana uses a wide variety of habitats. It is common in areas with trees (e.g. forests, parks, and urban parks; Table 7.14). In southern Chile, *G. nana* occupies mesophytic forests, montane forests, parklands, forest openings, forest/steppe ecotones, and shrublands (Vuilleumier 1985). In semi-desert areas, this owl species is commonly found in wooded ravines covered by clumps of *Maytenus boaria* (5 m in tall) and *Schinus polygamus* (<5 m) on south-facing slopes and *Acacia caven* on north-facing slopes (Jiménez and Jaksic 1989). In southern Chile, *G. nana* inhabits all forest type. *Glaucidium peruanum* inhabits oasis, savannahs, scrublands, and human-modified environments (Table 7.14). Moreover, it seems to be frequent in agricultural crops with trees, orchards, fruit trees, and city parks.

Table 7.14 Description of the habitat types used by Chilean owls

Habitat type	Condition	Orographic zone	Tfur	Bmag	Acun	Gnan	Gper	Afla	Sruf
Natural, open									
Sand and dune	sv, lv, tr	co, cv	–	–	+++	–	–	–	–
Scree and rockery[a]	sv, lv, ±tr	co, cv, an, hp	+	++	+	+	+	–	–
Oasis	sv, tr	cv	+	–	+	–	+	–	–
Pasturelands/grasslands[b]	dv, sv, tv, lv	co, cv, an, hp	+	–	++	+	–	+++	–
Humid prairies	dv, sv, tv, lv, tr	co, cv, an	+	–	++	+	–	++	–
Reedbed/cattails	dv, sv, tv, lv	co, cv	+	–	+	+	–	++	–
Peatlands	dv, lv	pp	–	–	–	+	–	+	–
Savannah[c]	dv, sv, tv, lv, tr	cv, an	++	+	+	++	+	+	–
Parklands	sv, tv, tr	cv, pp	++	+	+	+		+	–
Natural, closed									
Shrubland[d]	or, rg	co, cv, an, pp	++	++	++	++	+	+	–
Forest[e]	og	co, cv, an, pp	+	+	–	+++	–	–	+++
	sg	co, cv, an, pp	++	++	–	+++	–	–	+++
Forest sapling	dv, tv	co, cv, an, pp	+	+	–	+	–	–	+
Ecotone									
Forest-prairies		co, cv, an, pp	++	++	+	++	–	+	+
Forest-shrubland		co, cv, an, pp	++	++	+	++	–	+	+
Anthropogenic									
Lawns[f]	dv, lv, ±tr	cv	–	–	+	–	–	+	–
Grazing pastures[g]	dv, sv, lv, ±tr	co, cv, an, pp	++	–	++	+	–	+	–
Abandoned pastures[h]	dv, tv, ±tr	co, cv, an, pp	+++	–	++	+	–	+++	–
Brambles[i]	dv, tv	co, cv, an, pp	++	–	+	+	–	+++	–
Croplands[j]	dv, sv, tv, lv, ±tr	co, cv	++	–	+	++	++	++	–
Orchards[k]	dv, lv, tv	co, cv, an, pp	+	–	–	+	++	–	–

(continued)

Table 7.14 (continued)

Habitat type	Condition	Orographic zone	Tfur	Bmag	Acun	Gnan	Gper	Afla	Sruf
Vineyards	dv, tv	cv	+	–	–	+	–	–	–
Logged forests[l]	sv, tv	co, cv, an, pp	+	+	–	+	–	+	–
Recreational forests[m]	dv, tv, og, sg	co, cv, an, pp	+	+	–	+	–	–	+
Forestry plantations[n]	mp, dv, tv, nv	co, cv, an, pp	++	+	–	++	–	–	+
	yo, sv, ±nv	co, cv, an, pp	+	–	–	+	–	+	–
	ha, ±nv	co, cv, an, pp	+	–	–	+	–	+	–
Hamlets[ñ]	ol	cv, pp							
House gardens	sv, tv, nv, ev		+	–	–	+	+	–	–
Botanical gardens	sv, lv, tv, nv, ev		++	–	+	+	–	–	+
Urban parks[o]	sv, tv		++	–	+	++	++	–	–
Industrial buildings	±pp., ev, ac, ab		+	–	–	–	–	–	–
Buildings[p]	nv, ev, ac, ab		++	–	–	–	–	–	–

Tfur = *Tyto furcata*, Bmag = *Bubo magellanicus*, Acun = *Athene cunicularia*, Gnan = *Glaucidium nana*, Gper = *G. peruanum*, Afla = *Asio flammeus*, Sruf = *Strix rufipes*. Sign "plus" indicates the relative degree of use: + = occupied, but the use degree is unknown; ++ = frequently occupied; +++ = apparently preferred

Habitat types: [a]For example, stone/rock piles, ledges and rocky slopes, cliffs, and ravines. [b]For example, grasslands and pastures in Andean highlands, steppes, and Puna. [c]For example, tamarugal savannah (northern Chile), open thornscrub (central Chile). [d]For example, coastal shrubland, sclerophyllous shrubland, steppe-shrubland, and forest-originated shrubland. [e]For example, sclerophyllous forest, pure or mixed deciduous forest, evergreen forest, and coniferous forest. [f]Usually sports fields. [g]With livestock in low amount. [h]For example, extensive pastures along fences and road, inside of agricultural fields and airports. [i]For example, patches of blackberry (*Ribes ulmifolius*), rosehip (*Rosa moschata*), European buckthorn (*Ulex europaeus*), and gorse (*Diostea juncea*). [j]For example, cereal (wheat, oats, corn) and vegetable croplands. [k]Household and/or commercial use. [l]With grazing, mining, or firewood burning. [m]Protected forest remnants for educational and/or recreational use. [n]For example, commercial forestry plantations (e.g., pine, eucalyptus). [ñ]For example, farmhouses, sheds, barns, and farm machinery. [o]For example, tree-lined avenues and squares. [p]For example, residential neighborhoods with towers, chimneys, and dormers. Condition: dv = dense vegetation, sv = sparse vegetation, tv = tall vegetation, lv = low vegetation, tr = with trees, ±tr = with or without trees, or = original scrub little perturbed, rg = regenerated scrub, og = old growth (>200 years), sg = secondary growth (≤200 years), nv = with native vegetation, ±nv = with or without native vegetation, ev = non-native vegetation, mp = mature plantation (>20 years), yo = young (<20 years) or growing (<5 years) plantations, ha = harvested plantation, ±pp. =

with or without pollution, ac = active, ab = abandoned, ol = old. Orographic zone: co = coastal mountain range, cv = central valley, an = Andes mountain range, hp. = high plateau (altiplano), pp. = Patagonian plains. References: Molina 1782, Gay 1847, Cassin 1855, Germain 1860, Lataste (1895b), Lane and Sclater 1897, Crawshay 1907, Scott and Sharpe 1912, Bullock 1929, Hellmayr 1932, Barros 1945, 1950, 1963, Housse 1945, Olrog 1948, Goodall et al. 1951, Philippi et al. 1954, Bernath 1965, Johnson 1967, Cody 1970, Humphrey et al. 1970, Keith 1970, Meyer de Schauensee 1970, Texera 1973, Reise and Venegas 1974, Solar and Hoffmann 1975, Fulk 1976, Schlatter 1976, 1979, 2004, Péfaur et al. 1977, 1979, Jaksic et al. 1978, 1981, 1982, 1990, 1996, 1997, 1999, 2001, 2002, Venegas and Jory 1979, Núñez and Yáñez 1982, Schlatter et al. 1982, Rau et al. 1985, 1992, Vuilleumier 1985, 1998, Araya and Millie 1986, Jaksic and Jiménez 1986, Meserve et al. 1987, 1993, Jiménez and Jaksic 1989, Torres-Mura and Contreras 1989, Fjeldsa and Krabbe 1990, Iriarte et al. 1990, Muñoz-Pedreros and Murúa 1990, Ebensperger et al. 1991, Jaksic and Feinsinger 1991, Estades 1992, 1995, 1997, González 1993, Martínez 1993, 2005a, Ortiz et al. 1994, Torres-Contreras et al. 1994, Venegas 1994, Simeone 1995, Tala et al. 1995, Figueroa 1996, 1997, Martínez and Jaksic 1996, 1997, Rozzi et al. 1996, Torres-Mura et al. 1997, de la Peña and Rumboll 1998, Estades et al. 1998, Jorge et al. 1998, Martínez et al. 1998, Venegas and Sielfeld 1998, Couve and Vidal 1999, 2000, 2003, Díaz 1999, Gantz and Rau 1999, König et al. 1996, Marks et al. 1999, Pavez and Gabella 1999, Egli and Aguirre 2000, Figueroa et al. 2000, 2001a, b, c, 2004, 2006, 2009, Hoffmann and Lazo 2000, Jiménez 2000, Figueroa and Quintana 2001, Jaksic 2001, López and Domínguez 2002, Jaramillo 2003, González-Acuña et al. 2004, Ippi and Rozzi 2004, Lazo 2004, Pavez 2004a, Begall 2005, Correa and Roa 2005, Hertel et al. 2005, Martínez and González 2005, Mella 2005, Carmona and Rivadeneira 2006, Elgueta et al. 2006, Hahn et al. 2006, Vukasovic et al. 2006, Alvarado et al. 2007, Figueroa and Alvarado 2007, Hinojosa-Sáez et al. 2007, Rivas and Figueroa 2009, Bonacic and Ibarra 2010, Carevic 2011, Celis-Diez et al. 2011, Altamirano et al. 2012, Ibarra et al. 2012, Carevic et al. 2013, Cruz-Jofré and Vilina 2014, Ibarra et al. 2015, Rivas-Fuenzalida et al. 2015a, Mella et al. 2016, Muñoz-Pedreros et al. 2016a, authors, personal observation

Fig. 7.8 Adult *Strix rufipes* observed in the San Martin Experimental Forest, Valdivia, southern Chile (12 February 2013). This individual was observed at a site with a high concentration of monito del monte (*Dromiciops gliroides*), an endemic marsupial of the southern temperate forest that constitutes one of its main prey on the area (Photographs Ricardo A. Figueroa)

Asio flammeus occupies open grasslands, cattails, reed beds, wet meadows, open low shrublands, agricultural crops (e.g. wheat, oats), and young forest plantations (Table 7.14, Fig. 7.6). Eventually, some individuals explore forest edges and the interior of cleared or burned forests (Hahn et al. 2006; Figueroa and Alvarado 2007; R.A. Figueroa, personal observation). In the Juan Fernández Archipelago, *A. flammeus* inhabits extensive pastures on rocky, erosive, and slightly inclined terrains (Hahn et al. 2006).

Strix rufipes occupies diverse forest associations and successional stages of native forests (Jaksic and Feinsinger 1991; Martínez 2005a; Table 7.14). In southern Chile, this species inhabits temperate rainforests (Fig. 7.8), mesophytic forests, woodlands, forest openings, and forest/steppe ecotones (Vuilleumier 1985). In Valdivia city, southern Chile, *S. rufipes* is present in several small forest remnants inside the urban setting and in a highly degraded forest remnant in the peri-urban edge (R. P. Schlatter, personal observation; J. Ruiz, personal communication; R. A. Figueroa, personal observation).

7.3.8.2 Habitat Use

Although Chilean owls occupy different habitat types, they tend to use some more than others. González (2007) documented that in a rural area of southern Chile, the habitats most used by *Tyto furcata* were pasturelands, followed by shrublands and forests. Possibly, it is because pasturelands, especially those that are abandoned, concentrate a high abundance of rodents and contain old trees or buildings that provide shelter and nesting substrates (e.g. Jaksic and Jiménez 1986; R.A. Figueroa, unpublished data). Housse (1945) states that this species uses sites that assure three essential conditions: darkness during day, low disturbance, and high prey abundance. This is consistent with findings of Massa et al. (2015) who studied the behavioural patterns of an individual of *Tyto furcata* in Argentina using GPS technology.

In island hills of Metropolitan region, central Chile, *Athene cunicularia* appears to use more sites located in foothills (500–600 m a.s.l.) with a low slope (mean = 35%), bordering agricultural crops, and covered extensively by herbaceous vegetation (soil cover >80%; Roa 2011). Recently, Villagrán (2016) characterized the breeding habitat of *A. cunicularia* in suburban areas of Valdivia city at three different spatial scales: nest, breeding territory, and landscape. All nests were located on sandy ground and very close to a perch (range = 0.1–1.4 m). Breeding territories were characterized mainly by a high number of perches (mean = 57, range = 42–78) and a relatively herbaceous cover (mean = 66%, range = 28–90%). At landscape scale, no variable appeared to explain the presence of this owl species. Nests in coastal sandy areas may concentrate on sites with sparse shrubs of *Acacia caven*, with plenty of grass and relatively far from the beach (200–300 m, Pavez and Gabella 1999).

Across an altitude/disturbance gradient of temperate forest in an Andean area of southern Chile, Ibarra et al. (2012, 2015), based on vocal response, found that *Glaucidium nana* was more frequent in stands of old-growth evergreen forest (800–950 m a.s.l., 52.4% of vocal responses), less frequent in stands of old-growth forest of *Araucaria-Nothofagus* (1200–1300 m) and secondary forest (800–900 m a.s.l., 19.1% of vocal responses, respectively), and much less frequent in stands of forest disturbed by cattle and logging (600–800 m a.s.l.; 14.3% of vocal responses). According to Ibarra et al. (2012), the habitat variables that best explained the presence of *G. nana* were the presence of tall trees (height > 20 m, diameter at the breast height [dbh] > 50 cm) and high density of trees (170–190 individuals/ha). Such forest characteristics would provide greater availability of prey and shelter for a greater number of individuals of *G. nana* compared to younger forest remnants structurally less complex and with a higher degree of human disturbance.

In agricultural lands of southern Chile, *Asio flammeus* hunts and reproduces mainly on patches of abandoned pasture. In Osorno, southern Chile, individuals of *A. flammeus* do hunt with greater intensity on patches of abandoned pastures, which together constitute between 53% and 75% of the vegetation cover of their hunting ranges (Martínez et al. 1998). Within these pastures, there are small meadows and blackberry clumps used alternatively as roosting and shelter sites. Hunting areas also include fence lines, orchards, and villages. Fence lines are important elements of the habitat of *A. flammeus* as the fence posts are used as perches for hunting, resting, and preening (R. A. Figueroa, personal observation). In Traiguén, southern Chile, the breeding pairs establish their nests in very dense and extensive pastures near vegetable and cereal crops and with a wide network of fences (R.A. Figueroa and E.S. Corales, unpublished data). As was mentioned previously, abandoned pastures constitute sites with high concentration of prey, thus promoting their occupation by different species of owls.

A number of studies indicate that *Strix rufipes* tends to use more old-growth forest remnants. In Valdivian forests, Martínez and Jaksic (1996) found that *S. rufipes* prefers stands of secondary and old-growth forest, but independent of the successional stage, owls nest and roost in multilayered stands >100 years old, canopy cover >70%, >5 snags/ha (dbh ≥ 20 cm), and with evident signs of decay (woody debris and emergent trees with broken limbs and/or rotten core). Although hunting sites are structurally more variable, these also present a relatively closed canopy and a complex vegetation structure (Martínez 2005a; Rivas-Fuenzalida et al. 2015a). In an extensive remnant of the *Araucaria-Nothofagus* forest in the Tolhuaca National Park, sites occupied by *S. rufipes* contain large-sized trees (height = 20–30 m, dbh = 20–60 cm), closed canopy (coverage = 50–80%), large fallen logs on the ground, deep leaf litter, and dense understory composed of dense carpets of southern bamboo (*Chusquea* spp.), shrub clumps, and saplings (Figueroa et al. 2006). Somewhat more towards south, Ibarra et al. (2012) found that *S. rufipes* was more frequent in stands of old-growth *Araucaria-Nothofagus* forest (32.4% of vocal responses), less frequent in stands of secondary and old-growth evergreen forest (29.4% and 26.5% of vocal responses, respectively), and much less frequent in stands of multispecies forest disrupted by livestock and logging (11.8% of vocal response). Habitat variables that best explained the presence of *S. rufipes* in these forest stands were the presence of tall trees (height > 20 m, dbh > 50 cm), a relatively low density of trees, and high density of southern bamboo. In central Chile, the remnants of sclerophyllous forest occupied by *S. rufipes* tend to be more open, but they also contain features of old-growth forest, e.g. tall and old trees (height = 10 to 25 m, dbh = 0.1–1 m) with some of them being >200 years old, and canopy cover between 50% and 70% (Díaz 1999, Alvarado et al. 2007). *S. rufipes* also can use interior forest openings and forest edges as foraging sites (Díaz 1999).

Despite its strong association with extensive old-growth forest, *Strix rufipes* may also occupy forest remnants intermixed with or surrounded by pine plantations. Rivas-Fuenzalida et al. (2015a) surveyed the presence of this owl species across a mosaic of native forest/pine plantations in Nahuelbuta mountain range, southern Chile. The numbers of sites with the presence of *S. rufipes* were distributed almost equally between forests and plantations (20 and 18 sites, respectively). The sites in forest greatly varied in age (30–1000 years) and vegetation structure. Twelve sites were in remnants of mixed deciduous forest, seven in remnants of evergreen forest, and one in a remnant of old-growth *Araucaria araucana* forest. The remnant sizes were also quite variable (13 to >1000 ha). Although sites recorded in pine plantations had a homogeneous tree layer, many of them had native vegetation under the canopy or were close to streams bordered with strips of native trees and shrubs. The age and size of plantations was also variable (15–60 years old, 1–250 ha). The breeding sites of *S. rufipes* found across of forest/pine plantation mosaics show a variable structural complexity. Two nesting sites inside pine plantations had trees 15–25 years old, a canopy height of 11–16 m, a canopy cover of 55–63%, and a sparse understory composed of native plants and were located ≥100 m from the nearest forest stand (Estades et al. 1998). The other nesting site in a secondary forest patch of *Nothofagus* had presence of snags of native trees and was surrounded by an extensive commercial pine plantation (Vukasovic et al. 2006). Rivas-Fuenzalida et al. (2015a) recorded an adult individual of *S. rufipes* with a fledgling young in a small mature pine plantation (3.1 ha, 37 years old). In all the above cases, the structural habitat characteristics would ensure nesting substrates, prey abundance, and fresh and shady places during summer for *S. rufipes* (Martínez 2005a). Recently, however, one of us (R.A. Figueroa, personal observation) registered this owl species in highly degraded forest fragments across agricultural lands in southern Chile. Overall, evidence suggests that this owl species has some degree of plasticity regarding the use of forest habitats.

Summarizing, *T. furcata*, *B. magellanicus*, *G. nana*, and *G. peruanum* can be considered habitat generalists. However, the first three species are facultative users of forests. In contrast, *A. cunicularia*, *A. flammeus*, and *S. rufipes* are more specialized in their habitat requirements; the first two species use much more open habitats with grassland/shrubland, and the latter is strongly associated with mature forests.

7.3.9 Reproduction

Breeding aspects of Chilean owls have been documented by naturalists since the middle of the nineteenth century. However, many references contain only occasional observations. So far, the most detailed and original information has been provided by Housse (1945), Barros (1950, 1963) and Goodall et al. (1951). More recent observations have partially complemented the scarce information available (e.g. Estades et al. 1998; Vukasovic et al. 2006; Roa 2011; Ibarra et al. 2015, 2017).

7.3.9.1 Reproductive Phenology

In some regions of Chile, *Tyto furcata* breeds for much of the year (Table 7.15). In central Chile, individuals mate and prepare their nest during autumn, and pairs can raise their chicks until winter (Lataste 1895b; Barros 1963). According to Lataste (1895a), *T. furcata* could have up to three broods in the year. In southern Chile, incubation starts in spring (e.g. Bullock 1929; Barros 1963) usually lasting 30–33 days (Hoffmann and Lazo 2000; Pavez 2004a). Females lay their eggs at 1–2 days intervals, and the chicks hatch in staggered days (Lataste 1895b; Barros 1963). Housse (1945) states that in southern Chile, *T. furcata* delays 1 month the beginning

Table 7.15 Breeding phenology of Chilean owls

Species	Courtship	Nest preparation	Incubation	Chick rearing
Tyto furcata	Central Chile: autumn II (Apr), winter III (Sep), spring I (Sep)	Central Chile: autumn II (Apr)	Central Chile: spring I–III (Sep–Dec), summer I–III (Dec–Jan), autumn II (Apr). Southern Chile: spring II (Nov), summer II (Jan)	Central Chile: spring II–III (Oct–Dec), summer I (Jan), autumn II–III (Apr–Jun), winter (Jul)
Bubo magellanicus	Central Chile: spring I (Oct)	Central Chile: spring I (Sep)	Central Chile: spring I–II (Sep). Southern Chile: spring II (Nov). Southernmost Chile: spring II (Oct). Unspecified zone: winter III (Aug), spring (Sep)	Central Chile: spring II (Nov). Southern Chile: spring III (Dec)
Athene cunicularia			Central Chile: spring I–III (Sep–Dec). Southern Chile: spring II (Oct), summer I (Jan). Southernmost Chile: spring III (Nov)	Central Chile: spring II (Oct–Nov). Southern Chile: summer II (Feb)
Glaucidium nana	Central Chile: winter II (Jul). Southern Chile: winter III (Aug–Sep), spring II–III (Oct–Dec). Unspecified zone: winter III (Aug–Sep), spring II (Oct)	Unspecified zone: spring II (?)	Central Chile: spring I (Sep). Southern Chile: spring III (Dec), summer I–II (Dec–Jan)	Central Chile: spring II (Nov). Southern Chile: summer II (Jan–Feb)
Asio flammeus	Central Chile: spring I (Oct)	Central Chile: spring II (Oct)	Central Chile: spring II–III (Oct–Dec), summer I (Jan). Southern Chile: spring II (Oct), summer II (Jan–Feb), autumn I (Apr)	Southern Chile: summer II–III (Feb–Mar), autumn I (Mar–Apr)

(continued)

Table 7.15 (continued)

Species	Courtship	Nest preparation	Incubation	Chick rearing
Strix rufipes	Southernmost Chile: winter III (Aug), spring I (Sept–Oct)	Central Chile: spring I (Sep)	Central Chile: spring I–II (Sep–Nov). Southern Chile: spring III (Nov–Dec). Unspecified zone: autumn II (May)	Central Chile: spring II (Nov). Southern Chile: spring III–summer I (Dec–Jan), summer II (Feb), autumn I (Mar–Apr). Unspecified zone: autumn II (May)

The table contains only information that was available to the authors until the ending date of this review. Roman numerals indicate intra-seasonal stages: I = early season, II = mid-season, III = late season. Months are abbreviated into parentheses

References: *Tyto furcata* = 1–5, 6, 8, 9, 12, 13; *Bubo magellanicus* = 9, 12, 14, 18, 19, 22; *Athene cunicularia* = 8, 9, 10, 12, 20; *Glaucidium nana* = 2, 7, 9, 11, 12, 14, 15, 18, 19, 25, 26, 29; *Asio flammeus* = 4, 9, 12, 14, 16, 21, 24, 29; *Strix rufipes* = 9, 15, 17, 19, 23, 27, 28, 29. [1]Gay 1847, [2]Germain 1860, [3]Lataste 1895a, [4]1895b, [5]Raspail 1895, [6]D'Hamonville 1896, [7]Lane and Sclater 1897, [8]Bullock 1929, [9]Housse 1945, [10]Olrog 1948, [11]Barros 1950, [12]Goodall et al. 1951, [13]Barros 1963, [14]Johnson 1967, [15]Fjeldsa and Krabbe 1990, [16]González 1993, [17]Estades et al. 1998, [18]Couve and Vidal 1999, [19]König et al. 1999, [20]Egli and Aguirre 2000, [21]Figueroa et al. 2001a, [22]Pavez 2004a, [23]Vukasovic et al. 2006, [24]T. Rivas-Fuenzalida, personal communication, [25]Norambuena and Muñoz-Pedreros 2012, [26]Ibarra et al. 2014a, [27]Rivas-Fuenzalida et al. 2015a, [28]Ibarra et al. 2017, [29]authors, unpublished data

of the reproductive period. *Bubo magellanicus* appears to restrict its breeding period to spring and summer (Table 7.15). In Chile, the incubation time is not well delimited: 21 vs 35 days (Couve and Vidal 1999; Pavez 2004a). *Athene cunicularia* restricts its reproductive activity to spring and summer. Incubation starts from first egg and hatching is staggered (Goodall et al. 1951). In Chile, the incubation time is unclear: 18 vs 28 days (Housse 1945; Pavez 2004a). Following records of incubation dates, the reproductive period in southern Chile starts 1 month later than in central Chile (Table 7.15). *Glaucidium nana* breeds between late winter and spring in central Chile, between late winter and summer in southern Chile, and between spring and summer in southernmost Chile (Table 7.15). This owl species appear to have only one brood a year (Housse 1945; Barros 1950), but the incubation time is unclear in the literature: 15–17 vs 26–28 days (Housse 1945; König et al. 1999; Pavez 2004a). However, recently, Ibarra et al. (2014a) reported an incubation period of 15–17 days for a nest found in a stand of the Andean temperate forest in southern Chile. Findings of Ibarra et al. (2014a) confirm Housse (1945). In central Chile, *Asio flammeus* breeds between early spring and early summer and in southern Chile between spring and autumn (Table 7.15). It appears to have only one brood a year incubating for almost 3 weeks or 26 days (Housse 1945; Pavez 2004a). *Strix rufipes* reproduces during spring in central Chile and between summer and autumn in southern Chile (Table 7.15). Incubation time is unknown, but Ibarra et al. (2017) suggest an incubation period of at least 31 days.

7.3.9.2 Nest

Tyto furcata occupies a wide variety of nesting substrates, but the most used are elevated and well-protected cavities (Table 7.16). Occasionally, it may nest directly on substrates at the ground level (Housse 1945). Females lay their eggs directly on the ground and surround them with prey remains (Lataste 1895a; Housse 1945; Barros 1963; Rivas and Figueroa 2009).

Bubo magellanicus nests in natural cavities, crevices and high platforms (Couve and Vidal 2003; Rivas and Figueroa 2009) including abandoned nests of Falconiformes (Table 7.16). Barros (1945) documented a nest of *B. magellanicus* on a nesting platform of *Caracara plancus*, and Goodall et al. (1951) recorded a pair incubating on an abandoned nest of *Parabuteo unicinctus* on an *Acacia caven* tree. Housse (1945) found a nest on a woody platform (50-cm diameter, 10-cm high) composed of thick and dry sticks and placed on a tree at almost 3 m above ground. On cliffs, *B. magellanicus* can anchor its nests to plants or overhanging roots. This owl species also nests in shallow depressions on the ground in well-protected sites (Table 7.16). Housse (1945) described a nest made with aquatic plants on a small island in a lagoon covered with rushes.

Athene cunicularia typically nests in burrows below ground (Rivas and Figueroa 2009). Owl pairs may excavate a burrow themselves or readapt burrows of other animals (e.g. Gay 1847; Housse 1945; Barros 1963; Table 7.16). Occasionally, they nest in burrows under stumps or in the interior of hollow fallen trunks (Fernández et al. 1980, R.A. Figueroa, personal observation). In general, nest burrows consist of a single chamber running at ≥50 cm below ground (Housse 1945; Barros 1963). Chambers may reach 1.5–6 m in length (Housse 1945) and 16 ± 6.0 × 17 ± 4.0 cm in diameter (mean ± SD; Roa 2011) tending to be wider towards the distal extreme, just where the nest is placed. Depending on terrain, chambers can be zigzagging (Housse 1945). Pairs line the chamber's bottom with grass, dung and feathers. Around the burrow's entrance, there often are accumulations of excavated material and pellets. According to Housse (1945), *A. cunicularia* prefers to nest in inclined sites to avoid rain from entering the incubation chamber. Pairs can repair and reuse the same chamber over several years.

Glaucidium nana nests mainly in elevated tree cavities (Table 7.16), both natural and excavated by woodpeckers (e.g. *Colaptes pitius*, *Picoides lignarius*), at >1 m above ground (Goodall et al. 1957; Jiménez and Jaksic 1989; König et al. 1999; Ibarra et al. 2014a, 2015). It also accepts nest boxes placed between 3 and 5 m above ground (authors, unpublished data). Nest holes or crevices on cliffs can reach 30–50 cm in depth (T. Peddar, in Goodall et al. 1951). *G. nana* can occasionally nest below ground occupying holes above water level in talus of irrigation channels and cavities made by *Cinclodes patagonicus* on riverbanks (Housse 1945; Barros 1950). In remnants of Andean temperate forest, two nesting cavities were located in large

Table 7.16 Nesting substrates of Chilean owls

Substrate	Tfur	Bmag	Acun	Gnan	Gper	Afla	Sruf
Natural							
Elevated							
Tree holes	+	+	–	+	+	–	+
Cactus holes	–	–	–	–	+	–	–
Rocky crevices	+	+	–	+	–	–	–
Hollow stumps	–	–	–	–	–	–	+
Tree crown	–	+	–	–	–	–	–
Ravine cracks	+	+	–	+	+	–	–
On/among rocks	–	+	–	–	–	–	–
Abandoned hawk nests[a]	–	+	–	–	–	–	+
On ground							
Natural cavities[b]	–	–	–	+	–	–	–
Built cavities	–	–	+	–	–	–	–
Animal caves[c]	–	–	+	+	–	–	–
Among pastures/grassland	–	–	–	–	–	+	–
Among reedbed/cattail	–	–	–	–	–	+	–
Among shrubs	–	–	–	–	–	+	–
Fallen hollow logs	–	–	+	–	–	–	–
Low hollow stumps[d]	–	–	–	–	–	+	–
Forest ground[e]	–	+	–	–	–	–	¿
Anthropogenic							
Elevated							
Attics	+	–	–	+	–	–	–
Abandoned houses/buildings[f]	+	–	–	–	–	–	–
Bell towers	+	–	–	–	–	–	–
Wall crevices	+	–	–	+	+	–	–
Storehouses and warehouses	+	–	–	+	–	–	–
Unused chimneys	+	–	–	–	–	–	–
Grape hamper	+	–	–	–	–	–	–
Telephone poles	–	–	–	–	+	–	–
Nest boxes	+	–	–	+	–	–	–
On ground							
Ground in pine plantations[g]	–	–	–	–	–	–	+
Inside abandoned pastures	–	–	–	–	–	+	–
Inside croplands[h]	–	–	–	–	–	+	–
Among brambles[i]	–	–	–	–	–	+	–
Haystacks	+	–	–	–	–	–	–
Mine galleries	+	–	–	–	–	–	–
Abandoned wells	+	–	–	–	–	–	–

Tfur = *Tyto furcata*, Bmag = *Bubo magellanicus*, Acun = *Athene cunicularia*, Gnan = *Glaucidium nana*, Gper = *G. peruanum*, Afla = *Asio flammeus*, Sruf = *Strix rufipes*. The sign "plus" indicates it is used

(continued)

Table 7.16 (continued)

Substrates: [a]See text. [b]Cavities under tree roots or among rocks. [c]Caves and burrows built by foxes, lesser grisons (*Galictis cuja*), skunks (*Conepatus* spp.), rabbits, rodents (e.g., *Octodon degu*), and birds (e.g., *Cinclodes* spp., rinocríptidos; Des Murs 1847, Housse 1945, Mann 1978). [d]See text. [e]Depressions at the base of tree trunks or under broken branches and fallen trees (König et al. 1999). [f]Any protected room available. [g]On needle litter, under bushes, or among grass (Estades et al. 1998). [h]Crops of alfalfa, wheat, and oats. [i]European buckthorn and blackberry. *Tyto furcata* = 1–6, 9–12, 15–19, 25, 27, 29, 32–36, 39–43, 46, 47, 53; *Bubo magellanicus* = 7, 11, 12, 15, 18, 19, 24, 29–31, 34–36, 40, 41, 47; *Athene cunicularia* = 1, 2, 7, 9, 12, 13, 15, 17–22, 32, 34, 35, 38, 40–42; *Glaucidium nana* = 1, 3, 9, 11, 12, 14–16, 18–20, 23, 24, 27, 29–35, 40, 41, 47–49, 52; *Glaucidium peruanum* = 31, 32; *Asio flammeus* = 5, 8–10, 12, 15, 18, 19, 26, 27, 34, 35, 37, 38, 40, 41, 45, 53; *Strix rufipes* = 9, 12, 15, 24, 28, 30, 31, 34, 40–42, 44, 45, 47, 48, 50, 51. [1]Gay 1847, [2]Cassin 1855, [3]Germain 1860, [4]Lataste 1895a, [5]1895b, [6]Raspail 1895, [7]Lane and Sclater 1897, [8]Peters 1923, [9]Bullock 1929, [10]1935, [11]Barros 1945, [12]Housse 1945, [13]Olrog 1948, [14]Barros 1950, [15]Goodall et al. 1951, [16]1957, [17]Barros1963, [18]Johnson 1965, [19]1967, [20]Solar and Hoffmann 1975, [21]Mann 1978, [22]Schlatter et al. 1982, [23]Jiménez and Jaksic 1989, [24]Fjeldsa and Krabbe 1990, [25]Schlatter and Murúa 1992b, [26]González 1993, [27]Venegas 1994, [28]Estades et al. 1998, [29]Couve and Vidal 1999, [30]König et al. 1999, [31]Marks et al. 1999, [32]Egli and Aguirre 2000, [33]Hoffmann and Lazo 2000, [34]Figueroa et al. 2001a, [35]Jaksic et al. 2001, [36]2002, [37]Couve and Vidal 2003, [38]Jaramillo 2003, [39]Muñoz-Pedreros 2004, [40]Pavez 2004a, [41]Trejo 2004, [42]Martínez and González 2005, [43]Muñoz-Pedreros and Gil 2005, [44]Vukasovic et al. 2006, [45]Figueroa and Alvarado 2007, [46]Muñoz-Pedreros et al. 2010, [47]Celis-Diez et al. 2011, [48]Altamirano et al. 2012, [49]Ibarra et al. 2014a, [50]Rivas-Fuenzalida et al. 2015a, [51]Ibarra et al. 2017, [52]Figueroa et al. unpublished data, [53]authors, unpublished data

Nothofagus trees at 7–8 m tall. Cavities had a round-shaped entrance and were variable in diameter (6.0 × 6.8 cm and 9.5 × 12 cm). One cavity originated naturally in a standing dead southern beech (*Nothofagus dombeyi*), and the other was excavated by *C. pitius* in a live southern beech (*N. obliqua*; Ibarra et al. 2014a, 2015). A pair was observed nesting in a cavity at almost 20-m height in a large *Nothofagus* tree (>25 m tall) inside Valdivia city (J. Ruiz, personal communication). Pairs lightly line their nests with mosses, dry grass, leaf litter, dry sticks, and/or woody debris (Housse 1945; Barros 1950; R.A. Figueroa, personal observation). Some pairs can reuse a nest for several years (Housse 1945).

Asio flammeus lay its eggs in simple depressions on the ground in dense vegetation (Table 7.16). Usually, the pairs line in the bottom of the nest with stems of grass or other plants (Housse 1945). In agricultural lands, *A. flammeus* nests in abandoned pastures with tall and dense vegetation (Figueroa et al. 2001a). Two nests found in these pastures were similar in shape and structure. Each consisted of a "grot" of almost 30 cm diameter with a round bed composed of flattened stems of grass, stems of grass around the nest interlaced at their tips forming a roof over the nest (R.A. Figueroa and E.S. Corales, unpublished data). Such characteristics would ensure shade during the day, heat maintenance during the night, and protection against rain and aerial predators. The other nest consisted of a bed of grass on the ground under two bushes of blackberry (*Rubus constrictus*; 62-cm tall) which also served as a roof (R.A. Figueroa and E.S. Corales, unpublished data). Some pairs nest in the interior of hollow stumps (Housse 1945; R.A. Figueroa and E.S. Corales, unpublished data). González (1993) found a nest in a shrub patch of *Ulex europaeus*, but he did not describe its shape or structure.

Strix rufipes typically nest in tree cavities (Table 7.16). A nest found in the interior of a hollow trunk of an old tree was at almost 2 m in depth from the treetop opening (Housse 1945). Vukasovic et al. (2006) found a nest placed on the top of a dead broken *hualo* (*Nothofagus glauca*) at 3 m above ground; the nest was in an exposed cup-shaped cavity of almost 30 cm in diameter wrapped with woody material from the inner part of the trunk. Two nests found in pine plantations were in a simple depression on the ground (Estades et al. 1998). In Nahuelbuta, Rivas-Fuenzalida et al. (2015a) registered an adult female incubating on an abandoned nest of a pair of Chilean hawk (*Accipiter chilensis*) in the interior of an extensive secondary-growth deciduous forest (300 ha, 40 years old). It is possible that scarcity of natural cavities and fidelity to nesting sites compelled owl pairs to nest in highly vulnerable conditions. More recently, Ibarra et al. (2017) found a pair nesting in the interior of a large crevice (94-cm height × 33-cm width) at almost 5-m height in a large snag of southern beech (*Nothofagus obliqua*, dbh = 63.3 cm). Similarly, all nests of *S. rufipes* found in Andean forest remnants of southern Argentina were in natural tree cavities. Wallace (2010) found a nest on an inner ledge in the highest part of a southern beech (*Nothofagus dombeyi*) with its interior burned and hollow; the nest entrance was at almost 6-m height. Beaudoin and Ojeda (2011) reported finding four nests, three which were in tree cavities excavated by Magellanic woodpeckers (*Campephilus magellanicus*) and one was in a tree cavity created by internal rottenness of the trunk. All these nests were in large live southern beech (*N. dombeyi*; 1–2-m dbh, 18–32-m tall).

7.3.9.3 Eggs and Clutch Size

Egg characteristics and clutch sizes of the Chilean owls are summarized in Table 7.17. Whereas eggs of *Tyto furcata* are often described as white and glossy, Gay (1847) affirms that eggs are dull and opaque. It is possible that this author had observed slightly dirty eggs that were stained with bed material or debris. With respect to this, Raspail (1895) states that eggs of Chilean specimens are regularly oval, shinier, and larger compared to those of European specimens (length × width: 40–43 × 30.5–34.5 mm vs 37.3–39.6 × 28–30.4 mm). In the case of *Asio flammeus*, Chilean ornithologists have described clutches of up to seven eggs (Goodall et al. 1951). However, three clutches found in southern Chile had only 2–3 eggs (González 1993; R.A. Figueroa and E.S. Corales, unpublished data). Similarly, a nest found by P. Martens in the late 1800s had three chicks (Lataste 1895b). In addition, N. Piwonka photographed a nest with three eggs and a recently hatched chick (see Lazo et al. 2008). Johnson (1967) questioned the origin of the first eggs attributed to *Strix rufipes*. This author, recognizing that such eggs were of owls, doubted that these belonged to *S. rufipes* as they were found in a hawk's nest and without verifying presence of the nest's owner. Recent reliable records indicate that eggs of *S. rufipes* are white and round-shaped (Estades et al. 1998, A. Rivera in Altamirano et al. 2012; Ibarra et al. 2017). According to the literature, this owl species has the smallest clutch size among species of Chilean owls: 1–2 eggs (Table 7.17).

Table 7.17 Egg characteristics of Chilean owls

Species	Shape and texture	Color	Range Length (mm)	Width (mm)	Mass (g)	Eggs/Nest	N
Tyto furcata	rou/ova, smo	whi, bri[a]	33.9–44.4	29.4–34.5	und	2–7 [11][b]	36
Bubo magellanicus	rou, smo	whi, opa	48–53.3	41–45.9	und	2–5	12
Athene cunicularia	rou/ova, smo	whi, bri (opa)[c]	30.6–41	25.5–33.0	und	2–8	53
Glaucidium nana	rou	whi, opa	27–31.3	24–30	und	2–5 [1][d]	29
Asio flammeus	rou	whi	39.9–47	32.5–38	22.5–23.5	2–7	10
Strix rufipes	rou	whi	[41.9–43?[e]]	[31.5–32?[e]]	und	1–2 [2–3?[f]]	[3?[e]]

N = number of measured eggs

Shape/texture: rou = rounded, ova = oval, smo = smooth. Color: whi = white, opa = opaque, bri = bright, und = undescribed. [a]Unlike other authors, Gay (1847) indicates that the eggs of this species are matte and dull. [b]Goodall et al. (1951) indicate a maximum of 11 eggs. [c]While Housse (1945) describes eggs white bright, Goodall et al. (1951) state that eggs are white and opaque. [d]Hoffmann and Lazo (2000) indicate a minimum of 1 egg. [e]Measurements were given by Goodall et al. (1951), but the origin of eggs was questioned by Johnson (1967). [f]Several authors suggest 2–3 eggs, but the original source is unknown. The only reliable sources are Estades et al. (1998; 1–2 eggs), a photography of a nest with 1 egg and 1 chick taken by A. Rivera (see Altamirano et al. 2012), a photography of a nest with 2 eggs taken by M.A. Vukasovic (unpublished data), and Ibarra et al. (2017; 1 egg). References: *Tyto furcata* = 1–6, 8, 9, 11–14, 17, 20–24; *Bubo magellanicus* = 7, 9, 11, 13, 14, 17–19, 22, 23; *Athene cunicularia* = 1, 8, 9, 11–14, 20, 22, 23; *Glaucidium nana* = 1, 3, 7–11, 13, 14, 17–23, 27; *Asio flammeus* = 5, 9, 11, 13–15, 22, 29; *Strix rufipes* = 11, 16, 18, 19, 22, 23, 25, 26, 28. [1]Gay 1847, [2]Yarrell 1847, [3]Germain 1860, [4]Lataste 1895a, [5]1895b, [6]Raspail 1895, [7]Lane and Sclater 1897, [8]Bullock 1929, [9]Housse 1945, [10]Barros 1950, [11]Goodall et al. 1951, [12]Barros 1963, [13]Johnson 1965, [14]1967, [15]González 1993, [16]Estades et al. 1998, [17]Couve and Vidal 1999, [18]König et al. 1999, [19]Marks et al. 1999, [20]Egli and Aguirre 2000, [21]Hoffmann and Lazo 2000, [22]Pavez 2004a, [23]Trejo 2004, [24]Rivas and Figueroa 2009, [25]Altamirano et al. 2012, [26]A.M. Vukasovic, personal communication, [27]Ibarra et al. 2014a, [28]Ibarra et al. 2017, [29]authors, unpublished data

7.3.9.4 Productivity and Chick Growth

The productivity of Chilean owls may be inferred from their clutch sizes. However, because some eggs are infertile or are lost for some reason, clutch size is not always a good indicator of productivity (i.e. number of hatched and/or raised chicks).

Tyto furcata has a relatively high productivity. In Lake Peñuelas, productivity of 11 pairs was 26 young (22 alive + 4 dead) between 2002 and 2003 and 36 young (31 alive + 5 dead) between 2003 and 2004 (Muñoz-Pedreros et al. 2010). However, survival rate of juveniles in 1 year was only 35%. Chicks are born with white down which later turn yellow (Housse 1945; Rivas and Figueroa 2009). According to Barros (1963), the chicks leave the nest after 1 month. *Bubo magellanicus* breeds at least 2–3 chicks which hatch with yellowish-white down that later turns grey with a brown tint (Rivas and Figueroa 2009). *Athene cunicularia* breeds 2–5 chicks which hatch with white down that after 7 days turns bluish grey, acquiring later a creamy brown colour on the head and ventral region (Housse 1945; Barros 1963; Rivas and Figueroa 2009). In the third week, chicks show two dark strips on the throat and greyish-yellow on the belly and wings (Housse 1945; Rivas and Figueroa 2009). *Glaucidium nana* raises 3–4 chicks (Ibarra et al. 2014a; R. A. Figueroa and E. S. Corales, personal observation). Hatchlings are born with white down that after 10 days turns brown, except on ventral parts; at the start of the second week the chicks grow rapidly, and at almost 1 month in age they make their first flights (Housse 1945). *Asio flammeus* breeds at least three chicks (González 1993). Chicks hatch with white down which turns yellow at 15 days (Housse 1945). *Strix rufipes* breeds one to two chicks (Housse 1945; Vukasovic et al. 2006; Wallace 2010; J. Medel, personal observation; Ibarra et al. 2017). Depending on age, chicks have white or brown down (Fig. 7.7). When the chicks leave the nest, they exhibit a greyish-white plumage with ochre tint, showing already the typical barred pattern of adult owls (Martínez 1995; Figueroa et al. 2001a; P. Wallace, personal communication; J. Medel, personal communication).

7.3.9.5 Breeding Behaviour and Nestling Diet

In Chile *Tyto furcata* is known to be monogamous, although pairs may reconstitute after the death of one of its members (Housse 1945). During courtship, the male performs an "erotic dance" swinging her body back and forth, making a chattering like a "clic-clac …" or "clack-clack …" (Lataste 1895b). Barros (1963) mentions that the male flies around the perched female emitting short and repeated voices similar to "chi-chi-chi"; then, the female makes a clicking with her bill and takes flight, and the male immediately follows. According to Housse (1945), the female incubates all day, being fed by the male and briefly leaves the nest overnight. Chicks solicit food with insistent claims and parents feed them with insects and meat bits (Gay 1847). After the chicks leave the nest, they scatter across the field within a close radius, but returning during early morning to sleep in their nest (Gay 1847; Barros 1963).

Bubo magellanicus is an apparently monogamous species (Housse 1945). The pairs move to isolated sites at the beginning of the breeding season, and later during courtship both vocalize in duet (Housse 1945; König et al. 1999). Both pair members participate in the nest preparation and chick rearing (Housse 1945). According to Housse (1945), both sexes incubate eggs. However, König et al. (1999) point out that only females incubate while males provide them prey. Pairs of *B. magellanicus* are very aggressive when defending their nests (König et al. 1999; Martínez and González 2005). At the end of the breeding period, nestlings are expelled from their natal site by the parents (Housse 1945).

Athene cunicularia is a monogamous species with both members of the pair participating in nest preparation (Housse 1945). Although Gay (1847) states that males and females alternate in incubation, Housse (1945) points out that females incubate alone while males provide them food and defend the nest outside the burrow. The hatchlings are fed with pap of insects (Housse 1945). According to Housse (1945) and Barros (1963), pairs protect their nest against human intruders by executing strategies of distraction; one of the pair makes short flights issuing strident voices to attract attention and keep intruders away from the nest. When dogs or foxes come near nest sites, the adult owls scare them with wing touches, screaming and clicking while flying over (Scott and Sharpe 1912; Housse 1945; Barros 1963). When the chicks leave the nest, they remain near the burrow, moving away a little more in following days (Gay 1847). When self-sufficient, fledglings are expelled from nest sites by their parents (Gay 1847; Housse 1945). The adult female obstructs the burrow entrance with her body, while the adult male chases fledglings when they attempt to return; if the latter insist, the female repels them with flapping and pecking (Housse 1945).

Although pair fidelity is unknown in *Glaucidium nana*, this owl species appears to be monogamous. *G. nana* exhibits great vocal activity during courtship; the males call repeatedly at different times of the night, dawn and mid-morning (Barros 1950; Johnson 1967; Norambuena and Muñoz-Pedreros 2012). To attract the female, the male offers nest cavities in the following sequence: the male advertises a potential nesting cavity by singing near it, slipping into the hole and singing from the entrance; then the female approaches the male and poses beside him; the male emits a soft cooing "gl-gl-gl-clu-clu", raising repeatedly his tail and moving his body forward and back; the female inspects the hole and the male leaves; from the outside, the male looks towards the entrance; if she leaves the hole, the male flies with cooing notes to the hole and tries to attract her again; to convince her, male deposits prey inside the cavity, and if the female accepts the food, the pair-bond consolidates (Barros 1950; König et al. 1999). The female incubates alone while being fed by the male (Housse 1945; Barros 1950), and it may abandon the eggs if manipulated by humans (Barros 1950). The chicks are cared for and fed with pap of insects by both parents for 3–4 weeks (Housse 1945; König et al. 1999). In southern Chile, nestlings are fed with passerine birds (*Turdus falcklandii*, *Scelorchilus rubecula*, *Elaenia albiceps*, *Carduelis barbata*), marsupials (*Dromiciops gliroides*), rodents (*Rattus rattus*), and other small mammals (Ibarra et al. 2014a, 2015). Fledglings return to the nest to sleep during the first week (Housse 1945). During the breeding season, the fathers keep pair-bonds by issuing voices similar to a "t-r-r-r …" which are soft in the male, and quick, penetrating, and sharp in the female (Barros 1950).

According to some authors, *Asio flammeus* is vocally active during courtship emitting a series of calls like squeaks, whistles, and barks (Goodall et al. 1951; Johnson 1967). The female incubates alone while the male feeds her and remains near the nest when not hunting; if the female dies for some reason, the incubation may be continued by the male (Housse 1945). Both parents feed the chicks all day, even delivering them more prey than necessary (Housse 1945). González (1993) observed that adult owls brought a rodent every 45 min to chicks 2 weeks old. Nestlings demand food by emitting short, high-pitched voices like a "choo-choo" or "pi-pi-pi" (Housse 1945). In the presence of humans, adult owls protect eggs and chicks by performing distraction displays which consist in moving away ≥50 m from the nest, dropping down to the ground, and emitting squeaky agonizing voices (González 1993; R.A. Figueroa and E.S. Corales, unpublished data). Some pairs may abandon the eggs if the nest is disturbed (R. A. Figueroa, personal observation). The chicks remain with their parents for a few days after leaving the nest, and soon after, they hunt alone and sleep together (Housse 1945). In Chile, the degree of pair fidelity is unknown.

Strix rufipes is said to be a monogamous species (Martínez 1995). The males attract females by offering some profitable prey (Martínez 1995). Apparently, females incubate alone (Vukasovic et al. 2006). Estades et al. (1998) observed a single owl in a nest they visited on several occasions, which was supposedly a female. When the female incubates, she is fed by the male with small mammals, birds, and insects (Ibarra et al. 2017). Males transfer prey to female either inside or at the entrance of the cavity, and when males brought excess prey, they store it inside the nest for later consumption (Ibarra et al. 2017). Both parents feed nestlings (Vukasovic et al. 2006; Ibarra et al. 2017). While monitoring a pair with an owlet, Vukasovic et al. (2006) observed the following: the feeding activity started at early night and ended just before sunrise (21:00–06:00 h); adult owls visited the nest on many occasions (29–30 times) with a time interval of almost 16 min between visits with each visit lasting almost 2 min; adult owls brought food to the owlet on 19 out of 45 clearly recorded events (45% of all records); on most of the visits, the chick was fed directly by parents and sometimes the adults regurgitated food on the edge of the nest; the owlet activity was constant during the night and during the day; one parent covered the entire nest remaining attentive and vigilant, although dozing briefly. Similarly, Ibarra et al. (2017) observed that prey delivery along the incubation/nestling period occurred between early night and before sunrise (21:00–22:00 h and 05:00–06:00 h). Diet of incubating females and nestlings can consist of scansorial rodents (e.g. *Oligoryzomys longicaudatus*), arboreal marsupials (e.g. *Dromiciops gliroides*), birds, and forest-dwelling beetles (e.g. *Achantinodera cumingii*, *Chiasognathus grantii*; Vukasovic et al. 2006; Ibarra et al. 2017). Fledglings usually emit strident voices, possibly demanding food from parents (R. A. Figueroa, personal observation).

7.3.10 Longevity

A female *Tyto furcata* domesticated from a chick has lived almost 20 years to date (C. Barrientos, personal communication). An individual of *Bubo magellanicus* domesticated from chick reached an age of >40 years (J. Rottmann, personal communication). In the wild, *Glaucidium nana* can reach an age of 6–7 years (König et al. 1999).

7.3.11 Behaviour

Most of the information about the behaviour of Chilean owls comes from accounts of senior naturalists being the more outstanding works those of Gay (1847), Housse (1945), Barros (1950, 1963), and Goodall et al. (1951). More recent observations have partially complemented the scarce information available.

Table 7.18 Behavioral characteristics of Chilean owls

Character	Tfur	Bmag	Acun	Gnan	Gper	Afla	Sruf
Fidelity	++?[a, b, c]	+++[d]	+++	++[c]	+++	+?[b, e]	+++
Independence	++	++	++	+++	+++	+++	++
Aggressiveness	++	+++	+++	+++	+++	+?	+++
Tolerance	+++	++	++	+++	+++	++	++
Sociability	+	+	++	+	+	++	+
Sensitivity	+	?	?	+	+	+++	?
Secretiveness	++	++	+	+++	+++	++	+++
Territoriality	++	+++	+++	+++	+++	+++	+++
Envisage	+++	?	?	+++	?	?	+++
Voraciousness	?	?	?	+++	?	?	?
Tameness	+++	+++	+	+	?	++	++

Tfur = *Tyto furcata*, Bmag = *Bubo magellanicus*, Acun = *Athene cunicularia*, Gnan = *Glaucidium nana*, Gper = *G. peruanum*, Afla = *Asio flammeus*, Sruf = *Strix rufipes*. The sign "plus" indicate the relative strong of each character: + = low; ++ = moderate; +++ = high

Fidelity = permanent residence on a site; independence = hunts or rests solitarily; aggressiveness = reacts violently to a direct threat (e.g., predators, humans); tolerance = indifference to presence, activity, and/or residence of humans; sociability = ability to live and hunt in groups; sensitivity = abandonment of nest or residence site if it is perturbed; secretiveness = it moves silently and hiddenly; territoriality = vocal and physical defense of hunting or nesting areas; envisage = prey storage; voraciousness = kill and eat voraciously; tameness = habituation and dependence on humans.

Notes: [a]Barros (1963) mentions that *Tyto furcata* regularly change their nesting sites. [b]Rau and Jaksic (2004b) indicate that *Tyto furcata* and *Asio flammeus* are partial residents and local migrants. [c]Jaksic et al. (1996) found that these species can temporarily leave their residence area. [d]König et al. (1999) mention that birds of southernmost localities migrate to more protected sites during winter. [e]According to Pavez (2004a), this species is sedentary. References: *Tyto furcata* = 1, 3, 4, 7, 9, 10, 13, 14, 16, 17, 19–22, 24–27; *Bubo magellanicus* = 4, 6, 9, 10, 13, 14, 16, 19–21, 24, 26, 27; *Athene cunicularia* = 1–4, 7, 9, 10, 13, 17, 19–22, 27; *Glaucidium nana* = 4–6, 8–11, 13–17, 19–21, 24, 27, 28; *Glaucidium peruanum* = 15, 21, 24; *Asio flammeus* = 4, 8, 10, 12, 18–22, 24, 27; *Strix rufipes* = 4, 6, 10, 15, 16, 21, 23, 24, 27. [1]Lataste 1895b, [2]Scott and Sharpe 1912, [3]Bullock 1929, [4]Housse 1945, [5]Barros 1950, [6]Goodall et al. 1951, [7]Barros 1963, [8]Humphrey et al. 1970, [9]Schlatter 1979, [10]Jaksic and Jiménez 1986, [11]Jiménez and Jaksic 1989, [12]Venegas 1994, [13]Jaksic et al. 1996, [14]Couve and Vidal 1999, [15]König et al. 1999, [16]Couve and Vidal 2000, [17]Egli and Aguirre 2000, [18]Figueroa et al. 2001a, [19]Jaksic et al. 2001, [20]Couve and Vidal 2003, [21]Pavez 2004a, [22]Rau and Jaksic 2004b, [23]Martínez 2005a, [24]Martínez and González 2005, [25]Sanhueza 2007, [26]Bonacic and Ibarra 2010, [27]Ibarra et al. 2015, [28]authors, unpublished data

7.3.11.1 Individual Character and Social Behaviour

In general, the behaviour pattern of a species is influenced by the character or personality of its members. An outline of the individual character of Chilean owls is given in Table 7.18. However, because much of the information is circumstantial, our approach is subject to corrections.

According to the literature, three species of Chilean owls manifest some degree of social behaviour. *Athene cunicularia* is perhaps the most social species; several pairs may nest very close to each other. Housse (1945) counted 22 burrows in a single sand trench, even finding three burrows in a stretch of 5 m. Barros (1963)

Table 7.19 Circadian activity of Chilean owls during summer

Horary	Tfur[a]	Bmag[b]	Acun	Gnan	Afla	Sruf
00:00–02:00	hun, voc	voc	?	voc	?	voc, fee
02:00–04:00	hun, voc	voc	?	voc	?	voc, fee
04:00–06:00	res	voc	fly	voc	?	voc, fee
06:00–08:00	i	i	voc	voc	?[d]	i, vig, voc
08:00–10:00	i	i	res	res, voc[c]	res	i, vig
10:00–12:00	i	i	i, res	hun, voc[c]	fly, res	i, vig, fly[e]
12:00–14:00	i	i	i, res	hun, voc	?	i, vig, res[e], voc
14:00–16:00	i	i	voc, res	?, voc	hun	i, vig
16:00–18:00	i	i	i, res	hun, voc	?[d]	i, vig, res[e]
18:00–20:00	i	i	?	hun[c]	hun, res	i, vig, voc[e]
20:00–22:00	i	voc	voc	voc	hun, res	voc
22:00–24:00	hun, voc	voc	?	voc	?	voc, hun, fee

Tfur = *Tyto furcata*, Bmag = *Bubo magellanicus*, Acun = *Athene cunicularia*, Gnan = *Glaucidium nana*, Afla = *Asio flammeus*, Sruf = *Strix rufipes*. Activity: hun = hunts, voc = vocalizes, res = rests/alights, fli = flies, fee = feeds, vig = vigilant, i = inactive

[a]Venegas (1994) mentions that in Magallanes, southern Chile, it is possible to observe this owl during day. [b]Several authors indicate that this species can also hunt during the day, but without indicating a specific time (Mogensen 1930 en Humphrey et al. 1970; Pavez 2004a; Martínez and González 2005; Bonacic and Ibarra 2010). [c]Norambuena and Muñoz-Pedreros (2012) detected that vocal activity *G. nana* is much more prevalent during midmorning with a second peak toward late afternoon. In addition, these authors observed individuals of *G. nana* attacking small passerines between 11:00 and 12:00 h. [d]Páez-López et al. (2005) indicate that *Asio flammeus* is active at this time, but without specifying the type of activity. [e]See Ippi and Rozzi (2004). References: *Tyto furcata* = 1–3, 6, 7, 12; *Bubo magellanicus* = 7, 12; *Athene cunicularia* = 1, 7, 12; *Glaucidium nana* = 5, 12, 13; *Asio flammeus* = 4, 10, 12; *Strix rufipes* = 8, 9, 11, 12. [1]Gay 1847, [2]Lataste 1895a, [3]1895b, [4]Housse 1945, [5]Barros 1950, [6]1963, [7]Jaksic et al. 1981, [8]Ippi and Rozzi 2004, [9]Martínez 2005a, [10]Páez-López et al. 2005, [11]Vukasovic et al. 2006, [12]Norambuena and Muñoz-Pedreros 2012, [13]authors, personal observation

found small colonies formed by pairs which occupied their own burrow few metres from one another. Roa (2011) recorded several individuals congregated in specific sites, but their burrows were 100–150 m apart. Apparently, *Asio flammeus* can form hunting groups. Housse (1945) observed flocks of *A. flammeus* in sites with high prey availability and several individuals together during autumn possibly to displace in search of food. So far, we have not witnessed occurrences of large flocks, but we have observed small groups of 3–4 individuals searching for prey in the same hunting area (D. R. Martínez and R. A. Figueroa, personal observation). *Glaucidium nana* also could have some degree of sociability; Housse (1945) observed several pairs occupying cavities very close to each other in the same tree.

7.3.11.2 Circadian Activity

The circadian activity of Chilean owls is summarized in Tables 7.19 and 7.20. *Tyto furcata* and *Bubo magellanicus* are strictly nocturnal species, but occasionally these may be active during the day (e.g. Humphrey et al. 1970; Venegas 1994). In fact, in

Table 7.20 Circadian activity of Chilean owls during winter

Horary	Tfur[a]	Bmag[b]	Acun	Gnan	Afla	Sruf
00:00–02:00	hun	voc	?	voc	?	hun
02:00–04:00	?	voc	?	hun,voc	?	hun
04:00–06:00	?	voc	?	?	?	hun
06:00–08:00	i	voc	res	hun	?[c]	hun
08:00–10:00	i	i	voc	voc[b]	hun, res	i, res[d], hun[d]
10:00–12:00	i	i	res	res, voc[b]	hun	i, hun[d]
12:00–14:00	i	i	res	voc[b]	res	i, res[d], fli[d]
14:00–16:00	i	i	res	voc[b]	hun, res	i
16:00–18:00	i	i	voc	hun, voc[b]	hun, res	i, res[d]
18:00–20:00	voc, hun	voc	?	voc	hun, res	voc
20:00–22:00	hun	voc	?	voc	hun, res	voc
22:00–24:00	hun	voc	?	?	?	voc, fli

Tfur = *Tyto furcata*, Bmag = *Bubo magellanicus*, Acun = *Athene cunicularia*, Gnan = *Glaucidium nana*, Afla = *Asio flammeus*, Sruf = *Strix rufipes*. Activity: hun = hunts, voc = vocalizes, res = rests/alights, fli = flies, fee = feeds, vig = vigilant, i = inactive

[a]Venegas (1994) mentions that in Magallanes, southern Chile, it is possible to observe this owl during day. [b]Norambuena and Muñoz-Pedreros (2012) detected that vocal activity *G. nana* is much more prevalent during midmorning with a second peak toward late afternoon. [c]Páez-López et al. (2005) indicate that *Asio flammeus* is active at this time, but without specifying the type of activity. [d]See Ippi and Rozzi (2004). References: *Tyto furcata* = 1–3, 5, 7, 11; *Bubo magellanicus* = 5, 7, 11; *Athene cunicularia* = 7, 11; *Glaucidium nana* = 1, 5, 6, 11, 12; *Asio flammeus* = 4, 5, 10, 11; *Strix rufipes* = 8, 9, 11. [1]Gay 1847, [2]Lataste 1895a, [3]1895b, [4]Bullock 1929, [5]Housse 1945, [6]Barros 1950, [7]Jaksic et al. 1981, [8] Ippi and Rozzi 2004, [9]Martínez 2005a, [10]Páez-López et al. 2005, [11]Norambuena and Muñoz-Pedreros 2012, [12]authors, personal observation

central Chile Jaksic et al. (1981) found that between 11% and 14% of their prey were diurnal. The vocal activity of *T. furcata* seems to be more frequent during dusk or at early night (Barros 1963). *Athene cunicularia* and *Asio flammeus* exhibits continuous activity, but very little is known about their nocturnal activity. According to Barros (1963), *A. cunicularia* starts its activity during twilight, remaining inactive during the day. However, Jaksic et al. (1981) found that almost 83% of the prey consumed by this owl species were diurnal, suggesting an intense hunting activity during the day. Diurnal activity of *A. flammeus* seems to concentrate on certain hours of the day. In central Chile, Páez-López et al. (2005) recorded the most active individuals between 6:30–7:00, 11:30–12:00, 17:30–18:00 h, and 20:00–20:30 h. In southern Chile, the hunting activity during winter appears concentrated between 17:00 and 19:00 h and sometimes between 13:00 and 14:00 h (R.A. Figueroa, unpublished data). Housse (1945) observed individuals of *A. flammeus* overflying agricultural lands every afternoon between 15:00 and 16:00 h. This species tends to be more active during cloudy days of autumn and winter (Bullock 1929; Housse 1945; R. A. Figueroa, personal observation). *Glaucidium nana* has continuous activity, but vocal activity appears to be more intense during twilight and night (Barros 1950; Jiménez and Jaksic 1989). Recently, Norambuena and Muñoz-Pedreros (2012) documented that during the day *G. nana* tends to be more active at midmorning (09:00–11:00 h), apparently with a second activity peak towards late afternoon (18:00 h). A preliminary study suggests that during summer this owl may be

Table 7.21 Vocalizations of *Tyto furcata* described for Chile and adjacent areas

Vocalization (descriptor)		Characteristics	Sex	Period	Function
Type 1 (typical)	*cuiiii*...(1, 2)	Very sharp and strong, lugubrious	MF	N	H
	chuuuit/crrruit...(3, 4, 5)	lugubrious, plaintive squeal	MF	N	H
	chi-i-i-i-it (6)	Repeated squeal	MF	N	T
	chiiiii chiiiii (10)	–	?	N	H
	tschiit-tschiit (8)	Strong and strident	?	N	?
	sheeeerrr (7)	Harsh and tremulous (1–2 s)	?	?	?
	sshhhjj... (9)	Strong hissing	?	N	?
	shhhhhhhhhKKKK (11)	Penetrating whistle	?	?	?
	huishhhhhhhh (13)	Prolonged squeak	?	N	?
	quichchchchchk (12)	Strong and strident	?	?	?
Type 2	*clik-clak/clak-clak* (1, 2)	Soft chattering of beak	M?	N	R
Type 3	*Keu-keu-ke* (2, 3)	Soft chattering of beak	MF	N	WF
	Kec-kec-kec-kec (8)	Soft chattering	?	?	W
	kiik (11)	Single note	?	?	?
Type 4	*chi-chi-chi-chi-chi*...(6)	Metallic screech	M	N	R

Sex: M = male, F = female. Day period: N = night. Function: T = territorial, W = alarm, F = fear or fright, H = hunting, R = courtship and copulation,? = unspecified
[1]Lataste 1895a, [2]1895b, [3]Krahnass 1896a, [4]1896b, [5]Raspail 1896, [6]Barros 1963, [7]Fjeldsa and Krabbe 1990, [8]Couve and Vidal 1999, [9]Narosky and Babarskas 2000, [10]Figueroa et al. 2001a, [11]Jaramillo 2003, [12]Pavez 2004a, [13]Martínez and González 2005

active throughout the night, but exhibiting at least three vocal activity peaks: twilight and early night (21:00–23:00 h), midnight (01:00–03:00 h), and late-night and dawn (05:00–06:30 h) (unpublished data). According to variety of its prey species, this owl species hunts regularly during day, dusk, and night (Jiménez and Jaksic 1989). Circadian activity of *Glaucidium peruanum* is unknown, but could be similar to that of *G. nana*. *Strix rufipes* is a strictly nocturnal owl, but some individuals show diurnal activity. In Cabo de Hornos, Ippi and Rozzi (2004) recorded an active individual during the day on several occasions. In central Chile, the species has been recorded vocalizing regularly at 20:00 and 06:00 h and occasionally at 12:00 h (S. Alvarado, personal observation). In southern Chile, the vocal activity of *S. rufipes* begins 40 min after sunset and ends about 20 min before sunrise (Martínez 2005a). During summer, this owl appears to exhibit a pronounced vocal activity peak at early night (22:00–23:00 h) and two subsequent slight peaks at midnight (01:00–03:00 h) and late-night (04:00–06:00 h) (R.A. Figueroa and E.S. Corales, unpublished data).

7.3.11.3 Vocal Behaviour

The typical voice of *Tyto furcata* seems to be a territorial signal, and it is used mainly when hunting (Table 7.21). Interestingly, *T. furcata* utters metallic clicking sounds in flight, which might suggest echolocation as used by bats for orientation

Table 7.22 Vocalizations of *Bubo magellanicus* described for Chile and adjacent areas

Vocalization (descriptor)		Characteristics	Sex	Period	Function
Type 1 (typical)	*tuu-kukuru* (1)	–	?	N	?
	tucú-crrrrr tucú-crrrrr (7)	–	?	?	?
	tucu-tucu-tucúquerrrr...(13)	–	?	N	?
	ju-juju-ju-ju...(3)	More notes in Fs	MF	N	P
	huhOOh-urrrr...(9)	*urr* longer in Fs	MF	?	P
	hoo-hoo-Querrrrr (11)	Hollow and deep	?	?	?
	bu-boo-brr...(4)	Reverberant	MF	N	P
	wu-búh-worrrr(5)	Double; the *o* longer in Fs	MF	?	P
	bu-hóohworrrr (8)	Deep, *orr* longer in Fs	MF	?	P
	bubú-bu.-bubububu...(12)	Deep	?	?	?
	ñacurutú-úú...(2, 6)	–	?	N	H
	ñacurutú...(10)	Deep	?	?	?

Sex: M = male, F = female. Day period: N = night. Function: P = pair-bond, H = hunting,? = unspecified

[1]Crawshay 1907, [2]Housse 1945, [3]Goodall et al. 1951, [4]Humphrey et al. 1970, [5]König et al. 1996, [6]Plath 1996, [7]Couve and Vidal 1999, [8]König et al. 1999, [9]Marks et al. 1999, [10]Narosky and Babarskas 2000, [11]Jaramillo 2003, [12]Pavez 2004a, [13]Martínez and González 2005

(König and Weick 2008). The function of the typical voice of *Bubo magellanicus* is the pair-bond maintenance (Table 7.22). The typical voice of *Athene cunicularia* is a territorial and warning signal (Table 7.23). The two best-known voices of *Glaucidium nana* could have a dual role: territorial defence and pair-bond, or hunting (Table 7.24). A recent study shows that vocalizations associated with pair-bond is more frequent in spring (75% of spontaneous calls), and vocalization associated with territorial defence is more common during summer and autumn (75–100% of spontaneous calls; Norambuena and Muñoz-Pedreros 2012). The more common vocalization of *Glaucidium peruanum* apparently delimits territory (Table 7.25). So far, Chilean ornithologists have not identified a typical vocalization for *Asio flammeus*, but the three known voices appear to have more than one specific function (Table 7.26). According to our field experience, this species rarely vocalizes and its voices are only evident in situations of territorial defence, alarm or courtship. This would contribute to lessening its detectability in the field. The two typical vocalizations of *Strix rufipes* indicate territorial defence and pair-bond (Table 7.27).

7.3.11.4 Hunting Techniques

Almost all Chileans owls hunt by both active and passive searching (Table 7.28). *Tyto furcata* hunt actively by combining gliding and hovering flights (Table 7.29). In Osorno, southern Chile, one of us observed an individual of *T. furcata* during the night flying swiftly among trees with the intention of scaring and capturing Shiny Cowbirds (*Molothrus bonariensis*, R.A. Figueroa, personal observation). *Bubo*

Table 7.23 Vocalizations of *Athene cunicularia* described for Chile and adjacent areas

Vocalization (descriptor)		Characteristics	Sex	Period	Function
Type 1 (typical)	*Chiii...* (1, 5)	–	?	?	W?
	Chi-i-chi-i-i (10)	Repeated	?	ND	TW
	chiii-chi-chi-chi-chi (2, 8)	High-pitched squeal	?	ND	TW
	chrr-rriiii...tih, tih/ Prr-rrriiiiii...(7)	"h" ending is aspirate	?	N	TW
	zree tchichi chi chi....(9)	Squeaky chatter	?	?	?
	chhaaak-weet weet weet (13)	Strident claims	?	?	?
	cuichchchch-cuich-cuich-cuich-cuich (14)	Cry at intervals	?	ND	W
	kieee..kie..kié kié..(11)	–	?	?	?
	kiiiiiii-ki-ki-ki (15)	–	?	ND	W
	pip pip pip pip churrr (4)	Angry	FM	D	W
	PIIIIST-piist-piist-piist-piist (12)	Intermittent whistles	?	?	?
Type 2	*eep eep..*(9)	Rapid series	F	N	P
Type 3	*cu-co-ru* (6)	Mournful	?	N	?
	coo coo coo.....(9)	Long series, with end trill	M	N	P
	cuú-cuú-cuu/cuh-cuú-cuu... tíhtíh(13)	Guttural	?	?	?
	co-coquoi/coc-co-woy/ cocquoi-o (3)	–	?	A	?
Type 4	*Hú, hú, hú* (1, 10)	Hoot	?	SN	H
	woo whoWHO, oo, oo (9)	Flute voice	?	N	?
Type 5	*tih-tabac* (7)	Loud and sharp	?	?	?
Type 6	*rrtí, trr-rrtí.../ trr-rrtí-ti-ti-ti...*(7)	Lengthy	?	?	?
Type 7	*Piquí, piquí* (1)	–	?	?	?
	Peque-peque-peque (10)	–		SN	H

Sex: M = male, F = female. Day period: N = night, D = day, A = dawn, S = sunset. Function: T = territorial, W = alarm or warning, H = hunting, P = pair-bond,? = unspecified
[1]Gay 1847, [2]Crawshay 1907, [3]Scott and Sharpe 1912, [4]Peters 1923, [5]Housse 1945, [6]Goodall et al. 1951, [7]Barros 1963, [8]Humphrey et al. 1970, [9]Fjeldsa and Krabbe 1990, [10]Plath 1996, [11]Narosky and Babarskas 2000, [12]Figueroa et al. 2001a, [13]Jaramillo 2003, [14]Pavez 2004a, [15]Martínez and González 2005

magellanicus actively searches for prey primarily by using gliding (Table 7.29). *Athene cunicularia* hunts often by a sit-and-wait strategy (e.g. Housse 1945), but also can hunt actively by using several flight modes or running on the ground (Table 7.29). When individuals of *Glaucidium nana* actively hunt, they make fast, undulating straight flights (Table 7.29). In addition, taking advantage of their small body size, individuals of *G. nana* hunt through foliage either by directly attacking or approaching with stealthy steps and short flights to potential prey (Housse 1945; R. A. Figueroa, personal observation). *G. nana* also explores the inner of cavities searching for nestlings of other birds such as Striped woodpecker (*Picoides*

Table 7.24 Vocalizations of *Glaucidium nana* described for Chile and adjacent areas

Vocalization (Descriptor)		Characteristics	Sex	Period	Function
Type 1 (typical)	*cu-cu-cu-cu* (2)	Soft and monotone	F	NS	P
	huj huj huj huj huj huj (3, 10)	22–66 short whistles repeated	?	ND	P
	Chuuuu-chuuuu-chuuuu (4)	Soft and repeated	F	?	?
	kü-kü-kü-kü-kü-kü...(6)	Rapid, 20–30 short notes (3.5–5 notes/sec)	M	ND	T
		Similar, high-pitched, shorter song	F	ND	T
	u-u-u-u-u-u-u-u-...(5)	Monosyllable monotone whistle	?	?	?
	took took took took (7)	Monotone series (2 notes/sec)	?	?	?
	tu-tu-tu-tu-tu (8)	Acute, fluted, and repeated	?	SAD	T
	fi-fi-fi-fi-fi-fi-fi-fi...(9)	Monotonous succession (2–3 notes/sec)	?	N	T
Type 2 (typical)	*truí, truí, yi, yí ....yí-yí-yí-yí*(2, 10[a])	Rapid	M	NSD	TH
	ch-ch-ch (2)	Rapid, excited	M	NSD	TH
	chuí, chichuíchuíchuí (2)	Rapid, excited	M	NSD	TH
	trí--chíchíchíchí-chí (2)	Rapid, excited	M	NSD	TH
	tick-tick-ticktick-chrickchrick...(6)	Accelerated, metallic trill	M F	?	?
	trigigigirrr/trigigick (6)	Metallic trill	F(Y)	?	B
Type 3	*To-whit-to-whi* (1)[b]	Undertoned	?	ND	?
	Tue-tue-tue (4)[c]	–	?	?	?
Type 2	*t-r-r-r, t-r-r-r* (2)	Soft trill	M	N	P

Sex: M = male, F = female, (Y) = also fledgling young. Day period: N = night, D = day, A = dawn, S = sunset. Function: T = territorial, H = hunting, P = pair-bond, B = begging,? = unspecified
[a]Norambuena and Muñoz-Pedreros (2012) state that this type of vocalization has only one function of territorial delimitation. [b]Possibly corresponds to type 2. [c]Apparently corresponds to the type 1.
[1]Lane and Sclater 1897, [2]Barros 1950, [3]Jiménez and Jaksic 1989, [4]Plath 1996, [5]Couve and Vidal 1999, [6]König et al. 1999, [7]Jaramillo 2003, [8]Pavez 2004a, [9]Martínez and González 2005, [10]Norambuena and Muñoz-Pedreros 2012

lignarius; R. A. Figueroa, personal observation), American kestrel (*Falco sparverius*; Figueroa and Corales 2015), Thorn-tailed rayadito (*Aphrastura spinicauda*), and Chilean swallows (*Tachycineta meyeni*; Ibarra et al. 2014a). *Asio flammeus* is evidently an active hunter flying over extensive areas searching for prey (Bullock 1929; R. A. Figueroa, personal observation). When searching for prey, *A. flammeus* primarily use quartering flights in combination with gliding, hovering and soaring flights (Table 7.29, Fig.7.6). The only hunting mode documented for *Strix rufipes* is sit-and-wait (e.g. Martínez 2005a, Fig. 7.8). Possibly, many of the flying insects consumed by this owl are captured in the air.

Table 7.25 Vocalizations of *Glaucidium peruanum* described for Chile and adjacent areas

Vocalization (descriptor)		Characteristics	Sex	Period	Function
Type 1 (typical)	*toitoitoitoitoitoitoitoitoitoit*…(1)	Rapid irregular, 6–7 notes/sec	M	ND	T?
		Similar, tone somewhat higher	F	ND	T?
	took took took took (2)	Rapid trill (3–4 notes/sec)	?	?	?
	tu-tu-tu-tu (3)	Sharp, repeated, and rapid	?	AS	T
Type 2	*chirrp* (1)	Short, single or in series, high pitched	MF?	?	W?

Sex: M = male, F = female. Day period: N = night, D = day, A = dawn, S = sunset. Function: T = defense of territory, W = alarm or warning,? = unspecified
[1]König et al. 1999, [2]Jaramillo 2003, [3]Pavez 2004a

Table 7.26 Vocalizations of *Asio flammeus* described for Chile and adjacent areas

Vocalization (descriptor)		Characteristics	Sex	Period	Function
Type 1	*che che* (1)	Screeching	?	D	T
	chi-chi-chi/wheechiz (2)	Sharp	?	?	T
	shkkk-shhhhhh! (3)	Two-syllable harsh hissing	?	?	W
Type 2	*ree-you* (2)	Screamed voice	F	?	B
	nuuuuu/miuuu (4)	Screech, sharp	?	D	TW
Type 3	*boo-boo-boo* (2)	Hollow, deep	M	?	B

Sex: M = male, F = female. Day period: D = day. Function: T = defense of territory, W = alarm, B = pair-bond,? = unspecified
[1]Crawshay 1907, [2]Fjeldsa and Krabbe 1990, [3]Jaramillo 2003, [4]Figueroa and Alvarado 2007

7.3.11.5 Aggressive and Territorial Interactions

There is scarce information on aggressive and territorial interactions in Chilean owls. Some of their territorial behaviours can be inferred from vocalizations (Tables 7.21, 7.22, 7.23, 7.24, 7.25, 7.26 and 7.27) and flight modes (Table 7.29). *Tyto furcata* manifests territoriality vocally, but the occurrence of physical aggressions is unknown. According to Lataste (1895a, b), this species defends eggs and chicks until the last moment by taking an attack position, ruffling feathers, raising its wings and emitting a clicking similar to a "keu-keu …" (Table 7.21). *Athene cunicularia, Glaucidium nana,* and *Asio flammeus* manifest their territoriality by vocalizations and physical displays. The former emits loud vocalizations and displays fast undulating flights against intruders around its burrows or hunting sites. Individuals of *G. nana* often defend their nesting sites by directly attacking intruders with fast flights

Table 7.27 Vocalizations of *Strix rufipes* for Chile and adjacent areas

Vocalization (descriptor)		Characteristics	Sex	Period	Function
Type 1 (typical)	*hoo-hooo....crru crru* (1)	Copious and high-pitched scream	MF	N	?
	kru..kru..kru..(6)	Loud "growl"	?	?	?
	grr grooo grooo GROOO KOOO (7)	Muffling growl	?	?	?
	kokoko-kwowkwowkwowkwow…(4)	Growling (ko) and guttural (kwow)	MF	N	B
	cua-cua-cua-cua-cuo-cuo-cuo-coco…(8)	Like human laughter	MF	?	?
	ko-ko-ko-ko, ko-ko (10)	?	?	?	?
	cóo--cóo--cóo--cóo--cóo--cóo--cóo-…(9)	Usually 10 notes	MF	N	T
	có--có--có--có.-có--có--có—có (9)	Variation 1	MF	N	T
	có--coó--coó--coó—có (9)	Variation 2	MF	N	T
	¡có-có-có-có-có-có-có-có! (9)	Variation 3	MF	N	T
	brrr……ú AU ú ú ú ú ú ú ú ú ú ú ú ú…(5)	Nasal, high, reverberant (10–12 notes)	?	?	?
Type 2 (typical)	*kokokokwaihkwaikwkwowwkwow…*(4)	Guttural, high, "kwaih" explosive	F(M)	N	B?
	cóo-cóo-cóo-juaá-juaá-… cóo-cóocóo…(9)	"*juaá*" like laughter (4–6)	MF	N	TB
	juou…Juou…juo..jol…col…col…jol… juou (2)	?	MF	N	T
Type 3	*míííííííííooo…míííííííííooo… míííííííííooo* (9)	Like meow	F	N	B
Type 4	*poorr, poorr* (3)	Trilled	?	?	?
Type 5	*wraaaak!* (9)	Rough and lugubrious	?	N	?

Sex: M = male, F = female. Day period: N = day. Function: B = pair-bond, T = defense of territory,? = unspecified

[1]Fjeldsa and Krabbe 1990, [2]Straneck and Vidoz 1995, [3]de la Peña and Rumboll 1998, [4]König et al. 1999, [5]Marks et al. 1999, [6]Narosky and Babarskas 2000, [7]Jaramillo 2003, [8]Pavez 2004a, [9]Martínez 2005a, [10]Martínez and González 2005

and flapping their wings (R. A. Figueroa, personal observation). Individuals of *A. flammeus* defend hunting areas against conspecifics by making direct intimidating flights and emitting short, sharp vocalizations (Table 7.26). Pairs of *S. rufipes* delimitate their breeding territories by intense and prolonged vocal series (Table 7.27).

Table 7.28 Hunting modes of Chilean owls

Hunting mode	Tfur	Bmag	Acun	Gnan	Afla	Sruf
Active searching						
Aerial exploration among tree crown	+	+	–	+	–	–
Aerial exploration among bushes	+	+	–	+	–	–
Aerial exploration over ground	+	+	–	+	+	–
Flying and vocalizing	–	–	+	–	–	–
Flush flight	–	–	–	+	–	–
Aerial capture	–	–	+	–	–	?[c]
Terrestrial pursuit and fight	–	–	+[a]	–	–	–
Terrestrial pursuit and grip			+[b]			
Pasive searching (sit and wait)						
Perched on tree branches	+	–	+	+	+	+
Perched on shrub branches	–	+	+	+	–	–
Perched on fence posts or poles	+	–	+	+	+	–
Perched on ground	–	+	+	–	+	–

Tfur = *Tyto furcata*, Bmag = *Bubo magellanicus*, Acun = *Athene cunicularia*, Gnan = *Glaucidium nana*, Afla = *Asio flammeus*, Sruf = *Strix rufipes*
[a]According to Housse (1945), *Athene cunicularia* uses this technique to capture snakes moving on the ground. [b]Used for capturing insects. According to Scott and Sharpe (1912), owls grip so strongly insects that often lose their balance and stagger. [c]Possibly many of flying beetles consumed by *Strix rufipes* are captured on the air. References: Cassin 1855, Scott and Sharpe 1912, Housse 1945, Barros 1950, 1963, Venegas 1994, Couve and Vidal 2000, 2003, Ippi and Rozzi 2004, Martínez 2005a, Martínez and González 2005, Sanhueza 2007, authors, personal observation

Table 7.29 Flight modes of Chilean owls

Flight modes	Tfur	Bmag	Acun	Gnan	Afla	Sruf
Gliding, rectilinear	hun, car	hun, car	hun	def	hun	car
Gliding, undulated	–	–	dis, def	dis, car, hun	–	–
Gliding, curvilinear	–	?	esc	–	–	dis
Quartering	hun	–	–	–	hun	–
Cruising	dis, car	dis, car	dis, car	–	dis, def	–
Cruising, rapid	–	–	hun, def	hun	–	–
Cruising, slow	–	–	–	–	dis[c]	–
Hovering	hun	–	hun	–	hun	–
Soaring[b]	–	–	hun	–	hun	–
Diving	–	–	–	hun	–	–

Tfur = *Tyto furcata*, Bmag = *Bubo magellanicus*, Acun = *Athene cunicularia*, Gnan = *Glaucidium nana*, Sruf = *Strix rufipes*. Function[a]: hun = hunting, car = transport of prey, esc = escaping, dis = displacement, def = defense of territory
[a]Definitions are tentatives. [b]15–20 m height (Lataste 1895b; Housse 1945). [c]Like the flight of herons (Scott and Sharpe 1912). References: *Tyto furcata* = 2, 7, 9, 10, 13, 14; *Bubo magellanicus* = 13, 14; *Athene cunicularia* = 1, 4, 5, 11–14; *Glaucidium nana* = 6, 8, 13, 14, 17; *Asio flammeus* = 3, 4, 5, 8, 9, 12–17; *Strix rufipes* = 13, 14, 17. [1]Gould 1841, [2]Lataste 1895b, [3]Crawshay 1907, [4]Scott and Sharpe 1912, [5]Housse 1945, [6]Barros 1950, [7]1963, [8] Humphrey et al. 1970, [9]Jaksic 1985, [10]Couve and Vidal 2000, [11]Egli and Aguirre 2000, [12]Jaramillo 2003, [13]Muñoz-Pedreros and Ruiz 2004, [14]Pavez 2004a, [15]Páez-López et al. 2005, [16]Martínez and González 2005, [17]authors, personal observation

7.3.12 *Movements*

7.3.12.1 Foraging Movements

Tyto furcata and *Bubo magellanicus* tend to remain several months in the same place moving in an extensive area every night in search of prey (Housse 1945; Barros 1963; R.A. Figueroa, personal observation). *Athene cunicularia* and *Glaucidium nana* move during each night among distinct types of vegetation (e.g. grasslands, cattails, cropland, and pond edges) in search of prey (Barros 1950, 1963; R.A. Figueroa, personal observation). *Asio flammeus* usually moves through 2–3 foraging pastures at 0.5–3 km apart, sometimes flying high before moving to another pasture (>30 m; R.A. Figueroa, unpublished data). Judging from the accumulation of pellets under perches, individuals of *Strix rufipes* remain for a long time (15–60 days) in each of their foraging patches (Figueroa 1996).

7.3.12.2 Home Range

Judging from the prey carried to an attic, Housse (1945) speculated that the hunting area for a pair of *Tyto furcata* covers half a square league (800–1000 ha). In a rural area of southern Chile, the mean home range of two individuals of *T. furcata* released after being bred in captivity, reached 1051 ha (González 2007). This is consistent with estimates of home ranges of *T. furcata* in South and North America (almost 720–1746 ha, Marti 1992; Massa et al. 2015). Surprisingly, Muñoz-Pedreros et al. (2010), based on body size and by using allometric equations, estimated a considerably smaller hunting area for an individual of *T. furcata* (65 ha). However, it should be noted that this result is the product of a mathematical model and not a real estimate of the hunting area. In any case, we should consider that the home range of this species can reach an extension as large as 3174 ha and as small as 72 ha (Evans and Emlen 1947; Taberlet 1983; Marti 1992; Taylor 1994; Massa et al. 2015). König et al. (1999) indicate that the territory size of *Glaucidium nana* is almost 1 km^2 (100 ha), but without identifying the original source. This territory size is in agreement with estimates of home range of other species of *Glaucidium* (e.g. Strom and Sonerud 2001). In agricultural lands of southern Chile, the hunting range of *Asio flammeus* is 220–300 ha (N = 3 owl pairs; Martínez et al. 1998, R.A. Figueroa and E.S. Corales, unpublished data). Overall, this hunting range is more extensive than that estimated for *A. flammeus* in other regions (Village 1987) and could be due to patchy condition of pasturelands in southern Chile. In Valdivian forests, the home range of *S. rufipes* can reach between 180 and 1206 ha (median = 560 ha), tending to be more extensive in areas with fragmented forests (Martínez 2005a). However, in a highly fragmented forest landscape, Gantz and Rau (1999) detected *S. rufipes* only in a forest remnant of 350 ha (see also Rau et al. 2015). Nonetheless, this owl species could additively use small forest patches; members of a pair and their pellets were recorded in at least three forest fragments of 1–6 ha within an area of 200 ha (Figueroa et al. unpublished data). A most recent survey of *Strix rufipes* across agricultural lands in southern

Chile showed that one or more pairs inhabits several nearby forest fragments very variable in size (0.5–263 ha), shape (circular to highly irregular), and condition (highly degraded to secondary growth, or pure forest to intermixed with pine plantations) within a 1-km radius (R.A. Figueroa, unpublished data).

7.3.12.3 Migration and Residence

There is no evidence of long-distance migrations of Chilean owls. Instead, several authors mention the occurrence of local migrations and changes of residence. Housse (1945) affirms that *Tyto furcata* remains long in one place if there is an abundant and stable supply of prey. Barros (1963) states that this owl species changes residence after staying several months in the same place. Consistent with these conclusions, Rau and Jaksic (2004b) affirm that *T. furcata* is a partial migrant in southern Chile. *Bubo magellanicus* may move to sites with greater vegetation cover during winter (König et al. 1999). *Glaucidium nana* seems to be a winter visitor in northern Chile and a reproductive summer visitor in austral islands (Goodall et al. 1951; Humphrey et al. 1970). According to Vuilleumier (1985), *G. nana* is a partial migrant whose southernmost populations migrate to more northern localities during autumn. The observation of Housse (1945) about the sudden appearance of a "peregrine" flock of *Asio flammeus* during a rat plague in Capitan Pastene suggests that this owl species is nomadic. Rau (1994) states that *A. flammeus* is a partial migrant occupying agricultural pastures only in autumn and spring. However, it is possible that such observations reflect instead a numerical response to prey abundance (see Trophic Ecology) or an alternate use of foraging patches. Some authors think that southernmost individuals of *A. flammeus* (Aysen-Magallanes) are only summer residents (Couve and Vidal 2003; Jaramillo 2003).

7.3.13 Feeding and Trophic Ecology

Feeding is the more studied aspect of Chilean owls (Figueroa and Alvarado 2012). Description of the Chilean owls' prey was started by Gould (1841) who found remains of a crayfish in the stomach of an individual of *Strix rufipes* collected in an island of the Chonos Archipelago. A little later, Crawshay (1907) realized that *Athene cunicularia* changes its diet according to the season. This author observed that in Tierra del Fuego, this owl species preyed upon mainly rodents during winter and during summer consumed largely beetles, particularly brown "pololos" (*Aulacopalpus pilicollis*). Housse (1945) reported the first diet quantification of a Chilean owl. This author found 63 rodent skulls, 12 bird skulls, and remains of two bats in 78 pellets of *Tyto furcata*. From a historical point of view, information of Chilean owls' diet can be divided in two slightly overlapped periods: a descriptive period between 1841 and 1976 (Table 7.30) and a quantitative period started in 1973 and continuing up to date (see Jaksic 1997). As recent publications suggest, it is

Table 7.30 Prey types of Chilean owls reported in descriptive accounts until the 1970s

Prey type	Tfur	Bmag	Acun	Gnan	Afla	Sruf
Mammals						
Rodents	+	+	+	+	+	+
Marsupials	+	–	–	–	–	–
Bats	+	+	–	+	–	–
Hares/rabbits	–	+	+	–	+	–
Birds						
Passerines	+	–	+	+	+	+
Non-passerines	+	+	–	+	–	–
Reptiles						
Lizards	+	–	+	+	+	–
Snakes	–	+	+	–	–	–
Amphibian						
Toads/frogs	+	+	+	–	+	+
Crustaceans						
Crayfish	–	+	+	–	–	–
Crab	–	+	+	–	–	–
Small crabs	–	–	+	–	–	+
Insects						
Dragonflies/damselflies	–	–	+	+	–	–
Crickets	–	–	–	+	–	–
Locust	–	–	+	–	–	+
Beetles	+	+	+	+	+	+
Butterflies	–	–	+	+	–	–
Moths	+	–	–	–	–	–
Arachnid						
Tarantulas	–	–	+	–	–	–
Scorpions	–	–	+	+	–	–

Tfur = *Tyto furcata*, Bmag = *Bubo magellanicus*, Acun = *Athene cunicularia*, Gnan = *Glaucidium nana*, Afla = *Asio flammeus*, Sruf = *Strix rufipes*. The sign "plus" indicates that it is present, the sign "minus" indicates that it has not been registered

References: *Tyto furcata* = 2, 4, 5, 9, 12, 15–18, 21; *Bubo magellanicus* = 2, 9, 11, 12, 14–16, 18–21; *Athene cunicularia* = 1–3, 6–8, 10, 12, 15–18, 20–24; *Glaucidium nana* = 2, 6, 7, 12, 13, 15, 16, 18, 21, 22; *Asio flammeus* = 8, 11, 12, 15, 16, 18, 20; *Strix rufipes* = 1, 12, 18, 20. [1]Gould 1841, [2]Gay 1847, [3]Cunningham 1871, [4]Lataste 1895a, [5]1895b, [6]Lane and Sclater 1897, [7]Crawshay 1907, [8]Scott and Sharpe 1912, [9]Bullock 1929, [10]Housse 1935, [11]1939, [12]1945, [13]Barros 1950, [14]Olrog 1950, [15]Goodall et al. 1951, [16]1957, [17]Barros 1963, [18]Johnson 1965, [19]Greer and Bullock 1966, [20]Johnson 1967, [21]Humphrey et al. 1970, [22]Johnson 1972, [23]Solar and Hoffmann 1975, [24]Schlatter 1976

very probable that we are *ad portas* of a third period, one functional, which focuses on the structure and dynamic of food webs (Arim and Jaksic 2005; Arim et al. 2007; Farías and Jaksic 2011). Below, we discuss the diet patterns of adult owls. The scarce information on nestling/fledgling diet is provided in the sect. 7.3.9.5.

7.3.13.1 Diet Profile

Based on quantitative studies, we realized all Chilean owl species have a relatively conservative diet across very different bioclimatic zones (Fig. 7.9). However, the importance level of particular prey taxa may vary both seasonally and geographically.

Tyto furcata is characterized by having a rodent-based diet with small seasonal and geographical variation in the consumption of these preys (Table 7.31). Exceptionally, this owl may consume disproportionately passerine birds (González-Acuña et al. 2004; see sect. 7.3.13). *T. furcata* gets other prey taxa according to local availability in the field (e.g. reptiles or lagomorphs).

Bubo magellanicus has a strong consumption of mammalian prey, particularly rodents. Even though this owl species preys upon lagomorphs in relatively low numbers, these prey may equal the biomass provided by the rodent prey (Table 7.32). In central Chile, *B. magellanicus* may consume birds and invertebrates in relatively high numbers during spring-summer (Table 7.32).

Athene cunicularia shows a preeminent consumption of invertebrates all year round throughout Chile (Table 7.33). However, the energy provided by these prey is variable and depends on each particular prey taxa being consumed. Rodents are the more regular vertebrate prey, but consumption fluctuates seasonally. The lesser consumption of vertebrates relative to invertebrate prey is compensated by the greater biomass contribution of the former (Table 7.33). Among vertebrate prey, marine birds may become important prey of individuals of *A. cunicularia* that inhabit coastal areas or islands near the coastline. On the coast of Conception, southern Chile, Pavez and Gabella (1999) found remains of *Sterna* sp. and *Phalaropus* sp. (Charadriiformes) in 12% of all analysed pellets (N = 51). On islands of the coastal system of Coquimbo, northern Chile, Cruz-Jofré and Vilina (2014) found that Peruvian diving petrels (*Pelecanoides garnotii*) accounted for 40% of all consumed vertebrate prey (see also Contreras et al. 1999; Torres et al. 2011). An outstanding study is Carevic (2005, 2011, see also Carevic et al. 2013) who described for the first time the diet of *A. cunicularia* in the Atacama Desert, northern Chile, which also included an analysis of prey selection. Results of this study confirm the strong consistency in the diet composition of *A. cunicularia* among environmentally contrasting areas throughout Chile (Table 7.33).

Glaucidium nana exhibits an eclectic diet including rodents, birds and insects in similar proportions (Table 7.34). Individuals of *G. nana* from central and southern Chile show a very similar composition in diet (Table 7.34). In both cases, rodents and birds account for the highest contributions of biomass.

Asio flammeus is characterized by marked seasonal variations in diet composition (Rau et al. 1992; Martínez et al. 1998). Studies carried out in southern Chile showed that rodent consumption by this owl species increases from one third of all prey eaten during summer to almost 90% during winter (Table 7.35). In contrast, consumption of birds and insects increases towards spring and summer. During these latter seasons, avian preys make the greatest biomass contribution. The seasonal changes in diet composition appear to be influenced by temporal variations in

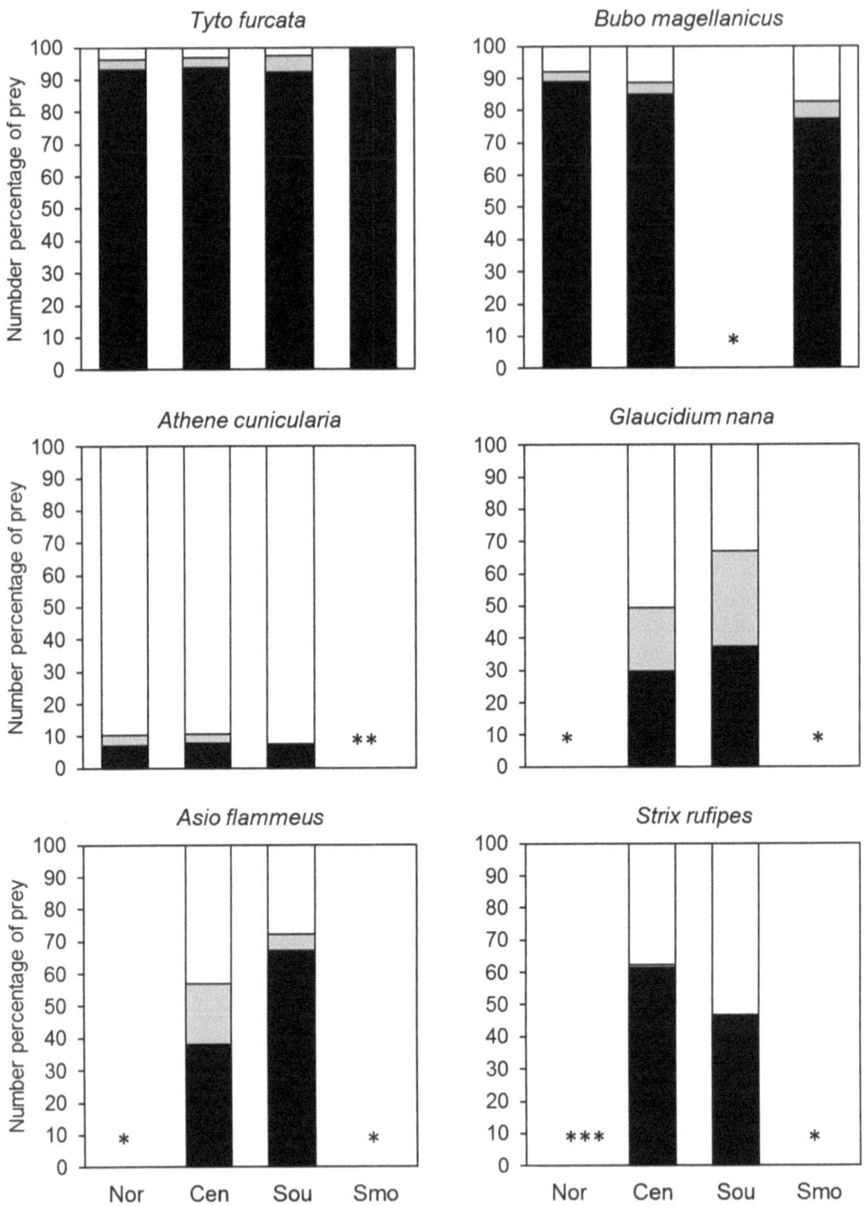

Fig. 7.9 Dietary profiles of Chilean owls built with information available in the literature and unpublished data of authors (see Appendix 7.1). Prey item: black bars = mammals, Gray bars = other vertebrates, white bars = invertebrates. Bioclimatic zones: Nor = northern Chile, Cen = central Chile, Sou = southern Chile, Smo = southernmost Chile (Notes: * the species is present, but no information, **extinct, *** the species is absent)

Table 7.31 Seasonal diet of *Tyto furcata* across bioclimatic zones of Chile

	Northern				Central				Southern				Southernmost	
	SS		FW		SS		FW		SS		FW		SS	
Prey item	F%	B%	F%	B%	F%	B%	F%	B%	F%	B%	F%	B%	F%	B%
Rodents	93.1	97.2	61.5	92.2	85.3	86.4	89.7	97.0	88.7	96.0	95.4	98.7	99.8	99.9
Marsupials	4.1	1.9	9	3.4	3.7	1.5	4.3	2.5	0.1	–[b]	–[b]	–[b]	0	0
Bats	0.3	0.1	0	0	0.1	–[b]	0	0	0.1	–[b]	0	0	0	0
Lagomorphs	0	0	0	0	2.5	10.7	0	0	0	0	0	0	0	0
Birds	1.0	0.4	5	2.6	4.1	1.2	1	0.4	8	3.8	2.6	1.2	0.2	0.1
Reptiles	0.7	0.3	4	1.3	0.1	–[b]	0	0	0.2	–[b]	0	0	0	0
Amphibians	0.3	–[b]	0	0	0	0	0	0	0	0	0	0	0	0
Invertebrates[a]	0.5	–[b]	20.5	0.5	4.2	–[b]	5.0	0.1	2.9	–[b]	1.9	–[b]	0	0
Total of prey items	391		78		2603		2428		1201		1880		531	
Total biomass (g)	18,991.7		3075.5		187,960.3		124,319.8		60,905.6		81,495.6		23,895.3	
N° of pellets	371		41		1789		1632		508		783		302	

SS = spring–summer, FW = fall–winter, F% = number frequency of prey item, B% = biomass percentage

[a]Mainly insects; only a scorpion was registered in northern Chile. [b]<0.1%. References: northern Chile = 15, 18, 22; central Chile = 1–5 8, 9, 12, 14, 17, 20; southern Chile = 6, 7, 11, 13, 16, 19, 21, 23; southernmost Chile = 10. [1]Housse 1945, [2]Reise 1970, [3]Schamberger and Fulk 1974, [4]Jaksic and Yáñez 1980a, [5]Cerpa and Yáñez 1981, [6]Rau et al. 1985, [7]Mayorga and Mansilla 1988, [8]Torres-Mura and Contreras 1989, [9]Zunino and Arcos 1989, [10]Iriarte et al. 1990, [11]Muñoz-Pedreros and Murúa 1990, [12]Ebensperger et al. 1991, [13]Simeone 1995, [14]Jaksic et al. 1997, [15]1999, [16]González-Acuña et al. 2004, [17]Begall 2005, [18]Carmona and Rivadeneira 2006, [19]Figueroa et al. 2009, [20]Muñoz-Pedreros et al. 2010, [21]Muñoz-Pedreros et al. 2016a, [22]Valladares et al. 2016, [23]authors, unpublished data

Table 7.32 Seasonal diet of *Bubo magellanicus* across bioclimatic zones of Chile

	Northern				Central				Southernmost			
	SS		FW		SS		FW		SS		FW	
Prey item	F%	B%	F%	B%	F%	B%	F%	B%	F%	B%	F%	B%
Rodents	89.8	96.9	93.2	99.0	41.4	28.3	87.2	56.5	80.4	81.6	85.9	92.7
Marsupials	3.7	2.5	1.6	0.6	1.6	0.3	0	0	0	0	0	0
Lagomorphs	0	0	0	0	13.6	66.7	12.8	43.5	0.7	10.9	0.3	5.8
Birds	0.9	0.4	0.5	0.3	16.0	4.5	0	0	1.5	7.1	2.3	1.2
Reptiles	0	0	0	0	0.1	–[b]	0	0	–[b]	–[b]	0	0
Insects	3.1	0.1	4.7	0.1	9.5	–[b]	0	0	17.3	0.3	11.5	0.3
Arachnids[a]	2.5	–[b]	0	0	17.7	0.1	0	0	0	0	0	0
Total of prey items	324		190		742		39		1625		732	
Total biomass (g)	13,287		7414.5		122,362		9188.3		81,821.2		29,044.5	
N° of pellets	161[c]		136		392		39		382		314	

SS = spring–summer, FW = fall–winter, F% = number frequency of prey item, B% = biomass percentage

[a]Spiders plus scorpions; [b]< 0.1%. [c]One of the studies did not report number of pellets. References: northern Chile = 9, 12; central Chile = 3, 4, 6, 10, 13, 15; southernmost Chile = 1–3, 5–8, 11, 14, 16, 17. [1]Reise and Venegas 1974, [2]Jaksic et al. 1978, [3]Yáñez et al. 1978,[4]Jaksic and Yáñez 1980a), [5]Rau and Yáñez 1981, [6]Jaksic et al. 1986, [7]Iriarte et al. 1990, [8]Tala et al. 1995, [9]Torres-Mura et al. 1997, [10]Mella 2002, [11]Torres-Mura and Lemus 2003, [12]Mella et al. 2016, [13]Muñoz-Pedreros et al. 2016b, [14]Álvarez 2016, [15]Corales et al., in preparation, [16]R.A. Figueroa and E.S. Corales, unpublished data, [17]D.R. Martínez, unpublished data

Table 7.33 Seasonal diet of *Athene cunicularia* across bioclimatic zones of Chile

	ISCC	Northern				Central				Southern			
	SS	SS		FW		SS		FW		SS		FW	
Prey items	F%	F%	B%[a]	F%	B%[a]	F%	B%	F%	B%	F%	B%	F%	B%
Rodents	1.7	10.2	65.3	1.7	35.9	10.4	79.9	5.4	72.6	4.5	67.3	10.5	77.9
Marsupials	1.9	3.1	10.4	0.5	5.3	0.9	3.3	0.2	1.7	0	0	0	0
Bats	0.1												
Lagomorphs	0.3	0	0	0	0	–[d]	2	0	0	0	0	0	0
Birds	5.6	0.3	0.6	0.7	5.3	0.7	1.8	0.4	1.8	0.1	0.7	–[d]	0.2
Reptiles	1	9.1	15.1	3.9	19.6	0.1	0.7	–[d]	0.3	0	0	0	0
Amphibians	0	0	0	0	0	3	1.8	2	2.3	0	0	0	0
Echinoid	–[d]	0	0	0	0	0	0	0	0	0	0	0	0
Insects	74	52.4	5.8	63.5	23.1	70.2	8.6	83.0	19.2	94.2	29.3	86.6	19.8
Chilopods	0	0	0	0	0	–[d]	–[d]	0.1	–[d]	0	0	0	0
Crustaceans[b]	0	0	0	0	0	0.2	–[d]	0	0	0.7	2.6	0.3	2.1
Arachnids[c]	15.4	24.9	2.8	29.7	10.8	14.5	1.8	8.8	2	0.5	0.1	2.6	–[d]
Mollusk	0	0	0	0	0	–[d]	–[d]	–[d]	–[d]	0	0	0	0
Total of prey items	3915	353		408		13,310		11,242		3396		3966	
Total biomass (g)	–[e]	3173		1122.5		108,682		48,587		10,918.6		17,364.2	
N° of pellets	468	86		90		2256		1550		443		489	
N° of prey remains	3455	0		0		0		0		0		0	

SS = spring–summer, FW = fall–winter, F% = number frequency of prey item, B% = biomass percentage. The islands of the Coastal System of Coquimbo (ISCC) are located at northern Chile, but these have been treated separately because of the insular condition of *A. cunicularia*

[a]Values lightly vary with those of the original source because we used distinct mean mass of prey. [b]Crayfish and/or small crabs. [c]Spiders plus scorpions. [d]< 0.1%. [e]It was not estimated due to the imprecise taxonomic identification of prey birds. References: ISCC = 8, 14; northern Chile = 11, 13; central Chile = 1, 2, 4, 5, 9, 10, 12; southern Chile = 3, 6, 7. [1]Péfaur et al. 1977, [2]Yáñez and Jaksic 1979, [3]Fernández et al. 1980, [4]Núñez and Yáñez 1982, [5]Schlatter et al. 1982, [6]Villagra and Mora 1988, [7]Pradenas 1991, [8]Zunino and Jofré 1999, [9]Torres-Contreras et al. 1994, [10]Silva et al. 1995, [11]Carevic 2005, [12]Roa 2011, [13]Carevic et al. 2013, [14]Cruz-Jofré and Vilina 2014

Table 7.34 Seasonal diet of *Glaucidium nana* across bioclimatic zones of Chile

	Central				Southern	
	SS		FW		SS	FW
Prey item	F%	B%	F%	B%	F%	F%
Rodents	24.1	50.7	42.3	79.5	27.0	42.7
Marsupials	1.7	2.0	0.4	0.4	0	0
Birds	14.4	40.3	13.7	18.1	13.4	29.6
Reptiles	2.8	4.0	0.9	0.3	3.2	2.0
Amphibians	0	0	0	0	3.2	2.9
Insects	55.0	2.9	40.6	1.6	53.2	22.8
Arachnids[a]	2.0	0.1	2.1	0.1	0	0
Total of prey items	535		234		156	307
Total biomass (g)	10,091.7		6002.1		–[b]	–[b]
N° of pellets	195		116		71	164

SS = spring–summer, FW = fall–winter, F% = number frequency of prey item, B% = biomass percentage

[a]Spiders plus scorpions. [b]Biomass was not estimated because the identification of some species is still in process. References: central Chile = 1, 2; southern Chile = 3. [1]Jiménez and Jaksic 1989, [2]Jiménez and Jaksic 1993, [3]R.A. Figueroa et al., unpublished data

Table 7.35 Seasonal diet of *Asio flammeus* in the Juan Fernández Archipelago (JFA) and in southern Chile

	JFA		Southern			
	SS		SS		WF	
Prey item	F%	B%	F%	B%	F%	B%
Rodents	9.8	1.9	35.5	36.8	89.6	95.0
Lagomorphs	60.8	74.3	0	0	0	0
Other mammals	5.9	3.1	0	0	0	0
Birds	19.6	20.6	10.5	61.9	1.4	4.8
Insects	3.9	–[a]	54	1.3	9	0.2
Total of prey items	51		420		585	
Total biomass (g)	22,273		17,136		33,167.5	
N° of pellets	39		125		321	

SS = spring–summer, FW = fall–winter, F% = number frequency of prey item, B% = biomass percentge

[a]<0.1%. References: JFA = 2, 5; southern Chile = 1, 3, 4. [1]Rau et al. 1992, [2]Fuentes et al. 1993, [3]Martínez et al. 1998, [4]Figueroa et al. 2009, [5]authors, unpublished data

the abundance of rodent prey in the field (Martínez et al. 1998). Although Páez-López et al. (2005) described the diet of *A. flammeus* in central Chile, the very low level of detail of their results prevented us from including it in our analysis. In contrast to mainland Chile, individuals of *A. flammeus* in Juan Fernández Archipelago prey mainly upon lagomorphs during summer (Table 7.35).

Table 7.36 Seasonal diet of *Strix rufipes* across bioclimatic zones of Chile

	Central				Southern			
	SS		FW		SS		FW	
Prey item	F%	B%	F%	B%	F%	B%	F%	B%
Rodents	50.3	93.3	60.0	91.6	18	81	35.9	88.6
Marsupials	3.8	1.7	3.8	1.4	2.0	7.1	2.9	6.1
Bats	0	0	0	0	0.1	0.3	0	0
Lagomorphs	0.5	4.3	0.5	3.7	0	0	0	0
Birds	0	0	1.4	2.8	1.2	1.8	0.8	1.1
Reptiles	0	0	0.3	–[d]	0	0	0	0
Amphibians	0	0	0	0	0.4	0.5	0.1	–[d]
Insects	35.3	0.5	32.6	0.4	77.4	9	59.6	4.1
Crustaceans[a]	0.3	–[d]	0.3	–[d]	0.6	0.2	0.2	–[d]
Arachnids[b]	9.8	0.1	1.1	–[d]	0	0	0.4	–[d]
Snails[c]	0	0	0	0	0.3	0.1	0.1	–[d]
Total of prey items	366		362		687		1065	
Total biomass (g)	24,558.5		29,020		5909.3		15,348.9	
N° of pellets	166		146		209		339	
N° of prey remains	17		54		0		0	

SS = spring–summer, FW = fall–winter, F% = number percentage of prey item, B% = biomass percentage

[a]Small crabs (*Aegla* spp.). [b]Only scorpions. [c]Forest snails. [d]<0.1%. References: central Chile = 4, 6–8; southern Chile = 1–3, 5. [1]Martínez 1993, [2]Martínez 1995, [3]Figueroa 1996, [4]Díaz 1999, [5]Figueroa et al. 2006. [6]Vukasovic et al. 2006, [7]Alvarado et al. 2007, [8]Ramírez 2008

The diet composition of *Strix rufipes* depends on bioclimatic zone (Table 7.36). In sclerophyllous forests, the seasonal changes in prey consumption are not so evident; small mammals (i.e. <500 g) and insects are the most consumed prey all year round. In contrast, in the southern temperate forest, *S. rufipes* shows strong seasonal changes in diet composition; while consumption of insect prey increases towards spring-summer, consumption of small mammals increases towards autumn-winter (Table 7.36). In addition, individuals inhabiting southern temperate forests prey much more upon arboreal and scansorial small mammals than do individuals in sclerophyllous forests (Figueroa et al. 2006). Independent of climatic zones, small mammals always account for greater biomass contribution.

7.3.13.2 Trophic Specialization

The degree of trophic specialization of Chilean owls can be inferred from Fig. 7.7. If we consider an arbitrary cut-off point of 50% of all eaten individual prey, *Tyto furcata* and *Bubo magellanicus* became mammal-specialist predators, *Athene cunicularia* qualifies as an arthropod-specialist predator, and *Glaucidium nana*, *Strix rufipes,* and *Asio flammeus* qualify as generalist predators (i.e. consume similar proportions of different types of prey). However, these definitions do not consider the biomass provided by the prey (e.g. Silva et al. 1995). Taking into account the biomass information in Tables 7.31, 7.32, 7.33, 7.34, 7.35 and 7.36, we tentatively redefine the degree of trophic specialization of Chilean owls. *T. furcata* may be considered rather as a predator super specialized on small mammals as these are the preeminent prey in its diet and provide most of the biomass. *B. magellanicus* may still be defined as a predator specialized in mammalian prey because these constitute the greater contribution to diet in term of both number and biomass. *A. cunicularia* seems to be rather a facultative generalist predator as it has the ability to consume typical local prey which can eventually make an important contribution in number and biomass to its diet. *G. nana* is redefined here as an absolute generalist predator. Although this species consumes more insects than mammals and birds, it obtains almost all the biomass from these latter. *A. flammeus* may be considered rather a predator specialized on vertebrates because birds and mammals are the basis of its diet in terms of both numbers and biomass. Finally, *S. rufipes* is redefined as a predator specialized on small mammals; although the contribution of small mammals to diet is variable in terms of number, small mammals always account for most of the biomass. The patterns of trophic specialization we describe for *T. furcata*, *B. magellanicus*, *A. cunicularia,* and *G. nana* have been corroborated by long-term studies conducted in central Chile (see below). In the case of *S. rufipes*, it was previously defined as a generalist predator taking a broad spectrum of prey (Martínez 1993). However, it should be noted that no prey, except for small mammals, contributes >5% in terms of biomass.

7.3.13.3 Latitudinal Trends in Diet

Assuming the diversity of prey species decreases towards southern latitudes, some authors have attempted to verify if this is reflected in the diet diversity of Chilean owls. Carmona and Rivadeneira (2006) compared the diet diversity of *Tyto furcata* among ten localities along 30° of latitude between Atacama Desert and southern Patagonia (22°–51°S). By using the H 'Shannon-Wiener index as an estimator of food diversity, these authors found that the total diversity in diet of *Tyto furcata* decreased towards southern latitudes. Carmona and Rivadeneira (2006) explained that it was because of the sustained decline in the consumption of arthropods towards the south and the localized consumption of reptiles in more northern localities (i.e. northern and central Chile). Surprisingly, however, the diversity of rodent prey in the diet increased markedly towards south. Recently, Valladares et al. (2016) described the diet of *T. furcata* in the Copiapó Valley, Atacama Desert. These authors detected a low total diversity of prey in the diet of this owl species on its study site compared with those of southern Chile where the number of prey increases to more than double. This contradicts Carmona and Rivadeneira (2006) who characterized the diet of *T. furcata* northern Chile as the most wide-ranging diet compared to other areas of Chile. This suggests that diet diversity of *T. furcata* in northern Chile may be locality-dependent, and further studies are needed to reliably establish diet patterns. Based on a higher number of localities (18), but covering a smaller range of latitudes (30–51°S), Muñoz-Pedreros et al. (2016a) found similar results. By using the standardized Levin's index, these authors obtained the highest values of total diversity in diet in localities of central Chile and lowest values in localities of southern Chile. In agreement with Carmona and Rivadeneira (2006), Muñoz-Pedreros et al. (2016a) also found that small-mammal diversity in diet at 51°S was as high as in central Chile.

Jaksic et al. (1986) analysed the diet *of Bubo magellanicus* in three localities along a latitudinal gradient between central and southernmost Chile (33–51°S). By using the Levin's standardized index, these authors also detected that diversity of prey in the owl's overall diet decreased steadily towards southern latitudes. In addition, Jaksic et al. (1986) found that whereas the proportion of avian prey in the diet decreased towards south, the opposite happened with insects. Recently, Muñoz-Pedreros et al. (2016b) re-analysed the latitudinal variations in diet of this owl species including a greater number of localities (N = 5). Although they found that overall prey diversity in diet among localities differed significantly, no clear latitudinal pattern was detected. In fact, overall diet diversities greatly differed among localities within a same bioclimatic zone. Like Jaksic et al. (1986) and Muñoz-Pedreros et al. (2016b) found that proportion of avian prey tends to decrease, and insect prey tends to increase towards southern latitudes. On the other hand, Mella et al. (2016) found that the diet diversity of *B. magellanicus* in northernmost Chile tends to be lower than in southernmost Chile. This suggests that the latitudinal trend in diet diversity of *B. magellanicus* throughout Chile is not linear but rather curvilinear. Results from all studies above may reflect the local availability and/or spectrum of prey at the time when diets were evaluated. However, in the case of *B. magellanicus*, results may also be an artefact of the low number of sampled localities.

7.3.13.4 Prey Selection

Jaksic (1989b) defined a "selective predator" as one that takes its prey in a different proportion relative to their field abundance and an "opportunistic predator" as one that takes their prey according to their field abundance. Several studies suggest that Chilean owls tend more to be selective than opportunist when preying upon vertebrate prey.

Tyto furcata seems to overconsume or under-consume some small-mammal species in relation to their field abundance (Jaksic 1979, 1986; Jaksic and Yáñez 1979; Simonetti and Walkowiak 1979; Jaksic et al. 1981, 1992, 1997; Iriarte et al. 1990; Muñoz and Murúa 1990). A most recent study conducted in Lago Peñuelas, central Chile, appears to confirm prey selection by this owl species. In this locality, at least three rodent species were apparently preferred (i.e. were consumed more than expected in relation to field abundance), and three were apparently rejected (i.e. were consumed less than expected in relation to field abundance) by *T. furcata* (Muñoz-Pedreros et al. 2016a). However, partially opposite results were found in agroecosystems of southern Chile, where the level of importance of rodent species in the diet of *T. furcata* and *Asio flammeus* roughly reflected the field abundance of prey species (Figueroa et al. 2009).

Bubo magellanicus also appears to have a selective consumption of certain species of small mammals (Jaksic 1979, 1986; Jaksic et al. 1977, 1981, 1992, 1997; Iriarte et al. 1990; Muñoz-Pedreros et al. 2016b), but in southernmost Chile, it may act as an opportunistic predator (Tala et al. 1995). In central Chile, *Athene cunicularia* and *Glaucidium nana* selectively prey upon an array of small-mammal species, overconsuming some and under-consuming others (Jaksic 1979; Jaksic et al. 1977, 1981, 1986, 1992, 1997; Torres-Contreras et al. 1994). Although Carevic (2011) denoted some degree of prey selection by *A. cunicularia* during spring in northern Chile, this owl tended to consume more rodent and insect species than were more often caught in live traps.

In Valdivian forests, Martínez and Jaksic (1997) found that *Strix rufipes* consumes scansorial/arboreal small mammals (e.g. *Oligoryzomys longicaudatus*, *Dromiciops gliroides*) more than expected relative to their field abundance. In a subsequent analysis, Alvarado and Figueroa (2006) point out that in the sclerophyllous forest, *S. rufipes* captures a higher number of terrestrial small mammals and/or with a mass > 66 g (e.g. *Abrocoma benetti*, *Octodon lunatus*, *Phyllotis darwini*), whereas in the southern temperate forest, it captures a higher number of climbers/arboreal small mammals and/or with a mass < 45 g (e.g. *O. longicaudatus*, *Abrothrix longipilis*, *Irenomy tarsalis*).

The seemingly selective predation on small mammals by Chilean owls could be a result of different factors, alone or combined, including circadian activity, prey vulnerability and abundance, microhabitat use, and prey size, along with hunting tactics, handling capacity, and energy requirement of owls (Jaksic et al. 1981; Jaksic 1986; Iriarte et al. 1990; Martínez and Jaksic 1996; Figueroa et al. 2009). However, in many cases it is also possible that the supposed prey selection is instead an artefact of the statistical procedures utilized and subsampling of the prey offer.

7.3.13.5 Prey Vulnerability

Jaksic et al. (1992) summarize the potential vulnerability of mammalian prey inhabiting central Chile. The two most frequent rodent species in the diet of local owls are *Abrocoma bennetti* and *Thylamys elegans*. *A. bennetti* is the second largest of the small-mammal assemblage, and *T. elegans* is the smallest one. The first species exhibits an unrefined escape response, combining the "freezing" of its movements with a slow trotting gait. This would make it more vulnerable to *Bubo magellanicus*, a predator able to hunt large prey with high efficiency. *T. elegans* runs for cover directly but slowly which makes it vulnerable to all local owls. *Octodon degus* is a colonial rodent that emits warning calls revealing its location to predators, thus explaining its high proportion in the diet of *Glaucidium nana*.

In southern Chile, *Oligoryzomys longicaudatus* and *Abrothrix olivaceus* are the rodent species most preyed upon by owls. In agricultural lands, the first species preferentially utilizes dense shrubs mixed with pasture on fence borders which increases the probability of predation because the fence posts are the favourite perches of owls (Martínez et al. 1998). In addition, this rodent species is highly mobile and vagrant (Murúa et al. 1986) making it easily detectable by predators. Within the forest, *O. longicaudatus* also climbs trees which make it vulnerable to owls (Martínez 1993; Martínez and Jaksic 1996). *A. olivaceus* occupies more open microhabitats which increase their vulnerability to *Tyto furcata*, *Asio flammeus*, *Athene cunicularia,* and *G. nana*. *Dromiciops gliroides* may constitute an important proportion of prey items in the diet of *S. rufipes* (Martínez 1993; Figueroa et al. 2006). This latter could be a consequence of the complete overlapping in circadian activity of both species and of arboreal habits of *D. gliroides*, involving vertical and horizontal displacements (R.A. Figueroa, unpublished data; Fig. 7.8).

7.3.13.6 Relationship Between Body Size of Owls and Dietary Diversity

In central Chile, Jaksic et al. (1981) found a positive correlation between the body sizes of *Bubo magellanicus*, *Tyto furcata,* and *Athene cunicularia* and their respective food-niche breadth (i.e. dietary diversity). However, the highest food diversity for the largest owl seemed to be due to a more even consumption of the same prey items taken by all three owls and not because it had incorporated more taxa prey in its diet (Jaksic and Yáñez 1980a). Furthermore, Jaksic et al. (1992) found that there is a temporally inconsistent relationship between body size and food diversity of owls. In southern Chile, up to now, there has not been found a positive relationship between the body sizes of *T. furcata*, *A. cunicularia,* and *Asio flammeus* and their respective food diversities; the food-niche breadth is always higher for the smaller owl species, i.e. *A. cunicularia* (Rau and Jaksic 2004b; Rau et al. 2005). Figueroa et al. (2009) found a positive relationship, although not significant, between the body size of *A. flammeus* and *T. furcata* and their respective dietary diversities (it should be noted that both species differ slightly in body size). In southernmost Chile, Iriarte et al. (1990) found that *B. magellanicus* had a higher food breadth than *T. furcata* because the former made a significant consumption of European hares, a prey item not consumed by *T. furcata*. The inconsistency of all these results was previously noticed by Jaksic (1989a) who concluded that the body size of Chilean raptors is a labile predictor of their food diversities.

7.3.13.7 Relationship Between Body Size of Owls and Their Prey

Chilean owls can capture vertebrate animals with a wide range of body size. Eventually, all owl species can access prey as large as its own size or even larger. In fact, *Glaucidium nana* is capable of capturing prey that significantly exceeds its body mass (Jiménez and Jaksic 1989, 1993; Figueroa and Corales 2015). The strength and size of its feet and talons allow this small owl to kill and easily handle large-sized prey (Jiménez and Jaksic 1989). Thus, the body size of owls by itself would not be a constraint to access large-sized prey.

Table 7.37 Mean mass of vertebrate prey consumed by Chilean owls across bioclimatic zones of Chile

		Mass range (g)				Mean mass (g)[c]			
Species[a]	Mass (g)[b]	N	C	S	SM	N	C	S	SM
Bubo magellanicus	886	15–200	16–534	–	17–1200	43	183	–	58.5
Strix rufipes	434	–	27–534	15–134	–	–	127	33.7	–
Asio flammeus	380	–	20–112	15–250	–	–	–	69.0	–
Tyto furcata	340	15–150	17–233	15–320	17–60	48.6	63.6	39.8	45
Athene cunicularia	291	20–80	20–230	24–250	–	34	42.0	35.5	–
Glaucidium nana	61–73	–	5–160	15–134	–	–	42.0	–	–

N = northern Chile, C = central Chile, S = southern Chile, SM = southernmost Chile
[a]Listed in decreasing body size to better visualize the relationship with the mean mass of their prey.
[b]From Table 7.5. [c]Mean mass = sum of the products between the number of individual prey by its mean mass divided by total number of prey included in the computation. References: Tables 7.31, 7.32, 7.33, 7.34, 7.35 and 7.36

Independent of the above, several studies have found a positive correlation between the mean body mass of Chilean owls and that of their vertebrate prey (Jaksic 1983; Iriarte et al. 1990; Figueroa et al. 2009). Some studies suggest that local owl assemblages tend to keep the mean body size of their prey over time (Jaksic et al. 1993). An interesting fact is that the relationship between the mean body size of some owls and that of their mammalian prey, commonly observed at the interspecific level, only is reflected at the intraspecific level when the prey species exceeds a certain body size. In Aucó, Castro and Jaksic (1995) found that *Bubo magellanicus* and *Tyto furcata*, despite the considerable difference in body size, showed no significant difference between the respective frequency distributions of the body size of *Phyllotis darwini* (mass 50–60 g), the rodent prey most consumed by both species. Subsequently, Santibáñez and Jaksic (1999) assessed the differences in body size in the consumption of *Abrocoma bennetti*, a rodent species almost 3–4 times larger than *P. darwini*. These authors found that *B. magellanicus* consumed those individuals almost 30% larger than those consumed by *T. furcata* (178 g and 145 g, respectively). In Table 7.37 we summarize estimates of body sizes of vertebrate prey made on a much broader database. It should be noted that for all owl species, regardless of their body size, the mean mass of vertebrate prey tends to vary geographically. Such variations could reflect the distribution of body size of local prey assemblages (Jaksic 1997; Figueroa et al. 2009).

An aspect directly related to the body size of prey is the energy return. Bozinovic and Medel (1988) computed the expected metabolic rate (KJ/day) and the theoretical daily food requirements (g/day) for *B. magellanicus*, *T. furcata*, and *A. cunicularia*, and they found that all three owl species consumed vertebrate prey of different body sizes according to their daily energetic requirements (Table 7.38). Thus, these authors postulated that Chilean owls should select vertebrate prey with a body size

Table 7.38 Estimates of the expected metabolic rate (EMR) and theoretical food requirement (TFR) of three Chilean owl species according to their body mass

Species	Mass (g)	EMR (KJ/day)[a]	TFR (g/day)[b]
Athene cunicularia	250	373.4	72.9
Tyto furcata	310	428.4	83.7
Bubo magellanicus	1250	1045.8	204.2

Reference: Bozinovic and Medel 1988
[a]EMR = $10.9(\text{mass})^{0.64}$. [b]TFR = EMR_{owl} (KJ/day)/6.65(KJ/g) × efficiency of assimilation (%)

that conforms to their energy requirements. However, Bozinovic and Medel (1988) were based on information from an area where the most consumed vertebrate prey was also the largest and most abundant in the field (*Octodon degus*; Jaksic et al. 1981), so their results were inconclusive. In other localities, Chilean owls tend to include among the most consumed prey those of larger size within its handling capabilities (e.g. Cerpa and Yáñez 1981; Meserve et al. 1987; Muñoz and Murúa 1990; Jaksic et al. 1992). In agricultural lands of southern Chile, the rodent species most preyed by *T. furcata* and *A. flammeus* were both those most abundant (*Abrothrix olivaceus*, *Oligoryzomys longicaudatus*) and the largest one (*Loxodontomys micropus*) within the local assemblage of small mammals (Figueroa et al. 2009). This suggests that Chilean owls can compensate between body size and abundance of prey to achieve their energy requirements.

7.3.13.8 Demographic Characteristics of Prey

In Fray Jorge, central Chile, Fulk (1976) found that juvenile individuals of *Abrothrix olivaceus* and *Phyllotis darwini* (the most frequent rodent prey species in the local predators' diet) constituted >50% of all prey taken by *Tyto furcata* and *Asio flammeus* for most of the year. Only in spring, both owl species took more adult than juvenile individuals of *P. darwini*. Fulk (1976) explained that such results were influenced by the availability and/or differential vulnerability of juvenile and adult rodents. In Las Cardas, central Chile, Péfaur et al. (1977) found that *Athene cunicularia* indistinctly preyed upon juvenile and adult individuals of *P. darwini*, but markedly on adult individuals of *A. olivaceus*, interpreting their results as evidence of prey selection on a prey size basis. In Aucó, Castro and Jaksic (1995) found that the proportion of adult individuals of *P. darwini* in the diet of *T. furcata* and *Bubo magellanicus* accounted for >70% of all prey items. These authors suggested that the highest consumption of adult mice would be associated with their increased vulnerability during breeding season when reproductive males in search of females are perhaps less cautious or move more within their home range and when pregnant females are slower at escaping from predators. On a coastal area of central Chile, Cerpa and Yáñez (1981) found that *T. furcata* disproportionately preyed upon older adult mice (almost 88% of all individual prey), suggesting that this would allow this owl species to achieve their optimal prey size to meet their energy requirements. In the Andean foothills of Santiago, central Chile, Schlatter et al. (1980) noted that *A.*

cunicularia consumed a small number of adult individuals of large-sized small-mammal species (>100 g), but a considerable amount of adult individuals of small-sized small-mammal species (<60 g). Apparently, *A. cunicularia* adjusted its diet according to a prey size it could handle. In southern Chile, Tillería and Rau (1991) found no evidence of a differential consumption on age classes of *A. olivaceus* by *T. furcata* and *A. cunicularia*, arguing that predation on this rodent species was rather influenced by its environmental availability. In Valdivia, Simeone (1995) found that *T. furcata* consumed almost three times more juvenile than adult mice, speculating that it may have been a consequence either of abundance, vulnerability, or easier prey handling.

Regarding sex of prey, Fulk (1976) found that both males and females of *A. olivaceus* and *P. darwini* were equitably represented in the pellets of *T. furcata* and *A. flammeus* for the most of year. Only during spring the males of *P. darwini* were disproportionately consumed in relation to their environmental abundance. As explained above, the vulnerability of male of *P. darwini* increases during the breeding season because they are more active and move more within their home range. Tillería and Rau (1991) found no difference between the proportion of males and females of *A. olivaceus* consumed by *T. furcata* and *A. cunicularia*. Since wild populations of *A. olivaceus* on agricultural lands tend to exhibit a sex ratio of 1: 1 (R.A. Figueroa, unpublished data), the results of Tillería and Rau (1991) suggest that both owl species take male and female mice in proportion to their respective abundances in the field.

7.3.13.9 Temporal and Spatial Segregation in the Use of Prey Resource

Jaksic et al. (1981) noted that among the small mammals preyed upon by two strictly nocturnal owls, *Bubo magellanicus* and *Tyto furcata*, almost 30% and 40% of individuals corresponded to species with crepuscular and nocturnal habits, respectively. In addition, almost 10% were diurnal species. On the other hand, the diet of *Athene cunicularia*, an owl with large diurnal activity, consisted of nearly 17% of nocturnal and crepuscular prey. This seems to be related to two factors operating singly or in combination: (i) diurnal and nocturnal owls may extend their hunting periods to crepuscular hours, thus having access to prey active at dusk and dawn; and (ii) prey species, whatever their activity patterns, may extend activities in the field exposing themselves to owls of different hunting periods (Jaksic et al. 1981; Muñoz-Pedreros et al. 1990; Figueroa et al. 2009). Differences in the hunting period of owls have an observable correlation at the level of prey species consumed, but they appear not to be sufficient to prevent the exploitation of prey with different activity rhythms and cause segregation in the use of prey resources (Jaksic et al. 1981; Jaksic 1982). Some owl species may even extremely extend their activity period which allows them to access to unusual prey. In Navarino Island, Ippi and Rozzi (2004) registered *Strix rufipes* (a typically nocturnal owl) capturing and eating small birds at midday (see Behavior).

Syntopic owl species with similar circadian rhythms appear to exhibit low segregation in the use of hunting habitats. Based on prey types, Jaksic et al. (1981) concluded that *B. magellanicus*, *T. furcata*, and *A. cunicularia* exploit the prey resources in a similar manner in open and dense shrubs. In agricultural lands of southern Chile, *T. furcata* and *Asio flammeus* showed a marginal segregation because both species converge on the same and most abundant prey species (Figueroa et al. 2009).

7.3.13.10 Response to Temporal Changes in Prey Abundance

The response of Chilean owls to temporal variations in prey abundance, particularly rodent prey, has been analysed from distinct perspectives and in attempts to answer different questions. Some questions have adapted to the theoretical interest of the moment, and others have emerged from the ocurrence of natural phenomenon that cause changes in biological productivity. Initially, ecologists sought to determine whether top predators manifest a functional and/or numerical response to temporal changes in the abundance of their prey. Subsequently, research focused on assessing how top predators respond to population explosions of rodent prey. Most recently, some studies tried to envision how top predators perceive the asynchrony in prey availability over the long term. Below we synthesize the main findings of these studies.

Functional Response Different studies show that Chilean owls respond functionally to increases in prey abundance (Jaksic and Simonetti 1987). Fulk (1976) found that *Tyto furcata* and *Asio flammeus* consumed much more *Abrothrix olivaceus* just when it was more abundant and much less *Phyllotis darwini* just when it was less abundant. Jaksic and Simonetti (1987) documented a similar trend for *Bubo magellanicus*, *T. furcata*, and *Athene cunicularia*. However, long-term studies have revealed that the intensity of response may vary among species of owls. In the semi-arid scrubland of central Chile, *A. cunicularia* does not appear to respond functionally to small-mammal outbreaks, *B. magellanicus* and *T. furcata* maintain preying upon small mammals until they reach their lowest abundance, and *G. nana* increases consumption of small mammals when these became most abundant (Jaksic et al. 1992, 1997). In southern Chile, *Asio flammeus* and *Strix rufipes* tend to adjust to the seasonal abundance cycle of their prey consuming many more rodents during winter and more insects during the summer, when these prey reach their maximum abundance, respectively (Figueroa 1996; Martínez et al. 1998; Figueroa and Martínez, in preparation). In northernmost Chile, Mella et al. (2016) detected a seasonal switching of prey for *B. magellanicus*: when *P. xanthopygus* increased in the diet, the consumption of *A. andinus* decreased, and vice versa. These authors suggest that alternation in the consumption of these two rodent species can be explained by the differential availability in the field across seasons, possibly as consequence of antagonist interactions. González-Acuña et al. (2004) documented an unusual case of functional response by *T. furcata* inside the city of Chillán. In the spring of 1999, this owl species made a considerably higher consumption of passerine birds (>50%

of all individual prey found in the pellets), coinciding with the recruitment period of these birds and the low availability of rodents.

Numerical Response Jaksic et al. (1992) observed that during a small-mammal irruption in Aucó, central Chile, three of the four owl species present in the area were active for all the study time, despite the dramatic decline in prey abundance after the outbreak. *Tyto furcata* was the only species that apparently abandoned the area immediately after that small-mammal abundance dramatically declined. In Fray Jorge, central Chile, *Athene cunicularia* responds strongly to the increase in prey availability, clearly co-varying the number of resident individuals with the density of small mammals in the field (Silva et al. 1995). Casual observations suggest that *A. flammeus* could numerically respond to fluctuations in prey abundance. Housse (1945) reported that in the winter of 1934 a flock of *A. flammeus* suddenly appeared in agricultural areas of Capitan Pastene, southern Chile, almost at the same time as a rat plague invaded croplands, and owls left the area after the end of the plague. Rau (1994) stated that in agricultural areas of Osorno, southern Chile, *A. flammeus* remains on grasslands during autumn and spring, doing local migrations for the rest of the year. This is consistent with the seasonal cycle in rodent abundance on agricultural grasslands of southern Chile (Martínez et al. 1998; Figueroa et al. 2009).

Response to Ratadas In Chile, the massive rodent irruptions or outbreaks are known as "ratadas" and are associated with two long-term cyclical events which increase several times the primary productivity: (i) "El Niño" Southern Oscillation and (ii) the bamboo bloom (Murúa et al. 1996; Jaksic and Lima 2003). Studies conducted in central Chile show that local owls respond differentially to ratadas caused by the "El Niño" event. In Aucó, *Bubo magellanicus* and *Tyto furcata* steadily consumed a high number of small mammals for 6 years, independent of a ratada that occurred between 1991 and 1992 (Jaksic et al. 1997). However, *B. magellanicus* consumed many more individuals of *Octodon degus* during the ratada, just when this rodent species became more abundant. Furthermore, *Athene cunicularia* and *Glaucidium nana* responded strongly to the ratada by increasing at several times the consumption of small mammals. In addition, *A. cunicularia* also responded numerically to the ratada; this owl increased in twice its spatial density regarding to the pre-ratada years, but returning to its original density during the post-ratada years (Silva et al. 1995; Jaksic et al. 1997). Circumstantial information suggests that *Strix rufipes* responds functionally to ratadas resulting from bamboo mast seeding. In Valdivian forest remnants, this owl species consumed many more *Oligoryzomys longicaudatus* (the species most numerically favoured by the bamboo bloom) during the year of ratada (1994–1995; 40–81% of all rodent prey found in pellets) than during the year when the bamboo bloom initiated (1993; 0–63% of all rodent prey found in pellets; R.A. Figueroa, unpublished data). In addition, during the year of ratada, the pellets of *S. rufipes* contained more than twice the number of *O. longicaudatus* regarding the year when began the bamboo bloom (R.A. Figueroa, unpublished data).

Perception to Asynchrony in the Prey Resource Dynamics In central Chile, the local owls seem to respond idiosyncratically to the temporal variations in the prey abundance; i.e. rather than to diverge in their diets in times of low prey abundance or to converge opportunistically in times of high prey abundance, local owls seem to perceive in a different manner the level of prey resource (Jaksic et al. 1993; Farías and Jaksic 2007). *Glaucidium nana* and *Bubo magellanicus* converge on several species of small mammals when these are abundant; they concentrate predation on a subset of rodent species as total abundance of small mammals decreases and converge on arthropods when rodent prey are too scarce. *Athene cunicularia* and *Tyto furcata* do not exhibit a clear response to changes in the abundance of small-mammal prey in the field, steadily consuming arthropods or rodents, respectively. However, *T. furcata* may abandon its hunting areas when the rodent density falls to its minimum. Thus, the first two species seem to perceive the entire spectrum of states in the level of prey resources. *A. cunicularia*, due to their lower metabolic rate (see sect. 7.3.18), appears to meet much of its energetic demands just by consuming insects which do not suffer marked numerical fluctuations. The sensitivity of *T. furcata* to sharp declines in density of rodent prey corroborates its trophic super-specialization on these preys.

7.3.13.11 Response to Introduced Prey

The consumption level of allochthonous vertebrates by Chilean owls has served as an indicator of acceptance of "foreign" prey within the original assemblage of prey. Even though three decades ago, Chilean raptors showed a low level of consumption of the European rabbit (*Oryctolagus cuniculus*, Jaksic and Yáñez 1980a); currently several raptor species consume this lagomorph in an important proportion. Initially, Jaksic and Yáñez (1980a) concluded that Chilean raptors are not efficient at hunting rabbits because they had not yet developed behavioural adjustments needed to capture an agile prey recently introduced. However, some owl species may make an important consumption of rabbits (at least of juveniles) suggesting they have gained ability in capturing this prey. In central Chile, rabbits can account for almost 13–16% of prey consumed by *Bubo magellanicus* (Jaksic and Yáñez 1980b; Jaksic and Soriguer 1981; Jaksic and Marti 1984; Jaksic et al. 1986). In Lago Peñuelas, Muñoz-Pedreros et al. (2016b; see also Gil-Cordero 2003) found that the proportion of rabbits in the diet of *Bubo magellanicus* reached 22% of all prey individuals (N = 81 individual), most of which were juveniles (534 g). A study conducted in Juan Fernández Archipelago revealed that nearly 61% of all prey items eaten by *Asio flammeus* were rabbits (Fuentes et al. 1993). Muñoz-Pedreros et al. (2010) documented an unusual case of high predation upon rabbits (17–28% of all prey items) by *Tyto furcata* in Lago Peñuelas. As in the case of*B. magellanicus*, the most of consumed rabbits were juvenile individuals, a fact which could have facilitated its capture and handling by *T. furcata*. *B. magellanicus,* and *A. flammeus* also prey upon European hare (Reise and Venegas 1974; Iriarte et al. 1990; R. A. Figueroa and E. S. Corales, personal observation). In Torres del Paine,

European hares can account for almost 17% of all prey items in the diet of *B. magellanicus* (Iriarte et al. 1990).

All Chilean owls prey at some degree on allochthonous synanthropic rodents (*Rattus* spp., *Mus domesticus*). A particularly interesting study is Simeone (1995) who evaluated the effect of urban sprawl on the diet of *T. furcata*. He found that the proportion of *Rattus* spp. in diet was only 3% when the urban edge was distant from the study site (1974–1979), but it increased up to 20% when the urban edge was very close to the study site (1990–1992). Simeone (1995) concluded that *T. furcata* is trophically sensitive to urban sprawl which can influence changes in its food profile at prey rodent level. A most recent study showed that consumption of *Rattus* spp. and *Mus domesticus* by *T. furcata* along Chile can vary <1 to 24% and <1 to 17%, respectively (Muñoz-Pedreros et al. 2016a). In 14 pellets of *Bubo magellanicus* found in a disturbed forest remnant of central Chile, one third of prey items were *R. rattus* (Corales et al. in prep.). This suggests that the level of predation on introduced rodents by Chilean owls depends on the intensity of land use. That is, many more introduced rodents will be eaten by owls in more human-modified areas.

7.3.13.12 Carrion Consumption

The consumption of carrion by owls is well documented in the Northern Hemisphere. In contrast, the consumption of carrion by Chilean owls is scarcely reported. Novoa et al. (2016) reported the record of a *Bubo magellanicus* consuming a carcass of a southern lapwing (*Vanellus chilensis*), but it was not so clear if this latter was captured or found dead.

7.3.14 Population Ecology

The population ecology of Chilean owls has been partially addressed by analysing their numerical responses to temporal variations in prey availability (Jaksic et al. 1992; Silva et al. 1995). The main results of such studies were already discussed in the previous section. More recently, Lima et al. (2002) conducted a more sophisticated empirical analysis of population dynamics and structure of food webs for two species of owls (*Tyto furcata* and *Bubo magellanicus*) and three species of small mammals (*Phyllotis darwini*, *Abrothrix olivaceus*, and *Thylamys elegans*) of the semiarid scrub of central Chile. Results of this analysis revealed that the population dynamics of *T. furcata* and *P. darwini* are closely coupled: the former has a negative effect only on *P. darwini*, and the latter has a positive effect on the dynamics of *T. furcata*. On the other hand, Lotka-Volterra models suggest that population dynamics of *B. magellanicus* is regulated by intraspecific competition and limitation of food. However, this owl species shows no mutual effect with the three rodent species considered in the analysis. Possibly, this is because *B. magellanicus* prefers consuming other larger rodent species (e.g. *Abrocoma bennetti*) which could be more closely associated to its population dynamics.

7.3.15 Community Ecology

7.3.15.1 Food Niche and Trophic Guild Structure

Community ecology studies of Chilean raptors started approximately three decades ago precisely by using as a model three species of sympatric owls: *Bubo magellanicus*, *Tyto furcata*, and *Athene cunicularia* (Jaksic et al. 1977). According to Jaksic (1997), such analysis marked the beginning of a tendency towards studies that were more sophisticated and of greater extent. Most of the community studies of Chilean raptors have been based on guilds analysis and focused on elucidating whether these emerge from the opportunist response of their members to any very abundant prey or from prey selection based on body size (Jaksic 1989b, 1997; Jaksic et al. 1997; Rau and Jaksic 2004b; Rau et al. 2005). Jaksic (1997) classified the community studies of Chilean predators – that in all cases included owls – in the following categories: (i) food-niche relationship among pairs of species and within collective assemblages, (ii) analysis of latitudinal gradients in trophic ecology, (iii) interannual comparisons of the trophic structure and dynamics of collective predator assemblages, and (iv) intercontinental comparisons in the trophic structure of diurnal and nocturnal raptor assemblages.

Most of the results of community studies have already been included in several of previous sections. Because there are already good syntheses about the community ecology of Chilean owls (Jaksic 1997; Rau and Jaksic 2004b), we do not discuss this subject further.

7.3.15.2 Intraguild Predation

The occurrence of intraguild predation is unknown within communities of Chilean birds of prey and owls. There are at least two reported supposed cases of intraguild predation involving Chilean owls. During spring of 1996, Figueroa and Corales (2015) observed an adult *Glaucidium nana* subduing by its talons a *Falco sparverius* nestling (15–20 days in age). The owl was observed just when he was leaving the nesting cavity of a pair of *F. sparverius*. After leaving the cavity, the owl dragged with difficulty its prey over a horizontal branch very close to the nest. Because *G. nana* is known to be a nest predator and it is able to capture and kill prey that greatly exceed its body mass (Barros 1950; Jiménez and Jaksic 1993), it is possible that it had really killed the kestrel nestling. During the summer of 2015, Ibarra et al. (2017) registered an adult chimango caracara (*Milvago chimango*) taking an owl nestling from a tree nest cavity of *Strix rufipes* in a secondary forest remnant in southern Chile. However, these authors could not verify if the chimango truly killed the owl nestling.

7.3.16 *Functional Ecology*

7.3.16.1 Ecological Redundancy

Ecological redundancy analyses are intended to identify if more than one species within an ecosystem has a similar ecological function (Walker 1992). In the semi-arid scrub of central Chile, Jaksic et al. (1996) evaluated the occurrence of ecological redundancy in an assembly of top predators, including four owl species. These authors used as an indicator of ecological redundancy the degree of food-niche overlap by setting a threshold value >50% in dietary similarity. The resident predators formed two different trophic guilds, although with some temporal variations. In the case of owls, *Tyto furcata* and *Bubo magellanicus* formed an almost strictly mammal-eating guild, but the level of diet similarity between the two varied widely. Moreover, *Athene cunicularia* and *Glaucidium nana* formed part of an omnivore guild, but only through the fourth year of the study. Later, *G. nana* and *A. cunicularia* temporarily left the area, and after both species returned, the original trophic guild structure was not evident in three of four seasons. According to Jaksic et al. (1996), these results show that ecological redundancy within assemblages of vertebrate predators may suffer temporal long-term changes as the food availability also changes, and therefore, short-term studies are insufficient to determine the diversity and the functional dynamics of such assemblages in ecosystems.

In a more recent study, Farías and Jaksic (2011) analysed the functional redundancy of an assemblage of vertebrate predators in temperate forest fragments of Chiloé Island, southern Chile. The authors found that there is a positive linear relationship between species richness and functional richness to the extent that forest fragments become more extensive and structurally more complex. This could be explained because predator assemblages tended to be taxonomically and functionally richer in larger fragments, which in turn are more diverse in terms of vegetation structure. According to these authors, their results predict a strong effect of deforestation on the taxonomic and functional richness of vertebrate predators in the southern temperate forest. However, the two species of owls present in the assemblage, *Strix rufipes* and *G. nana*, tended to maintain their functional traits despite forest fragmentation. Furthermore, the use of multiple patches of forest by an individual of *S. rufipes* could dilute potential correlations between forest patch size and functional uniqueness (R.A. Figueroa, unpublished data; see sect. 7.3.12).

7.3.16.2 Links Within Food Web Structure

The study of food web structure is of paramount importance to better understand the function and dynamics of ecosystems and determine its sensitivity to environmental changes. In Chile, Arim and Jaksic (2005) analysed the association between the number of trophic connections (i.e. species richness in diet) and ecosystem productivity based on a local assemblage of top predators. The richness of prey species in the diet of all predators varied idiosyncratically with precipitation ($\approx$ productivity) supporting the hypothesis that food web structure is dynamic. The richness of trophic connections of *Athene cunicularia* was positively associated with precipitation of the year and with that accumulated during the previous two years. Species richness in the diet of *Bubo magellanicus* showed a time lag in the functional relation to precipitation. The richness of trophic connections was higher at the end levels of accumulated rainfall during the previous two years which suggest a U-shaped functional relation. The richness of prey species in the diet of *Glaucidium nana* also presented a U-shaped functional relation to precipitation, but without time lags. With a 1-year time lag, *Tyto furcata* linearly reduced its richness of prey species as productivity increased. According to Arim and Jaksic (2005), the time lag displayed by all these owls regarding the association between productivity and the number of trophic connections suggests the involvement of population processes in determining the observed patterns.

In a further analysis, Arim et al. (2007) analysed the association between primary productivity and food chain length. Their analysis included the effect of primary productivity on predator species position and the predator assemblage food chain length. All predators showed an association between rainfall and the level of consumption of non-herbivorous prey (an index of food chain length), but this association varied functionally. In the case of owls, results were as follows. *T. furcata* and *A. cunicularia* displayed a delayed U-shaped functional response; i.e. the proportion of non-herbivorous prey in its diet was higher at the ends of precipitation level, but after a year. *B. magellanicus* showed a humped pattern (i.e. inverted U) in its immediate response to productivity; i.e. proportion of non-herbivorous prey increased with intermediate levels of precipitation. In the case of *G. nana*, there was a slight effect of productivity on incidence of non-herbivorous prey in its diet. The authors suggested that the observed response in *T. furcata*, *A. cunicularia*, and *G. nana* could originate from the combined effect of two or more processes dominating different levels of energy available or could be the result of a single mechanism not yet well known.

7.3.17 Spatial Ecology

Overall, owls are characterized by their low densities and secretive habits. Thus, knowing variations in detectability is pivotal to guide better monitoring programs. Moreover, understanding the relationship between habitat niche width and the ocurrence of owls may facilitate the development of appropriate management recommendations. Recently, at least two studies shed light on the detectability and spatial occurrence of Chilean forest owls.

7.3.17.1 Detectability and Occupancy

Ibarra et al. (2014b) examined the temporal and abiotic sources of variation associated with detection probabilities of *Strix rufipes* and *Glaucidium nana* in a mountainous forest area of southern Chile. They used a multi-season occupancy framework for modelling occupancy and detection. In the case of *S. rufipes*, occupancy was positively associated with elevation. Occupancy of *G. nana* was not associated with any covariate. Detectability for both owl species increased with greater moonlight and decreased with environmental noise, and for *G. nana* greater wind speed decreased its detectability. The probability of detecting *G. nana* increased non-linearly with the number of days from the start of surveys and peaked during mid-summer (see also Ibarra et al. 2015). The detection of both species was positively correlated with the detection of the other species. In order to improve monitoring protocols of *S. rufipes* and *G. nana*, Ibarra et al. (2014b) recommended both species should be surveyed simultaneously for a minimum of 3–4 times during a season, survey stations should be located away from noise, and observers should record the moon phase and weather conditions for each survey.

7.3.17.2 Occurrence Patterns and Niche Relationships

Based on the same aforementioned study, Ibarra et al. (2014c) assessed the ocurrence probability of *Strix rufipes* and *Glaucidium nana* across three spatial scales and tested whether they differed in resource utilization and peak ocurrence probability. The ocurrence probability of *S. rufipes* ranged between 0.05 and 1.0 across sites, and it was positively associated with the variability in stem diameter of trees and bamboo understory density. The ocurrence probability of *G. nana* ranged between 0.67 and 0.98, and it was positively associated with forest patch irregularity and edge effects, and forest cover at 180 ha. Relative to *G. nana*, *S. rufipes* had lower total resource utilization due to a lower ocurrence probability over gradients of all covariates, but achieved a similar peak ocurrence probability for resources related with stand-level forest complexity and forest stability at the landscape scale. Ibarra et al. (2014c) suggested that the occurrence of *S. rufipes* will be promoted if multiaged stands with a variety of tree sizes (dbh = 19.9 ± 9 cm [mean ± SD]), including large old-growth trees, with relatively high bamboo cover (34.2 ± 26.6% [mean ± SD]), are retained. Landscapes with forest cover >63.5% would also favour the occurrence of habitat-specialist owls. Nonetheless, in agricultural lands of southern Chile, individuals of *Strix rufipes* can subsist for several years in tiny (0.5–6 ha) and highly degraded (scarce presence of large trees) forest patches, although with dense bamboo carpets (R.A. Figueroa, unpublished data).

7.3.18 Ecophysiology

To our knowledge, there is only one study that partially addresses ecophysiological aspects of Chilean owls. Bozinovic and Medel (1988) calculated the expected metabolic rate and the theoretical energy requirement of three species of Chilean owls according to their body mass (Table 7.38). The results of this study have been useful to partially explain the apparent prey selection by the Chilean owls (see sect. 7.3.13).

7.3.19 Genetics

Apart from phylogenetic analyses, genetics of Chilean owls has been scarcely studied. Recently, Colihueque et al. (2015) compared mitochondrial cytochrome c oxidase subunit I (COI) sequences of specimens of *Tyto furcata* and *Asio flammeus* with sequences of specimens from other parts of the World. This study revealed a strong intraspecific genetic divergence for *T. furcata* from Chile/Argentina with respect to specimens of *T. alba* from Europe and Australasia (4.6–5.5%), and for *A. flammeus* from Chile/Argentina in comparison with specimens from North America, Europe, and Asia (3.1%). Interestingly, the level of genetic distance detected in both species exceeds the upper limit of intraspecific comparisons reported previously for Strigiformes. In the case of *T. furcata*, specimens from Chile showed low levels of sequence divergence (0–0.2%), and genetic divergence between specimens from Chile and those from elsewhere in the New World was also low (0.0–0.2%). Findings of Colihueque et al. (2015) support König and Weick (2008) and Wink et al. (2008) who consider *T. furcata* as a separate species from Old World barn owls. The divergence between specimens of *A. flammeus* from Chile/Argentina and those from elsewhere suggests that the taxonomic status of South American populations requires further analyses. On the other hand, it is possible that in many countries the genetic make-up of local populations of *T. furcata* may be influenced by hybridization between native owls and those introduced by European settlers (König and Weick 2008). The existence of intermediate morphs between *Glaucidium nana* and *G. peruanum* at northernmost Chile (Marín et al. 1989) suggests potential hybridization.

7.3.20 Biological Conservation

The necessity for protecting Chilean owls was raised early by several of the first natural historians in our country who argued that these birds were useful as biocontrollers of rodent pests, but they were unjustifiably persecuted by humans (e.g. Gay 1847; Reed 1905; Housse 1945; Barros 1950, 1963). It is remarkable that such arguments are the same on which the current biological conservation strategies and environmental laws are based to justify the protection of Chilean native predators. In Chile, the modern conservation focus is essentially based on the population status and in establishing conservation priorities for native species. More novel approaches have considered the analysis of ecological redundancy, bioindication, and human dimension. From a practical standpoint, habitat manipulation and private protection have emerged as attractive and promising strategies.

7.3.20.1 Population Status

Jaksic and Jiménez (1986) found that whereas some Chilean owl species are negatively affected by human activity, others are benefited. At the time the analysis of these authors was done, *Tyto furcata* seemed to increase in population size as a consequence of the increased habitat and prey availability generated by agricultural activities. *Bubo magellanicus* seemed to maintain stationary population sizes, except in southernmost Chile where it seemed to be increasing in number owing to forest clearance and the introduction of the European hare. *Athene cunicularia* seemed to be increasing in population size, except in southernmost Chile where its population size had decreased dramatically. In central and southern Chile, this owl species benefited from the increased habitat and prey availability as a result of agricultural development. On the other hand, in southernmost Chile, this owl was adversely affected by the intensive sheep raising maintained for several decades since the beginning of the twentieth century; the trampling of hundreds of sheep on their burrows and hunting areas forced the species to abandon many of their breeding sites (Venegas and Sielfeld 1998; Jaksic et al. 2002). According to Humphrey et al. (1970), *A. cunicularia* was extinct on the Isla Grande of Tierra del Fuego as early as 1920 because of the introduction of sheep. *Glaucidium nana* seemed to maintain a stationary population size, except in central Chile where its population seemed be increasing owing to the increased prey availability. In addition, this species had adapted well to urban expansion because the green spaces within cities attract many birds species, both native and introduced (e.g. *Passer domesticus*), which constitute a stable food supply. *Asio flammeus* seemed to be decreasing in population size in central and southern Chile as a result of the extensive loss of grasslands and wetlands (see also Pavez et al. 2010); the opposite seemed to occur in southernmost Chile where forest clearance increased the habitat availability. Finally, the population of *Strix rufipes* seemed be declining throughout its distribution range owing to forest loss and fragmentation (see also Martínez and Jaksic 1996; Gantz and Rau 1999; Omland et al. 2001; Martínez 2005a).

Subsequently, Jaksic et al. (2001) analysed the conservation status of raptors in the most urbanized area in Chile, the Metropolitan region. In the case of owls, these authors found the following tendencies. *T. furcata* and *G. nana* have been increasing in population size as a result of the increasing habitat and prey availability. In addition, *G. nana* is taking advantage of native bird flocks that take shelter in squares and city park trees. On the other hand, *A. cunicularia, A. flammeus,* and *S. rufipes* have been decreasing in population size as a direct consequence of the habitat loss. *B. magellanicus* is the only species exhibiting a stationary population size, but without an apparent cause.

From the above analyses, it is deducible that some owl species only benefit initially from human activity because if environmental changes become much too drastic, these may have rather negative effects. For example, excessive alteration of pasturelands or grasslands can cause the local extinction of *A. cunicularia* and *A. flammeus*.

7.3.20.2 Threats

There are a number of potential threatening factors for survival of Chilean owls (Table 7.39). As was mentioned earlier, habitat loss is the major threat to *Athene cunicularia*, *Asio flammeus,* and *Strix rufipes*. Illegal hunting is apparently not a major threat to Chilean owls (Jaksic and Jiménez 1986; Jaksic et al. 2001). Tala and

Table 7.39 Threats to Chilean owls

Threat	Tfur	Bmag	Acun	Gnan[g]	Afla	Sruf
Human pursuit[a]	+	++	+	+++	+	+
Habitat loss	+	+	+	+	+++	+++
Loss of prey	–	–	++	–	+?	++
Collisions with vehicles	+++	–	–	–	+?	–
Pesticides[b]	+?[e]	+?	+?	+?	+?[e]	+?
Electrocution[c]	+?	–	–	–	–	–
Domestic predators[d]	–	–	+[f]	–	+[f]	–

Tfur = *Tyto furcata*, Bmag = *Bubo magellanicus*, Acun = *Athene cunicularia*, Gnan = *Glaucidium nana*, Afla = *Asio flammeus*, Sruf = *Strix rufipes*. Degree of potential threat: + = low, ++ = intermediate, +++ = high,? = subject to confirmation

[a]Includes killing with firearms, by hand, by stoning, by trapping, and/or by poisoning. [b]Some rodenticides may indirectly cause the death of owls by consuming poisoned rodents. [c]Caused by collisions with high-voltage power lines. [d]Dogs, cats, and pigs. [e]These species frequently hunt in areas with young eucalyptus plantations where chemicals are applied against herbivorous rodents. [f]Because these species nest on the ground are more exposed to attacks of domestic predators. [g]Threats to *Glaucidium peruanum* are unknown but could be the same as for *G. nana*. References: *Tyto furcata* = 4, 11, 14, 15; *Bubo magellanicus* = 7, 9, 13–15; *Athene cunicularia* = 2, 6, 7, 11, 13, 15; *Glaucidium nana* = 1, 3, 7, 9, 13–15; *Asio flammeus* = 2, 6, 7, 8, 11, 15; *Strix rufipes* = 2, 5, 6, 8, 9, 10, 11, 12, 14, 15. [1]Barros 1950, [2]Jaksic and Jiménez 1986, [3]Jiménez and Jaksic 1989, [4]Schlatter 1992, [5]Martínez and Jaksic 1996, [6]Figueroa et al. 2001a, [7]Jaksic et al. 2001, [8]Vásquez and Simonetti 1999, [9]Konig et al. 1999, [10]Omland et al. 2001, [11]Tala and Iriarte 2004, [12]Martínez 2005a, [13]Bonacic and Ibarra 2010, [14]Celis-Diez et al. 2011, [15]authors, personal observation

Iriarte (2004) affirm that incidence of dead raptors in controls of hunting day is <0.03%. However, illegal hunting of owls has not been directly quantified and its impact could be underestimated. Owls of all species have come to rehabilitation centres with injuries caused either by firearms, traps, or collisions (S. Alvarado, unpublished data). However, there are no reports detailing the survival percentage of these owls. Although there has been no evaluation in our country, König et al. (1999) have suggested that rodenticides pose a threat to several species of Chilean owls. More information about the potential risk of rodenticides for owls is found in www.owlpage.com.

7.3.20.3 Conservation Priorities

The conservation priority of Chilean owls has been evaluated by various analyses at local and global scales. Results of these evaluations are summarized in Table 7.40. Despite some methodological differences, it should be noted that there is enough consistency between different local assessments. In general, species that are in the higher conservation priority are *Strix rufipes* and *Asio flammeus*. However, none of these owl species are considered urgent to preserve globally (see also Legislation and Protection). From an ecosystem perspective, ecological redundancy analysis suggests that all Chilean owls merit conservation priority (Jaksic et al. 1996; Farías and Jaksic 2011).

Table 7.40 Conservation priority of Chileans owls according to different criterions

References	Tfur	Bmag	Acun	Gnan	Gper	Afla	Sruf
Local assessments							
Jaksic and Jiménez 1986[a]	I	S	I	S[h]	I[h]	D	D
Glade 1988[b]	OD	OD	OD	OD	OD	IK	IK
Rottmann and López-Calleja 1992[b]	OD	OD	OD	OD	OD	IK	V
República de Chile 1996[b]	OD	OD	OD	OD	OD	IK	IK
Vásquez and Simonetti 1999[c]	ne	11	11	10	ne	14	14
Estades 2004[b]	OD	OD	OD	OD	OD	IK	IK
Pincheira-Ulbrich et al. 2008[d]	3	2	3	3	2	2	1
Global assessments							
Stotz et al. 1996 (S/CP/RP)[e]	L/L/L	L/L/L[i]	M/L/L	L/L/L[h]	M/L/L	L/M/M	M/L/L
Inskipp and Gillet 2005 (CITES)[f]	II	II	II	II	II	II	II
IUCN 2008[g]	LC	nr	LC	LC	nr	LC	LC

Tfur = *Tyto furcata*, Bmag = *Bubo magellanicus*, Acun = *Athene cunicularia*, Gnan = *Glaucidium nana*, Gper = *Glaucidium peruanum*, Afla = *Asio flammeus*, Sruf = *Strix rufipes*
Conservation categories according to the source: [a]I = increasing number, S = stationary number, D = decreasing number. [b]OD = out of danger, IC = inadequately known, V = vulnerable. [c]Values represent indices of sensitivity to changes in the landscape which range from 9 (low sensitivity) to 15 (high sensitivity). [d]Based on an index of conservation priority: 1 = maximum priority, 2 = special attention, 3 = no priority. [e]S = sensitivity to human disturbance, PC = conservation priority, PI = research priority, L = low, M = medium. [f]II = included in Appendix II of CITES (they may be traded under certain regulations). [g]LC = least concern. [h]Included into *G. brasilianum*. [i]Included into *B. virginianus*. ne = not evaluated, nr = not recognized as a valid species

7.3.20.4 Ecosystem Services

During the last years, the conservation of Chilean owls has been strongly supported by national agencies of public health and wildlife because the owls are recognized as effective biocontrollers of zoonosis rodent reservoirs (e.g. Figueroa et al. 2001a; Alvarado et al. 2015; Figueroa and Alvarado 2015). In addition, some national forestry companies have showed interest in the management of owls for biocontrol of forestry animal pests (Murúa and Rodríguez 1989). On the other hand, it is necessary to know the impact of predators on invader rats as these can negatively affect local biodiversity.

Biocontrol of Zoonosis Vectors Studies of the diet of *Tyto furcata* and *Strix rufipes* demonstrates that these species are important predators of leptospirosis/hantavirus rodent reservoirs (Figueroa et al. 2007). In fact, *T. furcata* captures a high proportion of adult individuals of *Oligoryzomys longicaudatus* (Figueroa et al. 2007), which are most likely to host hantaviruses. More recently, Muñoz-Pedreros et al. (2016a) evaluated the role of this owl species as a potential biocontroller of *O. longicaudatus* throughout the distribution area of the latter. These authors found that *T. furcata* can become an important harvester of *O. longicaudatus* which can account for almost 30% of all consumed prey items. In addition, in at least one locality, this owl species acted as a selective predator upon *O. longicaudatus*. Muñoz-Pedreros et al. (2016a) concluded that *T. furcata* is a potentially good biocontrol agent for the hantavirus reservoirs since it is a specialist predator on rodents with a selective intake of the *O. longicaudatus*. Eventually, *T. furcata* and *Bubo magellanicus* can consume a not negligible number of *Rattus norvegicus* (authors, unpublished data) which is a reservoir for Seoul hantavirus in Chile (Lobos et al. 2005).

Biocontrol of the Forestry/Agricultural Pests Studies of the role of Chilean predators as biocontrollers of forestry rodent pests have simultaneously implied habitat modifications (see below). Muñoz-Pedreros and Murúa (1990), after a habitat manipulation to attract birds of prey towards pine plantations by vegetation clearing and installing elevated perches, found that *T. furcata* was not an important harvester of *Octodon bridgesi* neither *Rattus rattus*, which are known to gnaw the bark of pines (Murúa and Rodríguez 1989). The relatively high consumption of introduced lagomorphs by *Bubo magellanicus* (see sect. 7.3.13.11) renders it as an important biocontroller of pine forestry pests. European rabbits and hares are known to sever twigs and stem of pine seedlings (Gajardo and Rodríguez 1985), browse sapling of native plants, and damage vegetable crops (Camus et al. 2008).

Biocontrol of Ecological Invaders The important consumption of allochthonous rats by *Tyto furcata* and *Asio flammeus* (see sect. 7.3.13.11) also renders them potential biocontrollers of ecological invaders in natural ecosystems. *Rattus norvegicus* and *R. rattus* have broadly invaded native forests, scrublands, and islands along Chile (Lobos et al. 2005). Up to our knowledge, the effects of these introduced rats in Chilean natural ecosystems have not been evaluated.

7.3.20.5 Bioindication

The bioindicator capacity of top predators is a strong argument for their conservation (e.g. Sergio et al. 2008). In Chile, the bioindicator capacity of birds of prey has only recently begun to be explored. Rau et al. (2005) subtly suggested that the dietary analysis of owls might be useful as a richness indicator of local prey species. A local biodiversity assessment in northern Chile showed that the pellets of *Tyto furcata* provide better information than did trappings regarding the local small-mammal richness (e.g. Jaksic et al. 1999).

Ibarra and Martin (2016) evaluated the surrogacy capacity of *Strix rufipes* and *Glaucidium nana* in Andean temperate forests. They specifically tested if there is a spatial correlation between owls and avian biodiversity by comparing probability of occurrence of owl species with avian taxonomic, endemism, and functional diversity. Results of these authors suggest that *S. rufipes* is a reliable surrogate for all avian biodiversity measures. On the contrary, *G. nana* did not function as a good surrogate. Ibarra and Martin (2016) associated the bioindicator capacity of *S. rufipes* to its aggregation across old-forest remnants structurally more complex and where a higher diversity of forest-dweller birds was likely. In fact, previous studies show that the mammalian prey diversity of *S. rufipes* tends to be higher in larger old-forest remnants (Figueroa et al. 1997). Further studies are needed to corroborate if findings of Ibarra and Martin (2016) are generalizable to the broadest spatial scale.

7.3.20.6 Human Perception

Chilean owls are perceived by rural communities in various ways, both positively and negatively. For example, the "pequén" (*Athene cunicularia*) and the "chuncho" (*Glaucidium nana*) are feature elements in peasant folklore, and popular poets have included them into their verses and songs (Plath 1996). Among all Chilean ornithological proverbs analysed by Ibarra et al. (2013; N = 89), six referred to owls (≈7% of all proverbs), with *G. nana* (N = 4) being the most mentioned species followed by *A. cunicularia* (N = 1) and *Tyto furcata* (N = 1). Interestingly, all these species are the most common owl species in rural, suburban, or urban areas, and thereby, it might increase inspiring encounters between owls and humans. In words of Ibarra et al. (2013), the presence of birds in proverbs and human experiences is an expression of the "landscape of human-bird encounter", connecting different expressions in the common legacy of biocultural heritage.

Recent studies suggest that positive or negative perceptions about owls by rural people depend on the species and its habits. Möller et al. (2004) evaluated the level of knowledge and attitudes towards birds of prey by schoolchildren, teachers, and villagers in a rural area of Valdivia, southern Chile. Their results indicated that most of the people had a very low level of knowledge and negative attitudes towards birds of prey (>50% of respondents). Interestingly, >80% of people thought that raptors are beneficial because they "eat mice". Silva-Rodríguez et al. (2006) evaluated the perception of subsistence farmers towards native birds in southern Chile. Surveys indicated that *Strix rufipes* is perceived as a bird of bad omen being persecuted by farmers when they sing near their homes. *G. nana* was widely perceived as harmful because it eats birds. The only species perceived as useful was *Tyto furcata* because it eats rodents. More recently, Piñones et al. (2016) assessed the level of knowledge and social appropriation of local avifauna by different social groups of an agricultural community in central-northern Chile. They found that *G. nana* is considered as a conflictive species because it may be harmful to productive activities and is said to be "bad omen bird". On the other hand, *T. furcata* is said to be a weather forecaster since its vocalizations indicate foggy nights.

In Chile, owls are salient elements in the cosmovision of indigenous peoples (Villagrán et al. 1999; Aillapán and Rozzi 2004; Martínez 2005b). In the narrative of Yagan people, the "owl" (*T. furcata*) is a symbol of wisdom (Rozzi 2004), and for Mapuche people, the "bad omen of owls" is feared, but accepted (Coña 1973;

Villagrán et al. 1999). In Andean highlands of northern Chile, the agro-pastoral communities' descendent from Quechua and Aymara ethnic groups perceive *T. furcata* and *Bubo magellanicus* as daunting birds (Castro and Rottmann 2016). The first species causes fear because it cries like a nursing baby at nightfall. *B. magellanicus* (juco tucucara) is considered a "bad omen bird" because its vocalizations warn of death. However, villagers show good knowledge about the identification and habits of these owls. For example, *B. magellanicus* is said to be a rodent and lizard eater (Castro and Rottmann 2016). All the above suggests that conservation of Chilean owls would be much more viable in indigenous lands. However, strengthening environmental education programs could help develop and establish a positive attitude towards owls in many more people in rural areas and, thereby, promote a more effective conservation. Such programs should be based essentially on ecosystem services provided by owls (e.g. biocontrol of rodent reservoirs of zoonosis).

7.3.20.7 Legislation and Protection

All Chilean owls are legally protected (Tala and Iriarte 2004). In its original text, the Law No. 4601 (República de Chile 1929), also known as Hunting Law, forbids indefinitely the hunting, transport, commercialization, and industrialization of any owl species, including subproducts such as eggs, chicks, and feathers (Title I, Article 2). Currently, the Hunting Law, with text modified by the Law No. 19473 (República de Chile 1996), forbids the hunting and capturing of any endangered, vulnerable, rare, and inadequately known wild animal species and of those that are beneficial for agricultural and forestry activity and the maintenance of the balance of natural ecosystems (Title II, Article 3). This includes all species of owls. In the case of *Strix rufipes*, its protection and habitat conservation is also promoted by the Law N° 20283 (República de Chile 2008), also known as Forest Law. In addition, our country has ratified several international conventions that promote the protection of owls: (i) Convention for the Protection of the Flora, Fauna, and Scenic Beauties of the Americas; (ii) Convention on International Trade in Endangered Species of Fauna and Flora, CITES; (iii) Convention on the Conservation of Migratory Species of Wild Animals; (iv) Convention on Wetlands of International Importance, RAMSAR; and (v) Convention on Biological Diversity (see Tala and Iriarte 2004).

It is important to indicate that whereas the Chilean legislation strictly forbids the owl trade, CITES permits its regulated trade (Appendix II). This apparent contraposition is explainable because in a global context no Chilean owl species is endangered, and its trade or transport could be possible if it is justified for scientific research.

7.3.20.8 Rehabilitation

Physical recovery of owls injured by humans is carried out in various rehabilitation centres throughout Chile (see www.sag.cl). In cases when rehabilitation is complete, the owls are released in areas where suitable habitat is available (Pavez

2004b; D. González-Acuña, unpublished data). Non-releasable owls are used in environmental education programs. With the aim of achieving significant and fascinating experiences, some centres perform activities promoting direct contact between non-releasable owls and children. Apart from the physical treatment to injured birds, some centres have achieved the breeding of captive pairs of *Tyto furcata* and *Bubo magellanicus* (Pavez 2004b; Tala and Iriarte 2004). Transplants of flight feathers (primary remiges) in *T. furcata* have also been successful (González-Acuña et al. 2011). We must also highlight the successful physical rehabilitation of *Strix rufipes* in one of these centres (Fig. 7.7). Although occasional, these experiences are promising for eventual programs of population recovery of these species.

7.3.20.9 Education and Outreach

Education and outreach programs about the ecological role of owls in Chile have been a powerful tool that has positively contributed to their conservation (Jaksic and Jiménez 1986; Simonetti et al. 1992; Figueroa 1995; Hidalgo 1999; Möller et al. 2004). In general, Chilean citizens seem to have increased their understanding of wildlife conservation through outreach and non-formal education campaigns conducted by media (radio, television, and newspapers), municipalities, government agencies and non-governmental organizations, rehabilitation centres, zoos, and some private companies (Jaksic and Jiménez 1986; Tala and Iriarte 2004; Figueroa and López 2007). In the particular case of owls, there are a number of books promoting knowledge of their natural history and ecological role as a foundation for conservation (Figueroa et al. 2001a; Figueroa and López 2007; Muñoz-Pedreros et al. 2004; Rivas and Figueroa 2009; Celis-Diez et al. 2011; Roa and Alvarado 2011; Altamirano et al. 2012). Recently, Rau (2014) and Alvarado et al. (2015) have emphasized the overall importance of birds of prey and owls for human well-being taking into account local studies carried out in Chile.

Government agencies playing an important role in education and conservation of raptors are the Agriculture and Livestock Service (www.sag.cl) and the National Forestry Corporation (www.conaf.cl). Among non-governmental organizations stand out the Ornithologists Union of Chile (www.unorch.cl) and the Center for Agricultural and Environmental Studies (www.ceachile.cl). In the context of the formal education, Figueroa (1995) proposed a strategy for using birds of prey, including owls, as models so that students more easily assimilate ecological concepts such as predation, competition, chains, and food webs, among others. This strategy includes field trips to allow close encounters with birds of prey, laboratory activities to identify their prey by pellets analysis, and review of scientific articles to enrich information. In the context of non-formal education, some recently created raptor exhibition centres are educating on the overall importance of birds of prey and owls through direct contact with live birds (Rivas-Fuenzalida et al. 2015b).

7.3.20.10 Management and Habitat Restoration

In Chile, there have been several experiences of habitat manipulation to attract or increase the number of owls. All these experiences have been more focused on the biocontrol of animals than the recovery of population size of endangered species. Independent of objectives, such experiences have enabled the assessment of the effectiveness of certain methods to increase the number of individuals of a particular species. As part of an integrated program of harmful rodent biocontrol in pine plantations, Murúa and Rodríguez (1989) installed elevated horizontal structures (perches) on free vegetation strips to attract birds of prey. This habitat modification was successful in attracting *Tyto furcata* to plantations. Subsequently, judging by the pronounced increase in the number of pellets, Muñoz-Pedreros and Murúa (1990) concluded that such habitat management allowed the increase in the effective occupation of *T. furcata* within plantations. The use of wooden nest boxes to attract birds of prey in forest plantations has also been successful. Schlatter and Murúa (1992b) documented the effective occupancy of two four nest boxes designed for *T. furcata*, and they even managed to reproduce inside the boxes. In addition, the smaller nest boxes were occupied by *Glaucidium nana* (see also Schlatter et al. 1991).

A habitat management program for biocontrol of hantavirus rodent reservoirs in rural areas of southern Chile included the implementation of wooden nest boxes with specific designs to attract *T. furcata*, *G. nana*, and *Strix rufipes* (Murúa et al. 2004, 2005; Ruiz et al. 2006; Figueroa et al. 2007). *Glaucidium nana* had the higher acceptation of boxes (occupancy success: agroecosystems = 53% [N = 36], pine plantations = 35% [N = 24), native forest = 17% [N = 23]), and at least three pairs raised their chicks inside boxes. The occupancy by *T. furcata* was remarkably low (2 of 60 boxes), and *Strix rufipes* did not occupy any of the boxes. Possibly, the availability of natural cavities was high enough around the sites where boxes for *T. furcata* and *S. rufipes* were installed. Muñoz-Pedreros et al. (2010) documented the results of a study on the effectiveness of wooden nest boxes to increase the population of *T. furcata* in order to improve a biocontrol method of hantavirus reservoir in central Chile. In 3 years, the occupancy success increased from 5% (1/20) to 55% (11/20), population density increased from 0.5 to 2 individuals/km^2, and productivity increased from 4 to 31 chicks. All these studies show that habitat modification by including artificial structures helps to effectively increase the population size of some owl species. Such experiences constitute a way for habitat restoration by increasing the supply of nesting and/or shelter spaces where the natural cavities and perches have been lost by anthropogenic activity.

Although in our country there have not been experiences of habitat restoration at the landscape scale, in some cases the purchase of land for establishing wildlife reserves has allowed the natural recovery of habitats that had been altered. The creation of a biological corridor at landscape scale (5600 km^2) in Nevados de Chillán is allowing the recovery of various types of original habitats that benefit several owl species (Figueroa and López 2007). The Biological Corridor Costa-Andes (www.

Table 7.41 External parasites of Chilean owls

Species	Tfur	Bmag	Acun	Gnan	Afla	Sruf
Lice						
Strigiphilus syrnii	–	–	–	–	–	+
Strigiphilus speotyti	–	–	+	–	–	–
Strigiphilus microgenitalis	–	–	–	+	–	–
Strigiphilus chilensis	–	+	–	–	–	–
Strigiphilus cursor	–	–	–	–	+	–
Kurodaia subpachygaster	+	–	–	–	–	–
Kurodaia caputonis	–	–	–	+	–	–
Strigiphilus (Tytoniella) aitkeni	+	–	–	–	–	–
Fleas	–	–	–	–	–	–
Hectopsylla psittaci	–	–	–	+	–	–

Tfur = *Tyto furcata*, Bmag = *Bubo magellanicus*, Acun = *Athene cunicularia*, Gnan = *Glaucidium nana*, Afla = *Asio flammeus*, Sruf = *Strix rufipes*
González-Acuña et al. 2006, Fuentes et al. 2011, Beaucornu et al. 2014, Moreno and González-Acuña 2015. The sign "plus" indicates presence

parquesparachile.cl), a supposedly ongoing project, is focused on protection of landscape-scale Valdivian forest, and therefore, it could benefit *Strix rufipes*.

7.3.21 Complementary Information

7.3.21.1 Parasites

González-Acuña et al. (2006) reported a first description of lice for Chilean owls. These authors found that each owl species hosts only one or two louse species. Subsequently, Orellana (2009) described a new louse species (*Kurodaia caputonis*) for *Glaucidium nana*. Recently, a flea species (*Hectopsylla psittaci*) was also described for *G. nana* by Beaucornu et al. (2014). Moreno and González-Acuña (2015) provided a more complete review on ectoparasites of Chilean owls. Their results are summarized in Table 7.41. Information on endoparasites is scarcer. Preliminary results indicate that *Tyto furcata* is internally parasitized by the nematode worms *Capillaria falconis* and *Porrocaecum depressum* (D. González-Acuña, unpublished data). Fuentes et al. (2011) reported presence of nematodes of the *Habronema* genus in individuals of *G. nana*.

It is important to mention that the parasite-host relationship in Chilean birds is an emerging and promising research line. As investigation of bird parasites advances, it will allow the better prevention and confrontation of problems of wildlife health, understanding of phylogenetic and co-evolutionary relationships among birds and parasites, complete information on diversity and taxonomy of bird parasites, and understanding of the role of parasite in food webs.

7.3.21.2 Taphonomic Usefulness

Taphonomy is a subdiscipline of palaeontology studying the post-mortem history and fossilization processes of bone remains. Accurate identification of bone deposition agents is crucial for reconstruction of palaeo-environments and resolves palaeoecological questions. Although the small-mammal skeletal elements constitute an important proportion of zoo-archaeological depositions, their origin is often uncertain. In general, owls are recognized as important agents of small-mammal bone accumulations. Owl pellets are characterized by survival of a higher proportion of bone material than pellets/scat produced by other avian and mammal predators, and thus they provide a more complete sample of the local fauna (Terry 2004). Identification of similar patterns among different small-mammal bone accumulations produced by a particular owl species would allow the determination of the origin of these accumulations.

Saavedra and Simonetti (1998) evaluated the taphonomic usefulness of *Tyto furcata* in central Chile and some North American localities. These authors tested the assumption that *T. furcata* generates bone depositions with specific recognizable patterns (e.g. degree of fragmentation of small-mammal bones), regardless of the composition of prey assemblages. Because there were important variations in the degree of bone fragmentation among local osseous accumulations, Saavedra and Simonetti (1998) concluded the use of one particular bone pattern, or an average pattern calculated from different *T. furcata* populations, is an inadequate tool for taphonomic study. Similar studies should be conducted for other Chilean owls to better determine their usefulness as taphonomic indicators (e.g., Montalvo et al. 2015).

7.4 Conclusions

We have summarized all available information on taxonomy, natural history, ecology, and biological conservation of Chilean owls. One of our results was the discovery and recovery of information that was totally unknown to many contemporary ornithologists, particularly for younger researchers. In addition, our review allowed us to realize that information provided by the first naturalists in our country has been cast into oblivion. On the other hand, we have also detected an increasing enthusiasm of young investigators for producing educational and outreach material on Chilean birds, which always include owls.

Even though our review is not analytical, we think that the wealth of information compiled here will allow the better understanding of the life history of Chilean owls and design of more robust conservation and management strategies. Regarding this, we should emphasize that there are still many insufficiently known aspects and important information gaps. The attaining of more complete knowledge of the biology of Chilean owls poses a number of challenges such as:

(i) To resolve definitively the validity of *Bubo magellanicus* as a separate species from *B. virginianus*
(ii) To define the geographical limits of *Glaucidium nana* and *G. peruanum*
(iii) To better determine their distribution in austral archipelagos
(iv) To know the actual extension of the distributional gap of *B. magellanicus*.
(v) To better estimate their population sizes and know critical thresholds of habitat loss at the landscape scale in order to better cope with potential numerical declines (Figueroa and Alvarado 2012); this is a priority for *Asio flammeus* and *Strix rufipes* which appear more sensitive to habitat loss
(vi) To know what variables are determining habitat use in order to better mitigate the negative effects of human activities at both local and regional scale
(vii) To know in detail their reproductive biology to lessen uncertainties about their population viability
(viii) To better know their home ranges and movement dynamics for refining of spatially based conservation strategies
(ix) To generate much more information on diet, feeding, and functional ecology in those less studied ecosystems (e.g. Patagonian steppe, agroecosystems, forests) in order to corroborate consistency in previously documented patterns and to determine their relevance in local food webs

Particular attention should be paid to those owl species residing on islands whose ecology is virtually unknown and is potentially more prone to local extinction. Genetic studies of Chilean owl are particularly important both to better evaluate (i) potential evolutionary divergences that could result from geographic isolation and (ii) genetic dilution or homogenization in owl populations inhabiting highly fragmented habitats (e.g. *Strix rufipes* in central Chile). Finally, we need to do much more education about the natural history of our wild species, including owls, and involve many more citizens in research and education projects to achieve our conservation goals more effectively. Most of these challenges imply long-term goals, but we hope these challenges will be addressed by young investigators and enthusiastic students with a great fascination not only for owls but also for all wildlife surround us.

As top predators, owls play a fundamental role in ecosystems by structuring the food webs, and thereby, they promote stability or an increase in local biodiversity. In addition, their ecological requirements make them useful as bio-indicators (i.e. allows for warning of the effects of environmental or ecological changes) and focal (i.e. allows the knowledge of which attributes should be present in the landscape to ensure both their own persistence and those of coexisting species; Lambeck 1997). On the other hand, many owl species are charismatic birds in virtue of which they can attract public support to conservation and education projects (Ducarme et al. 2013). Thus, the usefulness of owls as conservation tools should be not only addressed in academicals circles, but also we should make it possible that conservation concepts percolate across all human dimensions, including the economy. We think our review is a first step in this direction.

Acknowledgements We thank Álvaro García, Carlos Barrientos, Felipe Carevic, Franco Cruz-Jofré, Javier Medel, Jaime R. Rau, Jorge Ruiz, Makarena Roa, Jürgen Rottmann, Juan C. Torres-Mura, María A. Vukasovic, and Patricio Wallace who kindly permitted us access to their unpublished information; Ingo Hahn, Fabián M. Jaksic, Jaime R. Rau, Guillermo Riveros, Javier Simonetti, and Ana Trejo for providing literature; Roberto Murúa who allowed us to use his facilities and Rocío Sanhueza, Francisco Santander and Gustavo Escobar for valuable collaboration in searching information. The kindness and patience of Paula Enriquez is greatly appreciated. The accurate review of M.A. Martínez and editors substantially helped to improve presentation and quality of the final version of our manuscript. Finally, we thank Diane Haughney for helping us with the English translation. Part of the information on owls was also obtained by the authors' participation in the following projects: "Río Condor" (Forestal Trillium, 1994–95), "Búho de Bosque" (FONDECYT 92-327 y 1941256), "Ordenamiento Forestal R. N. Malleco" (CONAF Chile/ONF Francia, 1996, 2001), "Conservación del Huemul" (COD EFF Chile/SZF Alemania, 1997–2004), "Monitoreo de Reservorios de Hantavirus" (SAG/GORE Aysén, 1998–2003), "Estudio Faunístico P. N. Queulat" (CONAF, 2002–2003), "Rapaces Andinas" (2000–2004), and "Control Biológico de Hantavirus" (FONDEF D02I1005). We dedicate this review to Roberto P. Schlatter†, a great owl lover, and all naturalists who have greatly contributed to the knowledge of Chilean owls.

Appendix 7.1

References for Fig. 7.9: diet profile of Chilean owls

Species	Northern Chile	Central Chile	Southern Chile	Southernmost Chile
Tyto furcata	48, 58	1–3, 5, 12, 13, 15, 17, 23, 27, 28, 31, 33, 44, 54, 56, 64, 70	1, 21, 24, 30, 40, 53, 63, 67, 70	29
Bubo magellanicus	45, 71	8, 10, 15, 22, 33, 44, 50, 51, 69, 72, 73	–	4, 9, 10, 18, 22, 29, 41, 52, 74
Athene cunicularia	49, 51, 55, 65, 67, 68, 75	6, 7, 11, 16, 19, 20, 33, 37, 39, 44, 51, 66	14, 25, 32, 51	–
Glaucidium nana	–	26, 33, 35, 44	76	–
Asio flammeus	–	57	34, 46	–
Strix rufipes	–	47, 60, 61, 62	36, 38, 42, 43, 59, 67	–

References: [1]Housse 1945, [2]Reise 1970, [3]Capurro et al. 1971, [4]Reise and Venegas 1974, [5]Schamberger and Fulk 1974, [6]Jaksic and Yáñez 1977, [7]Péfaur et al. 1977, [8]Yáñez and Jaksic 1977, [9]Jaksic et al. 1978, [10]Yáñez et al. 1978, [11]Yáñez and Jaksic 1979, [12] Simonetti and Walkowiak 1979, [13] Jaksic and Yáñez 1979, [14] Fernández et al. 1980, [15] Jaksic and Yáñez 1980a, [16] Schlatter et al. 1980, [17] Cerpa and Yáñez 1981, [18] Rau and Yáñez 1981, [19]Núñez and Yáñez 1982, [20] Schlatter et al. 1982, [21] Rau et al. 1985, [22] Jaksic et al. 1986, [23] Maldonado 1986, [24] Mayorga and Mansilla 1988, [25] Villagra and Mora 1988, [26]Jiménez and Jaksic 1989, [27]Torres-Mura and Contreras 1989, [28] Zunino and Arcos 1989, [29]Iriarte et al. 1990, [30]Muñoz-Pedreros and Murúa 1990, [31]Ebensperger et al. 1991, [32] Pradenas 1991, [33] Jaksic et al. 1992, [34] Rau et al. 1992, [35] Jiménez and Jaksic 1993, [36] Martínez 1993, [37]Torres-Contreras et al. 1994, [38]Martínez 1995, [39] Silva et al. 1995, [40]Simeone 1995, [41]Tala et al. 1995, [42] Figueroa 1996, [43] Figueroa et al. 1997, [44] Jaksic et al. 1997, [45]Torres-Mura et al. 1997, [46] Martínez et al. 1998, [47]Díaz 1999, [48] Jaksic et al. 1999, [49] Zunino and Jofré 1999, [50] Mella 2002, [51] Gil-Cordero 2003, [52] Torres-Mura and Lemus 2003, [53]González-Acuña et al. 2004, [54]Begall 2005, [55] Carevic 2005, [56]Correa and Roa 2005, [57] Escobar et al. 2005, [58]Carmona and Rivadeneira 2006, [59] Figueroa et al. 2006, [60] Vukasovic et al. 2006, [61] Alvarado et al. 2007, [62] Ramírez 2008, [63] Figueroa et al. 2009, [64] Muñoz-Pedreros et al. 2010, [65]Carevic 2011, [66] Roa 2011, [67]Carevic et al. 2013, [68] Cruz-Jofré and Vilina 2014, [69]Mella et al. 2016, [70] Muñoz-Pedreros et al. 2016a, [71] Valladares et al. 2016, [72] Muñoz-Pedreros et al. 2016b, [73]E.S. Corales, in preparation, [74]R.A. Figueroa and E.S. Corales, unpublished data, [75]J.R. Rau, unpublished data, [76]R.A. Figueroa et al., unpublished data

Appendix 7.2 Overview of Some Methods Used in Ecological and Behavioural Studies of the Chilean Owls

The biological knowledge of wildlife increase as we improve study methods, develop new methods, and compare biases among different methods. Even though new technological advances are promising for owl research, these are not a panacea, and the efficiency and biases of some methods are still being evaluated. Below we briefly summarize some methodological advances in the ecological study of Chilean owls.

Diet Analysis

Pellet Identification For better identifying the pellets of Neotropical raptors, several authors have provided information about their morphometrics characteristics. In Chile, Muñoz-Pedreros and Rau (2004) detected a positive correlation between the body mass of birds of prey and owls and the size/mass of their pellets. However, when owls are analysed apart this correlation is not so evident. Although in heavier owls (*Bubo magellanicus*, *Strix rufipes,* and *Tyto furcata*), there is a positive linear correlation with the size of their pellets, this does not occur for lighter owls (*Asio flammeus*, *Athene cunicularia*). Even though a more adjusted correlation is observable between the mass of owls and that of their pellets, the pellets of *T. furcata* tend to be much heavier than those of the larger owl species, *B. magellanicus*. Previously, Trejo and Ojeda (2002) evaluated the morphometric characteristics of owl pellets in southern Argentina, including *B. magellanicus*, *T. furcata*, *S. rufipes,* and *Glaucidium nana*. In this case, there is a clear positive linear correlation between the mass of owls and that of their pellets. Based on these studies, it is possible to think that pellets may be attributed to a particular species of owl by its size or mass. However, correlations may be influenced by ecological or methodological factors. The size and mass of pellets may vary according to the prey spectrum and seasonal diet of each owl species. In general, those species that prey strongly upon small mammals tend to produce heaviest pellets (e.g. *T. furcata*; Muñoz-Pedreros and Rau 2004). Species such as *A. cunicularia* and *S. rufipes* produce heavier pellets during fall-winter when they consume more small mammals and lighter pellets during spring-summer when they consume more invertebrates (Schlatter et al. 1982; Figueroa 1996). On the other hand, pellet size may overlap among species, making it difficult to attribute a mean size or size range of pellets to a particular owl species (Trejo and Ojeda 2002). Because the mean size of pellets is influenced by sample size, in some cases the correlations observed can be instead a statistical artefact. Perhaps, the best manner to recognize pellets of a particular species of owl is a good familiarization with their appearance (i.e. shape, texture, and colour), know clearly where they were collected, and corroborate presence of owls in the collection site. For example, in agricultural lands the pellets of *T. furcata*, *A. flammeus*, and *A. cunicularia*, besides differing in size and appearance, may be discriminated by how they are distributed within habitats. The first species casts its pellets into pastures and grasslands widely. *A. cunicularia* typically accumulates its pellets at the entrance of burrows, and the latter species deposits its pellets inside tree hollows or under tree perches.

Prey Identification The early availability of keys for identification of small mammals and insects has greatly facilitated the study the diet of Chilean owls. However, identification of avian prey has been somewhat more difficult because of the unavailability of good identification keys and the difficulty of accessing voucher specimens deposited in reference collections. The latter is a limiting factor for investigators, particularly when they work in very distant places from museum or universities and personnel in charge hinder access to collections. Although in many cases, the shape and colour of feathers permit identification of avian prey, many passerine bird species exhibit similar colouration in plumage making it difficult to discriminate among species. In other cases, the presence of only colourless feathers in pellets impedes a good identification at a finer taxonomic level. Fortunately, in recent years several authors have provided identification keys of avian prey based on distinct anatomical elements. Reyes (1992) and Rau and Martínez (2004) developed an identification key based on the microscopic analysis of feather structures such as nodes and barbules. Although this key was developed at the order level, in many cases, it is possible to recognize feathers at genus level or even at species level. González-Acuña et al. (2010) and Seijas and Trejo (2011) developed detailed keys for identification of passerine prey based on the skull morphometrics. Lemus and Torres-Mura (2011) developed an identification key of avian prey based on sternum morphometrics. The virtue of this key is that it covers a wide spectrum of taxa (25 orders, 47 families, and 112 genera). No doubt, all these keys will be very useful to determine more reliably the diet of Chilean owls in future studies.

Biases in Diet Analysis To our knowledge, very little studies have evaluated the diet of Chilean owls by using complementary methods. Cruz-Jofré and Vilina (2014) characterized the diet of *Athene cunicularia* on several islands of the Coastal System of Coquimbo, northern Chile, based on pellets and prey remains. Unfortunately, because the prey remains were collected in only two of four islands, it is not possible to know well the biases of both methods for determining the diet of *A. cunicularia*. However, prey remains collected in Chañaral Island (N = 60) contained a high number of the marsupial *Thylamys elegans* (N = 20), three vampire bats (*Desmodus rotundus*), and a pair of lizards. The amount of *T. elegans* in prey remains was proportionately higher than in pellets. In addition, bats were not found in pellets. Findings of Cruz-Jofré and Vilina (2014) indicate that the combined analysis of pellets and prey remains permits a better determination of the diet of *A. cunicularia*. Ramírez (2008) evaluated the diet of *Strix rufipes* in a remnant of sclerophyllous forest in central Chile based on pellets and prey remains collected during 1 year. Interestingly, this author found no important differences between both methods regarding the importance level of each prey item. Perhaps, the use of complementary methods may be much more important in owl species having generalist food habits or in those that, due to their small size, dismember their prey much more (e.g. *Glaucidium nana*).

Evaluating Presence by Acoustical Detection

The success of conservation and management plans depends on a good knowledge about the distribution, habitat use, and activity of species. At the same time, this depends on our ability to better detect the presence of species. In the case of owls, detection may be particularly more difficult in comparison to other bird species because most of them have nocturnal activity and secretive habits and occur at low densities. However, owls are strongly territorial birds and many species manifest territoriality by vocalizations. Thus, many investigators are using vocal eliciting to detect presence of owls in particular areas or habitats (Contreras 2007).

Call Broadcasting Call broadcasting for detecting owls is being increasingly used in Chile. Up to our knowledge, the first investigators that used vocal eliciting to detect presence of a Chilean owl were Martínez and Jaksic (1996). To determine habitat use and abundance of *Strix rufipes* in Valdivian forest remnants, these authors vocally imitated their calls in forest areas after dusk or about midnight, when owls were more responsive. This incited resident owls to call back and reveal their presence. Surveys were conducted by calling from one spot for 60 s every 300 m as investigator walked along roads and permanent trails at each study site. If a response was received, the investigator moved 1000 m before calling again to assure that the same individual(s) was not following it. These authors did not survey for owls during nights with precipitation or wind.

Later, Trejo et al. (2011) carried out a study with the specific objectives to determine if *S. rufipes* responds to the playback of conspecific calls, determines if the playback calls increase detection rates of this owl species, and establishes the optimal duration of playbacks needed to detect owls present in a given area. Trejo et al. (2011) elicited vocal responses of owls by broadcasting of vocalization recordings by using the car digital audio system connected by a 10-m cable to a 120-watt megaphone and a 4 ohm. Each sampling period consisted of a 2-min pre-broadcast period (to detect spontaneous calls) and three playback bouts. Each bout consisted of a 1.5-min broadcast, with a 4-min post-broadcast period. The megaphone was pointed in the four cardinal directions, at chest height or overhead, for each call series, to maximize projection of the acoustic "lure". A second person, who always stood almost 50 m from the speaker, recorded the responses of owls. These authors did not survey on nights with precipitation or when winds exceeded 20 km/h. A greater number of owls were detected after the recorded calls were played than during the initial observation period (15% vs 85% of stations with a response). Trejo et al. (2011) concluded that use of playbacks is more efficient than passive detections for detecting *S. rufipes*. In addition, they also suggested that it may be more efficient to apply a longer sampling period with a sequential playback protocol to detect as many owls as are present near the station and recommended a 13-min protocol with two broadcast and listening periods for *S. rufipes*.

Subsequently, Ibarra et al. (2012, 2014b) used a somewhat more sophisticated method of vocal eliciting to detect presence of *Glaucidium nana* and *S. rufipes* in Andean forest remnants. They broadcast calls by using a portable amplifier with a volume adjusted to 100 db at 1 m in front of the speaker measured using a digital sound-level metre. Each survey started with a 1-min passive listening period, followed by playback of calls of both species played in a random sequence. For each species, the authors broadcast vocalizations for 30 s while rotating the amplifier 360°, and then they listened for 1 min so that they broadcast calls for each species twice and followed each time with 1 min of listening. In addition, Ibarra et al. (2014b) recorded a number of variables to evaluate factors associated with detectability: temperature, relative humidity, wind speed, cloud cover, environmental noise, and moonlight. Detectability for both owls increased with greater moonlight and decreased with environmental noise, and for *G. nana* greater wind speed decreased detectability. Detection of both species was positively correlated with the detection of the other species. The recommendation of Ibarra et al. (2014b) is that both species should be surveyed simultaneously for a minimum of 3–4 times during a season, survey stations should be located away from noise, and observers should record the moon phase and weather conditions for each survey.

Recently, Rivas-Fuenzalida et al. (2015a) evaluated the presence of *S. rufipes* in forest remnants and pine plantations of the Nahuelbuta mountain range, southern Chile. These authors determined presence of owls either by acoustic detections, direct observations, or indirect signs. Acoustical detections were made by both spontaneous calls and elicited calls. These authors elicited response of owls by broadcasting records of territorial vocalizations along roadways crossing forest remnants or pine plantations. In each station, the broadcast sessions lasted 15 min with three playback bouts at 3-min interval. Each playback bout consisted of a 30-s broadcast, with three continued repetitions of 10 s. Owls responded to vocal eliciting after 15 s to 6 min. Use of playbacks by Rivas-Fuenzalida et al. (2015a) was exploratory, but these preliminary experience and the protocols developed by Trejo et al. (2011) and Ibarra et al. (2012, 2014b) are being considered to improve ongoing owl surveys in Nahuelbuta and other localities of southern Chile. In fact, the first author of this review has preliminarily evaluated detectability of forest owls in southern Chile by using complementarily spontaneous and elicited calls. First results appear to partially support Trejo et al. (2011).

From 2007, the Agrarian and Environmental Studies Centre (Centro de Estudios Agrarios and Ambientales, CEA) is utilizing playbacks for monitoring of distinct raptor species. Investigators apply a standard protocol consisting of broadcasting of recorded calls by using a portable amplifier connected to a megaphone, repeating four times 1-min playbacks. By this method, investigators have registered response of a number of species including four owls: *T. furcata*, *B. magellanicus*, *G. nana*, and *S. rufipes* (Contreras 2007). However, investigators still have not published results of their studies.

The aforementioned studies suggest that call playback offers an efficient means to detect owls in large areas by shortening time involved in detection compared to methods involving passive detection (Trejo et al. 2011). However, we should take into account that vocal eliciting may also produce biased results. Exclusive use of playbacks may bias surveys towards paired owls (Martínez et al. 1992). In addition, when survey period is much extended, owls can be overcounted because individuals from several adjacent territories respond simultaneously to vocal eliciting (Trejo et al. 2011; R.A. Figueroa, personal observation). In contrast to Trejo et al. (2011), Martínez et al. (1992) suggest that the combined use of recording spontaneous calling owls and the broadcast of territorial voices may improve survey methods by revealing floaters. In some cases, depending on aspect investigated, spontaneous calls may be the only necessary method. For example, Norambuena and Muñoz-Pedreros (2012) evaluated efficiently the diurnal activity of *Glaucidium nana* using only records of spontaneous calls. Summarizing, the use of vocal eliciting by call broadcasting is an efficient method for detecting owls and, therefore, to estimate their distribution, abundance, and habitat use. On the other hand, call broadcasting is not recommended for evaluating the activity of owls because it may greatly alter natural behaviour. In this case, if telemetry is not possible, spontaneous calls are the most recommended method.

Passive Acoustic Records The passive recording of sound in a habitat can lead to better a understanding of the ecology and behaviour of species by revealing its acoustical interactions in a non-invasive manner (Merchant et al. 2015). The recent expansion of the study of acoustic interactions in the wild has been facilitated by the sophistication of autonomous acoustic recorder devices. The autonomous acoustic recorder devices are self-contained digital instruments that can be fixed to terrestrial structures to record the soundscape continuously on the scale of months (Mennill et al. 2012, Sousa-Lima et al. 2013). Each unit consists of battery-powered electronics for digital data acquisition and storage within a weather- or waterproof housing. The acoustic transducer (microphone) may be mounted on the device or attached via a cable. An autonomous acoustic recorder system should be chosen with a frequency range that encompasses the spectral content of all sounds of interest to the study (for much more detail see Merchant et al. 2015). To our knowledge, there are no published studies on the ecology or behaviour of Chilean owls using this technology. However, autonomous acoustic recorder devices have been occasionally used to detect owl presence in forest habitats in the context of environmental impact studies. Autonomous acoustic recorder devices may be a good alternative for studying the spontaneous vocal behaviour of owls, particularly under adverse environmental or climatic conditions for the researcher.

Evaluating Behaviour by Using Visual Recording

Direct observation by a biologist physically present in the field is the most traditional method for evaluating wildlife behaviour (O'Connell et al. 2011; Sagarin and Pauchard 2012). Depending on study species, focus, and investigator capabilities, this method can be quite efficient and perhaps the only necessary one. When possible, some investigators have complementarily used video infrared cameras for evaluating nest activity (Delaney and Grubb 1998). Remote automated photographic camera ("camera traps") is the newest tool for recording animal behaviour (O'Connell et al. 2011; Gallina and López-González 2011, de la Maza et al. 2013), and it is being used for studying the behaviour of owls. Below, we comment on use of these methods in the study of Chilean owls' behaviour.

Direct Visual Observation The extension of the daily activity towards the hours of the day of some species of Chilean owls has allowed researchers to study behavioural patterns with relative ease. For example, the patterns of movement and hunting behaviour of *Asio flammeus*, the diurnal activity of *Glaucidium nana,* and the chick-rearing behaviour of the *Strix rufipes* were studied simply by direct visual observation with the support of binoculars and/or telescopes (Martínez et al. 1998; Norambuena and Muñoz-Pedreros 2012; Vukasovic et al. 2006). In the particular case of *S. rufipes*, a strictly nocturnal species, Vukasovic et al. (2006) observed their behaviour by using binoculars taking advantage of the full moon during the time duration of their study.

Automated Photographic Record Recently, the use of automated photographic cameras has allowed the recording of the nocturnal behaviour of some Chilean owls. By using this camera type, Ibarra et al. (2014a, 2017) partially registered the hunting behaviour and feeding of nestlings of *G. nana* and breeding behaviour of *S. rufipes* during the night. Novoa et al. (2016) took the opportunity to use an automated camera for recording the consumption of a carcass of Chilean lapwing (*Vanellus chilensis*) by an individual of *Bubo magellanicus*. Even though this technology is efficient for recording behaviours difficult to observe directly by the researchers, the main limitation is that it only records events in one point in space, restricting the correct interpretation of the facts. For example, Novoa et al. (2016) could not determinate if consumption of a *V. chilensis* by a *B. magellanicus* was attributable to predation or scavenging.

Owl Tracking by Telemetry

Telemetry allows biologists to track animals across the landscape while generating information about their movement patterns. Animal movement is measured by consecutive triangulations in different locations following the signal strength of radio-transmitters or locations given by GPS devices (Kenward 1985; Meyburg and Fuller 2007; Walls and Kenward 2007). Despite the wide use of telemetry for tracking owls, this technology has just begun to be used in Chile. The only known study is that of González (2007) who radio-tracked an individual of *T. furcata* in southern Chile. However, this study has not formally been published.

Increasing technological innovation is allowing biologists to gain abundant information about the life history of wild animals, much of which cannot be easily obtained by direct observation. However, new technological tools should be only considered as a complementary support for field studies and not as a final solution. We should never lose sight that the presence of biologists in the field is pivotal for a better understanding and interpretation of our findings.

References

Aillapán L, Rozzi R (2004) Una etno-ornitología mapuche contemporánea: poemas alados de los bosques nativos de Chile. Ornitol Neotrop 15:419–434

Alaie K, Aliabadian M (2012) Mitochondrial DNA (CYTB) divergences in two distinct, Old World and New World barn owls. Iran J Anim Biosyst 8:47–55

Aliabadian M, Alaei-Kakhki N, Mirshamsi O et al (2016) Phylogeny, biogeography, and diversification of barn owls (Aves: Strigiformes). Biol J Linn Soc 119:904–918

Altamirano TA, Ibarra JT, Hernández F et al (2012) Hábitos de nidificación de las aves del bosque templado andino de Chile. Fondo de Protección Ambiental, Ministerio del Medio Ambiente, Santiago

Alvarado S, Figueroa RA (2006) A non-parametric discriminant analysis of dietary selectivity of the rufous-legged owl (*Strix rufipes*) based on habitat, taxa and prey size. In: Ellis C, Seipke SH (eds), Abstracts II Neotropical raptor conference, Iguazú, pp 19–20

Alvarado S, Figueroa RA, Shehadeh I et al (2007) Diet of the rufous-legged owl (*Strix rufipes*) at the northern limit of its distribution in Chile. Wilson J Ornithol 119:475–479

Alvarado S, Figueroa RA, Valladares P et al (2015) Aves rapaces de la Región Metropolitana de Santiago, Chile. Seremi del Medio Ambiente, Región Metropolitana de Santiago, Universidad Santo Tomás and Universidad de Chile

Álvarez GA (2016) Dieta del tucúquere (*Bubo magellanicus*, Gmelin, 1788) en la localidad de Cañadón Grande, Magallanes, Chile. Biology and Chemical Teacher Thesis. Universidad de los Lagos, Osorno.

Araya B, Bernal M (1995) Aves. In: Simonetti JA, Kalin-Arroyo MT, Spotorno A, Lozada E (eds) Diversidad biológica de Chile. Comisión Nacional de Ciencia y Tecnología, Santiago, pp 350–690

Araya B, Millie G (1986) Guía de campo de las aves de Chile. Editorial Universitaria, Santiago

Araya B, Bernal M, Schlatter RP, Salaberry M (1995) Lista patrón de las aves chilenas, 3ra edn. Editorial Universitaria, Santiago

Arim M, Jaksic FM (2005) Productivity and food web structure: association between productivity and link richness among top predators. Ibis 74:31–40

Arim M, Marquet PA, Jaksic FM (2007) On the relationship between productivity and food chain length at different ecological levels. Am Nat 169:62–72

Bangs O (1919) Notes on South American short-eared owls. Proc New Engl Zool Cl 6:95–98

Barros R (1945) Notas sobre nidificación de aves chilenas. Rev Univ 30:181–202

Barros R (1950) La historia del chucho, *Glaucidium nanum* (King). Rev Univ 35:17–30

Barros R (1963) El pequén y la lechuza. Rev Univ 48:87–96

Barros A (1976) Nuevas aves observadas en Islas Picton, Nueva, Lennox y Navarino Oriental. An Inst Patagonia (Chile) 7:189–193

Barros R, Díaz F (2008) Observando aves en Arica y Parinacota. Chiricoca 7:3–25

Beaucornu J-C, Moreno L, González-Acuña D (2014) New catalogue of fleas (Insecta-Siphonaptera) of Chile. Zootaxa 3900:151–203

Beaudoin F, Ojeda V (2011) Nesting of rufous-legged owls in evergreen *Nothofagus* forests. J Raptor Res 45:272–274

Beehler BM (2010) The forgotten science: a role for natural history in the twenty-first century? J Field Ornithol 81:1–4

Begall S (2005) The relationships of foraging habitat to the diet of barn owls (*Tyto alba*) from Central Chile. J Raptor Res 39:97–101

Bernath EL (1965) Observations in southern Chile in the southern hemisphere autumn. Auk 82:95–101

Block WM, Brennan LA (1993) The habitat concept in ornithology, theory and applications. In: Power DM (ed) Current ornithology, vol 11. Plenum Press, New York, pp 35–90

Bonacic C, Ibarra JT (2010) Fauna andina: historial natural y conservación. Pontificia Universidad Católica de Chile, Santiago

Bozinovic F, Medel R (1988) Body size, energetic and foraging mode of raptors in central Chile: an inference. Oecologia 75:456–458

Brito JL (1999) Vertebrados del humedal la Reserva Nacional El Yali y su costa, Santo Domingo, Chile central. An Mus Hist Nat Valp 24:121–126
Brito JL (2005) Concón (*Strix rufipes*) en la Reserva Nacional Peñuelas, Región de Valparaíso. Bol Chil Ornitol 11:33
Bruce MD (1999) Family Tytonidae (barn-owls). In: del Hoyo J, Elliott A, Sargatal J (eds) Handbook of birds of the world, vol 5. Lynx Edicions, Barcelona, pp 34–75
Bullock DS (1929) Aves observadas en los alrededores de Angol. Rev Chil Hist Nat 33:171–211
Bullock DS (1935) Las aves de la Isla Mocha. Rev Chil Hist Nat 39:232–252
Burton JA (ed) (1973) Owls of the world: their evolution, structure, and ecology. Peter Lowe/ Eurobook Ltd, New York
Camus P, Castro S, Jaksic FM (2008) El conejo europeo en Chile: historia de una invasión biológica. Historia 41:305–339
Capurro S, Castillo H, Ipinza J et al (1971) Requerimientos tróficos de *Tyto alba tuidara* (Gray) en la zona de tendencia mediterránea de Chile. Arch Biol Med Exp 8:12
Carevic FS (2005) Hábitos alimentarios de *Athene cunicularia* (Molina 1782) en la provincia de Iquique, Chile, y su relación con la abundancia de presas. Professional Biologist Thesis. Universidad Arturo Prat, Iquique
Carevic FS (2011) Rol del pequén (*Athene cunicularia*) como controlador biológico mediante el análisis de sus hábitos alimentarios en la provincia de Iquique, norte de Chile. Idesia 29:15–21
Carevic FS, Carmona ER, Muñoz-Pedreros A (2013) Seasonal diet of the burrowing owl *Athene cunicularia* Molina, 1782 (Strigidae) in a hyperarid ecosystem of the Atacama Desert in northern Chile. J Arid Environ 97:237–241
Carmona E, Rivadeneira M (2006) Food habits of the barn owl *Tyto alba* in the National Reserve Pampa del Tamarugal, Atacama Desert, North Chile. J Nat Hist 40:473–483
Cassin J (1855) Birds. In: Gilliss JM (ed) US Naval astronomical expedition to the southern hemisphere. AOP Nicholson, Washington, DC, pp 172–206
Castro SA, Jaksic FM (1995) Great horned and barn owls prey differentially according to the age/ size of a rodent in northcentral Chile. J Raptor Res 29:245–249
Castro V, Rottmann J (2016) Aspectos de la etno–ornitología de la provincia de El Loa, norte de Chile. Rev Chil Ornitol 22:64–78
Celis-Diez JL, Ippi S, Charrier A et al (2011) Fauna de los bosques templados de Chile. Guía de campo de los vertebrados terrestres. Ediciones Corporación Chilena de la Madera, Concepción
Cerpa C, Yañez J (1981) Variación estacional de la dieta de *Tyto alba* (Gray 1929) en la zona mediterránea de Chile central. Bol Mus Nac Hist Nat 38:137–145
Chapman FM (1922) Descriptions of apparently new birds from Colombia, Ecuador and Argentina. Am Mus Novit 31:1–8
Chester S (1995) Birds of Chile. Wandering Albatross, San Mateo
Chubb C (1910) On the birds of Paraguay. Ibis 52:53–78
Chubb C (1916) The birds of British Guiana. Collections FV McConnell
Clark R (1986) Aves de Tierra del Fuego y Cabo de Hornos, guía de campo. Editorial LOLA, Buenos Aires
Clark RJ, Smith DG, Kelso LH (1978) Working bibliography of owls of the World. Nat Wildl Fed Sci Tech Ser 1:1–319
Cody ML (1970) Chilean bird distribution. Ecology 51:455–464
Colihueque N, Gantz A, Rau JR et al (2015) Genetic divergence analysis of the common barn owl *Tyto alba* (Scopoli, 1769) and the short-eared owl *Asio flammeus* (Pontoppidan, 1763) from southern Chile using COI sequence. ZooKeys 534:135–146
Coña P (1973) Testimonio de un cacique mapuche. Pehuén Editores, Santiago
Contreras P (2007) Utilización de señuelos acústicos en el estudio de aves rapaces en el sur de Chile. Enlace (Soc Vida Silv Chile) 67:16–17
Contreras M, Garthe S, Luna-Jorquera G (1999) ¿El pequén de Isla Choros (*Athene cunicularia*) preda sobre el yunco (*Pelecanoides garnotii*)? In: Abstracts of the VI Congreso Chileno de Ornitología, Universidad de Antofagasta, Antofagasta, Chile, 11–13 noviembre 1999

Correa P, Roa A (2005) Relaciones tróficas entre *Oncifelis guigna*, *Lycalopex culpaeus*, *Lycalopex griseus* y *Tyto alba* en un ambiente fragmentado de la zona central de Chile. Mastozool Neotrop 12:57–60

Cory CB (1915) Descriptions of new birds from South America and adjacent islands. Field Mus Nat Hist Pub Ornithol 1:1–302

Cory CB (1918) Catalogue of birds of the Americas. Field Mus Nat Hist Pub Zool Ser 13:1–315

Couve E, Vidal C (1999) Donde observar aves en el Parque Nacional Torres del Paine, guía de identificación. Editorial Fantástico Sur, Punta Arenas

Couve E, Vidal C (2000) Aves del Canal Beagle y Cabo de Hornos. Editorial Fantástico Sur, Punta Arenas

Couve E, Vidal C (2003) Birds of Patagonia, Tierra del Fuego and Antarctic Peninsula, the Falkland Islands and South Georgia. Editorial Fantástico Sur, Punta Arenas

Crawshay R (1907) The birds of Tierra del Fuego. Bernard Quaritch, London

Cruces P, Cerda J (1999) Informe técnico Proyecto Liebres. Servicio Agrícola y Ganadero, Gobierno Regional de Aysén, Chile

Cruz-Jofré F, Vilina Y (2014) Ecología trófica de *Athene cunicularia* (Aves: Strigidae) en un sistema insular del norte de Chile: ¿Posible respuesta funcional y numérica frente a *Pelecanoides garnotii* (Aves: Pelecanoididae)? Gayana Zool 78:31–40

Cunningham RO (1871) Notes on the natural history of the Strait of Magellan and West Coast of Patagonian made during the voyage of HMS. Edmonston and Douglas, Edinburgh

de Lafresnaye F, D'Orbigny A (1837) Synopsis avium ab Alcide d'Orbigny in ejus per Americam meridionalem itinere, coilectarum et ab ipso viatore necnon A de Lafresnaye in ordine redactarum. Mag Zool 7:1–88

de la Maza M, Leichtle J, Beltrami E et al (2011) Técnicas de monitoreo de fauna. In: de la Maza M, Bonacic C (eds) Manual para el monitoreo de fauna silvestre en Chile. Pontificia Universidad Católica de Chile, Santiago, pp 55–130

Delaney DK, Grubb TG (1998) An infrared video camera system for monitoring diurnal and nocturnal raptors. J Raptor Res 32:290–296

De La Peña M, Rumboll M (1998) Birds of southern South America and Antarctica. HarperCollins Publish Ltd, London

Dellacasa VA, Figueroa RA, Alvarado SA et al (2011) Flight behavior of the Red-backed Hawk (*Buteo polyosoma*) during summer in Nevados de Chillán, Chile. Hornero 26:171–175

Des Murs O (1847) Aves. In: Gay C (ed) Historia Política y Física de Chile, Zoología. Tomo I. Supremo Gobierno de Chile, Santiago, pp 183–494

D'Hamonville B (1896) Sobre *Strix perlata*. Act Soc Sci Chili 6:18–25

Díaz I (1999) Food habits of the rufous-legged owl (*Strix rufipes*) in the Mediterranean sclerophyllous forest of central Chile. J Raptor Res 33:260–264

Díaz I, Sarmiento C, Ulloa L et al (2002) Vertebrados terrestres de la Reserva Nacional Río Clarillo, Chile central: representación y conservación. Rev Chil Hist Nat 75:433–448

Donázar JA, Travaini A, Ceballos O et al (1997) Food habits of the great horned owl in northwestern Argentine Patagonia: the role of introduced lagomorphs. J Raptor Res 31:364–369

D'Orbigny A (1847) Voyage dans l'Amerique Meridionale. Tomo IV, 3e Partie: Oiseaux. Paris

Ducarme F, Luque GM, Courchamp F (2013) What are "charismatic species" for conservation biologists? BioSci Master Rev:2–8

Ebensperger LA, Mella JE, Simonetti JA (1991) Trophic-niche relationships among *Galictis cuja*, *Dusicyon culpaeus*, and *Tyto alba* in central Chile. J Mammal 72:820–823

Egli G (2002) Voces de aves chilenas [audios en disco compacto]. Unión de Ornitólogos de Chile, Santiago

Egli G, Aguirre J (2000) Aves de Santiago. Unión de Ornitólogos de Chile, Santiago

Elgueta E, Reid S, Pliscoff P et al (2006) Catastro de vertebrados terrestres y análisis en seis hábitats presentes en la Reserva Nacional Futaleufu, provincia de Palena, X Región, Chile. Gayana Zool 70:195–205

Elortegui S, Torres-Mura JC (2002) Biodiversidad: el sentido del camino. In: Elortegui S, Moreira A (eds) Parque Nacional La Campana, origen de una Reserva de la Biosfera. Taller La Era, Santiago, pp 72–129
Enríquez PL, Eisermann K, Motta-Junior JC et al (2015) Una revisión de la taxonomía y sistemática de los búhos neotropicales. In: Enríquez P (ed) Los búhos neotropicales – diversidad y conservación. El Colegio de la Frontera Sur, Chiapas, pp 29–38
Escobar MAH, Vukasovic MA (2004) *Asio flammeus* en Parque Nacional Bosque Fray Jorge, Región de Coquimbo. Bol Chil Ornitol 10:39
Escobar MAH, Uribe SV, Vukasovic MA et al (2005). Hábitos alimenticios del nuco (*Asio flammeus*) en agroecosistemas de Chile central. In: Abstracts of the VIII Congreso Chileno de Ornitología, Universidad de Concepción, Chillán, 19–23 Octubre 2005
Estades CF (1992) Avifauna de la Segunda Región de Antofagasta. Bol Unión Ornitól Chile 14:12–18
Estades CF (1995) Aves y vegetación urbana: el caso de las plazas. Bol Chil Ornitol 2:7–13
Estades CF (1997) Bird-habitat relationships in a vegetational gradient in the Andes of central Chile. Condor 99:719–727
Estades CF (2004) Estrategia nacional para la conservación de aves 2004. Unión de Ornitólogos de Chile. Ediciones Universidad de Chile, Santiago
Estades CF, Temple SA, Gajardo A (1998) Unusual nesting of the rufous-legged owl? J Raptor Res 33:183
Evans FC, Emlen JT Jr (1947) Ecological notes on the prey selected by a barn owl. Condor 49:3–9
Farías AA, Jaksic FM (2007) El Niño events, the lean/fat scenario, and long-term guild dynamics of vertebrate predators in a south American semiarid ecosystem. Austral Ecol 32:225–238
Farías AA, Jaksic FM (2011) Low functional richness and redundancy of a predator assemblage in native forest fragments of Chiloe Island, Chile. J Anim Ecol 80:809–817
Ferguson-Lees J, Christie DA (2001) Raptors of the world. Christopher Helm, London
Fernández MI, Fernández MR, Vidal ME (1980) La alimentación del *Speotyto cunicularia cunicularia* (Molina 1782). Biology Teacher Thesis. Pontificia Universidad Católica de Chile, Temuco
Figueroa RA (1995) Las aves rapaces como un modelo para la enseñanza de la ecología y educación ambiental. In: Abstracts of the I Congreso Nacional de Profesores de Ciencia de Chile, Universidad de Chile, Santiago, 11–13 Oct 1995
Figueroa RA (1996) Fluctuación estacional e interanual en la dieta del concón (*Strix rufipes*) en el sector de Rupanco, sur de Chile. Biology and Natural Sciences Teacher Thesis. Universidad de los Lagos, Osorno, Chile
Figueroa RA (1997) Estudio faunístico de la Reserva Forestal Malleco: aves y mamíferos. Informe final. Corporación Nacional Forestal, Temuco
Figueroa RA, Alvarado S (2007) Aves Rapaces. In: Figueroa RA, López R (eds) Flora y Fauna Terrestre con Prioridad de Conservación del Corredor Biológico Nevados de Chillán-Laguna Laja. Comité pro Defensa de la Fauna y Flora, Comisión Nacional del Medio Ambiente and Sociedad Zoológica de Frankfurt, Concepción, pp 75–87
Figueroa RA, Alvarado S (2012) Ecología y conservación de los búhos chilenos: avances y desafíos de investigación. Ornitol Neotrop 23:209–219
Figueroa RA, Alvarado S (2015) Bioindicación y salud pública: rol de las aves rapaces como indicadores ambientales y de biodiversidad. In: Romero N, Sánchez HJ, Urrea M, Mateo MM (eds) Actas Científicas N°1 del Encuentro de Grupos de Investigación de América y África Latina (GRAAL). Bubok Publishing SL, España, pp 139–143
Figueroa RA, Corales ES (2015) Registros ocasionales de presas consumidas por aves rapaces en áreas boscosas del sur de Chile. Bol Chil Ornitol 21:150–154
Figueroa RA, López R (eds) (2007) Fauna y flora con prioridad de conservación del Corredor Biológico Nevados de Chillán-Laguna Laja. Comité pro Defensa de la Fauna y Flora. Comisión Nacional del Medio Ambiente and Sociedad Zoológica de Frankfurt, Concepción
Figueroa RA, Quintana V (2001) Comunidad invernal de aves en un paisaje agroforestal del centro-sur de Chile. Bol Chil Ornitol 8:31–35

Figueroa RA, Martínez DR, Corales ES (1997) Dieta de búhos de bosque (*Strix rufipes*) en fragmentos de bosque nativo del valle central del sur de Chile. In: Abstracts of the III Congreso Chileno de Ornitología, Universidad Santo Tomás, Santiago, 12–15 noviembre 1997

Figueroa RA, Bravo CA, Corales ES et al (2000) Avifauna del Santuario de la Naturaleza Los Huemules del Niblinto, Región del Bío Bío, Chile. Bol Chil Ornitol 7:2–14

Figueroa RA, Figueroa R, Corales ES et al (2001a) Caracterización y monitoreo de la fauna vertebrada de la Reserva Nacional Malleco, Chile. Informe final. Corporación Nacional Forestal, Temuco

Figueroa RA, Corales ES, Cerda J et al (2001b) Roedores, Rapaces y Carnívoros de Aysén. Servicio Agrícola y Ganadero, Gobierno Regional de Aysén, Coyhaique

Figueroa RA, Figueroa R, Corales ES et al (2001c) Fauna vertebrada del Parque Nacional Tolhuaca, Región de la Araucanía, Chile. Informe final. Corporación Nacional Forestal, Temuco

Figueroa RA, Figueroa R, Corales ES et al (2004) Fauna vertebrada del Parque Nacional Queulat, Chile. Informe final. Corporación Nacional Forestal, Coyhaique

Figueroa RA, Corales ES, Martínez DR et al (2006) Diet of the rufous-legged owl (*Strix rufipes*) in an Andean *Nothofagus-Araucaria* forest, southern Chile. Stud Neotrop Fauna E 41:179–182

Figueroa RA, Murúa RE, Schlatter RP et al (2007) Bio-control of hantavirus rodent reservoirs by raptorial birds in southern South America: implication for management in rural environments. In: Abstracts of the VII international conference HFRS, HPS and hantavirus, Hotel Sheraton, Buenos Aires, 13–15 Jun 2007

Figueroa RA, Rau JR, Corales ES et al (2009) Rodent prey of the common barn owl and short-eared owl during winter in agricultural areas of southern Chile. Wildlife Biol 15:129–136

Figueroa RA, Alvarado SA, Corales ES et al (2015) Los búhos de Chile. In: Enríquez-Rocha PL (ed), Búhos Neotropicales: diversidad y conservación. El Colegio de la Frontera Sur (Ecosur), San Cristóbal de las Casas, México, pp 173–273

Fjeldsa J, Krabbe N (1990) Birds of the high Andes. Apollo Books, Svendborg

Fuentes D, González-Acuña D (2011) Aberraciones cromáticas del plumaje en aves: nuevos reportes en Chile. Bol Chil Ornitol 17:113–121

Fuentes MA, Simonetti JA, Sepúlveda MS et al (1993) Diet of the red-backed buzzard (*Buteo polyosoma*) and the short-eared owl (*Asio flammeus suinda*) in the Juan Fernández Archipelago off Chile. J Raptor Res 27:167–169

Fuentes C, Orellana M, Ardiles K et al (2011) Parasitismo gastrointestinal y externo en chuncho (*Glaucidium nanum*) en la región del Bío Bío. In: Abstract of the X Congreso Chileno de Ornitología, Universidad Andrés Bello, Santiago, 27 al 30 de septiembre 2011

Fulk GW (1976) Owl predation and rodent mortality: a case study. Mammalia 40:423–427

Gajardo JF, Rodríguez J (1985) Utilización de repelentes en el control del daño ocasionado por lagomorfos en plantaciones de *Pinus radiata* de la V Region de Chile. In: Olivares B, Morales E (eds) Simposio *Pinus radiata* – investigación en Chile, vol 1. Universidad Austral de Chile, Valdivia, pp 292–297

Gallina S, López-González C (2011) Manual de técnicas para el estudio de la fauna, vol I. de Querétaro, Instituto de Ecología, Universidad Autónoma de Querétaro

Gantz A, Rau J (1999) Relación entre el tamaño mínimo de fragmentos boscosos y su riqueza de especies de aves en el sur de Chile. An Mus Hist Nat Valp 24:85–90

Gay C (1847) Aves. In: Gay C (ed), Historia Política y Física de Chile, Zoología. Tomo I. Supremo Gobierno de Chile, Santiago, pp 183–494

Germain MF (1860) Notes upon the mode and place of nidification of some of the birds of Chili. Proc Boston Soc Nat Hist 7:309–316

Girini JM, Palacio FX, Zelaya PV (2017) Predictive modeling for allopatric Strix (Strigiformes: Strigidae) owls in South America: determinants of their distribution and ecological niche-based processes. J Field Ornithol 88:1–15

Gil-Cordero C (2003) Ecología trófica de aves rapaces en un gradiente latitudinal de Chile. Master Thesis in Ecology. Universidad Austral de Chile, Valdivia

Glade A (ed) (1988) Red list of Chilean terrestrial vertebrates. Corporación Nacional Forestal, Santiago

Gmelin JF (1788) Caroli a Linné. Systema naturae per regna tria naturae: Tomo I. Impensis Georg Emanuel Beer Liptzig
González J (1980) Clave de identificación para Strigiformes de Chile (búhos y lechuzas). Not Mens Mus Nac Hist Nat (Chile) 25:7–11
González C (1993) Notas sobre el nuco (*Asio flammeus*). Bol Unión Ornitól Chile 15:13
González MC (2007) La radiotelemetría como instrumento de evaluación en el manejo de vida silvestre. Enlace (Soc Vida Silv Chile) 67:14–15
González-Acuña D, Ausset M, Skewes O et al (2004) Variación estacional en el consumo de roedores por la lechuza de campanario (*Tyto alba*) en un área suburbana de Chillán, centro-sur de Chile. Hornero 19:61–68
González-Acuña D, Muñoz R, Cicchino A et al (2006) Lice of Chilean owls: a first description. J Raptor Res 40:301–302
González-Acuña D, Zarate P, Donoso S et al (2010) Caracterización morfométrica de cráneos de aves passeriformes del centro-sur de Chile. Bol Chil Ornitol 16:74–90
González-Acuña D, Barrientos C, Muñoz S et al (2011) Trasplante de plumas en la lechuza blanca (*Tyto alba*). Reporte de caso clínico. Hospitales Veterinarios 3:88–96
Goodall JD, Johnson AW, Philippi RA (1951) Las aves de Chile, vol 2. Platt Establecimientos Gráficos SA, Buenos Aires
Goodall JD, Johnson AW, Philippi RA (1957) Las aves de Chile. Suplemento I. Platt Establecimientos Gráficos SA, Buenos Aires
Gould J (1841) Birds. In: Darwin C (ed) The zoology of the voyage of HMS Beagle. Smith Elder and Co, London, pp 1–156
Gray JE (1828) Class Aves. In: Griffith E, Pidgeon E (eds) The animal kingdom arranged in conformity with its organization by Cuvier, vol 6. Whittaker, Treacher and Co, London, pp 1–548
Greer J, Bullock D (1966) Notes on stomach contents and weights of some Chilean birds of prey. Auk 83:308–309
Griscom L, Greenway JC (1937) Critical notes on new neotropical birds. Bull Mus Comp Zool 81:417–437
Groothuis TGG, Carere C (2005) Avian personalities: characterization and epigenesis. Neurosci Biobehav Rev 29:137–150
Hahn I, Römer U (2002) Threatened avifauna of the Juan Fernández Archipelago, Chile: the impact of introduced mammals and conservation priorities. Cotinga 17:66–72
Hahn I, Römer U, Schlatter RP (2006) Population numbers and status of land birds of the Juan Fernández Archipelago, Chile. Senckenb Biol 86:1–17
Heidrich P, König C, Wink M (1995) Bioakustik, taxonomie and molekulare systematik amerikanischer sperlingskäuze. Stuttgarter Beitr Naturkd Ser A (Biol) 534:1–47
Hellmayr CE (1932) Birds of Chile. Publ Field Mus (Zool Ser) 19:1–472
Herrera CM, Jaksic FM (1980) Feeding ecology of the barn owl in central Chile and southern Spain: a comparative study. Auk 97:760–767
Hertel F, Martínez D, Lemus M et al (2005) Birds from Chungungo, Tilgo, and Pájaros Islands in north-central Chile. J Field Ornithol 76:197–203
Hidalgo H (1999) La pajarera de Samuel Encino. Arrayán Editores, Santiago
Hinojosa-Sáez A, Valenzuela-Dellarossa G, González-Acuña D (2007) Avifauna del barrio universitario de Concepción. Bol Chil Ornitol 13:42–46
Hoffmann A, Lazo I (2000) Aves de Chile – un libro también para niños. Universidad Andrés Bello. RIL Editores, Santiago
Housse RE (1925) Adición a los "apuntes sobre las aves de la isla La Mocha". Rev Chil Hist Nat 29:225–227
Housse RE (1933) Avifauna de Chiloé. Rev Chil Hist Nat 37:38–42
Housse RE (1935) Monografía del pequén (*Speotyto cunicularia cunicularia*, Molina). Rev Univ 20:575–584
Housse RE (1939) Aves nocturnas de Chile. Rev Chil Hist Nat 43:20–28
Housse RE (1945) Las aves de Chile en su clasificación moderna: su vida y sus costumbres. Ediciones Universidad de Chile, Santiago

Hudson WH (1874) On the habits of burrowing owls (*Pholeoptynx cunicularia*). Proc Zool Soc Lond 42:308–311
Humphrey P, Bridge D, Reynolds P, et al (1970) Birds of Isla Grande (Tierra del Fuego). Preliminary Smithsonian Manual, Museum of Natural History, University of Kansas, Lawrence
Ibarra JT, Martin K (2016) Beyond species richness: an empirical test of top predators as surrogates for functional diversity and endemism. Ecosphere 6:142. doi:10.1890/ES15-00207.1
Ibarra JT, Gálvez N, Gimona A et al (2012) Rufous-legged owl (*Strix rufipes*) and austral pygmy owl (*Glaucidium nanum*) stand use in a gradient of disrupted and old growth Andean temperate forests, Chile. Stud Neotrop Fauna E 47:33–40
Ibarra JT, Barreau A, Altamirano TA (2013) Sobre plumas y folclore: presencia de las aves en refranes populares de Chile. Bol Chil Ornitol 19:12–22
Ibarra JT, Altamirano TA, Martin K et al (2014a) Tree cavity-nesting of austral pygmy-owls (*Glaucidium nana*) in Andean temperate forests of southern Chile. J Raptor Res 48:82–85
Ibarra JT, Martin K, Altamirano TA et al (2014b) Factors associated with the detectability of owls in South American temperate forests: implications for nocturnal raptor monitoring. J Wildl Manag 78:1078–1086
Ibarra JT, Martin K, Drever MC et al (2014c) Occurrence patterns and niche relationships of sympatric owls in South American temperate forests: a multi-scale approach. Forest Ecol Manag 331:281–291
Ibarra JT, Altamirano TA, Vergara G et al (2015) Historia natural y autoecología del chuncho (*Glaucidium nana*) en el bosque templado andino de la Araucanía, sur de Chile. Bol Chil Ornitol 21:29–40
Ibarra JT, Altamirano TA, Vermehren A et al (2017) Observations of a tree-cavity nest of the rufous-legged owl and predation of an owl nestling by a chimango caracara in Andean temperate forests. J Raptor Res 51:85–88
Inskipp T, Gillett HJ (eds) (2005) Checklist of CITES species. United Nations Environment Programme (UNEP), World Conservation Monitoring Centre (WCMC), Cambridge
Ippi S, Rozzi R (2004) Actividad diurna y nocturna del concón (*Strix rufipes*) en los bosques del Cabo de Hornos. Bol Chil Ornitol 10:9–12
Ippi S, Anderson CB, Rozzi R et al (2009) Annual variation of abundance and composition in forest bird assemblages on Navarino Island, Cape Horn Biosphere Reserve, Chile. Ornitol Neotrop 20:231–245
Iriarte J, Franklin W, Johnson W (1990) Diet of sympatric raptors in southern Chile. J Raptor Res 24:41–46
IUCN (2008) IUCN red list of threatened species 2008. www.iucn.org
Jaksic FM (1979) Técnicas estadísticas simples para evaluar selectividad dietaria en Strigiformes. Medio Ambiente (Chile) 4:114–118
Jaksic FM (1982) Inadequacy of activity time as a niche difference: the case of diurnal and nocturnal raptors. Oecologia 52:171–175
Jaksic FM (1983) The trophic structure of sympatric assemblages of diurnal and nocturnal birds of prey. Am Midl Nat 109:152–162
Jaksic FM (1985) Toward raptor community ecology: behavior base of assemblage structure. J Raptor Res 19:107–112
Jaksic FM (1986) Predator uppon small mammals in shrublands and grasslands of southern South America: ecological correlates and presumable consequences. Rev Chil Hist Nat 59:209–221
Jaksic FM (1988) Trophic structure of some Neartic, Neotropical and Palearctic owl assemblages: potential roles of diet opportunism, interspecific interference and resource depression. J Raptor Res 22:44–52
Jaksic FM (1989a) Tamaño corporal de depredadores como predictor de atributos tróficos: el caso de las aves rapaces. Medio Ambiente (Chile) 10:23–26
Jaksic FM (1989b) What do carnivorous predators cue in on: size or abundance of mammalian prey? A crucial test in California, Chile and Spain. Rev Chil Hist Nat 62:237–249
Jaksic FM (1997) Ecología de los vertebrados de Chile. Ediciones Universidad Católica de Chile, Santiago

Jaksic FM (2001) Spatiotemporal variation patterns of plants and animals in San Carlos de Apoquindo, central Chile. Rev Chil Hist Nat 74:477–502

Jaksic FM, Carothers JH (1985) Ecological, morphological, and bionenergetic correlates of hunting mode in hawks and owls. Ornis Scand 16:165–172

Jaksic FM, Feinsinger P (1991) Bird assemblages in temperate forests of North and South America: a comparison of diversity, dynamics, guild structure and resource use. Rev Chil Hist Nat 64:491–510

Jaksic FM, Jiménez JE (1986) The conservation status of raptors in Chile. Bird of Prey Bull 3:95–104

Jaksic FM, Lima M (2003) Myths and facts on ratadas: bamboo blooms, rainfall peaks and rodent outbreaks in South America. Austral Ecol 28:227–236

Jaksic FM, Marti CD (1981) Trophic ecology of *Athene* owls in mediterranean-type ecosystems: a comparative analysis. Can J Zool 59:2331–2340

Jaksic FM, Marti CD (1984) Comparative food habits of *Bubo* owls in mediterranean-type ecosystems. Condor 86:288–296

Jaksic FM, Simonetti JA (1987) Predator/prey relationships among terrestrial vertebrates: an exhaustive review of studies conducted in southern South America. Rev Chil Hist Nat 60:221–244

Jaksic FM, Soriguer RC (1981) Predation upon the European Rabbit (*Oryctolagus cuniculus*) in mediterranean habitats of Chile and Spain: a comparative analysis. J Anim Ecol 50:269–281

Jaksic FM, Yáñez J (1977) Nuevos aportes al conocimiento de la dieta del *Speotyto cunicularia* en la Provincia de Coquimbo (IV Región). Bol Ornitológ (Chile) 9:9–10

Jaksic FM, Yáñez J (1979) The diet of the barn owl in central Chile and its relation to the availability of prey. Auk 96:619–629

Jaksic FM, Yáñez J (1980a) Differential utilization of prey resources by great horned owls and barn owl in central Chile. Auk 97:895–896

Jaksic FM, Yáñez J (1980b) ¿Quién controla las poblaciones de conejos introducidos? Medio Ambiente (Chile) 4:41–44

Jaksic FM, Yáñez J, Persico R et al (1977) Sobre la partición de recursos por las Strigiformes de Chile central. An Mus Hist Nat Valp 10:185–194

Jaksic FM, Rau JR, Yáñez J (1978) Oferta de presas y predación por *Bubo virginianus* (Strigidae) en el Parque Nacional Torres del Paine. An Inst Patagonia (Chile) 9:199–202

Jaksic FM, Greene HW, Yañez J (1981) The guild structure of a community of predatory vertebrates in central Chile. Oecologia 49:21–28

Jaksic FM, Seib RL, Herrera C (1982) Predation by the barn owl (*Tyto alba*) in mediterranean habitats of Chile, Spain and California: a comparative approach. Am Midl Nat 107:151–162

Jaksic FM, Yáñez J, Rau JR (1986) Prey and trophic ecology of great horned owls in western South America: an indication of latitudinal trends. J Raptor Res 20:113–116

Jaksic FM, Jiménez JE, Feinsinger P (1990) Dynamics of guild structure among avian predators: competition or opportunism. In: Bell BD (ed), Proceedings XX Congressus Internationalis Ornithologici, Christchurch

Jaksic FM, Jiménez JE, Castro S et al (1992) Numerical and functional response of predators to a long-term decline in mammalian prey at a semiarid Neotropical site. Oecologia 89:90–101

Jaksic FM, Feinsinger P, Jiménez JE (1993) A long-term study of the dynamics of guild structure among predatory vertebrates at a semi-arid Neotropical site. Oikos 67:87–96

Jaksic FM, Feinsinger PL, Jiménez JE (1996) Ecological redundancy and long-term dynamics of vertebrate predators in semiarid Chile. Conserv Biol 10:252–262

Jaksic FM, Silva SI, Meserve PL et al (1997) A long-term study of vertebrate predator responses to an El Niño (ENSO) disturbance in western South America. Oikos 78:341–354

Jaksic FM, Torres-Mura JC, Cornelius C et al (1999) Small mammals of the Atacama Desert (Chile). J Arid Environ 42:129–135

Jaksic FM, Pavez EF, Jiménez JE et al (2001) The conservation status of raptor in the Metropolitan Region, Chile. J Raptor Res 35:151–158

Jaksic FM, Iriarte JA, Jiménez JE (2002) The raptors of Torres del Paine National Park: species accounts, diversity, and niche relationships. Rev Chil Hist Nat 75:449–461

Jaksic FM, Silva-Aranguiz E, Silva SI (2004) Fauna del Parque Nacional Bosque de Fray Jorge: una revisión bibliográfica. In: Squeo FA, Gutiérrez JR, Hernández IR (eds) Historia Natural del Parque Nacional Bosque de Fray Jorge. Ediciones Universidad de La Serena, La Serena, pp 93–114

Jaramillo A (2003) Field guide to the birds of Chile. A and C Black Publishers, London

Jaramillo A, Johnson MTJ, Rothfels CJ et al (2008) The native and exotic avifauna of Easter island: then and now. Bol Chil Ornitol 14:8–21

Jiménez JE (2000) Effect of sample size, plot size and counting time on estimates of avian diversity and abundance in a Chilean rainforest. J Field Ornithol 71:66–87

Jiménez JE, Jaksic FM (1989) Biology of the austral pygmy-owl. Wilson Bull 101:377–389

Jiménez JE, Jaksic FM (1993) Variación estacional de la dieta del caburé grande (*Glaucidium nanun*) en Chile y su relación con la abundancia de presas. Hornero 13:265–312

Johnson AW (1965) The birds of Chile and adjacent regions of Argentina, Bolivia and Peru. Vol I. Platt Establecimientos Gráficos SA, Buenos Aires

Johnson AW (1967) The birds of Chile and adjacent regions of Argentina, Bolivia and Peru. Vol II. Platt Establecimientos Gráficos SA, Buenos Aires

Johnson AW (1972) Supplement to the birds of Chile and adjacent regions of Argentina, Bolivia and Peru. Platt Establecimientos Gráficos SA, Buenos Aires

Jorge R, Tabilo E, Mondaca V (1998) Avifauna de la Laguna de Punta Teatinos y ecosistemas adyacentes, bahía de Coquimbo, Chile. Bol Chil Ornitol 5:2–9

Kaup JJ (1862) Monograph of the Strigidae. Trans Zool Soc London 4:201–260

Keith AR (1970) Bird observations from Tierra del Fuego. Condor 73:361–363

Kelso EH (1934) A list of the owls of America. Biol Leaflet 4:43–45

Kelso L (1938) New owls from South America. Biol Leaflet 9: no pages

Kelso L (1941) Additional races of American owls. Biol Leaflet 13:2

Kenward RE (1985) Raptor radio-tracking and telemetry. ICBP Tech Publ 5:409–420

King PP (1828) Extracts from a letter addressed by Capt Philip Parker King to NA Vigors on the animals of the Straits of Magellan. Zool J 3:422–432

Kleinschmidt O (1940) Futnote zur Literatur. Alfred Laubmann, Die Vogel von Paraguay (Strecker u. Schrder, Stuttgart I. Band 1939, II. Band 1940). Falco 36:59–60

König C (1991) Zur taxonomie and ökologie der Sperlingskäuze (*Glaucidium* spp.) des Andenraumes. Ökol Vögel 13:15–76

König C, Heidrich P, Wink M (1996) Taxonomie des Uhus (*Bubo* spec.) im südlichen Südamerika. Stuttgarter Beitr Naturkd Ser A (Biol) 540:1–9

König C, Weick F, Becking JH (1999) Owls. A guide to the owls of the world. Pica Press, London

König C, Weick F (2008) Owls of the world, 2nd edn. Christopher Helm Publishers, London

Krahnass A (1896a) Les cris de l'Effraye du Chili et celui du nandou. Act Soc Sci Chili 5:143–144

Krahnass A (1896b) Note sur les cris de l'Effraye du Chili. Act Soc Sci Chili 7:87–89

Kusch A, Donoso R (2017) Registro de un tucúquere albino (*Bubo magellanicus*) en Tierra del Fuego, Chile. Bol Chil Ornitol 23 (in press)

Lambeck RJ (1997) Focal species: a multispecies umbrella for nature conservation. Conserv Biol 11:849–857

Lane AA, Sclater PL (1897) Field-notes on the birds of Chili. Ibis 7:177–195

Lataste F (1895a) La question de l'Effraye du Chili *Strix perlata* Lichtenstein ou *Strix flammea* Linné? Act Soc Sci Chili 4:165–176

Lataste F (1895b) Nouvelles observations sur les moeurs et les manifestations phonétiques de l'Effraye sudaméricaine (*Strix perlata* Lichstenstein). Act Soc Sci Chili 5:63–72

Lazo I (2004) Guía de aves más frecuentes en Chile. Imaginación Digital Ltda, Santiago

Lazo I, Ginocchio R, Cofré H et al (2008) Nuestra diversidad biológica. In: Rovira J, Ugalde J, Stutzin M (eds) Biodiversidad de Chile: patrimonio y desafíos. Comisión Nacional del Medio Ambiente, Santiago, pp 55–59

Lemus ML, Torres-Mura JC (2011) Esternón de aves chilenas: Struthioniformes a Piciformes. Bol Mus Nac Hist Nat (Chile) 60:67–102

Lesson RP (1828) Manuel d'Ornithologie, ou description des genres et des principales especes d'Oiseaux. Tomo I. Roret, Paris

Lichtenstein MHK (1823) Verzeichnis der Doubletten des Zoologischen Museums der Koeniglichen Universität zu Berlin nebst Beschreibung vieler bisher unbekannter Arten von Säugethieren, Voegeln, Amphibien und Fischen. Verzeichniss Doubletten Zoologischen Museums Universität, Berlin

Lima M, Stenseth NC, Jaksic FM (2002) Food web structure and climate effects on the dynamics of small mammals and owls in semiarid Chile. Ecol Lett 5:273–284

Lobos G, Ferres M, Palma RE (2005) Presence of the invasive genera *Mus* and *Rattus* in natural areas in Chile: an environmental and epidemiological risk. Rev Chil Hist Nat 78:113–124

López B, Domínguez E (2002) Presencia de *Strix rufipes* King 1828 (Strigiformes: Strigidae) en el Parque Histórico Rey Don Felipe, Península de Brunswick, Magallanes. An Inst Patagonia (Chile) 30:147–150

López R, Figueroa RA (1999) Censo de poblaciones de huemules y fauna asociada en el Parque Nacional Bernardo O'Higgins, Magallanes: Informe Técnico. Corporación Nacional Forestal, Punta Arenas

Luna-Jorquera G, Fernández CE, Rivadeneira MM (2012) Determinants of the diversity of plants, birds and mammals of coastal islands of the Humboldt current systems: implications for conservation. Biodivers Conserv 21:13–32

Maldonado S (1986) La dieta de la lechuza blanca (*Tyto alba*) en la Reserva Forestal Lago Peñuelas. Corporación Nacional Forestal, Valparaíso

Mann G (1978) Los pequeños mamíferos de Chile (marsupiales, quirópteros, edentados y roedores). Gayana Zool 40:1–342

Marín MA, Kiff LF, Peña L (1989) Notes on Chilean birds, with description of two new subspecies. Bull BOC 109:66–82

Marks JS, Cannings RJ, Mikkola H (1999) Family Strigidae (typical owls). In: del Hoyo J, Elliott A, Sargatal J (eds) Handbook of birds of the world, vol 5. Lynx Edicions, Barcelona, pp 76–242

Marti CD (1992) Barn owl. In: Poole A, Stettenheim P, Gill F (eds), The birds of North America, No 1. Academy of Natural Sciences/American Ornithologists' Union, Philadelphia/Washington, DC

Martínez DR (1993) Food habits of the rufous-legged owl (*Strix rufipes*) in temperate rainforests in southern Chile. J Raptor Res 27:214–216

Martínez DR (1995) Ecología trófica del concón (*Strix rufipes,* King 1828) y su respuesta a manipulaciones de fracción del espectro de presas en el bosque de San Martín. Master Thesis in Ecology. Universidad Austral de Chile, Valdivia, Chile

Martínez DR (2005a) El concón (*Strix rufipes*) y su hábitat en los bosques templados australes. In: Smith-Ramírez C, Armesto JJ, Valdovinos C (eds) Historia, biodiversidad y ecología de los bosques costeros de Chile. Editorial Universitaria, Santiago, pp 477–482

Martínez DR (2005b) Los Strigiformes en la mitología vernacular austral. In: Smith-Ramírez C, Armesto JJ, Valdovinos C (eds) Historia, biodiversidad y ecología de los bosques costeros de Chile. Editorial Universitaria, Santiago, pp 483–484

Martínez D, González G (2005) Las aves de Chile, nueva guía de campo. Ediciones del Naturalista, Santiago

Martínez DR, Jaksic FM (1996) Habitat, abundance and diet of rufous-legged owls (*Strix rufipes*) in temperate forest of southern Chile. Ecoscience 3:259–263

Martínez DR, Jaksic FM (1997) Selective predation on scansorial and arboreal mammals by rufous-legged owls (*Strix rufipes*) in southern Chilean rainforest. J Raptor Res 31:370–375

Martínez DR, Figueroa RA, Ocampo CL et al (1998) Food habits and hunting ranges of short-eared owls (*Asio flammeus*) in agricultural landscapes of southern Chile. J Raptor Res 32:111–115

Martínez JA, Serrano D, Zuberogoitia I (2003) Predictive models of habitat preferences for the Eurasian eagle owl *Bubo bubo*: a multiscale approach. Ecography 26:21–28

Massa C, Gabelli FM, Cueto GR (2015) Using GPS tracking to determine movement patterns and foraging habitat selection of the common barn-owl (*Tyto alba*). Hornero 30:7–12

Mathews GM (1916) Birds of Australia, vol 5. Witherby Co, London
Mayorga S, Mansilla A (1988) Estudio anual de la dieta de la lechuza blanca, *Tyto alba* (Gray 1829) en una localidad semiurbana de la Provincia de Osorno. Biology and Natural Sciences Teacher Thesis. Instituto Profesional de Osorno, Osorno
Mella JE (1999) Vertebrados terrestres posibles de encontrar en la XI Región de Aysén. Servicio Agrícola y Ganadero, Coyhaique
Mella JE (2002) Dieta del cernícalo (*Falco sparverius*) y del tucúquere (*Bubo magellanicus*) en un ambiente cordillerano de Chile central. Bol Chil Ornitol 9:34–37
Mella JE (2005) Cambios estacionales en la avifauna del Monumento Natural El Morado, cordillera de Santiago. Bol Chil Ornitol 11:2–10
Mella JE, Delgado A, Moya I et al (2016) Dieta estacional y alternancia en el consumo de presas por el tucúquere (*Bubo magellanicus*) en el altiplano del norte de Chile. Rev Chil Ornitol 21:3–10
Mennill DJ, Battiston M, Wilson DR et al (2012) Field test of an affordable portable wireless microphone array for spatial monitoring of animal ecology and behaviour. Methods Ecol Evol 3:704–712
Merchant ND, Fristrup KM, Johnson MP et al (2015) Measuring acoustic habitats. Methods Ecol Evol 6:257–265
Meserve PL, Shadrick EJ, Kelt DA (1987) Diets and selectivity of two Chilean predators in the northern semi-arid zone. Rev Chil Hist Nat 60:93–99
Meserve PL, Gutiérrez JR, Jaksic FM (1993) Effects of vertebrate predation on a caviomorph rodent, the degu (*Octodon degus*), in a semiarid thorn scrub community in Chile. Oecologia 94:153–158
Meyburg M-U, Fuller MR (2007) Spatial tracking. B. Satellite tracking. In: Bird DM, Bildstein K (eds) Raptor research and management techniques. Hancock House Publishers, Washington, USA, pp 242–248
Meyer de Schauensee R (1966) The species of birds of South America with their distribution. Livingston Publishing Co, Philadelphia
Meyer de Schauensee R (1970) A guide to the birds of South America. Academy of Natural Sciences of Philadelphia, Wynnwood
Meyer de Schauensee R (1982) A guide to the birds of South America. International Council Bird Preservation, Philadelphia
Molina GI (1782) Saggio sulla storia naturale del Chili di Gio. Stamp S Tommaso d'Aquino, Bologne
Möller P, Muñoz-Pedreros A, Gil C (2004) Educación ambiental y aves rapaces. In: Muñoz-Pedreros A, Rau JR, Yáñez J (eds) Aves rapaces de Chile. CEA Ediciones, Valdivia, pp 293–298
Montavo CI, Fernández FJ, Tallade PO (2015) The role of Bubo virginianus magellanicus as rodent bone accumulator in archaeological sites: a case study for the Atuel River (Mendoza, Argentina). Int J Osteoarchaeol 6. doi: 10.1002/oa.2509
Moreno L, González-Acuña D (2015) Los parásitos de las aves rapaces de chile: una revisión. Bol Chil Ornitol 21:93–102
Morgado E, Günther B, González U (1987) On the allometry of wings. Rev Chil Hist Nat 60:71–79
Morrison ML, Marcot BG, Mannan RW (2006) Wildlife-habitat relationships: concepts and applications. Island Press, Washington, DC
Muñoz-Pedreros A (2004) Aves rapaces y control biológico de plagas. In: Muñoz-Pedreros A, Rau JR, Yáñez J (eds) Aves rapaces de Chile. CEA Ediciones, Valdivia, pp 307–334
Muñoz-Pedreros A, Gil C (2005) Uso de cajas anideras para la lechuza blanca (*Tyto alba*) para control biológico y crianza campestre. Simposio: El uso de cajas nido en estudios de aves rapaces en Chile y Argentina: experiencias, logros y proyecciones. In: Abstracts of the VIII Congreso Chileno de Ornitología, Universidad de Concepción, Chillán, 19–23 Octubre 2005
Muñoz-Pedreros A, Murúa RE (1990) Control of small mammals in a pine plantation (central Chile) by modification of the habitat of predators (*Tyto alba*, Strigiformes, and *Pseudalopex* sp., Carnivora). Acta Oecol 11:251–261
Muñoz-Pedreros A, Norambuena H (2011) Dos siglos de conocimiento sobre las aves rapaces de Chile (1810–2010). Gestión Ambiental 21:69–93

Muñoz-Pedreros A, Rau JR (2004) Estudios de egagrópilas. In: Muñoz-Pedreros A, Rau JR, Yáñez J (eds) Aves rapaces de Chile. CEA Ediciones, Valdivia, pp 265–279

Muñoz-Pedreros A, Ruiz J (2004) Características y adaptaciones de las aves rapaces. In: Muñoz-Pedreros A, Rau JR, Yáñez J (eds) Aves rapaces de Chile. CEA Ediciones, Valdivia, pp 15–26

Muñoz-Pedreros A, Murúa RE, González L (1990) Nicho ecológico de micromamíferos en un agroecosistema forestal de Chile central. Rev Chil Hist Nat 63:267–277

Muñoz-Pedreros A, Rau JR, Yáñez J (eds) (2004) Aves rapaces de Chile. CEA Ediciones, Valdivia

Muñoz-Pedreros A, Gil C, Yáñez J et al (2010) Raptor habitat management and its implication on the biological control of the hantavirus. Eur J Wildlife Res 56:703–715

Muñoz-Pedreros A, Gil C, Yáñez J et al (2016a) Trophic ecology of two raptors, barn owl (*Tyto alba*) and white-tailed kite (*Elanus leucurus*), and possible implications for biological control of hantavirus reservoir in Chile. Wilson J Ornithol 128:385–397

Muñoz-Pedreros A, Yáñez J, Gil C et al (2016b) Spatial differences in the diet of the Magellanic horned owl *Bubo magellanicus* (Gmelin, 1788) in central Chile. New Zeal J Zool. doi:10.1080/03014223.2016.1249379

Murúa RE, Rodríguez J (1989) An integrated control system for rodent in pine plantations in central Chile. J Appl Ecol 26:81–88

Murúa RE, González LA, Meserve PL (1986) Population ecology of *Oryzomys longicaudatus philippii* (Rodentia: Cricetidae) in southern Chile. J Anim Ecol 55:281–293

Murúa RE, González LA, González M et al (1996) Efectos del florecimiento del arbusto *Chusquea quila* Kunth (Poaceae) sobre la demografía de poblaciones de roedores de los bosques templados fríos del sur chileno. Bol Soc Biol Concepción (Chile) 67:37–42

Murúa RE, Schlatter RP, Briones M et al (2004) Uso de cajas nido para potenciar el control biológico de roedores reservorios de Hantavirus en el sur de Chile. In: Abstracts of the II Simposio Argentino de Investigación y Conservación de Rapaces, Museo de Historia Natural de La Plata, 1–2 octubre 2004

Murúa RE, Schlatter RP, Briones M et al (2005) Resultados preliminares de un programa multiespecífico de cajas aniderас para aves rapaces enfocado a reducir las poblaciones de roedores reservorios de hantavirus. Simposio: El uso de cajas nido en estudios de aves rapaces en Chile y Argentina: experiencias, logros y proyecciones. In: Abstracts of the VIII Congreso Chileno de Ornitología, Universidad de Concepción, Chillán, 19–23 Octubre 2005

Narosky T, Babarskas M (2000) Aves de la Patagonia. Guía para su reconocimiento. Vázquez Mazzini Editores, Buenos Aires

Nijman V, Aliabadian M (2013) DNA barcoding as a tool for elucidating species delineation in wide-ranging species as illustrated by owls (Tytonidae and Strigidae). Zool Sci 30:1005–1009

Norambuena H, Muñoz-Pedreros A (2012) Diurnal activity of the austral pygmy owl (*Glaucidium nana*) in southern Chile. Wilson J Ornithol 124:634–636

Novoa F, Altamirano TA, Ibarra JT (2016) Consumo de queltehue (*Vanellus chilensis*) por tucúquere (*Bubo magellanicus*) en el bosque templado andino de la Araucanía: ¿depredación o carroñeo? Rev Chil Ornitol 22:46–48

Núñez H, Meriggio V (2004) Clave para la determinación de especies de aves rapaces de Chile. In: Muñoz-Pedreros A, Rau JR, Yáñez J (eds) Aves rapaces de Chile. CEA Ediciones, Valdivia, pp 221–227

Núñez H, Yáñez J (1982) Dieta de *Athene cunicularia* (Molina 1782) en la V Región (aves: Strigiformes). Not Mens Mus Nac Hist Nat (Chile) 26:6–7

O'Connell AF, Nichols JD, Karanth KU (2011) Camera traps in animal ecology – methods and analyses. www.springer.com

Olrog CC (1948) Observaciones sobre la avifauna de Tierra del Fuego y Chile. Act Zool Lilloana 5:437–531

Olrog CC (1950) Notas sobre mamíferos y aves del archipiélago de Cabo de Hornos. Act Zool Lilloana 9:505–532

Olrog CC (1963) Lista y distribución de las aves argentinas. Universidad Nacional de Tucumán, Tucumán

Omland K, Martínez DR, Gaillard D et al (2001) Conservación de depredadores: áreas extensas y bosques antiguos. In: Primack R, Rozzi R, Feinsinger P et al (eds) Fundamentos de

conservación biológica, perspectivas latinoamericanas. Fondo de Cultura Económica, México, pp 459–461
Orellana M (2009) Parasitismo gastrointestinal y externo en chuncho (*Glaucidium nanum* King, 1828) (Aves) en la Región del Bío Bío, Chile. Bachelor Thesis in Veterinary Medicine. Universidad de Concepción, Concepción
Ortiz J, Quintana V, Ibarra-Vidal H (1994) Vertebrados terrestres con problemas de conservación en la cuenca del Bío-Bío y mar adyacente. Ediciones Universidad de Concepción, Concepción
Páez-López M, Escobar MAH, Vukasovic MA, et al (2005) Patrones de actividad diurna de *Asio flammeus* en un agroecosistema de Chile central. In: Abstracts of the VIII Congreso Chileno de Ornitología, Universidad de Concepción, Chillán, 19–23 Octubre 2005
Pavez E (2004a) Descripción de las aves rapaces chilenas. In: Muñoz-Pedreros A, Rau JR, Yáñez J (eds) Aves rapaces de Chile. CEA Ediciones, Valdivia, pp 29–104
Pavez E (2004b) Centro para las aves rapaces chilenas. In: Muñoz-Pedreros A, Rau JR, Yáñez J (eds) Aves rapaces de Chile. CEA Ediciones, Valdivia, pp 335–351
Pavez E, Gabella JP (1999) Presencia de aves marinas en la dieta del pequén (*Athene cunicularia*) en la costa de la Octava Región. Bol Chil Ornitol 6:42–43
Pavez E, Lobos GA, Jaksic FM (2010) Cambios de largo plazo en el paisaje y los ensambles de micromamíferos y rapaces en Chile central. Rev Chil Hist Nat 83:99–111
Paynter RA (1988) Ornithological gazetteer of Chile. Harvard University Press, Cambridge, MA
Péfaur JE, Jaksic FM, Yañez J (1977) La alimentación de *Speotyto cunicularia* (Aves: Strigiformes) en la Provincia de Coquimbo. An Mus Hist Nat Valp 10:78–184
Péfaur JE, Jaksic FM, Yañez J (1979) Biological and environmental aspects of a mouse outbreak in the semi-arid region of Chile. Mammalia 43:313–322
Peredo R, Miranda L (2001) Nuevos registros para la avifauna del estuario del río Lluta (Arica, Región de Tarapacá). Bol Chil Ornitol 8:2–9
Peredo R, Knapton B, Jaramillo A et al (2007) Lista de las aves de la desembocadura del Río Lluta. Chiricoca 2:5–11
Peters JL (1923) Notes on some summer birds of northern Patagonia. Bull Mus Comp Zool 65:275–337
Peters JL (1940) Check-list of birds of the world, vol 4. Harvard University Press, Cambridge
Philippi RA (1868) Catálogo de las aves chilenas existentes en el Museo Nacional de Santiago. An Univ Chile 31:241–335
Philippi-B RA, Johnson AW, Goodall JD et al (1954) Notas sobre aves de Magallanes y Tierra del Fuego. Bol Mus Nac Hist Nat 26:1–53
Pincheira-Ulbrich J, Rodas-Trejo J, Almanza VP et al (2008) Estado de conservación de las aves rapaces de Chile. Hornero 23:5–13
Piñones C, Zuleta C, Alfaro L et al (2016) Diálogo intergeneracional en torno a las aves: análisis de su potencial para la educación ambiental y conservación del sitio Ramsar Las Salinas de Huentelauquén (Coquimbo, Chile). Rev Chil Ornitol 22:107–119
Plath O (1996) Lenguaje de los pájaros chilenos. Editorial Grijalbo S.A, Santiago
Poeppig E (1829) Schreiben des jetzt in Chile reisenden Hrn Dr Poppig
Pradenas G (1991) Dieta anual de *Athene cunicularia* (Molina 1782) en Colonia Loncoche, IX Región, Chile. Natural Sciences Teacher Thesis. Pontificia Universidad Católica de Chile, Temuco
Raimilla V, Rau JR, Muñoz-Pedreros A (2012) Estado de arte del conocimiento de las aves rapaces de Chile: situación actual y proyecciones futuras. Rev Chil Hist Nat 85:469–480
Ramírez MPA (2008) Ecología trófica del concón (*Strix rufipes rufipes*) en un fragmento de bosque esclerófilo de la V Región y su relación con especies plaga. Bachelor Thesis in Veterinary Medicine. Universidad Mayor, Santiago
Raspail X (1895) Examen comparatif de L'Oeuf de l'Effraye du Chili. Act Soc Sci Chili 5:55–62
Raspail X (1896) A proos de cris de l'Effraye du Chili. Act Soc Sci Chili 6:79
Rau JR (1994) ¿Residencia o migración local? Un compromiso entre la oferta de sitios de nidificación y la amplitud de nicho trófico en estrigiformes. In: Abstracts of the II Congreso Chileno de Ornitología, Valparaíso, septiembre de 1994
Rau JR (2014) Papel ecológico de las aves rapaces: del mito a su conocimiento y conservación en Chile. Universidad de los Lagos, Osorno

Rau JR, Jaksic FM (2004a) Diversidad de las aves rapaces de Chile. In: Muñoz-Pedreros A, Rau JR, Yáñez J (eds) Aves rapaces de Chile. CEA Ediciones, Valdivia, pp 121–127

Rau JR, Jaksic FM (2004b) Ecología de ensambles taxonómicos. In: Muñoz-Pedreros A, Rau JR, Yáñez J (eds) Aves rapaces de Chile. CEA Ediciones, Valdivia, pp 145–153

Rau JR, Martínez DR (2004) Identificación de los órdenes de aves chilenas a través de la microestructura de sus plumas. In: Muñoz-Pedreros A, Rau JR, Yáñez J (eds) Aves rapaces de Chile. CEA Ediciones, Valdivia, pp 229–234

Rau JR, Yañez J (1981) Nuevos antecedentes sobre la alimentación de *Bubo virginianus* Vieillot, en Magallanes (Aves, Strigiformes, Strigidae). Not Mens Mus Nac Hist Nat (Chile) 25:9–10

Rau JR, Martínez DR, Yáñez J (1985) Dieta de la lechuza blanca, *Tyto alba* (Strigiformes, Strigidae) en el sur de Chile. Bol Mus Reg Araucanía 2:134–135

Rau JR, Villagra M, Mora M et al (1992) Food habits of the short-eared owl (*Asio flammeus*) in southern South America. J Raptor Res 26:35–36

Rau JR, Muñoz-Pedreros A, Martínez DR (2005) Diversidad trófica de aves rapaces y mamíferos carnívoros en la Cordillera de la Costa. In: Smith-Ramírez C, Armesto JJ, Valdovinos C (eds) Historia, biodiversidad y ecología de los bosques costeros de Chile. Editorial Universitaria, Santiago, pp 537–538

Rau JR, Gantz A, Sade S et al (2015) Patrones de subconjuntos anidados de aves rapaces en fragmentos de bosque del sur de Chile. Bol Chil Ornitol 21:76–82

Reed E (1896) Catálogo de las aves chilenas. An Univ Chile 93:197–213

Reed CS (1905) Sobre las especies chilenas del suborden Striges y su utilidad. Bol Soc Agríc Sur, Concepción

Reid S, Cornelius C, Barbosa O et al (2002) Conservation of temperate forest birds in Chile: implications from the study of an isolated forest relict. Biodivers Conserv 11:1975–1990

Reise D (1970) Algunos datos sobre la alimentación de la lechuza blanca. Bol Ornitológ (Chile) 2:7

Reise D, Venegas W (1974) Observaciones sobre el comportamiento de la fauna de micromamíferos en la región de Puerto Ibáñez (Lago General Carrera), Aysén, Chile. Bol Soc Biol Concepción 47:71–85

Remsen JV Jr, Cadena CD, Jaramillo A, et al (2016) A classification of the bird species of South America. American Ornithological Union. http://www.museum.lsu.edu/~Remsen/SACCBaseline.html. Accessed 1 Oct 2016

República de Chile (1929) Ley de Caza N° 4601. Ministerio de Agricultura, Santiago

República de Chile (1996) Ley de Caza N° 19473. Diario Oficial, 4 Sept 1996

República de Chile (2008) Ley sobre recuperación del bosque nativo y fomento forestal N° 20283. Diario Oficial, 2 Jul 2008

Reyes S (1992) Clave para la identificación de los órdenes de aves chilenas: microestructura de los nodos de las bárbulas. Biology and Natural Sciences Teacher Thesis. Departamento de Educación, Instituto Profesional de Osorno, Osorno

Reyes R, Campos-Ellwanger P, Schlatter RP (2009) Avifauna de isla Guafo. Bol Chil Ornitol 15:35–43

Ridgway R (1876) The genus *Glaucidium*. Ibis 6:11–17

Ridgway R (1914) Birds of North and Middle America. Striges. US Nat Mus Bull 50:594–825

Rivas T, Figueroa RA (2009) Aves rapaces de la cordillera de Nahuelbuta y sus alrededores. Red Conservacionista del Patrimonio Natural de Contulmo y Comisión Nacional del Medio Ambiente, Chile

Rivas-Fuenzalida T, Asciones-Contreras N, Rivas F et al (2015a) Presencia del concón (*Strix rufipes*) en remanentes de bosque y plantaciones de pino de la cordillera de Nahuelbuta, sur de Chile. Bol Chil Ornitol 21:83–92

Rivas-Fuenzalida T, Asciones-Contreras N, González C (2015b) Centro de Aves Rapaces Ñankulafkén: una oportunidad para la conservación de las rapaces en Nahuelbuta, sur de Chile. Spizaetus 20:17–28

Roa M (2011) Selección de sitios de nidificación y ecología trófica del pequén (*Athene cunicularia,* Molina, 1782) en cerros islas de la eco-región mediterránea de Chile central. Master Thesis in Natural Resources. Pontificia Universidad Católica de Chile, Santiago

Roa M, Alvarado S (2011) Guía de aves rapaces. Características y atributos de las aves rapaces de Calera de Tango, Santiago de Chile

Robbins M (2011) Proposal (#328) to SACC: elevate *Bubo v. magellanicus* (Strigidae) to species level. http://www.museum.lsu.edu/~Remsen/SACCprop328.html. Accessed 17 Jul 2011

Rottmann J, López-Calleja MV (1992) Estrategia nacional de conservación de aves. Servicio Agrícola y Ganadero, Santiago

Roulin A (2003) Geographic variation in sexual dimorphism in the barn owl *Tyto alba*: a role for direct selection or genetic correlation? J Avian Biol 34:251–258

Rozzi R (2004) Implicaciones éticas de narrativas yaganes y mapuches sobre las aves de los bosques templados de Sudamérica austral. Ornitol Neotrop 15:435–444

Rozzi R, Martínez DR, Willson M et al (1996) Avifauna de los bosques templados de Sudamérica. In: Armesto JJ, Villagrán C, Kalin-Arroyo MT (eds) Ecología de los bosques nativos de Chile. Editorial Universitaria, Santiago, pp 135–152

Ruiz J, Figueroa RA, Schlatter RP et al (2006) Dieta y respuesta de *Glaucidium nanum* a cajas-nido: un análisis sobre su rol como biocontrolador de roedores reservorios de hantavirus. In: Ellis C, Seipke SH (eds), Abstracts II Neotropical raptor conference, Iguazú

Saavedra B, Simonetti JA (1998) Small mammal taphonomy: intraspecific bone assemblage comparison between South and North American barn owl, *Tyto alba*, populations. J Archaeol Sci:25:165–25:170

Sagarin R, Pauchard A (2012) Observation and ecology: broadening the scope of science to understand a complex world. Island Press, Washington, DC

Sanhueza R (2007) Lechuza blanca. Enlace (Soc Vida Silv Chile) 67:12

Sanhueza R, Muñoz-Pedreros P (2007) Clave de búhos y lechuzas. Enlace (Soc Vida Silv Chile) 67:11

Santibáñez DP, Jaksic FM (1999) Prey size matters at the upper tail of the distribution: a case study in northcentral Chile. J Raptor Res 33:170–172

Schalow H (1898) Verzeichniss der in Chile, Patagonien, Feuerland Und Auf Den Falklandinseln gesammelten Vogel. In: Plate L (ed), Fauna Chilensis. Zoologische Jahrbuecher, Jena Suppl 4(3):641–749

Schamberger M, Fulk G (1974) Mamíferos del Parque Nacional Fray Jorge. Idesia 3:167–179

Schlatter RP (1976) Aves observadas en un sector del Lago Riñihue, Provincia de Valdivia, con alcances sobre su ecología. Bol Soc Biol Concepción 50:133–143

Schlatter RP (1979) Avances de la ornitología en Chile. Arch Biol Med Exp 12:153–168

Schlatter RP (1992) Conducta extraña de la lechuza blanca (*Tyto alba*). Bol Unión Ornitól Chile 13:6

Schlatter RP (2004) Fauna de turberas de la XII Región y Tierra del Fuego. In: Blanco DE, de la Balze VM (eds), Los turbales de la Patagonia, bases para su inventario y la conservación de su biodiversidad. Wetlands International 19. Buenos Aires, pp 107–112

Schlatter RP, Murúa RE (1992a) Control biológico de plagas forestales: bosque artificial y biodiversidad. Ambiente y Desarrollo 8:66–70

Schlatter RP, Murúa RE (1992b) Monocultivos de pino insigne: cómo conciliar homogeneización de ambientes introducidos con recuperación de biodiversidad. In: Olivares B, Meneses M, Paredes G (eds) *Pinus radiata*: investigación en Chile (silvicultura, manejo y tecnología). Valdivia, Facultad de Ciencias Forestales, Universidad Austral de Chile, pp 19–28

Schlatter RP, Yáñez J, Núñez H et al (1980) The diet of the burrowing owl in central Chile and its relation to prey size. Auk 97:616–619

Schlatter RP, Yáñez J, Núñez H et al (1982) Estudio estacional de la dieta del Pequén, *Athene cunicularia* (Molina) (Aves, Strigidae) en la Precordillera de Santiago. Medio Ambiente (Chile) 6:9–18

Schlatter RP, Murúa RE, Reise D (1991) Rodales con monocultivos de pino insigne en Valdivia: casas anideras como contribución al control biológico integrado de plagas insectiles. In: Abstracts of the I Congreso Chileno de Ornitología, Universidad Austral de Chile, Valdivia, 15–19 Sept 1991

Schlegel H (1862) Oti. Revue methodique et critique de la collection des Oiseaux. Mus Pays-Bas 2:1–2

Sclater PL (1867) Notes on the birds of Chili. In: Proceedings of scientific meeting of Zoology Society of London, London, pp 319–340

Sclater PL (1891) On a second collection of birds from the province of Tarapaca, northern Chili. Proc Zool Soc Lond 59:131–137

Sclater PL, Salvin O (1873) Nomenclator avium neotropicalium. Sumptibus Auctorum, Londini

Scott WS, Sharpe RB (1912) Ornithology. Vol 2. Anatidae-Tytodidae. In: Scott WS (ed), Reports of the Princeton University expeditions to Patagonia, 1896–1899. Princeton University, Princeton, pp 505–718

Seijas S, Trejo A (2011) Clave para la identificación de los passeriformes del noroeste patagónico en base a la osteología craneal. Hornero 26:129–147

Sergio F, Caro T, Brown D et al (2008) Top predators as conservation tools: ecological rationale, assumptions, and efficacy. Ann Rev Ecol Evol Syst 39:1–19

Sharpe RB (1875a) Catalogue of the birds in the British Museum. Vol 2. Striges, or nocturnal birds of prey. Taylor and Francis, London

Sharpe RB (1875b) Contributions to a history of the Accipitres. The Genus Glaucidium. Ibis 5:35–59

Sharpe RB (1881) Birds. Account of the zoological collections made during the survey of HMS. "Alert" in the Straits of Magellan and on the coast of Patagonia. In: Proceedings of scientific meeting of Zoology Society of London, London, pp 6–18

Short LL (1975) A zoogeographic analysis of the South American Chaco avifauna. Bull Am Mus Nat Hist 154:163–352

Sibley CG, Monroe BL Jr (1990) Distribution and taxonomy of birds of the world. Yale University Press, New Haven

Sielfeld W (1977) Reconocimiento macrofaunístico terrestre en el área del Seno Ponsomby (Isla Hoste). An Inst Patagonia (Chile) 8:275–296

Sielfeld W, Amado N, Herreros J et al (1996) La avifauna del Salar del Huasco, Chile. Bol Chil Ornitol 3:17–24

Silva SI, Lazo I, Silva-Aránguiz E et al (1995) Numerical and functional response of burrowing owls to long-term mammal fluctuations in Chile. J Raptor Res 29:250–255

Silva-Aránguiz E (2012) Recopilación de la literatura ornitológica chilena 1810–2012. http://www.bio.puc.cl/auco/artic01/ornito01.htm#S. Accessed 6 Mar 2012 (discontinued webpage)

Silva-Rodríguez EA, Ortega-Solís GR, Jiménez JE (2006) Aves silvestres: actitudes, prácticas y mitos en una localidad rural del sur de Chile. Bol Chil Ornitol 12:2–14

Simeone A (1995) Ecología trófica del bailarín *Elanus leucurus* y la lechuza blanca *Tyto alba* y su relación con la intervención humana en el sur de Chile. Bachelor Thesis in Biological Sciences. Universidad Austral de Chile, Valdivia

Simonetti JA, Walkowiak A (1979) Presas de *Tyto alba* Gray, 1829 (Aves: Strigidae) en el Parque Nacional La Campana. An Mus Hist Nat Valp 12:89–91

Simonetti JA, Kalin-Arroyo MT, Spotorno A et al (1992) Hacia el conocimiento de la diversidad biológica de Chile. In: Halffter G (ed) La diversidad biológica de Iberoamérica. CYTED-D, Veracruz, pp 253–270

Solar V, Hoffmann R (1975) Las aves de la ciudad. Editorial Gabriela Mistral, Santiago

Sousa-Lima R, Norris TF, Oswald JN et al (2013) A review and inventory of fixed autonomous recorders for passive acoustic monitoring of marine mammals. Aquat Mamm 39:23–53

Steadman DW (1995) Prehistoric extinctions of Pacific island birds: biodiversity meets zooarchaeology. Science 267:1123–1131

Stotz DF, Fitzpatrick JW, Parker TA III et al (1996) Neotropical birds, ecology and conservation. University of Chicago Press, Chicago

Straneck RJ, Vidoz F (1995) Sobre el estado taxonómico de *Strix rufipes* (King) y de *Strix chacoensis* (Cherrie & Reichenberger) (Aves: Strigidae). Notulas Faunísticas 74:1–5

Strom H, Sonerud GA (2001) Home range and habitat selection in the pygmy owl *Glaucidium passerinum*. Ornis Fennica 78:145–158

Taberlet P (1983) An estimation of the average foraging radius of the barn owl *Tyto alba* based upon rejection pellets analysis. Terre et la Vie 38:171–178

Tala C, Iriarte A (2004) Conservación de aves rapaces en Chile. In: Muñoz-Pedreros A, Rau JR, Yáñez J (eds) Aves rapaces de Chile. CEA Ediciones, Valdivia, pp 281–294

Tala C, González B, Bonacic C (1995) Análisis de la dieta del tucúquere *Bubo virginianus* en el Valle del Río Ibáñez, Aysén. Bol Chil Ornitol 2:34–35

Taylor IR (1994) Barn owls. Predator-prey relationships and conservation. Cambridge University Press, Cambridge

Terry RC (2004) Owl pellet taphonomy: a preliminary study of the post-regurgitation – taphonomic history of pellets in a temperate forest. PALAIOS 19:497–506

Texera W (1973) Distribución y diversidad de mamíferos y aves en la Provincia de Magallanes. II: Algunas notas ecológicas sobre los canales patagónicos. An Inst Patagonia (Chile) 4:291–305

Tillería MS, Rau JR (1991) Determinación de clases de edad relativa y proporción de sexos en *Akodon olivaceus*: roedor-presa dominante de dos rapaces sintópicas del sur de Chile (*Tyto alba* y *Athene cunicularia*). In: Abstracts of the I Congreso Chileno de Ornitología, Universidad Austral de Chile, Valdivia, 15–19 Sept 1991

Torres MJ, Luna-Jorquera G, Quielfeldt P et al (2011) Dieta y conducta de depredación de *Athene cunicularia* sobre *Pelecanoides garnotii* en isla Choros. In: Abstracts of the X Congreso Chileno de Ornitología, Universidad Andrés Bello, Santiago, 27–30 Sept 2011

Torres-Contreras H, Silva-Aránguiz E, Jaksic FM (1994) Dieta y selectividad de presas de *Speotyto cunicularia* en una localidad semi-árida del norte de Chile a lo largo de siete años (1987–1993). Rev Chil Hist Nat 67:329–340

Torres-Mura JC (2004) Lista de las aves rapaces de Chile. In: Muñoz-Pedreros A, Rau JR, Yáñez J (eds) Aves rapaces de Chile. CEA Ediciones, Valdivia, pp 11–14

Torres-Mura JC, Contreras LC (1989) Ecología trófica de la lechuza blanca (*Tyto alba*) en los Andes de Chile central. Stud Neotrop Fauna E 24:97–103

Torres-Mura JC, Lemus ML (2003) Dieta del tucúquere (*Bubo magellanicus*) en la Reserva Nacional Lago Jeinimeni (XI Región). In: Abstracts of the VII Neotropical ornithological congress, Termas de Puyehue, Chile, 5–11 Oct 2003

Torres-Mura JC, Galaz JL, Rodríguez L (1997) Dieta del tucúquere (*Bubo virginianus*, Strigidae) en la Reserva Nacional Pampa del Tamarugal (I Región). In: Abstracts of the III Congreso Chileno de Ornitología, Universidad Santo Tomás, Santiago, 12–15 Nov 1997

Traylor MAJ (1958) Variation in South American great horned owls. Auk 75:143–149

Trejo A (2004) Reproducción de las aves rapaces de Chile. In: Muñoz-Pedreros A, Rau JR, Yáñez J (eds) Aves rapaces de Chile. CEA Ediciones, Valdivia, pp 155–168

Trejo A, Ojeda V (2002) Identificación de egagrópilas de aves rapaces en ambientes boscosos y ecotonales del noroeste de la Patagonia argentina. Ornitol Neotrop 13:313–317

Trejo A, Figueroa RA, Alvarado S (2006) Forest-specialist raptors of the temperate forests of southern South. Rev Bras Ornitol 14:317–330

Trejo A, Beaudoin F, Ojeda V (2011) Response of rufous-legged owls to broadcast of conspecific calls in southern temperate forests of Argentina. J Raptor Res 45:267–271

Valladares P, Urrutia N, Álvarez N et al (2016) Diet of the barn owl (*Tyto alba* Scopoli 1769) from the Copiapó valley, Atacama Desert, Chile. Interciencia 41:114–118

Vásquez RA, Simonetti JA (1999) Life history traits and sensitivity to landscape change: the case of birds and mammals of mediterranean Chile. Rev Chil Hist Nat 72:517–525

Venegas C (1994) Aves de Magallanes. Ediciones Universidad de Magallanes, Punta Arenas

Venegas C, Jory J (1979) Guía de campo para las aves de Magallanes. Publicaciones Instituto de la Patagonia, Serie Monografías 11, Punta Arenas, Chile

Venegas C, Sielfeld W (1998) Catálogo de los vertebrados de la Región de Magallanes y Antártica Chilena. Ediciones Universidad de Magallanes, Punta Arenas

Videler JJ (2005) Avian flight. Oxford University Press, London

Vieillot LJP (1817) Birds. Nouveau Dictionnaire d'Histoire Naturelie, Nouvelle edition. Vol 7. Paris

Vilina Y, Cofré H (2008) Aves terrestres. In: Rovira J, Ugalde J, Stutzin M (eds) Biodiversidad de Chile, patrimonio y desafíos. Comisión Nacional del Medio Ambiente, Santiago, pp 249–257

Vilina Y, Capella JJ, González J et al (1995) Apuntes para la conservación de la Reserva Nacional Pingüino de Humboldt. Bol Chil Ornitol 2:2–6
Village A (1987) Numbers, territory-size and turnover of short-eared owls *Asio flammeus* in relation to vole abundance. Ornis Scand 18:198–204
Villagra M, Mora M (1988) Dieta de *Athene cunicularia* Molina 1782 (Strigidae) en un agroecosistema de la provincia de Osorno. Biology Teacher Thesis. Instituto Profesional de Osorno, Osorno
Villagrán D (2016) Selección de hábitat por el pequén (*Athene cunicularia*) en la periferia de la ciudad de Valdivia, sur de Chile. Natural Resources Engineering Dissertation. Facultad de Ciencias Forestales y Recursos Naturales, Universidad Austral de Chile, Valdivia
Villagrán C, Villa R, Hinojosa LF et al (1999) Etnozoología Mapuche: un estudio preliminar. Rev Chil Hist Nat 72:595–628
von Berlepsch H, Stolzmann J (1892) On birds from Peru. In: Proceedings of scientific meeting of the Zoology Society of London, London, pp 387–388
Vuilleumier F (1985) Forest birds of Patagonia: ecological geography, speciation, endemism and faunal history. In: Buckley P, Foster M, Morton E et al (eds), Neotropical ornithology. American Ornithological Union (Ornithological Monographs 36), Washington DC, pp 255–304
Vuilleumier F (1998) Avian biodiversity in forest and steppe communities of Chilean Fuego-Patagonia. An Inst Patagonia (Chile) 26:41–57
Vukasovic MA, Escobar MA, Tomasevic JA et al (2006) Nesting record of rufous-legged owl (*Strix rufipes* King) in central Chile. J Raptor Res 40:172–174
Walker BH (1992) Biodiversity and ecological redundancy. Conserv Biol 6:18–23
Wallace P (2010) Primer registro de nidificación de lechuza bataraz austral (*Strix rufipes*) en Argentina. Nuestras Aves 55:3
Walls SS, Kenward RE (2007) Spatial tracking. A. Radio tracking. In: Bird DM, Bildstein KL (eds) Raptor research and management techniques. Hancock House Publishers, Washington, USA, pp 237–241
Weick F (2006) Owls (Strigiformes) – annotated and illustrated checklist. Springer, Berlin/Heidelberg/New York
Wetmore A (1922) New forms of Neotropical birds. J Wash Acad Sci 12:323–325
Wetmore A (1926) Observation on the birds of Argentina, Paraguay, Uruguay and Chile. US Nat Mus Bull 133:1–448
Wheeler L (1938) A new wood owl from Chile. Field Mus Nat Hist Pub Zool Ser 20:479–483
Wink M, Heidrich P, Sauer-Gürth H, El-Sayed A-A et al (2008) Molecular phylogeny and systematics of owls (Strigiformes). In: König C, Weick F (eds) Owls of the world, 2nd edn. Christopher Helm, London, pp 42–63
Wink M, El-Sayed A-A, Sauer-Gürth H et al (2009) Molecular phylogeny of owls (Strigiformes) inferred from DNA sequences of the mitochondrial cytochrome b and the nuclear RAG-1 gene. Ardea 97:581–591
Yáñez J, Jaksic FM (1977) Alimentación de *Bubo virginianus* en San Femando (VI Región). Bol Ornitológ (Chile) 9:8–9
Yáñez J, Jaksic FM (1979) Predación oportunista de *Speotyto cunicularia* en el norte chico (Aves: Strigiformes). Not Mens Mus Nac Hist Nat (Chile) 272:12
Yáñez J, Rau JR, Jaksic FM (1978) Estudio comparativo de la alimentación de *Bubo virginianus* (Strigidae) en dos regiones de Chile. An Mus Hist Nat Valp 11:97–104
Yarrell W (1847) Descriptions of the eggs of some of the birds of Chile. Proc Zool Soc London 15:51–55
Zimmer JT (1930) Birds of the Marshall Field Peruvian Expedition, 1922–1923. Field Mus Nat Hist Pub Zool Ser 17:1–480
Zunino S, Arcos M (1989) Nuevos antecedentes de la dieta de *Tyto alba* en el Parque Nacional La Campana. An Mus Hist Nat Valp 20:99–101
Zunino S, Jofré C (1999) Dieta de *Athene cunicularia* en isla Choros, Reserva Nacional Pingüino de Humboldt, IV Región. Bol Chil Ornitol 6:2–7

Chapter 8
The Owls of Costa Rica

Pablo Camacho-Varela and Randall Arguedas-Porras

Abstract Costa Rica is known worldwide for its high biodiversity concentrated in a rather small area, and the Strigiformes form no exception to this diversity. A total of 17 species have been recorded in the country during more than 150 years of ornithological research. Since 1964, owl species richness has not increased, and one species (*Asio stygius*) is considered in the category of "expected occurrence." Costa Rican owl species are classified in 11 genera, *Megascops* and *Glaucidium* being the more diverse in terms of species numbers. There are no endemic species, but two species are considered regional endemics, *Glaucidium costaricanum* and *Megascops clarkii*. Fourteen species are residents and three are accidental. For the Pacific slope, 13 species have been recorded, while ten species have been recorded for the Caribbean slope, and ten species are shared between them. One species is restricted to the Caribbean slope, while three species, previously considered as Pacific lowlands only, have recently been reported from the Caribbean slope. Three species are exclusively found in the highlands over 2000 m. All of the owl species are included in CITES, Appendix II. We have documented 58 existing published documents that include the 17 species of Costa Rican owls (37 corresponding to Strigidae, one to Tytonidae, and 20 in the order Strigiformes). Costa Rica's owls remain poorly studied, and there are considerable knowledge gaps for many species. It is essential to generate biological and ecological baseline data to know the actual status and population trends of Costa Rican owls, providing the Region with information that allows us to take conservation actions and further directions for these taxa.

Keywords AOU nomenclature • Expected occurrence • Resident species • Accidental species

---

P. Camacho-Varela (✉)
Fundación Rapaces de Costa Rica, Heredia, Costa Rica
e-mail: pcamacho@rapacesdecostarica.com

R. Arguedas-Porras
Fundación Rapaces de Costa Rica, Heredia, Costa Rica

Fundación Pro Zoológicos, San José, Costa Rica
e-mail: ranarg@gmail.com

P.L. Enríquez (ed.), *Neotropical Owls*, DOI 10.1007/978-3-319-57108-9_8

Crested Owl (*Lophostrix cristata*)

8.1 Introduction

Costa Rica is included among the 20 countries with the highest biodiversity of the world, in terms of the total of species, and despite that, the country has only 0.03% of the total continental world area. Costa Rica has one of the highest species densities (total species per area unit) in the world, harboring 3.6% of the planet's expected biodiversity (Obando 2007). Within this biological diversity, birds are no exception, with 918 species officially registered, and the Strigiformes certainly form part of this exceptional diversity (Garrigues et al. 2016).

The Costa Rican land mass is characterized by a highly irregular relief, which favors a wide variety of environments. A mountain range crosses the country in a northwest-southeast direction, dividing it in two slopes: the Atlantic or Caribbean and the Pacific (Herrera and Gómez 1993). Due to these geographic characteristics, climate, and geological history, five avifaunal zones are recognized: (1) North Pacific, (2) South Pacific, (3) Atlantic or Caribbean, (4) Highlands, and (5) Cocos Island (Fig. 8.1). Avifauna zones show very different ecological contexts, which markedly determines the distribution of many bird species. Cocos Island is located in the Pacific Ocean, 532 km from the continental coast (Slud 1964; Sánchez 2002; Calderón and May 2011; Montoya 2016). In this chapter, we discuss diversity, distribution, conservation, and status knowledge of Strigiformes in Costa Rica until 2016, based on a bibliographic review and information provided by the authors.

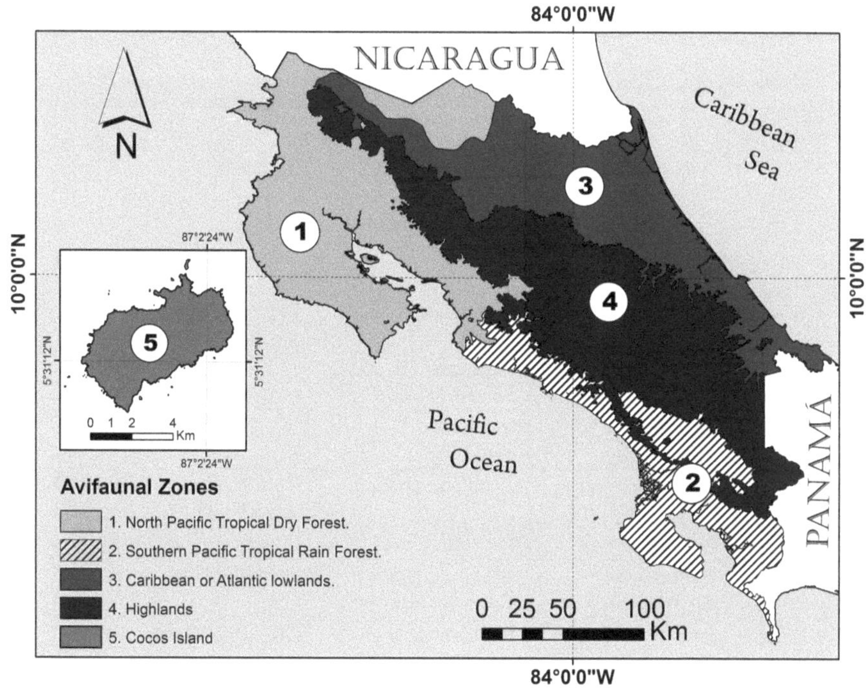

Fig. 8.1 Avifaunal zones of Costa Rica (Source: Adapted from Slud (1964) and Sánchez (2002). Datum: WGS84. Design and elaboration: P. Camacho 2016)

8.2 Taxonomic Diversity

Ornithological research in Costa Rica has 156 years of history and dates back to 1860 (Cabanis 1860; Obando-Calderón 2012). A total of 17 species of Strigiformes have been recorded for the country (Table 8.1). This number represents 1.85% of Costa Rican avian species (Stiles et al. 2003; Garrigues et al. 2016), 25.3% of the Neotropical owl species, and 23.2% of the Strigiformes species recorded on the American continent (Chesser et al. 2016; Remsen et al. 2016). It is important to consider that the number of species as well as the taxonomic classification can vary for the country depending of the consulted source, e.g., according to König et al. (2008), an additional species can be added to the list, *Megascops vermiculatus* (separating it from *M. guatemalae*), thus reaching 18 species, and the same happens with some nomenclature changes. In this chapter, we follow the North American Classification Committee (NACC) of the American Ornithologists' Union (AOU).

In 1910, 15 species were known for the country, according to Carriker (1910). However, the remaining two species (*Athene cunicularia* and *Asio flammeus*) had already been collected (10 and 43 years old, respectively) but were not documented (Slud 1964). Since 1964, all species of Strigiformes have been reported for the

Table 8.1 Owl species and status for Costa Rica (November 2016)

Family	Species	Spanish name/ (common name in Costa Rica)	English name	Distribution status
Tytonidae (1)				
	Tyto alba	Lechuza ratonera (Lechuza de campanario, Cara de gato, Búho)	Barn owl	R
Strigidae (16)				
	Megascops cooperi	Lechucita sabanera (Sorococa)	Pacific screech owl	R
	Megascops choliba	Lechucita neotropical (Estucurú)	Tropical screech owl	R
	Megascops guatemalae	Lechucita vermiculada	Vermiculated screech owl	R
	Megascops clarkia	Lechucita serranera	Bare-shanked screech owl	R-END
	Lophostrix cristata	Búho penachudo	Crested owl	R
	Pulsatrix perspicillata	Búho de anteojos (Oropopo, olopopo)	Spectacled owl	R
	Bubo virginianus	Búho grande	Great horned owl	R?
	Glaucidium costaricanum	Mochuelo montañero	Costa Rican pygmy owl	R-END
	Glaucidium griseiceps	Mochuelo enano	Central American pygmy owl	R
	Glaucidium brasilianum	Mochuelo común (Cuatro ojos, Majafierro)	Ferruginous pygmy owl	R
	Athene cunicularia	Lechuza llanera	Burrowing owl	A
	Ciccaba virgata	Lechuza café (Ju de León)	Mottled owl	R
	Ciccaba nigrolineata	Lechuza blanco y negro	Black-and-white owl	R
	Asio flammeus	Lechuza campestre	Short-eared owl	A
	Pseudoscops clamator	Búho listado	striped owl	R
	Aegolius ridgwayi	Lechucita parda (Lechucita de Alfaro)	Unspotted saw-whet owl	R

Source: Stiles et al. (2003), Garrigues and Dean (2014), Chesser et al. (2016), Garrigues et al. (2016), Remsen et al. (2016)
Estatus/status: *END* endemic, *R* resident, *R?* residence uncertain, *A* accidental

country, a list that has not increased to date (Morales 1978; Stiles and Lewis 1980; Stiles et al. 2003; Garrigues and Dean 2014; Garrigues et al. 2016).

The 17 species recorded in the country belong to two families: Tytonidae with one species and Strigidae with the remaining 16 (Stiles et al. 2003; Garrigues et al. 2016). Costa Rican owl species are classified in 11 different genera, of which

Table 8.2 Owl species with type locality for Costa Rica (November 2016)

Species	Type locality	Date/collector	Source
Megascops cooperi	Santa Ana, Costa Rica	Sept 4 1875; collected by José C. Zeledón	Ridgway (1878)
Megascops guatemalae (vermiculatus)	Costa Rica	No date; collected by General Lawrence	Ridgway (1887)
Aegolius ridgwayi	La Candelaria, Escazú, Costa Rica	March 29 1903; collected by Anastasio Alfaro	Alfaro (1905)
Glaucidium costaricanum	Costa Rica	No date; collected by Dr. van Patten	Kelso (1937)

Megascops and *Glaucidium* are the most diverse in terms of species numbers, with four and three species respectively, as well as *Ciccaba* that is represented by two species (Camacho-Varela 2014; Chesser et al. 2016; Garrigues et al. 2016).

Three species have their type locality in Costa Rica (Table 8.2) that means that they were collected for the first time in the country and that they were described based on the holotype (Ridgway 1878; Alfaro 1905; Kelso 1937). There is an additional specimen that some authors consider the holotype of *Megascops vermiculatus* (Ridgway 1887; König et al. 2008). Two species (*M. cooperi* and *A. ridgwayi)* were described as new species for science when they were collected; the rest of the species were posterior taxonomic reclassifications (Ridgway 1878; Alfaro 1905).

With respect to distribution status (Table 8.1), 14 species are recognized as residents (R), as they breed in Costa Rica and can be found year-round in the country (Stiles et al. 2003; Garrigues and Dean 2014; Garrigues et al. 2016). No endemic species of Strigiformes are found in Costa Rica; nevertheless two of them are considered regional endemics (R-END): *Glaucidium costaricanum* with a limited distribution to Costa Rica and northwestern Panama highlands (Fig. 8.2) and *Megascops clarkii*, a species restricted to Costa Rica, Panama, and northwestern Colombia, between 900 and 2500 m.a.s.l. (Kelso 1937; Enríquez and Rangel-Salazar 1997a; König et al. 2008).

On the other hand, there are three species whose status in the country is doubtful. Two of them are catalogued as accidental (A), i.e., a species without confirmation of breeding in Costa Rica and recorded just once or very few times, far from its normal distribution; finally one species is classified as an unknown resident (R?). None of the species are catalogued as migratory, yet the specific case of *Bubo virginianus* should be reviewed, this species is considered as unknown resident (R?), defining it as “a non-migratory species which breeds within Costa Rican territory but whose occurrence status is doubtful or unconfirmed” (Stiles et al. 2003; Garrigues et al. 2016) and, based on its actual context in the country, should be evaluated as migratory or accidental.

The only proved record of *B. virginianus* for Costa Rica, which incidentally was classified as a new subspecies for the country and later for Central America and southern Mexico (*B. v. mesembrinus*), is an adult female (No. 33218. from United States National Museum – USNM), undated, collected in San José by J. Carmiol (Oberholser 1904; Webster and Orr 1958). It is included in the lists of Costa Rica

Fig. 8.2 Costa Rican pygmy owl (*Glaucidium costaricanum*), San Gerardo de Dota, San José Costa Rica. 2 July 2012 (Photograph ©Mario Alberto Salazar Araya)

since 1869 (Frantzius 1869; Lawrence 1870). Carriker (1910) states that "evidently a rare bird in Costa Rica, and most likely found only in the higher portions of the country," while Slud (1964) mentions that he did not find it within the Costa Rican territory during his study. Before 1989, some observations were reported from the Central Valley (Cartago, Alajuela) and from Taboga, Guanacaste, but there are no precise dates indicated (Stiles et al. 2003). Much the same situation applies to Panama, which only has two historical records: a male specimen collected in 1968 (no month) in Chitra, Veraguas, and an observation in Isla Rancheria, Coiba, February 4th, 1956 (Salvin 1870; Wetmore 1968; Ridgely and Gwynne 1989; Jiménez-Ruiz et al. 2015).

Chris Artuso (with E. Carman), reported a *B. virginianus* singing at night, on December 27th, 2015 (eBird 2016), in the Volcano Irazú slopes, Cartago; previously, on April, 2004, another individual was reported resting during the day at Monteverde, Puntarenas (Garrigues 2004). Although, neither of those records were properly documented and they constitute the only two recent records for the country. Interestingly, all these record's dates, including the last two for Costa Rica, and the two from Panama, correspond with the Nearctic migratory season – September to April – according to Stiles et al. (2003), who classify this species as a very rare resident. Furthermore, they note that there are no nest records in the country. Similarly, based on the records for Honduras and Nicaragua, Olson (1997) points out that it is safe to assume that the species is a breeding resident, but its status for southern Central America seems to be uncertain, questioning its resident status for Costa Rica and Panama.

In the case of *Asio flammeus*, it is known from just two specimens (United States National Museum, USNM, No. 39735 and El Museo Nacional de Costa Rica – MNCR – No. 20119), both collected in San José, more than 99 years ago – December 18, 1863 and December 12, 1916 – according to Slud (1964), who also classified it as a rare winter visitant to Costa Rica. The other accidental species is *Athene cunicularia*, known in the country from just one specimen (male, American Museum Natural History – AMNH – 485,400) collected on December 20, 1900 in Los

Cuadros, southern slope of Volcán Irazú (Slud 1964; Stiles et al. 2003). Two possible records, one visual from northern Guanacaste in the 1980s and a picture from the surroundings of Rincon de la Vieja National Park in 2005, have not been confirmed (Stiles et al. 2003; Vargas 2012; Garrigues and Dean 2014). The species is accidental for Panama, and there is only one specimen recorded on December 13th, 1900, in Divala, Chiriqui (Ridgely and Gwynne 1989: Jiménez-Ruiz et al. 2015), so it can be inferred that an exceptional migration may have occurred around those dates (Stiles et al. 2003).

Finally, only one species is classified as "expected to occur"; Slud (1964) reports *Asio stygius* for Costa Rica in a "list of species not recorded" but expected for Costa Rica as occasional or accidental. *A. stygius* is a species with a wide but disjunct distribution, with some scarce, scattered, and poorly known geographical records (Rodríguez-Ruíz and Herrera-Herrera 2009). Because of its relative abundance, this owl is considered one of the rarest, most irregular, and hardest to observe species of Strigiformes (Marks et al. 1999; Monroy-Ojeda and Pedraza-Ruiz 2015). This species is reported from the neighboring countries, including Caribbean islands, but literally "skips" Costa Rica and Panama in its total distribution (Marks et al. 1999; König et al. 2008), and there are no current records that prove its presence in Costa Rica.

8.3 Distribution

Current knowledge about owl distribution in Costa Rica (14 resident and 3 accidental species), shows that the species are distributed in different ways in four of the five avifaunal zones, with Cocos Island the only region with no Strigiformes records (Table 8.3; Stiles et al. 2003; Garrigues and Dean 2014; Garrigues et al. 2016; Montoya 2016).

Ten of the 14 resident species (72%) are associated with tropical and subtropical regions, ranging between 0 and 1800 m of elevation above sea level. Only three species occupy exclusively cold temperate regions, above 2000 m (Stiles et al. 2003; Garrigues and Dean 2014). One species (*C. virgata*), which is the species with the widest distribution in the country, is found in almost every available habitat, from sea level to 2500 m approximately (Fig. 8.3 and Table 8.3).

There are 13 species recorded on the Pacific slope, while 11 are found on the Caribbean slope, including accidental species and isolated records. Ten species are shared between the two slopes (Table 8.3). The three accidental species were recorded on the Pacific slope and part of the highlands. Only one species, *G. griseiceps* (Fig. 8.4), is restricted to the Caribbean slope, while three additional species previously reported exclusively for the Pacific lowlands have since also been reported for Caribbean slope. *M. cooperi*, traditionally considered as occurring exclusively on the dry Pacific (Garrigues and Dean 2007), was also found on the northern Caribbean (Vargas 2012; Camacho-Varela 2014, Fig. 8.5). Likewise, *M. choliba* and *G. brasilianum,* previously thought to be confined to the Pacific slope, were recently found on the Caribbean slope (Camacho-Varela 2014, 2017, Fig. 8.6).

Table 8.3 Distribution of owl species by avifaunal zones for Costa Rica (November, 2016)

Species	Avifaunal zones of Costa Rica 1. Northern Pacific	2. Southern Pacific	3. Atlantic or Caribbean	4. Highlands	5. Cocos Island
Tyto alba	X	X	X	O	–
Megascops cooperi	X	O			–
Megascops choliba	O	X	X	O	–
Megascops guatemalae	X	X	X		–
Megascops clarkii				X	–
Lophostrix cristata	X	X	X	O	–
Pulsatrix perspicillata	X	X	X	O	–
Bubo virginianus	?			?	–
Glaucidium brasilianum	X	X	O	O	–
Glaucidium costaricanum				X	–
Glaucidium griseiceps			X		–
Athene cunicularia	?			?	–
Ciccaba virgata	X	X	X	X	–
Ciccaba nigrolineata	X	X	X		–
Asio flammeus				?	–
Pseudoscops clamator	X	X	X	O	–
Aegolius ridgwayi				X	

Source: Stiles et al. (2003), Camacho-Varela (2014), Garrigues and Dean (2014), Montoya (2016)
X normal distribution, *O* isolated or partial records, *?* accidental

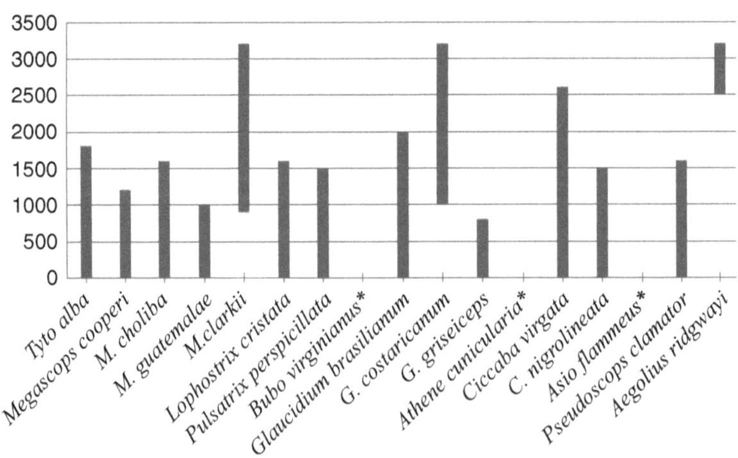

Fig. 8.3 Altitudinal distribution (m.a.s.l.) of the 17 recorded owl species for Costa Rica, November 2016 (*accidental species)

Fig. 8.4 Central American pygmy owl (*Glaucidium griseiceps*), Boca Tapada San Carlos, Alajuela Costa Rica. 22 de mayo del 2015 (Photograph ©Mario Alberto Salazar Araya)

The distribution of the genus *Megascops* is the best studied in the country (Fig. 8.5; Camacho-Varela 2014).

In a regional context, *M. cooperi* reaches its austral limit and three species (*M. choliba, M. clarkii*, and *G. costaricanum*) the septentrional limit of their distribution in Costa Rica (Marks et al. 1999; König et al. 2008; Camacho-Varela 2014). However, in the case of *M. choliba,* its limit is uncertain, because there is a record only 6.8 km away from the Nicaraguan border (Fig. 8.5), even though the species has never been recorded in that country (König et al. 2008; Camacho-Varela 2014, 2017; Trejo and Lezama-López 2015). On the other hand, the subspecies *Aegolius ridgwayi ridgwayi* also reaches its most northern limit in Costa Rica, separating it from *A. r. tacanensis*, Chiapas, Mexico, and from *A. R. rostratus*, Guatemala, Honduras, and El Salvador (Weidensaul 2015; Carman 2016).

8.4 Conservation Status

According to the Convention on International Trade in Endangered Species of Wild Fauna and Flora (CITES), all the owl species from Costa Rica are listed in Appendix II, not necessarily threatened, but whose trade have to be regulated in order to prevent any use that may be incompatible with its survival (CITES 2016). According to International Union for the Conservation of Nature (UICN 2016), all the species are listed as Least Concern (LC), with six species noted with decreasing populations (Table 8.4).

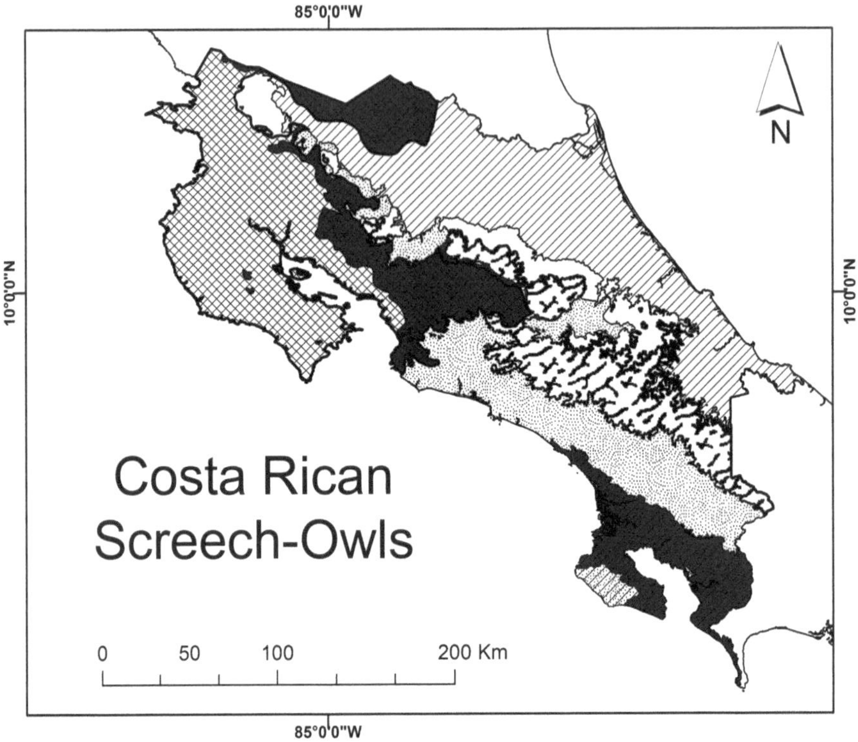

Elevation m	Species				Overlap zones
	M. choliba	M. cooperi	M. clarkii	M. guatemalae	
0					
500					
900					
1000					
1500					
1600					
2000					
3100					
3500					
3820					

Fig. 8.5 Actual distribution and altitudinal overlap strips of *Megascops* in Costa Rica (Taken from Camacho-Varela 2014)

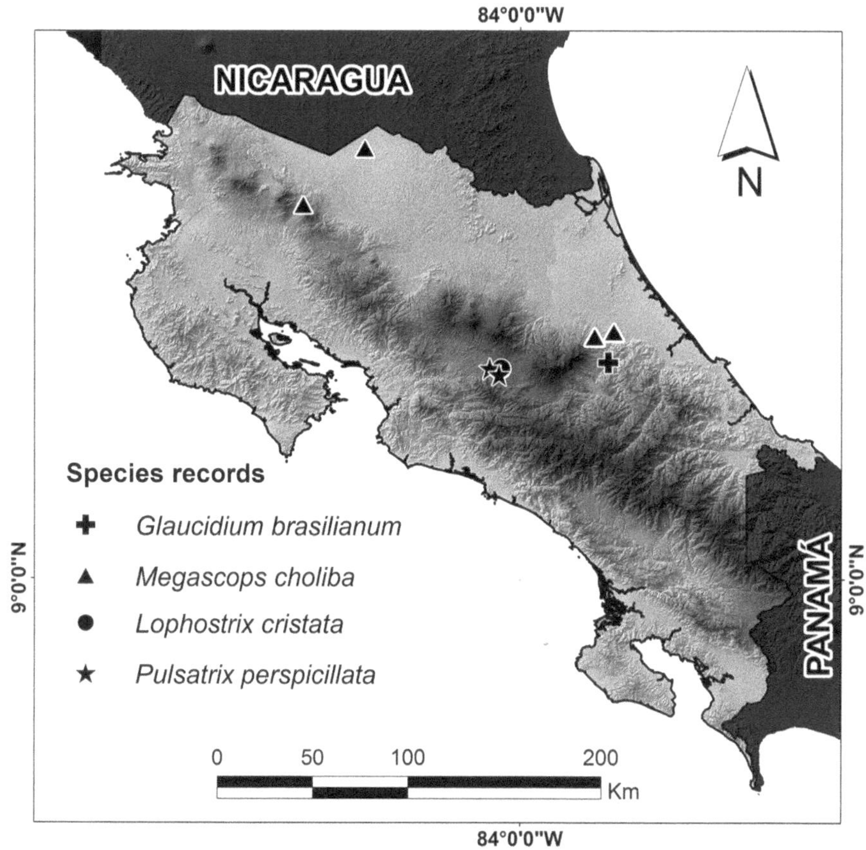

Fig. 8.6 New occurrence records of four owl species in Costa Rica (October 2016) (Source: Camacho-Varela (2017). Datum: WGS84. Design and elaboration: P. Camacho 2016)

On the other hand, at a national level, the Policies of Wildlife Conservation Law (Decretos 2005), article 26, considers three owl species with reduced or threatened populations (Table 8.4), as well as the species included in CITES Appendix II. None of the Strigiformes species are considered at risk of extinction according to the Wildlife Conservation Law.

Despite the above categorization, it is urgent to determine Costa Rican owls' conservation status more precisely and to update the information regarding this issue. The lack of information about species with reduced distribution ranges like *G. costaricanum* or *M. clarkii* is unacceptable. The same holds for the subspecies *A. r. ridgwayi*, for which no updated information exists. Nevertheless, there are some inconsistencies, for example, *L. cristata* is considered to have reduced or threatened populations but is one of the species with the widest distributions within the country, including records from San Jose city (Camacho-Varela 2017, Fig. 8.6). It is essential to generate biological and ecological baseline data in order to know the actual status and population trends of the owls of Costa Rica and the Region.

Table 8.4 Conservation status of owl species recorded for Costa Rica (November 2016)

Species	CITES appendix	IUCN Conservation status	IUCN Population trend	Wildlife conservation law
Tyto alba	II	LC	Stable	
Megascops cooperi	II	LC	Stable	
Megascops choliba	II	LC	Stable	
Megascops guatemalae	II	LC	Decreasing	
Megascops clarkii	II	LC	Decreasing	Threatened or reduced populations
Lophostrix cristata	II	LC	Stable	Threatened or reduced populations
Pulsatrix perspicillata	II	LC	Stable	
Bubo virginianus	II	LC	Stable	Threatened or reduced populations
Glaucidium brasilianum	II	LC	Decreasing	
Glaucidium costaricanum	II	LC	Stable	
Glaucidium griseiceps	II	LC	Stable	
Athene cunicularia	II	LC	Decreasing	
Ciccaba virgata	II	LC	Decreasing	
Ciccaba nigrolineata	II	LC	Stable	
Asio flammeus	II	LC	Decreasing	
Pseudoscops clamator	II	LC	Stable	
Aegolius ridgwayi	II	LC	Stable	

8.5 Conservation Strategies

There are no specific conservation strategies for any particular Costa Rican owl species. However, general strategies do exist, and they can contribute to their conservation; in this section we want to highlight two of them.

8.5.1 Protected Wild Areas

Costa Rica has been recognized worldwide as one of the leading countries in biodiversity conservation, and half a century ago, the country invested in the establishment of Wild Protected Areas, which nowadays comprise a total of 170, which represent 26.8% (1,340,872 ha) of the national continental area and 17.9% of the national marine area (SINAC 2010). Similarly, in May 2006, the National Program of Biological Corridors was established. Costa Rica has 37 official biological corridors with 1,753,822 ha, which represents 34% of the total continental area of the country. Likewise, it is important to highlight the strategy called Pay for Environmental Services (PSA – for its acronym in Spanish) that exists since 1996 and is powered by the Fondo Nacional de Financiamiento Forestal (FONAFIFO). This is an incentive given for forest areas protection, with the aims of mitigating the greenhouse effect, protecting water sources, conserving biodiversity, and maintaining scenic beauty. Since its implementation 20 years ago, the forest cover has increased (Asamblea Legislativa CR 1996; Barquero and Hernández 2015). Therefore, it will be important to link the ornithological component (including raptors) to the monitoring and assessment criteria used to be awarded to the PSA, which may generate important ecological information on the target species.

8.5.2 Important Bird Areas (IBA)

The objective of the IBA program from Costa Rica was to determine priority areas that allow bird conservation as well as their habitats (Sánchez et al. 2009; Sandoval and Sánchez 2011, Fig. 8.7). Of all the species included in this treaty, *M. clarkii* is the only one confirmed for IBAs under A2 criteria (restricted distribution area species) (Sandoval and Sánchez 2011).

A total of 21 IBAs have been declared for Costa Rica; they represent a total area of 3,070,976 ha (Fig. 8.7), almost 54% of the country's total area, and though they were not specifically meant to protect any Strigiformes species, such extensive areas will allow us to use IBAs information in other in situ conservation initiatives, as well as for the development of conservation actions for owl species included in those IBAs (Sánchez et al. 2009). Here we present the distribution of all owl species of Costa Rica (including accidental) for each of the 21 IBAs of Costa Rica (Table 8.5; Stiles et al. 2003; Garrigues and Dean 2014; eBird 2016).

8.6 Threats

The threats to Costa Rican owls have been evaluated using 20 years of data from wild rescued animals at the Simón Bolívar Zoo (Arguedas 2017 Fig. 8.8). Two main causes for rescue of incoming animals are recognized, the first includes neonates or recent fledglings collected by people that found them on the ground and the other are injured juveniles or adults.

Owl populations close to human environments are the most affected, including *Tyto alba, Megascops choliba, Pseudoscops clamator, Ciccaba virgata, Glaucidium brasilianum, and Pulsatrix perspicillata* (Figs. 8.9 and 8.10). Other species are less common to see but still reported, like *Ciccaba nigrolineata, Lophostrix cristata*, and

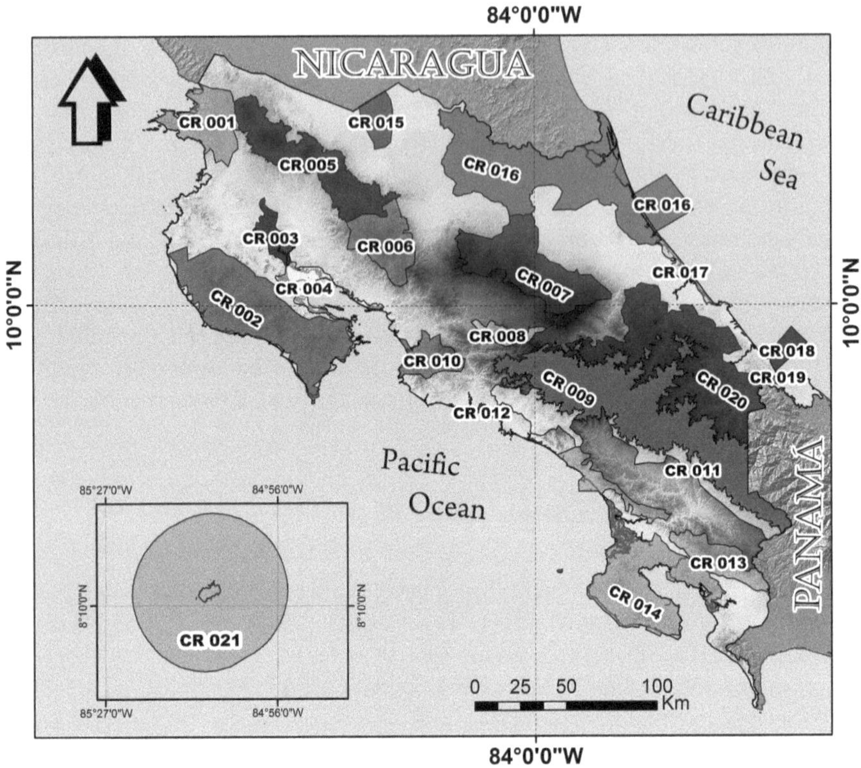

Fig. 8.7 Important bird areas (IBAs) in Costa Rica. CR-001, Guanacaste lowlands; CR-002, Nicoya Peninsula; CR-003, Palo Verde wetlands; CR-004, Mangroves and coastline of Nicoya; CR-005, Guanacaste cordillera; CR-006, Arenal-Monteverde cordillera; CR-007, Central Volcanic Belt; CR-008, El Rodeo, Hills of Escazú, and La Carpintera; CR-009, Talamanca Cordillera; CR-010, Tárcoles, Carara, La Cangreja; CR-011, Los Santos–Pacific La Amistad; CR-012, Central Pacific mangroves; CR-013, Coastal hills; CR-014, Sierpe and Osa Peninsula wetlands; CR-015, Maleku-Caño Negro; CR-016, Caribbean wetlands and grasslands; CR-017, Pacuare; CR-018, Cahuita and Gandoca; CR-019, Kéköldi; CR-020, Caribbean La Amistad; CR-021, Coco Island (Source: Sánchez et al. (2009), Sandoval and Sánchez (2011). Datum: WGS84. Design: Pablo Camacho 2016)

Megascops clarkii. The rest of the species within the Costa Rican diversity, are very rare or incidental; thus we do not have enough information about mortality or accidents.

Beside known individual cases, general threats affecting wildlife affect owl populations as well, with habitat loss, global warming, pollution, and possibly invasive exotic species and diseases being the major problems for this avian group (Obando-Calderon et al. 2014). Myths and legends of this group can surely mean a significant threat that should be considered (Enríquez and Rangel Salazar 2004).

Causes of incoming injured animals at rescue centers include car collisions, window collisions, fence wire collisions, and gunshots. From data from the retrospective study mentioned before, 54% of the owls that came from 1995 to 2015 were

Table 8.5 Distribution of Strigiformes species present in each IBA. November 2016

Species	CR-001	CR-002	CR-003	CR-004	CR-005	CR-006	CR-007	CR-008	CR-009	CR-010	CR-011	CR-012	CR-013	CR-014	CR-015	CR-016	CR-017	CR-018	CR-019	CR-020	CR-021
	0–500	0–1018	5–110	0	500–2020	400–200	50–3432	600–2480	1500–3820	0–1756	200–1500	0–10	0–1700	0	30–200	0–50	0	0–300	0–300	500–1500	0–575
Tyto alba	x	x	x	x	x	x	x	x	x	x	x	x	x	x	x	x	x	x	x	x	
Megascops cooperi	x	x	x	x	x	x	x			x		x			x						
Megascops choliba						x	x	x	x	x	x	x	x	x	x					x	
Megascops guatemalae		x			x	x	x							x	x	x	x	x	x	x	
Megascops clarkii						x	x	x	x		x									x	
Lophostrix cristata	x		x		x	x	x			x	x	x	x	x	x	x	x	x	x	x	
Pulsatrix perspicillata	x	x	x	x	x	x	x	x	x	x	x	x	x	x	x	x	x	x	x	x	
*Bubo virginianus**						x	x														
Glaucidium costaricanum						x	x		x		x									x	
Glaucidium griseiceps						x	x								x	x	x	x	x	x	
Glaucidium brasilianum	x	x	x	x	x	x	x	x		x	x	x	x							x	
*Athene cunicularia**							x														
Ciccaba virgata	x	x	x	x	x	x	x	x	x	x	x	x	x	x	x	x	x	x	x	x	
Ciccaba nigrolineata		x	x	x	x	x	x			x	x	x	x	x	x	x	x	x	x	x	
*Asio flammeus**							x														
Pseudoscops clamator		x	x	x		x	x	x		x	x	x	x	x	x	x	x	x	x	x	
Aegolius ridgwayi							x	x	x												

[x]Presence and records.

*Accidental species.

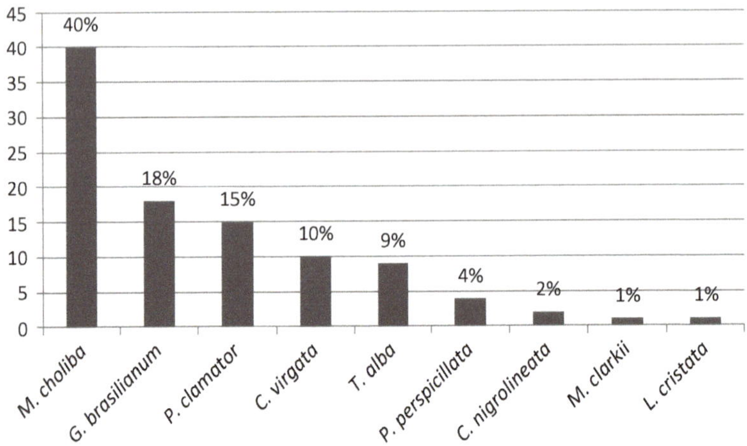

Fig. 8.8 Rescued owls at the Simón Bolívar Zoo in San José from 20 years (1995–2015) showing the percentage of incoming injured or orphaned owl species (n = 352) (Arguedas 2017)

Fig. 8.9 Barn owl (*Tyto alba)*, Carrizal de Alajuela, Alajuela Costa Rica. 3 de Enero del 2015 (Photograph ©Mario Alberto Salazar Araya)

Fig. 8.10 Striped Owl (*Pseudoscops clamator*), Barva de Heredia, Heredia, Costa Rica. 1 de Marzo del 2013 (Photograph ©Mario Alberto Salazar Araya)

because of trauma. Within trauma category, 78% were fractures (93% wing fracture and 7% leg fractures), 16% cranio-encephalic trauma, and the other 6% skin or general trauma lesion. 38% of the incoming owls were orphaned and the remaining 8% causes include intoxication, electrocution, or unknown causes (Arguedas 2017).

Understanding the causes of accidents and mortality is very important for taking conservation actions and for directing conservation and education efforts. Animals coming with lesions are all treated and then rehabilitated or kept in captivity, depending on the severity of the lesion, while others must be euthanized. Such efforts are useful for many species, especially when talking about reduced population species. However the main goal should be to mitigate the causes of the mortality events. That's why education and community divulgation programs should focus on preventing car and window collisions, as well as shooting animals (Arguedas 2017).

On the other hand, most of the orphaned animals are brought to rescue centers mainly because people find fledglings learning to fly on the ground and think they are true orphans, or they find them on their nests, in the case of species that naturally nest on the ground (*Pseudoscops clamator*) (Soto 2015). True orphans are not all that common but occur when trees are cut down with nests in them (personal observation). In both cases, education is probably the best tool we can use to make people learn about what to do when they find these birds (Obando-Calderón et al. 2014).

Economic costs of rehabilitation are expensive in terms of the medical and surgical procedures, personnel attending animals, food, and rehabilitation processes; thus there is an economic impact for countries that want to assume responsibility of their wildlife according to their environmental laws. That is why understanding and preventing underlying causes of accidents and mortality from stakeholders are vital for owl conservation (Arguedas 2017).

8.7 Ecological and Biological Knowledge Status

An analysis of the actual knowledge about Costa Rican Strigiformes was made based on an exhaustive bibliographic review online, as well as the online database Searchable Ornithological Research Archive (SORA 2016) and information from the BINABITROP database of the Organization for Tropical Studies – OTS (Fuentes-González 2013). A total of 58 published documents from the 17 owl species were found (no gray literature was considered). Of these publications, 37 correspond to Strigidae family, one to Tytonidae, and 20 are general data about the order (Appendix 8.1).

The species with the most information was *C. virgata*, it was found in 32 publications (55%); nevertheless, it should be considered that most of the studies correspond to distribution, and *C. virgata* is the most widely distributed species in the country. The least studied species was *A. flammeus*, with only five publications. Most of the papers (69%) dealt with Strigiformes records and distribution in the national territory, while a 27.6% corresponded to taxonomy issues (new species, reclassifications, others). No rigorous studies exist about diet analysis, which made up just 22.4% of the recorded articles, corresponding mainly to anecdotal or occasional reports instead of diet analysis per se. One of the few studies carried out with strict scientific principles in the country is Vaughan and McCoy (1982), who documented the diet of *T. alba* in a roosting place located in Cueva del Tigre, in Palo Verde National Park, Guanacaste. That study was made based on the pellet analysis, specifically on the skulls they contain. Similarly, reproductive biology data (nests, eggs, clutch and nestling size, phenology, and chick description) are limited, mainly corresponding to isolated reports. One of the most complete articles is about *M. clarkii* (Enríquez and Rangel-Salazar 1997a) and about *M. guatemalae* together with *G. costaricanum,* describing locality, habitat, nest, and eggs for both species (Marín and Schmitt 1991). Courtship and defense of a possible nest were documented for *A. ridgwayi*, but breeding was not verified (Carman 2016). In the case of the three accidental species, their nest and reproductive aspects are not known; the same holds for two of the residents: *G. griseiceps* and *C. nigrolineata* (Stiles et al. 2003).

8.8 Conclusions

In this chapter, we summarize the available information regarding distribution, taxonomy, natural history, ecology, and biological conservation for Costa Rican owls. In general terms, knowledge of the 17 Strigiformes species of the country is limited, even for the common ones. It is necessary to generate more information, especially on aspects like diet and reproductive biology of various species. Likewise, we consider that it is mandatory to study the taxonomic situation of *M. guatemalae* in the country, since there are three very well-recognized populations: Caribbean, middle elevations of Nicoya, and southern Pacific (Camacho-Varela 2014).

Finally, we agree that Strigiformes taxonomy and distribution are currently the two best known aspects for the country, in part because of Costa Rica's relatively small size, but also because it is a well-known destination for research and bird tourism. This activity has grown as a hobby in the last years and also the number of persons (tour guides, bird watchers, and environmental science professionals) which are constantly contributing with information about species sightings, natural history, and other information. In our country, we can also find online forms and social networks to report events as a citizen science contribution (Camacho-Varela et al. 2015; Soto 2016). As previously mentioned, a lot of information gaps remain, urging us to start generating baseline ecological and biological information to determine actual population trends of the Costa Rican and regional owls.

Acknowledgments To Paula Enríquez, without her enthusiasm and dedication, this chapter would not be possible. To John Van Dort, Mario A. Salazar, and Ricardo Pérez León for their contributions and support during the realization of this and A C. Sánchez for providing information about IBAs.

Mottled Owl (*Ciccaba virgata*)

Appendix 8.1

List of studies and references realized in owl of Costa Rica

Component	Family	Species	References
Behavior	Strigidae	*M. guatemalae, L. cristata, P. perspicillata, C. virgata, C. nigrolineata, G. griseiceps*	Enríquez and Rangel-Salazar (2004)
Behavior	Strigidae	*M. guatemalae, L. cristata, P. perspicillata, C. virgata, C. nigrolineata*	Enríquez and Rangel-Salazar (1997b)
Behavior	Strigidae	*G. brasilianum*	Sandoval and Wilson (2012)
Diet	Strigidae	*M. choliba*	Acosta-Chaves and Granados (2015)
Diet	Strigidae	*C. virgata*	Acosta-Chaves (2015)
Diet	Strigidae	*P. clamator*	De la (2015)

(continued)

Component	Family	Species	References
Diet	Strigidae	*G. griseiceps*	Janzen (1976)
Diet	Strigidae	*P. perspicillata*	Ramírez-Fernández and Solís-del Valle (2014)
Diet	Strigidae	*C. nigrolineata*	Sandoval-Vargas et al. (2008)
Diet	Tytonidae	*Tyto alba*	Vaughan and McCoy (1982)
Dieta, distribución	Strigidae	*C. nigrolineata*	Dickerman (1971)
Diet, distribution, reproduction	Strigidae	*A. ridgwayi*	Carman (2016)
Distribution	Strigidae	*A. ridgwayi*	Alfaro (1905)
Distribution	Strigidae	*G. brasilianum, C. virgata*	Alvarado Quesada and Durán Alvarado (2006)
Distribution	Strigidae	*G. brasilianum*	Alvarado-Quesada and Bolaños-Redondo (2011)
Distribution	Strigidae	*G. brasilianum, C. virgata, P. perspicillata, M. clarkii*	Alvarado-Quesada and Bolaños-Redondo (2012)
Distribution	Strigidae	*G. griseiceps*	Bangs (1909)
Distribution	Strigidae	*G. brasilianum*	Cabanis (1862)
Distribution	Strigidae	*M. cooperi, M. choliba, M. clarkii, M. guatemalae*	Camacho-Varela (2014)
Distribution	General	General	Carriker (1910)
Distribution	Strigidae	*L. cristata*	Cherrie (1893)
Distribution	Strigidae	General	Frantzius (1869)
Distribution	General	General	Garrigues and Dean (2007)
Distribution	General	General	Garrigues and Dean (2014)
Distribution	General	General	Garrigues et al. (2016)
Distribution	General	*T. alba, M. cooperi, M. choliba, G. brasilianum, C. virgata*	Guido-Granados and Rodríguez Arias (2011)
Distribution	General	General	Guido-Granados and Rodríguez Arias (2013)
Distribution	General	*T. alba, M. guatemalae, C. virgata, C. nigrolineata, G. griseiceps, L. cristata, P. perspicillata, P. clamator*	Harvey and Gonzalez Villalobos (2007)
Distribution	General	General	Morales (1978)
Distribution	Strigidae	*B. virginianus*	Oberholser (1904)
Distribution	Strigidae	*T. alba, L. cristata, C. nigrolineata, P. clamator*	Orians and Paulson (1969)
Distribution	Strigidae	*P. perspicillata, G. brasilianum*	Rodríguez Arias and Brenes-Cambronero (2010)
Distribution	Strigidae	General	Sánchez (2002)
Distribution	Strigidae	*G. costaricanum, M. guatemalae, M. choliba, M. clarkii, L. cristata, P. perspicillata, C. nigrolineata, C. virgata, A. ridgwayi*	Sánchez et al. (2004)

(continued)

Component	Family	Species	References
Distribution	General	*T. alba, P. perspicillata, G. brasilianum, P. clamator*	Slud (1980)
Distribution	Strigidae	*M. choliba*	Smith (1983)
Distribution	General	General	Stiles and Lewis (1980)
Distribution	General	General	Underwood (1899)
Distribution	Strigidae	*M. cooperi, M. choliba, P. perspicillata, A. cunicularia.*	Vargas (2012)
Distribution, Taxonomy	General	General	Chesser et al. (2016)
Distribution, Taxonomy	Strigidae	*M. trichopsis guatemalae*	Hekstra (1982)
Distribution, Taxonomy	Strigidae	*G. costaricanum*	Kelso (1937)
Distribution, Taxonomy	Strigidae	*Megascops*	Ridgway (1878)
General	General	General	König et al. (2008)
General	General	General	Lawrence (1870)
General	General	General	Marks et al. (1999)
General	General	General	Ridgely and Gwynne (1989)
General	General	General	Slud (1964)
General	General	General	Stiles and Skutch (2003)
General	General	General	Weidensaul (2015)
Parasites	Strigidae	*M. clarkii, P. perspicillata*	Marin (1988)
Perception	General	General	Enríquez and Mikkola (1997)
Population	Strigidae	*M. guatemalae, L. cristata, P. perspicillata, C. virgata, C. nigrolineata, G. griseiceps*	Enríquez and Rangel-Salazar (2001)
Reproduction	Strigidae	*M. clarkii*	Enríquez and Rangel-Salazar (1997a)
Reproduction	Strigidae	*M. guatemalae* and *G. costaricanum*	Marín and Schmitt (1991)
Reproduction	Strigidae	*G. brasilianum*	Valerio (2001)
Taxonomy	Strigidae	*M. guatemalae* and *G. costaricanum*	Ridgway (1887)
Taxonomy	Strigidae	*B. virginianus*	Webster and Orr (1958)

References

Acosta-Chaves VJ (2015) Opportunistic attack of mottled owl (*Ciccaba virgata*: Strigidae) on white-tipped dove (*Leptotila verreauxi*: Columbidae) in the El Copal Biological Reserve, Costa Rica. Spizaetus 20:19–31

Acosta-Chaves V, Granados G (2015) *Smilisca sordida* predation by a Tropical screech owl (*Megascops choliba*). Mesoamerican Herpetology 2(1):106–107

Alfaro A (1905) A new owl from Costa Rica. Proc Biol Soc Wash 18:217

Alvarado Quesada G, Durán Alvarado F (2006) Avifauna de los Cerros de Escazú, Costa Rica. Brenesia 66:37–47

Alvarado-Quesada G, Bolaños-Redondo S (2011) Avifauna del Parque La Libertad. Museo Nacional de Costa Rica, Departamento de Historia Natural

Alvarado-Quesada G, Bolaños-Redondo S (2012) Avifauna de El Rodeo, Mora, San José, Costa Rica. Brenesia 77:203–228

Arguedas R (2017) Causas de ingreso de Strigiformes al Zoológico Simón Bolívar durante 20 años (1995–2015). Revista Ciencias Veterinarias (in press)

Asamblea Legislativa CR (1996) Ley Forestal N° 7575. Asamblea Legislativa. Exp 11003. San José

Bangs O (1909) Notes on some rare or not well-known Costa Rican birds. Proc Biol Soc Wash 22:29–38

Barquero AI, Hernández G (2015) Bosques y Pago de Servicios Ambientales en Costa Rica. Ambientico 253:12–16

Cabanis J (1860) Üebersicht der im Berliner Mnseum befindlichen Vögel von Costa Rica. Jour für Orn 8(47): 321–336, 401–416

Cabanis JL (1862) Uebersicht der im Berliner Museum befindlichen Vögel von Costa Rica. Jour fur Ornith 10(5):321–336

Calderón P, May R (2011) Conozca las Aves. Introducción a la Ornitología. Asociación Ornitológica de Costa Rica, San José

Camacho-Varela P (2014) ¿Cómo identificar correctamente los búhos *Megascops* presentes en Costa Rica? Zeledonia 18(2):3–15

Camacho-Varela P (2017) Nuevos registros de la distribución de cuatro especies de búhos (Strigiformes: Strigidae) en Costa Rica. Brenesia (in press)

Camacho-Varela P, Hernández-Ugarte D, Salazar-Araya M, Jiménez-Córdoba C (2015) Primer registro de la anidación de *Buteogallus meridionalis* en Costa Rica y notas sobre su dieta reproductiva. Spizaetus 19:43–50

Carman E (2016) Nuevos datos de la historia natural del búho *Aegolius ridgwayi ridgwayi* en Costa Rica. IV Conferencia de Rapaces Neotropicales and II Simposio de Búhos Neotropicales Peregrine Fund-Fundación Rapaces de Costa Rica, La Fortuna, 10–13 Oct 2016

Carriker MA (1910) An annotated list of the birds of Costa Rica, including Cocos island. Ann Carnegie Mus 6:314–915

Cherrie GK (1893) Exploraciones zoológicas efectuadas en la parte meridional de Costa Rica por los años de 1891–92. I. Aves. Tipografía Nacional, San José

Chesser RT, Burns KJ, Cicero C, Dunn JL, Kratter AW, Lovette IJ, Rasmussen PC, Remsen JV, Rising JD, Stotz DF, Winker K (2016) Fifty-seventh supplement to the American ornithologists' union check-list of North American birds. Auk 133:544–560

CITES (2016) Checklist Of CITES Species http://checklist.cites.org. Accessed 4 Nov 2016

De la OJ (2015) Primer registro de depredación del Búho Listado (*Pseudoscops clamator*) sobre la Paloma Morada (*Patagioenas flavirostris*) en una zona urbana de la provincia de Alajuela, Costa Rica. Zeledonia 19(1):32–34

Decretos CR (2005) Reglamento a Ley de Conservación de la Vida Silvestre, Decreto N° 32633-MINAE. La Gaceta N° 180, (San José, Costa Rica) 20 de setiembre, 2005, pp 6–23

Dickerman RW (1971) Further notes on Costa Rican birds. Condor 73(2):252–253 eBird (2016) eBird: An online database of bird distribution and abundance. eBird, Cornell Lab of Ornithology, Ithaca. http://www.ebird.org. Accessed 2 Nov 2016

Enríquez PL, Mikkola H (1997) Comparative study of general public owl knowledge in Costa Rica, Central America and Malawi, Africa. In: Duncan JR, Johnson DH, Nicholls TH (eds) Biology and conservation of owls of the Northern Hemisphere: 2nd International symposium. Gen Tech Rep NC-190. U.S. Dept. of Agriculture, Forest Service, North Central Forest Experiment Station, St. Paul, pp 160–166

Enríquez PL, Rangel-Salazar JL (1997a) First nest record of the Bare-shanked screech-owl (*Otus clarki*). J Raptor Res 31(3):276–279

Enríquez PL, Rangel-Salazar JL (1997b) Intra-and interspecific calling in a tropical owl community. In: Duncan JR, Johnson DH, Nicholls TH (eds) Biology and conservation of owls of the Northern Hemisphere. Gem Tech. Rep. NC-190. USDA Forest Service, Denver, CO, pp 525–532

Enríquez PL, Rangel-Salazar JL (2001) Owl occurrence and calling behavior in a tropical rain forest. J Raptor Res 35:107–114

Enríquez PL, Rangel-Salazar JL (2004) Conocimiento popular sobre los búhos en los alrededores de un bosque húmedo tropical protegido en Costa Rica. Etnobiología 4:41–53

Frantzius AV (1869) Ueber die geographische Verbreitung der Vogel Costaricas und deren Lebensweise. Jour fur Ornith 17: 195–204, 289–318, 361–379

Fuentes-González G (2013) Aves de Costa Rica: una bibliografía anotada, 1st edn. Organización para Estudios Tropicales, San José

Garrigues R (2004) The Gone Birding Newsletter 5(2): www.angelfire.com/bc/gonebirding/news17.html. Accessed 2 Nov 2016

Garrigues R, Dean R (2007) The birds of Costa Rica. A field guide. Comstock Publishing Associates, Ithaca

Garrigues R, Dean R (2014) The birds of Costa Rica: A field guide, 2nd edn. Zona Tropical, Ithaca/Nueva York

Garrigues R, Araya Salas M, Camacho-Varela P, Montoya M, Obando-Calderón G, Ramírez-Alán O (2016) Lista Oficial de las Aves de Costa Rica Actualización 2016. Comité de Especies Raras y Registros Ornitológicos de Costa Rica-Comité Científico – Asociación Ornitológica de Costa Rica (AOCR). Zeledonia 20:2 (in press)

Guido-Granados I, Rodríguez Arias C (2011) Avifauna de la ciudad de San Ramón, Alajuela, Costa Rica y su área de expansión urbana. Brenesia 75–76:23–36

Guido-Granados I, Rodríguez Arias C (2013) Lista actualizada de las especies de aves de la Reserva Biológica Alberto Manuel Brenes y su área de amortiguamiento. Brenesia 80:40–58

Harvey CA, Gonzalez Villalobos JA (2007) Agroforestry systems conserve species-rich but modified assemblages of tropical birds and bats. Biodiver Conserv 16:2257–2292

Hekstra G (1982) Description of twenty-four new subspecies of American *Otus* (Aves Strigidae). Bull Zool Mus 9:49–63

Herrera W, Gómez LD (1993) Mapa de Unidades Bióticas de Costa Rica. 1: 685000. San José, INCAFO Costa Rica

Janzen D (1976) Food and feeding behavior of a captive Costa Rican least pigmy owl *Glaucidium minutissimum rarum* Griscom. (Aves: Strigidae). Brenesia 9:71–80

Jiménez-Ruiz B, Aparicio-Ubillúz K, Delgado-Botello F, et al. (2015) Los búhos de Panamá. In: Enríquez, PL (ed) Los Búhos Neotropicales:diversidad y conservación. El Colegio de la Frontera Sur, pp 522–558

Kelso L (1937) A Costa Rican race of Jardine's Pygmy owl. Auk 54:304

König C, Weick F, Becking JH (2008) Owls of the world, 2nd edn. Yale University Press, New Haven

Lawrence GN (1870) A catalogue of the birds found in Costa Rica. Ann Lyc Nat Hist New York 9:86–149

Marin M (1988) *Mallophagos* en algunas aves costarricenses. Brenesia 30:93–100

Marín M, Schmitt J (1991) Nests and eggs of some Costa Rican birds. Wilson Bull 103(3):506–509

Marks JS, Cannings RJ, Mikkola H (1999) Family Strigidae (typical owls). In: del Hoyo J, Elliot A, Sargatal J (eds) Handbook of the birds of the world, vol 5. Lynx Edicion, Barcelona, pp 76–242

Monroy-Ojeda A, Pedraza-Ruiz R (2015) El búho cara oscura (*Asio stygius*), nuevo registro para la avifauna del estado de Querétaro, México. Huitzil 16:70–73

Montoya M (2016) Lista Oficial de las Aves de la Isla del Coco. Comité de Especies Raras y Registros Ornitológicos (Comité Científico –AOCR) https://listaoficialavesdecostaricaword-press.com/isla-del-coco/ Accessed 2 Nov 2016

Morales R (1978) Las Lechuzas de Costa Rica. CATIE. Unidad de Áreas Silvestres, Costa Rica

Obando V (2007) Biodiversidad de Costa Rica en cifras. InBio-Sinac/Minae, San José

Obando-Calderón G (2012) Cronología: 152 años construyendo una lista de la avifauna de Costa Rica. Comité Científico, Asociación Ornitológica de Costa Rica (AOCR). Zeledonia 16(2):48–69

Obando-Calderón G, Acosta-Chávez V, Camacho-Varela P, Elizondo P, Montoya M, Oviedo S, Saénz Y (2014) Aves en problemas: las causas, cuándo y cómo ayudarlas. Zeledonia 18(1):28–53

Oberholser H (1904) A revision of the American great horned owls. Proc U S Natl Mus 27:177–192

Olson SL (1997) Avian biogeography in the islands of the Pacific coast of western Panamá. In: Phillips R (ed) A Festschrift. Horizon Communications, Albuquerque, pp 69–82

Orians GH, Paulson DR (1969) Notes on Costa Rican birds. Condor 71:426–431

Ramírez-Fernández J, Solís-del Valle M (2014) *Rhinophrynus dorsalis* (Mexican Burrowing Toad) Predation. Herpetological Review 45(3):480–481

Remsen JV, Areta JI, Cadena CD, et al (2016) A classification of the bird species of South America. American Ornithologists' Union http://wwwmuseumlsuedu/~Remsen/SACCBaselinehtm. Accessed 2 Nov 2016

Ridgely RS, Gwynne JA (1989) A Guide to the birds of Panama: with Costa Rica, Nicaragua, and Honduras, 2nd edn. Princeton University Press, Princeton

Ridgway R (1878) A review of the American species of the genus *Scops*. Proc U S Natl Mus 1:85–117

Ridgway R (1887) Description of two new species of Kaup's genus *Megascops* (*Megascops vermiculatus, Megascops hastatun, species nova*). Proc US Natl Mus 10:267–268

Rodríguez Arias C, Brenes-Cambronero L (2010) Fauna de vertebrados de la Reserva Madre Verde, Palmares, Costa Rica. Brenesia 73–74:108–120

Rodríguez-Ruíz ER, Herrera-Herrera JR (2009) A un siglo del registro del búho cara oscura (*Asio stygius*) en el centro de Tamaulipas y notas sobre su distribución en México y los Estados Unidos de América. Huitzil 10(2):56–60

Salvin O (1870) On some collections of birds of Veragua. Part II Proc Zool Soc Lond 1870:175–219

Sánchez J (2002) Aves del Parque Nacional Tapantí, Costa Rica / Birds of Tapantí National Park. Instituto Nacional de Biodiversidad (INBio). Santo Domingo de Heredia Costa Rica

Sánchez JE, Criado J, Sánchez C, Sandoval L (2009) Costa Rica. In: Devenish C, Díaz Fernández DF, Clay RP, Davidson I, Yépez Zabala I (eds) Important Bird Areas Americas - Priority sites for biodiversity conservation. BirdLife International (BirdLife Conservation Series No. 16), Quito, pp 149–156

Sánchez JE, Barrantes G, Durán F (2004) Distribución, ecología y conservación de la avifauna de la cuenca del río Savegre, Costa Rica. Brenesia 61:63–93

Sandoval L, Sánchez C (eds) (2011) Áreas Importantes para la Conservación de las Aves de Costa Rica. Unión de Ornitólogos de Costa Rica, San José

Sandoval L, Wilson D (2012) Local predation pressure predicts the strength of mobbing responses in tropical birds. Curr Zool 58(5):781–790

Sandoval-Vargas L, Biamonte E, Solano-Ugalde A (2008) Previously unknown food items in the diet of six Neotropical bird species. Wilson J Ornithol 120(1):214–217

SINAC- MINAET (2010). Políticas para las Áreas Silvestres Protegidas (ASP) del Sistema Nacional de Áreas de Conservación-SINAC 2011–2015. San José

Slud P (1964) The birds of Costa Rica, distribution and ecology. Bull Amer Mus Nat Hist 128:1–430

Slud P (1980) The birds of Hacienda Palo Verde, Guanacaste, Costa Rica. Smithson Contrib Zool 292:1–92
Smith SM (1983) *Otus choliba*. In: Janzen DH (ed) Costa Rican natural history. University of Chicago Press, Chicago, pp 592–593
SORA (2016) Searchable ornithological research Archive https://sora.unm.edu/. Accessed 4 Nov 2016
Soto M (2015) Expertos piden dejar de 'rescatar' pichones de búhos y lechuzas. La Nación, 21 Apr 2015. p 17A
Soto M (2016) 3000 ticos pajarean en Costa Rica. La Nación, 14 Nov 2016. p 16A
Stiles G, Lewis T (1980) Locational checklist of the birds of Costa Rica. Costa Rican Expeditions, San José
Stiles FG, Skutch AF (2003) Guía de aves de Costa Rica, 3rd edn. Editorial INBio, Santo Domingo de Heredia. Costa Rica
Stiles FG, Skutch AF, Gardner D (2003) GuÃa de aves de Costa Rica, tercera edn. INBio, Santo Domingo de Heredia
Trejo A, Lezama-López M (2015) Los búhos de Nicaragua. In: Enríquez, PL (ed) Los Búhos Neotropicales: diversidad y conservación. El Colegio de la Frontera Sur, pp 511–520
UICN (2016) The IUCN red list of threatened species. http://www.iucnredlist.org/. Accessed 4 Nov 2016
Underwood CF (1899) Avifauna costarriqueña: Lista revisada conforme a las últimas publicaciones. Museo Nacional de Costa Rica, San José
Valerio JP (2001) Anotaciones sobre el apareamiento de *Glaucidium brasilianum* (Mochuelo común). Zeledonia 5:1
Vargas M (2012) 4 especies de búhos cambian de casa y hábitos en el país. La Nación, 21 Oct 2012. p 18A
Vaughan C, McCoy M (1982) Barn owl food habits. Brenesia 19–20:614–615
Webster JD, Orr RT (1958) Variation in the great horned owls of middle America. Auk 75(2):134–142
Weidensaul S (2015) Owls of North America and the Caribbean. Peterson Reference Guides, New York
Wetmore A (1968) The birds of republic of Panamá. Part 2. – Columbidae (pigeons) to Picidae (woodpeckers). Smithsonian publication 4732, Washington, DC

Chapter 9
The Owls of Colombia

Sergio Chaparro-Herrera, Sergio Córdoba-Córdoba, Juan Pablo López-Ordoñez, Juan Sebastián Restrepo-Cardona, and Oswaldo Cortes-Herrera

Abstract We compiled existing information on the 28 owl species found in Colombia and evaluated information gaps. We found 23 species in the Andes and inter-Andean valleys, 15 in the Caribbean, 10 in the Pacific, 10 in the Orinoco region, 10 in the Amazon, and 4 in the isolated Sierra Nevada de Santa Marta mountain system, including one undescribed species. With regard to conservation status, and following the BirdLife International, one species (*Megascops colombianus*) is considered as Near Threatened (NT) and another (*Glaucidium nubicola*) as Vulnerable (VU), while the remaining owl species are considered as Least Concern (LC). There are no species-specific management or conservation plans for any owl in Colombia, but most species are present in areas with some degree of governmental or private protection. According to our analysis, knowledge of the Strigiformes in the country is very limited, which posits the necessity of developing research and understanding about such basic aspects as geographical variability, population status, reproductive ecology, and behavior, among others. Indeed, without this information, it will be very difficult to discern the current status, population trends, or ecological requirements of owls in Colombia in particular and in the Neotropical region in general.

Keywords Strigidae • Diversity • Conservation • Status of knowledge

---

S. Chaparro-Herrera (✉)
Asociación Bogotana de Ornitología (ABO), Bogotá, Colombia

Grupo de Ecología y Evolución de Vertebrados, Universidad de Antioquia, Medellín, Colombia
e-mail: sergioupn@gmail.com

S. Córdoba-Córdoba
Department of Ecology and Evolutionary Biology, Princeton University, Princeton, NJ, USA
e-mail: sergcordoba@yahoo.com

J.P. López-Ordoñez • O. Cortes-Herrera
Departamento de Biología, Universidad Nacional de Colombia, Bogotá, Colombia
e-mail: juanpablolopezo@gmail.com

J.S. Restrepo-Cardona
Facultad de Estudios Ambientales y Rurales, Pontificia Universidad Javeriana, Bogotá, Colombia

Fundación Rapaces de Colombia, Villamaría, Colombia
e-mail: restrepojuan@javeriana.edu.co

P.L. Enríquez (ed.), *Neotropical Owls*, DOI 10.1007/978-3-319-57108-9_9

Cloud-forest Pygmy Owl (*Glaucidium nubicola*)

9.1 Introduction

Colombia has the most diverse avifauna in the world with more than 1870 species of birds, of which 28 are owls. This makes Colombia one of the countries with the greatest number in the region along with Mexico, Peru, and Ecuador. In the Neotropical region, 84 species of Strigiformes have been described, representing 34% of the 250 known species in the world (König et al. 2008). This region contains 14 of the 26 genera of owls in the world, 11 of which are in Colombia; five of these are endemic to the region (*Megascops*, *Gymnoglaux*, *Lophostrix*, *Pulsatrix*, and *Xenoglaux*), four of them present in Colombia, and seven widespread species: barn owl (*Tyto alba*), burrowing owl (*Athene cunicularia*), striped owl (*Asio clamator*), great horned owl (*Bubo virginianus*), short-eared owl (*Asio flammeus*), mottled owl (*Ciccaba virgata*), and spectacled owl (*Pulsatrix perspicillata*) also reside in Colombia (see Enríquez et al. 2006).

The Republic of Colombia is located in the northwestern portion of South America, being 1,141,748 km^2 terrestrial and 928,660 km^2 marine (IGAC 1996). It borders Venezuela and Brazil to the east, Peru and Ecuador to the south, Pacific Ocean and Panama to the west, and the Caribbean Sea to the north. Due to its geographical location, the Colombian territory includes a variety of bioclimatic, hydrological, geological, and geomorphological logical components that make up a wide range of ecosystems. There are mountains, valleys, terraces, plateaus, and plains that support various plant formations; due to this it is possible to find pastures,

Andean forests, xeric vegetation, mature forests, savannas, swamps, and mangrove vegetation, among others. The country is composed of a complex topography dominated by the Andes, which crosses the territory from south to north. It reaches its greatest degree of structural complexity along three mountain ranges (Rangel-Ch et al. 1995a; Franco and Bravo 2005). Colombia's complex topography includes five major natural regions identified by their physical characteristics: relief, vegetation, climate, and soil conditions. These are the Andes region and inter-Andean valleys, Caribbean, Pacific, Orinoco, and Amazon (Fig. 9.1).

The species of owls are regionally represented as follows: Andes region and inter-Andean valleys (23 species), Caribbean region (15 species), Pacific region (10 species), Orinoco region (10 species), and Amazon region (10 species). The Sierra Nevada de Santa Marta (four species) is also considered an important subregion being an isolated mountain system.

The Andean region has a high physiographic and topographic complexity, consisting of three main branches, the Western, Central, and Eastern mountain ranges. Among the three ranges are two large valleys, the Cauca River between Western and Central Cordilleras and the Magdalena River between the Central and the East. The Western Cordillera has elevations up to 4200 m.a.s.l. At its northern end, it branches into three mountainous areas. The Cordillera Central is the highest elevation 5750 m.a.s.l. and extends southward in the East Cordillera of Ecuador; its northern end constitutes the Serrania de San Lucas.The Cordillera Oriental is the widest of the three and comparatively the lowest; Serranía del Perijá is the northern end on the border with Venezuela (Rangel-Ch et al. 1995a). The Andes are the most important factor determining regional climate in Colombia and the vegetation changes with altitude and humidity. This region contains 30% of the birds of the continent (Kattan et al. 2001). However, this diversity of birds is seriously threatened at the national level. The destruction and fragmentation of vegetation, pollution, and hunting have increased the number of species in all risk categories (Renjifo et al. 2002). The largest urban centers are concentrated in this region with almost 70% of the population and much of the agricultural and livestock activities of the country; their natural ecosystems have lost about 60% of its original extension (Villarreal 2006).

The Caribbean region of Colombia is mainly a plain with some low-rise mountains stretching from the coast of the Caribbean Sea near its border with Panama to Venezuela; the mountainous portion of the Sierra Nevada de Santa Marta (ca. 12,000 km^2) which is independent of the Andean system and geographically isolated by the valleys of the rivers Magdalena and Cesar is excluded. The vegetation in this area is constituted of bushes and cacti covering wide portions of the region; remnants of tropical dry forest and residual forests in these semiarid zones create a transition to more humid regions. Except the foothills of the Sierra Nevada de Santa Marta, this vast region is the driest in the country; in some places (the peninsula of La Guajira) annual precipitation does not exceed 500 mm. It is the second most densely populated region of the country and also the province with the highest degree of alteration. This is largely due to deforestation and the suitability of land for cattle ranching and agriculture, as well as for logging, mining, and tourism (Rangel-Ch et al. 1995c; Negret 2001).

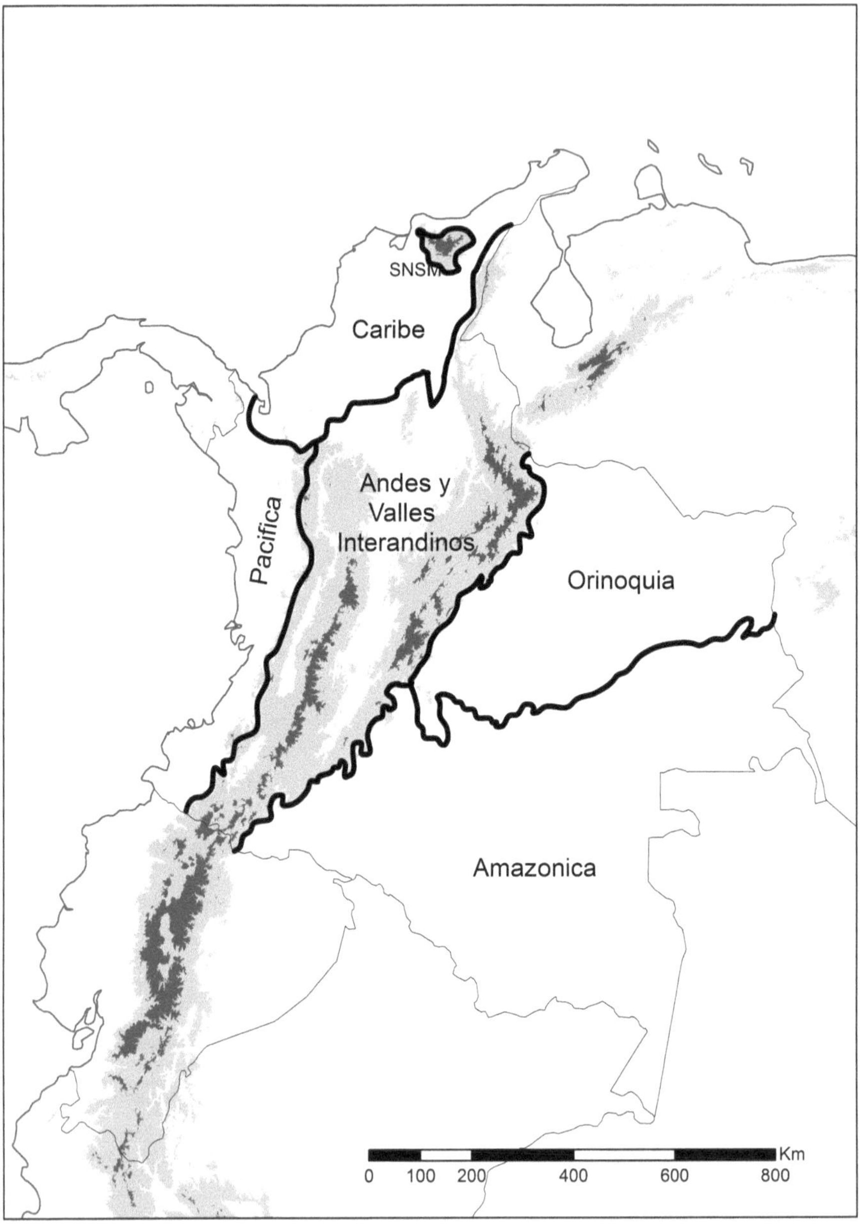

Fig. 9.1 Natural regions of Colombia (excluding the Insular Region). Andes Region, inter-Andean valleys, Caribbean region, Pacific region, Orinoco region, Amazon region, and Sierra Nevada de Santa Marta (SNSM)

The Pacific region is located west of the western mountain range and is characterized as a region covered with very wet rainforest (Negret 2001). It is a region extremely rich in species and has a high degree endemism; in terms of avian diversity, it is estimated that there are 778 species of birds, which correspond to 43% of all species in Colombia (Rangel-Ch 2004). The main natural hazards are the volcanic eruptions and tsunamis. Extensive areas recently planted with African oil palms have also altered the landform. In several localities lagoons and marshes have been drained so the land can be used for agriculture (Rangel-Ch et al. 1995b).

The region comprises the Orinoquía Colombian tropical savanna or Eastern Plains stretching from the foothills of the Cordillera Oriental, north of the Guaviare River to the limits with Venezuela. It is a vast plain where natural savanna vegetation predominates, covered with grasses, shrubs, palms, and gallery forests accompanying the course of numerous rivers (Negret 2001). The topography of the plains varies between 500 m.a.s.l. in the vicinity of the Andes, and 200 m.a.s.l. along the Orinoco River is interrupted by several rivers flowing toward the east. The most important rivers are the Arauca, Casanare, Meta, and Guaviare; they all originate at high altitudes in the Cordillera Oriental and run almost straight line to the Orinoco. This region of Colombia is one of the most poorly studied and faces strong transformations of the landscape because of the expansion of the agricultural frontier, livestock pastures establishment of oil palm plantations and rice, prospecting and oil exploitation, permanent gas flaring in these holdings, and alteration by frequent burning, often caused by livestock farmers and crop farmers (Rangel-Ch et al. 1995d; Peñuela et al. 2011).

The Amazon region in Colombia represents 30% of the national territory and includes the Sierra de la Macarena, the Serranía de Chiribiquete, and the plains of Yarí-Caguán. The Colombian Amazon covers the vast rainforests of the Amazon River basin, forming an almost continuous green carpet, from the foothills of the eastern cordillera, south of the Guaviare River, passing to the Amazon plain and the mountains, hills, plateaus, hills, and isolated hills, to the border with Venezuela, Brazil, Peru, and Ecuador (Hurtado 1992).

Numerous factors, including soil and climate, can produce significant differences in the formation of the rainforest, consisting of several formations. Permanently or periodically flooded forests by white-water rivers (*varzeas*) or black water (*igapos*) result in the predominance of Amazon savannas sparse vegetation with natural or seasonal savannas composed of grasses and rocky outcrops. Finally on the eastern slopes of the Cordillera Oriental foothill, rainforest and Andean forests (Hurtado 1992; Negret 2001) are present. Among the activities that most affected this region are exploration and oil exploitation, logging, expansion of the agricultural frontier, exploitation of commercially valuable species of fish in major rivers, hunting, contribution of nondegradable waste chemicals employed in the processing of coca, and deforestation by cutting down and burning the rainforest formations in the mountain foot and alluvial valleys (Castaño 1993; Rangel-Ch et al. 1995e).

For the nomenclature to species level of each species of owl, we follow Remsen et al. (2016). However, we did comments for some species such as *G. brasilianum*, *Megascops guatemalae*, and *Ciccaba virgata* according to the classification of

König et al. (2008). The altitudinal distribution of owls in Colombia is divided into four sections according to the heights where several species predominate. Nine species mainly occur in low-lying areas below 1000 m.a.s.l. (e.g., tawny-bellied screech owl *Megascops watsonii*); six use both low-lying areas as intermediate altitudes in the mountains below 2400 m.a.s.l. (e.g., black-and-white owl *Ciccaba nigrolineata*); ten are schematic mountain species, usually above 800–1000 m.a.s.l.

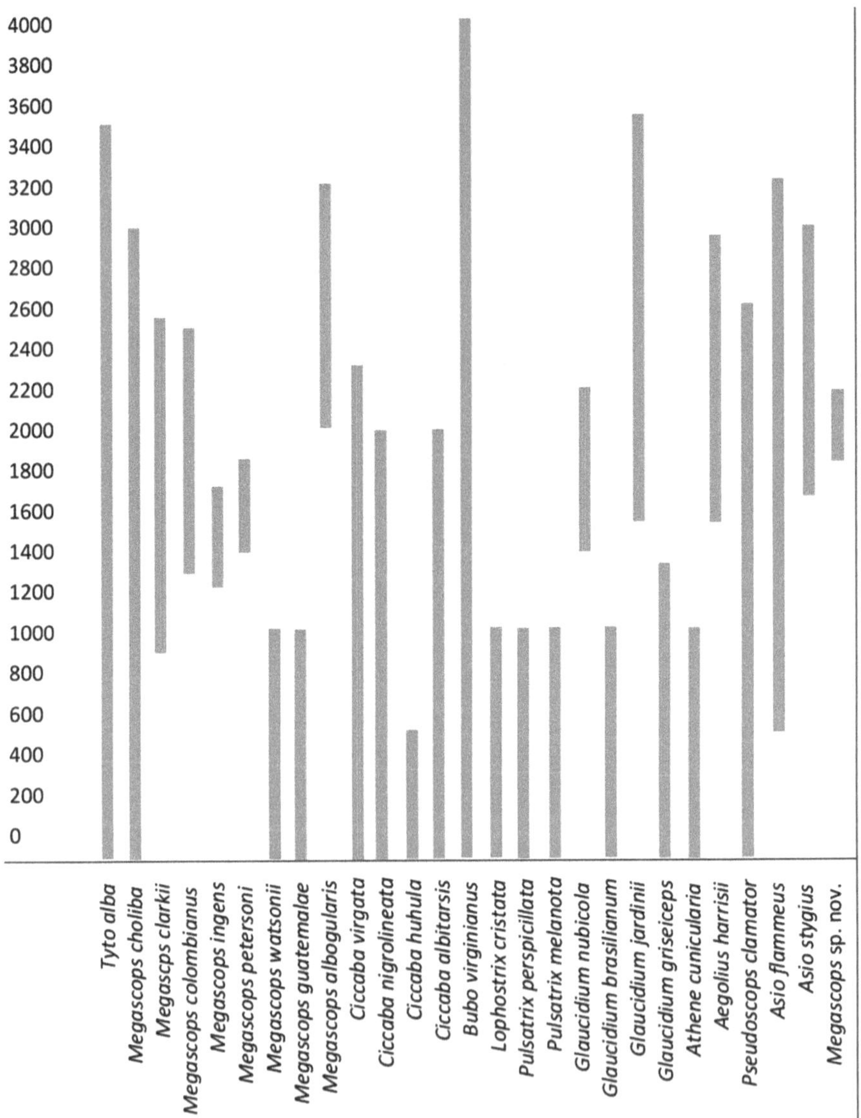

Fig. 9.2 Altitudinal ranges of the 28 species of owls registered in Colombia

(e.g., Colombian screech owl *Megascops colombianus*, buff-fronted owl *Aegolius harrisii*); and three are over all the altitudinal gradients (e.g., tropical screech owl *Megascops choliba*, great horned owl *Bubo virginianus*; Fig. 9.2).

9.2 State and National Conservation

Of the 28 species of owls present in Colombia, a species is considered Near Threatened (NT) (Colombian screech owl) and another as Vulnerable (VU) (cloud-forest pygmy owl *Glaucidium nubicola*), and the remaining species have been assessed as Least Concern (LC) (BirdLife International 2016). Buhito Santa Marta (*Megascops* sp. nov.) being a new species without formal description has not been officially evaluated.

Its restriction to the Sierra Nevada de Santa Marta and due to the destruction of their habitat throughout its distribution, it could be placed in some category of threat; further information to help establish the degree of vulnerability of the species is required. In Colombia there is no management and conservation plans specifically designed for owls; however, many species are present in areas with some degree of protection, whether governmental areas (Natural National Parks, Regional Parks) or private reserves. For four species, bare-shanked screech owl (*Megascops clarkii*), rufescent screech owl (*M. ingens*), great horned owl (*Bubo virginianus*), and subtropical pygmy owl (*Glaucidium parkeri*), there are no conservation strategies within the national territory. Three other species, screech owl (*Megascops* sp. nov.), band-bellied owl (*Pulsatrix melanota*), and cloud-forest pygmy owl, are known only from their presence in a Natural National Park (NNP), the Sierra Nevada de Santa Marta, the Serranía of Churumbelos Auka-Wasi, and Tatamá, respectively. Although some species of owls are among the birds found in various IBA (Important Bird Area), this does not guarantee protection, since they are not legally protected at the federal level. Finally the species with greatest protection because of its wide geographical distribution are tropical screech owl (41 protected areas), Andean pygmy owl (*Glaucidium jardinii*) in 19, rufous-backed owl (*Ciccaba albitarsis*) in 17, ferruginous pygmy owl (*Glaucidium brasilianum*) in 16, white-throated screech owl (*Megascops albogularis*) in 15, and mottled owl (*Ciccaba virgata*) in 14 (Appendix 9.1). The tawny-bellied screech owl, despite having a wide distribution, is known only from two protected areas: PNN Mountainous area of the Macarena and PNN Mountainous Chiribiquete. However, it is believed that it is also found in the PNN Amacayacu, and it has been registered near and has a wide distribution in the Amazon region.

Despite such high species richness of owls in the country and their wide distribution, and ecological value within the ecosystems where they live, overall knowledge of Strigiformes within the country is extremely limited. While a good number of documents confirm the presence and distribution of the various species of owls in Colombia, the lack of local and regional information on some basic aspects as population, reproductive ecology, and behavior make it difficult to understanding the ecological requirements of these species. Without this ecological information, it is very difficult to elucidate the present state and population trends of Neotropical owls (Enríquez et al. 2006). Most existing information on owls in Colombia is limited to records of distribution of different species, mainly of specimens held in biological collections (Appendix 9.1), and in some cases direct observations, vocalizations, and recordings. The species with the highest number of specimens in scientific collections is the tropical screech owl with 335 individuals, followed by the barn owl (163) and mottled owl (141) (Appendix 9.1), while for the screech owl (*Megascops* sp. nov) and subtropical pygmy owl, no specimens have been deposited in a scientific collection. Bare-shanked screech owl, cinnamon screech owl (*M. petersoni*), band-bellied owl (*Pulsatrix melanota*), cloud-forest pygmy owl, and Central American pygmy owl (*G. griseiceps*) are represented by one, two, or three specimens; however, the vast majority of these specimens are in collections outside Colombia (Appendix 9.1).

In most cases lack of information and specimens available for study limits our ability to investigate their morphological and behavioral variability, much less understand relationships between one or more species. As for the plumage coloration, different species show different dark phases ("rufous") as *Megascops* species or dark and gray phases in some *Glaucidium* species; it is not known if these characteristics are linked to populations or only are individual variations. Many of these plumage and morphology features, which have traditionally made it difficult to interpret and classify kinship relations within the group, are being reevaluated using the vocalizations of species. Vocalizations are very important for communication in owls and generally in nocturnal birds, being even more relevant and informative to understanding their relationships, behavior, and ecology. However, very few vocalizations have been recorded in Colombia. This limited data set makes it difficult to study the variability between populations and species.

Also, several species still have many taxonomic problems, including species of the genus *Megascops*, mainly the group of *M. guatemalae (centralis/vermiculatus, napensis)* and the complex *M. ingens/colombianus*, *M. watsonii,* and *M. petersonii*, as in the genus *Glaucidium* with *G. griceiceps* (*hardyi* and Santa Marta*),* among others, where the lack of geographical and ecological variability, as the boundaries between subpopulations and species in several groups that are found in Colombia normally is very large, demonstrates the need to consolidate and strengthen national collections improving taxonomic representation, geographically and temporally throughout the collections (Cuervo et al. 2006). Many species require molecular analysis that is then considered along with information on their ecology, abundance, and natural history.

There is little information on the ecology of many species of owls in South America, and only in some cases there is information of population densities or habitat use (e.g., Terborgh et al. 1990). However, in Colombia researchers have barely begun to try to estimate the population densities (Fierro-Calderón and Córdoba-Córdoba 2014; S. Córdoba-Córdoba and O. Marín-Gómez pers. obs.), with inconclusive preliminary results due to the lack of records and the absence of long-term monitoring. Not only we lack information on habitat requirements of different species, but we also have no information about those with restricted range to one or a few types of habitat or even for those most widely distributed in different types of vegetation. We need to understand the habitat use or habitat selection of species and how the effects of fragmentation and habitat degradation influence in their populations. König et al. (2008) mentioned that among the main causes of threat faced by owls, which are also present in Colombia, are road accidents, illegal traffic, the use of pesticides, and transforming landscapes (both grassland and forest) in agricultural areas loosing habitats for owls.

In some cases information about diet comes from ecologists visiting roosts and collecting pellets. We know of some species that feed on large vertebrates, small vertebrates (including bats and rodents, birds, frogs, others), invertebrates, and many insects. For the majority of species, no information is known regarding diet or how it changes over time, so the diet of 77% of owls present in Colombia requires study (Appendix 9.1 and information for each species). Also we understand little regarding the natural history of owls in the Neotropics. We have very little information from the breeding season of most species, either because very few nests have been described (e.g., Borrero 1962; Freeman and Julio 2010; Appendix 9.1 information for each species) or because it has not been possible to track individuals over time. We do not have good knowledge of reproductive biology, parental care, age of first reproduction, or aspects of longevity and behavior.

After a subjective approximation of the degree of knowledge of 28 species of Colombia's owls showed that no species has "good" information for Colombia (Appendix 9.1). We can consider that for two species, there is an average degree of knowledge; for ten species, there is a low degree of knowledge; and for 16 species, our knowledge is deficient (Appendix 9.1). We hope this chapter will be a useful guide as to the state of knowledge of the species of owls in Colombia.

9.3 Species in Colombia

Tyto alba **(Scopoli, 1769) o** *T. furcata* **according König et al. (2008) and Alibadian et al. (2016)**

Common Name: Spanish, Lechuza Común; English, barn owl

Description: Medium-sized to large nocturnal bird (38 cm); colors cream whitish to light brown, without "ears" and long legs. A whitish-shaped facial disk, more or less chestnut, and the areas in front of the eyes are black with fuscous border; gray back, vermiculated with black and spattered with whitish and fuscous; underneath it varies from white, lightly streaked with tan in the chest and flanks, even before intense ocher, sprinkled with black, dark brown iris, pink beak, brown legs, similar immature. In flight at night, it has a ghostly appearance with a lower surface of the whitish wings and pale body (Hilty and Brown 1986; ABO 2000). Weight of the bird, between 311 and 573 g for *T. a. pranticola*/*guatemalae*, but *T. a. contempta* has no known weight and is smaller in size than many other subspecies of the group (König et al. 2008).

Voice: A variety of loud hisses, snoring, and scraping sounds, but no hoots. More often a harsh cry or hissing *chrrrrrriii*; in the nest, strong hiss but often a squeak note (Hilty and Brown 1986). The vocalization is produced from perch or in flight. It may be a special form of echolocation (König et al. 2008).

Distribution: A cosmopolitan species; in the New World is resident from S of Canada to South America (ABO 2000). In Colombia it is found up to 3500 m.a.s.l. Most records are from the lowlands of the Caribbean region, in the three cordilleras (although not for the W Cordillera Occidental), near Cúcuta in N Santander and east of the Andes from Meta (Villavicencio, Macarena, Carimagua) and the Orinoco region. Also in the S throughout the entire Amazon region to Putumayo (Hilty and Brown 1986; Salaman et al. 2009). See Alibadian et al. (2016) for complete analysis in phylogeny, biogeography, and diversification.

Ecology: Despite its widespread distribution in the New World, its dietary habits and other aspects of its natural history in North/South America have not yet been adequately documented (Delgado-V and Calderón-F 2007). It is a strictly nocturnal bird that flies during the day only when it is disturbed in its resting place. Local in semi-open regions and around human settlements where it often rests and nests (Hilty and Brown 1986; König et al. 2008). It has been observed in oil palm plantations in the municipality of Orocué (Chaparro-Herrera pers. obs.). It feeds on vertebrates (mainly rodents and amphibians), marsupials, and bats and also consumes insects in lesser proportions (Delgado-V and Cataño-B 2004; Delgado-V and Calderón-F 2007; Delgado-V and Ramírez 2009). In general, hunting from a perched site but also catching its prey in flight (König et al. 2008). The presence of *T. alba* in many regions helps to keep rodent populations low, which become agricultural pests and affect public health (Fuentes et al. 2009). They show great fidelity to nesting sites, usually a hollow or a quiet spot in some building, steeples, cavities in old trees in the woods, or sometimes in caves; nest contains a minimal amount of feathers, branches, and litter, barely enough so that the 3–4 round white eggs do not fall to the ground. The size of the clutch and the success of the reproduction depend on the supply of food (Hilty and Brown 1986; ABO 2000; König et al. 2008).

Conservation Status: It has a very wide range of distribution, and its population is more than 10,000 mature individuals. Therefore, it is considered a species of Least Concern (LC) worldwide (BirdLife International 2016).

Conservation Strategies: It is present in the Sanctuary of Fauna and Flora Guanentá Alto Río Fonce; Forest Reserve Bogotá eastern; Nature Reserves in Bojonawi, La Ventana, Refugio Nimajay, Pitalito, Wakuinali y La Pedregoza; Natural Reserve of Birds El Colibrí del Sol, Chincherry, Ranita Dorada and Halcón Dorado (Palacios et al. 2005; CAR 2009; Salaman et al. 2009; Peñuela et al. 2011; GeoSIB 2012).

Megascops choliba, *Otus choliba* **(Vieillot, 1817)**
Common Name: Spanish, Currucutú Común; English, tropical screech owl

Description: Length between 21 and 25 cm. A gray brown (clear phase) or rufous (rufous phase) plumage, the belly is white with black stripes in under parts; vermiculated with black and brown in the abdomen. White spots in the scapular feathers; it always shows a relatively distinctive blackish edge in the face, yellow eyes, and beak pale gray. Inmatures: clear gray barred with dark brown or rufous according to plumage phase (ABO 2000). Weight of the bird: between 100 and 160 g and average of 132 g (Dunning 2008; König et al. 2008).

Voice: It is more active vocally after sunset or before dawn (Hilty and Brown 1986). It shows two vocalizations, type A is a short and purring trill and follows two or three notes accented: *gurrrrku-kúk* or *gurrrrku-kúkúk*. When it is excited after playback, final notes are more numerous and pronounced in a rhythm stuttered *gurrrrku-kúk-gukúk-gugukoohk*. Female calls similar as the male, but its vocalizations are higher and less frequent. Type B call showed for both sexes; it is a bubble *bububúbubu* or *curr cutú tutu* during courtship or when owls start to sing in dawn. It vocalization is a preamble to vocalization Type A and is a response to playback when there are interference in its territory. When it is alarmed, the calls are "empty": *hahahahahahaha hahaha…* in a scale descendent. A soft: *wook* seems to be a contact function (König et al. 2008).

Distribution: It shows wide distribution in South America, from Costa Rica in Central America to N Argentina, and is absent in W Ecuador, W Peru, and Chile (Hilty and Brown 1986; Sick 1997; Holt et al. 1999; König et al. 2008). This species has the highest distribution area in Colombia since 3000 m.a.s.l. It is found in lowlands in the Caribbean region; in Valle del Río Magdalena and Valle del Río Cauca; in La Guajira Peninsula; in San Andrés and Providencia islands; in Cordillera Oriental, Central, and Occidental; in la Serranía de Perijá; in el Macizo Colombiano; and in Amazon and Orinoquia region; it is absent in Pacific region (Hilty and Brown 1986; Salaman et al. 2009).

Ecology: Although it is a common species and is distributed widely in different ecosystems (even in urban areas), it is difficult to see, and few ecological information exist (Sick 1997; Holt et al. 1999). It lives in different habitats (savannas with dispersal trees, open forests, crops areas, forest edges, dry forest, open fields, areas with bushes, urban parks, crops) (König et al. 2008). It is strictly nocturnal and starts its activity during sunset and generally in pairs (Hilty and Brown 1986; König et al. 2008). It nests in almost any kind of cavity, holes in the ground with subterranean galleries, in fence posts, even in clay pits in house yards (D. Ortega pers. com.), or in abandoned nest of woodpeckers between 2 and 10 m high where it lays 2–4 white eggs (Hilty and Brown 1986; ABO 2000). Its breeding biology and the diet are few known in Colombia. Delgado-V (2007) described the diet of this species in Medellín City and similar to Motta-Junior (2002) in Brazil and found that this species feeds of invertebrates (mainly insects) and eventually feeds small vertebrates.

Conservation Status: The distribution of this species is wide and therefore cannot be considered as V ulnerable, under the criterion size of the distribution area. The species is Least Concern globally (BirdLife International 2016).

Conservation Strategies: It is distributed in Natural National Parks, Chiribiquete, El Tuparro, La Paya, Araracuara, Macuira, and Los Katios; Sanctuaries of Fauna and Flora Iguaque and Otún Quimbaya; Forest Reserves Río Blanco and Quebrada Olivares, Yotoco and La Romera, Los Cristales, and Guaymaral; Natural Reserves Nechí-Bajo Cauca, Nukak, Río Barroso and San Juan, and El Ático; Biological

Reserve Carpanta; Regional Natural Park El Vínculo; Natural Reserves of Civil Society Tambito, La Casa de la Abuela, Las Unamas, Rey Zamuro, Mata Redonda, Manaco6, El Caduceo, Agua Verde, Bojonawi, La Ventana, Refugio Nimajai, Pitalito, Wakuinali and La Pedregoza, and Natual Reserves of Birds Loro Orejiamarillo, El Paujil, Reinita Cielo Azul, El Dorado, Arrierito Antioqueño, *Pauxi pauxi*, Ranita Dorada and el Halcón Colorado (Donegan and Dávalos 1999; Salaman et al. 2009; Peñuela et al. 2011; GeoSIB 2012).

Megascops clarkii, *Otus clarkii* **(Kelso, L. and E. H. Kelso, 1935)**
Common Name: Spanish, Currucutú Manchado; English, bare-shanked screech owl

Description: Length between 20 and 25 cm. A pale brown to gray brown plumage, it has a rufous or clear gray phase identical in both sexes. Plumage dotted and with "ears," yellow iris, rufous coffee on the back, striated crown and spathe with black, winged, and scapular coverings dotted with white, cinnamon face, without well-defined facial border, and throat and brown chest dotted with white and blackish streak; the rest of white lower parts are irregularly streaked. Barred with narrowly striated fawn of black (dotted), most of the naked tarsus (not visible in the field) 180 g, being one of the heaviest species in the genus (Holt et al. 1999).

Voice: Vocalization is usually a deep *wook wook wook* repeated with intervals of a few seconds, and it also has a deep *hu-hu* whistle, *Hoo Hoo hoo*. The third and fourth notes are the most notorious, which is heard at a distance. The female has a slightly keener call, plus an additional call described as a rather loud howling sound *coo, coo-coo* (Holt et al. 1999).

Distribution: It is a monotypic species; its area of distribution is restricted to Costa Rica, Panama, and NW Colombia, found between 900 and 2500 m.a.s.l., and considered a rare species in its range. It is a resident species, without seasonal or altitudinal movements. In Colombia, a specimen was collected between the border of Panama and Colombia in Cerro de Tacarcuna and another specimen collected in the department of Chocó, Unguia municipality, Cuchilla del Lago (Hilty and Brown 1986; Holt et al. 1999).

Ecology: This species inhabits very dense montane forests, edges of cloud forest, as well as small forest relics and is frequently found in family groups, even in the reproductive age. It is a strictly nocturnal species; however, it can be sighted hunting in the late afternoon on forest edges, on clearings, or in the canopy, feeding mainly on insects such as crickets, grasshoppers, beetles, and spiders along with some shrews and small rodents (Holt et al. 1999). Little is known about the reproductive period, based on the description of a nest found in Costa Rica, in April 1994, where a female was nesting in a natural cavity at 3.3 m in the trunk of an oak (*Quercus copeyensis*) and subsequent observed feeding chicks orthopteran and coleopteran insects (Enríquez et al. 1997).

Conservation Status: Although this species may have a small area of distribution and despite the fact that the population trend appears to be declining, it is not believed to approachVulnerable thresholds; however, the size of the population has not been officially quantified. For these reasons, the species is evaluated for Least Concern (LC) (BirdLife International 2016).

Fig. 9.3 Colombian screech owl (*Megascops colombianus*) endemic species of Colombia. Municipality of Pereira, Risaralda department, 09 February 2012 (Photograph ©Diego Calderón Franco; www.colombiabirding.com)

Conservation Strategies: Because of their restricted distribution, little is known about their potential geographic distribution; however, it is present in several protected areas in Costa Rica and Panama. In Colombia, the Cerro de Tacarcuna has not yet been declared a protected area, this being the only area where this owl species is known for Colombia (J.P. López-Ordoñez pers. obs.).

Megascops colombianus **(Traylor, 1952)**
Common Name: Spanish, Currucutú Colombiano and Autillo Colombiano; English, Colombian screech owl

Taxonomy: It was initially described as a subspecies of *Megascops [Otus] ingens* (Traylor 1952). Later it was elevated to a species based on its larger size, the tarsos proportionally longer and almost totally naked, as well as to present the coloration of plumage similar to *Otus petersoni* (Fitzpatrick and O'Neill 1986). He was considered part of the group of South American brown-eyed screech owls, made up of *M. petersoni*, *M. ingens*, *M. watsonii*, and *M. marshalli* (Fitzpatrick and O'Neill 1986).

Description: Length between 26 and 28 cm. There are two known morphs, one brown gray and the other reddish brown. The plumage is almost uniform in color, with short "ears," slightly pronounced facial disk of the same color as the back. Relatively long and strong heels, partially with feathers in the upper part of the tarsus only. Dark brown iris and pinkish grayish eyelids. Wings and tail barred dark and clear. Below light colored with dark brown (König et al. 2008). Weight of the bird: female 210 g ($n = 1$) (Hilty 1977) and males 150–156 g ($n = 2$) (Fitzpatrick and O'Neill 1986) (Fig. 9.3).

Voice: Its vocalization presents several short notes repeated and constant like those of a flute, where the first notes are many times more *tu tua tu tua* and then if constant *tu-tu-tu-tu-tu*... between 18 and 50 s. It emits them mainly in the sunset and after the dusk, although also infrequently heard at other times (O. H. Marín-Gómez pers. comm.).

Distribution: Endemic species of Colombia. It is distributed mainly by the W slope of the western Andes range in Colombia and N Ecuador. There are recent records for this species in three locations on the W slope of the Central Cordillera in Colombia. Its altitudinal range is between 1250 and 2450 m.a.s.l. It has been registered in Cordillera Occidental (Department of Chocó: Cerro del Torra); Valle del Cauca, in the watershed of the Anchicayá and Verde River basins in cloud forest, and 9 km N–W of the Dagua, in the Yotoco Forest Reserve, the W Cordillera Occidental; Cauca (El Tambo, San Antonio-Guapi, and Munchique National Natural Park); and Nariño: Ricaurte, Planada Natural Reserve, Barbacoas-Attaquer, Natural Reserve Río Ñambí (P. Florez pers.comm., S. Córdoba-Córdoba pers. obs.), andTumaco-La Guayacana (Traylor 1952; Hilty 1977; Fitzpatrick and O'Neill 1986; Negret 1994; Biomap 2006; Álvarez-Rebolledo et al. 2007). Cordillera Central-department of Quindío: Cedro Rosado, Armenia, Alto Quindío -Salento, Bremen-La Popa Nature Reserve and Barbas River Canyon Nature Reserve Forest and Research (Arbeláez-Cortés et al. 2011; O. H. Marín-Gómez pers. comm.); Risaralda, Sanctuary of Fauna and Flora Otún -Quimbaya (S. Córdoba-Córdoba pers. obs., D. Calderón-Franco http://www.flickr.com/photos/colombia_birding_diego/6787377548/ -12 febrero 2012).

Ecology: Colombian screech owl is a rare and discontinuous ranging species, restricted to montane forests (Fitzpatrick and O'Neill 1986; Hilty and Brown 1986; Holt et al. 1999; Freile and Chaves 1999; Renjifo et al. 2002; BirdLife International 2016). There visual and sound recordings of individuals belonging to those recognized for this species, but it is necessary to confirm if this species is only in the Cordillera Central (Arbeláez-Cortés et al. 2011; O. H. Marín-Gómez pers. comm., S. Córdoba-Córdoba pers.obs.), which would indicate a wider distribution, and could extend toward the Nof the central mountain range and the S toward the Colombian Massif. The known habitats for the species are mature, continuous forests, secondary forest, thinned forests, forest edges, and high stubble (S. Córdoba-Córdoba and O. H. Marín-Gómez pers. obs.). The species could use some fragmented habitats, as it has been observed perched on solitary tall trees in grasslands and forest edges, although there are extensive forests nearby (Freile and Chaves 1999).

Little is known of its ecology, so it is not known if this species is displaced or if it can coexist with other species of owls that are also distributed to the same forests, as with *Megascops albogularis*, which is larger. Aggressive and possibly more adapted to forest edges than *M. colombianus*, although *M. albogularis* tends to be found at higher elevations (S. Córdoba-Córdoba pers. obs.). This species has been little studied and only recently has been recorded more frequently because it is possible to differentiate it partially by its song of *Megascops ingens* or *M. petersoni*, which present a similar pattern of vocalization. It feeds on small mammals and large and small insects (Freile and Chaves 1999). There is no general information on its reproduction, although its breeding season could include several months between December and March, due to the collection of a female specimen with developed gonads and abundant fat by January (Hilty 1977), a juvenile specimen collected, and the registration of a female with a juvenile at the end of February and early March in the Andean forest (Estela et al. 2004). There are no population estimates for this species. From the records in Barbas-Bremen by vocalization counting points (Marín-Gómez and Ospina 2008), a density of 0.004 ind/h was estimated (O. H. Marín-Gómez pers. comm. and S. Córdoba-Córdoba and O. H. Marín-Gómez pers. obs.).

Conservation Status: Considered to be almost Near Threatened (NT) (Renjifo et al. 2002; BirdLife International 2016), the species is mainly threatened by the loss of habitat, which continues throughout the area of distribution (Renjifo et al. 2002), as well as in the new localities in the central mountain range. The loss and fragmentation of forests throughout their distribution indicate that they have lost 44% of their habitat (Renjifo et al. 2002). Due to its restricted distribution and habitat degradation processes, it has been considered almost threatened (Renjifo et al. 2002).

Conservation Strategies: There are records in the Munchique Otún Quimbaya Natural National Park, Sanctuary of Fauna and Flora, Yotoco and La Planada Forest Reserve, Natural Reserve Río Ñambí, Regional Natural Park Páramo del Duende, the Bremen-La Popa Forestry and Research Natural Reserve, and the AICA Barbas River Canyon (however, the latter is not guaranteed to be protected). This species has been registered at the Farallones of Cali Natural National Park. However, it is necessary to know its actual distribution, to carry out studies on its ecological requirements and to make more accurate population estimates (BirdLife International 2016, S. Córdoba-Córdoba and O. H. Marín- Gómez pers. obs.).

Megascops ingens, *Otus ingens*, *Scops ingens* (**Salvin, 1897**)
Common Name: Spanish, Autillo Pálido; English, rufescent screech owl

Taxonomy: The species presents unresolved taxonomic issues and frequently has been included with *M. colombianus* as a subspecies, as well as *M. petersoni* as cospecific. Currently two subspecies are recognized: *venezuelanus* (Phelps and Phelps Jr., 1954) -N Colombia and NW Venezuela, and *ingens* (Salvin, 1897) - Andes from NE Ecuador to W Bolivia (Remsen et al. 2016).

Description: Length between 24 and 28 cm. It is a large owl with small "ears," coffee iris. Plumage brown on the back, vermiculated blackish; partially concealed nuchal collar whitening to scapulae and winged white coverings; dark brown and cinnamon barred rump and tail; facial disk coffee colored with no sharp edge, throat before finely barred coffee, and the rest of the underside white with a few narrow and prominent black striate, crossed by fine scattered brown lines (Hilty and Brown 1986). Weight of the bird: between 134 and 180 g in males; between 140 and 223 g in females (Holt et al. 1999).

Voice: The primary vocalization is a series of quick notes *tu*, almost 50 in 10 s. The secondary is a series of abruptly accelerated pitch-like vocalizations after the fourth note *hu hu hu tu tuututu* (Holt et al. 1999).

Distribution: This species is distributed from Venezuela to N Bolivia. In Colomb+ia, it is distributed between 1200 and 2250 m.a.s.l., on the W flank of the Serranía del Perijá, La Guajira (Hiroca), Santander (Florida and Encino-Biological Reserve Cachalú), Boyacá (Pajarito), and Cundinamarca (Bojacá), subspecies *venezuelensis* (Hilty and Brown 1986; Córdoba-Córdoba and Álvarez-Rebolledo 2003; GeoSIB 2012; A. Marín pers. comm.).

Ecology: It is a rare species that occurs in moist montane forests; it is distributed between 1200 and 2250 m.a.s.l. It feeds primarily on insects and small vertebrates and is strictly nocturnal, apparently foraging in the understory and canopy. No reproductive information is available. No seasonal or migratory movements have been observed (Holt et al. 1999; König et al. 2008).

Conservation Status: This species has a wide distribution range and the population trend appears to be stable. However, the size of the population has not been estimated. The species is considered of Least Concern (LC) (BirdLife International 2016).

Conservation Strategies: Forest destruction threatens some areas of its distribution. In Colombia, it has not been registered in protected areas throughout its distribution, while for Venezuela it is located toward the Sierra del Perijá National Park (J.P. López- Ordoñez pers.obs.).

Megascops petersoni **(Fitzpatrick and O'Neill, 1986)**
Common Name: Spanish, Currucutú Canela, Lechicita Canela, and Autillo de Peterson; English, cinnamon screech owl

Taxonomy: Possibly forms a superspecies with *Megascops marshalli* and alternatively would be considered as conspecific with *Megascops ingens* or *Megascops colombianus* (König et al. 2008; Remsen et al. 2016).

Description: Length between 18 and 22 cm. Cinnamon brown dorsal coloration. It has feathers of the crown and back with thin double, wavy bars, alternating between dark brown and pale. No pale forehead color. Crown slightly brown darker than back. Tufts of moderate length (29 mm). Tarsus almost fully feathered, iris is brownish brown, darkest facial disk with more conspicuous blackish edges, and slight striation vermiculated in the chest. It differs from *M. ingens* because of its smaller size and the total absence of white on any part of the body (Fitzpatrick and O'Neill 1986). The weight of the bird: between 88 and 119 g in males, average of 97 g; females 92–105 g and average 98 g (Fitzpatrick and O'Neill 1986; König et al. 2008).

Voice: Series of equally spaced notes *Buu-buu-buu*. In rapid succession, it raises its tone at the beginning and then slowly decays (König et al. 2008).

Distribution: This species was considered restricted to the mountain forests of Peru and Ecuador (Fitzpatrick and O'Neill 1986). Populations are now known in Colombian territory, located in the foothills of the Cordillera Central and Cordillera Oriental between 1400 and 1820 m.a.s.l. Cordillera Central: sound recorded in La Forzosa Reserve, Roble Arriba, Anorí municipality (Antioquia) (http://www.xeno-canto.org/57504), and two specimens collected in Alto de Chaquiral, Anorí municipality (Cuervo et al. 2008). Cordillera Oriental: San Vicente de Chucurí en Santander (www.xeno-canto.org/143842), one specimen was deposited in the collection of the Institute of Natural Sciences (ICN) unlisted and may be in San Antonio del Tequendama municipality, but it requires confirmation.

Ecology: Nocturnal. Its diet is not known, but it is believed that it consumes insects and small vertebrates. In Ecuador and Peru, the species inhabits cloud forests, between 1650 and 2500 m.a.s.l. (Fitzpatrick and O'Neill 1986; Holt et al. 1999). In Colombia, the records correspond to very humid montane forests (bmh-PM) with elevations between 1550 and 1850 m.a.s.l. (Cuervo et al. 2008; Freeman and Julio 2010). Only reproductive information from a nest found in a natural cavity low in the trunk of a tree, contained a white egg placed in a shallow depression on soft wood, which was abandoned. Likely incubate period at least 32 days (Freeman and

Julio 2010). There are specimens collected in the month of July in Peru with not very developed gonads although a male with tests of 7 × 4 mm and a female with follicles not greater than 2 mm. It could be concluded that the breeding season is prior to July and August, which are dry months in Peru, due to the size of the gonads found and the molt, suggesting that they were close to the end of the breeding season. This is supported by the existence of a first-year juvenile individual with adult plumage of 28 July 1976. All the specimens collected showed body molting and some molting on wings or tail in July (Fitzpatrick and O'Neill 1986; König et al. 2008).

Conservation Status: Although this species has a small area of distribution, it is not believed to meet the criteria to be considered Vulnerable by the area criterion. Likewise, although it seems that its population is decreasing, this does not seem sufficient to be considered vulnerable by the criterion of population tendency. It has been considered a species of Least Concern (LC) worldwide (BirdLife International 2016).

Conservation Strategies: In the Cordillera Central, *M. petersoni* is found in the La Forzosa Nature Reserve and the Arrierito Antioqueño Natural Reserve of Birds (Cuervo et al. 2008; Freeman and Julio 2010). In the Cordillera Oriental, it is located in the Natural Reserve Las Palmeras in Cubarral, Meta (O. Cortes-Herrera pers. obs.); all these correspond to private areas.

Megascops watsonii **(Cassin, 1848)**
Common Name: Spanish. Currututú Selvático, Buhito Selvático, and Autillo Orejudo; English, tawny-bellied screech owl

Taxonomy: The subspecies *M. w. usta*, (Southern tawny-bellied screech owl) registered S of the Amazon is considered, by some authors, to be a different species (Remsen et al. 2016). In this account we do not separate the two subspecies.

Description: Length between 19 and 23 cm. It has tufts like elongated "ears" and presents eyebrows and blackish facial ring. The plumage of the upper parts is of different shades of brown, which helps camouflage it. There are forms of grayish plumage and reddish brown. The coloration of the lower parts varies according to the subspecies: in *M. w. watsonii* (N Amazon), the background color is tawny reddish to ocher, with black lines. The subspecies S of the Amazon *M. w. usta*, sometimes considered as a different species (Remsen et al. 2016), is generally pale to pale with shades of grayish or yellowish grayish, with whitish markings from the chest and few thin black lines (König et al. 2008). Weight of the bird: males and females between 114 and 172 g, average 129 g (Dunning 2008).

Voice: Vocalization consists of a long phrase with many notes, such as a *buubububu* sequence, which begins with low volume and then increases and weakens toward the end, emitted mainly during twilight (Hilty and Brown 1986; König et al. 2008).

Distribution: It is found in Bolivia, Brazil, Colombia, Ecuador, French Guiana, Guayaquil, Peru, Suriname, and Venezuela (König et al. 2008). For Colombia it is distributed throughout the Amazon and Orinoquia region (Hilty and Brown 1986). Orinoquia: specimens in the municipality of Vista Hermosa and La Macarena,

Guapaya River (Meta), and La Morelia (Caquetá) (Chapman 1917; Blake 1962; Instituto de Ciencias Naturales 2004). In the Amazon region: on the Amu River, Puerto Abeja, and Serranía del Chiribiquete (Caquetá), Calderón River 35 km N of Leticia (Amazonas), and Apaporis River (Vaupés-Amazona) (Álvarez et al. 2003; Instituto de Ciencias Naturales 2004; Biomap 2006; Stiles 2010).

Ecology: Mainly nocturnal, although it vocalizes at twilight. This species is found up to 600 m.a.s.l. in forests of the Amazon Basin and eventually in the foothills of the Cordillera Oriental (Hilty and Brown 1986; König et al. 2008). It is more frequent in zones with continuous forests, less depth of litter in the ground, and high canopy. It seems that the fragmentation and alteration of the forest structure could affect the presence and abundance of this species in the Amazon jungle (Barros and Cintra 2009). Census of flooded forests in the Amazon recorded 0.4 ind/h (Borges et al. 2004). Even so, this species is described as locally "common" in the localities where it has been recorded (Stotz et al. 1996; König et al. 2008). Their nest and their reproductive biology are not known, just as their diet is unknown though it is likely similar to that of other *Megascops*.

Conservation Status: This species has a wide distribution area, and although it seems that its population is declining, it does not reach the thresholds to be considered as Vulnerable; therefore, it has been considered Least Concern (LC) worldwide (BirdLife International 2016).

Conservation Strategies: This species is present in the Macarena and Serranía de Chiribiquete Natural National Parks (Blake 1962; Álvarez et al. 2003; Instituto de Ciencias Naturales 2004; Biomap 2006); it is presumed in the Amacayacu Natural National Park since it has been collected in its vicinity and due to its wide distribution in the Amazon (Hilty and Brown 1986; Stiles 2010).

Megascops guatemalae **(Sharpe, 1875) (includes** *M. g. vermiculatus* **and** *M. g. centralis*, *M. napensis*)

Common Name: Spanish, Currucutú Vermiculado, LechucitaVermiculada, Buhito Guatemalteco, and Autillo Guatemalteco; English, vermiculated screech owl

Taxonomy: May form a superspecies with *Megascops vermiculatus*, *M. hoyi*, and *M. sanctaecatarinae*. Sometimes considered co-specimens with *M. vermiculatus*, it presents vocal differences with *M. guatemalae*, which has sometimes been considered as a separate species (König et al. 2008). A more extensive analysis is needed that includes genetic information and vocalizations to clarify the species and subspecies in this complex group. A division is made between *M. guatemalae* (Mexico to N of Costa Rica), *M. vermiculatus* (Costa Rica to NE of South America NE of Colombia), *M. napensis* (E of Ecuador and Colombia along the E slope of the Andes to Perú and N Bolivia), and *M. roraimae* (mountains N of Venezuela and adjacent areas of Brazil) (Dickinson 2003; König et al. 2008). According König et al. (2008), Colombia has two of those species (*M. vermiculatus* and *M. napensis*). Ridgely and Greenfield (2001), based on vocal records, propose that the subspecies *M. g. centralis* be a separate species that is distributed in the geographical Chocó and would be in Colombia; König et al. (2008) synonimize the central taxis with *vermiculatus*, proposing that this species is distributed in Colombia. Here we treat records in Colombia as *M. guatemalae*, according Remsen et al. (2016), and we included *M. g. vermiculatus*, *M. g. centralis*, and *M. g. napensis* as subspecies.

Description: Length between 20 and 23 cm; its plumage is reddish gray, and unlike the other members of the same genus, it has feathers that cover the feet. The tail is relatively long; the bottom has longitudinal and some horizontal stripes. His face is bordered by dark feathers and has relatively short tufts. The eyes are yellow. The peak is olive green (Hilty and Brown 1986; König et al. 2008). Weight of the bird: for *M. g. vermiculatus* average weight is 118 g (Dunning 2008).

Voice: Its vocalization appears to be a trembling trill very fast 5–8 s long o'o'o'o'o'..., with many fast notes similar to the calls of *Megascops asio* (König et al. 2008).

Distribution: This species is widely distributed in Central and South America, Mexico, Guatemala, Belize, Honduras, Costa Rica, Nicaragua, Panama, Venezuela, Guyana, Ecuador, Peru, Bolivia, and Brazil. For Colombia, it is distributed in Valle del Río Magdalena, Chocó biogeographic, and footprints of the Cordillera Oriental and Central. Chocó geographic: Buenaventura (Valle del Cauca), Serranía del Baudó, Alto del Buey (Chocó) (Meyer de Schauensee 1945; Biomap 2006). Cordillera Occidental: Valle del Cauca, Restrepo, Bravo River, Calima River and Risaralda districts, municipality of Mistrató, 8 km NO Jeguadas, Pisones (Instituto de Ciencias Naturales 2004). Cordillera Central: an old specimen possibly from the surroundings of Medellín (Antioquia) (Biomap 2006). Cordillera Oriental and Valle del Río Magdalena: Serranía de los Yariguíes, under Simacota, municipality of Simacota and Cerro de la Paz (Santander) (Donegan et al. 2010) and El Paujil Natural Reserve of Birds, municipality of Puerto Boyacá (Boyacá and Santander) (Freeman et al. 2011, http://www.xeno-canto.org/10835). Sierra Nevada de Santa Marta: a registry for Las Taguas, municipality of Santa Marta (Magdalena) (Todd and Carriker 1922; Biomap 2006), could be another species so requires verification.

Ecology: Very little is known about its biology, it is considered as local and rare (Freeman et al. 2011), and the size of its population is unknown. Its natural habitat appears to be dry tropical or subtropical forests or moist, lowlands and tropical forests up to 1200 m.a.s.l. (König et al. 2008). Their reproductive biology and diet are not known.

Conservation Status: This species is widely distributed; its population seems stable, should be large, and is therefore considered of Least Concern (LC) (BirdLife International 2016).

Conservation Strategies: It is present in the Serranía de los Yariguies National Natural Park (Donegan et al. 2010) and in the El Paujil Natural Reserve of Birds. It is likely in the Tatamá National Natural Park, since the specimen collected in Mistrato, Alto de Pisones, is very close there (Instituto de Ciencias Naturales 2004).

Megascops albogularis, *Syrnium algobularis* (**Cassin, 1849**)
Common Name: Spanish, Currucutú Gorgiblanco; English, white-throated screech owl

Taxonomy: Some authors suggest including it within the genus *Macabra* by the absence of "erectile ears" but by its vocal pattern must be included within the genus *Megascops*. The *obscurus* and *aequatorialis* subspecies are possible morphs as a result of individual variation. Six subspecies are recognized: subspecies *obscurus* (Phelps and Phelps Jr., 1953b), Serranía del Perijá, NW Venezuela; subspecies *meridensis* (Chapman 1923), Andes W of Venezuela; subspecies *macabrum* (Bonaparte 1850), central and W Andes from Colombia, S of Ecuador to the N of Peru; subspecies *albogularis* (Cassin 1849), Andes E of Colombia to N Ecuador; subspecies *aequatorialis* (Chapman 1922) eastern Ecuador; and subspecies *remotus* (Bond and Meyer de Schauensee 1941), E of Ecuador.

Description: Length between 20 and 27 cm. Very dark with contrasting white throat and "ears" barely noticeable. Iris yellow head and chest, dark brown finely dotted with white and tan, border of black and diffuse facial disk, conspicuous and narrow white throat, belly and lower parts with tawny scattered dark brown streaks. Immature: with finely buffed white and uniformly blackish barred (Hilty and Brown 1986). Weight of the bird: 185 g (Holt et al. 1999).

Voice: The vocalization of the male which presents from seven to 30 rough notes *churrochurro-churo-chu-chu-chu-chu* can be heard uninterrupted for 1 min, this being his primary vocalization. Occasionally there are heard shouts spaced almost five per second, gradually descending, and also has reported a descending trill of 10–14 shouts every 5–10 s, this being his secondary vocalization. Frequently, both sexes refer to the synchronous duet, the slightly higher and more acute female (Holt et al. 1999).

Distribution: This species is distributed in South America from Colombia to Bolivia; for Colombia three subspecies are known between 2000 and 3200 m.a.s.l.: *M. a. obscurus* (Serranía del Perijá), *M. a. macabrum* central and W Andes of Colombia (W Andes from Antioquia (Paramo of Frontino and Farallones de Citará), Risaralda (Tatamá Natural National Park), and Cauca (Munchique Natural National Park, Serranía del Pinche) (Negret 1994; Pulgarín and Múnera 2006; Casas and Ayerbe-Quiñones 2006; Krabbe et al. 2006; Echeverry-Galvis and Córdoba-Córdoba 2007) and Central Andes from Antioquia (Angostura), La Lana (San Pedro de los Milagros, Antioquia) (http://www.xeno-canto.org/39524, www.xeno-canto.org/77200), Caldas (Manizales- Río Blanco Reserve), Risaralda (Laguneta), Tolima (Toche) (Meyer de Schauensee1954, www.xeno-canto.org/96101), Cauca (Purace, Popayan, Quintana), and Nariño (Doña Juana Cascabel Volcanic Complex) (Ayerbe-Quiñones 2006; Ayerbe-Quiñones et al. 2008)), and *M. a. albogularis* Eastern Andes (Santander (Serranía de los Yariguíes, Alto Cantagallos and La Aurora), Boyacá (Soatá, Iguaque) (Córdoba-Córdoba and Echeverry-Galvis 2006; Donegan et al. 2007, www.xeno-canto.org/94531), and Cundinamarca (Bogota-Quebrada La Vieja, Bojacá, Chingaza Natural National Park) (Hilty and Brown 1986; Echeverry-Galvis et al. 2006; Córdoba-Córdoba et al. 2007; GeoSIB 2012).

Ecology: A nocturnal and occasionally crepuscular, rare, and local species, known from scattered localities in the mountains, mainly on jungle edges, open forest, and semi-open areas with trees. With respect to reproduction, very little is known (T. K. Salmon in Sclater and Salvin 1879). The nests are generally found at the level of the soil between ferns or grass. A white egg was discovered and incubated, in an abandoned cup-shaped nest on a bush near the ground (Sclater and Salvin 1879). In Peru, a female was recorded with ovaries developed in July, nesting in October (Ecuador). It is a resident species, apparently does not have seasonal or altitudinal movements (Holt et al. 1999).

Conservation Status: This species has a wide distribution area; however, the population status has not been quantified. Therefore, the species is considered of Least Concern (LC) (Bird Life International 2016).

Conservation Strategies: It is found in the Natural National Parks Las Orquíadas Tatamá, Munchique, Doña Juana Casacabel Volcanic Complex, Puracé, Chingaza, and Serranía of the Yariguíes; Forest Reserve Protector Rivers Black and White, Río Blanco and Oriental Forest of Bogotá; Protective Forest Reserve Producer Laguna de Pedro Palo; Natural Reserve of the Civil Society Chicaque; Nature Reserve Carpanta; Biological Reserve Encenillo and Integrated Management District Salto del Tequendama-Cerro Manjuí (ABO 2000; Moreno and Camargo 2008; CAR 2009, S. Chaparro-Herrera and J. P. López-Ordoñez pers. obs.).

Megascops sp. nov. **Spanish name: Buhito de Santa Marta**
This species has not yet been formally described and therefore has not been evaluated by BirdLife International. Current records suggest a restricted distribution, and due to the destruction of the habitat (agriculture, illegal crops, felling, and burning of forests) that still continues in the Sierra Nevada de Santa Marta, this species could be considered in some category of threat. Recent records in El Dorado Natural Reserve of Birds by birdwatchers describe it as common, but the population size has not been estimated. However, it is likely that its population is rapidly declining as a result of the loss of habitat (O. Cortes-Herrera pers. obs.). This species has been recorded only toward the Cuchilla de San Lorenzo, municipality of Minca in the Sierra Nevada of Santa Marta, Department of Magdalena, Colombia. The vegetation types in the area are cloud forest, humid forest, forest edges, and grassland, in elevations between 1800 and 2500 m.a.s.l. The species has been recorded in the protected area of the Sierra Nevada National Natural Park of Santa Marta. In spite of this, it is necessary to elucidate urgently the ecological requirements, population densities, and demographic characteristics of the species (O. Cortes-Herrera pers. obs.) (Fig. 9.4).

Lophostrix cristata **(Daudin, 1800)**
Common Name: Spanish, Búho Crestado; English, crested owl

Description: Length between 38 and 43 cm. Unmistakably large with prominent white superciliary that continues with large, erect, and partially white "ears." Variable tops, dark brown to light buff coffee, somewhat mottled with ruffle; wing and scapular coverlets with large white spots; dark brown and black barred primates; rufous facial disk; lower parts ante finely barred griffon and brownish vermiculate (Hilty and Brown 1986). Weight of the bird: very little information, two males of 425 and 510 g and two females of 545 and 620 g (Dunning 2008).

Voice: Vocalization is similar to the croaking of a toad. It begins with a stuttering sound that accelerates to deeper, a guttural purring and harsh *k-k-kkkk-krrrrraou*.

Fig. 9.4 *Megascops* sp. nov. *Common Name*: Buhito de Santa Marta. Species endemic of Colombia has not yet been formally described. Sector Minca, Municipality of Santa Marta, Magdalena Department, 30 November 2011 (Photograph ©Diego Calderón Franco; www.colombiabirding.com)

In the distance the introduction is slow and the notes are inaudible, so the vocalization sounds like *krrrrr*. This vocalization is pronounced at intervals of several seconds (König et al. 2008).

Distribution: It is distributed locally from S Mexico through Central America to Colombia, Ecuador, Venezuela, Suriname, Guianas, Brazil, and Peru to Bolivia (Hilty and Brown 1986; König et al. 2008). In Colombia, this species is distributed up to 1000 m. in lowland areas of the Pacific region, from the border with Panama to the S to Barbacoas in Nariño, in the lowlands of the Caribbean region, to the E along the base of the N of the Andes to the average Valle del Río Magdalena in Santander and in Puerto Triunfo (Antioquia), E of the Andes from W Caquetá to the S in the Amazon region (Hilty and Brown 1986; Salaman et al. 2009, Chaparro-Herrera pers. obs.).

Ecology: A strictly nocturnal owl, apparently rare and associated with lowland rainforests and rainforests, it can be found in primary forests, forest edges, or secondary growth forests. It is often seen in pairs in the undergrowth or middle heights. During the day it rests in dense vegetation, especially in thickets along the rivers. It prefers the proximity to water (Hilty and Brown 1986; Stiles and Skutch 1995; König et al. 2008). This species feeds on invertebrates (mainly large insects) but probably also consumes small vertebrates (Hekstra 1973; König et al. 2008). Apparently it nests in natural holes of mature trees (König et al. 2008); however, aspects of their reproductive and hunting behavior have not yet been studied in Colombia.

Conservation Status: Species has a very wide distribution and therefore does not approach the thresholds to be considered as a vulnerable species under the range size criterion. It is considered a species of Least Concern (LC) (BirdLife International 2016).

Conservation Strategies: It is present in the Natural National Park Ensenada de Utria, Pangan and El Paujil Natural Reserve of Birds (Salaman et al. 2009; GeoSIB 2012), and Natural Reserve Río Claro Cañón-El Refugio (Chaparro-Herrera pers. obs.). König et al. (2008) mention that their conservation status is unknown.

Pulsatrix perspicillata **(Latham, 1790)**
Common Name: Spanish, Búho de Anteojos and Lechuzón de Anteojos; English, spectacled owl

Description: Length between 43 and 52 cm. It is a round-headed bird and has yellow eyes, dark brown head and lower parts, superciliary spreads that extend into white glasses, white throat, and chest with a brown band. Immature: almost totally white anteriorly, with black face (in a heart shaped form) and brown wings. It may take up to 5 years to reach full adult plumage (Hilty and Brown 1986; König et al. 2008). Weight of the bird: average of 750 g, range 571–980 g (Stiles and Skutch 1995; König et al. 2008) (Fig. 9.5).

Voice: Six-to-eight low-resonance bass-calling series *boo-boo-boo-boo-boo-boo,* also a descending series of notes low *woof.* More vocal on moonlit nights (König et al. 2008).

Distribution: Its range extends from S Mexico to the N of Argentina (König et al. 2008). In Colombia it is distributed up to 1000 m.a.s.l. Present in the geographic Chocó, Caribbean zone, Valle alto del Río Magdalena, Orinoco, and Amazon region. *P. p. chapmani*: geographic Chocó, Chocó municipalities of Unguia-Tanela River, and Juradó, Juradó River, in Nariño Corregidor of Altaquer, Río Ñambi Natural Reserve, in Valle del Cauca, municipality of Cali-Cauquita River (Biomap 2006). *P .p. perspicillata*: Amazon region (Caquetá department, municipality of Solano-Tres Esquinas, Orteguaza River) and Cairo (Municipality of Florence, Morelia (Instituto de Ciencias Naturales 2004)). Caribbean region: Department of Bolívar, Carmen de Bolívar 10 km from Arroyo Playón; Department of Magdalena, municipalities of Pivijay, Santa Marta, La Tigrera; and Department of Córdoba, municipality of Tierra Alta-Cerro Murrucucu. Valle alto del Río Magdalena: Department of Huila, municipality of Villavieja (Biomap 2006).

Ecology: Its natural habitat corresponds to tropical forests, humid lowland forests up to 1800 m.a.s.l. (König et al. 2008). It is mostly nocturnal but occasionally active on cloudy days. It is a relatively common owl but has rarely been observed. Noted to rest at variable heights between the 2 m and the canopy (Hilty and Brown 1986). Normally the activity begins after sunset and continues until dawn. In the day it is

Fig. 9.5 Spectacled owl (*Pulsatrix perspicillata*). Municipality of Manizales, Caldas department, June 2012. (Photograph ©Juan Pablo López-Ordoñez)

often sheltered in trees with dense foliage and bamboo patches; after sunset it flies with a gentle flapping of a perch on a perch facing the ground (König et al. 2008). This species consumes a variety of animals (insects, mammals, birds, and reptiles) and because of their size, can feed on large preys such as opossums (*Didelphys*) and sloths (*Bradypus variegatus*) (Bryson et al. 2009). Records of pellets found in Cerro de Oro-Oaxaca (Mexico) contained: *Marmosa* sp. bats, birds (*Momotus momota*, *Leptotila* sp.), insects (*Melanototus globosa*, *Pseudomonas dophyllinae*, *Ericlus spiniger*, *Golofa* sp.), and small crustaceans (Gómez de Silva 1997). Although it does not seem to require a large area of continuous forest for its reproductive success (Hume 1991), Stiles and Skutch (1995) mention that it depends on wooded areas for nesting. The reproductive behavior of this species is little known (König et al. 2008). Nests have been recorded in large cavities located in trees of about 15 m high (Stiles and Skutch 1995). The female incubates two white eggs for 5 weeks and biparental care, where the male brings food for the female and chicks. Almost always only one of the chicks survives, which they care for about a year (König et al. 2008). This species tends to increase its vocal behavior in days near the summer solstice, possibly not to overlap its vocalization with other species of owls (Enríquez and Rangel-Salazar 2001; König et al. 2008).

Conservation Status: This species has a wide distribution area, and although it seems that its population is decreasing, it is estimated that its population size is not so small; it is considered as a species of Least Concern (LC) (BirdLife International 2016).

Conservation Strategies: It is located in the Natural National Parks of Ensenada de Utría, the Katios, Sierra de La Macarena, Tinigua, and Amacayacu (Instituto de Ciencias Naturales 2004; Biomap 2006).

Pulsatrix melanota **(Tschudi, 1844)**
Common Name: Spanish, Búho Ventribandeado, Búho de Vientre Bandeado, and Lechuzón Barreado; English, band-bellied owl

Taxonomy: Some authorities join *Pulsatrix melanota* and *P. koeniswaldiana* into a superspecies, suggesting that they could be treated as conspecific (Remsen et al. 2016).

Description: Length between 44 and 48 cm, round head, and white head. Chest with a broad dark brown band, the ventral part has dark brown crosslines with light brown belly bottom, short tail dark brown with white lines. Plumage of chicks is pale but has a black mask (Hilty and Brown 1986; König et al. 2008). Weight of bird is not known. But it could be similar to that of *P. perspicillata* with an average of 750 g (Stiles and Skutch 1995).

Voice: Not well known, its vocalization is several notes followed *bu-bu-ubububububu* (König et al. 2008).

Distribution: It is distributed on the E slope of the Andes from S Colombia along E Ecuador, Peru, and N and W Bolivia (Ridgely and Greenfield 2001; Hennessey et al. 2003; Schulenberg et al. 2007). In Colombia, it is rather a rare species; there is only one especimen in the Museum of Chicago (FMNH of Chicago) without sex or local data (Biomap 2006). There are recent observations in the Serranía de los Churumbelos-Río Nabúeno (700 m.a.s.l.) and Alto Hornoyaco River in the department of Cauca; Rumiyaco River and Guamuez Putumayo Pumping Station possibly further north (Salaman et al. 1999, 2002a, b, 2007).

Ecology: Nocturnal habitats are located in tropical forests, in moist forests and forest edges, and locally in open forests (Holt et al. 1999; Restall et al. 2006). König et al. (2008) suggested that the diet of this species could be similar to that of the other two species of the genus *Pulsatrix*, the tawny-browed owl (*P. koeniswaldiana*) and the spectacled owl (*P. perspicillata*). In the stomach contents of an individual in Ecuador contained Orthoptera and Coleoptera remains, small spiders, seeds, plant, and mineral material; there were no signs of bones, feathers, claws, or other structures suggesting vertebrates (Cadena et al. 2011).

Conservation Status: This species has a wide distribution area, and although it seems that its population is declining, it is estimated that its population size is not declining rapidly; it is considered Least Concern (LC) (BirdLife International 2016).

Conservation Strategies: This species is present in the Serranía de los Churumbelos Auka-Wasi Natural National Park (Salaman et al. 2007).

Bubo virginianus, Strix virginiana **(J. F. Gmelin, 1788)**
Common Name: Spanish, Búho Americano; English, great horned owl

Taxonomy: The species has numerous subspecies, poorly differentiable or apparently as a result of individual variation. Twelve subspecies are distributed and recognized in North, Central, and South America (Holt et al. 1999).

Description: Length in males of 51 cm and females of 60 cm. Large, robust owl, "ears" long and erect. Yellow iris, dark brown on the back, speckled gray forehead, bluish white facial disk fringed with black, white throat, lower parts narrowly barred dark brown and whitish, chest with few broad black streaks (Hilty and Brown 1986; Holt et al. 1999). Weight of the bird can vary between subspecies; in males the weight is between 680 and 1585 g; females are generally heavier between 1000 and 2500 g (Holt et al. 1999; König et al. 2008).

Voice: During mating, males perform a series of so-called *hu-huhoooo, hooh hooh* for 3 s, while the females perform the same call but an additional note at the beginning, *huhuhuhoo hooh hooh* (Holt et al. 1999).

Distribution: Distributed from North America to Tierra de Fuego (South America). It is in an altitudinal gradient up to 4000 m.a.s.l. Two subspecies are reported for Colombia: Andes from Colombia to NW Peru (subspecies *nigrescens*) and in the lowlands of the E region of Colombia, to the Guayanos shield (subspecies *ñacurutu*). In Colombia, it has been reported in the municipalities of Caldas and Belmira (Antioquia), Mompós, San Fernando, San Jacinto del Cauca (Bolívar), Ayapel, Santa Cruz de Lorica, Best Corner (Córdoba), Tolú Old, Santiago de Tolú (Sucre), Riohacha (La Guajira), Bucaramanga (Santander), Manizales, Río Blanco Reserve (Caldas), Santiago de Cali (Valle del Cauca), Piamonte (Cauca), Baraya (Huila), and El Porvenir (Meta) (Hilty and Brown 1986; GeoSIB 2012).

Ecology: It is a widely distributed species found in a wide variety of habitats, from forest areas, secondary growth, pastures, and open areas. It is a strictly nocturnal species, which feeds on various preys, including small mammals (rabbits, mice), terrestrial birds (but also water birds), reptiles, amphibians, fish, and insects, and other invertebrates have been reported in their diet. Their breeding season is from December to July, but little is known for Colombia. Nest in large cavities of logs or near the ground in depressions or in the base of trees; incubate two to three eggs between 28 and 35 days. Longevity is known an individual of more than 28 years. There are reports of migration in North America, with movements of more than 250 km from the banding station (Holt et al. 1999; König et al. 2008).

Conservation Status: This species has a wide distribution, and the population trend seems to be stable; however, the size of the population in Colombia has not been quantified. The species is categorized as Least Concern (LC) interntionally (BirdLife International 2016).

Conservation Strategy: Despite its wide distribution in Colombia, there is no information on the presence of this species in protected areas.

Ciccaba virgata **(Cassin 1843) o** *Strix virgata* **(König et al. 2008)**
Common Name: Spanish, Búho Moteado; English, mottled owl

Taxonomy: König et al. (2008) separated this species into two different taxa: *C. v. virgata* (from N Colombia and E of the Andes to Argentina (Misiones), absent from Pacific slopes of the Andes and W of Santa Marta and Perijá Mountains), and *C. v. squamulata* (fom Mexico, Central America and probably to N Colombia, Santa Marta Mountains, and W Cordillera of the Andes to SW Ecuador), both present in Colombia. However, this species is considered here as a single species according Remsen et al. (2016) (Fig. 9.6).

Description: Without "ears." Brown eyes, brown back mottled with black (seems blackish at a distance), superciliary and narrow edge of facial white disk, remiges and tail barred dark brown and gray dyed, chest before strongly mottled and striated blackish coffee, and lower limbs broadly striated dark brown (Hilty and Brown 1986).

Fig. 9.6 Mottled owl (*Ciccaba virgata*). Serranía de San Lucas, Bolívar department, September 2010 (Photograph ©Juan Pablo López-Ordoñez)

Weight of the bird: males between 220 and 256 g, average of 240 g; females between 307 and 366 g, mean of 336 g (Gerhardt et al. 1994a; Dunning 2008; König et al. 2008).

Voice: The male's vocalizations are series of about five clear, rather spaced cries, equally deep and resonant, with emphasis in the fourth, where the fifth is weaker and somewhat slow after a short *who-who-who-whóho*. This phrase repeated at intervals of several seconds. The female has a similar but more acute vocalization. In the beginning of breeding season, male and female sing in duet. The female emits a slight and timid call (König et al. 2008). Also, female rarely emit a cat-like scream (Hilty and Brown 1986).

Distribution: This species is distributed from Mexico and Central America to NE Argentina and E Paraguay in South America (Hilty and Brown 1986; Infonatura 2007; König et al. 2008). In Colombia, it is distributed up to 2300 m.a.s.l. in the lowlands of the Pacific and Caribbean region, in the lowlands of the N of the Andes, Serranía del Perijá, and Sierra Nevada de Santa Marta, to the S in the Valle del Río Cauca and Magdalena (Huila) and to the E of the Andes from W of Meta and in the subtropical Amazon (Amazon region), also present in Vichada, Guainía, and Serranía de La Macarena (Hilty and Brown 1986; McNish 2007; Salaman et al. 2009).

Ecology: Strictly a nocturnal bird that begins its activity at dusk and inhabits between the middle level and the canopy in the forests. During the day it rests between the dense foliage or inside natural orifices in the trunks of the trees and is one of the most vocal Neotropical owls in the region (Hilty and Brown 1986; König et al. 2008). It appears that this species is tolerant to deforestation and can be found in forest clearings and edges, as well as semi-open areas or secondary forest forests (Stiles and Skutch 1995). It is associated with riparian vegetation (1100 m.a.s.l.) in La Unión, municipality of Quipile, Cundinamarca (Sua and Chaparro 2010). Gerhardt et al. (1994b) mention that this owl species is the most numerous in the neotropic forests, and its range is very wide. It consumes small vertebrates such as rodents, reptiles (including snakes), and amphibians; it also feeds on arthropods and also on small birds (Gerhardt et al. 1994a; König et al. 2008). This species can enter urban areas for food (Enríquez 1995). There is no detailed information on reproductive biology in Colombia, but breeding birds have been found between February and May, one in April and one in July (Hilty and Brown 1986). They nest in cavities or thick branches between epiphytes or in abandoned nests of other larger birds where it lays two white eggs (Gerhardt and Bonilla 1991; König et al. 2008).

Conservation Status: It has a wide distribution and apparently the population is stable. Therefore, it is considered as Least Concern (LC) (BirdLife International 2016).

Conservation Strategies: It is present in the National Natural Parks, Munchique, Tatamá, and Tinigua, Sanctuary of Fauna and Flora Otún Quimbaya, Protected Forestry Reserve Río Blanco and Yotoco, Natural Reserve Río Barroso and San Juan, Natural Reserve of the Civil Society La Reseda, and in the Natural Reserves of the Birds Pangan, El Paujil, El Dorado, *Pauxi pauxi*, Ranita Dorada, and Halcón Colorado (Salaman et al. 2009; Peñuela et al. 2011; GeoSIB 2012).

Ciccaba nigrolineata **(Sclater, 1859),** *Strix nigrolineata* **(Sclater, 1859, König et al. 2008)**
Common Name: Spanish, Búho Carinegro; English, black-and-white owl

Description: Length between 35 and 40 cm; yellowish brown eyes, black back finely barred with white on the back, black facial disk, superciliary, indistinct band bordering white-spotted facial disk, white below, and narrowly and uniformly barred black tail with narrow white bars (Hilty and Brown 1986). Weight of the bird: males between 404 and 436 g, average of 435 g; female between 468 and 535 g, average of 482 g (Dunning 2008; König et al. 2008) (Figs. 9.7 and 9.8).

Voice: Vocalization is more like a phrase of quick, low, and deep guttural notes that gradually grows in volume and tone, followed after a short pause of 0.25 s by a strong, explosive, and loud *wów* as if it were a crying or moaning and then followed after a rest of 0.2 s by a weak and short or slightly lower pitch. In the field recorded as: *wobobobobobo wów ho*. These phrases are repeated in intervals of several seconds. The female has a similar but light vocalization. Prolonged vocalization is only described in young birds (König et al. 2008).

Distribution: It is distributed from Central Mexico through Central America to Venezuela and Colombia and Occidental of Ecuador and NW of Peru (König et al. 2008). In Colombia it is distributed until 2000 m.a.s.l. In the lowlands of the Caribbean region and the Pacifica region, in the Valle del Río Cauca and Magdalena, records to NE of the Andes, Sierra Nevada de Santa Marta, Serranía de Perijá, and the E and W Cordilleras (Hilty and Brown 1986; Salaman et al. 2009).

Ecology: Apparently very local in the humid forest, edges of forest, secondary forest, and clearings of forest with associated trees, sometimes near human settlements (Hilty and Brown 1986). It is strictly nocturnal. During the day it remains well hidden in the dense foliage, between vines or on a branch near the main trunk, usually high above the ground, resting on the branches of the middle or upper levels of the forest, males and females. They may find resting together (König et al. 2008). It feeds mainly especially insects, especially beetles and orthoptera, but also consumes

Fig. 9.7 Black-and-white owl (*Ciccaba nigrolineata* or *Strix nigrolineata*). Municipality of Santa Marta, Magdalena department, 29 November 2011 (Photograph ©Diego Calderón Franco; www.colombiabirding.com)

Fig. 9.8 Black-and-white owl (*Ciccaba nigrolineata* or *Strix nigrolineata*). Municipality of Santo Domingo, Antioquia department, 08 June 2011 (Photograph ©Santiago David-R)

small mammals such as bats and other small vertebrates (Ibáñez et al. 1992; König et al. 2008). Aspects about their behavior, reproductive biology, and eating habits are limited and unknown in Colombia. In Central America they nest in natural cavities, where they laid two white eggs (König et al. 2008).

Status of Conservation: It has a very wide distribution area, and its population size is very large, with more than 10,000 mature individuals. Therefore, it is considered a species of Least Concern (LC) (BirdLife International 2016).

Conservation Strategies: Although considered a fairly common species in 12 Latin American countries, its conservation status is uncertain (König et al. 2008). It is present in the Natural National Parks Ensenada de Utria, Tatamá and Munchique, and Natural Reserves of the Birds El Paujil and El Dorado (Salaman et al. 2009; GeoSIB 2012).

Ciccaba huhula **(Daudin, 1800)**
Common Name: Spanish, Carabo Negro; English, black-banded owl

Taxonomy: This species is often classified in the genus *Ciccaba* along with other species such as *Ciccaba virgata*, *Ciccaba nigrolineata*, *S. albigularis*, and sometimes *S. woodfordii* (Africa) based on differences in the structure of the external ear – not compared to other species in the genus *Strix*. However, DNA studies suggest that all of these are related and would be better maintained in a single *Strix* genus (Holt et al. 1999; König et al. 2008).

Description: Length between 30 and 36 cm. No "ears," iris coffee and sometimes yellow orange, and blackish facial disk. Completely black with very white lines on the back and belly, superciliary and margins of white discolored facial disks, and black tail with several narrow white bars (Holt et al. 1999; König et al. 2008). Weight of the bird: 370 g (Dunning 2008).

Voice: The vocalization is characterized by an ascending series of four deep *bhú* notes, followed by descending, sharp, and slight notes *buhu* (Holt et al. 1999).

Distribution: *C. h. huhula* is distributed in Colombia and Venezuela until 500 m.a.s.l. E of the Andes in Meta, Caquetá, and S of the Guaviare River, Amazonas (Hilty and Brown 1986; GeoSIB 2012).

Ecology: Known in Colombia from four specimens but often goes unnoticed throughout its distribution. It species is reported in clearings and forest interior, especially in Araucaria forest. It has been adapted to sites with anthropic disturbance, especially in areas with banana and coffee crops. There is no information about their reproductive period and no seasonal movements. It feeds on insects, especially beetles, mantids, and small vertebrates (Holt et al. 1999).

Conservation Status: This species has a wide distribution area and the population trend seems stable; however, the population size has not been quantified. For these reasons, the species is assessed as Least Concern (LC) (BirdLife International 2016).

Conservation Strategy: In Colombia it has been reported in the Amacayacu National Natural Park (Holt et al. 1999).

Ciccaba albitarsis, *Syrnium albitarse* **(Bonaparte, 1850)**
Common Name: Spanish, Búho Ocelado; English, rufous-banded owl

Taxonomy: This species has also been classified in the genus *Ciccaba* along with other species such as *Ciccaba virgata*, *Ciccaba nigrolineata*, *S. huhula*, and sometimes *S. woodfordii* (Africa) based on differences in external ear structure compared to other species in the *Ciccaba* genus. But DNA studies suggest that they are all related and would be better maintained in the *Strix* genus (Holt et al. 1999; König et al. 2008).

Description: Length between 30 and 36 cm. Without "ears," coffee iris, head and upper parts blackish brown heavily barred and stained with buffy rufous, superciliar and white bridal area, white throat, chestnut barred and stained fawn and whitish (indefinite pectral band), and feathers of the rest of the inferior parts marginalized platinum white and centrally divided with rufous coffee, forming large white and square "ocellate" spots; the plumage color of the immature is uniform (Holt et al. 1999). Weight of the bird: between 265 and 350 g ($n = 2$) (Museum of the Alexander von Humboldt Institute) (Fig. 9.9).

Voice: It has a short, deep series of deliberate notation *hu, hu-hu-hu HOOa* with a pause after the first note, the next three a little faster, and the long and strongly emphasized final note in rhythm differs from that of *C. huhula* and *C. nigrolineata*. This species repeats at intervals of 8–11 s (Holt et al. 1999).

Distribution: Andes from South America, from N Venezuela to central and W Bolivia. In Colombia: Cerro Bravo, Jericó, Piedras Blancas Hydroelectric Power Station Miraflores Municipal Reserve, La Estrella, Fredonia, Ebejico (Antioquia). Pajarito (Boyacá), Manizales, Natural Reserve Río Blanco (Caldas), El Tambo, National Natural Park Munchique (Cauca), Ensenada de Utria (Chocó), Cáqueza, Choachí, Zipacón, La Calera, Guasca, Junín (Cundinamarca) Junín (Nariño), Salento (Quindío), and Yotoco (Valle del Cauca) (Hilty and Brown 1986; GeoSIB 2012).

Ecology: It is a common species but with restricted distribution. It inhabits humid montane forest and cloud forest between 1700 and 3700 m.a.s.l. but also in open areas and patches of forest. Little is known about its diet, probably feeding on

Fig. 9.9 Rufous-banded owl (*Ciccaba albitarsis*) (Photograph ©Andrea Beltrán)

insects in the canopy, and it is an owl of nocturnal habits but active toward dusk and dawn. With regard to their reproduction, an individual was recently emerged from the nest toward the third week of June, and a juvenile was observed in August. It is a resident species with no migratory movements (Holt et al. 1999).

Conservation Status: This species has a wide distribution area and the population trend seems to be stable; however, the size of the population has not been quantified. The species is evaluated as Least Concern (LC) (BirdLife International 2016).

Conservation Strategy: It has been reported in the National Natural Parks Tamá, Munchique, Farallones de Cali, Tatamá, Puracé, Nevado del Huila, Las Hermosa, Nevado del Ruíz, Ensenada de Utria and Chingaza; Protected Forestry Reserve La Planada, Río Blanco and Yotoco; Nature Reserve of the Civil Society of the Alto Quindío Acaime; Natural Reserves Carpanta, La Montana del Oso and Miraflores (J. P. López- Ordoñez pers. obs.).

Glaucidium nubicola **(Robbins and Stiles, 1999)**
Common Name: Spanish, Buhito Nubícola; English, cloud-forest pygmy owl

Description: Small owl (c.14–16 cm) compact and without "ears." Yellow eyes, yellow green beak, and yellow legs. Dark chocolate coffee on the head and back, with some white spots on the crown and sides of the head, no face ring with conspicuous edge and coffee with whitish concentric marks, with "false eyes" on the nape of the neck, blacks bordered with white. Wings with white spots, very short tail in general, black and with five white bands (Robbins and Stiles 1999; König et al. 2008). Weight of the bird: males weigh 76.1 g ($n = 3$) up to 80 g; Female 79 g ($n = 1$) (Robbins and Stiles 1999).

Voice: Vocalization is generally a long sequence of whistles sung as pairs *tu-tu tu-tu tu-tu...* equidistant, sometimes with single notes or triplets (Robbins and Stiles 1999).

Distribution: Endemic species of Colombia. It is distributed along the Pacific slope of the Andes in the Cordillera Occidental of Colombia and to the S to the SW of Ecuador. In Colombia, there are few records of this species, mainly between 1400 and 2200 m.a.s.l., although there is an auditory record of Robbins at 900 m.a.s.l. in Ecuador (Freile et al. 2003). Distributed at Alto de Pisones (Mistrato), Tatamá National Natural Park (Pueblo Rico-Risaralda), San Antonio (Cali and Chicoral), La Cumbre (Valle del Cauca), Reserve La Palnada (Ricaurte), Río Nambí Reserve, Barbacoas, and the El Pangan (Junín-Nariño) (Robbins and Stiles 1999; Stiles et al. 2002; Echeverry et al. 2008; Fierro-Calderón and Montealegre 2010; Fierro-Calderón and Córdoba-Córdoba 2014).

Ecology: It is a rare species with very low densities (Fierro-Calderón and Montealegre 2010; Fierro-Calderón and Córdoba-Córdoba 2014) and apparently seldom located when heard. It has been found in mature or cloud forests on slopes, young and mature secondary forests, and on forest edges with dense soils (Robbins and Stiles 1999; Freile et al. 2003; Fierro-Calderón and Córdoba-Córdoba 2014). Little is known of its ecology, but its habits are presumed to be similar to those of other species of the genus *Glaucidium*. It has been heard mainly in the morning but is active throughout the day. It is found in the canopy and middle stratum of the forest, occasionally low to understory (Stiles et al. 2002). Their diet consists mainly of invertebrates and small vertebrates, such as crickets, cicadas, and bedbugs, as well as lizards and birds (Miller 1963; Robbins and Stiles 1999). It nests in hollows of trees and old woodpecker cavities (König et al. 2008; Olmedo 2011b). The breeding season is not known, but it has been suggested that it would be between February and June (Robbins and Stiles 1999). It could be longer because a juvenile has also been observed in August (Greeney and Nunnery 2006).

Conservation Strategies: It is present in the Tatamá National Natural Park and in the Important Bird Area (IBA)) La Planada, Río Ñambí, El Pangán, Chicoral and San Antonio (Echeverry et al. 2008). However, this does not ensure the protection of their populations; its presence has not been confirmed in some protected areas that are in their potential distribution.

Glaucidium jardinii **(Bonaparte, 1855)**
Common Name: Spanish, Buhito Andino, Mochuelo Andino, and Mochuelo Montañero; English, Andean pygmy owl

Description: Length between 15 and 16 cm; round head without "ears." Yellow eyes and greenish yellow beak and legs. Dark brown on the back; stippled crown of gray coffee. Half necked with "false eyes"; black tail with four white bands. Rufous or clear phase: head, upper region, and chest rufous with little or no streaks. Immature: similar to the adults of its phase but the crown and nape with clear lists instead of points or without spots (Hilty and Brown 1986; ABO 2000). Weight of the bird: males between 54 and 77 g, average 62 g; females between 54 and 69 g, average 61.3 g (general average 66.9 g) (Echeverry-Galvis et al. 2006) (Fig. 9.10).

Voice: Vocalization is a series of 4–20 or more clear whistles emitted with regular rhythm, about two per second *tut tut tut tut*.... When it is very excited, it mixes a purr between the series of whistles (ABO 2000). It calls with a long series of whistled *puup*, emitted in pairs (Hilty and Brown 1986).

Distribution: It is distributed in the Andean region from W Venezuela to Central Bolivia. In Colombia between 1500 and 3500 m.a.s.l. (Krabbe et al. 2006; Arbeláez-

Fig. 9.10 Andean pygmy owl (*Glaucidium jardinii*). Rufous morph. Cerro Montezuma, Valle del Cauca department, 21 May 2012 (Photograph ©Diego Calderón Franco; www.colombiabirding.com)

Cortés et al. 2011). Distributed in the Serranía del Perijá (Cerro Pintado, La Paz and Manaure Balcón del Cesar, Cesar) and locally in the Cordillera Oriental from the Tamá Natural National Park (Norte de Santander) to the E of Bogotá (Choachí) and La Plata and Palermo (Huila) (Hilty and Brown 1986; Biomap 2006; Ardila et al. 2007; IUCN 2016); Cordillera Central from E Medellín (Antioquia) to the mountains of Quindío (Hilty and Brown 1986; Arbeláez-Cortés et al. 2011), Cajamarca (Tolima) (http://www.xeno-canto.org/128292), and W flank and elevated areas of the Central and E Cordillera in the Cauca (Ayerbe-Quiñones et al. 2008); in the Cordillera Occidental in the Paramo of Frontino, municipality of Urrao (Antioquia) (Krabbe et al. 2006), Tatamá Natural National Park, Pueblo Rico municipality (Risaralda) (Echeverry-Galvis and Córdoba-Córdoba 2007), visual records in Cerro Torrá in the S of Chocó and San Antonio, Km 18, La Cumbre (Valle del Cauca) and W flank in the Cauca (Hilty and Brown 1986; Ayerbe-Quiñones et al. 2008; IUCN 2016); Cauca Natural Reserve Tambito, El Tambo municipality (Donegan and Dávalos 1999), the NW flank of the Serranía de los Chrumbelos (Salaman et al. 1999, 2002a, b), and Magdalena River (Ayerbe-Quiñones et al. 2008). Nariño in the municipalities of Tumaco and Barbacoas (Biomap 2006).

Ecology: Known in a few dispersed localities in humid mountain forests and edges (Hilty and Brown 1986), also in oak forests (*Quercus*) with dense epiphytic growth (Salaman et al. 2002a, b). It is present in secondary forest, edges of secondary forest, grazed or dwarf forest, and alder (*Alnus acuminata*) forest plantations with low and dense understory (Verhelst et al. 2001; Ardila et al. 2007). There is also premontane forest and montane forest (Arbeláez-Cortés et al. 2011). It goes unnoticed by remaining in the canopy or average levels of vegetation, sometimes leaving large trees in adjacent paddocks or clearings (Hilty and Brown 1986; ABO 2000). Their diet is based on the consumption of insects, large invertebrates, and very small vertebrates through stalking or persecution (Stiles and Roselli 1998; Ardila et al. 2007). It nests in hollows of trees, often in an old nest of a woodpecker, where it lays 2–3 white and round eggs (ABO 2000).

Conservation Status: This species has a wide distribution range, and its population size may be moderately small to large and the trend of the population seems to be stable. It is not believed to be approaching the Vulnerable thresholds. However, the size of the population has not been quantified, but it is considered as Least Concern (LC) (BirdLife International 2016).

Conservation Strategies: It is present in the Protected Forest Reserves of the E forest of Bogotá, Blanco and Negro Rivers, and Peñas del Aserradero (Stiles and Roselli 1998; Álvarez-Rebolledo et al. 2007; CAR 2009; S. Chaparro-Hererra pers. obs.); Biological Reserve Encenillo; Chicaque Civil Society Natural Reserve; Nature Reserve Carpanta and Sanctuary of Fauna and Flora Iguaque (Moreno and Camargo 2008; GeoSIB 2012; S. Chaparro-Herrera pers. obs.); Protected Forest Reserve Páramo de Urrao, Regional Natural Park Ucumari (Naranjo 1994; Krabbe et al. 2006; GeoSIB 2012), Nature Reserve of the Alto Quindío Acaime Civil Society (GeoSIB 2012), Sanctuary of Fauna and Flora Otún Quimbaya (GeoSIB 2012), National Natural Parks Chingaza, Tatamá, Munchique, Purace, Tamá and Serranía de Los Churumbelos Auka-Wasi; Reserva Natural Tambito y Forest

Reserve La Planada (Donegan and Dávalos 1999; Salaman et al. 1999; Salaman et al. 2002a, b; Echeverry-Galvis and Córdoba-Córdoba 2007; Ayerbe-Quiñones et al. 2008; IUCN 2016; GeoSIB 2012).

Glaucidium griseiceps **(Sharpe, 1875)**
Common Name: Spanish, Mochuelo Centroamericano and Buhito Enano; English, Central American pygmy owl

Taxonomy: Formerly considered part of the *Glaucidium minutissimum* group (Howell and Robbins 1995; König et al. 2008).

Description: Length between 14 and 16 cm; with the crown, nape and back olive brown to reddish brown. Spotted whitish on the face.The iris and feet are yellow; the beak and yellow green wax. It has black spots as eyes in the nape of the neck. Scapular feathers and winged wings covered with white. Blackish tail with white bars; the sides of the chest between coffee and cinnamon.Immature: the crown has no spots and no ocular spots on the nape (Howell and Robbins 1995; Stiles and Skutch 1995). Weight of the bird is between 49 and 57 g, and Colombian specimens weigh 54.5 g (König et al. 2008; Moreno-Palacios and Rodríguez-Ortíz 2008).

Voice: It begins with 2–4 equally spaced screams, followed by a very brief pause and then a series of 6–18 very similar notes *huu-huu, huu-huu-huu*...; trills may precede the series of cries (Holt et al. 1999). Also a series of c. 6–15 short and fast whistles (Donegan et al. 2007; Moreno-Palacios and Rodríguez-Ortíz 2008).

Distribution: From SE Mexico to NWof Ecuador. In Colombia the subspecies *griseiceps* is present, until 1350 m.a.s.l. (Howell and Robbins 1995; Donegan et al. 2010). Distributed in Monte Libano (Córdoba); Antioquia: Serranía de San Lucas municipality of Anorí; Chocó: Natural National Park Los Katios, Cerro Cuchillo municipality of Río Sucio and Bahia Solano; Valle del Cauca: municipality of Darién. Santander: Serranía de los Yariguíes and surrounding lowlands, municipalities of Zapatoca and Simacota and the Natural Reserve of Birds El Paujil located in the piedmont of the Serranía de las Quinchas on the W slope of the Cordillera Oriental, municipalities of Puerto Boyacá, Boyacá, and Cimitarra (Robbins and Howell 1995; Salaman et al. 2002a, b; Biomap 2006; Moreno-Palacios and Rodríguez-Ortíz 2008; Donegan et al. 2010; eBird 2011).

Ecology: Present in tropical humid forest and premontane forest (Howell and Robbins 1995; Donegan et al. 2010). It inhabits the canopy of primary forest with semi-open undergrowth and large trees including *Clathrotropis brachypetala* (Fabaceae), *Cavanillesia* sp. and *Catostemma alstonii* (Bombacaceae), *Virola sebifera* (Myristicaceae), and *Pseudolmedia laevis* (Moraceae), with high canopy at c.30 m and presence of lianas and epiphytes (Salaman et al. 2002a, b; Moreno-Palacios and Rodríguez-Ortíz 2008). It is believed that it feeds mainly on insects, but there is not much information. Moreno-Palacios and Rodríguez-Ortíz (2008) observed an individual who had captured a pigeon (*Claravis pretiosa*). Their reproductive biology is not known. There are no nests described for this species but could be in abandoned termite or woodpeckerholes (König et al. 2008). A juvenile was recorded on the ground accompanied by an adult in March (Moreno-Palacios and Rodríguez-Ortíz 2008).

Conservation Status: This species has a wide distribution area, and its population size may be moderately small to large, and the population trend seems to be stable, so it is not believed to be approaching Vulnerable thresholds. For these reasons, the species is evaluated for Least Concern (LC) (BirdLife International 2016).

Conservation Strategies: Present in the Katios and Serranía de los Yariguíes Natural National Parks; Forest Reserve Protector Río León; Nature Reserve Nechí-Bajo Cauca; Natural Reserve of Birds *Pauxi pauxi* and El Paujil (Robbins and Howell 1995; Salaman et al. 2002a, b; Moreno-Palacios and Rodríguez-Ortíz 2008; Donegan et al. 2010; eBird 2011).

Glaucidium brasilianum **(Gmelin, 1788)**
Common Name: Spanish, Buhito Ferrugíneo and Mochuelo Común; English, ferruginous pygmy owl

Taxonomy: This species has sometimes been considered different from the forms of Central America and S United States (*G. ridgwayi*), which include NW South America by the Pacific and part of the Atlantic Coast of Colombia; *G. brasilianum* would only be distributed in South America according to this proposal (König et al. 2008). Here we include both *G. brasilianum* and *G. ridgwayi* forms in Colombia as a single species according to Remsen et al. (2016).

Description: Length between 17 and 20 cm. No "ears," bright yellow eyes and legs, beak and greenish yellow wax, black claws. Two phases of coloration, grayish brown or rufous on the back; crown and face finely striated gray brown or whitish; "false eyes" black on each side of the nape; eyebrows white; blackish tail with 5–6 visible white bars. Immature: the crown with spotts, and poorly defined the ocular spots (Hilty and Brown 1986; Stiles and Skutch 1995). Weight of the bird: males between 46 and 74 g, average 64.3 g; females between 62 and 95, average 77.2 g (Dunning 2008; König et al. 2008) (Fig. 9.11).

Voice: Its called is a long series (up to several minutes) of *tuut* or *poip* sounds, with 2.5 notes per second, usually with a small tail shake with each note produced; when it is excited produces a series of short purrs or chirrups trills. It calls more frequently at dusk or before sunset, sometimes during the day (Hilty and Brown 1986; Holt et al. 2017).

Distribution: It is found from the extreme SWof the United States to the N of Chile, central part of Argentina and Uruguay (Hilty and Brown 1986). In Colombia it is present until 1000 m.a.s.l. mainly in the N Caribbean Coast (subspecies *medianum*) from Colosó (Sucre), Magdalena, and La Jagua de Ibiríco (Cesar) to Dibulla and Maicao in the La Guajira and in East Andes (subspecies *ucayalae*) from Arauca and Vichada to Ipilaes (Nariño) and Leticia (Amazonas). There are isolated records in Remedios and Santa Fe de Antioquia (Antioquia) (Hilty and Brown 1986; Strewe and Navarro 2003; Biomap 2006; McNish 2007; Ardila et al. 2007, http://www.xenocanto.org/86123).

Ecology: Moderately common to locally common in lowlands and mountain foot (especially in drier areas) from the dry forest and semi-open areas of scrub with trees and weeds to edges of humid forest *terra firme* and *várzea* forest, (it is a seasonal floodplain forest inundated by white-water rivers), and also in gallery forests

Fig. 9.11 Ferruginous pygmy owl (*Glaucidium brasilianum*). Municipality of Puerto Concordia, Meta department, 03 January 2017 (Photograph © Wilmer Andrés Ramírez Riaño)

and open areas with isolated trees in the Orinoquia (Hilty and Brown 1986; McNish 2007). It is very active during the day (McNish 2007; König et al. 2008). It uses hollows of trees (3–12 m) or cavities in termites for their reproduction and puts of two to five white eggs (Hilty and Brown 1986). It consumes insects, large invertebrates, and very small vertebrates (Ardila et al. 2007). In Guaviare, Santa María (Boyacá) and Vista Hermosa (Meta, Sierra de la Macarena National Natural Park) were observed and heard on the following day mixed flocks of birds, while in Dibulla (La Guajira) listened to the sunset (J.P. López-Ordoñez pers. obs.).

Conservation Status: This species has a range of distribution and large population size, and despite the fact that the population trend appears to be declining, it is not believed to be approaching the Vulnerable thresholds. For these reasons, the species is evaluated for Least Concern (LC) (BirdLife International 2016).

Conservation Strategies: Present at the Sanctuaries of Fauna and Flora Los Flamencos and Ciénaga Grande de Santa Marta; Protected Forestry Reserve Serranía de Coraza and Montes de María; Natural Reserves Nukak, Mamancana, Playa Güio (Guaviare) (A. Sua pers.comm.); National Natural Parks Sierra de la Macarena, El Tuparro and Amacayacu; Natural Reserves of Civil Society Bojonawi, Hato Corozal, La Ventana, Nimajay Refuge, Wakuinali, Pitalito, and Buena Vista Natural Reserve (Hilty and Brown 1986; Moreno-Bejarano and Álvarez-León 2003; Strewe and Navarro 2003; Biomap 2006; Peñuela et al. 2011; GeoSIB 2012, http://www.xeno-canto.org/18116, http://www.xeno-canto.org/86123).

Glaucidium parkeri **(Robbins and Howell, 1995)**
Common Name: Spanish,, Mochuelo de Parker

It is a species whose distribution was known only on the eastern slope of the Andes of Ecuador, Peru and Bolivia, but no for Colombia (Hennessey et al. 2003; König et al. 2008; Holt et al. 2017). In January 2014, was confirmed its presence for Colombia, in the locality El Valle de Sibundoy, between the municipalities of San Francisco and Mocoa, department of Putumayo to 1800 m.a.s.l. (Acevedo-Charry et al. 2015). It is important to study the distribution of this species in Colombia, especially in foot of mountains of E mountain range, but also its biology and ecology.

Description: Length 14 cm; yellow green beak, yellow eyes, and fingers. It has brown and whitish mottled facial disk. Above brown, crown and sides of the head with gray stained and white dots bordered with blackish. It has “false eyes” black on each side of the nape; primary and secondary feathers dark brown with conspicuous and irregularly shaped white spots; black tail with five white bands. Weight of the bird: males between 60 and 64 g (n = 3) (Robbins and Howell 1995; König et al. 2008; Holt et al. 2017).

Voice: Calls with three or four short phrases (normally 2–4 or 6 en intervals of several seconds) of low tone increasing when ending the two notes, *hu-hu, hu*, and having a hesitation before the last note (König et al. 2008; Holt et al. 2017).

Distribution: It is little known. It is distributed in eastern slope of Andes from Putumayo in Colombia to S of Ecuador (Condor mountain range, Cutucú mountain range) and Perú and N Bolivia between 1450 and 1975 m.a.s.l. (König et al. 2008). But there are records to 2050 m.a.s.l. S in Perú and between 700 and 1300 m.a.s.l. in NE Bolivia and under 1000 m.a.s.l. in Ecuador.

Ecology: It is distributed in mountain forest and cloud forest with epiphytes in andinas slopes. No information exists about its diet but it may feed on big insects, small reptiles, and birds. Also no imformation is known on reproduction but could breed in nest cavities abandoned for woodpeakers (König et al. 2008; Holt et al. 2017).

Conservation Status: This species is not considered Vulnerable with any criterion (distribution, demography, population size). It is considered Least Concern (LC) (BirdLife International 2016).

Conservation Strategies: Only has been recorded in one locality for the country, but it is present in the Important Bird Area Valle de Sibundoy-Putumayo (Acevedo-Charry et al. 2015).

Athene cunicularia **(Molina, 1782)**
Common Name: Spanish, Mochuelo Conejo, Mochuelo Terrero, and Búho Llanero; English, burrowing owl

Description: Length between 19 and 26 cm. Terrestrial owl, diurnal, without “ears,” yellow eyes, with long legs and short tail. Back rufous and densely dotted; superciliary and white forehead; mostly white lower parts with blackish chest band (often interrupted in the center); chest and sides barred and dotted with coffee. Immature: buffy white underneath (Hilty and Brown 1986). Bird weight: one specimen of the Colombian subspecies weigh 155 g; the species average is 150 g (Dunning 2008; König et al. 2008) (Fig. 9.12).

Voice: Vocally is very active, with a varied repertoire. The male vocalizes *cu-cuhooh*, repeated at intervals of several seconds. It varies individually according to the state of excitation of the bird. The female pronounces a similar but softer vocalization. Both sexes have a call for communication *kwekwekwekweeh* when they are alarmed, increasing the volume to make it very strong. A squeak *chreeh-ketketket* is pronounced

Fig. 9.12 Burrowing owl (*Athene cunicularia*). Municipality of Tauramena, Casanare department, 23 November 2012 (Photograph ©Jennifer del Río)

in similar situations. They also emit a *chee-gugugugugug* apparently with contact function. Young owls emit dry rattle sounds especially when there are disturbances at the nest (König et al. 2008).

Distribution: It is distributed from the plains of W North America to the S of Central America mainly to the Pacific coast; also some islands of the Caribbean like Hispaniola and very local in the west of Cuba. Local in the NW of South America and the Andean region. It presents a wide distribution in the east of South America from Pará in Brazil to Patagonia and Tierra del Fuego in Argentina, where it is very rare (König et al. 2008). In Colombia it is distributed up to 1000 m.a.s.l. in the lowlands of the Caribbean region, in the Cesar valley (Casacará) to the E to the La Guajira Peninsula; in arid portions of the Valle Alto del Río Magdalena in Tolima and Huila; in the high valley of the Patía River; in the E part of the Andes in the Llanos to the S to the River Guaviare and the S of Meta (La Macarena) in the Orinoquia region (Hilty and Brown 1986; Salaman et al. 2009; Ayerbe-Quiñones and López-Ordóñez 2011).

Ecology: It is a fairly common local owl, which is expanding its occupancy in the face of the destruction of forest habitats in some regions, because it inhabits dry and open environments with few trees, such as sheets, deserts, and grasslands (Sick 1997). This species is largely diurnal but more active at dusk and sometimes active during the night. Very terrestrial, during the day it is observed perched on the ground or on a rock, mounds of earth, poles of near, or other places of rest of low height (Hilty and Brown 1986; König et al. 2008). It has been considered a kind of opportunistic trophic habit, with a diet mostly composed of arthropods, such as beetles and other insects, spiders, and scorpions, but eventually consumes small vertebrates such as rodents, reptiles, and amphibians, occasionally small birds (Haug et al. 1993; Holt et al. 1999; König et al. 2008).

Conservation Status: Presents a wide distribution area. In spite of the fact that the population trend appears to be declining, it is not believed to be rapid enough to be considered a vulnerable species. Therefore, it is classified as a species of Least Concern (LC) (BirdLife International 2016).

Conservation Strategies: It is present in the Natural Reserve Palmarito, El Tuparro National Natural Park, in the Natural Reserves of the Civil Society, La Casa de la Abuela, Las Unamas, Rey Zamuro, Mata Redonda, Manacáo, El Caduceus, La Reseda, La Esperanza 1 and 2, La Gloria and Agua Verde (Omacha Foundation et al. 2008; Peñuela et al. 2011; GeoSIB 2012).

Aegolius harrisii **(Cassin, 1849)**
Common Name: Spanish, Búho Bicolor and Buhito Frentianteado; English, buff-fronted owl

Description: Small owl (about 20 cm) with large round head without "ears." Yellow green eyes, dark beak with whitish tip, and pink yellow legs. Dark chocolate coffee on the back; head with crown front and cream face ring with narrow black outer edge; spots above the eyes, such as narrow "eyebrows" brown to black, blackish barb; conspicuous cream nuchal band; wings and tail with white spots and cream; and a band on the feathers along the shoulder. Bottom color light cream. Short tail and generally dark brown to black. Immature: similar to the adults of its phase but the crown and nape with clear lists instead of spots or without clear spots (Hilty and Brown 1986; Córdoba-Córdoba and Ahumada 2005; König et al. 2008). Weight of bird: subspecies *A. h. harrisii* has an average weight of 122.6 g (Córdoba-Córdoba and Ahumada 2005).

Voice: Little known. Vocalization is a rather sharp and short-lived vibrating trio of 3–10 s that sometimes trembling *turrrrrr'urrrr'urrrrrrr....* Many times toward the end of a vocalization the volume drops. An alarm call "*stacato*" your *tu-tu tutututututu....* It also emits short corners *tut tut* possibly for contact (Hilty and Brown 1986; König et al. 2008).

Distribution: It is distributed in the Andean region from Venezuela to N Argentina with widely dispersed records throughout the Andes (subspecies *harrisii*) and to the Atlantic coast of S Brazil, Uruguay, N of Argentina, and Paraguay (subspecies *iheringi* and *dabbenei*) and isolated specimens in Cerro Neblina, Tepuyes of Venezuela. In Colombia there are very few records of this species between 1500 and 2900 m.a.s.l. Present in the Tambo, Cerro Munchique (Cauca) and Llorente (Nariño) in the W Cordillera; Ibagué (Tolima) and Bosque de Florencia, Samaná (Caldas) in the central mountain range; Bojacá and Zipacón (Cundinamarca) and the Tamá National Natural Park in Herrán (North Santander) for the Cordillera Oriental (von Sneidern 1954; Fitzpatrick and Willard 1982; Fjeldså and Krabbe 1990; Córdoba-Córdoba and Ahumada 2005; Parra-Hernández et al. 2007; GeoSIB 2012, A. Betancourt pers. comm.).

Ecology: It is a rare and apparently very localized species, although possibly unseen in many areas because it is not obvious. Apparently activity is mainly nocturnal, although has been observed active during the day (Ribas and Santos 2007, S. Córdoba-Córdoba pers. obs.). There is little information on its ecology, and this seems to differ between known subspecies. In the Andes, the types of vegetation that it uses include mountain forests, Andean forests bordering the páramo, where it also uses forest clearings and semi-open areas, even fragmented forests and dry groves (Hilty and Brown 1986; Córdoba-Córdoba and Ahumada 2005; König et al. 2008; Olmedo 2011a). Their known diet are insects and small vertebrates (mainly rodents) from prey remains found in a cavity nest and stomach contents (Parker et al. 1985; Córdoba-Córdoba and Ahumada 2005; Olmedo 2011a, S. Córdoba-Córdoba pers. obs.), although it could try to capture birds or bats (Córdoba-Córdoba and Ahumada 2005, S. Córdoba-Córdoba pers. obs.). It nests in hollows in tree trunks, in natural cavities or in old woodpecter nests at different heights of the ground; nests with three eggs have been observed (König et al. 2008).

Conservation Status: It presents a wide distribution area and therefore does not approach the thresholds to be considered as vulnerable species under the range size criterion. Population size appears to be stable, and although its population size has not been quantified, its population is not thought to be less than 10,000 individuals. Because of this, it is considered a kind of Least Concern (LC) (BirdLife International 2016).

Conservation Strategies: It can be found in the Natural National Parks of Muchique, Florencia Forest and Tama (North Santander) (Negret 1994; Córdoba-Córdoba and Ahumada 2005; GeoSIB 2012).

Asio clamator, Pseudoscops clamator, Rhinoptynx clamator **(Vieillot, 1807)**
Common Name: Spanish, Búho Rayado and Búho Listado; English, striped owl

Taxonomy: Formerly called *Rhinoptynx clamator*. Genetic information indicates that this species is closer to *Asio otus* and in turn to *Asio flammeus*, indicating that it should belong to the genus *Asio* (König et al. 2008).

Description: Length between 30 and 38 cm; with long "ears" (4.5 cm) cafes with blunt rim. Eyes light brown, beak black. Dorso before tawny and striated blackish; remiges and tail barred from dark to blackish brown; whitish facial disk with fuscous ridge; white throat; rest of white underside very distinctly striated from dark to blackish brown. In flight, from below, it shows prominent black patches on the "mufiecas" (part where the wing is bent halfway between the man and the wing tip). Immature: they have a cinnamon face with white, black, and tan border; pale white head slightly lined with black; plumage of the ante body with gray bars (Hilty and Brown 1986; ABO 2000). Weight of the bird: males between 335 and 485 g (average 406 gr); females between 400 and 556 g (average 484 g) (Dunning 2008; König et al. 2008) (Fig. 9.13).

Voice: Series of approximately seven notes equidistant *hu hu hu*... low and in the same tone; also a single nasal *purr* of almost a second, taller and stronger in the middle: *junnNNNnnj*; also 7–8 sharp barks like dog *ju-jou*! *Jou*! *Jou*! ... *o or, or, or* (Hilty and Brown 1986; ABO 2000).

Distribution: From SE Mexico to N Argentina and Uruguay. In Colombia the *clamator* subspecies is distributed to 2600 m.a.s.l. (more frequent below 500 m). Present on the Caribbean coast from E Atlantic to the Santa Marta region (Hilty and Brown 1986; Biomap 2006; Strewe et al. 2009) and a register in the lower Sinu River (Córdoba) (Estela and López-Victoria 2005); the middle and Valle alto del Río Magdalena from San Agustín (Huila) to El Espinal (Tolima) and a registry in La Dorada (Caldas) (Hilty and Brown 1986; Biomap 2006; CorpoCaldas and Asociación Calidris 2010); Cordillera Oriental in Bogotá, Tena, Anapoima (Cundinamarca) (http://www.xeno-canto.org/54441, GeoSIB 2012); Mesa de los Santos and Simacota (Santander), Pamplona-Bucaramanga (Paramo de Santurbán)

Fig. 9.13 Striped owl (*Asio clamator*). Municipality of Chinchina, Caldas department, 11 November 2010 (Photograph © María Teresa Jaramillo)

and Cúcuta (Norte de Santander) (Biomap 2006; Donegan et al. 2010; eBird 2011); E of Andes in Sacamá (Casanare), Villavicencio (Meta) and Florencia (Caquetá), the distribution of these species in Orinoquia is not well known (Biomap 2006; McNish 2007). The central mountain range in Medellin (Antioquia), Palestine, Chinchia and Salamina (Caldas), Dos Quebradas and Pereira (Risaralda), Armenia (Quindío), and E flank in Ibagué and Lérida (Tolima) (Delgado-V et al. 2005; Biomap 2006; Parra-Hernández et al. 2007; CorpoCaldas and Calidris Association 2010; Arbeláez-Cortés et al. 2011; Botero and Jaramillo 2011; Losada-Prado and Molina-Martínez 2011; eBird 2011); Cordillera Occidental in Cali (Valle del Cauca) and Caloto, Santander de Quilichao, Popayan plateau and Valle Alto del Río Patia (Cauca) (Ayerbe-Quiñones et al. 2008; eBird 2011).

Ecology: It is a common species locally of open areas with isolated trees like clear savannas, pastures, and wetlands. It also frequents secondary forest, forest edges, agricultural and upland areas, and sometimes plantations. During the day, it sleeps in low bushes on the ground; hunting on the fly or from exposed perches, swooping on prey including small mammals, large insects, and sometimes birds and lizards (Hilty and Brown 1986; ABO 2000; McNish 2007). In the city of Medellín, the main component (58%) in the diet is represented by introduced rodents such as *Rattus rattus*, *R. norvegicus*, and *Mus musculus*. In addition to the rodents, several insects of the orders Blattaria and Orthoptera were found (Delgado-V et al. 2005). There are some records of nests and eggs in Bogotá city and surrundings between March and July; nests are on the ground or close to the base of a tree or bush (Riaño et al. 2017; S. Chaparro-Herrera pers.obs.). Juveniles are recorded on December to January. In July a nest was found with two white eggs in Córdoba Ecological District Park (https://flic.kr/p/BbA7WU) (S. Chaparro-Herrera pers. obs.). This species lay 2–4 eggs (generally three) in nests on the ground, in low cavities, or over dead leaves in trunks of palms (König et al. 2008). On December there are records of fledgling in the University of Antioquia (Medellín) aboce palms (J. L. Parra pers. obs.).

Conservation Status: This species has a wide range and population size, and its population trend appears to be stable. This species is considered Least Concern (LC) (BirdLife International 2016).

Conservation Strategies: Present in the Ecological District Park of Córdoba, La Conejera, Techo, Torca and Guaymaral and Meandro del Say (S. Chaparro-Herrera pers. obs.). Forest Protector Reserve Producer Laguna de Pedro Palo (http://www.xeno-canto.org/54441). Tayrona National Natural Park (W. Naranjo pers. comm.), Via Parque Isla de Salamanca and Sanctuary of Fauna and Flora Ciénaga Grande de Santa Marta (Moreno-Bejarano and Álvarez-León 2003; Biomap 2006).

Asio stygius **(Wagler, 1832)**
Common Name: Spanish, Búho Orejudo; English, stygian owl

Taxonomy: Possibly two *stygius* and *robustus* subspecies may be found in Colombia (Borrero 1967; König et al. 2008). However, the subspecies *robustus* has traditionally been recognized, but see distribution of subspecies in König et al. (2008), who classify the present form in Colombia with subspecies *stygius*. A taxonomic revision of the group is necessary.

Description: Length between 38 and 46 cm. Robust and dark; prominent "ears" close together, with a pale patch on the forehead. Yellow eyes, gray beak to light greenish yellow, blackish legs. Blackish brown back with a few blotches; facial disk with central part blackish and the peripheral part before dark listed with black; a light gray speckle on the rim of the disk above and a striking triangular spot on the forehead between the "ears." Back blackish somewhat stained with opaque. Immature: general color lighter gray, more or less mottled, barred with listed with opaque. It is distinguished from other large owls by the combination of their dark colors, long "ears," and yellow eyes (Hilty and Brown 1986; ABO 2000). Weight of the bird: 408–675 g, average of 565 g (Echeverry-Galvis et al. 2006; Dunning 2008) (Fig. 9.14).

Voice: A strong, low-pitched *ju* or *ju ju*, repeated at intervals of 6–10 s; the female a *miah* short as a cat's meow (Hilty and Brown 1986; Holt et al. 1999).

Distribution: Resident in the Greater Antilles, from N Mexico to Nicaragua, and in the Andes locally from Venezuela to N Argentina. In Colombia it is present mainly between the 1700 and 3000 m.a.s.l. (occasionally below) in the Cordillera Occidental in El Tambo, Popayán, Totoró (Cauca), and a registry in Cali (Valle del Cauca); Cordillera Central in Angostura (Antioquia) to the E of the Andes in Ipiales (Nariño); Cordillera Oriental from Choachí (Cundinamarca) to Suratá (Santander), San Agustin and near Villavieja (Huila). In the E plains (a register in Villavicencio-Meta) (Borrero 1967; Hilty and Brown 1986; ABO 2000; Verhelst et al. 2001; Losada-Prado et al. 2005; Biomap 2006; Parra-Hernández et al. 2007; Ayerbe-Quiñones et al. 2008; Donegan et al. 2010; Arbeláez-Cortés et al. 2011, http://www.xeno-canto.org/77201, http://www.xeno-canto.org/331805, http://www.xeno-canto.org/25458). Recently there are records in the N of Colombia: Dibulla (La Guajira), Aguaclara (Norte de Santander), and the Magdalena River basin: Garzón (Huila) (J.P. López-Ordoñez pers. obs).

Fig. 9.14 Stygian owl (*Asio stygius*). Municipality of Medellín, Antioquia department, 29 September 2012 (Photograph © Sebastián Vieira)

Ecology: Strictly nocturnal and of few known habits, rare and local in its distribution. It inhabits humid mountain forests, primary and secondary forest, forest edges, open areas, or shrubs with high dense tree patches in rural areas or some well-wooded urban parks (Hilty and Brown 1986; ABO 2000; Verhelst et al. 2001; Losada-Prado et al. 2005; Parra-Hernández et al. 2007). Once observed in thorny scrub (435 m.a.s.l.) in the Valle alto del Río Magdalena. In the Serranía de los Churumbelos seen in oak forest (*Quercus*), 15–20 m high, with many epiphytes in the canopy and 10 m understory dominated by Ericaceae, epiphytes, and shrubs (Salaman et al. 1999). It appears that in the savanna of Bogotá, it consumes both birds (*Zenaida auriculata*, *Porphyrio martinica*, *Coccyzus americanus*, *Sturnella magna*, and *Pyrocephalus rubinus*) and rodents and other small mammals, including bats; in Popayán, ten individuals were captured in the central plaza when they arrived at night to hunt the *Zenaida auriculata* roosting on woodcuts overnight, as well as large beetles (Borrero 1967; Hilty and Brown 1986; ABO 2000, S. Chaparro-Herrera pers.obs.). The nest has not been described in Colombia. In other countries it nests on the ground or takes advantage of old nests of other birds in trees or shrubs (Hilty and Brown 1986). A juvenil found to early July in Medellín (Santa Elena) (I. Mesa pers. comm.).

Conservation Status: This species has a wide range and population size, and although the trend of the population seems to be decreasing, it is not considered as Vulnerable but as Least Concern (LC) (BirdLife International 2016).

Conservation Strategies: Present in the National Parks Munchique, Purace, Serranía de los Churumbelos Auka-Wasi, Serranía de los Yariguíes; Natural Reserve Tambito; District of Integrated Management Serranía de los Yariguíes and Natural Reserve of

Birds Reina Cielo Azul; Protective Forest Reserve Río Blanco and Quebrada Olivares (Salaman et al. 1999; Ayerbe-Quiñones et al. 2008; Donegan et al. 2010); Ecological District Park Córdoba, La Conejera, Burro (P. Camargo pers.comm., S. Chaparro-Herrera pers. obs.).

Asio flammeus **(Pontoppidan, 1763)**
Common Name: Spanish, Búho Campestre and Búho Orejicorto; English, short-eared owl

Description: Length between 33 and 42 cm; mainly diurnal; with very short "ears." Yellow eyes and beak and black claws.Wings and feet moderately long, legs fully feathered. Body almost completely gray brown, striated on the back and underneath brown; showing in flight a clear area at the base of primaries and conspicuous black patches on the "wrist"; facial disk gray brown dark turning black around the eyes (Hilty and Brown 1986; ABO 2000). Weight of the bird: no known weight for this species in Colombia. For other subspecies, the average weight of males is 315 g and for females 378 g (Dunning 2008).

Voice: Generally silent, but during the courtship male issues a series of 13–16 notes in flight *hoo-hoo-hoo-hoo-hoo-hoo* (Holt et al. 1999). It also produces a meow like a cat; a *cri cri cri...* sharp in defense of the nest; more rarely a series of 6–10 *tuut* or a long, rough and buzzing note *dziiiiaaa* or *yyyyyyiiiaa* (Hilty and Brown 1986; ABO 2000).

Distribution: Widely distributed both in the world; populations in North America are partially migratory (winter to Costa Rica), locally in South America from Colombia to Argentina. In Colombia, the subspecies *bogotensis* is found between 500 and 3700 m.a.s.l. (probably much higher), in Popayán, Puracé, Totoró, and Alto Patia valleys (Cauca) (Biomap 2006; Ayerbe-Quiñones et al. 2008); a register on the E flank of Central Ibagué (Tolima) (Parra-Hernández et al. 2007); and Cordillera Oriental with records in Tunja (Boyacá), Bogotá, Funza, Mosquera, Gachancipa and Sopó (Cundinamarca) (Biomap 2006) and in E Andes in La Macarena, Carimagua (Meta) and Orocué (Casanare) (Hilty and Brown 1986; McNish 2007, S. Chaparro-Herrera pers. obs.).

Ecology: Owl of open areas with partially diurnal and crepuscular habits, often seen by its characteristic irregular flight with slightly crooked wings. Solitary or in small groups; lies on the ground and fence posts (Hilty and Brown 1986). It inhabits pastures, savannas, and paramo; originally numerous in the savanna of Bogota and abundant in marshy areas as in cultivated ones; possibly the replacement of native pastures by the "kikuyo" grass has affected it, because it makes it difficult to catch prey. In any case, the subspecies bogotensis seems in danger of local extinction in the savanna of Bogotá (Borrero 1962; Hilty and Brown 1986; ABO 2000). Usually it flies very low, to no more than two to four meters of height; always in form buoyant but erratic. For hunting (often in the afternoon, especially on cloudy days), it descends to the ground in search of rodents and other small animals detected visually or by sound (Borrero 1962; ABO 2000). During most of the day, it rests in dense vegetation or near the ground, rarely in taller trees (ABO 2000). In Orinoquia it also inhabits in the estuary (McNish 2007) and was observed in oil palm plantations in Orocué, Casanare (S. Chaparro-Herrera pers. obs.). It is a monogamous and territorial species (König et al. 2008). It lays three white eggs on the ground among grasslands; two nests found in September in the savanna of Bogotá (one with eggs, another with chicks); the young leave the nest before flying (Borrero 1962; Hilty and Brown 1986).

Conservation Status: This species has a wide distribution area and a stable population size, and although the population may be declining, it is not considered Vulnerable; the species is a Least Concern (LC) (BirdLife International 2016).

Conservation Strategies: Present in the Ariari-Guayabero Integrated Management District, National Natural Park La Macarena in Meta; La Herrera Lagoon Water Reservation (Hilty and Brown 1986; Biomap 2006), National Natural Park Sumapaz (P. Camargo pers. comm.), and Jaboque and Guaymaral Ecological District Park.

Acknowledgments We want to especially thank to Paula Enríquez who has been the manager of this book and to promote the creation of the group that organized this chapter in the Congress of Neotropical Ornithology in Cuzco, Peru in 2011. We also thank the Alexander von Humboldt Institute (IAvH) for the information provided. SCH thanks F. G. Stiles and N. Ocampo-Peñuela for the information provided. A Diego Calderón, Jennifer del Río, María Teresa Jaramillo y Sebastian Viera through pictures. Thanks to SCC and María Ángela Echeverry for their support to attend the Congress in Peru. SCC would like to thank Oscar Humberto Marín-Gómez and Karolina Fierro-Calderón shared and Department of Ecology and Evolutionary Biology at the University of Princeton, NJ for their support for the Congress in Peru. JSR thanks J. A. Salazar curator of the Museum of Natural History at the University of Caldas. We would like to thank deeply to Jack Clinton Eitniear for his help in the English translation of this chapter.

Mottled Owl (*Ciccaba virgata*)

Spectacled Owl (*Pulsatrix perspicillata*)

Appendix 9.1

The Owls of Colombia

Species	Conservation status	Conservation lstrategies	Number of specimens in collections	Natural history	Diet	Knowledge state
Tyto alba	LC	SFF,RFP,RNSC, RNA (12)	163 (60/103)	Yes/Yes	Yes	Medium
Megascops choliba	LC	PNN,SFF,RFP,RN,RB, PNR,RNSC,RNA (41)	335 (161/174)	No/No	Yes	Medium
Megascops clarkii	LC	–	1 (1/0)	No/No	No	Deficient
Megascops colombianus	NT	PNN,SFF,RFP,RN, PNR,RNFI (7)	16 (16/0)	No/No	No	Deficient
Megascops ingens	LC	–	7 (4/3)	No/No	No	Deficient
Megascops petersoni	LC	RN,RNA (3)	2 (1/1)	Yes/No	No	Low

Species	Conservation status	Conservation lstrategies	Number of specimens in collections	Natural history	Diet	Knowledge state
Megascops watsonii	LC	PNN (2)	16 (9/7)	No/No	No	Deficient
Megascops guatemalae	LC	PNN,RNA (2)	7 (6/1)	No/No	No	Deficient
Megascops albogularis	LC	PNN,RFP,RFPP,RNSC, RN,RB,DMI (15)	122 (79/43)	No/No	No	Low
Megascops sp.nov.	–	PNN (1)	0	No/No	No	Deficient
Lophostrix cristata	LC	PNN,RNA (3)	37 (30/7)	No/No	No	Deficient
Pulsatrix perspicillata	LC	PNN (5)	25 (4/21)	No/No	No	Deficient
Pulsatrix melanota	LC	PNN (1)	1 (1/0)	No/No	No	Deficient
Bubo virginianus	LC	–	35 (18/17)	No/No	Yes	Deficient
Ciccaba virgata	LC	PNN,SFF,RFP,RN, RNSC,RNA (14)	141 (104/37)	No/No	No	Low
Ciccaba nigrolineata	LC	PNN,RNA (5)	32 (24/8)	No/No	No	Deficient
Ciccaba huhula	LC	PNN (1)	5 (2/3)	No/No	No	Deficient
Ciccaba albitarsis	LC	PNN,RFP,RNSC, RN (17)	68 (43/25)	No/No	No	Low
Glaucidium nubicola	VU	PNN (1)	2 (1/1)	No/No	No	Deficient
Glaucidium jardinii	LC	PNN,SFF,RFP,RNSC, RN,RB,PRN (19)	56 (46/10)	No/No	No	Low
Glaucidium griseiceps	LC	PNN,RFP,RN,RNA (6)	3 (2/1)	No/No	No	Deficient
Glaucidium brasilianum	LC	PNN,SFF,RFP,RNSC, RN (16)	119 (82/37)	No/No	No	Low
Glaucidium parkeri	LC	–	0	No/No	No	Deficient
Athene cunicularia	LC	PNN,RNSC,RN (12)	81 (43/38)	No/Yes	No	Low
Aegolius harrisii	LC	PNN (3)	12 (8/3)	No/No	No	Deficient
Pseudoscops clamator	LC	PNN,SFF,RFP,PED (8)	19 (5/14)	No/No	Yes	Low

Species	Conservation status	Conservation lstrategies	Number of specimens in collections	Natural history	Diet	Knowledge state
Asio stygius	LC	PNN,RFP,RN,RNA, DMI,PED (10)	46 (18/28)	No/No	Yes	Low
Asio flammeus	LC	PNN,DMI,PED,RH (4)	35 (12/23)	Yes/No	Yes	Low

Conservation Status according to BirdLife International 2016: Vulnerable (VU), Near Threatened (NT), Least Concern (LC). Conservation Strategies: Natural National Park (PNN), Sanctuary of Fauna and Flora (SFF), Forest Reserve (RFP), Natural Reserve (RN), Biological Reserve (RB), Regional Natural Park (PNR), Natural Reserve of Civil Society (RNSC), Natural Reserve of Birds (RNA), Ecological District Park (PED), Integrated Management District (DMI), Water Reservation (RH), and Nature Reserve Forest and Research (RNFI), total of localities in parenthesis Diet: information in Colombia. Knowledge state: High, Medium, Low, Deficient (base on the information, categories and value scores: Conservation Strategies –Number of localities (maximum 4): No presence in protected areas = 0, between 1-10 = 1, between 11-20 = 2, more of 20 = 4 points; Number of specimens – general knowledge of distribution (maximum 4): No specimens = 0 points, less of 40 = 1, between 41-200 = 2, between 201-500 = 3, more than 500 = 4 points; Lifecycle (maximum 4): No/No = 0, Yes/No = 2, No/Yes = 2, Yes/Yes = 4; Diet (Maximum 2): No = 0, Yes = 2; therefore Knowledge is high: proportion greater than 0.85; knowledge medium: between 0.50 and 0.85; knowledge low: between 25 and 50, and knowledge deficient: between 0 and 0.25

References

ABO (2000) Aves de la Sabana de Bogotá: guía de campo. ABO-CAR, Bogotá

Acevedo-Charry O, Cárdenas Á, Coral-Jaramillo B et al (2015) First record of subtropical pygmy owl *Glaucidium parkeri* in the Colombian Andes. Bull BOC 135(1):77–79

Aliabadian M, Alaei-Kakhki N, Mirshamsi O et al (2016) Phylogeny, biogeography, and diversification of barn owls (Aves: Strigiformes). Biol J Linnean Soc. doi:10.1111/bij.12824

Álvarez MA, Umaña M, Mejía GD et al (2003) Aves del Parque Nacional Natural Serranía de Chiribiquete. Amazonía-Provincia de la Guyana. Colombia Biota Colombiana 4(1):49–63

Álvarez-Rebolledo M, Caro V, Laverde O et al (2007) Guía sonora de las aves de los Andes colombianos. Instituto de Recursos Biológicos Alexander von Humboldt y Cornell Laboratory of Ornithology, Bogotá

Arbeláez-Cortés E, Marín-Gómez OH, Duque-Montoya D et al (2011) Birds, Quindío Department, Central Andes of Colombia. Check List 7(3):227–247

Ardila ME, Rangel-Ch JO, Rodríguez JC (2007) Avifauna de la alta montaña del Perijá. In: Rangel-Ch O (ed) Colombia Diversidad Biótica V: La alta montaña de la Serranía de Perijá. Instituto de Ciencias Naturales, Bogotá

Ayerbe-Quiñones F (2006) Avifauna del complejo volcánico Doña Juana-Cascabel: riqueza, endemismos y especies amenazadas. Informe Técnico convenio marco UAESPNN-WWF. Presentado a WWF-CORPODOÑAJUANA, La Cruz, Nariño

Ayerbe-Quiñones F, López-Ordóñez JP (2011) Adiciones a la avifauna del valle alto del río Patía, un área interandina del suroccidente colombiano. Boletín SAO 20(2):1–17

Ayerbe-Quiñones F, López-Ordóñez JP, González-Rojas MF et al (2008) Aves del Departamento del Cauca-Colombia. Biota Colombiana 9:77–132

Banks RC, Cicero C, Dunn JL et al (2003) Forty-fourth supplement to the American ornithologists' union check-list of North American birds. Auk. doi:10.1642/0004-8038(2003)120[0923:FSTTAO]2.0.CO;2
Barros OG, Cintra R (2009) The effects of forest structure on occurrence and abundance of three owl species (Aves: Strigidae) in the Central Amazon forest. Fortschr Zool. doi:10.1590/S1984-46702009000100014
Biomap (2006) Base de datos Darwin: Proyecto BioMap base de datos de distribución de la avifauna Colombiana. http://www.biomap.net . Accessed 01 Apr 2012
eBird (2011) eBird: an online database of bird distribution and abundance. Versión 2, Ithaca. http://www.ebird.org. Accessed 26 Mar 2012
Birdlife International (2016) IUCN red list. http://www.birdlife.org. Accessed 22 Dec 2016
Blake ER (1962) Birds of the Sierra Macarena, eastern Colombia. Chicago Natural History Museum 44(11):69–112
Borges SH, Henriques LM, Carvalhães A (2004) Density and habitat use by owls in two Amazonian forest types. J Field Ornithol. doi: http://dx.doi.org/10.1648/0273-8570-75.2.176
Borrero JI (1962) Notas varias sobre *Asio flammeus bogotensis* en Colombia. Rev Biol Tropical 10(1):45–59
Borrero JI (1967) Notas sobre hábitos alimenticios de *Asio stygius robustus*. El Hornero 10:445–447
Botero JE, Jaramillo MT (2011) *Pseudoscops clamator*. Boletín SAO 20(2):87–88
Bryson VJ, Roland K, Lowman MD et al (2009) Evidence for three-toed sloth (*Bradypus variegatus*) predation by spectacled owl (*Pulsatrix perspicillata*). Edentata Journal. doi:10.1896/020.010.0113
Cadena H, De Vinueza FB, Bonaccorso E (2011) Notas sobre la dieta del búho ventribandeado (*Pulsatrix melanota*) en Ecuador. Ornitol Neotrop 22:471–475
CAR (2009) Forest reserve bosque oriental de Bogotá. Corporación Autónoma Regional y Conservación Internacional, Bogotá
Casas CF, Ayerbe-Quiñones (2006) Caracterización y diversidad de la avifauna en el sector del Diviso-Mandur, Vereda Naranjal, Municipio de Argelia, Cauca. Informe Técnico, proyecto Corredor de Conservación Biológico y Multi-cultural Sector Munchique-Pinche, Cordillera Occidental Colombiana. Presentado a Fundación Proselva, Popayán
Castaño C (1993) Situación general de la conservación de la biodiversidad en la región Amazónica: Evaluación de las áreas protegidas, propuestas y estrategias. Proyecto FAO/TCP/RLA/0160. Quito, Ecuador
Chapman FM (1917) The distribution of bird-life in Colombia: a contribution to a biological survey of South American. Bull of the Mus of Nat History 36:1–729
Córdoba-Córdoba S, Ahumada JA (2005) Confirmation of buff-fronted owl *Aegolius harrisii* for the Cordillera Oriental of Colombia. Bull BOC 125(1):56–58
Córdoba-Córdoba S, Álvarez-Rebolledo M (2003) Guía sonora de las aves del departamento de Norte de Santander-Colombia. Cuchilla, Toledo y PNN Tama. Serie CARS, Banco de Sonidos Animales. Instituto de Investigación de Recursos Biológicos Alexander von Humboldt, Villa de Leyva, Boyacá
Córdoba-Córdoba S, Echeverry-Galvis MA (2006) Diversidad de aves de los bosques mixtos y de roble del Santuario de Flora y Fauna de Iguaque, Boyacá. In: Solano C, Vargas N (eds) Memorias del I Simposio Internacional de Robles y Ecosistemas Asociados. Fundación Natura-Pontificia Universidad Javeriana, Bogotá
Córdoba-Córdoba S, Echeverry-Galvis MA, Umaña AM et al (2007) Caracterización de la biodiversidad de cuatro áreas del Parque Nacional Natural Chingaza, Cundinamarca, Colombia. Programa de Inventarios de Biodiversidad. Grupo GEMA, Proyecto Andes. Instituto de Investigación de Recursos Biológicos Alexander von Humboldt, Bogotá, Colombia
CorpoCaldas, Asociación Calidris (2010) Estado de conocimiento de las aves en el departamento de Caldas: prioridades de conservación y vacíos de información. Corporación Autónoma Regional de Caldas. Manizales, Colombia
Cuervo AM, Cadena CD, Parra JL (2006) Seguir colectando aves en Colombia es imprescindible: un llamado a fortalecer las colecciones ornitológicas. Ornitol Colombiana 4:51–58

Cuervo AM, Pulgarín P, Calderón D (2008) New distributional bird data from the cordillera central of the Colombian Andes, with implications for the biogeography of northwestern South America. Condor. doi:10.1525/cond.2008.8555
Delgado-V CA (2007) Dieta del Currucutú *Megascops choliba* (Strigidae) en la ciudad de Medellín, Colombia. Boletín SAO 17(2):114–117
Delgado-V CA, Calderón-F D (2007) La dieta de la lechuza común *Tyto alba* (Tytonidae) en una localidad urbana de Urabá, Colombia. Boletín SAO 16(2):94–97
Delgado-V CA, Cataño-B EJF (2004) Diet of the barn owl (*Tyto alba*) in the lowlands of Antioquia, Colombia. Ornitol Neotrop 15:413–415
Delgado-V CA, Ramírez JD (2009) Presas de la lechuza común (*Tyto alba*) en Jardín, Antioquía, Colombia. Ornitol Colombiana 8:88–93
Delgado-V CA, Pulgarín-R PC, Calderón-F D (2005) Análisis de egagrópilas del Búho Rayado (*Asio clamator*) en la ciudad de Medellín. Ornitol Colombiana 3:100–103
Dickinson EC (ed) (2003) The Howard and Moore complete checklist of the birds of the world, Revised and enlarged, 3rd edn. Christopher Helm, London
Donegan TM, Dávalos LM (1999) Ornithological observations from Reserva Natural Tambito, Cauca, south-west Colombia. Cotinga 12:48–55
Donegan TM, Avendaño JE, Briceño ER et al (2007) Range extensions, taxonomic and ecological notes from Serranía de los Yariguíes, Colombia's new national park. Bull BOC 127(3):172–213
Donegan TM, Avendaño JE, Briceño-L ER et al (2010) Aves de la Serranía de los Yariguíes y tierras bajas circundantes, Santander, Colombia. Cotinga 32:72–89
Dunning JB (2008) CRC handbook of avian body masses, 2nd edn. CRC Press, Boca Raton
Echeverry MA, Córdoba S, Castellanos J (2008) In: Devenish, C, Franco AM. Directorio Nacional de AICAS de Colombia. http://aicas.humboldt.org.co. Edición cartográfica: Unidad de SIG - IAvH - Instituto Humboldt-Cartografía y análisis SIG: Directorio Nacional de AICAS de Colombia
Echeverry-Galvis MA, Córdoba-Córdoba S (2007) New distributional and other bird records from Tatamá Massif, West Andes, Colombia. Bull BOC 127(3):213–224
Echeverry-Galvis MÁ, Córdoba-Córdoba S, Peraza CA et al (2006) Body weights of 98 species of Andean cloud-forest birds. Bull BOC 126:291–298
Enríquez PL (1995) Abundancia relativa, uso de hábitat y conocimiento popular de los Strigiformes en un bosque húmedo tropical en Costa Rica. Tesis de Maestría. Programa Regional en Manejo de Vida Silvestre para Mesoamérica y el Caribe. Universidad Nacional de Costa Rica, Heredia
Enríquez PL, Rangel-Salazar L (2001) Owl occurrence and calling behavior in a tropical rain forest. J Raptor Res 35(2):107–114
Enríquez PL, Rangel-Salazar L, Marshall JT (1997) First nest record of the bare-shanked sreech-owl (*Otus clarkii*). J Raptor Res 31(3):276–279
Enríquez PL, Johnson DH, Rangel-Salazar JL (2006) Taxonomy, distribution and conservation of owls in the neotropics: a review. In: Rodríguez-Estrella R (ed) Current raptor studies in México. Centro de Investigaciones Biológicas del Noroeste y CONABIO, La Paz
Estela FA, López-Victoria M (2005) Aves de la parte baja del río Sinú, Caribe colombiano; inventario y ampliaciones de distribución. Bol Inv Marinas Costeras 34:7–42
Estela FA, Arbeláez D, Fajardo D et al (2004) Caracterización ornitológica del Páramo del Duende y su zona de amortiguación. Asociación para el Estudio y Conservación de las Aves Acuáticas de. Colombia -Calidris, Cali
Fierro-Calderón K, Córdoba-Córdoba S (2014) *Glaucidium nubicola*. In: Renjifo LM, Gómez MF, Velásquez-Tibatá J et al (eds) Libro rojo de aves de Colombia, Volumen I: bosques húmedos de los Andes y la costa Pacífica. Editorial Pontificia Universidad Javeriana e Instituto Alexander von Humboldt, Bogotá
Fierro-Calderón K, Montealegre C (2010) Nuevo registro del buhito nubícola (*Glaucidium nubicola*) en la cordillera Occidental de Colombia. Boletín SAO 20(1):29–33
Fitzpatrick JW, O'Neill JP (1986) *Otus petersoni*, a new species of screech-owl from the eastern Andes, with systematic notes on *O. colombianus* and *O. ingens*. Wilson Bull 98:1–14

Fitzpatrick JW, Willard DE (1982) Twenty-one bird species new or little known from the Republic of Colombia. Bull BOC 102:153–158

Fjeldså J, Krabbe N (1990) Birds of the high Andes. Zool. Mus., Univ. of Copenhagen & Apollo Books, Svendborg

Franco AM, Bravo G (2005) Áreas importantes para la conservación de las aves en Colombia. In: BirdLife International Y Conservación International. Áreas importantes para la conservación de las aves en los Andes Tropicales: sitios prioritarios para la conservación de la biodiversidad. BirdLife International y Conservación International (Serie de Conservación de BirdLife No. 14), Quito, pp 117–132

Freeman BG, Julio C (2010) The nest and egg of cinnamon screech owl *Megascops petersoni* in central Colombia. Cotinga 32:107

Freeman BG, Hilty SL, Calderón-F E et al (2011) New and noteworthy bird records from central and northern Colombia. Cotinga 34:12–25

Freile JF, Chaves JA (1999) Colombian screech-owl *Otus colombianus*. Cotinga 12:95–96

Freile JF, Chaves JA, Iturralde G et al (2003) Notes on the distribution, habitat and conservation of the cloud-forest pygmy-owl (*Glaucidium nubicola*) in Ecuador. Ornitol Neotrop 14:275–278

Fuentes L, Poleo J, Gonzales Y (2009) Lechuza de campanario, principal depredador de roedores en el cultivo de arroz en el estado de Guárico. INIA HOY 4:97–101

Fundación Omacha, Fundación Palmarito, WWF Colombia (2008) Caracterizaciones biológicas de la Reserva Natural Palmarito, Casanare, Colombia

GeoSIB (2012) Strigidae. http://hermes.humboldt.org.co/visoruniversal2010/bin/Visor.html. Accessed 25 May 2012

Gerhardt RP, Bonilla N (1991) Nest descriptions, food habits, and home range size of black and white owls (*Ciccaba nigrolineata*). In: Whitacre DF, Burnham WA, Jenny JP (eds) Progress report IV: Maya project: use of raptors and other fauna as environmental indicators for design and management of protected areas and for building local capacity for conservation in Latin America. The Peregrine Fund, Boise, pp 149–157

Gerhardt RP, Gerhardt DM, Flatten CJ et al (1994a) The food habits of the sympatric *Ciccaba* owls in northern Guatemala. J Field Ornithol 65:258–264

Gerhardt RP, Gerhardt DM, Flatten CJ et al (1994b) Breeding biology and home range of two *Ciccaba* owls. Wilson Bull 106:629–639

Gómez de Silva G (1997) Diet of the spectacled owl (*Pulsatrix perspicillata*) during the rainy season in northern Oaxaca, México. J Raptor Res 31(4):385–387

Greeney HF, Nunnery T (2006) Notes on the breeding of north-west Ecuadorian birds. Bull BOC 126:38–45

Haug EA, Millsap BA, Martell MS (1993) Burrowing owl. Birds North Am 61:1–20

Hekstra GP (1973) Scops and screech owls. In: Burton JA (ed) Owls of the world. EP Dutton & Co, Inc., New York, pp 94–115

Hennessey AB, Herzog SK, Sagot E (2003) Lista anotada de las aves de Bolivia, 5th edn. Asociación Armonía/BirdLife International, Santa Cruz de la Sierra

Hilty SL (1977) *Chlorospingus flavovirens* rediscovered, with notes on other Pacific Colombian and Cauca Valley birds. Auk 94:44–49

Hilty SL, Brown WL (1986) A guide to the birds of Colombia. Princeton University Press, Princeton

Holt DW, Berkley R, Deppe C et al (1999) Barn-owls to hummingbirds. In: del Hoyo J, Elliott A, Sargatal J et al (eds) Handbook of the birds of the world, vol 5. Lynx Edicions, Barcelona

Holt DW, Berkley R, Deppe C et al (2017) Subtropical pygmy-owl (*Glaucidium parkeri*). In: del Hoyo J, Elliott A, Sargatal J et al (eds) Handbook of the birds of the world alive. Lynx Edicions, Barcelona. http://www.hbw.com/node/55071. Accessed 19 Jan 2017

Howell SN, Robbins MB (1995) Species limits of the least pygmy-owl (*Glaucidium minutissimum*) complex. Wilson Bull 107:7–25

Hume R (1991) Owls of the world. Illustrated by T. Boyer. Running Press Book Publishers, Philadelphia

Hurtado A (1992) Amazonia colombiana: una puesta en lugar. In: Andrade G, Hurtado A, Torres R (eds) Amazonia colombiana, diversidad y conflicto. Centro de Estudios Ganaderos y Agrícolas, Bogotá, pp 13–24

Ibáñez C, Ramo C, Busto B (1992) Notes on food habits of the black and white owl. Condor. doi:10.2307/1369226

IGAC (1996) Diccionario Geográfico de Colombia, Tomo I-IV. Instituto Geográfico Agustín Codazzi, Bogotá

Infonatura: Animals and Ecosystems of Latin America (2007) Version 5.0. Arlington. http://www.natureserve.org/infonatura. Accessed 14 May 2012

Instituto de Ciencias Naturales (2004) Facultad de Ciencias, Universidad Nacional de Colombia. http://www.biovirtual.unal.edu.co. Accessed 10 May 2012

IUCN (2016) IUCN red list of threatened species. Versión 2016-2. www.iucnredlist.org. Accessed 26 Dec 2016

Kattan G, Morales G, Padu F (2001) Patrones de distribución geográfica de la diversidad de aves en los Andes Colombianos. Boletín SAO 12(22–23):5–11

König C, Weick F, Becking J-H (2008) Owls of the world, 2nd edn. Christopher Helm, Londres

Krabbe N, Flórez P, Suárez G et al (2006) The birds of Paramo de Frontino, western Andes of Colombia. Ornitol Colombiana 4:39–50

Losada-Prado S, Molina-Martínez G (2011) Avifauna del bosque seco tropical en el departamento del Tolima (Colombia): Análisis de la comunidad. Caldasia 33(1):271–294

Losada-Prado S, Carvajal-Lozano AM, Molina-Martínez YG (2005) Listado de especies de aves de la cuenca del río Coello (Tolima, Colombia). Biota Colombiana 6(1):101–116

Marín-Gómez OH, Ospina JC (2008) Monitoreo de la avifauna del AICA Barbas-Bremen. Informe Técnico. Fundación Ornitológica del Quindío y Corporación Autónoma del Quindío, Armenia

Mcnish T (2007) Las aves de los llanos de la Orinoquia. M&b Producciones Y Servicio, Riqueza Natural

Meyer de Schauensee R (1945) Notes on Colombian birds. J Acad Nat Sci Phila 97:1–16

Miller AH (1963) Seasonal activity and ecology of the avifauna of an American equatorial cloud forest. UC Publ Zool 66:1–78

Moreno N, Camargo P (2008) Estado actual de la avifauna en la Reserva Biológica Encenillo (RBE) Guasca-Cundinamarca. Tesis, Universidad INCCA de Colombia, Bogotá

Moreno-Bejarano LM, Álvarez-León R (2003) Fauna asociada a los manglares y otros humedales en el Delta-Estuario del río Magdalena, Colombia. Revista Acad Colomb Ci Exact 27(105):517–534

Moreno-Palacios M, Rodríguez-Ortíz E (2008) Nuevo registro del mochuelo cabecigris (*Glaucidium griseiceps*) en el valle medio del Río Magdalena, Colombia. Ornitol Colombiana 6:92–95

Motta-Junior JC (2002) Diet of breeding tropical screech-owl (*Otus choliba*) in Southeastern Brazil. J Raptor Res 36(4):332–334

Naranjo LG (1994) Inventario y estudio de la comunidad aviaria del Parque Regional Ucumarí. Corporación Autónoma Regional de Risaralda, Pereira

Negret AJ (1994) Lista de aves registradas en el Parque Nacional Natural Munchique, Cauca. Novedades Colombianas: Nueva Época 6:69–83

Negret AJ (2001) Aves en Colombia amenazadas de extinción. Serie de Estudios de la Naturaleza. Universidad del Cauca, Editorial Universidad del Cauca, Colombia

Olmedo I (2011a) *Aegolius harrisii*. In: Freile JF, Bonaccorso E (eds) Aves de Ecuador, Ver. 3.0. Museo de Zoología, Pontificia Universidad Católica del Ecuador, Quito. http://zoologia.puce.edu.ec/vertebrados/anfibios/FichaEspecie.aspx?Id=1116. Accessed 19 Dec 2011

Olmedo I (2011b) *Glaucidium nubicola*. In: Freile JF, Bonaccorso E (eds) Aves de Ecuador, Ver. 3.0. Museo de Zoología, Pontificia Universidad Católica del Ecuador, Quito. http://zoologia.puce.edu.ec/vertebrados/anfibios/FichaEspecie.aspx?Id=1310. Accessed 19 Dec 2011

Palacios H L, Velasco JM, Rivero R et al (2005) Plan de manejo 2005–2009 del Sanctuary of Fauna and Flora Guanenta Alto Río Fonce, Boyacá, Colombia

Parker TA, Schulenberg TS, Graves GR et al (1985) The avifauna of the Huancabamba region, northern Peru. In: Buckley PA, Foster MS, Morton ES, Ridgely RS et al (eds) Neotropical ornithology. Ornithology Monographs. doi:10.2307/40168282
Parra-Hernández RM, Carantón-Ayala DA, Sanabria-Mejía JS et al (2007) Aves del municipio de Ibagué - Tolima, Colombia. Biota Colombiana 8(2):199–220
Peñuela L, Castro F, Ocampo-Peñuela N (2011) Reservas Naturales del nodo Orinoquia en su rol de Conservación de la Biodiversidad. Fundación Horizonte Verde y Resnatur, Meta, Colombia
Pulgarín P, Múnera PW (2006) New bird records from Farallones del Citará, Colombian Western Cordillera. Boletín SAO 16(1):44–53
Rangel-Ch JO (ed) (2004) Colombia. Diversidad Biótica IV. El Chocó Biogeográfico/Costa Pacífica. Instituto de Ciencias Naturales, Universidad Nacional de Colombia, Bogotá
Rangel-Ch JO, Aguilar-P M, Sánchez-C H et al (1995a) Región costa Pacífica.In: Rangel-Ch JO (ed) Colombia, Diversidad Biótica I. Instituto de Ciencias Naturales. Convenio Inderena-Universidad Nacional de Colombia, Bogota pp 121–139
Rangel-Ch JO, Aguilar-P M, Sánchez-C H et al (1995b) Región de la Amazonia. In: Rangel-Ch JO (ed) Colombia, Diversidad Biótica I. Instituto de Ciencias Naturales. Convenio Inderena-Universidad Nacional de Colombia, Bogota pp 82–103
Rangel-Ch JO, Lowy-C P, Aguilar-P M (1995c) Marco general y alcances del estudio. In: Rangel-Ch JO (ed) Colombia, Diversidad Biótica I. Instituto de Ciencias Naturales. Convenio Inderena-Universidad Nacional de Colombia, pp 17–25
Rangel-Ch JO, Lowy-C P, Sánchez-C H (1995d) Región Caribe. In: Rangel-Ch JO (ed) Colombia, Diversidad Biótica I. Instituto de Ciencias Naturales. Convenio Inderena-Universidad Nacional de Colombia, pp 217–232
Rangel-Ch JO, Sánchez-C H, Lowy-C P et al (1995e) Región de la Orinoquia. In: Rangel-Ch JO (ed) Colombia, Diversidad Biótica I. Instituto de Ciencias Naturales. Convenio Inderena-Universidad Nacional de Colombia, pp 239–254
Remsen J V, Cadena CD, Jaramillo A et al (2012) A classification of the bird species of South America. American Ornithologists' Union http://www.museum.lsu.edu/~Remsen/SACCBaseline.html. Accessed 29 May 2012
Remsen J V, Cadena CD, Jaramillo A, et al. (2016) A classification of the bird species of South America. American Ornithologists' Union. http://www.museum.lsu.edu/~Remsen/SACCBaseline.html. Accessed 29 Jan 2016
Renjifo LM, Franco-Maya AM, Amaya-Espinel JD (eds) (2002) Libro rojo de aves de Colombia. Serie Libros Rojos de Especies Amenazadas de Colombia. Instituto de Investigación de Recursos Biológicos Alexander von Humboldt y Ministerio del Medio Ambiente, Bogotá
Restall R, Rodner C, Lentino M (2006) Birds of Northern South America: an identification guide, volume 1: species accounts. Yale University Press, New Haven
Riaño J, Paqui MF, Córdoba-Córdoba S et al. (2017) Nest and chicks of *Pseudoscops clamator* (Aves: Strigidae) in the highland plateau of the Sabana de Bogotá, Colombia. Acta biol. Colomb. doi: http://dx.doi.org/10.15446/abc.v22n1.54380
Ribas CF, Santos REF (2007) Novo registro documentado do cabure-acanalado *Aegolius harrisii* (Cassin, 1849) para o estado do Paraná. Actual Ornitol 140(4):4–5
Ridgely RS, Greenfield PJ (2001) The birds of Ecuador: status, distribution, and taxonomy. Cornell University Press, New York
Robbins MB, Howell SN (1995) A new species of pygmy-owl (Strigidae: *Glaucidium*) from the eastern Andes. Wilson Bull. doi:10.2307/4089365
Robbins MB, Stiles FG (1999) A new species of pygmy-owl (Stringidae: *Glaucidium*) from the Pacific slope of the Northern Andes. Auk. doi:10.2307/40168282
Salaman PGW, Donegan TM, Cuervo A (1999) Ornithological surveys in Serranía de los Churumbelos, southern Colombia. Cotinga 12:29–39
Salaman PGW, Donegan TM, Cuervo AM (2002a) New distributional bird records from Serranía de San Lucas and adjacent Central Cordillera of Colombia. Bull BOC 122:285–304

Salaman PGW, Stiles FG, Bohórquez CI et al (2002b) New and noteworthy bird records from the east slope of the Andes of Colombia. Caldasia 24(1):157–189

Salaman P, Donegan T, Davison D et al (2007) Birds of Serranía Churumbelos, their conservation and elevational distribution. Conservación Colombiana 3:4–93

Salaman P, Donegan T, Caro D (2009) Listado de las aves de Colombia. Conserv Colombiana 8:1–89

Schulenberg TS, Stotz DF, Lane DF et al (2007) Birds of Peru. Princeton University Press, Princeton

Sclater PL, Salvin O (1879) On the birds collected by the late Mr. T.K. Salmon in the State of Antioquia, United States of Colombia. Proc Zool Soc Lond. doi:10.1111/j.1096-3642.1879.tb02684.x

Sick H (1997) Ornitologia Brasileira. Ed. Nova Fronteira SA, Rio de Janeiro

Stiles FG (2010) La avifauna de la parte media del río Apaporis, departamentos de Vaupés y Amazonas, Colombia. Revista Acad Colomb Ci Exact 34(132):381–396

Stiles FG, Roselli L (1998) Inventario de las aves del bosque altoandino: Comparación de dos métodos. Caldasia 20:29–43

Stiles FG, Skutch AF (1995) Guía de aves de Costa Rica. Translated into Spanish by: Loretta Rosselli. Illustrated by Dana Gardner. Instituto Nacional de Biodiversidad (INBio), Heredia

Stiles FG, López-Lanus B, Renjifo L M (2002) *Glaucidium nubícola*. In: Renjifo L M., Franco-Maya AM, Amaya-Espinel JD et al (eds) Libro rojo de aves de Colombia. Serie Libros Rojos de Especies Amenazadas de Colombia. Instituto de Investigación de Recursos Biológicos Alexander von Humboldt y Ministerio del Medio Ambiente, Bogotá, pp 238–240

Stotz DF, Fitzpatrick JW, Parker TA et al (1996) Neotropical birds: ecology and conservation. University of Chicago Press, Chicago

Strewe R, Navarro C (2003) New distributional records and conservation importance of the San Salvador Valley, Sierra Nevada de Santa Marta, northern Colombia. Ornitología Colombiana 1:29–41

Strewe R, Villa-de León C, Álzate J et al (2009) Las aves del campus de la Universidad del Magdalena, Santa Marta, Colombia. Revista Intrópica. doi:http://dx.doi.org/10.21676/23897864.141

Sua A, Chaparro S (2010) Composición y estructura gremial del ensamblaje de aves de la vereda La Unión en el Municipio de Quipile, Departamento de Cundinamarca, Colombia. Tesis, Universidad Pedagógica Nacional, Bogotá

Terborgh J, Robinson SK, Parker TA et al (1990) Structure and organization of an Amazonian forest bird community. Ecol Monogr. doi: 10.2307/1943045

Todd WE, Carriker MA (1922) The birds of the Santa Marta region of Colombia: a study in altitudinal distribution. Ann Carnegie Museum. doi:http://dx.doi.org/10.5962/bhl.title.56257

Traylor MA (1952) A new race of *Otus ingens* (Salvin) from Colombia. Nat Hist Misc Chicago Acad Sci 99:1–4

Verhelst JC, Rodríguez JC, Orrego O et al (2001) Aves del Municipio de Manizales- Caldas, Colombia. Biota Colombiana 2(3):265–284

Villarreal H (2006) Ecosistemas terrestres naturales. In: Chávez ME y Santamaría M (eds) Informe sobre el avance en el conocimiento y la información de la biodiversidad 1998–2004, Tomo II. Instituto de Investigación de Recursos Biológicos Alexander von Humboldt y Ministerio del Medio Ambiente. Bogotá, p 1998–2004

Von Sneidern K (1954) Notas sobre algunas aves del Museo de Historia Natural de la Universidad del Cauca, Popayán, Colombia. Nov Colombianas 1:3–13

Chapter 10
The Owls of Ecuador

Juan F. Freile, Esteban A. Guevara, Cecilia Pacheco, and Tatiana Santander

Abstract Species richness of owls (Strigiformes) in Ecuador is remarkably high (28 species) despite its small territory (c. 280,000 km^2). This diversity is not evenly distributed across mainland Ecuador, with higher species richness in tropical areas, humid forests, and pristine habitats. Only two species occur in the Galapagos Islands. Although there are no species endemic to Ecuador, the distribution ranges of at least five are primarily confined to Ecuador. The knowledge about Ecuadorian Strigiformes is insufficient. The natural history and distribution of the two Galapagos endemic taxa (*Tyto alba punctatissima* and *Asio flammeus galapagoensis*) have been fairly accurately documented. In contrast, little has been published on continental owl species. Currently, only two species are considered as globally threatened or near threatened and five as threatened at a national level. Nonetheless, poor knowledge about populations, habitat use, natural history, distribution, and vulnerability to extinction might have precluded accurate assessments of the conservation status of several species. We suggest further investigating the basic ecology, distribution, populations, and relationships with human of Ecuadorian owls, in order to better understand their current conservation status.

Keywords Endangered species • Owl distribution • Natural history • Strigiformes

J.F. Freile (✉)
Comité Ecuatoriano de Registros Ornitológicos,
Pasaje El Moro E4–216 y Norberto Salazar, Tumbaco, Ecuador
e-mail: jfreileo@yahoo.com

E.A. Guevara • T. Santander
Aves y Conservación, Joaquín Tinajero E3-05 y Jorge Drom, Quito, Ecuador

C. Pacheco
Secretaría de Ambiente DMQ, Río Coca E6-85 y Genovesa, Quito, Ecuador

P.L. Enríquez (ed.), *Neotropical Owls*, DOI 10.1007/978-3-319-57108-9_10

Buff-fronted Owl (*Aegolius harrisii*)

10.1 Introduction

Ecuador is one of the most diverse countries on Earth (Mittermeier et al. 1997). It has the fourth richest bird fauna in the Neotropical region, with more than 1670 species (Freile et al. 2015c; Remsen et al. 2016). Diversity of nocturnal birds is equally noteworthy. Nonetheless, knowledge about Ecuadorian birds is still insufficient (Freile et al. 2006, 2014). Information about the basic ecology of most nocturnal birds is scanty, resulting in underappreciating their vulnerability to extinction in a scenario of escalating land use changes. In this manuscript, we assess the diversity, distribution, conservation, and current knowledge of nocturnal birds of prey (Strigiformes) occurring in Ecuador, aiming to understand their current status and delineate research priorities.

10.2 Taxonomic Diversity

Ecuador has 28 species of owls according to the most recent taxonomic revision of South American birds (Freile et al. 2015c; Remsen et al. 2016; Table 10.1). *Tyto alba* is the only representative of the family Tytonidae; however, the taxonomy of this widespread species needs a thorough revision (König et al. 2008) that might result in the validation as full species of the Galapagos Islands endemic taxon (*T. a. punctatissima*).

The remaining 27 species belong to the family Strigidae and are classified in 10 genera (Table 10.1); *Megascops* and *Glaucidium* are the most diverse genera, with eight and six species, respectively. The taxonomy of *Megascops* is not well resolved, so the total number of Strigidae species in Ecuador might increase once the taxonomic status of at least three subspecies becomes clear (König et al. 2008; Dantas et al. 2015; N. Krabbe, pers. comm. 2015). Two of these subspecies occur in humid forests: *M. guatemalae centralis* (northwestern lowlands and foothills) and *M. g. napensis* (eastern foothills and Andean slopes). The third subspecies, *M. roboratus pacificus*, of southwestern dry forests, possibly differs from the nominal form (*M. r. roboratus*), which occurs in the Marañon river valley of northeast Peru and extreme southeast Ecuador (König et al. 2008).

The taxonomy of the *M. guatemalae* superspecies (including *M. vermiculatus*) needs a comprehensive revision. Molecular, vocal, morphological, and zoogeographic evidences suggest that several taxa currently ranked as subspecies actually represent valid species (König et al. 2008). Conservatively, Remsen et al. (2016) treat all South American forms of this complex as subspecies of *M. guatemalae*. Two of them, *M. g. roraimae* (including *M. g. napensis*) and *M. g. centralis*, have been given species status by some authors (Hardy et al. 1999; Ridgely and Greenfield 2001). Yet, König et al. (2008) separate *M. g. napensis* from *M. g. roraimae*. In a more recent paper, Dantas et al. (2015) support the separation of the Central American *M. guatemalae* from *M. vermiculatus* of Central and South America and support a sister relationship between *M. v. napensis* and *M. v. roraimae*. Nonetheless, these authors do not upgrade the latter two taxa to full species status. They did not sample *M. v. centralis* but conclude that *M. vermiculatus* from eastern Panama is sister to *M. v. napensis* and *M. v. roraimae*. A more detailed systematic assessment, with a broader geographic sampling, is needed to resolve this intricate taxonomic case. Another poorly resolved case is that of *M. ingens* and *M. colombianus*. Vocal, plumage, and morphology differences are subtle and less marked than between populations of *M. ingens ingens* from Ecuador and Peru, suggesting that *M. colombianus* might not merit species status (Freile and Castro 2013; N. Krabbe, pers. comm. 2016).

While the taxonomy of other owl genera occurring in Ecuador is more resolved than that of the genus *Megascops* (Remsen et al. 2016), a few noteworthy examples need attention. Two *Glaucidium* species (*G. parkeri* and *G. nubicola*) were only recently described (Robbins and Howell 1995; Robbins and Stiles 1999), whereas the subspecific identity of *G. griseiceps* in Ecuador is still unclear. Further, the Andean populations of *G. peruanum* might also represent a separate taxon (Ridgely and Greenfield 2001). Another interesting case is that of *C. virgata*, which is divided into two species by König et al. (2008): *C. virgata* in the Amazon and *C. squamulata* in the Pacific lowlands, a split not followed by subsequent authors (Mikkola 2013; Remsen et al. 2016). Finally, another mystery is the status of the "San Isidro" owl (*Ciccaba* sp.). This "form" was discovered in 1999–2000 at a very limited geographic area in the upper Napo province, at 2000 m.a.s.l. (M. Lysinger, unpublished

Table 10.1 Strigiformes of Ecuador, with common names in English, Spanish (Ridgely and Greenfield 2001, 2006), and vernacular names given in several regions within Ecuador (Valarezo-Delgado 1984), with modifications adapted to local knowledge. Taxonomy follows the South American Classification Committee (SACC; Remsen et al. 2016)

Scientific name	English name	Spanish name	Vernacular name
Tyto alba	Barn owl	Lechuza campanaria	Chusig, Lechuza
Megascops choliba	Tropical screech owl	Autillo tropical	Búho
Megascops roboratus	Peruvian screech owl	Autillo roborado	Búho
Megascops colombianus	Colombian screech owl	Autillo colombiano	Búho
Megascops ingens	Rufescent screech owl	Autillo rojizo	Búho
Megascops petersoni	Cinnamon screech owl	Autillo canelo	Búho
Megascops watsonii	Tawny-bellied screech owl	Autillo ventrileonado	Búho
Megascops guatemalae	Vermiculated screech owl	Autillo vermiculado	Búho
Megascops albogularis	White-throated screech owl	Autillo goliblanco	Búho
Lophostrix cristata	Crested owl	Búho penachudo	Búho copetudo
Pulsatrix perspicillata	Spectacled owl	Búho de anteojos	Búho de anteojos
Pulsatrix melanota	Band-bellied owl	Búho ventribandeado	Búho
Bubo virginianus	Great horned owl	Búho coronado americano	Cuscungo
Ciccaba virgata	Mottled owl	Búho moteado	Búho
Ciccaba nigrolineata	Black-and-white owl	Búho blanquinegro	Búho
Ciccaba huhula	Black-banded owl	Búho negribandeado	Búho
Ciccaba albitarsis	Rufous-banded owl	Búho rufibandeado	Búho
Glaucidium nubicola	Cloud-forest pygmy owl	Mochuelo nuboselvático	Buhito
Glaucidium jardinii	Andean pygmy owl	Mochuelo andino	Cuscunguito
Glaucidium parkeri	Subtropical pygmy owl	Mochuelo subtropical	Buhito
Glaucidium griseiceps	Central American pygmy owl	Mochuelo cabecigrís	Buhito
Glaucidium brasilianum	Ferruginous pygmy owl	Mochuelo ferruginoso	Buhito
Glaucidium peruanum	Peruvian pygmy owl	Mochuelo del Pacífico	Buhito
Athene cunicularia	Burrowing owl	Búho terrestre	Pigpiga
Aegolius harrisii	Buff-fronted owl	Buhito frentianteado	Buhito
Pseudoscops clamator	Striped owl	Búho listado	Búho
Asio stygius	Stygian owl	Búho estigio	Búho grande
Asio flammeus	Short-eared owl	Búho orejicorto	Búho grande

data, 2011). Its vocalizations closely resemble those of *C. nigrolineata* and *C. huhula*, whereas its plumage is rather intermediate between *C. nigrolineata* and *C. huhula* (Freile et al. 2015a). It is isolated from *C. nigrolineata* by the Andes, as the latter occurs in the Pacific lowlands and foothills. On the other hand, *C. huhula* ranges mostly below 1000 m.a.s.l. in the Andean-Amazonian foothills (i.e., 1000 m lower in elevation). It has been suggested that it represents a valid taxon or simply a local variation of the rather variable *C. huhula* (Freile et al. 2015a; N. Krabbe, pers. comm. 2011), but further information is needed. A comprehensive study is currently underway.

Even though Ecuador does not have a country endemic species of owl, at least five species are regional endemics (Stattersfield et al. 1998; Ridgely and Greenfield 2001): *M. roboratus*, *M. colombianus*, *M. petersoni*, *G. nubicola*, and *G. parkeri*. The global distribution ranges of these species are mainly confined to Ecuador's current political boundaries, barely reaching adjacent areas of southern and eastern Colombia and northern Peru.

Species richness of owls is not even across mainland Ecuador. The Andes (above 2000 m.a.s.l.) is the richest region, followed by the eastern Andean slopes (1000–2000 m.a.s.l.) and the Amazonian lowlands (below 1000 m.a.s.l.) (Figs. 10.1 and 10.2; Table 10.2). However, the Pacific lowlands harbor more species than the Amazon lowlands altogether. The rather homogenous forests of the Amazon lowlands are home to nine species mostly distributed throughout the entire region. On the contrary, the western lowlands are home to 11 species, some of them ranging only in the wet northwest and others only in the dry southwest. Owl diversity is higher in forested ecosystems than in open areas and in humid forests than in dry forests (Table 10.2). Restricted-range or semi-endemic species (see above) are confined to humid forests, excepting *M. roboratus* of deciduous and semi-deciduous forests. Likewise, these regional endemics tend to prefer forests and woodland rather than open or degraded habitats (Ridgely and Greenfield 2001).

10.3 Distribution

Ecuador can be broadly divided in the following biogeographic regions (Sierra 1999; Ridgely and Greenfield 2001; Freile and Santander 2005): (1) Chocó, (2) Manabí-Tumbes, (3) Northwestern Andes, (4) Eastern Andes, (5) Southern Andes, (6) Amazonian ridges, (7) Amazon, and (8) Galapagos (Fig. 10.3).

The Chocó region is located in the northwestern lowlands and lower Andean slopes, extending southward along the Andean foothills to the southern provinces of El Oro and Loja. Six species occur in this region, including two species that, in Ecuador, are confined to the Chocó (*C. nigrolineata* and *G. griseiceps*). Further, three species are shared with the Amazon lowlands (*Lophostrix cristata*, *Pulsatrix perspicillata*, and *C. virgata*), and one species is shared with the eastern Andean foothills and Amazonian ridges: *M. guatemalae* (but see previous comments of taxonomy; König et al. 2008; Dantas et al. 2015).

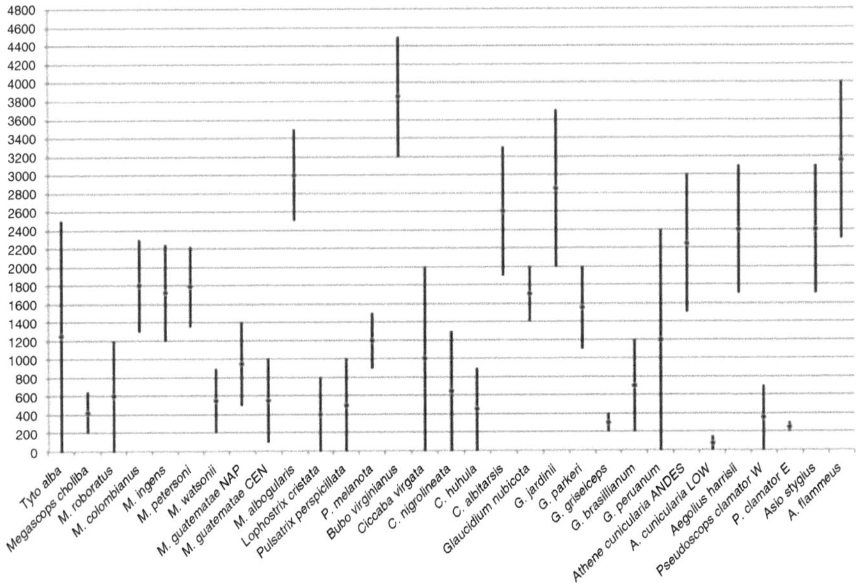

Fig. 10.1 Elevational range of Ecuadorian Strigiformes. *Athene cunicularia* has two figures: one for the subspecies *pichinchae* (ANDES) and one for the lowlands' subspecies (*punensis* in the W) and (possibly *carrikeri* in the E) (LOW); *Pseudoscops clamator* has two figures: one for the population in the western lowlands (W) and one for the eastern lowlands (E); *Megascops guatemalae* has two figures: one (NAP) for the subspecies *roraimae* (*napensis*) of the east Andean foothills and one (CEN) for the subspecies *centralis* of the Chocó lowlands

The Manabí-Tumbes region was already remarked for its high avian endemism by Chapman (1926). Five owl species occur in this region, including two semi-endemics (*M. roboratus* and *G. peruanum*) and one species nearly confined to the region in Ecuador (*Pseudoscops clamator*); one species is shared with dry Andean valleys (*Athene cunicularia*), which was also recently found in scattered river islands in Rio Napo, Amazonian Ecuador (Mischler and Hicks 2015), and one species is distributed throughout most of Ecuador (*T. alba*). The distribution of *M. roboratus* and *G. peruanum* comprises the Manabí-Tumbes region and the dry Marañon valley of northern Peru. In Ecuador, the Marañon region extends through the small Rio Mayo valley, in the southernmost part of Zamora Chinchipe province. Both regions share endemic species, often represented by different subspecies, indicating a common biogeographic history (Cracraft 1985; Stattersfield et al. 1998). The distribution of *P. clamator* (nominate subspecies) is puzzling, as it has recently been discovered in river islands along major Amazonian rivers, as well as in the southeastern foothills at the Bombuscaro region of Podocarpus National Park (C. Vits and A. Solano-Ugalde, pers. comm. 2009).

The Andes region as a whole has more species of Strigiformes, with 19 species. However, the northwestern, eastern, and southern Andes of Ecuador have different biogeographical affinities that result in differing patterns of diversity and endemism

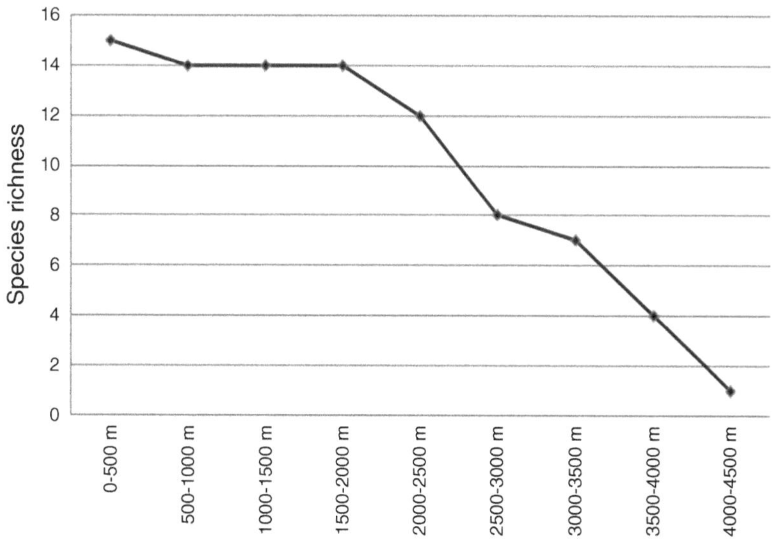

Fig. 10.2 Elevational distribution patterns of Ecuadorian Strigiformes. The *Y-axis* shows species richness by elevational band, and the *X-axis* shows elevation bands. Some species occur in more than one elevation band, so numbers are not cumulative

(Fjeldså 1992). The Northwestern Andes has two regional endemics (*M. colombianus* and *G. nubicola*), while four other species are confined, in Ecuador, to the Eastern Andes (*P. melanota*, *M. ingens*, *M. petersoni*, and *G. parkeri*). The Amazonian ridges of Cutucú and Cóndor are separated from the Andes by the low valleys of the Santiago and Zamora rivers, respectively, and have a distinct geological history from the Andes. They do not have endemic owls, but endemism in other avian families is noteworthy (Robbins et al. 1987; Schulenberg and Awbrey 1997; Stattersfield et al. 2008). The subspecies *M. guatemalae napensis* is nearly confined to these cordilleras in Ecuador, but it occurs locally along the Andean foothills.

Four species are restricted in Ecuador to the Amazon lowlands (*M. watsonii*, *M. choliba*, *G. brasilianum*, and *C. huhula*); all of them have a widespread distribution throughout the Amazon basin (König et al. 2008). The remaining five Amazonian species also occur in the western lowlands and the Andean slopes (e.g., *T. alba*, *L. cristata*, *P. perspicillata*, *C. virgata*, *P. clamator*; Table 10.2).

The Galapagos Islands, being isolated geographically for thousands of years, have a remarkable endemic avifauna (Wiedenfeld 2006). Even though the two taxa occurring in Galapagos are currently considered as subspecies of two owls of global distribution (*T. alba* and *A. flammeus*), a more thorough taxonomic assessment will likely result in upgrading them to species status (König et al. 2008). The Galapagos owls share a fairly long history of geographical isolation, resulting in significant morphological and behavioral traits that substantiate treating them as valid species (De Groot 1983; Mikkola 2013).

Table 10.2 Owl diversity by biogeographic regions and major habitat types in Ecuador. Biogeographic zones broadly correspond to Chapman (1926), Sierra (1999), Ridgely and Greenfield (2001), and Freile and Santander (2005) (see Fig. 10.3)

	Tropical NW	Tropical SW	Subtropical W	Andes W	Andes E	Subtropical E	Tropical E	Galapagos	Humid forest	Dry forest	Open areas	Paramo
Tyto alba		1		1	1		1	1			1	
Megascops choliba						1	1		1			
Megascops roboratus		1				1				1		
Megascops colombianus			1						1			
Megascops ingens						1			1			
Megascops petersoni						1			1			
Megascops watsonii							1		1			
Megascops guatemalae	1					1			1			
Megascops albogularis				1	1				1			
Lophostrix cristata	1						1		1			
Pulsatrix perspicillata	1		1				1		1	1		
Pulsatrix melanota						1			1			
Bubo virginianus				1	1						1	1
Ciccaba virgata	1		1				1		1			

Ciccaba nigrolineata	1		1						1			
Ciccaba huhula						1	1		1			
Ciccaba albitarsis				1	1				1			
Glaucidium nubicola			1						1			
Glaucidium jardinii				1	1				1			
Glaucidium parkeri						1			1			
Glaucidium griseiceps	1								1			
Glaucidium brasilianum							1		1		1	
Glaucidium peruanum		1	1							1	1	
Athene cunicularia		1		1						1	1	
Aegolius harrisii				1	1				1	1		
Pseudoscops clamator		1					1			1	1	
Asio stygius				1	1				1		1	
Asio flammeus				1	1			1			1	1
Total	6	5	6	9	8	8	9	2	21	6	8	2

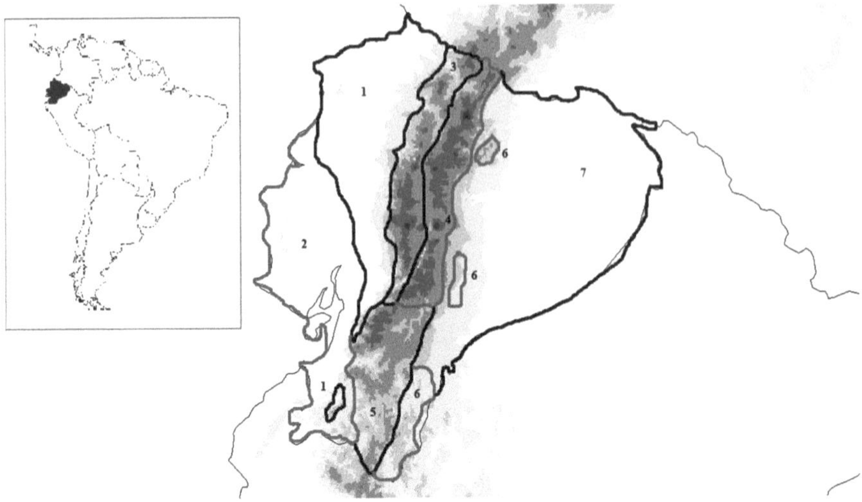

Fig. 10.3 Biogeographic division of Ecuador in eight regions determined following Sierra (1999), Ridgely and Greenfield (2001), and Freile and Santander (2005): (1) Chocó, (2) Manabí-Tumbes, (3) Northwestern Andes, (4) Eastern Andes, (5) Southern Andes, (6) Amazonian ridges, (7) Amazon, (8) Galapagos (not shown)

The distribution of eleven species in Ecuador is poorly known, but they might have fragmented or restricted ranges. The distribution ranges of *G. griseiceps* and *Aegolius harrisii* are limited to small regions in the Chocó lowlands and Andean valleys, respectively. Likewise, species like *M. ingens* and *P. melanota* are known from scattered localities despite their presumed continuous distribution along the entire Andean-Amazonian foothills.

Generally, the distribution of owls in the Ecuadorian Chocó is continuous from the Colombian boundary to southern Esmeraldas and northern Pichincha provinces, but the distribution of some species extends southward along the Andean base to El Oro province, near Peru; some Chocoan species also occur locally in the summit of coastal ridges of Mache Chindul (Carrasco et al. 2008) and Chongón Colonche (Becker et al. 2000). Tumbesian species, on the other hand, extend their distribution northward to central-north Manabí province and locally up along dry Andean valleys (like the Jubones River) in Azuay, El Oro, and Loja provinces. The distribution of owl species like *M. colombianus* and *C. albitarsis* in the Northwestern Andes ends in northern Cotopaxi province, as it occurs in several avian taxa endemic to the Andean portion of the Chocó biogeographic region (Krabbe et al. 1998). Owl species ranges in the Eastern Andes tend to be more continuous, not being interrupted by well-known distribution barriers like the Zamora River (Krabbe 2009).

10.4 Conservation Status

Only two owl species occurring in Ecuador are currently ranked as threatened or near threatened (NT) globally (BirdLife International 2016c), and five species are included in Ecuador's Red Data Book (Granizo et al. 2002).

Glaucidium nubicola is considered as vulnerable at global and national scales following IUCN (2001, 2003) categories and criteria [criterion B1a+b (i,ii,iii,iv,v): extent of occurrence smaller than 20,000 km^2, severely fragmented, less than ten localities, in continuous decline]. Its global population is likely small (less than 10,000 individuals) and declining, and its geographic range is small (less than 7000 km^2). Deforestation in this restricted range is intense and rampant (Granizo et al. 2002; Freile et al. 2003b; BirdLife International 2016a).

Megascops colombianus is considered near threatened at global and national scales under similar considerations as *G. nubicola* [criterion B1a+b (i,ii,iii,v) and also criterion C1: population estimated at less than 10,000 individuals and declining in at least 10% during the last 10 years or three generations]. Even though its populations are likely declining, its total numbers are likely fairly large, and its distribution range covers more than 13,700 km^2 (Granizo et al. 2002; BirdLife International 2016b).

At a national scale, *G. griseiceps* and *A. harrisii* are ranked as vulnerable (VU), both due to habitat loss (Granizo et al. 2002). Poor knowledge of the distribution and natural history of *A. harrisii* precludes an accurate assessment of its conservation status, but its range in Ecuador is likely fragmented, small, and possibly declining [criterion B1+2abcd: extent of occurrence smaller than 20,000 km^2, area of occupancy smaller than 2000 km^2, habitat severely fragmented and in continuous decline] (Granizo 2002). Similarly, *G. griseiceps* seemingly has a small population in Ecuador (fewer than 5000 adult individuals) and occurs in a small geographic range (smaller than 2500 km^2), wherein deforestation is extensive and rampant (Jahn and Mena-Valenzuela 2002a). These authors estimated that the Ecuadorian population of *G. griseiceps* would reduce it at least 10% in the next three generations [criteria A3c, C1: population reduction projected or suspected in area of occupancy, extent of occurrence, and/or habitat quality; population smaller than 10,000 adult individuals and declining].

Lastly, *M. guatemalae centralis* is ranked as near threatened in Ecuador because its population and distribution range are small and declining but not in the extent needed to reach the thresholds for IUCN threatened categories. Jahn and Mena-Valenzuela (2002b) estimated a population of over 12,000 individuals, which is likely declining in 20–40% in the next three generations. This decline rate is high and might result in a vast reduction of this owl and other Chocó endemic birds in the near future. We consider it likely that *M. guatemalae centralis* should be upgraded to a threat category (VU).

10.5 Threats

Possibly, the main threats for owl species in Ecuador are habitat loss and fragmentation resulting from deforestation for timber extraction and the expansion of large-scale agriculture and cattle raising (Granizo et al. 2002), given that most species prefer forested habitats. The Chocó lowlands are facing a dramatic deforestation rate, possibly the worst in the country (Sierra et al. 1999; Lepers et al. 2005; Peralvo et al. 2015). Estimates of the natural vegetation loss in this region range from 65% to 95% (Dodson and Gentry 1991; MAE 2010; Peralvo et al. 2015). The high deforestation rate in the Chocó region is mostly due to massive timber extraction for commercial purposes and land use change from forest to monoculture of oil palm and banana (Freile and Santander 2005). Further, road building has facilitated the expansion of the agricultural frontier, resulting in new human settlements that increase pressure over primary forests.

Similarly, the high Andes has suffered an extensive transformation of landscapes, from natural habitats to pastureland and monoculture crops over the last five centuries (Hidalgo Nistri 1998; Peralvo et al. 2015). Very little forest remains in the Andean slopes that face to the central valley, and timber extraction and fires are still frequent in these remnant forests (Freile and Santander 2005). Additional, open-pit mining is a serious and mounting threat to Andean forests, with number and extent of mining concessions escalating noticeably (López et al. 2003). More than 1326 mining concessions exist in Ecuador; the largest ones are located in the southern provinces of Morona Santiago and Zamora Chinchipe and in the northern provinces of Carchi, Imbabura, Esmeraldas, and Pichincha.

The Cordillera del Cóndor in southeast Ecuador is seriously imperiled by mining exploitation, threatening a number of endemic species (Freile and Santander 2005). Large-scale mining provokes fragmentation and alteration of natural habitats, air and water pollution, and a number of socio-environmental conflicts (Miranda et al. 2003).

Although habitat loss is apparently the main threat for most owl species, other risks exist but are poorly documented and not well understood. A frequently used method to control agricultural pests like insects and rodents is the application of anticoagulant pesticides (Nebel and Wright 1999). These pesticides might provoke poisoning of predators that feed on this kind of prey, including owls (Madden 2002). Some studies show that owls—both common and rare species—largely feed on rodents and insects (König et al. 2008).

The effects of global climate change on owl species populations are not well known but might result in drastic changes of avian communities and biodiversity patterns in general (Peterson et al. 2002; Dawson et al. 2011). Furthermore, illegal hunting and prosecution of some owl species are another threat inadequately documented and likely in need of more research (Enríquez et al. 2006). In several urban and rural areas, nocturnal birds are considered bad omen beings and are linked to supernatural phenomena (Enríquez and Rangel-Salazar 2004). As a consequence, they are prosecuted, chased away from their habitats, and even killed (Enríquez and Rangel-Salazar 2004; Charpentier and Martínez 2007). Owls are also exterminated because they are considered predators of domestic poultry (P. Sánchez, pers. comm. 2010).

10.6 Conservation Actions

There are no specific conservation measures for owls in Ecuador, but the following general actions contribute to their conservation:

10.6.1 Protected Areas

The Ecuadorian government acknowledges the national heritage of natural areas (Patrimonio de Areas Naturales Protegidas, PANE) as its main tool to conserve biodiversity in Ecuador and to promote sustainability in natural areas through ecotourism and preservation of genetic diversity (MAE 2007). The national bylaw of forestry and conservation of natural areas and wildlife, in force since 1981, regulates protection of forests and natural vegetation and controls the national heritage of natural areas. Planning, management, development, protection, and control of the state-run areas of the national heritage of natural areas are responsibility of the Ministry of the Environment (MAE).

The PANE currently comprises 49 protected areas, covering approximately 15% of Ecuador's territory and most natural ecosystems occurring in Ecuador. Forty-seven protected areas are located in continental Ecuador, one is insular (Galapagos National Park) and one exclusively marine (Galapagos Marine Reserve) (Ecolap and MAE 2007; MAE 2015; www.ambiente.gob.ec).

Complete inventories of the avifauna of the PANE areas are nonexistent, but we estimate that most Strigiformes species are protected in at least one conservation area, excepting three species that are not positively recorded inside the PANE (*G. nubicola*, *P. clamator*, and *A. stygius*); further, some species are deficiently represented in the PANE (e.g., *G. griseiceps*, *A. cunicularia*, *M. roboratus*, *A. harrisii*). Modeling the geographic distribution of all species after a comprehensive compilation of museum and literature data (J. F. Freile, unpublished data, 2016) is favoring a more accurate assessment of species ranges' protection provided by the PANE (Freile and Castro 2013).

Another land protection initiative embraces protection forests, either governmental or private. Protection forests are part of the National Forest Heritage, but their actual conservation level varies considerably. Up to 160 protection forests were registered up to 2002 (Ayala 2002), 88 being private (Arias and Tobar 2003). Alongside, there are a good number of privately protected areas, as well as areas protected by local communities. Several private reserves are part of the National Network of Private Conservation Areas (RBPE, in Spanish), an organization with more than 60 members and over 70,000 ha throughout Ecuador (Freile and Santander 2005). Some private organizations, particularly Jocotoco Foundation, Mindo CloudForest Foundation, EcoMinga Foundation, and others, manage several reserves across the country, especially in areas poorly covered by the PANE. Overall, more than 210,000 ha were privately protected in Ecuador by 2003, a figure that needs updating. Moreover, land owned by indigenous and Afro-Ecuadorian organizations covers roughly 7.5 million ha (Añazco et al. 2010). Other protected land not yet embraced by the PANE includes two very large areas of continuous and mostly pristine forests, home of indigenous nationalities: the Intangible Zones of Cuyabeno-Imuya (603,380 ha) and Yasuni (982,000 ha).

The RBPE has established important guidelines for managing private and communal protected areas, but implementing these guidelines is constrained by a weak legal framework for these protected areas. National laws protect communal territories, but not all communities are conservation oriented. There is a protracted discussion in Ecuador to expand the PANE to four subsystems, aiming to strengthen the engagement of civil society and local governments in conservation and sustainability (MAE 2007). These subsystems include national areas protected (the PANE itself), plus areas protected by local governments; land protected by indigenous, local, and Afro-Ecuadorian communities; and private conservation areas.

The PANE would include national areas managed by the National Environmental Authority, whereas the remaining three subsystems would include areas of local interest technically and legally supported by the National Environmental Authority but managed by local governments, communities, or private landowners. Areas under these subsystems might be part of the PANE in particular cases, after a thorough management study, but management and administration would remain in local governments, communities, and landowners. Currently existing protection forests, on the other hand, need a careful assessment of actual conservation status and of their importance for nature conservation.

Ecuador has 108 Important Bird Areas (IBAs; Freile and Santander 2005), which were identified on the basis of the ornithological knowledge. Even though IBAs are not a land protection category, they represent key areas where conservation measures might be implemented. Some ecological corridors between protected areas are also identified in Ecuador; in these corridors, conservation is promoted by cultural landscape management (Freile and Santander 2005).

The Ecuadorian government supports the implementation of national conservation strategies or action plans for threatened species, as tools for prioritizing scientific research, technical procedures, and administrative decisions (Freile and Rodas 2008). Strategies and action plans are developed in concerted processes based on actual knowledge about the species and effective conservation actions, identifying threats, stakeholders, and potential solutions. Several national strategies and action plans have been developed in Ecuador, with the active contribution of the wildlife department of the MAE. These initiatives include the National Conservation Strategies for the Andean condor (*Vultur gryphus*) and the harpy eagle (*Harpia harpyja*) and the Conservation Action Plans for the waved albatross (*Phoebastria irrorata*) and the black-breasted puffleg (*Eriocnemis nigrivestis*; Jahn and Santander 2008). Regrettably, most documents are archived in the MAE but are not formally published nor fully implemented. There are no similar initiatives for threatened owl species, but their research and conservation interest have increased notably in recent years.

10.6.2 Red Data Book

Red Data Books are tools for prioritizing conservation actions on those species considered extinction prone in the short term. Ecuador's latest version of the Red Data Book (Granizo et al. 2002) includes 161 threatened species: 16 critically endangered (CR), 47 endangered (EN), and 98 vulnerable (VU). Two owls are ranked as VU (*G. griseiceps* and *A. harrisii*), whereas *M. colombianus* and *M. "centralis"* are classified as near threatened (NT) in Ecuador. Another species also ranked as VU in Ecuador was assessed by Freile et al. (2003b) but not included in Granizo et al. (2002) because it was not evaluated. Species included in the Red Data Book are legally protected according to the most recent environmental laws released by the Ministry of the Environment Ecuador (Texto Unificado, Libro IV de la Biodiversidad, Título II, Art. 61).

10.7 Current Knowledge

Knowledge about the general ecology and even the distribution of owls in Ecuador is deficient (Freile et al. 2012, 2015b). Only 36 published studies about the 28 owls of Ecuador existed up to 2004 (i.e., a ratio of 1.28 publication per species) according to an analysis of the state of knowledge about the birds of Ecuador (Freile et al. 2006). Of these 36 publications, 22 belong to Strigidae (9 to the genus *Megascops*, 5 to *Glaucidium*) and 3 to Tytonidae (Freile et al. 2005). This analysis was limited to field information generated only in Ecuador or to publications that included Ecuadorian species and excluded technical reports not formally published.

Although some avian orders (e.g., Tinamiformes, Caprimulgiformes, Galbuliformes) have less published information, knowledge about Ecuadorian owls is also insufficient. A more detailed analysis of the 36 publications on Strigiformes revealed that 13 are general publications (e.g., field guides or genus/family/order taxonomic revisions; Hekstra 1982; König 1991; Olson 1995; Hardy et al. 1999; König et al. 2008). Meanwhile, nine studies are focused on the two taxa endemic to Galapagos (nine about *T. alba punctatissima* only and seven about both *T. a. punctatissima* and *Asio flammeus galapagoensis*, e.g., De Groot 1983). Three publications deal with *Bubo virginianus* (e.g., De Vries 1981) and *M. roboratus* (e.g., Williams and Tobias 1996), two with *M. colombianus* (e.g., Salvadori et al. 2004) and *G. nubicola* (e.g., Freile et al. 2003b), and one for the following four species: *M. choliba*, *C. albitarsis*, *Athene cunicularia*, and *Asio flammeus*. Most publications present basic, anecdotic information on the natural history and distribution of these species.

There is a recent upsurge of publications about owls in Ecuador, particularly distribution surveys (Freile and Castro 2013) and diet descriptions (e.g., Cadena-Ortiz et al. 2011, 2016), and several studies are currently being undertaken on the feeding ecology, population density, habitat use, and the effects of forest fragmentation on owl species (e.g., Freile et al. 2016; Pozo-Zamora et al. 2017). Recent research on owls has modified figures presented by Freile et al. (2006, 2015b). For instance, we presently have new field data on species that previously lacked published information (e.g., Cadena-Ortiz et al. 2011) or had only scarce information (e.g., Cadena-Ortiz et al. 2016). Nonetheless, the general pattern discussed by Freile et al. (2006, 2012) still remains. Most information published in recent years results from anecdotal observations or from limited samples—temporally and spatially—and we are still documenting the general aspects of natural history and distribution but not yet assessing patterns and processes.

There is no published information, generated in Ecuador, for 13 owl species (Table 10.3). The breeding biology of seven species is completely unknown (Greeney et al. 2009), including basic descriptions of nests. Nests of only four species

Table 10.3 Current knowledge about Strigiformes occurring in Ecuador

Species	Breeding	Diet	Roost	Terr.	Pop.	Out	Nest
Tyto alba	1	2	2	1	2	3	Y
Megascops choliba	0	0	1	0	1	2	Y
Megascops roboratus	1	1	0	0	1	1	Y
Megascops colombianus	1	1	0	1	2	1	Y
Megascops ingens	0	0	0	0	1	0	N
Megascops petersoni	0	1	0	0	1	0	N
Megascops watsonii	0	1	0	0	1	1	N
Megascops guatemalae	0	0	0	0	1	1	Y
Megascops albogularis	1	1	0	0	1	1	N
Lophostrix cristata	0	0	1	0	1	1	Y
Pulsatrix perspicillata	0	1	1	0	1	2	Y
Pulsatrix melanota	0	1	1	0	1	1	N
Bubo virginianus	0	1	1	0	1	3	Y
Ciccaba virgata	0	0	0	0	1	2	Y
Ciccaba nigrolineata	0	0	1	0	1	2	Y
Ciccaba huhula	0	0	1	0	1	2	Y
Ciccaba albitarsis	0	1	1	0	1	1	N
Glaucidium nubicola	1	0	1	1	2	1	Y
Glaucidium jardinii	1	0	1	0	1	1	Y
Glaucidium parkeri	0	0	0	0	1	1	N
Glaucidium griseiceps	0	0	1	0	1	2	Y
Glaucidium brasilianum	0	1	1	0	1	2	Y
Glaucidium peruanum	1	1	1	0	1	1	Y
Athene cunicularia	1	1	1	1	1	3	Y
Aegolius harrisii	0	0	0	0	1	1	Y
Pseudoscops clamator	0	1	0	0	1	2	Y
Asio stygius	0	0	1	0	1	2	Y
Asio flammeus	2	2	2	2	1	3	Y

Data from Holt et al. (1999), Freile et al. (2005, 2006), and a revision of recent literature of studies carried out in Ecuador. *terr.* territories, *Pop.* populations, *Out* studies carried out in other countries, *Nest* nests reported or studied in the entire global distribution of the species
0 unknown, *1* anecdotal reports, *2* studied, *3* several studies

have been documented in Ecuador (e.g., Freile et al. 2003a; Salvadori et al. 2004). Diet in Ecuador has been documented for *Tyto alba* (De Groot 1983; Moreno 2010; Moreno and Román 2013; Brito et al. 2015): *Athene cunicularia* (Cadena-Ortiz et al. 2016), *Pulsatrix melanota* (Cadena-Ortiz et al. 2011), and *Asio flammeus* (De Groot 1983; Pozo-Zamora et al. 2017). Further, Cadena-Ortiz et al. (2013) reported stomach contents of museum specimens of *Pulsatrix perspicillata*, *Glaucidium brasilianum*, *G. peruanum*, and *Pseudoscops clamator*. Only the information of both Galapagos taxa was compiled during an entire year (De Groot 1983), but this study is the only published data to date—there is a current update of information for *T. alba* (Wagner et al. 2016).

Some natural history information generated in Ecuador is not yet formally published. For example, Charpentier and Martínez (2007) studied the diet, roosts, and threats of *T. alba* inside Cuenca city, Azuay province. They document a diet comprised mostly of rodents but also including beetles and birds. They found a relative abundance of 0.95 individuals/km^2. This study demonstrates that owl research in urban environments is feasible and important from an educational perspective. Likewise, Varela (2012) undertook a year-round census of populations and habitat use of the owl community at an Andean dry forest. He found six species, which were not evenly distributed in the study area. Two species (*T. alba* and *A. cunicularia*) occupied more habitats and were more abundant than the remaining species, which were confined to more mature habitats within the study site (*C. albitarsis* and *A. harrisii*) or to specific areas with tall trees (*A. stygius*). Information compiled by Varela (2012) will be published in due course.

The distribution of at least 11 species is not well documented. One of the least known species in terms of distribution is *A. harrisii*, currently known from less than 12 localities (Rodas et al. 2005; Freile et al. unpublished data, 2016), and *G. griseiceps*, which is known for less than 10 sites in a very limited area. Besides, the potential seasonal or altitudinal movements of species are unknown. Even though all species are presumed to be sedentary, *B. virginianus*, *A. stygius*, and *P. clamator* apparently tend to establish territories for a given timespan and vacate them, moving to a new territory.

An accurate assessment of the species' conservation status is likely hampered by deficient knowledge about their ecology and distribution, as is reflected in the fact that only two species are presently considered as threatened or near threatened at global scale (BirdLife International 2016c) and five at national level (Granizo et al. 2002; Freile et al. 2003b; see "Conservation Status"). *Glaucidium nubicola* was not included in the national Red Data Book but later was suggested as vulnerable in the country (Freile et al. 2003b). The conservation status of most Strigiformes in Ecuador is probably less favorable than currently considered, particularly in species like *A. harrisii*, *M. ingens*, *M. petersoni*, *P. melanota*, *G. parkeri*, and *A. stygius*. From a conservation perspective, studying and documenting owls' ecology, distribution, and vulnerability to extinction are imperative.

Present trends in owl research in Ecuador are promising. For example, in the V Reunion Ecuatoriana de Ornitología (V Ecuadorian Ornithology Meeting), held in August 2016, 5 out of 42 presentations were about owls, whereas in the IV REO

(August 2014), no presentations were related to owls, and in the III REO (August 2012), there were just two. Future—or currently undergoing—owl studies in Ecuador include dietary descriptions, censuses in several ecosystems, breeding biology, assisted breeding, human perceptions, distribution assessments, and cultural importance.

10.8 Conclusions

Diversity of Strigiformes in Ecuador is barely lower than that of Peru (30 species) and slightly higher than in Colombia (27 species) (Remsen et al. 2016); notably, these countries are much larger than Ecuador. This situation makes Ecuador a top priority country for research and conservation of owls. The distribution of at least four species is mostly confined to Ecuador (*Megascops roboratus*, *M. colombianus*, *M. petersoni*, and *Glaucidium nubicola*). There are at least two studies published about two of these semi-endemic species (*M. roboratus* and *G. nubicola*), while information about *M. colombianus* and *M. petersoni* is scarcer (Table 10.3). More field studies in general ecology and distribution of owls in Ecuador are needed in order to accurately assess their vulnerability to extinction.

Knowledge about the 28 Strigiformes of Ecuador is very limited, even about widespread and common species that tolerate high levels of anthropogenic disturbance and which are well studied elsewhere (e.g., *Bubo virginianus*, *Athene cunicularia*). Likewise, the taxonomic status of some species/subspecies in Ecuador is still unresolved (e.g., *Asio flammeus galapagoensis*, *Megascops guatemalae centralis* and *M. g. napensis*, *Glaucidium griseiceps*).

Current conservation status of most Strigiformes is apparently stable, as estimated by BirdLife International (2016c) and Granizo et al. (2002). However, it seems plausible that poor knowledge is hindering a more accurate assessment and that species are more vulnerable to extinction than presently feared. Seven species occur in open, degraded habitats, while eight forest species tolerate moderate habitat degradation and occur in forest borders and adjacent clearings (Table 10.2). The remaining 13 species are forest dependent and might be highly susceptible to severe alterations in their habitats.

It is essential to generate and publish information on the owls of Ecuador so as to understand their current conservation status and implement education and awareness campaigns aiming to prevent intentional killing driven by unproven beliefs (Enríquez and Rangel-Salazar 2004). Further knowledge about owls and any educational activities implemented will ultimately contribute to biodiversity conservation.

Striped Owl (*Asio clamator*)

Stygian Owl (*Asio stygius*)

Acknowledgements Thanks to Paula Enríquez for inviting us to prepare this chapter and for her help with bibliography; to Ana Charpentier, Pablo Sánchez, Mitch Lysinger, Alejandro Solano, Catherine Vits, Héctor Cadena, Paolo Piedrahita, and Niels Krabbe for sharing unpublished data; and to Tjitte de Vries, Diego Cisneros-Heredia, Alejandro Solano-Ugalde, José Luis Rangel-Salazar, and two anonymous referees for their comments on this manuscript. J. F. Freile thanks Fundación Numashir, Percy Sladen Memorial Fund (Linnean Society of London), Gobierno Provincial de Pichincha, Heimo Mikkola, and Neotropical Bird Club for supporting owl research in Ecuador; to John Blake, Jordan Karubian, and David Johnson (Global Owl Project) for their contributions to owl research; and to Paola Moscoso, Santiago Varela, Diego Castro, Scott Walter, Beto González, Mike Darkes, Luke Browne, and Julio Loor for fieldwork. T. Santander and E. Guevara thank the support and facilities provided by Aves & Conservación to prepare this manuscript.

References

Añazco M et al (2010) Sector forestal ecuatoriano: propuestas para una gestión forestal sostenible. Serie Investigación y Sistematización 8, Programa Regional Probona, Intercooperation, Quito

Arias V, Tobar M (2003) VII. Ecuador In: Swift B et al (eds) Conservación privada en Latinoamérica: herramientas legales y modelos para el éxito, Environmental Law Institute, Pronatura

Ayala M (2002) Conservación en manos privadas. Ecuador Terra Incognita 16:20–23

Becker CD et al (2000) Interesting bird records from the Colonche hills, western Ecuador. Cotinga 13:55–58

Birdlife International (2016a) Species factsheet: *Glaucidium nubicola*. http://www.birdlife.org. Accessed 15 Sept 2016

Birdlife International (2016b) Species factsheet: *Megascops colombianus*. http://www.birdlife.org. Accessed 15 Sept 2016

Birdlife International (2016c) IUCN Red list for birds. http://www.birdlife.net/datazone. Accessed 18 Jun 2016

Brito J et al (2015) Mamíferos pequeños en la dieta de la lechuza *Tyto alba* (Strigiformes: Tytonidae) en dos localidades del occidente de Ecuador, con ampliación distribucional de *Ichthyomys hydrobates* (Rodentia: Cricetidae). Pap Avulsos Zool (São Paulo) 55(19):261–268

Cadena-Ortiz H, Bahamonde-Vinueza D, Bonaccorso E (2011) Notas sobre la dieta del búho ventribandeado (*Pulsatrix melanota*) en Ecuador. Ornitol Neotrop 22(3):471–475

Cadena-Ortiz H, Freile JF, Bahamonde-Vinueza D (2013) Información sobre la dieta de algunos búhos (Strigidae) del Ecuador. Ornitol Neotrop 24(4):469–474

Cadena-Ortiz H et al (2016) Diet of the burrowing owl *Athene cunicularia* in two locations of the inter-Andean valley Ecuador. Rev Bras Ornitol 24(2):122–128

Carrasco L, Cook A, Karubian J (2008) Extensión del rango de distribución de ocho especies de aves en las montañas de Mache-Chindul, Ecuador. Cotinga 29:72–76

Chapman FM (1926) The distribution of bird-life in Ecuador. Bull Amer Mus Nat Hist 55:1–784

Charpentier AL, Martínez JD (2007) Abundancia y dieta de *Tyto alba*, la lechuza del campanario, en la ciudad de Cuenca. Dissertation, Universidad del Azuay

Cracraft J (1985) Historical biogeography and patterns of differentiation within the South American avifauna: areas of endemism. In: Buckley PA, et al (eds) Neotropical ornithology. Ornithological monographs 36, American Ornithologists Union, Washington, pp 49–84

Dantas SD et al (2015) Molecular systematics of the new world screech-owls (*Megascops*: Aves, Strigidae): biogeography and taxonomic implications. Mol Phylogenet Evol 94:626–634

Dawson TP et al (2011) Beyond predictions: biodiversity conservation in a changing climate. Science 32:53–58

De Groot RS (1983) Origin, status and ecology of the owls in Galapagos (Ecuador). Ardea 71:167–182

De Vries T (1981) Presas y periodo de reproducción del Cuscungo (*Bubo virginianus*) en el páramo del Cotopaxi, Ecuador. In: Anon (ed) Memorias de las V Jornadas Ecuatorianas de Biología. Sociedad Ecuatoriana de Biología, Quito, pp 34–35

Dodson CH, Gentry AH (1991) Biological extinction in western Ecuador. Ann Mo Bot Gard 78:273–295

ECOLAP, MAE (2007) Guía del Patrimonio de Áreas Naturales Protegidas del Ecuador. ECOFUND, FAN, DarwinNet, IGM, Quito

Enríquez PL, Rangel-Salazar JL (2004) Conocimiento popular sobre los búhos en los alrededores de un bosque húmedo tropical protegido en Costa Rica. Etnobiología 4:41–53

Enríquez PL, Johnson DH, Rangel-Salazar JL (2006) Taxonomy, distribution and conservation of owls in the neotropics: a review. In: Rodríguez-Estrella R (ed) Current raptor studies in México. Conabio, Mexico, pp 254–307

Fjeldså J (1992) Biogeographic patterns and evolution of the avifauna of relict high altitude woodlands of the Andes. Streenstrupia 18:9–62

Freile JF, Santander T (2005) Áreas importantes para la conservación de las aves en Ecuador. In: Boyla K, Estrada A (eds) Áreas importantes para la conservación de las aves en los Andes tropicales. Sitios prioritarios para la conservación de la biodiversidad. BirdLife International Conservation Series No. 14, Quito, pp 283–470

Freile JF, Rodas F (2008) Conservación de aves en Ecuador: ¿cómo estamos y qué necesitamos hacer? Cotinga 29:48–55

Freile JF, Castro DF (2013) New records of rare screech owls (*Megascops*) and pygmy owls (*Glaucidium*), with taxonomic notes and a conservation assessment of two globally imperilled species in Ecuador. Cotinga 35:5–10

Freile JF, Bonaccorso E, Santander T (2003a) First nesting report of the West Peruvian Screech-owl (*Otus roboratus*). Ornitol Neotrop 14(1):107–111

Freile JF et al (2003b) Notes on the distribution, habitat and conservation of the Cloud-forest Pygmy-owl (*Glaucidium nubicola*) in Ecuador. Ornitol Neotrop 14(2):275–278

Freile JF et al (2005) Listado bibliográfico de las aves del Ecuador. EcoCiencia, Fundación Numashir, Quito

Freile JF et al (2006) La ornitología en Ecuador: un análisis del estado actual del conocimiento y sugerencias para prioridades de investigación. Ornitol Neotrop 17(2):183–202

Freile JF, Castro DF, Varela S (2012) Estado del conocimiento, distribución y conservación de aves rapaces nocturnas en Ecuador. Ornitol Neotrop 23(Suppl):241–250

Freile JF, Greeney HF, Bonaccorso E (2014) Current neotropical ornithology: research progress 1996–2011. Condor: Ornithol Appl 116:84–96

Freile JF et al (2015a) An attempt to solve the enigmatic taxonomy of the 'San Isidro Owl' (*Ciccaba* aff. *huhula*). In: Abstracts of the X Neotropical Ornithology Congress, Neotropical Ornithological Society, Sociedade Brasileira de Ornitologia, Manaus, 19–24 Jul 2015

Freile JF et al (2015b) Los búhos de Ecuador. In: Enríquez PL (ed) Los Búhos Neotropicales: diversidad y conservación. El Colegio de la Frontera Sur, San Cristóbal de Las Casas, pp 333–353

Freile JF et al (2015c). Lista oficial de las aves del Ecuador, Comité Ecuatoriano de Registros Ornitológicos. http://www.ceroecuador.wordpress.com. Accessed 12 Dec 2016

Freile JF et al (2016) Estudios de diversidad, poblaciones, hábitat y ecología de búhos en dos localidades de Ecuador. In: Abstracts of the IV Neotropical Raptor Conference, II Neotropical Owls Symposium, Neotropical Raptor Network and Grupo de Especialistas en Búhos Neotropicales, La Fortuna, 10–13 Oct 2016

Granizo T (2002) *Aegolius harrisii*. In: Granizo, T et al (eds). Libro rojo de las aves del Ecuador. Simbioe, Conservación Internacional, EcoCiencia, Ministerio del Ambiente, UICN, Quito, pp 248–249

Granizo T et al (eds) (2002) Libro rojo de las aves del Ecuador. Simbioe, Conservación Internacional, EcoCiencia, Ministerio del Ambiente, UICN, Quito

Greeney HF, Solano-Ugalde A, Gelis RA (2009) Proyecto Ecuador Nidícola, Yanayacu Natural History Research Group. http://www.yanayacu.org/Ecuador_nests/paginas_principales/nhema_home.html. Accessed 8 Mar 2010

Hardy JW, Coffey BB, Reynard GB (1999) Voices of the new world owls. ARA Records, Gainesville
Hekstra GP (1982) Description of twenty-four new subspecies of American *Otus* (Aves, Strigidae). Bull Zool Mus Univ Amsterdam 9:49–63
Hidalgo-Nistri F (1998) Los antiguos paisajes forestales del Ecuador: una reconstrucción de sus primitivos ecosistemas. Hombre y Ambiente 46, Editorial Abya-Yala, Quito
Holt DW et al (1999) Strigidae species accounts. In: Del Hoyo J, Elliot A, Sargatal J (eds). Handbook of the birds of the world, vol 5: Barn-owls to Hummingbirds. Lynx Edicions, Barcelona, pp 153–242
International Union for Conservation of Nature (2001) IUCN Red List categories and criteria: version 3.1. IUCN Species Survival Commission, Gland and Cambridge
International Union for Conservation of Nature (2003) Guidelines for application of IUCN Red List criteria at regional levels: version 3.0. IUCN Species Survival Commission, Gland and Cambridge
Jahn O, Mena-Valenzuela P (2002a) *Glaucidium griseiceps*. In: Granizo T et al (eds) Libro rojo de las aves del Ecuador. Simbioe, Conservación Internacional, EcoCiencia, Ministerio del Ambiente, UICN, Quito, pp 246–247
Jahn O, Mena-Valenzuela P (2002b) *Otus centralis*. In: Granizo T et al (eds) Libro rojo de las aves del Ecuador. Simbioe, Conservación Internacional, EcoCiencia, Ministerio del Ambiente,UICN, Quito, p 361
Jahn O, Santander T (2008) Species conservation action plan for the Black-breasted Puffleg *Eriocnemis nigrivestis*. Aves and Conservación, BirdLife International, Quito
König C (1991) Zur taxonomie und ökologie der Sperlingskause (*Glaucidium* spp.) des Andenraumes. Oekol Vögel 13:15–76
König C, Weick F, Becking J-H (2008) Owls of the world, 2nd edn. Christopher Helm, London
Krabbe N (2009) Arid valleys as dispersal barriers to high-Andean forest birds in Ecuador. Cotinga 29:28–30
Krabbe N et al (1998) Avian diversity in the Ecuadorian Andes. An atlas of distribution of Andean forest birds and conservation priorities. Centre for Research on Cultural and Biological Diversity of Andean Rainforests (DIVA), DIVA Technical Report No. 4. Rønde
Lepers E et al (2005) A synthesis of information on rapid land-cover change for the period 1981–2000. Bioscience 55:115–124
López F, Torres M, Beltrán R (2003) La minería en el Parque Nacional Podocarpus. Fundación Ecológica Arcoíris, Loja
Madden W (2002) Racumine rodenticide – potential environmental impact on birds. In: Newton I et al (eds) Ecology and conservation of owls. CSIRO Publishing, Collingwood, pp 296–301
Mikkola H (2013) Owls of the world, a photographic guide. Bloomsbury Publishing, London
Ministerio del Ambiente del Ecuador-MAE (2007) Plan estratégico del Sistema Nacional de Áreas Protegidas del Ecuador 2007–2016. Proyecto Sistema Nacional de Áreas Protegidas (SNAP-GEF), REGAL-ECOLEX, Quito
Ministerio del Ambiente del Ecuador-MAE (2010) Mapa de deforestación histórica del Ecuador: 1990–2000–2008. Ministerio del Ambiente, Quito
Ministerio del Ambiente del Ecuador-MAE (2015) Estrategia nacional de biodiversidad del Ecuador y plan de acción 2015–2020. Ministerio del Ambiente, Quito
Miranda M et al (2003) Mining and critical ecosystems: mapping the risks. World Resources Institute, Washington, DC
Mittermeier RA, Robles-Gil P, Mittermeier CG (1997) Megadiversity. Earth's biologically wealthiest nations. Conservation International, Cemex S.A., Washington, DC
Mischler TC, Hicks DJ (2015) First record of Burrowing Owl *Athene cunicularia* in Amazonian Ecuador. Cotinga 37:118–119
Moreno P (2010) Mamíferos presentes en la dieta de la lechuza de campanario (*Tyto alba*) en Valdivia, provincia de Guayas, Ecuador. Av Cienc Ing (Quito) 3:B87–B90

Moreno P, Román JL (2013) Clasificación del género *Reithrodontomys* en el Ecuador y comentarios sobre la alimentación de la lechuza de campanario (*Tyto alba*) en los alrededores de Quito. Bol Técnico, Serie Zool 8–9:16–23
Nebel BJ, Wright RT (1999) Ciencias ambientales. Ecología y desarrollo sostenible. Prentice Hall, México
Olson SL (1995) The genera of owls in the Asioninae. Bull Brit Ornithol Club 115(1):35–39
Peterson AT et al (2002) Future projections for Mexican faunas under global climate change scenarios. Nature 416:626–629
Peralvo M et al (2015) Áreas prioritarias para la conservación del Ecuador continental. Ministerio de Ambiente, Condesan, Pontificia Universidad Católica del Ecuador, GIZ, Quito
Pozo-Zamora G et al (2017) Primer reporte de la dieta del Búho Orejicorto *Asio flammeus* (Strigiformes: Strigidae) en Pichincha, Ecuador. Rev Ecuatoriana Ornitol 1 (in press)
Remsen JV et al (2016) A classification of the bird species of South America. American Ornithologists' Union, South American Classifications Committee. http://www.museum.lsu.edu/~Remsen/SACCBaseline.html. Accessed 19 Jun 2016
Ridgely RS, Greenfield PJ (2001) The birds of Ecuador. Cornell University Press, Ithaca
Ridgely RS, Greenfield PJ (2006) Aves del Ecuador. Academia de Ciencias de Philadelphia, Fundación Jocotoco, Quito
Robbins MB, Howell SNG (1995) A new species of pygmy-owl (Strigidae: *Glaucidium*) from the eastern Andes. Wilson Bull 107(1):1–6
Robbins MB, Stiles FG (1999) A new species of pygmy-owl (Strigidae: *Glaucidium*) from the Pacific slope of the northern Andes. Auk 116(2):305–315
Robbins MB et al (1987) The avifauna of the Cordillera de Cutucú, Ecuador, with comparisons to other Andean localities. P Acad Nat Sci Phila 139:243–259
Rodas F, Contreras X, Tinoco B (2005) Aviturismo: rutas del austro. Ministerio de Turismo, Naturaleza & Cultura Internacional, Cuenca
Salvadori A, Harvey R, McBain W (2004) A nest of Rufescent (Colombian) Screech-Owl *Otus ingens colombianus*, at Mindo, Ecuador. Cotinga 21:77–78
Schulenberg TS, Awbrey K (eds) (1997) The Cordillera del Cóndor region of Ecuador and Peru: a biological assessment. Conservation International RAP Working Papers No. 7, Washington, DC
Sierra R (1999) Vegetación remanente del Ecuador continental, circa 1996. 1:1'000.000. Proyecto INEFAN/GEF-BIRF, Wildlife Conservation Society, Quito
Sierra R, Campos F, Chamberlin J (1999) Áreas prioritarias para la conservación de la biodiversidad en el Ecuador continental. Un estudio basado en la diversidad de ecosistemas y su ornitofauna. Ministerio del Ambiente/Proyecto INEFAN, GEF-BIRF, EcoCiencia, Wildlife Conservation Society, Quito
Stattersfield AJ et al (1998) Endemic bird areas of the world. Priorities for biodiversity conservation. BirdLife International Conservation Series No. 7, Cambridge
Valarezo-Delgado S (1984) Aves del Ecuador: sus nombres vulgares. Tomos II–III. Museo Ecuatoriano de Ciencias Naturales, Quito
Varela S (2012) Distribución espacial, diversidad y uso de hábitat de las aves nocturnas (Strigiformes) en el Bosque Protector Jerusalem, Pichincha, Ecuador. Dissertation, Universidad Central del Ecuador, Quito
Wagner H, Bairlein F, Piedrahita P (2016) Foraging of Barn Owls on Santa Cruz (Galapagos, Ecuador): a preliminary study. In: Abstracts of the 149th Jahresversammlung der Deutschen Ornithologen-Gesellschaft, Stralsund, 28 Sept–3 Oct 2016
Wiedenfeld DA (2006) Aves, The Galápagos Islands. Checklist 2:1–27
Williams RSR, Tobias JA (1996) West Peruvian Screech-Owl, *Otus roboratus*. Cotinga 6:76–77

Chapter 11
The Owls of El Salvador

Ricardo Pérez-León, Iselda Vega, and Néstor Herrera

Abstract Information on 13 nocturnal raptors (Strigiformes) distributed in El Salvador was organized using 900 records from 1912 to 2008. Two hundred and fifty-eight bird studies from El Salvador were revised, of which 72 contain information on the group or on individual owl species (50 unpublished papers and 22 published papers). In addition, 18 habitat types were evaluated and classified as natural ecosystems (costal vegetation, mangroves, secondary vegetation, pastures, alluvial forest, deciduous forest, semi-deciduous forest, riparian forest, pine forest, pine-oak forest, oak forest, cloud forest), agrosystems (agricultural land, livestock areas, coffee plantation, cypress tree plantations), and human settlements. The natural ecosystems with greatest diversity were identified in terms of species evenness. Distribution maps of owl species in the country were obtained and used to define classification categories for the species according to their habitat. We determined that five species are habitat generalist, two are habitat specialist of open habitats, three species are generalists of forest areas (they use diverse type of natural forest), and three are specialists of forest areas (they use specific types of natural forest). The main threats to nocturnal raptors are capture for the illegal pet trade and deforestation caused by changes in land use. In the short term, specific information on nocturnal raptors will be necessary in order to define their distribution status in fragmented landscape and the habitat requirements of each owl species.

Keywords El Salvador • Nocturnal raptors • Owl abundance • Fragmentation

---

R. Pérez-León (✉)
GIS and landscape ecology specialist, El Salvador raptor conservation group, Tres Ríos, Cartago, Costa Rica
e-mail: perez_leon@yahoo.com

I. Vega
El Salvador raptor conservation group, San Salvador, El Salvador

N. Herrera
Environmental consultant, San Salvador, El Salvador

P.L. Enríquez (ed.), *Neotropical Owls*, DOI 10.1007/978-3-319-57108-9_11

Pacific Screech Owl (*Megascops cooperi*)

11.1 Introduction

The first studies regarding the avifauna in El Salvador date from 1912, when foreign researchers visited the country to collect information about biodiversity present in the region. Since then, the studies focused on birds were sporadic, until the 1970s when formal investigations took place in the country (Dickey and Van Rossem 1938; Thurber et al. 1987). Also, during this period the ecosystems' conservation improved with the building of the first national parks and the natural protected areas' system in El Salvador (MARN 2006a).

During the last two decades (1990 and 2000), the increasing biodiversity research and conservation actions have been notorious countrywide, providing the opportunity to obtain information about bird species distribution, threats, and habitat needs. Despite these efforts, the studies focused on nocturnal birds of prey (so-called owls) are still incipient (Komar and Herrera 1995b, Vega 2006).

According to the official El Salvador's birds' list (MARN 2009), 551 species have been recorded in the country, including 13 owl species (order Strigiformes, families Tytonidae and Strigidae). The owls (known as "tecolotes" and "lechuzas" in El Salvador) are an endangered group of birds in the country (Vega 2006).

This chapter systemizes information about 900 owl records countrywide, including data from 1912 to 2008, and other important owl observations during 2016, helping to know more about owls' actual distribution and conservation status, as well as providing statements regarding their habitat use, threats, and important conservation strategies to foster their preservation in El Salvador.

11.2 Area Description

El Salvador is the smallest country in Central America, located at the coordinates 13° 50'N y 88° 50'O, with 20,000 km^2, is the most populated, with at least 5,700,000 people living in the territory (about 336 habitants/km^2; CIA 2009). The weather is tropical, determined by an alternation of dry and wet seasons and transitions among them (SNET 2007). The topography includes volcanoes, mountain chains, and coastal and internal valleys (Jiménez et al. 2004; PNODT 2004; MARN 2006a).

In the country exist at least 18 natural formations, including mangroves, mountain evergreen forest (cloud forest), riparian forest, alluvial forest, deciduous and semi-deciduous forests, morro savanna (*Crescentia* spp.), swamp areas, sand dune vegetation, and floating vegetation over water bodies. Also, there are secondary growing areas, mainly over volcanic lava and natural grasslands (MARN 2006b). Nevertheless, El Salvador is the Central American country with less natural land cover (13% approx.), mainly because the natural ecosystems have suffered dramatic degradation by human activities, including land use shifts to crop and pasture lands, coffee plantations, sugar cane, and cotton plantations (MARN 2006a, b). In addition, about 5% of the territory is covered by cities and towns (Fig. 11.1).

11.3 Taxonomy

The ecogeography of El Salvador, determined by altitude levels ranging from 0 to 2800 m over sea level, includes different ecosystem types in the landscape, influencing the distribution of avian communities (Jiménez et al. 2004). At that time, 13 species of owls have been recorded in the country, mainly observed during 1912–1927 by researchers Dickey and Van Rossem (1938), who collected specimens of 12 species in 19 localities countrywide (Table 11.1, Appendix 11.1).

In the next years, the information collected by other bird watchers and biologists has contributed to update the owl species for the national bird list. This information was not published formally in the past until this effort to compile and analyze the owl data available.

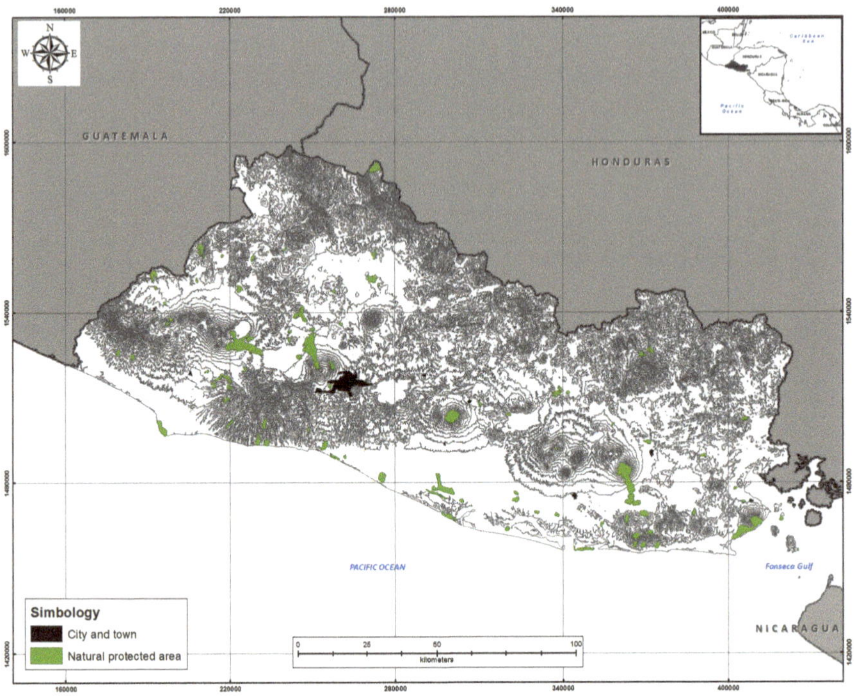

Fig. 11.1 El Salvador location and its natural protected areas (Source: shapefiles in LAMBERTCC-ESA-NAD27, Environmental and Natural Resources Minister of El Salvador)

11.4 Habitat Distribution

El Salvador, despite its ecosystem damaged, still harbors a diversity of habitat types, mainly due to its important ecogeography and volcanic landscape. The habitat fragmentation has altered avian communities and species richness, mostly to those species of mid to great size needing of large extensions of continuous forest for breeding needs (Thurber et al. 1987).

Despite the actual landscape arrangement in the country, there are species still surviving probably because their habitat needs are resilient to land use shifts. Nevertheless, there are species whose populations are isolated to forest fragments, endangering their survival in the country.

To analyze the owl species distribution, the land cover of El Salvador was the baseline to define habitat categories, considering the following criteria: (i) the habitat type described in each owl record and (ii) the habitat type classification used in El Salvador (MARN 2006b). This classification process allowed identifying 18 habitat types used by owls, among natural ecosystems, agroecosystems, and human towns and cities (Table 11.2, Appendix 11.2).

To analyze the distribution of each species among ecosystems and habitat types, each owl record was tabulated in a database. This data allowed to obtain the species

Table 11.1 Species and subspecies of owls recorded in El Salvador

Scientific name	Subspecies	English name	Spanish name
Tytonidae			
Tyto alba	*guatemalae*	Barn owl	Lechuza, Lechuza común
Strigidae			
Megascops cooperi	*cooperi*	Pacific screech owl	Tecolote de Cooper o tecolote
Megascops trichopsis	*mesoamericanus*	Whiskered screech owl	Tecolote de montaña o tecolote
Lophostrix cristata	*stricklandi*	Crested owl	Tecolote de cuernos
Pulsatrix perspicillata	*saturata*	Spectacled owl	Búho de anteojos
Bubo virginianus	*melancerus*	Great horned owl	Mistiricuco, búho de cuernos
Glaucidium brasilianum	*ridgwayi*	Ferruginous pygmy owl	Aurorita
Athene cunicularia	*hypugaea*	Burrowing owl	Búho de praderas o lechuza de suelo
Ciccaba virgata	*centralis*	Mottled owl	Búho café o pájaro león
Ciccaba nigrolineata	Monotypic	Black-and-white owl	Búho blanquinegro
Strix fulvescens	Monotypic	Fulvous owl	Búho fulvo
Pseudoscops clamator	*clamator*	Striped owl	Búho cornudo cariblanca
Aegolius ridgwayi	*rostratus*	Unspotted saw-whet owl	Tecolote abetero sureño

English name obtained from the Checklist of North American Birds (AOU 2010); Spanish names based on local knowledge (Howell and Webb 1995; MARN 2009)

richness and diversity (by using the Shannon index relating the species richness and the quantity of records for each species, which values range from 0 to the logarithm of the richness – Magurran 1988 cited by Moreno 2001).

Following this analysis, the habitats of natural ecosystems were the most diverse (Fig. 11.2), where the highest values of species richness and diversity were observed in cloud forest, deciduous forest, riparian forest, and pine-oak forest (Figs. 11.3 and 11.4, Appendix 11.2). The habitat types with less species richness and diversity values were the oak forest and grasslands. The agroecosystems and human towns show the same evenness in diversity values (Fig. 11.2). Despite this, the coffee plantations represent the highest owl species richness and diversity, defining this plantation as the most important for owl communities' survival (at least 12% of the countries' land cover is coffee plantations (MARN 2006b)). The human towns located at rural areas were most diverse (Fig. 11.4), probably due to their association to natural protected areas, secondary growing areas, and agroecological farms (including coffee plantations).

A habitat distribution analysis was also developed to categorize each species into habitat preferences. Each category was obtained using the following criteria: (1)

Table 11.2 Habitat types and land use description used to analyze owl species distribution

Habitat types and land use	Altitude over sea level	Description
Coastal vegetation (VC)	0–10 m	This vegetation is located over sand dunes in the coastline, it is mainly shrubby, but it can be associated to scattered trees planted by humans. Sometimes it can be associated to secondary growth and deciduous forest
Mangrove (MN)	0–10 m	This habitat type comprehends a dense to open wetland forest, located at estuaries and large river deltas approaching the Pacific Ocean coast
Alluvial forest (SA)	10–30 m 400 m	This habitat type is categorized as ombrophile forest, seasonally flooded during the rainy season. It is associated to riparian forest of the largest rivers in the country, lagoons, and lakes. Actually, this habitat type is isolated, surrounded by croplands and pastures, so it is considered of great importance for wildlife movement across the landscape countrywide
Grasslands (PA)	Various altitudes	It refers to natural savanna-type lands, associated to scattered trees and shrubs of the gender *Crescentia* spp. and the species *Curatella americana*
Secondary growth vegetation (VS)	Various altitudes	It includes shrub areas associated to scattered trees of secondary growth. It can be located over volcanic lava and abandoned farmlands
Riparian forest (SR)	Various altitudes	It is an evergreen forest located at riverbanks and river sides, as well as surrounding lagoons and lakes. Sometimes it can be associated to swamps. Actually, this habitat type is isolated, surrounded by croplands and pastures, so it is considered of great importance for wildlife movement across the landscape countrywide
Deciduous forest (SD)	10–1000 m	This forest is characterized by the total fall of leaves during the dry season. It is categorized as closed tropical deciduous forest and is located mostly at low lands and foothills
Semi-deciduous forest (SS)	30–1000 m	This habitat is categorized as closed ombrophile forest, which leaves fall during dry season partially, due to the association of deciduous and evergreen species. It is mostly found at foothills
Pine forest (PI)	1000–2000 m	This is categorized as coniferous forest dominated by trees of genus *Pinus*, found in the northern mountain chains, near the border with Honduras. Understory vegetation is scattered
Pine-oak forest (PR)	1000–2000 m	This type of habitat is semi-open evergreen forest. It is commonly found between pine forest and oak forest. Sometimes it can be found in association to Liquidambar trees (*Liquidambar styraciflua*). Understory vegetation is abundant
Oak forest (RO)	1500–2000 m	Mainly evergreen forest dominated by trees of genus *Quercus*, mostly found before the cloud forest at mountains. Understory vegetation is abundant

(continued)

Table 11.2 (continued)

Habitat types and land use	Altitude over sea level	Description
Cloud forest (SN)	2000–2800 m	Mainly continuous evergreen forest. It is represented in patches of different sizes, located at top of mountains and volcanoes' hillsides across the country
Croplands (AG)	Various altitudes	These lands are used for annual and seasonal crops (e.g., corn, been, sugarcane, cotton). Sometimes they can be associated to scattered trees and secondary growth vegetation
Coffee plantations (CA)	600–1600 m	Farms of coffee plantations are exposed or not to sun. The coffee associated to trees (used as shade) is considered as an opportunity for biodiversity conservation. Sometimes it is associated to small natural habitat patches and secondary growth vegetation
Cypress forest (CP)	1600–2200 m	These are monospecific plantations of cypress (*Cupressus lusitanica*) used for foresting in the past. It actually remains associated to coffee plantations and secondary growth patches
Pastures (GA)	Mainly at 10/50 m, also at mountain slopes up to 2000 m	These lands are common at the Pacific slope and internal valleys in the country, used for cattle grounds. It could be associated to secondary growth, scattered trees, and live fences
Town – rural areas (AR)	Various altitudes	It refers to human towns not densely populated. Infrastructure is scattered and cut by farms of different sizes (croplands, pastures, and coffee plantations). It can be associated to natural protected areas or secondary growth lands
Urban areas – city (CI)	Various altitudes	It refers to dense urban areas, with high presence of building infrastructure and intense human activity. It includes few forested areas as parks and botanical gardens

Obtained from the land cover map of El Salvador (MARN 2006b)

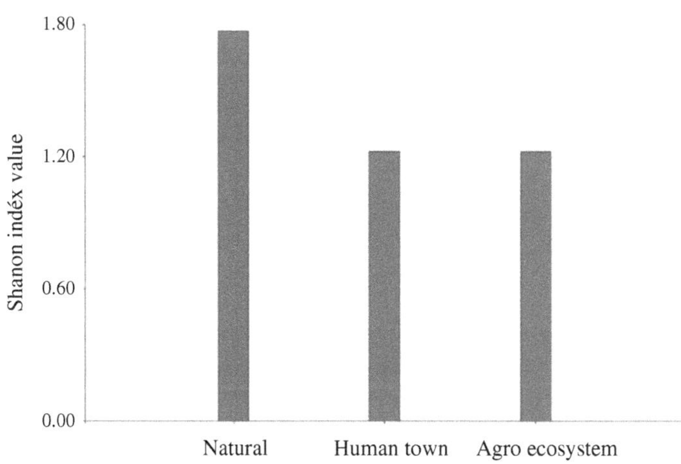

Fig. 11.2 Owl species diversity (using Shannon index) for ecosystem types (natural, agroecosystems, and human towns) – January 2009

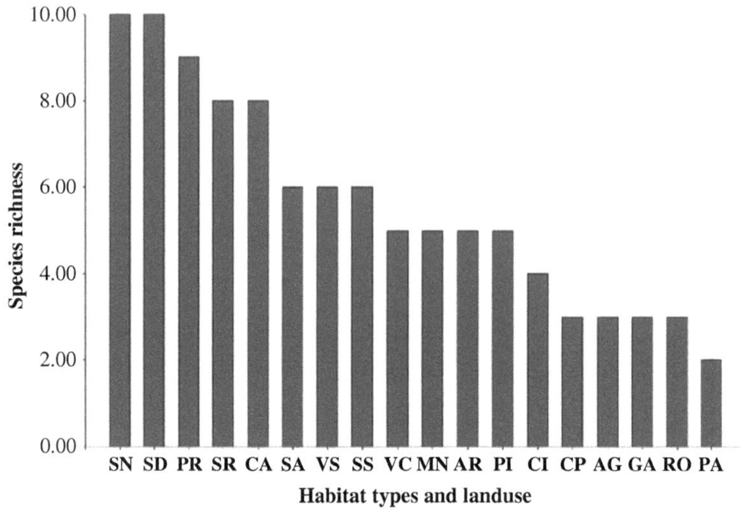

Fig. 11.3 Owl species richness distribution among 18 habitat types and land uses – January 2009 (acronyms' meaning is in Table 11.2)

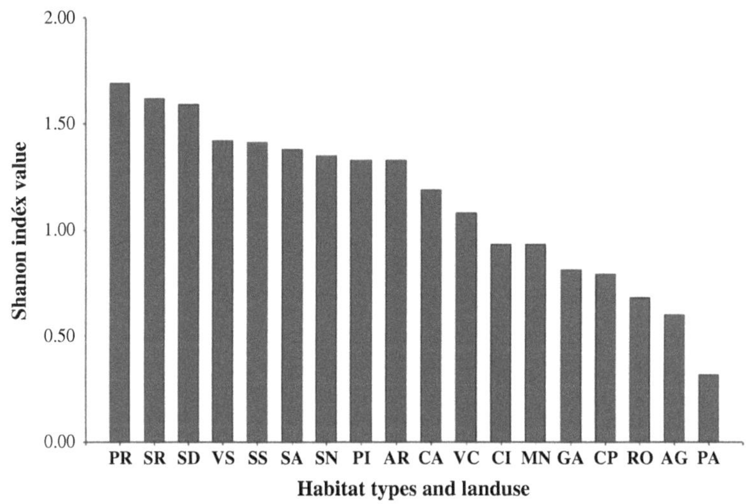

Fig. 11.4 Owl diversity (using Shannon index) among 18 habitat types and land uses – January 2009 (acronyms' meaning is in Table 11.2)

habitat generalist (HaGe) for species occurring in 10–17 habitat types (including agroecosystems and human towns), (2) habitat specialists (HaEs) for species occurring just in natural habitats and ecosystems, (3) forest generalists (FoGe) for species occurring in more than four natural habitat types (4) forest specialists (FoEs) for species occurring in just specific natural habitat types.

Five species were defined as habitat generalists: barn owl (*Tyto alba*), Pacific screech owl (*Megascops cooperi*), great horned owl (*Bubo virginianus*), ferruginous pygmy owl (*Glaucidium brasilianum*), and mottled owl (*Ciccaba virgata*); two species as habitat specialists: burrowing owl (*Athene cunicularia*) and striped owl (*Pseudoscops clamator*); three as forest generalists: whiskered screech owl (*Megascops trichopsis*), spectacled owl (*Pulsatrix perspicillata*), and fulvous owl (*Strix fulvescens*); and three as forest specialists: crested owl (*Lophostrix cristata*), black-and-white owl (*Ciccaba nigrolineata*), and unspotted saw-whet owl (*Aegolius ridgwayi*).

According to the analysis, the highest owl species richness (13) was found in forested habitats. In this sense, it is important to notice these types of habitats are of great relevance for owl communities' conservation in the actual landscape arrangement in El Salvador.

A set of maps for each species was built using the geographic coordinates of each owl record (Figs. 11.5 and 11.6) to identify easily those species more widespread in the country versus those most specialized. These maps are more helpful to owl conservation as they are updated periodically, providing evidence of any shifts for each species distribution map.

11.5 Owl Biology and Ecology in El Salvador

The first report about owl distribution status in El Salvador was published in 1938, as part of the monograph *The Birds of El Salvador* by Dickey and Van Rossem (1938), who traveled across the country looking for birds from 1912 to 1927. In this document, *T. alba*, *M. cooperi*, *P. perspicillata*, and *G. brasilianum* were categorized as common in low tropical lands (0–1000 m), whereas *M. trichopsis* was found fairly common in pine-oak forest at the northern mountain chains of the country. *L. cristata* was categorized as a rare species found at highest altitudes in the country (above 2000 m). *B. virginianus*, *A. clamator*, *C. nigrolineata*, and *A. cunicularia* were also categorized as rare species inhabiting low lands. *C. virgata* was the only species found widespread in the country, from 0 to 2500 m, while *S. fulvescens* was very common in cloud forest.

Aegolius ridgwayi was documented by J.T. Marshall at a cloud forest area (Marshall 1943). This researcher also found *M. trichopsis* in woodlands and coffee plantations, *B. virginianus* in cloud forest, *P. clamator* at open forested areas, and *C. nigrolineata* in deciduous forest.

Between 1979 and 1980, J.N. West studied the birds of prey at the El Imposible National Park, being the first study to be done in the country regarding birds of praycommunity (West 1988). West documented information about distribution, feeding habits, nesting, and behavior of *T. alba*, *M. cooperi*, *P. perspicillata*, *G. brasilianum*, *C. virgata*, and *C. nigrolineata* at a dry forest and secondary growing area, in the western coast mountains of El Salvador.

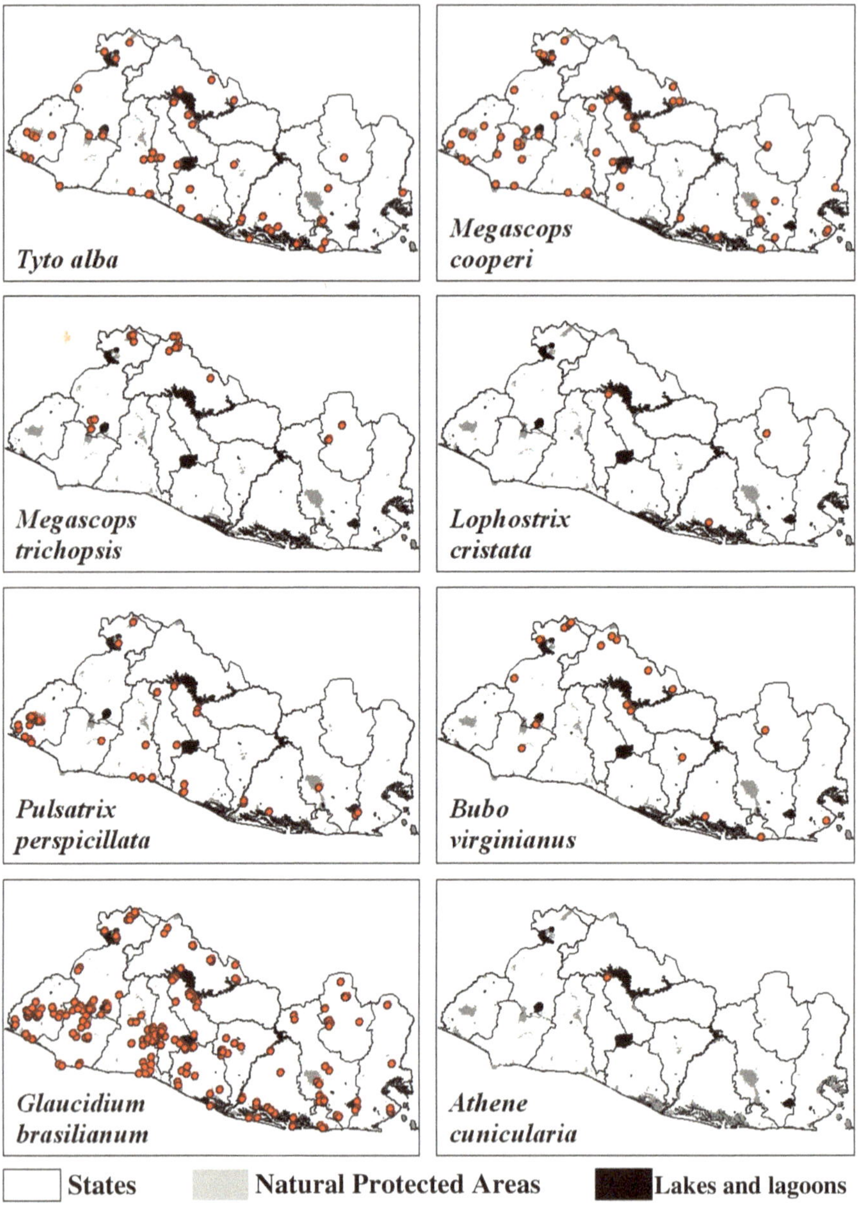

Fig. 11.5 Owl species occurrence maps for El Salvador (January 2009)

The conservation status for the avifauna in the country published at the end of the 1980s provided information about distribution and nesting records for *M. cooperi*, *P. perspicillata*, *B. virginianus*, *S. fulvescens*, and *P. clamator*, based on observations of avifauna led from 1966 to 1982 (Thurber et al. 1987).

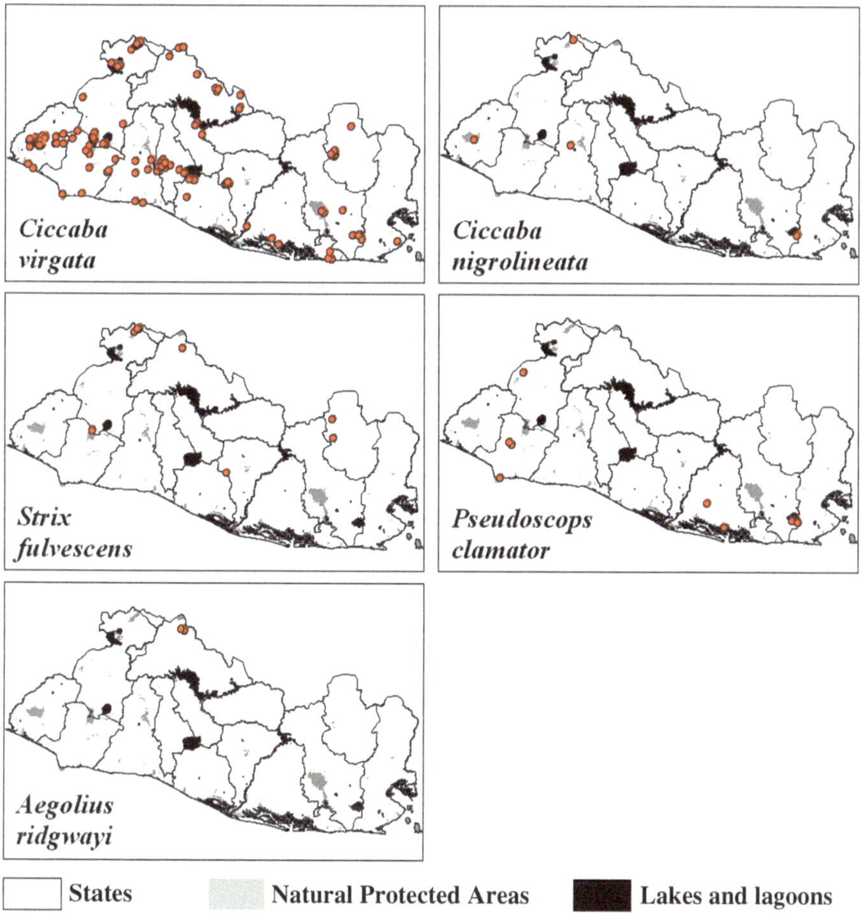

Fig. 11.6 Owl species occurrence maps for El Salvador (January 2009)

From 2002 to 2008, West researched and monitored owls in three natural protected areas, visiting different forested habitat types, including alluvial forest (at Nancuchiname), deciduous forest (at El Imposible National Park), and cloud and pine-oak forests (at Montecristo National Park). The monitoring sessions were led during dry and early rainy seasons, by walking transects and playing sound decoys searching for owls. The species observed include *T. alba*, *G. brasilianum*, *M. cooperi*, *M. trichopsis*, *P. perspicillata*, and *C. virgata*. Specific information about West findings was not obtained for this paper.

From October to December 2005, I. Vega (2006) studied owls in a deciduous forest, looking for birds by walking transects. She determined the abundance of five species. *Glaucidium brasilianum* was the most abundant with 53 owls/km; *M. cooperi* had 30 owls/km. The lowest abundant species were *B. virginianus* with 9 owls/km, *C. virgata* 7 owls/km, and *P. perspicillata* with 2 owls/km.

Table 11.3 Biological and ecological status of owl species in El Salvador until 2008

Scientific name	Breeding status	Feeding information	Relative abundance information
Tytonidae			
Tyto alba	Breeder	Complete	
Strigidae			
Megascops cooperi	Breeder	Poor information	
Megascops trichopsis	Breeder	Poor information	
Lophostrix cristata	Non breeder	No information	
Pulsatrix perspicillata	Breeder	Poor information	Minimum information
Bubo virginianus	Indirect evidence	No information	Minimum information
Glaucidium brasilianum	Breeder	Complete	Minimum information
Athene cunicularia	Unknown	No information	
Ciccaba virgata	Breeder	Complete	Minimum information
Ciccaba nigrolineata	Breeder	No information	
Strix fulvescens	Unknown	Poor information	Minimum information
Pseudoscops clamator	Breeder	No information	
Aegolius ridgwayi	Unknown	Poor information	

Data obtained from Dickey and Van Rossem (1938), Marshall (1943), Thurber et al. (1987), West (1988), Komar (2002), and Vega (2006)

We reviewed 258 avifauna papers to find owl information available for the country, but only 72 of them included data regarding owl species. From this total, only 32 are published in formal documents. A total of 900 individual owl records correspond to the data available in those 72 documents. Nevertheless, this information was sufficient just to approximate a conservation status for each species.

The status of distribution in the country corresponds to assumptions inferred from the data available. For example, the breeding status is well known for only nine species, but others, including *B. virginianus*, have just juvenile sightings and no nesting details. *L. cristata*, *A. cunicularia*, and *A. ridgwayi* have no breeding records so far. For *L. cristata* (observed in 1925 by Dickey and Van Rossem) and *A. ridgwayi* (registered in 1942 by Marshall 1943) resent records are not available, and *A. cunicularia* was not observed since 1925 until a recent bird observed in 2008 in the eastern El Salvador, at La Union area, near the border with Honduras (J. Fagan and L. Lara pers. comm. 2008).[1]

Information about feeding habits is available for only three species: *T. alba*, *G brasilianum*, and *C. virgata*, detailing prey species. Other five species (*M. cooperi*, *M. trichopsis*, *P. perspicillata*, *S. fulvescens*, and *A. ridgwayi*) have little information about feeding habits (Table 11.3).

Relative abundance information is available for few species. Cortez de Galán et al. (1994) found a low abundance in an urban area (0.004 birds/ha), while Komar and Herrera (1995a) registered an observation frequency of 2% for *C. virgata* and

[1] Jesse Fagan and Karla Lara, Associate Researchers of SalvaNATURA, Fundación Ecológica de El Salvador.

1% for *P. perspicillata* in a forest interior area at El Imposible National Park (surveying 48 point counts). Komar and Herrera (1995a) obtained an observation frequency of 3% for *G. brasilianum* at the same national park and a density of five birds of *C. virgata* per km surveyed. Komar (2002) estimated a frequency of 0.1 individuals for *S. fulvescens*, during 7 days of transect surveys in a cloud forest in the country.

Vega (2006) studied the relative abundance of owls at a secondary grown deciduous forest area, by surveying three transects of 8 km long (each). Per transect, this study registered 52.64 owls/km of *G. brasilianum*, 29.82 owls/km of *M. cooperi*, 8.77 owls/km of *B. virginianus*, 7.02 owls/km of *C. virgata*, and 1.75 owls/km *P. perspicillata*.

11.6 Conservation Status in El Salvador

Little information is known about the conservation status of owls in the country, mostly because of the lack of studies dedicated to this group of birds. A recent effort was led to classify the avifauna conservation status in El Salvador, considering local distribution criteria used by IUCN (Komar et al. unpublished data)[2]; this document details breeding and feeding habits recorded for each avian species to evaluate the conservation status, including owl species (Table 11.3, Appendix 11.1).

According to this classification, *L. cristata*, *C. nigrolineata*, and *A. ridgwayi* are considered species in critical danger of extinction, and *S. fulvescens* (an endemic species of the Northern Pacific slope of Central America; Howell and Webb 1995) is considered as a threatened species, leading all these four species to be of most importance for conservation in El Salvador. Nevertheless, it is important to include in these conservation categories those species who are habitat specialists and whom distribution is restricted to specific habitat types. In this sense, *A. cunicularia* and *P. clamator* are candidates to be in critical danger for the country, due to the poor information available for these species.

11.7 Threats for Owl Conservation in the Country

In El Salvador, the ecosystem degradation and land shifts to agriculture and pastures (occurred mostly during the last century) have led to a process of species substitution/adaptation, where those species known as habitat generalists are dominant over other less resilient species (Thurber et al. 1987). This habitat loss has restricted the distribution of several owl species, but populations of *T. alba*, *M cooperi*, *C. virgata*, and *G. brasilianum* have survived to this fragmentation process (Figs. 11.5 and 11.6).

[2] Komar, O., N. Herrera, L. Girón, R. Ibarra Portillo. Unpublished data. La lista roja de aves de El Salvador. San Salvador, El Salvador. SalvaNATURA (Biodiversity series No. 3).

There are other human activities affecting owl populations in the country, including illegal wildlife trade and illegal hunting (Rivas pers. comm. 2008)[3]. The owl species suffer of intense illegal trade in the local market. Dominguez (1994) registered that 4.4% of the total illegal wildlife trade belongs to owl species in El Salvador. In addition, the National Police Corp (Environmental Unit) between 1995 and 2008 registered 1.1% owls used for illegal trading. A total of eight species are commonly found in the local market, but the most frequent are *G. brasilianum* and *C. virgata*.

Among other threats to owl species, the myths and rituals are considered of less frequency and have not been documented properly in the country. Nevertheless, this causes owl persecution and illegal trapping, and also some are killed (Ramos and Mendoza 2000).

11.8 Conservation Strategies

In the past, specific conservation strategies regarding owl species have not been developed in the country. Probably, the Naional BiodiversityStrategy has been the only initiative to conserve avifauna, including owls, due to its main goal to consolidate natural protected areas as a conservation network in the landscape and their recognition as human heritage.

Among the natural protected area categories are Wildlife Refuges, National Parks, and Private Reserves. Some of these areas are managed by nonprofit organizations using a coadministration figure with the National Environmental Ministry. This allows the participation of social organizations and the private sector in conservation. At the same time, the environmental ministry has developed geographical land units, named "conservation areas," which are conjunctions of large territories harboring natural protected areas and biological corridors. This provides the opportunity to develop territorial arrangement and land use planning (MARN 2006a, MARN-UICN 2006).

Despite there are no specific efforts to protect particular fauna species, there are local wildlife monitoring programs in some natural protected areas, generating important information regarding the occurrence of vertebrate species in different habitat types.

The design and development of conservation strategies implies to gather information about distribution status of owl species. This duty, although has started incipiently, needs the participation of the Salvadoran society, as a whole, providing opportunities to develop research and conservation projects. At the time, the Environmental Unit of the National Police Corp is an opportunity to prevent habitat loss and wildlife persecution, as well as illegal trade. Also, particular conservation efforts have included land owners, mostly those owning coffee plantations and agroforest farms, which provide habitat for fauna survival. In the near future, the land use management and planning has to consider biodiversity conservation priorities and the environmental services provided by natural protected areas (Table 11.4).

[3] Rivas, I. O. 2008. National Police Corp, Environmental Unit.

Table 11.4 Owl species confiscated from the local market, from 1995 to 2008

Species	1996–1997	1998–1999	2000–2001	2003	2004–2008	Total
Tyto alba	7	15	1	9	11	43
Megascops cooperi	1	5		1	3	10
Megascops trichopsis	2					2
Pulsatrix perspicillata	3					3
Bubo virginianus	1	1		4	1	7
Glaucidium brasilianum	12	30	1	4	30	77
Ciccaba virgata	14	24	2	7	28	75
Pseudoscops clamator				2		2
Total	40	75	4	27	73	219

Data obtained from Mendoza and Ramos (1999), Peña de López and Herrera (1999), Ramos and Ricord de Mendoza (2000), Ibarra Portillo and Vega (2005), Rivas pers. comm. (2008), and Ramírez pers. comm. (2008)

11.9 Expected Owl Species in the Country

The initiatives to study owl biology and ecology are few, and no conservation efforts have been developed specifically focused on this group of birds. Unlikely, the owl information mainly has been recorded in casual sightings or has been collected in projects led for avifauna. Also, there is a lack of capacitated personnel to identify owls properly, as well as no enough budget and equipment, all necessary to develop a systematic and continuous research dedicated to owl species.

Despite this situation, in the past 20 years, birdwatching has increased unexpectedly in El Salvador, but mostly as a hobby. This is important, considering most of the recent and new species records for the country have been obtained by birdwatchers. For owl species, considering its conspicuous habits, the detection of new species has been not possible; besides there are six species expected to occur in the country (Table 11.5).

In the near future, studies focused on owl biology and ecology are needed, allowing to document habitat use and the specific avifauna communities owls belong to, as a way of determining these birds' needs for surviving and achieving conservation actions. This also will document the existence of expected species for the country.

11.10 Owl Observations During 2016

During the last years, owls' records have been sporadic in the country. Few interesting records include the most updated observation of *Aegolius ridgwayi*, in a northern cloud forest patch near the border of Honduras (Aguirre and Cardoza 2016), and *Bubo virginianus* at Suchitoto, near the Suchitlan lake (Vega 2016 pers. comm.).

Table 11.5 Expected owl species for El Salvador

Species	English common name	Spanish name	Distribution status	Conservation status
Otus barbarus	Bearded screech owl	Tecolote barbudo	FoEs	Endangered
Otus flammeolus	Flammulated screech owl	Tecolote flameado	FoEs	
Megascops guatemalae	Middle American screech owl	Tecolote vermiculado	FoGe	
Glaucidium gnoma	Northern pygmy owl	Tecolotito Serrano	FoEs	
Glaucidium griseiceps	Central American pygmy owl	Tecolotito Centroamericano	FoGe	
Asio stygius	Stygian owl	Búho cornudo oscuro	FoGe	

FoEs forest specialist, *FoGe* forest generalist (Obtained from König et al. 1999)

11.11 Conclusions

According to the information analysis developed in this chapter, the highest owl diversity is found in forested habitats. Considering these habitat types are mostly isolated in the country, there is an important need to promote the creation of agro-forest farmlands allowing owl species to survive in the landscape.

The knowledge about the conservation status of owls is poor. This implies a need of research efforts to document habitat needs for each species, helping to design adequate conservation strategies and actions, considering human welfare and owls' preservation as important goals to be integrated.

In the near future, research is needed to update owl species information, helping to determine birds' fluctuations and habitat use in the landscape. This, also, will help to determine how important natural protected areas are for owl preservation. In this sense, there is a need to encourage birdwatchers and biologist to continue doing owl observation and to develop correct identification skills.

Unspotted saw-whet Owl (*Aegolius ridgway*)

Acknowledgments We want to thank the Science Program from SalvaNATURA (a Salvadoran nonprofit organization) for allowing us to use the avian database of El Salvador, which was the baseline to obtain owl data needed to develop diversity analysis and to build distribution maps. Special thanks to Dra. Jane Nol West for contributing information about owl research at cloud forest areas and also for providing data about expected owl species in El Salvador. Also we would like to thank all people for contributing to owl records which were not included in the SalvaNATURA's avian database, especially Leticia Andino, Oscar Bolaños, Alicia Díaz, Jesse Fagan, Emerson Flores, Carlos Funes, Vicky Galán, Néstor Geovanni García, Ricardo Ibarra Portillo, Tom Jenner, Roselvy Juárez, Oliver Komar, Romaní Martin, Esmeralda Martínez, Álvaro Moisés, Luis Pineda, Karla Pérez León, Zoila Pérez, Jorge Ramos, Roberto Rivera, Marvín Rivas, Sofía Trujillo, and Carlos Zaldaña. Also, special thanks to Vladlen Henríquez, who did geographic and coordinate adjustments in the avian database. Data of confiscated owls from the local market was provided by Sada Francoisa Amaya, Ana Cecilia Peña de López, Ada Ramírez, and Ivan Orlando Rivas.

Appendix 11.1

Owl species recorded in El Salvador until 2008 (habitat distribution information was determined using the data collected in this chapter; conservation status was obtained from Komar et al. unpublished data)

Scientific name	Subspecies	English name	Spanish name	Habitat distribution[a]	Conservation status[b]
Tytonidae					
Tyto alba	*guatemalae*	Barn owl	Lechuza	HaGe	VU D1
Strigidae					
Megascops cooperi	*cooperi*	Pacific screech owl	Tecolote de Cooper	HaGe	NT
Megascops trichopsis	*mesoamericanus*	Whiskered screech owl	Tecolote de montaña	FoGe	EN B1 ab(iii) + 2 ab(iii)
Lophostrix cristata	*stricklandi*	Crested owl	Tecolote de cuernos	FoEs	DD
Pulsatrix perspicillata	*saturata*	Spectacled owl	Búho de anteojos	FoEs	VU D1
Bubo virginianus	*melanocerus*	Great horned owl	Mistiricuco, Búho de cuernos	HaGe	VU D inmig
Glaucidium brasilianum	*ridgwayi*	Ferruginous pygmy owl	Aurorita	HaGe	LC
Athene cunicularia	*pypugaea*	Burrowing owl	Búho de praderas	HaEs	NT
Ciccaba virgata	*centralis*	Mottled owl	Búho café, Pájaro León	HaGe	LC
Ciccaba nigrolineata	*(monotípica)*	Black-and-white owl	Búho blanquinegro	FoEs	CR B1 ab(iii) D
Strix fulvescens	*(monotípica)*	Fulvous owl	Búho fulvo	FoGe	CR B1 ab(iii) D
Pseudoscops clamator	*clamator*	Striped owl	Búho Cornudo Cariblanca	HaEs	CR D; EN B2 ab(iii)
Aegolius ridgwayi	*rostratus*	Unspotted saw-whet owl	Tecolote abetero Sureño	FoEs	DD

[a]Habitat distribution: *HaGe* habitat generalist, *HaEs* habitat specialist, *FoGe* forest generalist, *FoEs* forest specialist

[b]Conservation status (Komar et al. unpublished data – IUCN criteria): *VU D1* vulnerable, population is restricted, population size < 1000 mature individuals, *NT* near threat; *EN B1* ab(iii) endanger, (B1) distribution range < 5000 km^2, (iii) area of extension and habitat quality – (a) severely fragmented or less than five localities of occurrence and (b) declining continuously, *DD* insufficient data or not adequate to evaluate the risk of extinction, *VU D inmig* vulnerable, population to small or restricted, immigrant, *LC* least concern, a species evaluated not classified in other status, *CR B1* ab(iii) *D* in critical danger, high risk of population declining, distribution range very small (B1) < 100 km^2 – (a) habitat severely fragmented or just one locality of occurrence and (b) declining continuously, *CR D, EN B2* ab(iii) in critical danger, population size <50 mature individuals

Appendix 11.2

Habitat distribution of owl species in El Salvador

Species	Croplands	Coffee plantation	Cypress forest	Pastures	Towns (rural areas)	Urban cities	Alluvial forest	Deciduous forest	Cloud forest
Tyto alba	+	+		+	+	+	+	+	+
Megascops cooperi		+		+	+	+	+	+	+
Megascops trichopsis		+							+
Lophostrix cristata								+	+
Pulsatrix perspicillata		+						+	+
Bubo virginianus		+	+		+		+	+	+
Glaucidium brasilianum	+	+	+	+	+	+	+	+	+
Athene cunicularia								+	
Strix virgata	+	+	+		+	+	+	+	+
Strix nigrolineata								+	+
Strix fulvescens		+							+
Asio clamator							+	+	
Aegolius ridgwayi									

Species	Secondary growth vegetation	Semi-deciduous forest	Mangrove	Grasslands	Pine forest	Pine-oak forest	Oak forest	Coastal vegetation	Riparian forest
Tyto alba	+	+	+			+		+	+
Megascops cooperi	+	+	+			+	+	+	+
Megascops trichopsis	+				+	+			
Lophostrix cristata									
Pulsatrix perspicillata	+	+	+			+		+	+
Bubo virginianus		+			+	+		+	+
Glaucidium brasilianum	+	+	+		+	+	+	+	+
Athene cunicularia									
Strix virgata	+	+	+			+	+	+	+
Strix nigrolineata									+
Strix fulvescens					+	+			
Asio clamator				+				+	
Aegolius ridgwayi					+	+			

References

Aguirre Alas C, Cardoza J (2016) Primer registro fotográfico del búho moreno (*Aegolius ridgwayi*) en Cerro El Pital, San Ignacio, Chalatenango, El Salvador. Spizaetus 22:2–5

American Ornithologist Union (2010). North America and Middle America Bird Checklist. Available at http://checklist.aou.org

CIA (Central Intelligence Agency) (2009) The on-line world factbook. https://www.cia.gov/library/publications/the-world-factbook/. Accessed 24 Jan 2009

Cortez de Galán M, Herrera de Benítez D, Komar N, Menjivar N, Rodríguez W (1994) Estudio de la Avifauna de la Ciudad Universitaria (Universidad de El Salvador), durante la estación transitoria seco-lluviosa en 1991. Rev de Flora y Fauna 9:1–8

Dickey DR, Van Rossem AJ (1938) The birds of El Salvador. Zool Series 23:1–609, Chicago, Field Museum of Natural History

Domínguez JP (1994) Estado actual del comercio local de fauna silvestre vertebrada en el área metropolitana. Universidad de El Salvador, Escuela de Biología

Howell SNG, Webb S (1995) A guide to the birds of Mexico and Central America. Oxford University Press, New York

Ibarra Portillo R, Vega CM (2005) Informe anual de actividades en el trabajo coordinado sobre disposición de fauna silvestre 2003, MARN-FUNZEL. Ministerio de Medio Ambiente, San Salvador

Jiménez I, Sánchez-Marmol L, Herrera N (2004) Inventario nacional y diagnóstico de los humedales de El Salvador. MARN/AECI, San Salvador

Komar O (2002) Birds of Montecristo National Park, El Salvador. Ornitol Neotrop 13:167–193

Komar O, Herrera N. (1995a) Avian inventory of El Imposible National Park, San Benito and Río Guayapa sectors. In: Avian diversity at El Imposible National Park and San Marcelino Wildlife refuge. Working Paper No. 4, Wildl Conserv Soc. pp 6–31

Komar O, Herrera N (1995b) Avian inventory of Bosque Las Lajas, Complejo San Marcelino Wildlife refuge. In: Avian diversity at El Imposible National Park and San Marcelino Wildlife Refuge. Working Paper No. 4, Wildl Conserv Soc. pp 32–55

König C, Weick F, Becking J-H (1999) Owls – a guide to the owls of the world. Yale University Press, New Haven

MARN (2006a) Áreas de conservación (on line). http://www.marn.gob.sv/areas_de_conservacion.htm. Accessed 1 Oct 2006

MARN (2006b) Mapa de uso del suelo de El Salvador: formato shape para ARCVIEW. San Salvador

MARN (Ministerio de Medio Ambiente y Recursos Naturales) (2009) Lista oficial de aves de El Salvador www.marn.gob.sv/listados. Accessed 1 Oct 2006

MARN-UICN (Ministerio de Medio Ambiente y Recursos Naturales-Unión Mundial para la Naturaleza) (2006) In: Hernández G (ed) Estado de la gestión compartida de las áreas protegidas de El Salvador. San Salvador

Marshall JT Jr (1943) Additional information concerning the birds of El Salvador. Condor 45:21–33

Mendoza R, Ramos LA (1999) El Salvador: rescate y rehabilitación de fauna silvestre. In: Drews C (ed) Rescate de fauna en el Neotropico. Editorial Universidad Nacional, Heredia, pp 285–298

Moreno CE (2001) Métodos para medir la biodiversidad. MyT–Manuales y Tesis SEA, vol. 1, Zaragoza

Peña de López AC, Herrera N (1999) El Salvador: disposición de animales por el Servicio de Parques Nacionales y Vida Silvestre en 1995. In: Drews C (ed) Rescate de fauna en el Neotropico. Editorial Universidad Nacional, Heredia, pp 299–309

PNODT (Plan Nacional de Ordenamiento Territorial) (2004) Catálogo de espacios naturales (on line). http://www.marn.gob.sv. Accessed 1 Oct 2006

Ramos L, Ricord de Mendoza Z (2000) Tenencia de fauna silvestre en El Salvador. In: Nassar-Montoya F, Crane R (eds) Actitudes hacia la fauna en Latinoamerica. Humane Society Press, Washington, DC, pp 85–95

SNET (Sistema Nacional de Estudios Territoriales de El Salvador) (2007) El clima en El Salvador. http://www.snet.gob.sv. Accessed 26 Sep 2007

Thurber WA, Serrano JF, Sermeño A, Benítez M (1987) Status of uncommon and previously unreported birds of El Salvador. Proc West Found Vertebr Zool 3:109–293

Vega I (2006) Abundancia relativa y distribución de aves rapaces nocturnas en el área natural protegida Montaña de Cinquera, El Salvador. Mesoamericana 10:50

West JN (1988) The raptors of El Imposible National Park, El Salvador. Unpublished thesis submitted for the Masters of Science Degree. Central Washington University

Chapter 12
The Owls of French Guiana

Olivier Claessens, Nyls de Pracontal, and Johan Ingels

Abstract A total of 13 owl species, 1 Tytonidae and 12 Strigidae, have been recorded in French Guiana. Twelve of them are resident species, and one, the short-eared owl *Asio flammeus*, is a rare vagrant, probably from north-western South America. Two species have been recorded for the first time in recent years: the Roraiman (vermiculated) screech owl *Megascops guatemalae roraimae* probably remained undetected before due to its rarity and to lack of knowledge of its voice. On the contrary, the burrowing owl *Athene cunicularia* is a recent addition to the French Guianan avifauna, benefiting from local deforestation for agricultural projects. Seven owl species are forest species, and four of them are restricted to primary or old contiguous secondary forest. The remaining six species live in semiopen to open habitats and forest edges. At present, French Guianan owl populations do not seem at risk, although destruction of coastal natural habitats may constitute a danger in future. For that reason and because of the small size of populations, three species are listed as Vulnerable and one as Endangered at the national level. Our knowledge of the distribution and biology of many owl species is greatly deficient and needs urgent research for conservation purposes.

Keywords Breeding • Feeding habits • Owl distribution • Habitat

O. Claessens (✉)
Clos de Bénédicte, 2510 route de Montabo, F-97300 Cayenne,, French Guiana,, France
e-mail: oclaessens@wanadoo.fr

N. de Pracontal
GEPOG, 15 avenue Pasteur, F-97300 Cayenne, French Guiana, France
e-mail: association@gepog.org

J. Ingels
Galgenberglaan 9, B-9070 Destelbergen, Belgium
e-mail: johan.ingels@skynet.be

P.L. Enríquez (ed.), *Neotropical Owls*, DOI 10.1007/978-3-319-57108-9_12

Tawny-bellied Screech Owl (*Megascops watsonii*)

12.1 Introduction

French Guiana with its 84,000 km^2 of land forms part of what is geographically called the Guiana Shield, situated along the Atlantic coast of South America between the Orinoco and Amazon rivers. It is a Département français d'Outre-Mer (DOM), i.e. a French overseas department and therefore part of the European Union. In 2013, the population density was three inhabitants per km^2, with 90% of the population living along the coast (Wikipédia 2016).

The climate is humid and tropical with a mean humidity level of 80% and a rainfall up to 400 cm per year for certain inland areas (Barret 2006). The primary rainforest covering 90% of French Guiana is one of the five remaining vast and undisturbed tropical humid rainforests in South America (Barret 2006). This vast rainforest together with a mosaic of biotopes in the coastal region, such as mangroves, savannas, marshes, sand ridges, *terra firme* forest and white-sand forest, and a shoreline of 350 km contribute to a high level of biodiversity. The Sommet Tabulaire with an altitude of 830 m is the highest hill range in French Guiana.

A national park (33,900 km^2) and seven nature reserves (approximately 3000 km^2) (Appendix 12.1) form vast protected areas spread over the entire department (Fig. 12.1). Twelve Important Bird Areas (IBAs) supplement these protected areas (Boyé et al. 2009), but they do not all benefit from a protection status.

Approximately 730 bird species are found in French Guiana, of which 13 are owls (Comité d'Homologation de Guyane 2016), which represent 1.8% of the avifauna. Seven owl species are forest-dwelling species; four of them are restricted to primary or old contiguous secondary forest, tawny-bellied screech owl (*Megascops*

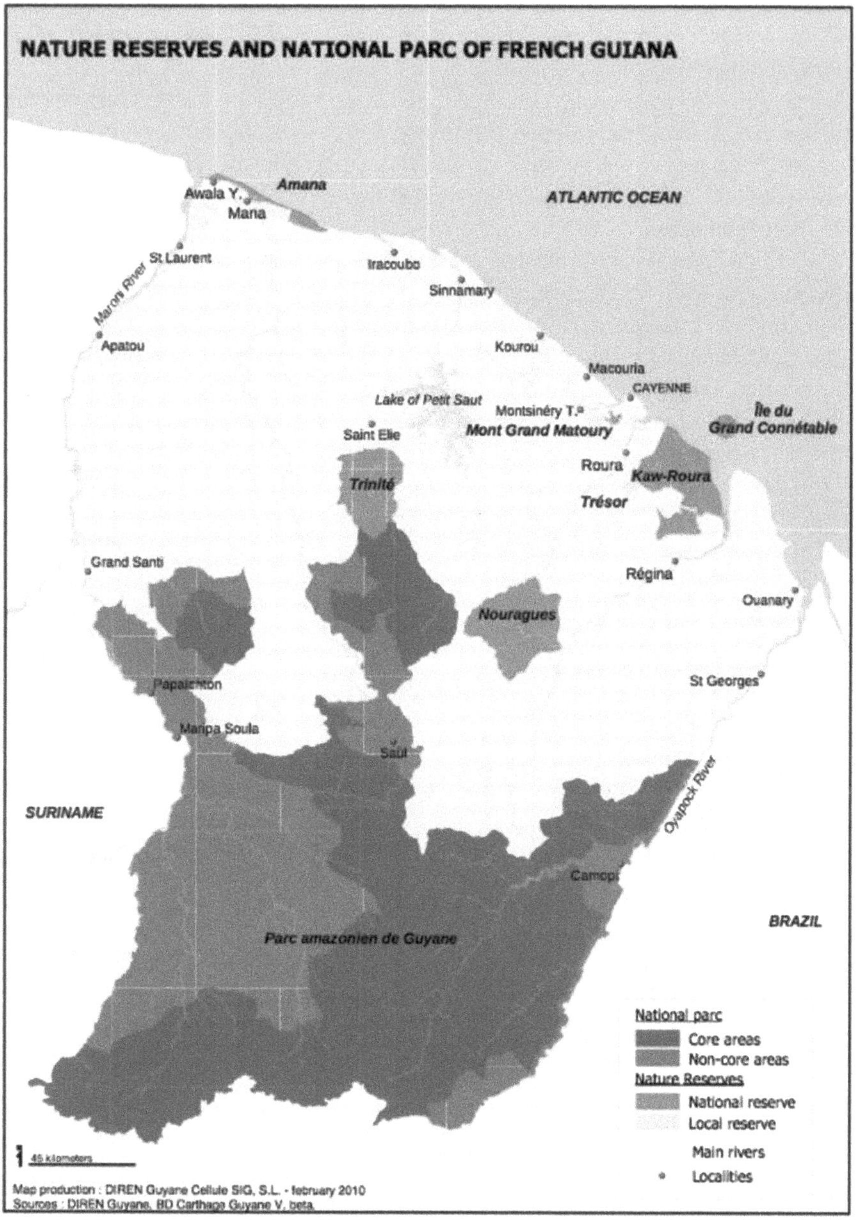

Fig. 12.1 Map with national park, nature reserves and main localities in French Guiana (© DIREN Guyane Cellule SIG, S.L., Feb 2010, BD Carthage Guyane, version beta)

watsonii), Roraiman (vermiculated) screech owl (*M. guatemalae roraimae*), crested owl (*Lophostrix cristata*) and Amazonian pygmy owl (*Glaucidium hardyi*), while spectacled owl (*Pulsatrix perspicillata*), mottled owl (*Ciccaba virgata*) and black-banded owl (*Ciccaba huhula*) are found in a wide variety of wooded habitats. Six species live in semiopen to open habitats or at forest edges: barn owl (*Tyto alba*), tropical screech owl (*Megascops choliba*), great horned owl (*Bubo virginianus*), burrowing owl (*Athene cunicularia*), striped owl (*Asio clamator*) and short-eared owl (*Asio flammeus*).

All 13 owl species are protected in French Guiana since 1986 by a decree of the French Prime Minister, updated on 4 April 2015. At present, French Guianan owl populations do not seem at risk, although destruction of coastal natural habitats may constitute a danger in future. Coastal forested ridges, savannas and white-sand forests are rare, restricted and highly threatened by the development of urbanisation, mining, infrastructures and agriculture (CTG 2016). For that reason and because of the small size of their populations, three species are listed as Vulnerable and one as Endangered at the national level by UICN-France et al. (2017). Our knowledge of the distribution and biology of many owl species is greatly deficient and needs urgent research for conservation purposes.

12.2 Species Accounts

Barn owl *Tyto alba*

Local names English, also common barn owl; French, Effraie des clochers, Chouette effraie; Creole (French Guiana), Chwèt; Kali'na, Prikyo:ko; Wayana, Pehpe; Wayãpi, Tõlõwõwõ; Sranan-tongo (Surinam Creole), Puspusi-owrukuku.

Taxonomy Represented by the subspecies *Tyto alba hellmayri*, occurring from eastern Venezuela, including Margarita Island through the Guianas to northern Brazil above the Amazon River, and also in Trinidad and Tobago (König et al. 2008). These authors consider the barn owls of the entire American continent as a separate species, *Tyto furcata*, the American barn owl, with the subspecies *Tyto furcata hellmayri* for French Guiana (Fig. 12.2).

Status and conservation Not Threatened, Least Concern (IUCN-BirdLife International 2010), CITES II (Bruce 1999). This species is a widespread resident in the coastal region. The habit of hunting along roads makes it vulnerable to collisions with vehicles. This owl often perches on electricity wires, which may involve a risk for electrocution. On 18 January 2015, an adult great black hawk *Buteogallus urubitinga* was seen preying on a barn owl near Kourou (D. Faure in GEPOG 2016).

Distribution The barn owl is rather common but restricted to the coastal region (Tostain et al. 1992; GEPOG 2016). It is regularly observed in villages and towns such as Saint-Laurent-du-Maroni, Mana, Sinnamary, Kourou, Macouria, Cayenne and its suburbs Matoury and Rémire-Montjoly, Kaw and Saint-Georges. It was

Fig. 12.2 Barn owl (*Tyto alba*) with prey, on an electricity wire along the road D8 near Mana, 14 April 2009 (Photograph Patrick Ingremeau)

found in several places in the centre of Cayenne. It was also seen at places distant from human settlements, like the marshes of Matiti along road D15 to the Dégrad Guatemala west of Kourou, the savannas of the Trésor Nature Reserve south of Roura or the Montagne Bruyère near Ouanary. This owl is absent from the forested interior, and there are no records from remote villages in the southern part of the country.

In French Guiana, this owl seems more common in the western part of its distribution range, although this may simply reflect an unequal observation effort. It is regularly observed in the marshes of Panato and in the Amana Nature Reserve near Awala-Yalimapo, along road D22 from Mana to Awala-Yalimapo, and at the Crique Irakompapi between Mana and Organabo. It is especially common in the ricefields around Mana. On 22 December 2008 during the rice harvesting period, 17 barn owls were counted around the silos of the Compagnie Rizicole de l'Ouest Guyane along road D8 (P. Ingremeau, pers. comm. 2009).

Habitat Barn owls occur in a great variety of habitats according to availability of prey. However, its occurrence is usually closely related to the presence of man. It is mostly found in villages and towns where it often roosts and nests in buildings. It is also found in partially open areas such as agricultural land, pastureland and savannas with trees, and patches of light woodland around human settlements where it hunts. There is one record of a barn owl in an isolated savanna within the forest in the Trésor Nature Reserve on 27 April 2012 (B. Villette, *fide* A. Hauselmann in GEPOG 2016); however, whether this bird was established there or just a vagrant individual is uncertain.

As rodents form an important part of its diet, it is not surprising that this owl is more numerous in the rice-growing area around Mana.

Behaviour and breeding Barn owls roost by day in tree holes or buildings. They hunt at night, starting at or slightly before dusk usually around 7.00 p.m., often from a low, open perch or sometimes by flying low over open fields. They may occasionally hunt in daylight, as seen in the marshes of Panato near Awala-Yalimapo (O. Claessens in GEPOG 2016) and suggested by an individual seen sitting at midday on the ground in the grassy savannas of Corossony near Sinnamary (A. Renaudier in GEPOG 2016).

The diet of barn owls in French Guiana was studied by analysing prey remains in pellets collected in three localities in western (Awala-Yalimapo), central (Sinnamary) and eastern (Kaw marshes) coastal French Guiana (F. Catzeflis 2015, unpublished). Small rodents, mostly *Holochilus sciureus*, followed by *Oligoryzomys fulvescens*, *Zygodontomys brevicauda* and *Nectomys rattus*, constituted 71% of 823 identified preys, bats represented 15%, birds 10%, and opossums 4%. Proportions of taxa varied among localities, with bats making up to 38% of preys at Awala-Yalimapo, birds 15% of preys at Sinnamary, while *Holochilus sciureus* made up to 86% of preys at Kaw. More occasional preys included a few juvenile individuals of the large opossums *Didelphis marsupialis* and *Philander opossum,* as well as the large Brazilian spiny tree rat *Makalata didelphoides*, with 4 individuals out of 134 rodents at Awala.

Another set of pellets collected at La Bordelaise near Macouria in 2013 (F. Catzeflis 2015, unpublished) contained remains of 26 small mammals, including as many as eight *Cryptonanus* nov. sp., a rare and still undescribed gracile mouse opossum (Catzeflis et al. 2014), besides four bats, one *Dermanura* sp. and three *Molossus* sp., four *Marmosa murina*, one *Holochilus sciureus*, one *Nectomys rattus*, three *Oligoryzomys fulvescens*, four *Zygodontomys brevicauda* and one *Rattus rattus*. Analysing prey remains in barn owl pellets gave also invaluable information on small mammal communities and lead to the discovery of new opossum and rodent species for French Guiana, including the savanna specialist *Sigmodon alstoni* (Baglan and Catzeflis 2016).

The barn owl can be also a regular predator of birds. At a nesting site at Kaw, many legs of wattled *jacana Jacana jacana* were found among prey remains. These jacanas are abundant in the surrounding marshes and were probably preyed on while roosting at night (N. de Pracontal, pers. obs. 2004). And pellets collected in the savanna of Montsinéry in June 2016 contained 16 small birds and only one common cane mouse *Zygodontomys brevicaudata* (F. Catzeflis, 2016, unpublished). In the city of Cayenne, barn owls were seen hunting by flying consistently along and just below the overhang of the roof of buildings, presumably in search of sleeping birds (O. Claessens in GEPOG 2016). An individual was observed killing a bat *Artibeus* sp. caught in a mist net during a bat census in the nature reserve of Mont Grand Matoury (O. Vrignaud in GEPOG 2016). Another one was observed preying on large insects and on a frog in the headlights of a car at the Guiana Space Centre near Kourou (A. Vinot in GEPOG 2016).

Little is known about its reproduction in French Guiana (Tostain et al. 1992). At the end of May 2005 in Kourou and at the end of October 2008 in Mana, barn owls were seen flying with strong wingbeats at some height above the houses, sometimes two individuals together, uttering long series of staccato or 'kleak-kleak' calls. We interpret this performance as courtship behaviour, although Marti et al. (2005) give this call no sexual signification.

Although it is sometimes found nesting in tree holes or in natural cavities in cliffs, this owl depends largely on human constructions, such as houses and barns, for nest sites (König et al. 2008). Local houses are constructed with a ceiling isolating the living rooms from the roof of corrugated iron sheets, which creates an attic with suitable nesting possibilities for this owl. At the beginning of November 2001, barn owls entering regularly under a roof were observed in Kourou, suggesting nesting (J.-P. Policard in GEPOG 2016). This owl also nests under roofs at Awala-Yalimapo. A pair resides and probably nests since years under the bridge over the Mana River near Mana (A. Renaudier, 2009, unpublished). At Kaw, a nest is located since at least 7 years in a corner on the floor of the attic of the local school. On 29 February 2009, i.e. in the first half of the rainy season, the nest contained two 15-day-old chicks (N. de Pracontal, pers. obs. 2009). In December 2008, a nest with five nestlings was found in a ruinous building of the Compagnie Rizicole de l'Ouest Guyane near Mana (P. Ingremeau, pers. comm. 2009). Copulation was observed in Cayenne on 28 June 2012, i.e. at the end of the rainy season (L. Kelle in GEPOG 2016). In May 2012, a pair was found nesting in a cavity in the wall of a quarry near Sinnamary (S. Uriot in GEPOG 2016). Nest sites are usually occupied for years, if birds are not prevented from entering their nesting place by human intervention.

Supposed or confirmed breeding attempts in French Guiana, as well as our observations of courtship flights, are all from February to June, i.e. during the rainy season, except a pair engaged in courtship in October. This agrees with data in neighbouring Suriname, where this owl breeds almost all year-round, with a peak between November and April at the onset of the long rainy season (Ribot 2011).

This owl does not hoot. Its most frequent call is a rough cat-like shriek, which gave it its name in Surinam Creole, puspusi-owrukuku, which means 'cat-owl'. The staccato call described above might be a particular vocalisation of the American subspecies and deserve more thorough investigations in a taxonomical perspective.

Tropical screech owl *Megascops choliba*

Local names French, Petit-duc choliba; Sranan-tongo (Surinam Creole), Owrukuku.

Taxonomy Also listed as *Otus choliba* (Sibley and Monroe 1990; Howard and Moore 1991; CINFO 1993; Fossé 2011). Represented by the subspecies *Megascops choliba crucigerus*, occurring in Trinidad and the eastern parts of Colombia, Ecuador and Peru, southern Venezuela, the Guianas and in north-eastern Brazil (König et al. 2008).

Status and conservation Not Threatened, Least Concern (IUCN-BirdLife International 2010), CITES II (Holt et al. 1999). A widespread but uncommon resident in the coastal region. The population is estimated to be less than 1000 individuals (UICN-France et al. 2017). The habit of hunting along roadsides makes it vulnerable to collisions with vehicles. Seven out of 81 records (8.6%) in the database Faune-Guyane of the Groupe d'Étude et de Protection des Oiseaux en Guyane (GEPOG 2016) are road kills. A more important threat is the destruction of its habitat for agriculture, mining, roads or urbanisation (CTG 2016). For these reasons added to its limited distribution, it is listed Vulnerable in the National Red List (UICN-France et al. 2017).

Distribution The tropical screech owl is uncommon in the coastal region in semiopen rural areas, around habitations and even in and around villages. All observations were made in the coastal region (GEPOG 2016). Being strictly nocturnal, it is more often heard than seen.

This owl is most probably absent from the forested interior, e.g. it has not been heard or seen around Saül, a well-birded area in central French Guiana with suitable habitat (V. Pelletier, pers. comm. 2009). The innermost locality is Auberge de l'Approuague near Régina, a partially deforested area c.50 km inland (A. Vinot in GEPOG 2016).

Habitat Tropical screech owls prefer bushes and shrubbery in semiopen and rural areas and light second growth forest, but also near habitations, in gardens in open villages and in suburbs and parks of towns. There are two records from the urban centre of Cayenne (GEPOG 2016). Individuals were also seen along the roads D8 and D22 between Organabo and Awala, both through old forested sand ridges in subcoastal lagoons, also in old mangrove along the river of Kaw, and in a swampy and low forest patch surrounded by open marshes at Savane Angélique, in northwestern French Guiana. In French Guiana, this screech owl avoids the dense primary forest of the interior.

Behaviour and breeding Tropical screech owls are normally not shy and can be observed at close range. It is often heard and seen in gardens around houses. At night, it may be found perched in bushes or trees and on electricity and telephone wires along roads from which it forages typically alone, mostly at lower levels (A. Renaudier, pers. comm. 2009). It can also be seen snatching insects in air along roadsides, especially at street lights. An individual was seen sitting next to a small snake on a dirt road through the marshes of Panato near Awala-Yalimapo (A. Renaudier and F. Deroussen, pers. comm. 2009). We have no other information on the diet of this owl in French Guiana.

Its song was tape-recorded at dawn in the village of Awala-Yalimapo by Renaudier and Deroussen (2008).

This owl nests mostly in cavities in trees. On 23 June 2006, a nest was discovered at Awala-Yalimapo (Renaudier et al. 2007). The nest cavity was situated on top of a dead beheaded awara palm tree *Astrocaryum vulgare*. The palm tree with a height of c.5 m and a diameter of c.15 cm stood in the middle of a small group of bushes, low trees and awara palms next to the parking of the primary school. Human habitations were only c.100 m away. The palm tree was decaying which did not allow checking the nest contents. However, the continuous presence of an adult, visible in the opening of the nest cavity, suggested incubation. About a week after the nest was discovered, vegetation below the nest tree was burnt-over and the nest abandoned. In 2013, a pair raised youngs in a hole of a wall at the agriculture school of Matiti, near Macouria. Breeding was also confirmed by the presence of fledglings along the road to Stoupan near Matoury in 2007 and at the Guiana Space Centre near Kourou in 2016. Two nestlings were seen at Matiti and at Stoupan, and only one at the Guiana Space Centre. All four nesting attempts were between April and August, during the main rainy season.

Tawny-bellied screech owl ***Megascops watsonii***

Local names French, Petit-duc de Watson; Wayãpi, Tõlõwõwõ; Sranan-tongo (Surinam Creole), Owrukuku.

Taxonomy Also listed as *Otus watsonii* (Sibley and Monroe 1990; Howard and Moore 1991; CINFO 1993; König et al. 2008; Fossé 2011). Represented by the subspecies *Megascops watsonii watsonii,* occurring from the lowlands in eastern Colombia south to north-eastern Peru and east to the Guianas and Amazonian Brazil, north of the Amazon River (König et al. 2008). These authors split the tawny-bellied screech owls of South America into the Northern tawny-bellied screech owl (*Megascops watsonii*) above the Amazon River and the Southern tawny-bellied screech owl (*Megascops usta*) below that river, on the basis of vocal differences. However, Remsen et al. (2016a) reject this split.

Status and conservation Not Globally Threatened, Least Concern (IUCN-BirdLife International 2010), CITES II (Holt et al. 1999). A common resident, mainly of the forested interior, and rare in white-sand forests in the coastal region. The population is estimated to be 130,000–300,000 pairs (UICN-France et al. 2017). Elsewhere in its range, destruction of its habitat is a major threat (König et al. 2008). However large-scale deforestation does not occur in French Guiana. The species occurs in good numbers in protected areas in the interior, i.e. in the Trésor, Nouragues and Trinité Nature Reserves and in the Parc Amazonien de Guyane. It is listed Least Concern at the national level (UICN-France et al. 2017).

Distribution The tawny-bellied screech owl is widespread and quite common in the forested interior, from sea level up to 600 m.a.s.l. (GEPOG 2016). It has been found in most of surveyed areas of primary forest throughout French Guiana, e.g. in the Mana River basin in north-western French Guiana, along the Route de Saint-Elie, around the lake of Petit Saut, in the Trinité and Nouragues Nature Reserves, around Saül in the centre, at Trois-Sauts in the extreme south-east of the country, in the area of the Crique Armontabo and on the Montagne de Kaw in the north-east. It was not found on the Massif Lucifer, a plateau 550 m.a.s.l., within its normal range in western French Guiana, although it was present at the Crique Tobie (120 m.a.s.l.) a few kilometres away. It was also found up to an altitude of 600 m on the Mont Itoupé (Sommet Tabulaire), in southern French Guiana. Singing males separated from 350 to 800 m were heard on several occasions at two localities near Saint-Georges in the period 2014–2016 (O. Claessens in GEPOG 2016). On 12 November 2007, three singing individuals were heard along 1 km of forest trail in the area of the Crique Limonade near Saül, in central French Guiana (O. Claessens in GEPOG 2016). White-sand forests, in the western part of the coastal region, only provided very few observations of this screech owl near Mana and Organabo (Tostain et al. 1992, A. Renaudier, pers. comm. 2009). No singing individuals were heard there, thus suggesting that these localities on the margins of the species' range lie outside its normal habitat and may not support stable breeding populations.

Habitat This owl prefers old-growth primary and mature secondary lowland forest in the interior, where it lives in the understorey and at middle levels. It is seldom seen in clearings and at forest edges. The scarcity of records in the coastal white-sand forests confirms that this dry, low and clear forest type does not constitute an optimal habitat for this owl.

Behaviour and breeding Little is known about general habits and breeding of the tawny-bellied screech owl, which are probably similar to those of other screech owls (Holt et al. 1999; König et al. 2008). It is nocturnal but may start calling just before dusk. It is less usual to hear it singing in the middle of the day, for example, at 10.00 a.m. on 1 July 2014 near Saül (O. Claessens in GEPOG 2016).

We have no information on the diet of this screech owl in French Guiana. Elsewhere within its range, it mainly feeds on insects but probably also occasionally on small vertebrates (König et al. 2008). This owl is regularly caught at night, less than 2 m above the ground, in mist nets raised along forest trails in search for bats. An individual was seen hunting along such a trail, making short and fast flights from perch to perch (O. Claessens, pers. obs. 1994).

Although there is no doubt about the breeding of this screech owl in French Guiana, we still have no confirmed breeding record. Tawny-bellied screech owls occupied in July 1987 a nest hole in a tree along the Route de Saint-Elie. This nest was abandoned in March 1987 by red-necked woodpeckers *Campephilus rubricollis* (Tostain et al. 1992). However, whether the nest hole was used for breeding or simply for roosting could not be verified. Conversely, an individual was seen on 7 May 2008 at the entrance of a large hole 10 m up in a tree, at km 194 on road N1 near Mana (A. Vinot in GEPOG 2016).

Vermiculated Screech Owl *Megascops guatemalae*

Local names Petit-duc vermiculé.

Taxonomy Also listed as *Otus guatemalae* (Sibley and Monroe 1990; Howard and Moore 1991; CINFO 1993; Fossé 2011). Represented by the subspecies *Megascops guatemalae roraimae* (Roraiman screech owl, Petit-duc de Roraima), occurring in the mountains of northern Venezuela and mountainous regions in southern Venezuela and adjacent Brazil and in Guyana and Suriname (Holt et al. 1999; Braun et al. 2007; Ottema et al. 2009).

Sibley and Monroe (1990) and others (see Remsen et al. 2016a) split *Megascops guatemalae* into two species, with *M. guatemalae* (Mexico to Costa Rica) and *M. vermiculatus* for all South American taxa, named Guatemalan and vermiculated screech owl, respectively. König et al. (2008) consider *Megascops g. vermiculatus*, *M. g. roraimae* and *M. g. guatemalae* (Howard and Moore 1991) as full species, on vocal, morphological and zoogeographical grounds, named vermiculated, foothill or Roraima and Guatemalan screech owl, respectively. However, Remsen et al. (2016a) reject these splits.

Status and conservation Not Globally Threatened, Least Concern (IUCN-BirdLife International 2010), CITES II (Holt et al. 1999). The vermiculated screech owl is a recent addition to the avifauna of French Guiana. First discovered in 2009, it is now known from eight localities, with new localities being added to its known distribution each year. An apparently rare resident, it is obviously more widespread and more common than previously thought. Since its forested habitat extends over most of the country and is still mostly intact, the vermiculated screech owl does not seem at risk in French Guiana.

Distribution The vermiculated screech owl was only recently discovered in French Guiana, and its distribution is still incompletely known. The first records were from the Montagne de Kaw, south-east of Cayenne, where the species was mist netted and photographed, or tape-recorded, at three different sites (Claessens and CHG 2015). Most subsequent records were on the Petites Montagnes Tortue west of Régina (Claessens and CHG 2015), at the Savane-roche Virginie south of Régina, at the Crique Nancibo near Cacao, at the Crique Maweyo and at the Crique Gabaret near Saint-Georges, and again on the Montagne de Kaw (Claessens et al. 2014, and unpublished records). These localities are all in the north-eastern part of French Guiana and represent an extension of the known species' range of 450 km eastwards (O'Shea and Ramcharan 2012). The species was also heard on Mont 414 in the Trinité Nature Reserve and photographed and tape-recorded at Borne 4 near the border with the Brazilian state of Amapá, in extreme southern French Guiana. Thanks to the knowledge of its song, this screech owl will undoubtedly be found in more localities throughout the country in future.

Habitat Rainforest on slopes of foothills of tepuis and other mountains, locally at lower elevations (König et al. 2008). This species needs almost solid forest (Holt et al. 1999). All the localities where it is known in French Guiana are in primary forest, either in lowland forest or on forested hills. Altitudes for the French Guianan localities range from 5 m at the Crique Nancibo up to 400 m.a.s.l. on Mount 414 in the Trinité Nature Reserve.

Behaviour and breeding Nothing is known about the biology of the vermiculated screech owl in French Guiana.

Crested Owl ***Lophostrix cristata***

Local names French, Duc à aigrettes; Sranan-tongo (Surinam Creole), Owrukuku.

Taxonomy Represented by the nominate subspecies *Lophostrix cristata cristata*, occurring in South America east of the Andes, from Venezuela, the Guianas to Colombia and the Amazon region south to northern Bolivia, west to eastern Ecuador and eastern Peru (König et al. 2008).

Status and conservation Not Globally Threatened, Least Concern (IUCN-BirdLife International 2010), CITES II (Holt et al. 1999). This owl is a widespread and common resident, mainly in the forested interior where it is the commonest owl. Local densities of 4 pairs per km^2, with singing individuals 300–500 m apart, were found in the Dékou-Dékou Biological Reserve in north-western French Guiana in January 2011, at Mont Itoupé (Sommet Tabulaire) in the south in October 2010, and near Saint-Georges in the north-east in December 2014 and 2015 and March 2016 (O. Claessens in GEPOG 2016). The total population is estimated to be above 200,000 pairs (UICN-France et al. 2017). In Saül, in the first half of October, up to six birds were sometimes heard singing each night in the forest around the village (J. Ingels, pers. obs. 2008, 2009). Planned deforestation in the coastal area (CTG 2016) will cause the distributional limit to recede towards the interior; however, this projected loss is supposed to be negligible in the short term with regard to the large numbers and the remaining large distribution over the country of this owl. It is listed Least Concern in the National Red List (UICN-France et al. 2017).

Distribution The crested owl is widespread in French Guiana. It was found in almost all surveyed localities within the forest block throughout the entire country. In the coastal region, it is present as long as there are large bouts of forest, for example, at the Savane des Pères near Kourou, only 3.5 km from the coast and near Macouria south of Cayenne. It was heard on 3 November 2011 on Mont Baduel, an isolated forested hill in Cayenne (T. Luglia in GEPOG 2016).

Habitat Crested owls are mainly found in lowland forest, either in pristine or in old secondary forest, where they prefer areas of dense vegetation like vine tangles and edges of treefalls (Tostain et al. 1992).

Behaviour and breeding Crested owls are nocturnal. They hunt and call from perches at mid- to subcanopy levels, but rest often lower in dense undergrowth. We have no information on its diet and hunting behaviour in French Guiana. Crested owls are difficult to watch and they are more often heard than seen. The typical song is a short, low-pitched, toadlike 'grrroow', sporadically brought throughout the night. Nothing is known about the reproduction of this owl in French Guiana. Fledged juveniles were seen in the Trinité Nature Reserve on 5 November 2016 and near Sinnamary on 26 November 2016, suggesting hatching in the middle of dry season (O. Claessens in GEPOG 2016, J.-C. Varlez in GEPOG 2016).

Spectacled Owl *Pulsatrix perspicillata*

Local names French, Chouette à lunettes; Creole (French), Chwèt-linèt; Kali'na, Po:po:po, Mutu; Wayana, Kulëu, Pupuli; Wayãpi, Tapupu; Sranan-tongo (Suriname Creole) Krabu-owrukuku.

Taxonomy Represented by the nominate subspecies *Pulsatrix perspicillata perspicillata,* occurring from Venezuela, the Guianas, Brazil and from eastern Colombia south to northern Bolivia (König et al. 2008).

Status and conservation Not Globally Threatened, Least Concern (IUCN-BirdLife International 2010), CITES II (Holt et al. 1999). This owl is a local resident, fairly common in the coastal region. Densities can be high in suburban and rural areas, e.g. pairs are spaced by c.500 m in the suburbs of Cayenne (GEPOG 2016). It also occurs with lower densities in the forested interior. It is listed Least Concern in the National Red List (UICN-France et al. 2017).

Distribution The spectacled owl is widespread in French Guiana (GEPOG 2016). In the interior, it has been found in more than 30 localities spread from north to extreme south, e.g. at Akouba Booko Gô Soula on the upper Malani River, in the Mitaraka Massif in the Tumuc-Humacs, at Borne 4 on the Brazilian border and at Trois-Sauts, including the Trinité and Nouragues Nature Reserves, in a wide area around Saül, on the Pic du Croissant and on the Piton de l'Armontabo near Saint-Georges, and in the Trésor Nature Reserve on the Montagne de Kaw.

In the coastal region, the owl is present from west to east although it is rare near Mana in the extreme western part. It looks more common from Kourou to Cayenne, Rémire-Montjoly and Matoury, i.e. in the most populated and man-altered area.

Habitat Spectacled owls occur in a wide variety of wooded habitats, though most commonly in disturbed, semiopen habitats. In undisturbed primary forest, it is usually found along creeks and small streams, in vine tangles or in treefall edges. It is also found in old mangroves along the coast, in secondary forest in man-altered areas and in rural and urbanised areas as long as large trees are present. This owl is rare in secondary and semiopen white-sand forests of north-western French Guiana (A. Renaudier, pers. comm. 2009).

Behaviour and breeding Spectacled owls are essentially nocturnal. They start activities after dusk and continue to dawn. Spectacled owls are more often heard than seen. The song is a typical series of deep, resonant hoots, descending, growing softer and faster and resembling the sound produced by a vibrating metal sheet. Male and female give antiphonal duets. This owl sings year-round and at any hour of the night; however, it is most vocal during calm, moonlit nights. At Saül, individuals were heard singing and pairs duetting every night in the first half of October (J. Ingels, pers. obs. 2008, 2009).

Despite being common, little is known of the nesting of spectacled owls in French Guiana. A pair nested in 2015 and 2016 in a tree hole in Cayenne. The nest was 7 m up in the cavity left by a broken main branch of a mango tree. This pair raised two young each year, which fledged in the first half of September and stayed for at least 1 month with the adults in neighbouring trees. In addition, recently fledged young have been found on 9 December 1987 on the Montagne du Mahury in Rémire-Montjoly, on 3 March 1988 along the Route de l'Anse near Sinnamary (Tostain et al. 1992), on 11 September 1993 at the lake of Petit Saut (J. Judas in GEPOG 2016) and at the end of May 2009, on 3 July 2010, on 5 April 2012 and on 28 and 30 September 2013 in Cayenne (T. Réquillart, pers. comm. 2009, M. Cobigo and O. Claessens in GEPOG 2016). Thus, breeding seems to occur year-round. On 26 February 2014, a pair was photographed perched side by side, singing and preening one another, at Stoupan, Matoury (S. Uriot in GEPOG 2016).

Little is known of the diet of spectacled owls in French Guiana. On 26 February 2012 on the Montagne de Kaw, an adult spectacled owl was observed preying on a rusty tree frog *Hypsiboas boans* (M. Dewynter, pers. comm. 2012). Of 34 prey remains identified in pellets collected in September and October 2015 and 2016 in Cayenne from a breeding pair and their offspring, half were Norway rats *Rattus rattus*, the remaining were seven flat-faced fruit-eating bat *Artibeus planirostris*, six brown rats *R. norvegicus*, a young common opossum *Didelphis marsupialis*, two small birds and a lizard (F. Catzeflis, 2016, unpublished).

Great Horned Owl ***Bubo virginianus***

Local names French, Grand-duc d'Amérique, also Grand-duc de Virginie; Creole (French Guiana), Grochwèt; Kali'na, Urukureya, Abilikiki, Pilitoko; Wayãpi, Ulukulea; Wayana, Pehpe; Sranan-tongo (Surinam Creole), Owrukukugranman.

Taxonomy Represented by the subspecies *Bubo virginianus nacurutu* from northern South America (König et al. 2008).

Status and conservation Not Threatened, Least Concern (IUCN-BirdLife International 2010), CITES II (Holt et al. 1999). A rare resident in coastal French Guiana, though considered relatively common by Tostain et al. (1992). It is listed Endangered at the national level, with a population estimated to be less than 100 pairs (UICN-France et al. 2017). The major threat is loss of habitat due to urbanisation, industry or mining.

Distribution The great horned owl has only been heard and seen in the coastal region, within 5 km from the coast, with the exception of the marshes of Kaw, c.10 km from the estuary of the Approuague River. Records in the interior of the country are at least very rare, if not erroneous. A sight record by C. Érard in 1991 along the Sinnamary River near Petit Saut in the forested interior is mentioned by Tostain et al. (1992).

Observations were all made in the coastal region of French Guiana, near Awala-Yalimapo, in the ricefields near Mana, along the Route de l'Anse and in the Savane Jojo near Sinnamary, in the marshes of Matiti west of Kourou, and at Concorde and Stoupan near Macouria (GEPOG 2016). Most of these localities are favourite birding sites, so we believe that this owl is more evenly distributed along the coast of French Guiana.

Habitat Although it can be found in a wide range of semiopen to open habitats, this owl typically inhabits edges of old mangrove in French Guiana. Territories must include woody habitats as roosting and nesting sites, as well as open areas for hunting. Great horned owls are common in Awala-Yalimapo and in low bushes and small trees on sand ridges along the beach near this Amerindian village, where they often use lady-of-the-night cactuses *Cereus hexagonus* as perches, and in old mangrove near Mana and along road D15 east of Kourou. Hunting areas are beaches, ricefields, swamps, grassland savannas or grazing lands at both sides of these forested ridges. There are a few records inland at the edge of secondary forest in man-disturbed areas near Macouria.

Behaviour and breeding The great horned owl is a discrete owl, roosting by day in dense foliage of bushes or trees. At dusk when becoming active, it often utters a few calls from its roost, before flying off. It was seen hunting in the marshes of Panato near Awala-Yalimapo as early as 6.00 p.m., i.e. about 1 h before nightfall, and it was heard singing in the marshes of Matiti at 6.00 p.m. when a great black hawk *Buteogallus urubitinga* perched in a nearby tree (A. Vinot in GEPOG 2016) and in the ricefields near Mana around 7.20 p.m.

Few data on feeding are available for French Guiana. On beaches used as laying sites by sea turtles, great horned owls are regular predators of young turtles emerging after hatching.

Great horned owls have not yet been found nesting in French Guiana, although pairs accompanied by a presumed juvenile were observed on three occasions along road D15 in the marshes of Matiti in May and November (V. Rufray, pers. comm. 2011). Songs were heard in April (2), June (1), September (2) and October (1).

Mottled Owl ***Ciccaba virgata***

Local names English, also American wood owl (Holt et al. 1999); French, Chouette mouchetée; Sranan-tongo (Surinam Creole), Owrukuku.

Taxonomy Also listed as *Strix virgata* (Sibley and Monroe 1990; CINFO 1993; Howard and Moore 1991; König et al. 2008; IOC 2010). This species is represented by the subspecies *Ciccaba virgata macconnelli*, occurring in the Guianas only (König et al. 2008).

Status and conservation Not Threatened, Least Concern (IUCN-BirdLife International 2010), CITES II (Holt et al. 1999). This owl is rare, poorly known but probably widespread resident.

Distribution The mottled owl is only known from the northern half of the country, south to Saül in central French Guiana, although probably also present further south. Because of our poor knowledge of its vocalisations and risks of confusions with calls of other species, especially of the black-banded owl, auditory records must be considered with caution. In the interior of French Guiana, confirmed records, i.e. sightings or captures, have been reported from the Saül area (V. Pelletier, pers. comm. 2009), from the camp Inselberg and Saut Pararé in the Nouragues Nature Reserve, from the road N2 near Cacao and from several localities near Roura, from Saint-Eugène at the lake of Petit Saut, from the Mont Tabulaire and the area of the Crique Aya in the Trinité Nature Reserve, and from the Massif Lucifer. In the coastal region, it has been seen in the marshes of Matiti, at the Guiana Space Centre and on the Montagne des Singes near Kourou, at Paracou and in the Savane Terre Blanche near Sinnamary (A. Renaudier, pers.comm. 2011) and along the road N1 east of Iracoubo. Its song was registered near Montsinéry-Tonnégrande along the road of Saut Léodate (O. Claessens in GEPOG 2016). In the region of Mana at the western end of the coastal region, it has been seen in fragmented forest at Bassin Mine d'Or, between Mana and Awala-Yalimapo, and on the Montagne de Fer (A. Renaudier in GEPOG 2016, J. Ingels, pers. comm. 2009). Calls or songs attributed to this owl were also heard in the region of Saint-Georges.

Habitat The mottled owl is most common in lowland primary forest, both inside and at forest edges, often in humid forests along rivers or creeks. However, it is also found in secondary woodland. It has been seen at several occasions in forest patches and in pastures in the coastal region. It was mist netted at 560 m.a.s.l. on the Massif Lucifer on 7 November 2005 (K. Pineau in GEPOG 2016).

Behaviour and breeding Little is known of the behaviour of the mottled owl. It is strictly nocturnal, hunting only at night. It rests during day in vine tangles, in tree falls or in thick vegetation along creeks, where it roosts in the understorey as low as two metres above the ground. An individual was mist netted 17 m up in the canopy of primary forest on the Massif Lucifer (K. Pineau in GEPOG 2016). It is a shy owl, leaving its roost when approached and moving with agility through dense understorey vegetation (A. Renaudier, pers.comm. 2009, O. Claessens, pers. obs. 1996). We have no information on its diet and nesting in French Guiana.

Black-banded Owl ***Ciccaba huhula***

Local names French, Chouette huhul; Sranan-tongo (Surinam Creole), Peniblaka-owrukuku.

Taxonomy Also listed as *Strix huhula* (Sibley and Monroe 1990; CINFO 1993; König et al. 2008; IOC 2010; Fossé 2011). Represented in French Guiana by the nominate subspecies *Ciccaba huhula huhula*, occurring in eastern Colombia, southern Venezuela and the Guianas to north-eastern Brazil, south to eastern Peru, north-western Argentina, northern Paraguay and north-eastern Argentina (König et al. 2008).

Status and conservation Not Threatened, Least Concern (IUCN-BirdLife International 2010), CITES II (Holt et al. 1999). This owl is a scarce and poorly known resident. The scarcity of records may be due to its true rarity and/or to its secretive behaviour.

Distribution The black-banded owl was found mainly in the coastal region, i.e. near Saint-Laurent-du-Maroni, at the edge of the Trou Poisson Savanna near Iracoubo, around Sinnamary, at Wayabo and at the Guiana Space Centre near Kourou, at Stoupan and on the Montagne de Kaw near Roura. In the interior, it was sighted or its song was heard at Saut Tamanoir on the Mana River, on the Mont Belvédère in central French Guiana, in the Trinité and Nouragues Nature Reserves, around Saül (V. Pelletier, pers. comm. 2009) and at Petit Saut on the Sinnamary River.

Habitat This owl is found in a variety of forested habitats, though primarily in lowland forest, both inside and at edges of forest.

Behaviour and breeding Almost nothing is known of the behaviour of the black-banded owl in French Guiana. It is strictly nocturnal, and it seems to rest and hunt higher in trees than the mottled owl. On 14 November 2014 at the end of the dry season, a fledgling was fed by an adult at the Guiana Space Centre near Sinnamary (S. Uriot in GEPOG 2016).

Amazonian Pygmy Owl ***Glaucidium hardyi***

Local names French, Chevêchette d'Amazonie; Sranan-tongo (Surinam Creole), Owrukuku.

Taxonomy Also listed as Hardy's pygmy owl (Sibley and Monroe 1990). Monotypic, *Glaucidium hardyi*, occurring in Amazonian South America from north-eastern Brazil through the Guianas to south-eastern Venezuela, north-eastern Bolivia, eastern Ecuador and south-eastern Peru (Holt et al. 1999; König et al. 2008).

Status and conservation Not Globally Threatened, Least Concern (IUCN-BirdLife International 2010), CITES II (Holt et al. 1999). This owl species is widespread and a common resident throughout French Guiana. The local density may be high, e.g. seven individuals were heard singing on less than 1 km^2, a mean 300 m of each other, sometimes even less than 200 m, near Saint-Georges in March 2016 (O. Claessens in GEPOG 2016).

Distribution The Amazonian pygmy owl is widespread throughout the forested interior of French Guiana (Fig. 12.3), from the northern limit of the forest block in the coastal area, south to the border with the Brazilian state of Amapá. In the coastal area, it is present in localities connected to the interior continuous forest, as close as 6 km from the coast at Bassin Mine d'Or near Mana, but absent from disjunct forest patches nearer to the coast, e.g. around Cayenne.

Habitat This owl lives in the upper strata of tall humid lowland forest, from sea level up to 560 m on the Lucifer plateau. It is sometimes found at forest edges around clearings. It was heard in 1995 on a 2.5 ha-island of the lake of Petit Saut (O. Claessens in GEPOG 2016); however, it is absent from isolated forest fragments in the coastal area.

Fig. 12.3 Amazonian pygmy owl (*Glaucidium hardyi*) along the Piste du Dégrad Florian, near Mana, 11 October 2008 (Photograph Tanguy Deville)

Fig. 12.4 Amazonian pygmy owl (*Glaucidium hardyi*) at Angoulème near Mana, 2 June 2005 (Photograph Alexandre Renaudier)

Fig. 12.5 Amazonian pygmy owl, *Glaucidium hardyi* near Maripasoula, 25 September 2011 (Photograph Tanguy Deville)

Behaviour and breeding Amazonian pygmy owls are active and vocal both at night and partly during the day (Figs. 12.4 and 12.5). It is often heard, however, difficult to see due to its tiny size. It usually lives in mid-level and subcanopy strata, where its small size and fast flight make it an elusive bird. This owl is easily attracted by playback or imitation of its song. On 4 June 2005 at Angoulême near Mana, a pair was attracted and excited by the whistling of an observer and copulated (A. Renaudier in GEPOG 2016). A breeding pair was observed near Maripasoula from 24 to 26 September 2011, in the dry season (Deville and Ingels 2012). The nest which contained a full-grown juvenile was an old woodpecker hole, 25 m up in a vertical branch of an isolated dead tree, in a clearing along a dirt road. The diameters of the branch and nest hole entrance were c.20 cm and c.7.5 cm, respectively. The young was supplied by the adult with a small, half-eaten, unidentified passerine at 2.00 p.m. and the day after with a cicada at 8.50 p.m. During observation periods in the morning and in the afternoon, one adult stayed near the nest all the time, leaving only for short foraging bouts. It sometimes sang softly from inside the nest or at the nest entrance. We have no other information about the diet of this pygmy owl or about its breeding biology in French Guiana. On 4 July 2014 at Saül, an Amazonian pygmy owl was observed perching at the entrance of a natural hole created by a broken branch, at least 15 m up in a tall, isolated dead tree at the edge of a large clearing (O. Claessens and S. Scellier in GEPOG 2016). That the hole was an active nest could not be ascertained.

Burrowing Owl ***Athene cunicularia***

Local name French, Chevêche des terriers.

Taxonomy Also listed as *Speotyto cunicularia* (Sibley and Monroe 1990). Probably represented by subspecies *Athene cunicularia minor*, occurring in the savannas of the upper Rio Branco in Brazil and adjacent parts of Guyana and Suriname.

Status and conservation Not Threatened, Least Concern (IUCN-BirdLife International 2010), CITES II (Holt et al. 1999). The burrowing owl is a recent addition to the avifauna of French Guiana, first discovered in the west of Kourou in 2006 (Giraud-Audine et al. 2007). Successive nesting attempts were observed near Sinnamary in 2006 and 2007 (Ackermann et al. 2008); however, local farmers said they knew the species there since 2004. New breeding sites or individual birds are found each year. This owl is a locally uncommon resident and breeding species. Main dangers for this owl are road kills and destructions of nesting sites, but these risks are low in French Guiana. Despite the small population in French Guiana, the burrowing owl does not seem exposed to extinction risk. Numbers are rather increasing. Local deforestation for agriculture, mining or industrial projects (CTG 2016) may favour further spreading of the burrowing owl in French Guiana, as is observed in the Amazonian department Pando in Bolivia (Miserendino 2007). On the other hand, natural savannas are seriously threatened by similar projects; thus, the availability of natural habitat may be reduced in future. This owl is listed as Near Threatened in the National Red List, because of its small, though increasing, population (UICN-France et al. 2017).

Distribution All records of burrowing owls in French Guiana are located in a narrow coastal region, due to the uniform forest cover of much of French Guiana. Most of them are concentrated in the Kourou–Sinnamary area (GEPOG 2016).

The first single individual was seen and photographed in 2006 on recently deforested land, ca.20 km south-west of Kourou (Giraud-Audine et al. 2007).

In the western part of the coastal area, a road kill was found in 2008 at km 189 on the road N1, near the crossing with road D8 to Mana. A single individual was seen at Mana in 2011. And a breeding pair was observed at Iracoubo at the end of 2015 and in 2016.

Near Sinnamary, successive nesting attempts were observed in 2006 and 2007 in the savannas of Corossony (Ackermann et al. 2008). And a breeding pair was photographed in a nearby pasture in 2015. It was also known there by farmers since several years (C. Bergère, pers. comm. 2015).

Around Kourou, after the discovery of the species in 2006, burrowing owls were seen in a savanna at the Guiana Space Centre west of Kourou in 2007 and in the marshes of Matiti, east from Kourou in 2009, 2010 and 2015. A breeding pair was photographed in 2015–2016 in the Aubanèle Savanna south-west from Kourou, where it was known at least since 2014. And a single individual was photographed on a building of the hospital of Kourou in December 2012.

A solitary burrowing owl was found in the city of Cayenne for the first time in May 2012. It was joined by a mate only in November 2015, and the pair successfully bred in 2016. An individual was seen at Cotonnières between Cayenne and Matoury in April 2016.

East of Cayenne, a burrowing owl was seen in the nature reserve of the Grand Connétable Island, 18 km off the estuary of the Approuage River, on 4 July 2008 and again in September 2009 (Ackermann et al. 2008; Renaudier et al. 2010).

The burrowing owl is a recent arrival, expanding its distribution range on the Guiana Shield (Giraud-Audine et al. 2007; Ackermann et al. 2008; Ottema et al. 2009). The above observations are indicative of a rapid though probably still underestimated colonisation of non-forested coastal areas of French Guiana. Thanks to vast, private, inaccessible pasturelands and to the species' ability to settle also in small lawns or in waste areas in cities, there are undoubtedly more localities with pioneer individuals or established pairs, as suggested by the discovery of breeding pairs by birdwatchers many years after their first observation by local farmers.

Habitat Most long-staying individuals and nesting pairs were established in pastures, in savannas or in grassy areas in villages and cities. These habitats can be described as the natural habitats of this species in French Guiana. The first bird, however, was seen in a recently cleared land along a dirt road, still covered with burnt and unburnt tree remains and situated in a largely forested area. Savannas and grazing lands, more natural habitats for this owl, were only found at a distance of several km. The Grand Connétable Island, where a burrowing owl was found, is an 8 ha rocky island, eroded by an earlier exploitation of guano, now locally covered with short bushes and cactuses and still occupied by large colonies of seabirds.

Behaviour and breeding Little is known about the biology and diet of the burrowing owl in French Guiana. Although burrowing owls prefer drier habitats, its presence in more humid savanna areas of the Guiana Space Centre is not surprising. This bird was seen foraging on frogs and toads. Batrachians and crustaceans have been recorded elsewhere as food items (Holt et al. 1999). Another bird was photographed while preying on a toad, eating earthworms and a common cane mouse *Zygodontomys brevicauda* on an urban lawn (O. Tostain and M. Giraud-Audine in GEPOG 2016). Three burrows found were excavated in the sandy soil of an embankment of a drainage ditch, in grazing lands and in a burnt-over savanna. A pair successfully bred in a drainage pipe of a wasteland, an opening into a ditch along an urban street. Fence posts and termite hills are frequently used as lookouts, where the owls are easily observed during day. Urban birds rest on the ground, on small buildings and under overhangs. The record of an individual on the Grand Connétable Island highlights the species' ability to disperse over unsuitable habitats, which may facilitate its colonization of new suitable sites.

Striped Owl ***Asio clamator***

Local name French, Hibou strié; Wayana, Kulëu; Wayãpi, Ulukuleayawa; Sranantongo (Surinam Creole), Owrukuku.

Taxonomy Previously listed as *Pseudoscops clamator* by the South American Classification Committee (SACC) who finally merged it into *Asio* in order to maintain the latter genus monophyletic, following results from genetic analysis by Wink et al. (2009) (Remsen et al. 2016b). Represented by the nominate subspecies *Asio clamator clamator,* occurring from Colombia and Venezuela, south to eastern Peru and central to north-eastern Brazil outside the Amazon forest (König et al. 2008).

Status and conservation Not Threatened, Least Concern (IUCN-BirdLife International 2010), CITES II (Holt et al. 1999). This species is fairly common in coastal French Guiana (Tostain et al. 1992), although its status is poorly known because of its strictly nocturnal habits. The habit of hunting along roads makes it vulnerable to collisions with vehicles. One to three individuals per year are found dead along an 8 km-long section of road D8 near Mana (A. Renaudier, pers. comm. 2009). This owl often perches on electricity wires, what may involve a risk for electrocution. It is listed as Vulnerable in the National Red List, due to the small size of the population (UICN-France et al. 2017).

Distribution Restricted to the coastal region. Most observations of striped owls were made around Mana and around Kourou. A total of six individuals in July 2008 and five individuals in April 2009 were counted after sunset sitting on a telephone wire along the 8 km-long road section mentioned above (P. Ingremeau, pers. comm. 2008, J. Ingels, pers. obs. 2009). It is also present near Sinnamary and Matoury and in and around Cayenne. Tostain et al. (1992) mention an observation on the inselberg Piton Baron in the Massif des Émerillons, in southern French Guiana. An individual settled during a few days in April 2013 on the Grand Connétable Island, a rocky island 18 km off the estuary of the Approuague River east of Cayenne.

Habitat Striped owls are found in a variety of semiopen to open habitats, such as light woodland with open areas, savannas and marshes with scattered trees, plantations and ricefields, agricultural and suburban and urban areas, e.g. in the cities of Cayenne, Rémire-Montjoly and Kourou, with wastelands, gardens, hedges or patchy forest, including along mangrove.

Behaviour and breeding This owl is nocturnal, becoming active at dusk. Most observations of individuals sitting on fence posts, traffic signs and electricity and telephone wires along roads are made by drivers at night. It hunts over open countryside, flying low and swooping down on prey. On 19 March 2006, an adult striped

Fig. 12.6 Striped owl (*Pseudoscops clamator*) with prey probably *Nectomys rattus* on a telephone wire along the road D8 near rice silos near Mana, 14 April 2009 (Photograph Patrick Ingremeau)

owl was seen hunting by flying low over the bushes and papyrus vegetation of the marshes of Panato near Awala-Yalimapo, in the way barn owls hunt (A. Renaudier in GEPOG 2016). It also perches on fence posts, bare branches and wires looking out for prey and then dives to the ground. A striped owl sitting on a telephone wire along road D8 with a rodent, probably *Nectomys rattus* in its claws, was photographed on 14 April 2009 (Fig. 12.6; P. Ingremeau, pers. comm. 2009). On 13 April 2016, a dispersing individual which stayed on the Grand Connétable Island was observed eating a ruddy turnstone *Arenaria interpres*. One pellet collected on the site contained legs of shorebirds, apparently semipalmated sandpiper *Calidris pusilla (*K. Pineau/Nature Reserve of the Grand Connétable Island in GEPOG 2016, S. Uriot, pers. comm. 2016).

A pair bred in August 2015 in a wasteland prior to building in Cayenne. Although they were seen only after the young had fledged, they had obviously nested on the ground on a knoll covered with herbaceous and bushy vegetation, from where the birds were seen flying out on evenings. An adult worried by the presence of the observer yelled at him a series of barking calls while ruffling its plumage (O. Claessens in GEPOG 2016). According to observations of fledged young, brood size is most often one or two; however, three recently fledged juveniles with an adult were observed near Mana on 21 September 2008 (Fig. 12.7; A. Renaudier, pers.

Fig. 12.7 Striped Owl (*Pseudoscops clamator*) two juveniles on a telephone wire along the road D8 near Mana, 20 September 2008 (Photograph Alexandre Renaudier)

comm. 2008). Recently fledged young were seen or heard year-round (13 independent records; GEPOG 2016, also Renaudier and Deroussen 2008, and A. Renaudier, pers. comm. 2010).

Stygian Owl ***Asio stygius***

Local name French, Hibou maître-bois.

This owl was once cited for French Guiana on the basis of a tape-recording (Renaudier and Comité d'Homologation de Guyane 2010). It was the only record for the country. However, this record was later revised and rejected by the French Guiana Rarities Committee. Thus, this owl no longer belongs to the bird list of French Guiana (Claessens and Comité d'Homologation de Guyane 2015).

Short-Eared Owl ***Asio flammeus***

Local name French, Hibou des marais.

Taxonomy Probably represented by the subspecies *Asio flammeus pallidicaudus* occurring in northern South America (Restall et al. 2006). This subspecies is not recognised by König et al. (2008) who maintain *Asio flammeus bogotensis*. The dates of the French Guianan records suggest that they may involve austral migrants belonging to the subspecies *Asio flammeus suinda*.

Status and conservation Not Threatened, Least Concern (IUCN-BirdLife International 2010), CITES II (Holt et al. 1999). A rare vagrant which has been seen only twice in French Guiana, in 1978 and in 1984. There are no recent observations of this owl in French Guiana.

Distribution A short-eared owl was seen on lawns near the Atlantic Ocean in Kourou from 26 to 28 July 1978. And another individual was seen in grassland at the Rochambeau (now Félix Eboué) airport near Cayenne on 28 and 29 June 1984 (Tostain et al. 1992).

Habitat Open areas such as savannas, grassland, pastures and ricefields.

Behaviour and breeding Nothing is known about the biology of the short-eared owl in French Guiana. Both sightings mentioned were done by day in open grasslands. Short-eared owls are known to perform long-distance dispersing movements in response to fluctuations in prey populations, i.e. small rodents (Marks et al. 1999). Such wide-scale and irruptive behaviour may facilitate the occurrence outside the normal range. French Guiana is obviously a marginal destination for the species.

12.3 Conclusion

Thirteen owl species are recorded in French Guiana, 12 are resident species, and 1, the short-eared owl, is a rare vagrant either from northern or southern South America. Two species, the burrowing owl and the vermiculated screech owl, were only recently observed for the first time in the country. The ferruginous pygmy owl (*Glaucidium brasilianum*) has not yet been recorded in the country but is a potential species to be found in French Guiana.

Neither direct prosecution nor hunting occurs for these nocturnal birds in French Guiana. In the interior, deforestation is low and mainly restricted to the edges of the forest block along roads. However, in recent years, increasing large-scale illegal gold mining and accompanying deforestation far in the interior are a worrying phenomenon. For those species living in the coastal region, especially in semiopen or open habitats, the main risks appear to be road kills and habitat loss. All are near threatened or threatened at the national level, and one, the great horned owl, is listed Endangered, with respect to IUCN criteria (UICN-France et al. 2017).

Thanks to the growing ornithological community and to the efficiency of the participative database Faune-Guyane (www.faune-guyane.fr) in collecting reliable data, the status of most owls in French Guiana is fairly well known. However, some of them are only rarely recorded and are obviously easily overlooked in the field, e.g. both *Ciccaba* species.

Our knowledge of the biology of all 13 species is, however, still largely deficient. More studies from a scientific point of view, e.g. on food, general behaviour, breeding biology and ecology, are urgently needed and are critical for conservation purposes.

Acknowledgements We thank the GEPOG (Groupe d'Étude et de Protection des Oiseaux en Guyane) for the permission to use data from the participative database Faune-Guyane (www.faune-guyane.fr), Jean-Pierre Policard for giving us the names of the owls in the main indigenous languages of French Guiana, François Catzeflis for identifying preys of a striped owl and of a burrowing owl and prey remains in spectacled and barn owls pellets and Sylvain Uriot for identifying preys in a striped owl pellet and for frequent fruitful discussions about French Guianan birds. We also thank Charles Bergère, Rachel Berzins, Clément Cambrezy, François Catzeflis, Ciao J. Carlos, Fernand Deroussen, Antoine Hauselmann, Patrick Ingremeau, Thomas Luglia, Sophie Maillé, Vincent Pelletier, Kévin Pineau, Jean-Pierre Policard, Alexandre Renaudier, Thomas Réquillart,

Vincent Rufray, Jean-Luc Sibille, Marc Thibault, Olivier Tostain, Sylvain and Quentin Uriot and Alexandre Vinot, and all the contributors to the database Faune-Guyane, for sharing their observations. We are grateful to the late Alexandre Renaudier for his critical reading of the first draft of the manuscript and for his many useful comments and additional data. And we mention the generous help with photos by Tanguy Deville, Patrick Ingremeau and Alexandre Renaudier. We thank DEAL (Direction de l'Environnement, de l'Aménagement et du Logement) and especially Sébastien Linares, for the map of the nature reserves.

Appendix 12.1 Surfaces of the national park (Parc Amazonien de Guyane) and seven nature reserves (Réserves Naturelles, RN) in French Guiana

Parc Amazonien de Guyane	33,900 km^2
RN du Mont Grand Matoury	21 km^2
RN régionale Trésor	25 km^2
RN de l'Île du Grand Connétable	79 km^2
RN de l'Amana	144 km^2
RN de la Trinité	760 km^2
RN de Kaw-Roura	947 km^2
RN des Nouragues	1000 km^2

References

Ackermann L, Chrétien M, Giraud-Audine M et al (2008) Further observations on the burrowing owl *Athene cunicularia* in French Guiana. Alauda 76:254–256

Baglan A, Catzeflis F (2016) Barn owl pellets collected in coastal savannas yield two additional species of small mammals for French Guiana. Mammalia 80:91–95

Barret J (ed) (2006) Atlas illustré de la Guyane. IRD (Institut de Recherche pour le Développement), Marseille

Boyé A, Brown A, Collier N et al (2009) French overseas departments and territories. In: Devenish C, Díaz Fernández DF, Clay RP et al (eds) Important bird areas Americas – priority sites for biodiversity conservation, BirdLife conservation series no. 16. BirdLife International, Quito, pp 213–228

Braun MJ, Finch DW, Robbins MB et al (2007) A field checklist of the birds of Guyana, 2nd edn. Smithsonian Institution, Washington, DC

Bruce MD (1999) Family Tytonidae. In: del Hoyo J, Elliott A, Sargatal J (eds) Handbook of the birds of the world, Barn-owls to Hummingbirds, vol 5. Lynx Edicions, Barcelona, pp 34–75

Catzeflis F, Barrioz S, Szpigel JF et al (2014) Marsupiaux et Rongeurs de Guyane. Edition Institut Pasteur de la Guyane, Cayenne

CINFO (Commission Internationale des Noms Français des Oiseaux, consisting of Devillers P, Ouellet H, Benito-Espinal E, et al (1993) Noms français des Oiseaux du Monde, 1st edn. Éditions MultiMondes Inc./Édition Chabaud, Sainte-Foy/Bayonne

Claessens O, Comité d'Homologation de Guyane (2015) Rare birds in French Guiana in 2008–10. Cotinga 37:66–78

Claessens O, Rufray V, Tostain O et al (2014) Les oiseaux rares en Guyane en 2010 et 2011. Report of the Comité d'Homologation de Guyane. www.chn-france.org/upload_content/rapport%20chg%202010-2011.pdf

Comité d'Homologation de Guyane (2016) Liste des oiseaux de Guyane. Version March 2016. www.gepog.org/sites/default/files/fichiers/photos_chants/listes/CHG_2016.pdf. Accessed 31 May 2016

CTG (Collectivité Territoriale de la Guyane) (2016) Schéma d'Aménagement Régional de la Guyane. https://www.ctguyane.fr/sar-approuve/. Accessed 1 Dec 2016

Deville T, Ingels J (2012) Observations at a nest of Amazonian pygmy owl *Glaucidium hardyi* in French Guiana. Neotrop Birding 11:23–26

Fossé A (2011) Liste française des oiseaux du monde. Version 4 April 2011. http://www.digimages.info/listeoiseauxmonde/present.html. Accessed 31 May 2011

GEPOG (Groupe d'Étude et de Protection des Oiseaux en Guyane) (2016) Faune-Guyane. www.faune-guyane.fr. Accessed 15 Oct 2016

Giraud-Audine M, Ackermann L, Tostain O et al (2007) The burrowing owl *Athene cunicularia*, a new species for French Guiana. Alauda 75:421–422

Holt DW, Berkley R, Deppe C et al (1999) Family Strigidae. In: del Hoyo J, Elliott A, Sargatal J (eds) Handbook of the birds of the world, Barn-owls to hummingbirds, vol 5. Lynx Edicions, Barcelona, pp 185–193

Howard R, Moore A (1991) A complete checklist of the birds of the world, 2nd edn. Academic Press, London

IOC (International Ornithological Congress) World Bird Names (2010) IOC world birds list. Version 2010–4. http://www.worldbirdnames.org/. Accessed 31 May 2011

IUCN-BirdLife International (International Union for Conservation of Nature-BirdLife International) (2010) The IUCN Red List of threatened species. http://www.iucnredlist.org/. Accessed 31 May 2011

König C, Weick F, Becking JH (2008) Owls of the world, 2nd edn. Christopher Helm, London

Marks JS, Cannings RJ, Mikkola H (1999) Typical owls (Strigidae). In: del Hoyo J, Elliott A, Sargatal J (eds) Handbook of the birds of the world, Barn-owls to Hummingbirds, vol 5. Lynx Edicions, Barcelona, pp 76–242

Marti CG, Poole AF, Bevier LR (2005) Barn Owl (*Tyto alba*). The birds of North America. Rodeward PG (ed). Ithaca: Cornell Lab of Ornithology. Retrieved from the Birds of North America: https://birdsna.org/Species-Account/bna/species/barnowl. doi:10.2173/bna.1

Miserendino R (2007) Registro del Chiñi (*Athene cunicularia*) para la Amazonia boliviana. Kempffiana 3:23–24

O'Shea BJ, Ramcharan S (2012) Appendix 9. Cumulative bird list for SE Suriname RAP. In: Conservation International-Suriname. Southeast Suriname RAP survey. Grensgebergte and Kasikasima. March 9–29, 2012. Preliminary Results. Conservation International-Suriname, Paramaribo, Suriname, p 105–117

Ottema O, Ramcharan S, Ouboter P et al (2009) Four bird species new to Surinam. Cotinga 31:140–141

Remsen JV Jr., Cadena CD, Jamarillo A, et al (2016a). A classification of the bird species of South America. Version 30 Nov 2016. American Ornithologists' Union. www.museum.lsu.edu/~Remsen/SACCBaseline.html. Accessed 1 Dec 2016

Remsen JV Jr., Cadena CD, Jamarillo A, et al (2016b) Proposal (713) to South American Classification Committee: Merge *Pseudoscops clamator* into *Asio*. www.museum.lsu.edu/~Remsen/SACCprop713.htm. Accessed 30 Oct 2016

Renaudier A, Claessens O, Comité d'Homologation de Guyane (2010) Les oiseaux rares en Guyane en 2008 et 2009. Rapport du Comité d'Homologation de Guyane. http://www.chn-france.org/upload_content/rapport%20chg%202008-2009.pdf. Accessed 30 Nov 2010

Renaudier A, Comité d'Homologation de Guyane (2010) Rare birds in French Guiana in 2005–07. Cotinga 32:OL 95–OL104
Renaudier A, Deroussen F (2008) Bird voices from French Guiana. 3 CDs set. Naturophonia/Groupe d'Étude et de Protection des Oiseaux en Guyane/Muséum National d'Histoire Naturelle, Paris
Renaudier A, Rodet M, Noizet S (2007) Neotropical notebook, French Guiana. Cotinga 28:88
Restall R, Rodner C, Lentino M (2006) Birds of northern South America. Christopher Helm, London
Ribot JH (2011) Birds of Suriname, South America. Version May 2011. www.1.nhl.nl/~ribot/. Accessed 31 May 2011
Sibley CG, Monroe BL Jr (1990) Distribution and taxonomy of birds of the world. Yale University Press, New Haven
Tostain O, Dujardin J-L, Érard C et al (1992) Oiseaux de Guyane. Société d'Études Ornithologiques, Brunoy
UICN-France (Union internationale pour la Conservation de la Nature-France), MNHN (Muséum National d'Histoire Naturelle, Paris) and GEPOG (Groupe d'Étude et de Protection des Oiseaux en Guyane) (2017) La Liste rouge des Espèces menacées en France, chapitre Oiseaux de Guyane. Paris
Wink M, El-Sayed A-A, Sauer-Gürth H et al (2009) Molecular phylogeny of owls (Strigiformes) inferred from DNA sequences of the mitochondrial cytochrome b and the nuclear RAG-1 gene. Ardea 97:581–591
Wikipédia (2016) Démographie de la Guyane. https://fr.wikipedia.org/wiki/D%C3%A9mographie_de_la_Guyane. Accessed 10 Dec 2016

Chapter 13
The Owls of Guatemala

Knut Eisermann and Claudia Avendaño

Abstract This compilation of recent data on the distribution, ecology, and conservation status of owls (Strigiformes) in Guatemala is based on an extensive literature review and numerous unpublished observations. Twenty species of owls have been recorded in Guatemala, of which 18 are resident. Breeding has been reported for 17 species, and it is assumed for one species. Two species are considered rare or accidental nonbreeding visitors to Guatemala during the northern winter. Guatemala's region with the highest species richness in owls is the highlands, where 17 species have been recorded. Twelve species have been recorded in the Pacific slope lowlands and 13 species in the Atlantic slope lowlands. We analyzed the data in the presence and relative abundance of owls from 105 sites from 1989 to 2016. According to the weighted mean value of the relative abundance index across three ornithogeographic regions, the most common owls in the country are (abundance ranking in descending order): Mexican wood owl (*Strix squamulata*), Ridgway's pygmy owl (*Glaucidium ridgwayi*), Guatemalan screech owl (*Megascops guatemalae*), black-and-white owl (*Strix nigrolineata*), American barn owl (*Tyto furcata*), Central American pygmy owl (*Glaucidium griseiceps*), Guatemalan pygmy owl (*Glaucidium cobanense*), great horned owl (*Bubo virginianus*), fulvous owl (*Strix fulvescens*), unspotted saw-whet owl (*Aegolius ridgwayi*), whiskered screech owl (*Megascops trichopsis*), crested owl (*Lophostrix cristata*), and Pacific screech owl (*Megascops cooperi*). Guatemala has an adequate legal framework to protect owl species (32% of the country is legally protected), but the conservation is not efficient, causing threats to owl populations. Of 18 resident owl species, 12 are forest specialists. In a vulnerability assessment applying IUCN Red List criteria on a national level, one species has been evaluated as Critically Endangered (CR), one as Endangered (EN), nine as Vulnerable (VU), five as Near Threatened (NT), two as Least Concern (LC), and two as not applicable. Habitat alterations through agriculture, mining, and oil drilling are the main threats. Of the remaining forests, 14% (5500 km^2) were lost from 2000 to 2010, and the pressure on natural habitat will further increase. In addition, owls in Guatemala are threatened by direct persecution because of popular superstitions. The network of 21 Important Bird Areas (IBA) in Guatemala includes populations of all owl species. Three species have been recorded

K. Eisermann (✉) • C. Avendaño
PROEVAL RAXMU Bird Monitoring Program, Cobán, Alta Verapaz, Guatemala
e-mail: knut.eisermann@proeval-raxmu.org; claudia@proeval-raxmu.org

P.L. Enríquez (ed.), *Neotropical Owls*, DOI 10.1007/978-3-319-57108-9_13

in at least 10 IBAs, 12 species in 5–9 IBAs, 4 in 2–4 IBAs, and 1 species in only one IBA. We consider the increase of the education level among the Guatemalan society the main key to protect habitats within the IBAs. A higher level of education would help to slow down population growth, increase environmental awareness, and consequently diminish pressure on natural areas.

Keywords Conservation status • Distribution • Relative abundance • Strigiformes

Fulvous Owl (*Strix fulvescens*)

13.1 Introduction

Although owls (Strigiformes) belong to the most popular birds, they also belong to the least known. Because of the nocturnal activity of most of the species, much of their natural history remains to be discovered by us humans adapted to diurnal activity. In Guatemala, more than 740 bird species have been reported (Eisermann and Avendaño 2007, KE unpub. data). The study of owls in the country began with the first specimen collections in the nineteenth century, compiled in the *Biologia Centrali-Americana* (Salvin and Godman 1897–1904), the first thorough regional biodiversity inventory for Middle America. Seventeen of the currently known 20 owl species of Guatemala were reported therein. Ridgway (1914) added stygian owl (*Asio stygius*), and Griscom (1932) added striped owl (*Asio clamator*) to the list of Guatemalan owls. Land (1970) included 19 species in his field guide to Guatemalan birds and suggested that Pacific screech owl (*Megascops cooperi*) should occur in the country, which was also mentioned earlier by Marshall (1967). Dickerman (1975) reported this species for the first time in Guatemala. Thus, recent compilations on the bird diversity of Guatemala contain 20 species of owls (Howell and Webb 1995; Eisermann and Avendaño 2007, 2015). This chapter is a translated, revised, and updated version of the Spanish original work which was based on data until 2013 (Eisermann and Avendaño 2015). This update contains data until 2016, enhancing the information on the distribution and residency status of several species. The objective is to provide a compilation of current data on the distribution and natural history of owls in Guatemala, to identify threats, and to recommend conservation strategies. Based on a thorough literature review and own data from 1989 to 2016, we provide here an updated classification of the relative abundance and residency status of all owl species recorded in Guatemala.

13.2 Study Area and Methods

13.2.1 Study Area

Guatemala covers an area of 108,900 km^2, bordering to the Mexican states of Chiapas, Tabasco, Campeche, and Quintana Roo and the Central American countries Belize, Honduras, and El Salvador. The Guatemalan Caribbean coastline is approximately 150 km long and the Pacific coastline 250 km. The elevation of the land area of Guatemala ranges from sea level to 4220 m. In ornithology, a reasonable biogeographic classification divides the country in three regions (Fig. 13.1). The highlands (1000–4220 m) cover approximately 37,500 km^2 (34% of the country), the lowlands (<1000 m) of the Pacific slope and interior valleys cover 19,000 km^2 (18%), and the lowlands of the Atlantic slope cover 52,400 km^2 (48%). According to a nationwide land cover mapping at the scale 1:50,000 (Ministerio de Agricultura, Ganadería y Alimentación 2006), 38% (41,580 km^2) of Guatemala is covered with forests, of which 31,554 km^2 are broadleaf forests, 2496 km^2 coniferous forests, 6316 km^2 mixed forests, 206 km^2 mangroves, and 1007 km^2 forest-covered wetlands. A more recent forest cover mapping identified 34% of the country covered with forest, with annual deforestation rate of 1% from 2006 to 2010 (Regalado et al. 2012). Almost half of the country is used for agriculture (see Sect. 13.4). Geographic coordinates of all sites mentioned in the text are listed in Appendix 13.1; sites and limits of departments are mapped in Fig. 13.2.

13.2.2 Data Source

Historic and current distribution of owls was compiled based on a thorough literature review, our own data, and unpublished museum specimen records. We reviewed publications and "gray literature" (i.e., unpublished reports, thesis) based on a

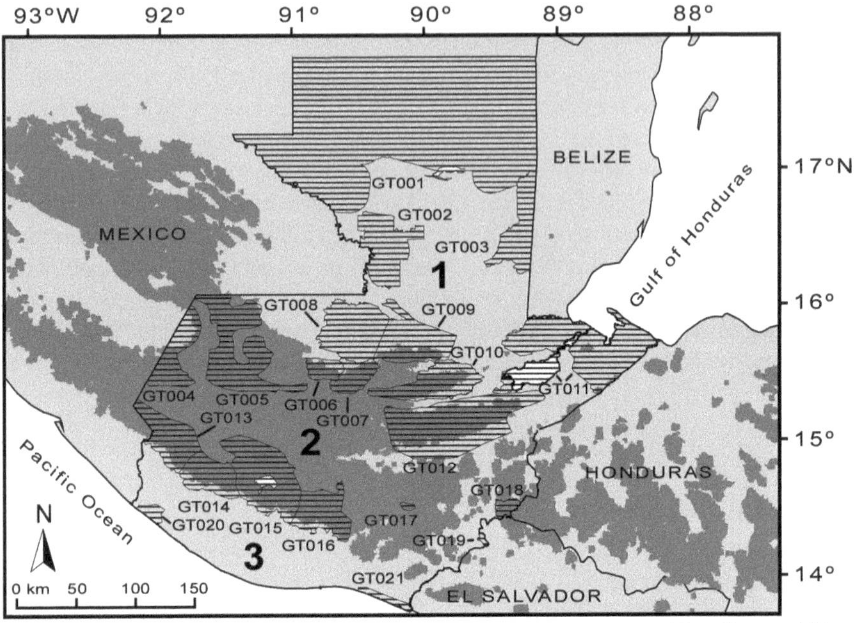

IBA GT001 Maya-Lacandón (9 species of Strigiformes)
IBA GT002 Río La Pasión (5)
IBA GT003 Chiquibul (2)
IBA GT004 Cuilco (1)
IBA GT005 Cuchumatanes (9)
IBA GT006 Cerro El Amay (7)
IBA GT007 Sacranix (8)
IBA GT008 Lachua-Ik'bolay (8)
IBA GT009 Candelaria-Campur (2)
IBA GT010 Yalijux (11)
IBA GT011 Caribe de Guatemala (9)
IBA GT012 Sierra de las Minas-Motagua (12)
IBA GT013 Tacaná-Tajumulco (5)
IBA GT014 Volcán Santiaguito (11)
IBA GT015 Atitlán (14)
IBA GT016 Antigua Guatemala (13)
IBA GT017 Cerro Miramundo (0)
IBA GT018 Montecristo (7)
IBA GT019 Lago de Güija (5)
IBA GT020 Manchón-Guamuchal (5)
IBA GT021 Monterrico-Río La Paz (5)

Fig. 13.1 Ornithogeographic regions of Guatemala: 1, Atlantic slope lowlands; 2, highlands; 3, Pacific slope lowlands. *Light-gray shade*, elevation ≤1000 m; *dark-gray shade*, elevation >1000 m; *horizontal hatching*, Important Bird Areas (IBAs) according to Eisermann and Avendaño (2009b). The list includes the international code and the name of the IBAs, and in parenthesis, the number of owl species recorded

bibliography of ornithological literature (Eisermann and Avendaño 2006) and more current literature up to 2016. Data for some sites were enhanced by recent observations published in eBird (Sullivan et al. 2009; eBird 2016). We used data of our own observations from 1989 to 2016. We reviewed specimens in the Zoological Museum Berlin, Germany, Zoological Collection of the Universidad Del Valle de Guatemala, and Natural History Museum of Universidad de San Carlos de Guatemala and reviewed databases of specimens accessible through VertNet (Constable et al. 2010; VertNet 2016). Acronyms of museums mentioned in the text are AMNH, American Museum of Natural History, New York; LACM, Natural History Museum of Los Angeles County, Los Angeles, California; MVZ, Museum of Vertebrate Zoology, Berkeley, California; and ZMB, Zoological Museum Berlin, Germany.

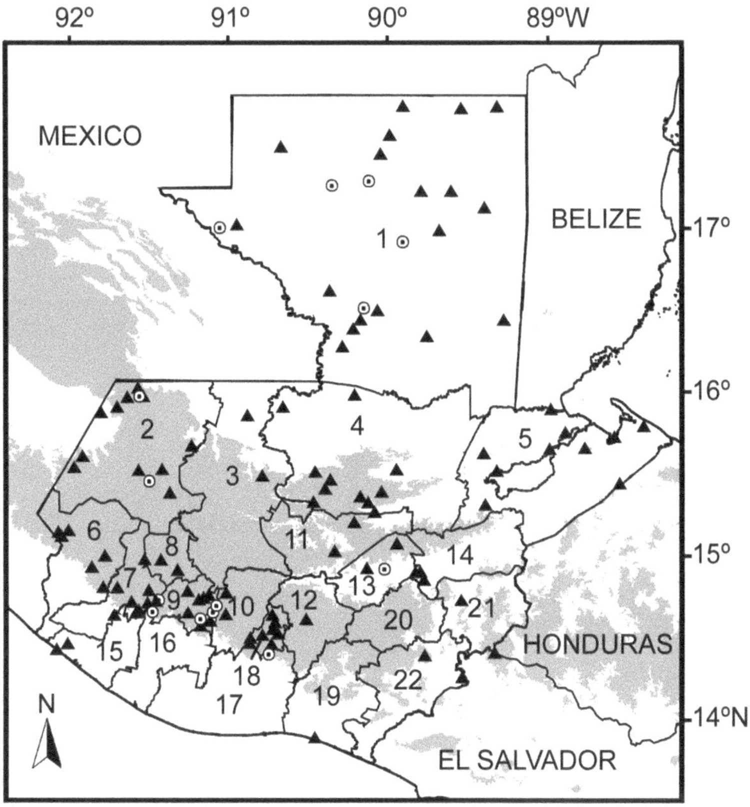

Fig. 13.2 Localization of sites with recent owl records (1989–2016) in Guatemala. *Triangles* mark sites used for relative abundance estimate; *circles* mark other sites with recent records. *Gray shade* marks highlands >1000 m. Departments of Guatemala: 1, Petén; 2, Huehuetenango; 3, Quiché; 4, Alta Verapaz; 5, Izabal; 6, San Marcos; 7, Quetzaltenango; 8, Totonicapán; 9, Sololá; 10, Chimaltenango; 11, Baja Verapaz; 12, Guatemala; 13, El Progreso; 14, Zacapa; 15, Retalhuleu; 16, Suchitepéquez; 17, Escuintla; 18, Sacatepéquez; 19, Santa Rosa; 20, Jalapa; 21, Chiquimula; and 22, Jutiapa

13.2.3 Estimation and Classification of Relative Abundance

To estimate current relative abundance, we considered data from 1989 to 2016 for 105 sites (Fig. 13.2): 55 sites from the highlands, 40 sites from the Atlantic slope lowlands, and 10 sites from the Pacific slope lowlands. Because standardized abundance data have been published for only a few sites, we apply a simplified abundance index divided into four classes for each site: common (numerical value 4), high probability to detect several individuals or territories in a day of observation; fairly common (numerical value 3), high probability to detect an individual or territory in a day of observation; uncommon (numerical value 2), high probability to detect the species in a week of observation; and rare (numerical value 1), few records, low probability to detect the species in a week of observation, or single record for a site. We used a numeric value 0 for unrecorded species. To determine a ranking of relative abundance in each region, we calculated the arithmetic mean of index values across all sites. To rank relative abundance across the entire country, we used weighted mean for each region according to the percentage area (Atlantic slope lowlands 48%, highlands 34%, Pacific slope lowlands 18%), hence $I_c = 0.48*I_a + 0.34*I_h + 0.18*I_p$, with I_c relative abundance index for the country, I_a relative abundance index for Atlantic slope lowlands, I_h relative abundance index for highlands, and I_p relative abundance index for Pacific slope lowlands.

13.2.4 Taxonomy and Nomenclature

The taxonomy of Neotropical owls has not yet been fully resolved (Enríquez et al. 2015); thus, nomenclatural changes are expected in the future. In this chapter, we follow the taxonomy and nomenclature of common and scientific names of owls by König et al. (2008). Scientific names of all species mentioned in the text are listed in Table 13.1.

13.3 Results

13.3.1 Owl Diversity

Two families of owls (Strigiformes) occur in Guatemala: barn owls (Tytonidae), in Spanish locally known as *lechuzas*, and true owls (Strigidae), locally known as *tecolotes* or *búhos*. According to modern taxonomy (König et al. 2008), 20 species of Strigiformes have been recorded in Guatemala (Eisermann and Avendaño 2006, 2007): one species in the genus *Tyto*, one *Psiloscops*, four *Megascops*, one *Bubo*, one *Pulsatrix*, three *Strix*, one *Lophostrix*, three *Glaucidium*, one *Aegolius*, one *Athene*, and three *Asio* (Table 13.1). Guatemala covers part of the distribution range of 25% of the approximately 80 Neotropical owl species and 8% of the approximately 250 owl species of the world.

13.3.2 Spatial Distribution

Of the three ornithogeographic regions of Guatemala, species richness of owls is highest in the highlands (17 species). In the lowlands, 12 species have been recorded on the Pacific slope and 13 species on the Atlantic slope (Table 13.1). Endemism in birds is sometimes defined to areas <50,000 km^2 (Terborgh and Winter 1983; Bibby et al. 1992; Wege and Long 1995; Stattersfield et al. 1998). Two of the owl species of Guatemala have a restricted range: bearded screech owl (*Megascops barbarus*), restricted to the Atlantic slope highlands of Guatemala and the Mexican state of

Table 13.1 Residency status, distribution, and vulnerability of owls in Guatemala

Taxa[a]	Subspecies in Guatemala[b]	Residency status in Guatemala[c]	Distribution in ornithogeographic regions[d]	Habitat[e]	Vulnerability on a national level[f]	Distribution in IBAs[g]
TYTONIDAE						
American barn owl *Tyto furcata* (Temminck) 1827	*T. f. guatemalae* (Ridgway) 1874, *T. f. pratincola* (Bonaparte) 1838	R (*T. a. guatemalae*), r (*T. a. pratincola*)	A, H, P	S, A, P, U	NT	GT001, GT007, GT008, GT010, GT011, GT015, GT016, GT019, GT021
STRIGIDAE						
Flammulated owl *Psiloscops flammeolus* (Kaup) 1852		R	H	Fc, Fm	CR D	GT005, GT015, GT016
Pacific screech owl *Megascops cooperi* (Ridgway) 1878	*M. c. cooperi* (Ridgway) 1878	R	P	S, A	VU A3c	GT012, GT019, GT020, GT021
Whiskered screech owl *Megascops trichopsis* (Wagler) 1832	*M. t. mesamericanus* (van Rossem) 1932	R	H	Fm, Fc, U	VU A3c	GT005, GT007, GT014, GT015, GT016, GT018
Bearded screech owl *Megascops barbarus* (Sclater & Salvin) 1868		R	H	Fm, Fb, Fc	VU A3c	GT005, GT006, GT010, GT012
Guatemalan screech owl *Megascops guatemalae* (Sharpe) 1875	*M. g. guatemalae* (Sharpe) 1875	R	A, H, P	Fb, Fm, S, P	VU A3c	GT001, GT002, GT006, GT007, GT008, GT010, GT011, GT012, GT014
Great horned owl *Bubo virginianus* (Gmelin) 1788	*B. v. mesembrinus* (Oberholser) 1904	R	A, H, P	Fc, Fm, S, A, U	NT	GT004, GT005, GT007, GT012, GT013, GT014, GT015, GT016, GT018, GT019

(continued)

Table 13.1 (continued)

Taxa[a]	Subspecies in Guatemala[b]	Residency status in Guatemala[c]	Distribution in ornithogeographic regions[d]	Habitat[e]	Vulnerability on a national level[f]	Distribution in IBAs[g]
Spectacled owl *Pulsatrix perspicillata* (Latham) 1790	*P. p. saturata* Ridgway 1914	R	A, H, P	Fb, P	VU A3c	GT001, GT008, GT011, GT014, GT015, GT018
Mexican wood owl *Strix squamulata* (Bonaparte) 1850	*S. s. centralis* (Griscom) 1929	R	A, H, P	Fb, Fm, Fc, S, A, P, U	LC	GT001, GT002, GT003, GT005, GT006, GT007, GT008, GT010, GT011, GT012, GT014, GT015, GT016, GT018, GT019, GT020, GT021
Black-and-white owl *Strix nigrolineata* (Sclater) 1859		R	A, H, P	Fb, Fm, P	VU A3c	GT001, GT002, GT008, GT010, GT011, GT014, GT015, GT016
Fulvous owl *Strix fulvescens* (Sclater & Salvin) 1868		R	H	Fb, Fm	VU A3c	GT006, GT010, GT012, GT013, GT014, GT015, GT016, GT018
Crested owl *Lophostrix cristata* (Daudin) 1800	*L. c. stricklandi* Sclater & Salvin 1859	R	A, H, P	Fb	VU A3c	GT001, GT002, GT008, GT010, GT011, GT012, GT018, GT020
Guatemalan pygmy owl *Glaucidium cobanense* Shape 1875		R	H	Fc, Fm, Fb, S, A	NT	GT005, GT006, GT007, GT010, GT012, GT013, GT014, GT015, GT016
Central American pygmy owl *Glaucidium griseiceps* Sharpe 1875		r	A, P	Fb, P	VU A3c	GT001, GT006, GT007, GT008, GT011, GT015

Ridgway's pygmy owl *Glaucidium ridgwayi* Sharpe 1875		R	A, H, P	S, A, P, U	LC	GT001, GT002, GT003, GT005, GT007, GT008, GT009, GT010, GT011, GT012, GT013, GT014, GT015, GT016, GT018, GT019, GT020, GT021
Burrowing owl *Athene cunicularia* (Molina) 1782	*A. c. hypugaea* (Bonaparte) 1825	vagM	A, H, P	A	NA	GT005, GT009, GT012, GT015, GT016
Unspotted saw-whet owl *Aegolius ridgwayi* (Alfaro) 1905		R	H	Fm, Fb, Fc, S, A	NT	GT005, GT006, GT010, GT012, GT013, GT014, GT015, GT016
Stygian owl *Asio stygius* (Wagler) 1832	*A. s. robustus* Kelso 1934	R	H, A	Fc, Fm, Fb, P	EN D	GT010, GT012, GT014, GT015, GT016
Striped owl *Asio clamator* (Vieillot) 1807	*A. c. forbesi* (Lowery & Dahlquest) 1951	R	A, P	S, A	NT	GT001, GT011, GT020, GT021
Short-eared owl *Asio flammeus* (Pontoppidan) 1763	*A. f. flammeus* (Pontoppidan) 1763	vagM	H	A	NA	GT016

[a]Nomenclature according to König et al. (2008)
[b]Subspecies in Guatemala: see species accounts in the text, nomenclature according to Dickinson and Remsen (2013)
[c]Residency status according to Eisermann and Avendaño (2007) and recent data: R–breeding resident, r–resident, breeding presumed, V–non-breeding visitor, vagM–migratory vagrant
[d]Ornithogeographic region: A–Atlantic slope lowlands, H–highlands >1,000 m, P–Pacific slope lowlands
[e]Habitat: A–open and agricultural area, Fc–coniferous forest (including pine plantations), Fm–mixed coniferous/broadleaf forest, Fb–broadleaf forest, S–scrub (including arid forest, thorn scrub, and secondary growth scrub), P–permanent plantations with shade trees (coffee and cardamom plantations), U–urban area
[f]Vulnerability according to Eisermann and Avendaño (2006) and recent data. Categories and criteria according to IUCN (2003, 2012a, b): LC–Least Concern: Common and widely distributed species which does not qualify under the Critically Endangered, Endangered, Vulnerable, or Near Threatened categories. NT–Near Threatened: Species not yet qualifying for the categories Critically Endangered, Endangered, or Vulnerable, but likely to qualify in the near future. VU A3c–Vulnerable: A reduction in population size $\geq 30\%$ projected or presumed within the next 10 years or three generations, whichever is longer (up to a maximum of 100 years); based on a reduction of area of occupancy, area of occurrence, or quality of habitat. EN D–Endangered: Total population size in Guatemala estimated to be <250 mature individuals. CR D–Critically Endangered: Total population size in Guatemala estimated to be <50 mature individuals. NA–Not applicable: Species occurring only as vagrants in Guatemala were not evaluated
[g]Delimitation of Important Bird Areas (IBA) according to Eisermann and Avendaño (2009a, b). See Fig. 13.1 for localization of IBAs

Chiapas, and Guatemalan pygmy owl (*Glaucidium cobanense*), restricted to the highlands of Guatemala, Honduras, and the Mexican state of Chiapas. Fulvous owl (*Strix fulvescens*) occurs mainly in the highlands of Guatemala, Honduras, El Salvador, and the Mexican state of Chiapas, but it has also been recorded in the highlands west of the Isthmus of Tehuantepec in the Mexican state of Oaxaca (Gómez de Silva 2010; Ramírez-Julián et al. 2011).

Of the 20 species of owls in Guatemala, five have been recorded mainly in the lowlands below 1000 m and seven mainly in the highlands above 1000 m. The records of eight species range from the lowlands to the highlands (Fig. 13.3).

13.3.3 Temporal Distribution

Of the 20 owl species in Guatemala, 18 are residents. Breeding has been confirmed for 17, and it is presumed for one species: Central American pygmy owl (*Glaucidium griseiceps*). Flammulated owl (*Psiloscops flammeolus*) has been considered a non-breeding visitor previously (Eisermann and Avendaño 2015), but nesting has been evidenced recently (Eisermann et al. 2017). Burrowing owl (*Athene cunicularia*) and short-eared owl (*Asio flammeus*) have been recorded as vagrants (Table 13.1). Short-distance migrations, including elevational migrations, have not been reported from owls in Guatemala.

13.3.4 Habitat Associations

Most of the owl species of Guatemala occur at least partially in forest habitats. Of the 18 resident species, 12 are forest specialists, including forest-like plantations such as coffee *Coffea arabica* and cardamom *Elettaria cardamomum* plantations shaded by a canopy of trees. Some of these owl species also occur in small forest patches in urban areas, such as the whiskered screech owl (*Megascops trichopsis*) (Table 13.1).

13.3.5 Relative Abundance

Only five species, Guatemalan screech owl (*Megascops guatemalae*), bearded screech owl, Mexican wood owl (*Strix squamulata*), fulvous owl (*Strix fulvescens*), and Ridgway's pygmy owl (*Glaucidium ridgwayi*), have been classified as locally common at some of the 105 sites. At most sites, however, these species were recorded as rare, or not recorded at all (Appendix 13.1); thus, the mean relative abundance indices per region are low. Mexican wood owl and Ridgway's pygmy owl are the species with the highest mean values (1.9 and 1.7) in some of the ornithogeographic regions, thus classifying as uncommon (Table 13.2). The mean

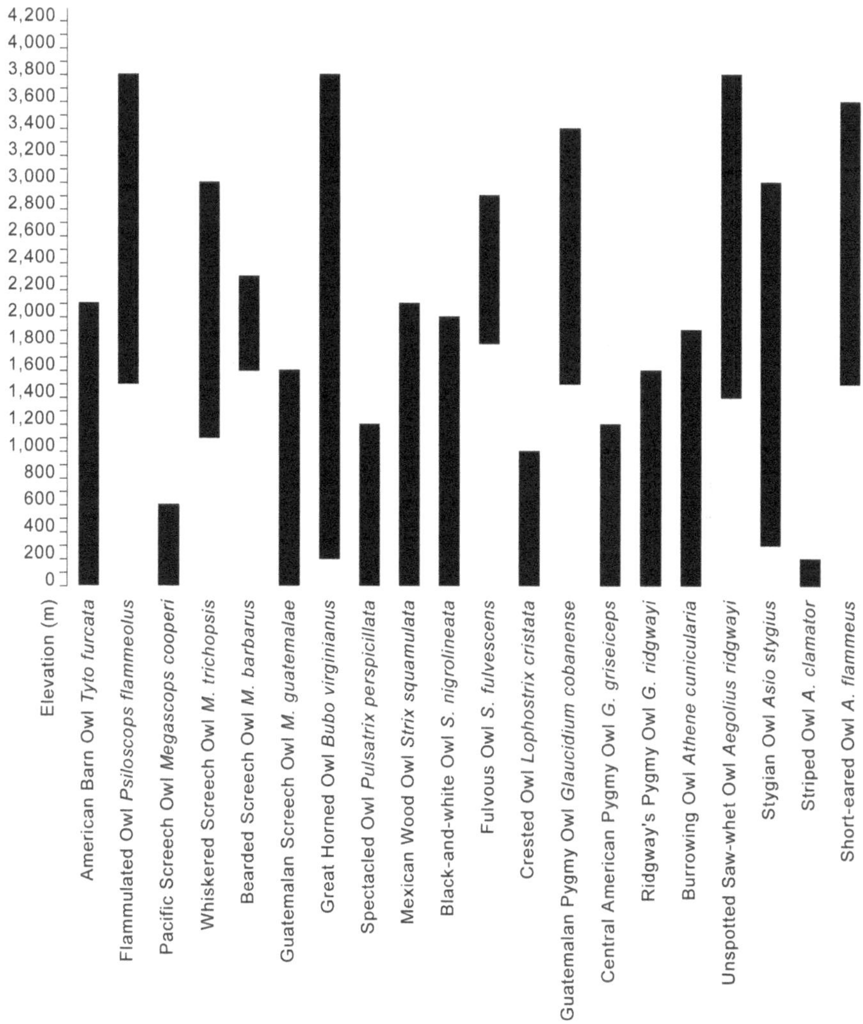

Fig. 13.3 Elevational range of owl records in Guatemala

relative abundance index for most species is below 1.5, indicating they are rare, which is not surprising for carnivore species. This index provides a comparison of the set of most common species among the three regions. It may, however, be biased because the information is based on data which does not consider detection probability. Currently, no higher-quality data are available.

Species with the highest relative abundance index in the Atlantic slope lowlands were (in descending order of abundance): Mexican wood owl, Guatemalan screech owl, Ridgway's pygmy owl, black-and-white owl (*Strix nigrolineata*), Central American pygmy owl, American barn owl (*Tyto furcata*), crested owl (*Lophostrix cristata*), and spectacled owl (*Pulsatrix perspicillata*). In the Pacific slope lowlands,

Table 13.2 Index of relative abundance (A) of owl species and number of sites with records (B) among a total of 105 sites in three ornithogeographic regions of Guatemala, according to data from 1989 to 2016

	Atlantic slope lowlands $n = 40$ sites		Highlands (>1,000 m) $n = 55$ sites		Pacific slope lowlands $n = 10$ sites	
Species	A[a]	B	A[a]	B	A[a]	B
American barn owl *Tyto furcata*	0.28	9	0.18	10	0.30	3
Flammulated owl *Psiloscops flammeolus*	0	0	0.02	1	0	0
Pacific screech owl *Megascops cooperi*	0.03	1	0	0	0.60	3
Whiskered screech owl *Megascops trichopsis*	0	0	0.38	17	0	0
Bearded screech owl *Megascops barbarous*	0	0	0.20	5	0	0
Guatemalan screech owl *Megascops guatemalae*	1.25	23	0.20	6	0.10	1
Great horned owl *Bubo virginianus*	0.08	3	0.47	24	0.10	1
Spectacled owl *Pulsatrix perspicillata*	0.18	6	0.02	1	0.10	1
Mexican wood owl *Strix squamulata*	1.85	32	1.15	31	1.90	7
Black-and-white owl *Strix nigrolineata*	0.68	18	0.11	5	0.20	1
Fulvous owl *Strix fulvescens*	0	0	0.51	14	0	0
Crested owl *Lophostrix cristata*	0.20	7	0.02	1	0.10	1
Guatemalan pygmy owl *Glaucidium cobanense*	0	0	0.64	28	0	0
Central American pygmy owl *Glaucidium griseiceps*	0.43	9	0.02	1	0.10	1
Ridgway's pygmy owl *Glaucidium ridgwayi*	0.93	23	0.45	16	1.70	8
Burrowing owl *Athene cunicularia*	0	0	0	0	0	0
Unspotted saw-whet owl *Aegolius ridgwayi*	0	0	0.44	19	0	0
Stygian owl *Asio stygius*	0.03	1	0.13	7	0	0
Striped owl *Asio clamator*	0.08	3	0	0	0.20	2
Short-eared owl *Asio flammeus*	0	0	0	0	0	0

[a]Mean of the numeric index of relative abundance at each site: 0–not recorded, 1–rare (few records, low probability to detect the species in a week of observation, or single record for a site), 2–uncommon (high probability to detect the species in a week of observation), 3–fairly common (high probability to detect an individual or territory in a day of observation), 4–common (high probability to detect several individuals or territories in a day of observation). Data in Appendix 3.1

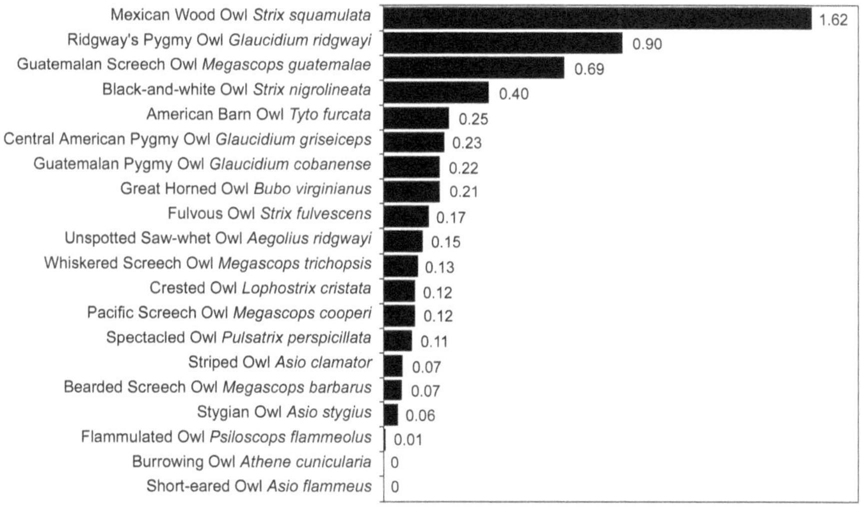

Fig. 13.4 Relative abundance index of owls in Guatemala based on data from 1989 to 2016. The index values result from weighted means across three ornithogeographic regions (Atlantic slope lowlands: 40 sites, highlands: 55 sites, Pacific slope lowlands: 10 sites). At each site, species were categorized in five abundance classes (0–4, see Methods)

most common owls were Mexican wood owl, Ridgway's pygmy owl, Pacific screech owl, American barn owl, black-and-white owl, and striped owl (*Asio clamator*). Most common owls in the highlands >1000 m were Mexican wood owl, Guatemalan pygmy owl, fulvous owl, great horned owl (*Bubo virginianus*), Ridgway's pygmy owl, unspotted saw-whet owl (*Aegolius ridgwayi*), whiskered screech owl, bearded screech owl, Guatemalan screech owl, and American barn owl (Table 13.2). According to the weighted mean value of the relative abundance index across the three ornithogeographic regions, the most common owls in the country were (abundance ranking in descending order): Mexican wood owl, Ridgway's pygmy owl, Guatemalan screech owl, black-and-white owl, American barn owl, Central American pygmy owl, Guatemalan pygmy owl, great horned owl, fulvous owl, unspotted saw-whet owl, whiskered screech owl, crested owl, and Pacific screech owl (Fig. 13.4).

It follows an annotated list of all owl species recorded in Guatemala, with information on distribution and research history in Guatemala.

13.3.6 Species Accounts

American barn owl (*Tyto furcata*) is widely distributed throughout the Americas (König et al. 2008). Two subspecies were recorded in Guatemala, *T. f. guatemalae* in the south of the country (Griscom 1932) and *T. f. pratincola* in the north of dpto. Petén (van Tyne 1935). The limit of distribution between both subspecies remains uncertain (Marti 1992; Bruce 1999). König et al. (2008) did not recognize these subspecies. American barn owl belongs to the most widespread owls in Guatemala and has been recorded in all ornithogeographic regions (Table 13.2, Fig. 13.4, and Appendix 3.1), including urban areas.

Flammulated owl (*Psiloscops flammeolus*) ranges from western North America to northern Central America (AOU 1998). Currently no subspecies is recognized (Dickinson and Remsen 2013). Griscom (1935) described the subspecies *P. f. guatemalae* based on specimens from San Miguel Dueñas (Salvin and Godman 1897–1904), later corrected to *P. f. rarus* (Griscom 1937; Dickerman 1987). We consider flammulated owl as one of the rarest owls in Guatemala (Table 13.2, Fig. 13.4, and Appendix 13.1). The residency status of flammulated owl in Guatemala remained unknown for long. Based on presumed immature specimens from southern Mexico, Kaup (1859) assumed there was a resident population in the south of the species' range. Consequently, Land (1970) classified the species as resident in Guatemala. Phillips (1942) assumed that flammulated owl might occur only as a nonbreeding visitor in Guatemala. Linkhart and McCallum (2013) mentioned if breeding evidence would be found in Oaxaca, Mexico, it might also breed in Guatemala and probably in El Salvador (indirect record based on feathers; Marshall 1978). Because of the few historic records in Guatemala, all during the northern winter, flammulated owl was until recently classified as a migratory vagrant (Eisermann and Avendaño 2007, 2015). The nesting of flammulated owl in Guatemala was recorded for the first time in 2016 in the Sierra Los Cuchumatanes, dpto. Huehuetenango (Eisermann et al. 2017) (Fig. 13.5). The nearest known nesting site is located in the Mexican state of Veracruz, at a distance of 750 km to the northwest (Eisermann et al. 2017). Nesting has not been reported from southern Mexico and El Salvador (Marshall 1978; Enríquez-Rocha et al. 1993; Howell and Webb 1995; AOU 1998; Komar and Domínguez 2001; Pérez León et al. 2015). In addition to the recent nesting record, there are only historic specimens from three sites in Guatemala: San Miguel Dueñas, dpto. Sacatepéquez; near Parramos, dpto. Chimaltenango; and at Cerro Tecpán, dpto. Chimaltenango (Sharpe 1875c; Salvin and Godman 1897–1904; Dearborn 1907; Eisermann et al. 2017). It remains unknown if migratory flammulated owls reach Guatemala and how regular they occur as resident breeding birds in this country (Eisermann et al. 2017). The new status as breeding bird southeast of the Isthmus of Tehuantepec justifies a taxonomic reevaluation of the subspecies *P. f. rarus*.

Pacific screech owl (*Megascops cooperi*) ranges mainly along the Pacific slope lowlands from southern Mexico to Costa Rica (AOU 1998). In Guatemala, it occurs locally (Table 13.2, Fig. 13.4, and Appendix 13.1). It was recorded for the first time by Dickerman (1975) and was the last owl species added to the avifauna of the country. Nesting has been reported from La Avellana, dpto. Santa Rosa, in 1974 (Dickerman 2007). An active nest with two eggs in Manchón-Guamuchal, dpto. San Marcos, on 19 March 2015 (R. Esquivel, Personal Communication), where the female was also seen on 25 March 2015 (KE and CA, photograph), is only the second nesting record for Guatemala. In Manchón-Guamuchal, dptos. San Marcos and Retalhuleu, Pacific screech owl was regularly seen during several visits

Fig. 13.5 Female flammulated owl (*Psiloscops flammeolus*) shortly before leaving the nest hole at dusk. Sierra Los Cuchumatanes, dpto. Huehuetenango, 29 April 2016 (Photograph © Knut Eisermann)

from 2000 to 2002 (J. Berry in Eisermann and Avendaño 2015). It was recently also recorded in Monterrico, dpto. Santa Rosa, in May 2009 (O. Barden in Eisermann and Avendaño 2015), on 25 November 2014 (KE and CA, photograph), and on 14 January 2015 (KE and CA, pers. obs.), on the Salvadoran side of Lake Güija (Herrera 2005), and in Reserva Heloderma, dpto. Zacapa, in the Motagua Valley (J. Berry, J. P. Cahill in Jones and Komar 2013; site therein erroneously reported as Niño Dormido Regional Municipal Park). Reserva Heloderma is located in a transition zone between the arid lowlands of the Pacific slope and the Atlantic slope. This site is located approximately 60 km north of Lake Güija. Only Guatemalan screech owl has been recorded at other sites in the dry scrub of Motagua Valley (Eisermann and Avendaño 2015), suggesting that Pacific screech owl may have colonized this valley until recently or that it occurs only very locally, probably due to competition with Guatemalan screech owl. Pacific screech owl occurs also in Costa Rica locally in the Caribbean slope lowlands (Camacho-Varela 2014). Records of Pacific screech owl in Guatemala range from sea level to 600 m.

Whiskered screech owl (*Megascops trichopsis*) ranges from the southwestern United States to northern Central America (AOU 1998; Fig. 13.6). In Guatemala, it occurs mainly in the highlands at 1000–3000 m (Table 13.2, Fig. 13.4, and Appendix 13.1) but rarely also below 1000 m (San Bernardo, dpto. Guatemala; Griscom 1935). Eisermann and Avendaño (2015) listed recent records at Reserva Pachuj, dpto. Sololá, in November 2007 and December 2008; at Cerro Tecpán, dpto. Chimaltenango, 2009–2011; in Finca El Pilar, dpto. Sacatepéquez, in December 2009; in Finca Filadelfia, dpto. Sacatepéquez, in December 2010; at Volcán de Agua, dpto. Sacatepéquez, in December 2010 and December 2012; in Laguna

Fig. 13.6 Whiskered screech owl (*Megascops trichopsis*), rufous morph, Volcán de Agua, dpto. Sacatepéquez, 19 December 2012 (Photograph © Knut Eisermann)

Lodge Eco-Resort and Nature Reserve, Santa Cruz La Laguna, dpto. Sololá, in August 2008; in Novillero (Parque Corazón del Bosque), dpto. Sololá, in December 2012; at Montaña Sacranix, dpto. Alta Verapaz, in October 2001; at Finca San Joaquín, dpto. Alta Verapaz, in July 2007; in Guatemala City, dpto. Guatemala, in July 2010; in Parque Regional Municipal Los Altos de San Miguel Totonicapán, dpto. Totonicapán, in 2008 (K. Cleary in Eisermann and Avendaño 2015); at Volcán Candelaria, dpto. Quetzaltenango, 2001–2003 (J. Berry in Eisermann and Avendaño 2015); and at Volcán San Pedro, dpto. Sololá, in November 2008 (J. Duerr in Eisermann and Avendaño 2015). It was also recorded at Finca Chaculá, dpto. Huehuetenango (an individual on 14 April 2012, KE, pers. obs.), 6 km southeast of Huehuetenango (road kill, 20 November 2016, KE and CA, pers. obs.), and at Volcán Suchitán, dpto. Jutiapa (MVZ 188302, Museum of Vertebrate Zoology Berkeley 2015). Whiskered screech owl lives in Guatemala in pine-oak and coniferous forests and woodlands. Two juveniles collected near San Pedro Sacatepéquez, dpto. Guatemala, on 19 April 1973 (AMNH 813288–813,289; American Museum of Natural History 2013a, b) confirm breeding in Guatemala.

Bearded screech owl (*Megascops barbarus*) is distributed in a small area of approximately 9800 km^2 (KE, unpub. data; Fig. 13.7a) in the Atlantic slope highlands of southeastern Chiapas, Mexico, and Guatemala (AOU 1998) (Fig. 13.7b).

Fig. 13.7 Bearded screech owl (*Megascops barbarus*). (**a**) Pair of a rufous morph male and brown morph female at Montaña Yalijux, dpto. Alta Verapaz, 29 May 2012. (**b**) Distribution in Guatemala. Historic records are marked with *triangles*, recent records with *circles*. (**c**) Two rufous morph nestlings at Montaña Yalijux, dpto. Alta Verapaz, 21 May 2014 (Photographs © Knut Eisermann)

The holotype was collected near Santa Bárbara, dpto. Baja Verapaz (Sclater and Salvin 1868). Later it was recorded near Cobán, dpto. Alta Verapaz (Ridgway 1914), near Uspantán, dpto. Quiché (Griscom 1932). Recently it has been recorded at Montaña Guaxac, dpto. Alta Verapaz (Eisermann and Avendaño 2007); in Reserva Chelemhá, dpto. Alta Verapaz (Eisermann and Avendaño 2015); at Cerro El Amay, dpto. Quiché (Eisermann et al. 2013; Eisermann and Avendaño 2015);

and in the western part of Sierra de las Minas, dpto. El Progreso, in December 2011 (Eisermann and Avendaño 2015; KE, voice recording). A specimen of a male collected 2 km south of Yalambojoch, dpto. Huehuetenango, on 8 January 2009 (Museum of Vertebrate Zoology, Berkeley, MVZ 184214; R. A. Jiménez, photographs) is the first record for the Sierra Los Cuchumatanes mountain range. A nest at Cerro El Amay, dpto. Quiché, in April 2010 (M. V. Hernández in Eisermann and Avendaño 2015), and several broods at Montaña Yalijux 2012–2016 (Eisermann and Avendaño 2015; KE and CA, pers. obs.) (Fig. 13.7c) are the first nesting records for Guatemala. Undocumented reports from Santa Rosa, Cuilco, dpto. Huehuetenango (Pérez 2006), and from Volcán Atitlán, dpto. Suchitepéquez (Nájera 2010), require verification. Although the bearded screech owl is locally common in Guatemala, we consider it one of the less common owls in the country because of its small range and local distribution (Table 13.2, Fig. 13.4, and Appendix 13.1). The elevational range of all recent and exactly localized records is 1700–2300 m. The bearded screech owl lives in Guatemala along the edge of humid broadleaf forest, pine-oak forest, and pine plantations.

Guatemalan screech owl (*Megascops guatemalae*) occurs on the Atlantic and Pacific slope of Mexico, the Atlantic slope of Guatemala, Honduras, and Nicaragua (AOU 1998; Marks et al. 1999). Reports from northern Costa Rica (König et al. 2008; Mikkola 2014) are apparently erroneous, because records south of the Lake of Nicaragua are attributed to the closely related vermiculated screech owl (*Megascops vermiculatus*) (Marks et al. 1999). The holotype of Guatemalan screech owl was collected in Guatemala (Sharpe 1875c, Dickerman 1987). In this country, it is mainly restricted to the Atlantic slope lowlands and foothills, where it occurs in different types of habitats from sea level to 1600 m, including lowland rainforest, cloud forest, pine-oak forest, and arid scrub (KE and CA, pers. obs.). Once it was recorded in the Guatemalan Pacific slope foothills, where a probably strolling bird was heard at Finca Patrocinio, dpto. Quetzaltenango in 2002 (J. Berry in Eisermann and Avendaño 2006). Guatemalan screech owl belongs to the most common owls in Guatemala (Table 13.2, Fig. 13.4, and Appendix 13.1).

Great horned owl (*Bubo virginianus*) is widespread in North, Middle, and South America (AOU 1998). Dickinson and Remsen (2013) recognized the subspecies *B. v. mesembrinus* for Guatemala, described by Oberholser (1904), and *B. v. mayensis* (Nelson) 1901 restricted to the Yucatán peninsula. Griscom (1935) proposed to use the name *B. v. mayensis* for populations in Middle America, applied by Johnsgard (2002), Weick (2006), and König et al. (2008). In Guatemala, the great horned owl is widely distributed in the semiarid Pacific slope highlands (pine-oak forest, coniferous forest) and arid interior valleys (thorn scrub) (Table 13.2, Fig. 13.4, and Appendix 13.1). It is rare in the humid Atlantic slope highlands, where it has been recorded in Cobán, dpto. Alta Verapaz, in 2010 (J. P. Cahill in Eisermann and Avendaño 2015), and later in each year until 2014 (KE, pers. obs.). Breeding was confirmed in Cobán by observation of a pair with a fledgling in February 2014 (KE, pers. obs., voice recording). It was also recorded in urban area of Guatemala City, dpto. Guatemala, in September 2005 (Eisermann and Avendaño 2015). In the Atlantic slope lowlands, it has been reported from the arid Motagua

Fig. 13.8 Pair of great horned owl (*Bubo virginianus*) at the day roost on thorn scrub in Motagua Valley, Parque Regional Municipal Lo de China, El Jícaro, dpto. El Progreso, 7 September 2010 (Photograph © Knut Eisermann)

Valley near Usumatlán, dpto. Zacapa (Land 1962a); in Parque Regional Municipal Lo de China, El Jícaro, dpto. El Progreso, in September 2010 (Eisermann and Avendaño 2015; Fig. 13.8); in Sabana Grande, dpto. Chiquimula, in September 2010 (Eisermann and Avendaño 2015); and in Reserva Heloderma, El Arenal, dpto. Zacapa, in January 2014 (J. P. Cahill, eBird, S16366246). The great horned owl has not been documented from dpto. Petén. The species can be expected at all elevations in Guatemala, where all records range from 200 to 3730 m. The highest record is from Parque Regional Municipal Todos Santos Cuchumatán, dpto. Huehuetenango, where an individual was seen at 3730 m on 27 August 2016 (KE and CA, photo).

Spectacled owl (*Pulsatrix perspicillata*) ranges from southern Mexico to northern Argentina (AOU 1998). We consider it one of Guatemala's less common owls (Table 13.2, Fig. 13.4, and Appendix 13.1), where it has been recorded in lowland and foothill rainforests on both slopes, from sea level to 1200 m. On the Atlantic slope, it has been recorded 50 km east of Tikal, dpto. Petén (van Tyne 1935), in Parque Nacional Tikal (Beavers 1992); Parque Nacional Sierra del Lacandón, dpto. Petén (near La Pasadita archaeological site in February 2002; R. B. McNab in Eisermann and Avendaño 2015); Parque Nacional Laguna del Tigre, dpto. Petén (near Buena Vista in 2006; M. Córdova in Eisermann and Avendaño 2015); Parque Nacional El Rosario, dpto. Petén (11 October 2013, J. P. Cahill, pers. comm.); Parque Nacional Laguna Lachuá, dpto. Alta Verapaz (Avendaño 2001); Cerro San Gil, dpto. Izabal (Cerezo et al. 2005); and Refugio de Vida Silvestre Punta de Manabique, dpto. Izabal (17 October 2013, J. P. Cahill, Personal Communication). Historically, spectacled owl has been reported from several sites on the Guatemalan Pacific slope: near Escuintla, dpto. Escuintla (Salvin and Godman 1897–1904); near Antigua Guatemala, dpto. Sacatepéquez (Ridgway 1914); at Hacienda California near Ocós, dpto. San Marcos (Griscom 1932); and in El Cacahuito near Taxisco, dpto. Santa Rosa (Tashian 1953). A specimen was collected in 1864 in Costa Cuca (ZMB 18081, Zoological Museum Berlin, Germany; Eisermann and Avendaño 2015). The exact locality is unknown, because Costa Cuca was in the nineteenth century a region covering the southern part of the modern dpto. Quetzaltenango (Eisermann 2011b). Four individuals of spectacled owl were liberated at Finca El Faro, dpto. Quetzaltenango, southeast of Volcán Santa María in 1989 (Vannini and Morales Cajas 1989). The only recent records from the Pacific slope of Guatemala are from Volcán Atitlán and were in Reserva Los Tarrales, dpto. Suchitepéquez; an individual was seen on 1 April 2009 (A. A. Anzueto in Eisermann and Avendaño 2015), two on 21 July 2009 (J. de León Lux in Eisermann and Avendaño 2015), one on 16 December 2009 (Eisermann 2010b), and one on 14 December 2015 (L. de León Lux, photograph; Eisermann 2016). In El Salvador, the spectacled owl has been recorded near the Guatemalan border at Cerro Montecristo, dpto. Santa Ana (Komar 2000), and in Parque Nacional El Imposible, dpto. Ahuachapán (Komar 2003).

Mexican wood owl (*Strix squamulata*) ranges from Mexico to northwestern South America (König et al. 2008, Fig. 13.9). This is the most common owl species in Guatemala, from sea level to 1800 m, locally to 2100 m (Table 13.2, Fig. 13.4, and Appendix 13.1). It is rare above this elevation. It lives mainly in broadleaf for-

Fig. 13.9 Mexican wood owl (*Strix squamulata*) is the most common owl in Guatemala. Reserva Los Tarrales, dpto. Suchitepéquez, 15 December 2012 (Photograph © Knut Eisermann)

ests but occurs also in mixed broadleaf and coniferous forests, plantations and adjacent open habitats, and small forest fragments in urban areas.

Black-and-white owl (*Strix nigrolineata*) ranges from southern Mexico to northeastern South America (AOU 1998). On the Guatemalan Atlantic slope, it has been recorded in dpto. Petén at the following sites: Parque Nacional Tikal (Smithe and Paynter 1963; Gerhardt et al. 1994a, 2012); Biotopo Naachtún-Dos Lagunas and Biotopo San Miguel La Palotada-El Zotz (Whitacre et al. 1991; Jones and Sutter 1992); Reserva Biológica San Román, Monumento Cultural Aguateca, and Monumento Cultural Ceibal (AHT 2000); Parque Nacional El Rosario (11 October 2013, J. P. Cahill, eBird S15381975); Parque Nacional Sierra del Lacandón (Tenez 2007); Parque Nacional Laguna del Tigre (Baumgarten 1998; Ordóñez 1998; Castillo Villeda 2001); Parque Nacional Mirador-Río Azul (Radachowsky et al. 2004; Budney et al. 2008); and archaeological site El Tintal (Budney et al. 2008). In dpto. Alta Verapaz, it has been recorded in Cahabón (Salvin and Godman 1897–1904), Panzós (Land 1963), and Parque Nacional Laguna Lachuá (Avendaño 2001; Eisermann 2001b) and in dpto. Izabal in the foothills of Sierra de las Minas in Selempín (Seglund and Conner 1997), Cerro San Gil (Cerezo et al. 2005), Refugio de Vida Silvestre Punta de Manabique (Eisermann 2001a), Área de Uso Múltiple Río Sarstún (Jones and Komar 2010a), and Sierra del Merendón (9 July 2012, J. P. Cahill, eBird S11305278). On the Pacific slope of Guatemala, it has been recorded historically in dpto. Suchitepéquez near Mazatenango (Salvin and Godman 1897–1904). Recent records on the Pacific slope are located in Reserva Los Tarrales, dpto. Suchitepéquez (Eisermann and Avendaño 2015), including a juvenile (Fig. 13.10); at Finca Las Nubes on Volcán Santo Tomás, dpto. Suchitepéquez (22 March 2011,

Fig. 13.10 Juvenile black-and-white owl (*Strix nigrolineata*) in Reserva Los Tarrales, dpto. Suchitepéquez, 26 June 2008 (Photograph © Susanne Arbeiter)

Fig. 13.11 (**a**) Fulvous owl (*Strix fulvescens*), here an adult, is in Guatemala the most obvious owl species in humid broadleaf forests above 2000 m. Reserva Chelemhá, dpto. Alta Verapaz, 16 April 2016. (**b**) This fledgling is the first documented breeding record for fulvous owl in Guatemala, Reserva Chelemhá, 6 May 2013 (Photographs © Knut Eisermann)

M. Retter, eBird S31493329); in Loma Linda, dpto. Quetzaltenango (Eisermann and Avendaño 2015); on the southern slope of Volcán de Agua (one individual at Finca El Zur, dpto. Escuintla, on 15 November 2015; D. Aldana, eBird S25925893); and at Finca El Pilar, dpto. Suchitepéquez (one individual on 23 January 2016, KE, voice recording; two individuals on 7 May 2016, KE and CA, pers. obs.). This owl has been reported from Parque Nacional El Imposible, dpto. Ahuachapán, El Salvador (Komar 2003), 15 km from the Guatemalan border. The black-and-white owl has been reported in Guatemala in the elevational range from sea level to 2000 m. We consider it one of the most common owls of Guatemala (Table 13.2, Fig.13.4, and Appendix 13.1).

Fulvous owl (*Strix fulvescens*) is restricted to the highlands of the Mexican state of Chiapas and the Central American countries Guatemala, Honduras, and El Salvador (AOU 1998; Fig. 13.11a) and has recently also been recorded in the highlands of the Mexican state of Oaxaca, west of the Isthmus of Tehuantepec (Gómez de Silva 2010; Ramírez-Julián et al. 2011). The species was described based on syntypes from Guatemala (Sclater and Salvin 1868; Dickerman 1987). In Guatemala, fulvous owl replaces Mexican wood owl in humid broadleaf forests above 2000 m, where it is the most obvious owl species (Table 13.2, Fig. 13.4, and Appendix 13.1). Guatemalan records range in elevations from 1800 to 2900 m. It has recently been reported at the following sites in dpto. Alta Verapaz, Montaña Caquipec (Eisermann and Schulz 2005) and Montaña Yalijux (Renner et al. 2006; Eisermann and Avendaño 2015); dpto. Quiché, Cerro El Amay (Eisermann et al. 2013; Eisermann and Avendaño 2015); dpto. El Progreso, Sierra de las Minas (Eisermann 1999; Eisermann and Avendaño

2015); dpto. Chiquimula, Cerro Montecristo (Eisermann 2006); dpto. Sacatepéquez, Finca El Pilar (Eisermann and Avendaño 2015), Finca Filadelfia (J. Fagan in Jones and Komar 2010a), and Volcán Agua (Eisermann and Avendaño 2015); dpto. Sololá, Volcán Atitlán (Eisermann and Avendaño 2015), Volcán San Pedro (J. Rivas in Eisermann and Avendaño 2015), and Volcán Tolimán (24 February 2001, P. Kaestner, pers. comm.); dpto. Quetzaltenango, Volcán Santa María (Vannini 1989) and Volcán Santo Tomás-Zunil (Brooks and Gee 2006, J. Berry in Eisermann and Avendaño 2015); and dpto. San Marcos, Refugio del Quetzal, San Rafael Pie de la Cuesta (Eisermann and Avendaño 2015), and Vega del Volcán (two individuals on 31 March 2015; KE, voice recording). A fledgling in Reserva Chelemhá, dpto. Alta Verapaz, in May 2013 (Fig. 13.11b), is the first documented breeding record for Guatemala. Previously, an immature, killed by local people in Chicacnab, Montaña Caquipec, dpto. Alta Verapaz, was seen in August 1998 (Eisermann and Avendaño 2015).

Crested owl (*Lophostrix cristata*) ranges from southern Mexico to Brazil (AOU 1998). The subspecies *L. c. stricklandi* was described based on syntypes from dpto. Alta Verapaz (Sclater and Salvin 1859; Dickerman 1987). In Guatemala, this owl has been reported from humid broadleaf forests from sea level to mainly below 1000 m. We consider it one of the less common owls in Guatemala (Table 13.2, Fig.13.4, and Appendix 13.1). On the Guatemalan Atlantic slope, it has been reported at the following sites in dpto. Petén, Parque Nacional Sierra del Lacandón (Tenez 2007), an individual on a day roost near Temple IV in Parque Nacional Tikal in 2004 (A. E. Hernández in Eisermann and Avendaño 2015), and Parque Nacional Laguna del Tigre, where listed without documentation by Pérez and Castillo Villeda (2000) and documented for the first time with a photograph of an injured individual near the archaeological site El Perú in February 2010 (M. Rivera Mejía in Eisermann and Avendaño 2015) (Fig. 13.12); dpto. Alta Verapaz, Parque Nacional Laguna Lachuá (Eisermann and Avendaño 2015), an undocumented record of a bird at Montaña Yalijux (Renner et al. 2006) at over 2000 m that was presumably a strolling individual; dpto. Izabal, Sierra Santa Cruz (Pérez 1998), Cerro San Gil where recorded for the first time in September 2010 (Eisermann and Avendaño 2015), and Sierra del Merendón (one individual on 14 May 2016, M. Ramírez, eBird S29676182); and dpto. El Progreso, a female collected near Tulumaje in 1932 (LACM 17710, Natural History Museum of Los Angeles County 2016). On the Guatemalan Pacific slope, crested owl has been reported historically from near Escuintla, dpto. Escuintla (Salvin 1874), near San Pedro Mártir (6 km northeast of Escuintla), and near San Diego, dpto. Escuintla, on Volcán de Agua (Salvin and Godman 1897–1904), which probably corresponds to Finca San Diego, 10 km north of Escuintla. The only recent record from the Pacific slope is from Finca Cataluña, dpto. Retalhuleu, 18 km east of Ocós, in December 2010 (J. Berry in Eisermann and Avendaño 2015), and on Cerro Montecristo on the Salvadoran-Guatemalan border (Herrera et al. 1998).

Guatemalan pygmy owl (*Glaucidium cobanense*) is restricted to the highlands of southern Mexico and northern Central America (Chiapas in Mexico, Guatemala, and Honduras). The description of the species is based on syntypes collected in dpto. Alta Verapaz (Sharpe 1875b; Dickerman 1987). The recent descrip-

Fig. 13.12 This photograph represents the first documented record of crested owl (*Lophostrix cristata*) in Parque Nacional Laguna del Tigre, dpto. Petén. Archaeological site El Perú, 10 February 2010 (Photograph © Melvin Rivera Mejía)

tion of vocalizations (Eisermann and Howell 2011; Howell and Eisermann 2011) supports the specific separation of Guatemalan pygmy owl from mountain pygmy owl (*Glaucidium gnoma*) in the Mexican highlands northwest of the Isthmus of Tehuantepec. In Guatemala, Guatemalan pygmy owl is widespread in the highlands at 1500–3400 m (mainly above 1800 m) (Table 13.2, Fig. 13.4, and Appendix 13.1). Nesting records at Montaña Yalijux, dpto. Alta Verapaz, in March 2010, March 2012, May 2013, and May 2014 (Eisermann and Avendaño 2015) (Fig. 13.13), as well as April 2015 and April 2016 (J. Mez, pers. comm., R. Rax, pers. comm., KE, pers. obs.), are the first nest records for Guatemala.

Central American pygmy owl (*Glaucidium griseiceps*) ranges from southern Mexico to northwestern South America (AOU 1998). The subspecies *G. g. griseiceps* has been described based on syntypes from Guatemala (Sharpe 1875a, Dickerman 1987). In Guatemala, this owl lives mainly in humid broadleaf forest in the Atlantic slope lowlands at 0–600 m, locally up to 1200 m (Table 13.2, Fig. 13.4, and Appendix 13.1). It is rare in the north of dpto. Petén (Beavers 1992), where it has recently been recorded in El Remate in May 2009 (O. Barden in Eisermann and Avendaño 2015), and in Parque Nacional Tikal in February 2013 (L. Oliveros in Eisermann and Avendaño 2015). Two records from the Guatemalan Pacific slope, in the western part of this zone (Howell and Webb 1995; voice recording by B. Whitney, S. Howell, Personal Communication) and in Reserva Los Tarrales, dpto. Suchitepéquez (Eisermann and Avendaño 2006) are presumably of strolling birds. No resident population is known from the Guatemalan Pacific slope. The Central American pygmy owl presumably breeds in Guatemala, but nesting has not been reported yet.

Ridgway's pygmy owl (*Glaucidium ridgwayi*) ranges from the southern United States throughout Middle America to the northwest of Colombia (König et al. 2008). In Guatemala, this owl is widespread and locally common (Table 13.2, Fig. 13.4, and Appendix 13.1) in scrub, dry forests, woodlands, plantations, and open habitats including urban areas from sea level to 1600 m.

Fig. 13.13 Guatemalan pygmy owl (*Glaucidium cobanense*). (**a**) Nestlings few hours before fledging. Montaña Yalijux, dpto. Alta Verapaz, 24 May 2014. (**b**) Juvenile 2 days after fledging, Montaña Yalijux, 27 May 2012 (Photographs © Knut Eisermann)

Burrowing owl (*Athene cunicularia*) lives in breeding, partially migratory populations from southwestern Canada to northern Mexico and in resident populations in South America. Records in southern Mexico and Central America are associated with migratory birds (König et al. 2008). In Guatemala, the burrowing owl has been reported historically on the Pacific coast near Champerico, dpto. Retalhuleu; near Puerto San José, dpto. Escuintla (Salvin and Godman 1897–1904); in the highlands in San Miguel Dueñas, dpto. Sacatepéquez (Salvin and Godman 1897–1904); in San Lucas Tolimán, dpto. Sololá, and Huehuetenango, dpto. Huehuetenango (Griscom 1932); in Alotenango, dpto. Sacatepéquez (Wetmore 1941); in interior valleys near San Jerónimo, dpto. Baja Verapaz (Salvin and Godman 1897–1904); in Gualán, dpto. Zacapa (Dearborn 1907); and also in the Atlantic slope lowlands in Lanquín, dpto. Alta Verapaz (Salvin and Sclater 1860; Salvin and Godman 1897–1904). Griscom (1932) reported this owl as locally common; subsequently it was reported as such by Land (1970), although he did not report this species in his collections (Land and Wolf 1961; Land 1962a, b, 1963). The burrowing owl has not been reported recently in Guatemala; it was recorded for the last time more than 80 years ago. Thus, we consider this owl a migratory vagrant in Guatemala, similar to the status in Honduras (Bonta and Anderson 2002; Gallardo 2014) and El Salvador (Dickey and van Rossem 1938; Komar 1998).

Unspotted saw-whet owl (*Aegolius ridgwayi*) ranges from southern Mexico to western Panama (AOU 1998; Fig. 13.14a). Until recently, the distribution of this owl in Guatemala was little known. Few historic records have been published for Guatemala. Salvin and Godman (1897–1904) reported it from Quetzaltenango, dpto. Quetzaltenango; Griscom (1930) from Sacapulas, dpto. Quiché; and Baepler (1962) from Soloma, dpto. Huehuetenango. Based on recent observations from 1989 to

Fig. 13.14 Unspotted saw-whet owl (*Aegolius ridgwayi*). (**a**) Adult at Volcán de Agua, dpto. Sacatepéquez, December 2010. (**b**) Distribution in Guatemala. *Circles* mark the localization of all records; most of them are recent observations 2000–2016. (**c**) Nestlings at Montaña Yalijux, dpto. Alta Verapaz, February 2013 (Photographs © Knut Eisermann)

2016, we consider this owl widespread throughout the Guatemalan highlands (Fig. 13.14b, Table 13.2, Fig. 13.4, and Appendix 13.1), where it is now known from 22 topographic units. It was recorded at the following sites in dpto. Alta Verapaz, Montaña Caquipec (Eisermann and Schulz 2005), Montaña Yalijux including Reserva Chelemhá (Eisermann and Avendaño 2015), and Tzalamilá (one individual at 1400 m on 17 November 2012, KE and CA, voice recording); dpto. Quiché, Cerro

El Amay (Eisermann et al. 2013; Eisermann and Avendaño 2015); dpto. El Progreso, western Sierra de las Minas (Eisermann and Avendaño 2015), Los Albores, and western Sierra de las Minas (J. P. Cahill, eBird, S11536505); dpto. Jutiapa, Volcán Suchitán (Valdez et al. 1999); dpto. Huehuetenango, two territories 5 km southeast of Todos Santos Cuchumatán on 4 December 2014 (KE and CA, voice recording), one in Chiabal on 27 January 2016 (KE, pers. obs.), and a pair in the Parque Regional Municipal Todos Santos Cuchumatán at 3720 m on 18 November 2016 (KE and CA, photo); dpto. San Marcos, Parque Regional Municipal Canjulá, Sibinal (Eisermann and Avendaño 2015), and Parque Regional Municipal San Pedro Sacatepéquez (Eisermann and Avendaño 2015); dpto. Quetzaltenango, Volcán Candelaria (J. Berry in Eisermann and Avendaño 2007), Volcán Santo Tomás-Zunil (J. Berry in Eisermann and Avendaño 2015), and San Carlos Sija (Eisermann and Avendaño 2015); dpto. Totonicapán, Parque Regional Municipal Los Altos de San Miguel Totonicapán (J. P. Cahill, eBird, S11242340; two individuals on 10 September 2014, KE and CA, pers. obs.); dpto. Sololá, Volcán Atitlán (one individual on 15 December 2010 and on 14 December 2015, two individuals on 18 December 2016; KE, photo, voice recording), Volcán Tolimán (one on 24 February 2001, P. Kaestner, pers. comm.), Los Robles (one heard several times 2013–2015) (E. Buchán, pers. comm.), near Agua Escondida (one heard in December 2015) (G. López, pers.comm.); dpto. Chimaltenango, Cerro Tecpán (Eisermann and Avendaño 2015) and Finca Patoquer (Eisermann and Avendaño 2015); and dpto. Sacatepéquez, Volcán Agua (Eisermann 2013; Eisermann and Avendaño 2015), Finca Filadelfia (Eisermann and Avendaño 2015), Volcán Acatenango (Eisermann and Avendaño 2015), Finca El Pilar (one individual on 2 January 2016, KE, pers. obs.), and probably Volcán Fuego (Tenez 2005a; D. Tenez, pers. comm.). A fledgling in dpto. Chimaltenango, in February 2006, and a fledgling in dpto. San Marcos, in January 2011, and a successful brood with fledged juveniles in March 2013 (Fig. 13.14c) (Eisermann and Avendaño 2015), as well as a clutch in October 2015 in Montaña Yalijux, dpto. Alta Verapaz (KE, pers. obs.), are the first nesting records for Guatemala. According to these observations, the nesting season in Guatemala seems to range from September to March. Records of the unspotted saw-whet owl in Guatemala range in elevation from 1400 to 3730 m (suitable habitat available up to 3800 m), with most records above 1900 m. There are only two records at 1400 m. The unspotted saw-whet owl inhabits in Guatemala coniferous forest and woodlands, coniferous plantations, edges of pine-oak and montane humid broadleaf forest, and montane scrub with scattered trees. Two subspecies have been described in Guatemala, *A. r. rostrata* from Sacapulas (Griscom 1932) and *A. r. tacanensis* in Soloma (Baepler 1962). Validity of subspecies has been doubted (Marks et al. 1999), and König et al. (2008) did not recognize any subspecies. Eisermann (2013) described vocalizations of the unspotted saw-whet owl and compared it to vocalizations of the Guatemalan pygmy owl, which can be a field identification challenge especially during dusk and dawn.

Stygian owl (*Asio stygius*) ranges from northern Mexico to northern Argentina and the Caribbean (AOU 1998). In Guatemala, we consider it one of the rarest owls, with only few records from 11 sites on 9 topographic units, mainly from the highlands (Fig. 13.15a, Table 13.2, Fig. 13.4, and Appendix 13.1), ranging in elevation from 300 to 3000 m. Stygian owl has been recorded in a variety of habitats in

Fig. 13.15 Stygian owl (*Asio stygius*). (**a**) Distribution in Guatemala. *Circles* mark the localization of all records. (**b**) Adult in Loma Linda, dpto. Quetzaltenango, 12 January 2011 (Photograph © Knut Eisermann)

Guatemala, including pine-oak forest, humid broadleaf forest at low and mid elevations, coniferous forest, and coffee plantations. A historic record was reported from Cobán, dpto. Alta Verapaz (Ridgway 1914). Recently it was recorded in Reserva Posada Montaña del Quetzal near Biotopo Mario Dary, dpto. Baja Verapaz (P. Hubbell in Eisermann and Avendaño 2007); at Volcán Candelaria, dpto. Quetzaltenango (J. Berry in Eisermann and Avendaño 2007); Reserva Los Andes, dpto. Suchitepéquez (KE and CA in Jones and Komar 2009, J. L. Yuxón in Eisermann and Avendaño 2015); Volcán San Pedro, dpto. Sololá (J. Duerr in Eisermann and Avendaño 2015); Semuc Champey, dpto. Alta Verapaz (A. Monroy Ojeda in Jones and Komar 2010b); Loma Linda, dpto. Quetzaltenango (Eisermann and Avendaño 2015) (Fig. 15b); Reserva Los Tarrales, dpto. Suchitepéquez (one individual on 13 December 2014 and on 14 December 2015, G. López, pers. comm.); and Finca El Pilar, dpto. Sacatepéquez (one individual on 16 December 2014, KE, pers. obs.; one individual on 3 January 2016, KE, voice recording). Single injured birds were encountered in Guatemala City (O. Ericastilla in Eisermann and Avendaño 2015) and in Finca Rubelchaim, dpto. Alta Verapaz (10 April 2015, J. P. Cahill, eBird, S22820198). A fledged juvenile was seen in Reserva Los Andes, dpto. Suchitepéquez, in February 2011 (Holt et al. 2014), representing the first breeding record for Guatemala.

Striped owl (*Asio clamator*) ranges in open, savannah-like habitats and forest edges from southern Mexico to northern Argentina (AOU 1998; Marks et al. 1999). This owl has been rarely reported in Guatemala (Table 13.2, Fig. 13.4, and Appendix 13.1). Boucard (1878) first reported it for Guatemala without locality. Salvin and Godman (1897–1904) ignored this publication because of several inaccuracies

Fig. 13.16 This juvenile striped owl (*Asio clamator*) in Hacienda Tijax, dpto. Izabal, represents the first breeding record for Guatemala, 18 April 1998 (Photograph © Eugenio Gobbato)

(Griscom 1932). This owl was documented for the first time in Guatemala with several specimens near La Avellana, dpto. Santa Rosa in 1973–1976 (Dickerman 2007). Recently it was also recorded near Cebollito, dpto. Santa Rosa, in 2013 (A. Chávez, pers. comm.). At another site near the Pacific coast, in Manchón-Guamuchal, dpto. San Marcos, it was recorded on 26 October 2014 (J. P. Cahill and J. de León Lux, eBird S20373838). Based on observations in Belize (Howell et al. 1992), Howell and Webb (1995) presumed occurrence in the Guatemalan Atlantic slope lowlands. The first record for this area, and the first breeding record for the country, was a nest with two juveniles in a cattle ranch near Río Dulce Fronteras, dpto. Izabal, in March 1998 (E. Gobbato in Eisermann and Avendaño 2015; Fig. 13.16). Later an adult was photographed near Puerto Barrios, dpto. Izabal, in 2006 (J.-L. Betoulle and N. Komar in Eisermann and Avendaño 2015). Advancing deforestation in Guatemala augments available habitat for striped owl. It has recently been reported from northwestern dpto. Petén, where a bird was heard in the southeastern part of Parque Nacional Laguna del Tigre (J. P. Cahill in Jones and Komar 2015), and repeated records were photo documented in the surroundings of San Benito since October 2015 (C. Echeverría, eBird, S25372846, S27173179). Two birds were recorded at Las Guacamayas, Chiapas, Mexico (Gómez de Silva 2012), 20 km from the Guatemalan border of dpto. Quiché. These records indicate that the known range from the Mexican states of Tabasco and northern Chiapas (Howell and Webb 1995) extends more than 200 km to the southeast. Because of large-scale deforestation, this owl may be expected throughout the southern dpto. Petén and northern part of the dptos. Quiché, Alta Verapaz, and Izabal.

Short-eared owl (*Asio flammeus*) is a nearly cosmopolitan species. It breeds in northern North America, northeastern Europe, northern Asia, as well as in the Caribbean, northern and southern South America. This owl occurs as a nonbreeding visitor in southern Mexico and Central America (AOU 1998). In Guatemala, this species has been reported only once through the collection of two specimens on the slopes of Volcán de Agua, dpto. Sacatepéquez (Salvin 1866), more than 140 years ago. We consider that owl a migratory vagrant in Guatemala.

13.4 Vulnerability on a National Level

13.4.1 Classification of Vulnerability

Eisermann and Avendaño (2006) evaluated the vulnerability of all bird species of Guatemala applying IUCN Red List criteria on a national level. Criteria are quantitative, considering population size and area of distribution. We updated this information for the 20 owl species in Guatemala based on IUCN criteria (2003, 2012a, b). We classified one species as Critically Endangered (CR), one as Endangered (EN), nine as Vulnerable (VU), five as Near Threatened (NT), two as Least Concern (LC) on a national level, and two as not applicable (NA) (Table 13.1). Flammulated owl is classified as Critically Endangered, because it was recently found as a resident breeding bird in Guatemala, with a breeding population probably less than 50 mature birds. We classified stygian owl as Endangered, because the total population in Guatemala is presumed to be less than 250 mature individuals. We classified other species specialized in forest habitats as Vulnerable, considering that populations are at risk to decrease 30% or more within 10 years or three generations because of a decline of the area of occupancy due to deforestation. Annual deforestation rate in Guatemala was an estimated 1.4% or 550 km^2 from 1990 to 2000 (FAO 2011), equaling a forest loss of 5500 km^2 in 10 years. A more recent evaluation estimated an annual deforestation rate of 1% from 2000 to 2006 (Regalado et al. 2012). Both deforestation rates did not discriminate between primary forest and secondary forests or plantations. Thus, we expect the loss of primary forest to be higher. Extensive forest fires, invasions in protected areas, and management deficiencies cause deforestation even within protected areas (ParksWatch 2005). It appears unlikely that the situation will improve in the midterm, because the human population is growing rapidly in Guatemala. It increased 35% from 1994 to 2002 (INE 2002), and from 2010 to 2050 it is estimated to double to 27.9 million (CEPAL 2010). It follows a discussion of the main threats to owl populations in Guatemala.

13.4.2 Habitat Alteration

The alteration of habitat caused by human activity is the principal threat to most owl species in Guatemala. One of the most drastic habitat alterations is the conversion from forest to open agricultural land. Because of the ongoing deforestation in

Guatemala, we classified all breeding species specialized in forest habitats as Vulnerable, Endangered, or Critically Endangered on a national level. Considering that Guatemala has a primarily agricultural economy, the growing population implies an increasing demand for agricultural land. Agriculture is a main cause of deforestation. Population growth causes migrations into forested areas, converting them into agricultural land (Loening and Markussen 2003; Carr 2004, 2005, 2008a, b; Carr et al. 2006). According to a land use mapping in 2003, a total of 29,979.6 km^2 (27.5% of Guatemala) were used for agriculture (Ministerio de Agricultura, Ganadería y Alimentación 2006). Scrub covered 23,925 km^2, of which the majority is part of agricultural crop rotation systems. Consequently, more than 40% of the country is used for agriculture. Annual crops (mainly corn *Zea mays*) covered 13,579.7 km^2 (12.5% of the country); perennial and semi-perennial crops (mainly coffee *Coffea arabica*, sugarcane *Saccharum officinarum*, cardamom *Elettaria cardamomum*, banana *Musa* spp., rubber tree *Hevea brasiliensis*, African oil palm *Elaeis guineensis*, cacao *Theobroma cacao*, and fruit trees) covered 11,454.9 km^2 (10.5% of the country), pasture 4381.7 km^2 (4% of the country), and gardens, nurseries, and vegetable crops 563.3 km^2 (0.5% of the country) (Ministerio de Agricultura, Ganadería y Alimentación 2006). Areas for fuel crops such as African oil palm and Jatropha (*Jatropha curcas*) have been expanded recently (Ribeiro Gallo 2007), causing additional loss of primary forest, because especially Jatropha can be grown on poor soils not adequate for traditional crops.

Increasing fragmentation of forest areas by agricultural plots augments the spread of pesticides used in crops (herbicides, fungicides, insecticides, rodenticides), which can affect owls directly by poisoning or indirectly through alterations in food availability (Blus 1996; Marks et al. 1999). Little information about pesticide impact on owls is available on a worldwide scale (Marks et al. 1999), and no data is available for Guatemala.

Habitat alteration caused by petrol and opencast mineral exploitation is another threat for owl populations in Guatemala, because it is planned in extensive areas. A total of 36,785 km^2 (34% of the country) is used or planned for mineral and petrol exploration and exploitation (Ministerio de Energía y Minas 2011a, b). This area includes 6960 km^2 of forest (20% of the remnant forest area in the country), which is principal habitat for owls (Fig. 13.17).

Guatemala's road density has been increasing from 14.51 km/100 km^2 in 2008 to 15.36 km/100 km^2 in 2013 (CEPAL 2015). Road construction causes direct loss of natural habitat because of the area covered by roads, and in addition, it facilitates the transport of illegally harvested timber and the foundation of new settlements. Consequently, the agricultural border advances. In the Maya Biosphere Reserve (IBA Maya-Lacandón), which is part of the largest Neotropical forest north of the Amazon basin, several new roads are planned. In case all of these are constructed, a loss of 183,000 ha of forest can be expected (Ramos et al. 2007), which equals to 10% of the area of this biosphere reserve. An increasingly denser road network may also elevate the number of road kills among owls because of collision with vehicles.

Other communication infrastructures alter owl habitat in less obvious ways. Communication towers, power lines, fence lines, and wind turbines harbor potential hazards for owls because of risk of collision, displacement, and indirect effects on

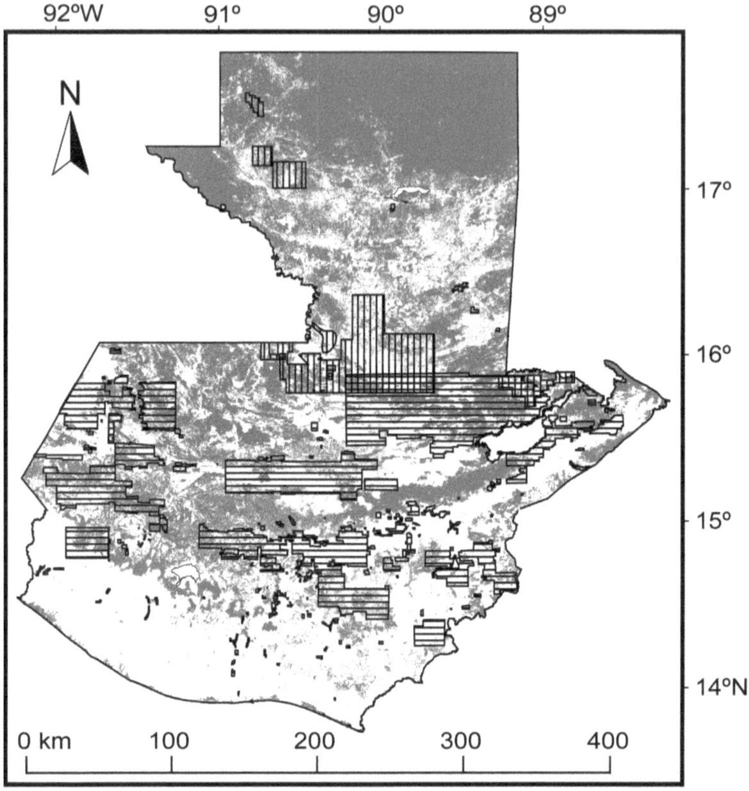

Fig. 13.17 Distribution of remaining forest (*gray shade*) in Guatemala based on a mapping by Ministerio de Agricultura, Ganadería y Alimentación (2006), and areas of mineral exploitations and exploration (*horizontal hatching*) and petrol drilling (*vertical hatching*) according to Ministerio de Energía y Minas (2011a, b)

prey availability (Gove et al. 2013). Cellular telephone networks have been developed rapidly in Guatemala in the past two decades through the installation of communication towers throughout the country. Initial studies in Europe suggest that electromagnetic contamination through the antennas has negative impact on wildlife (Balmori 2005, 2006; Balmori and Hallberg 2007). The impact on owls has not been studied. In Europe and North America, several owl species have been reported being killed by collision or electrocution on power lines, communication towers, and wind turbines (Fitzner 1975; Sergio et al. 2004; Smallwood et al. 2007). The overall impact on populations is difficult to assess but can be locally significant for rare and sensitive species (Horch and Keller 2005; Gove et al. 2013; Loss et al. 2014; Marques et al. 2014). In Guatemala, the impact of these infrastructures on owls has not been studied.

Global atmospheric contaminations cause climate changes, which alter the water cycle regime (Wigley et al. 1997; Karl and Trenberth 2003). An increasing temperature of ocean surfaces may enhance the occurrence of tropical storms (Trenberth

2005; Solomon et al. 2007; Parry et al. 2007), which can impact bird populations in tropical forests (Tejeda-Cruz and Sutherland 2005). Christensen et al. (2007) predict higher temperatures and less precipitation in Guatemala at the end of the twenty-first century. More pronounced dry seasons increase the risk of forest fires. Possible effects on tropical forests caused by climatic changes are still little studied (Clark 2007; Fischlin et al. 2007), but it appears possible that humid broadleaf forests convert gradually into coniferous and mixed forests, challenging habitat specialists of broadleaf forests. Climate changes will alter the distribution pattern of species. Peterson et al. (2001) modeled distribution changes for cracids in Mexico caused by climate changes, resulting in an increased area of distribution for some species and a decreased area of distribution for other species, with a high risk of extinction. Thomas et al. (2004) estimated the extinction of 15–37% of species with restricted range of distribution until 2050, using data for mammals, birds, amphibians, reptiles, butterflies, and plants in 20% of Earth's surface.

Part of the habitat destruction in Guatemala is caused by ignorance and violation of legal frames of nature conservation, including the destruction of protected areas, documented exemplarily for the Parque Nacional Laguna del Tigre (ParksWatch 2005).

Volcanism is a natural cause of habitat alteration. Eruptions cause local destruction of vegetation. Guatemala has currently three active volcanoes: Volcán Pacaya, dptos. Escuintla and Guatemala; Volcán de Fuego, dptos. Escuintla, Sacatepéquez, and Chimaltenango; and Volcán Santiaguito, dpto. Quetzaltenango.

13.4.3 Direct Persecution and Disturbance on Roost and Nest Sites

Direct persecution and disturbances on roost and nest sites are threats to owls, although the impact on populations has not been quantified. It is a common belief among the rural population of Guatemala, that owls attract death and illness (Eisermann and Avendaño 2015). This belief is common in many cultures (Enríquez and Mikkola 1997; Marks et al. 1999), including the Mayan culture, where owls are symbols of death and destruction (Tozzer and Allen 1910; Sharer 1994). Common names for American barn owl and for owls (involving all species) exist in all Mayan languages and in Garífuna (Appendix 13.2). It occurs that people kill owls when encountered and try to avoid that they vocalize near houses. For instance, in a village at Montaña Yalijux, dpto. Alta Verapaz, people cut an avocado tree because an owl was calling from it (Eisermann and Avendaño 2015). Villagers of Chicacnab (Montaña Caquipec) killed a young fulvous owl encountered along a forest trail (Eisermann and Avendaño 2015).

Owls are especially vulnerable on diurnal roost sites and on nest sites. Disturbances on these sites occur in part based on bad intentions, probably driven by superstitions, and in part based on ignorance. At Montaña Yalijux, Maya Q'eqchi' farmers destroyed intentionally broods of Guatemalan pygmy owl, bearded screech owl, and unspotted saw-whet owl (KE, pers. obs.; R. Rax, pers. comm.). Birdwatching is a rapidly developing pastime activity (La Rouche 2003), and Guatemala is being promoted as a destination for traveling bird-watchers (Bland 2007; Eisermann 2007a, b, 2011a, Cocker 2008). Watching birds can cause negative impact upon populations (Sekercioglu 2002). Repeated broadcast of recorded vocalizations or vocalization imitations to attract owls, use of strong flashlights for seeing owls at night, flash photography, and knocking nest trees causing brooding birds to appear at the nest entrance may cause behavioral alterations and decreasing nesting success (Gehlbach and Gehlbach 2000). Negative impacts by bird-watchers in Guatemala have not been quantified.

Illegal trade is another cause of direct persecution. Vannini and Morales Cajas (1989) mentioned the capture of spectacled owl, Mexican wood owl, and pygmy owl (*Glaucidium* spp.). Local state authorities confiscated owls repeatedly in Guatemala. Rescue stations of the association ARCAS received a total of 63 owls from 2002 to 2015: 24 Mexican wood owls, 18 pygmy owls (*Glaucidium* spp.), 6 American barn owls, 5 striped owls, 5 great horned owls, 3 Guatemalan screech owls,1 black-and-white owl, and 1 unspotted saw-whet owl (ARCAS 2002, 2003, 2004, 2005, 2006, 2007, 2008, 2009, 2010, 2011, 2012, 2013, 2014, 2015). This equals 2.2% of the 2813 birds received, with parrots and parakeets (Psittacidae) being the most common.

The general lack of environmental awareness among the Guatemalan society, caused by low education level, lack of political will, social injustice, and poverty, is a major obstacle in reducing threats to owls in this country. Education deficiencies allow that superstitions of owls persist. According to an estimate by the Economic Commission for Latin America and the Caribbean, illiteracy rate was 25% of the Guatemalan population aged 15 years and older in 2010, and 55% lived in poverty in 2002, having an income below the basic food basket (CEPAL 2010). This indicates that improving education should be a key factor in a conservation strategy.

13.5 Conservation Strategies

The legal conditions in Guatemala are favorable for nature conservation. The country has an extensive network of protected areas based on a national law on protected areas. Guatemala has signed several international treaties. A network of Important Bird Areas has been identified, and Guatemala is part of the group of megadiverse countries since 2010. Conservation implementation, however, is still inefficient in Guatemala.

13.5.1 Protected Areas

The network of protected areas in Guatemala covers 32% of the country (CONAP 2011), based on a national law on protected areas (Ley de Áreas Protegidas, Decreto 4-89 and reforms Decretos 18-89, 110-96, 117-97 del Congreso de la República de

Guatemala). The National Council for Protected Areas (CONAP) is the government authority responsible for the management of protected areas, of which many are co-managed by other government agencies (e.g., National Forest Institute INAB and Institute for Archaeology and History IDAEH) and nongovernment organizations or private entities. Of the 192 protected areas declared from 2001 to 2010, 141 are private reserves (CONAP 2011).

13.5.2 Important Bird Areas

BirdLife International developed the Important Bird Area program as a prioritization tool to protect the world's birds as umbrella species for biodiversity conservation. These sites are identified based on populations of globally threatened species, species with restricted areas of distribution, biome-restricted species, and large bird aggregations (Devenish et al. 2009). The list of triggered species for the identification of IBAs in Guatemala included four owls: Pacific screech owl, whiskered screech owl, bearded screech owl, and fulvous owl (Eisermann and Avendaño 2009a, b). According to modern taxonomy, it should also include Guatemalan pygmy owl. The network of IBAs includes populations of all owl species of Guatemala. Three species (Ridgway's pygmy owl, Mexican wood owl, and great horned owl) have been recorded in at least 10 IBAs, 12 species in 5–9 IBAs, 4 species in 2–4 IBAs, and 1 species in a single IBA (Table 13.1).

The IBAs of Guatemala cover an area of 51,884 km^2 (48% of the country), and the size of the IBAs ranges from 43.6 to 20,950.9 km^2 (Fig. 13.1). Of the total area within IBAs, 61.2% (31,770 km^2) are covered with unaltered habitat (mainly primary forest, natural scrub, and wetlands). Habitat altered through human activity (mainly agricultural area and secondary growth scrub) covers 38.3% (19,885 km^2) of the IBA, and urban area covers 0.5% (229 km^2). Of the total area within IBAs in Guatemala, 60% (31,000 km^2) are located within protected areas, and 40% (20,884 km^2) lack legal protection. Seven of the 21 IBAs are legally protected in 0–5% of the area, five IBAs in 6–20% of the area, three IBAs in 21–50%, two IBAs in 51–90%, and four IBAs in 91–100% of the area (Eisermann and Avendaño 2009a).

13.5.3 International Conventions

The government of Guatemala signed several international conventions for the conservation of biodiversity, including the (1) Ramsar Convention for the conservation of wetlands, (2) Convention on International Trade in Endangered Species of Wild Fauna and Flora (CITES), (3) United Nations Convention on Biological Diversity, (4) Convention on Climate Change, and (5) Convention for the Conservation of the Biodiversity and the Protection of Wilderness Areas in Central America.

13.5.4 Education

International conventions, declaration of protected areas, and the designation of Important Bird Areas are tools for defining the legal frame and priorities for conservation. Conservation, however, happens with and through the local people. Improving the education level is key factor for successful conservation in situ. Raising the basic education level would lead to a lower growth rate of the population and consequently reduce the pressure toward natural areas. Coverage of schools, also in remote villages, has improved considerably after signing the peace accords in 1996, which ended a 36-year civil war. Statistical indicators on school participation, however, reveal deficiencies of the Guatemalan education system. Only 46% of the population aged 13–15 years were attending high school in 2015 (Ministerio de Educación 2016). This number reflects the reality in many rural communities in Guatemala, where parents do not consider the education of their children a priority. Raising the education level and standard of living is a complex task, to be completed through local efforts by families, communities, national and international nongovernment organizations, and government agencies.

To reduce direct persecution of owls, it is necessary to improve the image of how Guatemalans perceive these birds. We recommend reinforcing environmental education in schools with dedicated information about the natural history and ecological role of owls. Involving public media, especially radio, which is still the most common media used in remote villages in Guatemala, but also TV, press, and social Internet media, may help to reach the adult population.

13.5.5 Alternative Land Use

Tourism is one of the main sectors of the Guatemalan economy, and nature tourism, especially bird-watching, is developing in the country. The income from visitors to state, community, or private protected areas can be considerable (Naidoo and Adamowicz 2005). In order to keep negative impacts by bird-watchers on owl populations as low as possible, we recommend following ethical principles of bird-watching (American Birding Association 2016).

13.6 Conclusions and Priorities for Future Research

Owls belong to the least studied birds in Guatemala (Eisermann and Avendaño 2006; Enríquez et al. 2012), for one resident species nesting has not even been reported (Table 13.1). In this chapter, we update information on distribution, habitat use, and relative abundance of previous compilations (Salvin and Godman 1897–1904; Ridgway 1914; Griscom 1932; Land 1970; Howell and Webb 1995; Enríquez et al. 2006; Eisermann and Avendaño 2007, 2015). Areas in Guatemala with few or no data on the distribution of owls are located principally in the south of the country (dptos. Jalapa, Jutiapa, Chiquimula, and Santa Rosa; southern part of dptos. Escuintla, Suchitepéquez, and Retalhuleu) and in the semiarid highlands in the interior (southern part of dpto. Quiché, northern part of dptos. Totonicapán and Chimaltenango, western part of dpto. Baja Verapaz, dpto. Guatemala) (Fig. 13.2).

The majority of bird studies in Guatemala involving some species of owls are limited to data on presence. The quality of Neotropical bird studies has been criticized (Winker 1998; Vuilleumier 2004). Eisermann and Avendaño (2006) criticized that many of the ornithological studies in Guatemala remain unpublished; reports and digital data often become lost after some years if not deposited in public archives. To improve this situation, (1) we recommended to improve the study design, especially of rapid assessments carried out by nongovernment organizations, government agencies, and universities. Rapid assessments should include adequate methodologies to detect nocturnal species and determine their abundance. Fast technological development has made digital photography and digital voice recording widely accessible, which facilitate the documentation of field records. (2) We recommended including publication costs in study budgets. Many studies remain unpublished because of lack of funding. (3) If studies are not published formally, we recommended providing access to the data using public databases, such as eBird (Sullivan et al. 2009). Vocalizations can be deposited online via eBird or at xeno-canto (Planqué and Vellinga 2005).

The natural history of owls in Guatemala remains widely unstudied. No species has been studied thoroughly, including relatively common and widespread species such as Guatemalan screech owl and Ridgway's pygmy owl. The only owls with some natural history data published from Guatemala are Mexican wood owl, black-and-white owl, Guatemalan pygmy owl, bearded screech owl, unspotted saw-whet owl, stygian owl, and flammulated owl. During Peregrine Fund's Maya Project in northern dpto. Petén, Whitacre et al. (1991), Jones, and Sutter (1992) provided abundance data for Mexican wood owl, black-and-white owl, and Guatemalan screech owl. Census methodology was described by Whitacre et al. (1992). Gerhardt (1991) examined the reaction of Mexican wood owl to broadcasted vocalizations. Gerhardt et al. (1994a, b), Gerhardt and Gerhardt (2012), and Gerhardt et al. (2012) described breeding biology, home range, and food of Mexican wood owl and black-and-white owl in the Parque Nacional Tikal, and Gerhardt and Gerhardt (1997) published biometric data for both species. Eisermann and Howell (2011) described four different vocalizations of Guatemalan pygmy owl from the Guatemalan highlands, and Eisermann (2013) described vocalizations of unspotted saw-whet owl. Eisermann et al. (2017) described first nesting and habitat of flammulated owl in Guatemala. Holt et al. (2014) published the first evidence of breeding of stygian owl in Guatemala and Eisermann and Avendaño (2015) the first breeding records for Guatemalan pygmy owl, unspotted saw-whet owl, bearded screech owl, fulvous owl, and striped owl. The first nest site, roost sites, and biometric data for bearded screech owl were provided by Enríquez and Cheng (2008) and Enríquez et al. (2010) from Chiapas, Mexico. Because the area of distribution of this owl is small, including only the highlands of southeastern Chiapas and the Atlantic slope highlands of Guatemala, the ecology is presumably similar throughout the range. The ecology of other species with more extensive distribution patterns may vary between regions. The knowledge on the natural history from other regions should be considered a starting point for studies in Guatemala. Current data on Neotropical owls were published in 18 country chapters by Enríquez (2015). The biology and ecology of species which occur also in North America and Europe, have been compiled elsewhere. The series *Birds of North America* covers the following owls also occurring in Guatemala: American barn owl (Marti 1992), flammulated owl (Linkhart and McCallum 2013), whiskered screech owl (Gehlbach and Gehlbach 2000), great horned owl (Houston et al. 1998), Ridgway's pygmy owl (Proudfoot and Johnson 2000), burrowing owl (Haug et al. 1993), and short-eared owl (Holt and Leasure 1993). Johnsgard (2002) and Weidensaul (2015) complied information for North American owls. One species, short-eared owl, occurs also in Europe, whose biology and ecology in this area has been described by Mikkola (1983) and Glutz von Blotzheim and Bauer (1994).

The phylogeny of owls is not yet fully resolved (Wink et al. 2004, 2008). Of the owls in Guatemala, the taxonomic status of the following taxa remains uncertain (Enríquez et al. 2015): American barn owl (relation between *Tyto furcata* in the New World and *T. alba* in the Old World), status of subspecies *Psiloscops flammeolus rarus* of flammulated owl, Pacific screech owl (relation between *M. cooperi* from southern Mexico to Costa Rica and *M. lambi* in Oaxaca, Mexico), Guatemalan screech owl (relation between *Megascops guatemalae* in Mexico and northern Central America and *M. vermiculatus* in southern Central America and northern South America), great horned owl (relation between the widespread *Bubo virginianus* and *B. magellanicus*, restricted to the Andes and southern South America) (Enríquez et al. 2015), Mexican wood owl (relation between *Strix squamulata* in Mexico, Central America, and northern South America and *S. virgata* in South America), genus *Ciccaba* (relation between genera *Ciccaba* and *Strix*), fulvous owl (relation between *Strix fulvescens* of northern Central America and southern Mexico and *Strix varia* in North America and northern Mexico), Guatemalan pygmy owl (relation between *Glaucidium cobanense* in southern Mexico and northern Central America, *G. gnoma* of northern Mexico, and *G. californicum* in western North America), Central American pygmy owl (relation between *Glaucidium griseiceps* and *G. minutissimum* s.l. in South America), unspotted saw-whet owl (relation between *Aegolius ridgwayi* and *A. acadicus*), and striped owl (status of the genus *Pseudoscops*). Further molecular and vocal analyses could provide new insight into the taxonomic status of owls in Guatemala.

The human population of Guatemala is growing rapidly, causing an increasing pressure on natural areas and owl habitat. Most owls of Guatemala live primarily in forests. Hence, the populations of most species are threatened. The conservation of Important Bird Areas (IBAs) in Guatemala could help to protect populations in the long term. Guatemalan IBAs support populations of all owl species recorded in the country. The key to efficient conservation in Guatemala is an improvement of the education.

We hope this compilation encourages (1) more studies on owls in Guatemala to fill gaps in our knowledge on abundance, habitat use, breeding biology, demography, and sensitivity toward human impacts and (2) more efforts to educate the Guatemalan society about owls. Guatemala carries a key responsibility for the conservation of some of Middle America's owl species.

Acknowledgments This compilation of the current knowledge on owls of Guatemala is based on a thorough ornithological bibliography for the country, which was supported by a multitude of persons and institutions (see acknowledgments in Eisermann and Avendaño 2006). We thank the following communities, institutions, and people, who have supported our field research on owls in Guatemala under the roof of PROEVAL RAXMU Bird Monitoring Program: communities of Rocjá Pomtilá, Sanimtacá, Xalabé, Samac, Loma Linda, Santa María Tzejá, Vega del Volcán, Unión Reforma, Sibinal, and Chiabal; Markus Reinhard, Vera Hurschler de Reinhard, and Armin Schumacher of Reserva Chelemhá; Andy Burge, Mónica Barrios, and staff of the Reserva Los Tarrales; Olga Hazard, Jim Hazard, Jaime Freire, and staff of Reserva Los Andes; Francisco Falla of Finca San Sebastián; Mario Aguilar, Lorena de Aguilar, Paulino Chuvac, and staff of Finca Patrocinio; Mayah and Jeffro of Reserva Laguna Lodge Atitlán; Carlos Mazariegos, Jamie Staples, Marcelino Orozco, and Saúl Sandoval (San Rafael Pie de la Cuesta); Misael Baltazar (Parque Regional Municipal San Pedro Sacatepéquez, dpto. San Marcos), Elías Barrera López (CONAP, Uspantán), Kyle Richard (Peace Corps, Sibinal), David Unger (PROEVAL RAXMU), Esteban Matías (Chiabal); Vinicio Bravo, Otilio Bravo, and Román Bartolón (Asociación de Desarrollo Agroforestal Integral Sostenible – ADAFIS); and Rafael Esquivel and José Roberto Jarquín. The National Council of Protected Areas (CONAP) provided research permits.

We thank the following persons for contributing unpublished data: Rogelio Rax Xo, Ernesto Chalib Rax, Juan Mez Saquí, Ernesto Col, Efraín Pop Chub, Efraín Caal, and Roberto Caal (PROEVAL RAXMU Bird Monitoring Program); Susanne Arbeiter (PROEVAL RAXMU Bird Monitoring Program and Fachhochschule Eberswalde, Germany), Roan B. McNab, and Marcial Córdova (Wildlife Conservation Society, Guatemala); Andy Burge, Josué de León Lux, and Gerardo López (Lalo); Everilda Buchán, Léster de León Lux, and Aaron de León Lux (Reserva Los Tarrales); Jesús Lucas Yuxón (Reserva Los Andes); Colum Muccio, Oda Ericastilla, and Luisa María Ortiz Samayoa (ARCAS); Steve Howell (Point Reyes Bird Observatory); Kurt Duchez, Franklin Herrera, and Julio Romero (CONAP Central); Eugenio Gobbato (Hacienda Tijax); and Rafael Esquivel, Jason Berry, Kate Cleary, Jean-Luc Betoulle, Nicholas Komar, Aquiles Estuardo Hernández, Javier Rivas, Robin Schiele, Claudia Múnera, Roger Wilson, Daniel Tenez, Jorge Antonio Jarquín, Abel Alejandro Anzueto, Janet S. Duerr, Olivier Barden, María Vicente Hernández, Carol Anderson, John Paul Cahill, Melvin Rivera Mejía, Luis Oliveros, Alfonso Chávez, and participants of the Christmas Bird Counts in Tikal, Antigua Guatemala, and at Volcán Atitlán organized by CAYAYA BIRDING. Thanks to Eugenio Gobbato, Susanne Arbeiter, and Melvin Rivera Mejía for the contribution of photographs. We thank Sylke Frahnert and Pascal Eckhoff for access and assistance during reviewing specimens in Zoological Museum Berlin (ZMB), Germany; Mayra Maldonado (Zoological collection of the Universidad Del Valle de Guatemala) and Brenda Chávez (Natural History Museum of Universidad de San Carlos de Guatemala) for facilitating access to the collection under their responsibility; and Claudia Múnera for reviewing specimens. Betsy Sánchez of Garífuna Cultural Center, Livingston, and Tomás Sánchez translated common names into Garífuna.

We appreciate comments on an earlier version of the manuscript by three anonymous reviewers, improvements on the English usage provided by an anonymous reviewer, and editorial efforts by Paula L. Enríquez. This contribution of the PROEVAL RAXMU Bird Monitoring Program includes data collected during projects supported by US Fish and Wildlife Service, Stiftung Artenschutz, Germany; Verein Sächsischer Ornithologen, Germany; Unión para Proteger el Bosque Nuboso (UPROBON), Guatemala; Reserva Los Tarrales, Heifer International Guatemala, Austrian Development Agency, Asociación de Desarrollo Agroforestal Integral Sostenible (ADAFIS), and CAYAYA BIRDING, Guatemala.

Appendix 13.1

Geographic coordinates of all sites mentioned in the text and relative abundance of owls at 105 sites

Site	Geographic coordinates	References[a]	Region[b]	Relative abundance index for each species[c]																			
				TYT FUR	PSI FLA	MEG COO	MEG TRI	MEG BAR	MEG GUA	BUB VIR	PUL PER	STR SQU	CIC NIG	STR FUL	LOP CRI	GLA COB	GLA GRI	GLA RID	ATH CUN	AEG RID	ASI STY	ASI CLA	ASI FLA
Agua Escondida, dpto. Sololá	14.6641°N 91.1072°W																						
Alotenango, dpto. Sacatepéquez	14.4869°N 90.8050°W																						
Antigua Guatemala, dpto. Sacatepéquez	14.5586°N 90.7331°W	R. Wilson, Eisermann and Avendaño (2015)	H	1	0	0	0	0	0	0	0	0	0	0	0	0	0	1	0	0	0	0	0
Área de Uso Múltiple Monterrico, dpto. Santa Rosa	13.8986°N 90.4703°W	Dickerman and Brash (1980), Robbins and Dowell (1992), Dickerman (2007); O. Barden in Eisermann and Avendaño (2015); KE and CA, pers. obs.; A. Chávez, Personal Communication	P	1	0	2	0	0	0	0	0	3	0	0	0	0	0	3	0	0	0	1	0
Área de Uso Múltiple Río Sarstún, dpto. Izabal	15.9029°N 88.9986°W	Arrivillaga et al. (1992), J. Fagan and L. Andino in Jones and Komar (2010a)	A	0	0	0	0	0	0	0	0	1	1	0	0	0	0	0	0	0	0	0	0
Archaeological site El Mirador, dpto. Petén	17.7531°N 89.9174°W	Budney et al. 2008	A	0	0	0	0	0	3	0	0	4	0	0	0	0	0	0	0	0	0	0	0
Archaeological site El Perú, dpto. Petén	17.2717°N 90.3656°W		See data for Parque Nacional Laguna del Tigre																				
Archaeological site El Tintal, dpto. Petén	17.5761°N 89.9991°W	Budney et al. (2008)	A	2	0	0	0	0	3	0	0	4	3	0	0	0	0	0	0	0	0	0	0

Biotopo Chocón Machacas, dpto. Izabal	15.7628°N 88.9048°W	Pérez Consuegra et al. (2001)	A	0	0	0	0	0	0	0	0	1	0	0	0	0	0	0	0	0	0	0	0
Biotopo Mario Dary, dpto. Baja Verapaz	15.2134°N 90.2203°W	Bardolf and Bauer (1992), Eisermann and Avendaño (2006, 2015)	H	0	0	0	0	0	0	0	0	4	0	0	0	0	0	0	0	0	1	0	0
Biotopo Naachtún - Dos Lagunas, dpto. Petén	17.7384°N 89.5566°W	Whitacre et al. (1991), Jones and Sutter (1992)	A	0	0	0	0	0	4	0	0	4	2	0	0	0	0	0	0	0	0	0	0
Biotopo San Miguel La Palotada - El Zotz, dpto. Petén	17.2340°N 89.8069°W	Whitacre et al. (1991), Jones and Sutter (1992)	A	0	0	0	0	0	4	0	0	4	2	0	0	0	0	0	0	0	0	0	0
Buena Vista, Parque Nacional Laguna del Tigre, dpto. Petén	17.2961°N 90.1267°W		See data for Parque Nacional Laguna del Tigre																				
Cahabón, dpto. Alta Verapaz	15.6059°N 89.8123°W																						
Carmelita, dpto. Petén	17.4618°N 90.0556°W	Molina (1998)	A	0	0	0	0	0	1	0	0	1	0	0	0	0	0	0	0	0	0	0	0
Cebollito, dpto. Santa Rosa	13.8748°N 90.4281°W		See data for Área de Uso Múltiple Monterrico																				
Cerro Cruz Maltín, dpto. Huehuetenango	15.6803°N 91.2402°W	Garcia Barrientos and Gaitán González (2003)	H	0	0	0	0	0	0	1	0	0	0	0	0	0	0	0	0	0	0	0	0
Cerro El Amay, dpto. Quiché	15.4945°N 90.7989°W	Eisermann et al. (2013), Eisermann and Avendaño (2015)	H	0	0	0	0	4	3	0	0	4	0	3	0	2	1	0	0	1	0	0	0

(continued)

Site	Geographic coordinates	References[a]	Region[b]	Relative abundance index for each species[c]																			
				TYT FUR	PSI FLA	MEG COO	MEG TRI	MEG BAR	MEG GUA	BUB VIR	PUL PER	STR SQU	CIC NIG	STR FUL	LOP CRI	GLA COB	GLA GRI	GLA RID	ATH CUN	AEG RID	ASI STY	ASI CLA	ASI FLA
Cerro Montecristo, dpto. Chiquimula	14.4167°N 89.3500°W	Herrera et al. (1998), Komar (2000), Eisermann (2006)	H	0	0	0	1	0	0	1	1	1	0	1	0	0	0	1	0	0	0	0	0
Cerro San Gil, dpto. Izabal	15.6667°N 88.7833°W	Robbins and Dowell (1992, 1993, 1995); Dowell et al. (1994), Robbins (1996), Cerezo (2001), Cerezo et al. (2005), Eisermann and Avendaño (2015)	A	0	0	0	0	0	4	0	2	4	2	0	1	0	3	1	0	0	0	0	0
Cerro Tecpán, dpto. Chimaltenango	14.7828°N 91.0268°W	Eisermann and Avendaño (2015)	H	0	0	0	2	0	0	2	0	0	0	0	0	1	0	0	0	2	0	0	0
Cobán, dpto. Alta Verapaz	15.4705°N 90.3701°W	Eisermann and Avendaño (2015)	H	1	0	0	0	0	1	1	0	1	0	0	0	0	0	1	0	0	0	0	0
Champerico, dpto. Retalhuleu	14.2954°N 91.9107°W																						
El Cacahuito, dpto. Santa Rosa	14.0904°N 90.4469°W																						
El Estor, dpto. Izabal	15.5249°N 89.3354°W	Dowell et al. (1994)	A	0	0	0	0	0	0	0	0	0	0	0	0	0	0	1	0	0	0	0	0
El Jobal, dpto. Huehuetenango	15.6167°N 91.9333°W	Sandoval (2000)	P	0	0	0	0	0	0	0	0	0	0	0	0	0	0	1	0	0	0	0	0
El Rancho, dpto. El Progreso	14.9158°N 90.0074°W																						
El Remate, dpto. Petén	16.9943°N 89.6922°W	KE, O. Barden, J. P. Cahill in Eisermann and Avendaño (2015)	A	1	0	0	0	0	0	0	0	1	0	0	0	0	1	0	0	0	0	0	0

Escuintla, dpto. Escuintla	14.3057°N 90.7840°W																						
Finca Cataluña, dpto. Retalhuleu	14.4727°N 92.0241°W	J. Berry in Eisermann and Avendaño (2015)	P	0	0	0	0	0	0	0	0	0	0	0	1	0	0	0	0	0	0	0	0
Finca Chaculá and Yalambojoch, dpto. Huehuetenango	15.9754°N 91.6517°W	KE and CA, pers. obs., R. A. Jiménez, Personal Communication	H	0	0	0	1	1	0	1	0	0	0	0	0	0	0	1	0	0	0	0	0
Finca Dos Marias, dpto. San Marcos	14.8174°N 91.8054°W	D.S. Cooper in Eisermann and Avendaño (2015)	H	0	0	0	0	0	0	0	0	0	0	0	0	0	0	1	0	0	0	0	0
Finca El Carmen, dpto. Huehuetenango	15.9167°N 91.7167°W	Sandoval (2000)	H	0	0	0	0	0	0	0	0	1	0	0	0	0	0	0	0	0	0	0	0
Finca El Faro Volcán Santa María (<1000 m), dpto. Quetzaltenango	14.6755°N 91.5819°W	Vannini (1989), Morales Cajas (1991)	P	0	0	0	0	0	0	0	1	1	0	0	0	0	0	0	0	0	0	0	0
Finca El Faro, Volcán Santa María (>1000 m), dpto. Quetzaltenango	14.6755°N 91.5819°W	Vannini (1989), Morales Cajas (1991)	H	0	0	0	0	0	0	1	0	0	0	1	0	1	0	0	0	0	0	0	0
Finca El Pilar, dpto. Sacatepéquez	14.5333°N 90.7000°W	Eisermann (2010a); Eisermann and Avendaño (2015); KE, pers. obs.	H	0	0	0	1	0	0	1	0	2	1	3	0	2	0	0	0	1	1	0	0
Finca El Zur, dpto. Escuintla	14.4072°N 90.7564°W		See data for Volcán Agua																				
Finca Filadelfia, dpto. Sacatepéquez	14.6016°N 90.7196°W	Eisermann (2010a), J. Fagan in Jones and Komar 2010a, Eisermann and Avendaño (2015)	H	0	0	0	3	0	0	2	0	3	0	2	0	0	0	2	0	1	0	0	0

(continued)

Site	Geographic coordinates	References[a]	Region[b]	Relative abundance index for each species[c]																			
				TYT FUR	PSI FLA	MEG COO	MEG TRI	MEG BAR	MEG GUA	BUB VIR	PUL PER	STR SQU	CIC NIG	STR FUL	LOP CRI	GLA COB	GLA GRI	GLA RID	ATH CUN	AEG RID	ASI STY	ASI CLA	ASI FLA
Finca Las Nubes, dpto. Suchitepéquez	14.6668°N 91.4914°W																						
Finca Los Cimientos, dpto. Huehuetenango	15.8833°N 91.8167°W	Sandoval (2000)	P	0	0	0	0	0	0	0	0	0	0	0	0	0	0	1	0	0	0	0	0
Finca Patoquer, dpto. Chimaltenango	14.6505°N 91.0297°W	C. Prahl in Eisermann and Avendaño (2015)	H	0	0	0	0	0	0	0	0	0	0	0	0	0	0	0	0	1	0	0	0
Finca Patrocinio, dpto. Quetzaltenango	14.6600°N 91.6000°W	J. Berry in Eisermann and Avendaño (2015), Eisermann and Avendaño (2015)	P	0	0	0	0	0	1	0	0	4	0	0	0	0	0	2	0	0	0	0	0
Finca San Diego, dpto. Escuintla	14.4142°N 90.7867°W																						
Finca San Francisco, dpto. Huehuetenango	15.9833°N 91.5500°W	Sandoval (2000)	H	1	0	0	0	0	0	0	0	1	0	0	0	0	0	0	0	0	0	0	0
Finca Santa Victoria, dpto. Sololá	14.7692°N 91.1340°W	Eisermann and Avendaño (2015)	H	0	0	0	0	0	0	1	0	0	0	0	0	1	0	0	0	0	0	0	0
Gualán, dpto. Zacapa	15.1136°N 89.3588°W																						
Guatemala City, dpto. Guatemala	14.6190°N 90.5246°W	Eisermann and Avendaño (2015), C. Múnera in Eisermann and Avendaño (2015), ARCAS, CONAP, MUSHNAT	H	0	0	0	1	0	0	1	0	1	0	0	0	0	0	1	0	0	1	0	0

Hacienda California, dpto. San Marcos	14.5592°N 92.1753°W																						
Huehuetenango, dpto. Huehuetenango	15.3201°N 91.4702°W																						
La Avellana, dpto. Santa Rosa	13.9212°N 90.4675°W			See data for Área de Uso Múltiple Monterrico																			
La Cumbre, El Paraíso, Cuilco, dpto. Huehuetenango	15.5500°N 91.9833°W	Sandoval (2000)	H	0	0	0	0	0	0	1	0	0	0	0	0	0	0	0	0	0	0	0	0
La Pasadita (archaeological site), dpto. Petén	17.0095°N 91.0621°W																						
Laguna Lodge Eco-Resort and Nature Reserve, dpto. Sololá	14.7433°N 91.1972°W	Eisermann and Avendaño (2015)	H	0	0	0	1	0	0	1	0	0	0	0	0	1	0	0	0	0	0	0	0
Laguna Yolnabaj, dpto. Huehuetenango	16.0333°N 91.5833°W	Sandoval (2000)	H	0	0	0	0	0	1	0	0	1	0	0	0	0	0	0	0	0	0	0	0
Lake Güija, dpto. Jutiapa	14.2667°N 89.5500°W	Herrera (2005)	P	1	0	1	0	0	0	1	0	1	0	0	0	0	0	1	0	0	0	0	0
Lanquín, dpto. Alta Verapaz	15.5761°N 89.9803°W																						
Loma Linda, dpto. Quetzaltenango	14.7287°N 91.6270°W	Eisermann and Avendaño (2015)	H	0	0	0	0	0	0	0	0	4	1	0	0	0	0	0	0	0	1	0	0
Los Robles, dpto. Sololá	14.7048°N 91.0853°W																						

(continued)

Site	Geographic coordinates	References[a]	Region[b]	Relative abundance index for each species[c]																			
				TYT FUR	PSI FLA	MEG COO	MEG TRI	MEG BAR	MEG GUA	BUB VIR	PUL PER	STR SQU	CIC NIG	STR FUL	LOP CRI	GLA COB	GLA GRI	GLA RID	ATH CUN	AEG RID	ASI STY	ASI CLA	ASI FLA
Manchón-Guamuchal, dptos. Retalhuleu and San Marcos	14.4368°N 92.0914°W	J. Berry in Eisermann and Avendaño (2015), K. Eisermann and C. Avendaño, pers. obs., J. de León Lux, eBird, S20373838	P	0	0	3	0	0	0	0	0	3	0	0	0	0	0	3	0	0	0	1	0
Mazatenango, dpto. Suchitepéquez	14.5339°N 91.5042°W																						
Momostenango, dpto. Totonicapán, 7 km south of town	14.9825°N 91.4370°W	Eisermann and Avendaño (2015)	H	0	0	0	0	0	0	0	0	0	0	0	0	1	0	0	0	0	0	0	0
Montaña Caquipec, dpto. Alta Verapaz	15.3667°N 90.1833°W	Eisermann and Schulz (2005)	H	0	0	0	0	0	0	0	0	1	0	4	0	2	0	0	0	1	0	0	0
Montaña Guaxac (<1000 m), dpto. Alta Verapaz	15.3333°N 90.1333°W	Eisermann (2001b)	A	0	0	0	0	0	1	0	0	4	0	0	0	0	0	3	0	0	0	0	0
Montaña Guaxac (>1000 m), dpto. Alta Verapaz	15.3333°N 90.1333°W	Eisermann (2001b)	H	0	0	0	0	1	2	0	0	4	0	0	0	0	0	3	0	0	0	0	0
Montaña Sacranix (<1000 m), Alta Verapaz	15.5203°N 90.4658°W	Eisermann (2001b)	A	1	0	0	0	0	4	0	0	4	0	0	0	0	2	3	0	0	0	0	0
Montaña Sacranix (>1000 m), dpto. Alta Verapaz	15.5203°N 90.4658°W	Eisermann (2001b)	H	1	0	0	1	0	4	0	0	4	0	0	0	2	0	3	0	0	0	0	0

Montaña Yalijux (>1000 m), dpto. Alta Verapaz	15.3971°N 90.0496°W	Renner et al. (2006), Eisermann and Avendaño (2015); KE, pers. obs.	H	1	0	0	0	4	0	0	0	1	0	4	1	2	0	1	0	2	0	0	0
Monterrico, dpto. Santa Rosa	13.8926°N 90.4800°W		See data for Área de Uso Múltiple Monterrico																				
Monumento Cultural Aguateca, dpto. Petén	16.3914°N 90.2284°W	AHT International (2000)	A	0	0	0	0	0	1	0	0	1	1	0	0	0	0	1	0	0	0	0	0
Monumento Cultural Ceibal, dpto. Petén	16.5067°N 90.0739°W	AHT International (2000)	A	0	0	0	0	0	1	0	0	0	1	0	0	0	0	0	0	0	0	0	0
Morazán, dpto. El Progreso	14.9335°N 90.1423°W	Eisermann and Avendaño (2015)	A	0	0	0	0	0	2	0	0	0	0	0	0	0	0	4	0	0	0	0	0
Novillero (Parque Corazón del Bosque), dpto. Sololá	14.7941°N 91.2678°W	Eisermann and Avendaño (2015)	H	0	0	0	1	0	0	1	0	0	0	0	0	1	0	0	0	0	0	0	0
Ocós, dpto. San Marcos	14.5080°N 92.1946°W																						
Panajachel, dpto. Sololá	14.7396°N 91.1595°W																						
Panzós, dpto. Alta Verapaz	15.3996°N 89.6407°W																						
Parque Nacional El Rosario, dpto. Petén	16.5198°N 90.1584°W																						

(continued)

Site	Geographic coordinates	References[a]	Region[b]	Relative abundance index for each species[c]																			
				TYT FUR	PSI FLA	MEG COO	MEG TRI	MEG BAR	MEG GUA	BUB VIR	PUL PER	STR SQU	CIC NIG	STR FUL	LOP CRI	GLA COB	GLA GRI	GLA RID	ATH CUN	AEG RID	ASI STY	ASI CLA	ASI FLA
Parque Nacional Laguna del Tigre, dpto. Petén	17.5045°N 90.6780°W	Baumgarten (1998), Ordoñez (1998), Castillo Villeda (2001); M. Córdova, M. Rivera Mejía in Eisermann and Avendaño (2015), J. P. Cahill in Jones and Komar (2015)	A	1	0	0	0	0	1	0	1	4	2	0	1	0	0	1	0	0	0	1	0
Parque Nacional Laguna Lachuá, dpto. Alta Verapaz	15.9167°N 90.6667°W	Avendaño (2001), Eisermann (2001b), Eisermann and Avendaño (2015)	A	1	0	0	0	0	3	0	1	4	2	0	2	0	2	1	0	0	0	0	0
Parque Nacional Mirador-Río Azul, dpto. Petén	17.7500°N 89.3333°W	Radachowsky et al. (2004)	A	0	0	0	0	0	3	0	0	4	1	0	0	0	0	0	0	0	0	0	0
Parque Nacional Sierra del Lacandón, dpto. Petén	17.0298°N 90.9571°W	Morales (2001), Puebla-Olivares et al. (2002), Tenez (2007), R.B. McNab in Eisermann and Avendaño (2015)	A	1	0	0	0	0	1	0	1	3	1	0	1	0	1	1	0	0	0	0	0
Parque Nacional Tikal, dpto. Petén	17.2333°N 89.6167°W	Beavers et al. (1991), Beavers (1992), Gerhardt et al. (1994a), Eisermann (2007c, 2008b, 2009b, 2010c), L. Oliveros in Eisermann and Avendaño (2015)	A	0	0	0	0	0	2	0	1	3	2	0	1	0	1	1	0	0	0	0	0
Parque Nacional Yaxhá-Nakum-Naranjo, dpto. Petén	17.1327°N 89.4124°W	Seavy et al. (1995), J. P. Cahill (eBird)	A	0	0	0	0	0	1	0	0	1	1	0	0	0	0	0	0	0	0	0	0

Parque Regional Municipal Canjulá, Sibinal, dpto. San Marcos	15.1295°N 92.0619°W	Eisermann and Avendaño (2015)	H	0	0	0	0	0	0	0	0	0	0	0	0	1	0	0	0	1	0	0	0
Parque Regional Municipal Lo de China, El Jícaro, dpto. El Progreso	14.8959°N 89.8860°W	Eisermann and Avendaño (2015)	A	0	0	0	0	0	0	1	0	0	0	0	0	0	0	3	0	0	0	0	0
Parque Regional Municipal Los Altos de San Miguel Totonicapán, dpto. Totonicapán	14.9223°N 91.3323°W	Eisermann and Avendaño (2015), J. Berry, K. Cleary, J. P. Cahill in Eisermann and Avendaño (2015)	H	1	0	0	1	0	0	1	0	0	0	0	0	1	0	0	0	1	0	0	0
Parque Regional Municipal Niño Dormido, Cabañas, dpto. Zacapa	14.9099°N 89.8109°W	Eisermann and Avendaño (2015)	A	0	0	1	0	0	2	0	0	0	0	0	0	0	0	3	0	0	0	0	0
Parque Regional Municipal San Pedro Sacatepéquez, dpto. San Marcos	15.0085°N 91.7939°W	Eisermann and Avendaño (2015)	H	0	0	0	0	0	0	0	0	0	0	0	0	1	0	0	0	2	0	0	0
Parque Regional Municipal Todos Santos Cuchumatán, dpto. Huehuetenango	15.5313°N 91.5823°W	García Barrientos and Gaitán González (2003), Eisermann et al. (2017), KE and CA, pers. obs.	H	0	1	0	0	0	0	1	0	0	0	0	0	2	0	0	0	2	0	0	0
Parramos, dpto. Chimaltenango	14.6095°N 90.8030°W																						
Patzún, dpto. Chimaltenango	14.6834°N 91.0148°W																						

(continued)

Site	Geographic coordinates	References[a]	Region[b]	Relative abundance index for each species[c]																			
				TYT FUR	PSI FLA	MEG COO	MEG TRI	MEG BAR	MEG GUA	BUB VIR	PUL PER	STR SQU	CIC NIG	STR FUL	LOP CRI	GLA COB	GLA GRI	GLA RID	ATH CUN	AEG RID	ASI STY	ASI CLA	ASI FLA
Pepajau, dpto. Huehuetenango	15.5373°N 91.4279°W	Garcia Barrientos and Gaitán González (2003)	H	0	0	0	0	0	0	1	0	0	0	0	0	1	0	0	0	0	0	0	0
Puerto Barrios, dpto. Izabal	15.7351°N 88.5980°W	J.-L. Betoulle in Eisermann and Avendaño (2015)	A	2	0	0	0	0	0	0	0	2	0	0	0	0	0	3	0	0	0	1	0
Puerto San José, dpto. Escuintla	13.9246°N 90.8210°W																						
Quetzaltenango, dpto. Quetzaltenango	14.8333°N 91.5167°W																						
Refugio de Vida Silvestre El Pucté, dpto. Petén	16.6277°N 90.3749°W	AHT International (2000)	A	0	0	0	0	0	0	0	0	1	0	0	0	0	0	0	0	0	0	0	0
Refugio de Vida Silvestre Petexbatún, dpto. Petén	16.4507°N 90.1814°W	AHT International (2000)	A	0	0	0	0	0	0	0	0	1	0	0	0	0	0	0	0	0	0	0	0
Refugio de Vida Silvestre Punta de Manabique, dpto. Izabal	15.8000°N 88.4167°W	Eisermann (2001a), J. P. Cahill, eBird, S15494859	A	0	0	0	0	0	0	0	1	4	1	0	0	0	0	2	0	0	0	0	0
Refugio de Vida Silvestre Xutilhá, dpto. Petén	16.3460°N 89.7714°W	AHT International (2000)	A	0	0	0	0	0	0	0	0	1	0	0	0	0	0	0	0	0	0	0	0
Refugio del Quetzal (municipal reserve of San Rafael Pie de la Cuesta), dpto. San Marcos	14.9398°N 91.8748°W	Eisermann and Avendaño (2015)	H	0	0	0	0	0	0	0	0	1	0	1	0	1	0	0	0	0	0	0	0

Reserva Atitlán, dpto. Sololá	14.7524°N 91.1673°W	Eisermann and Avendaño (2015)	H	0	0	0	0	0	0	0	0	1	0	0	0	0	0	0	0	0	0	0	0
Reserva Biológica San Román, dpto. Petén	16.2826°N 90.2950°W	AHT International (2000)	A	0	0	0	0	0	0	0	0	1	1	0	0	0	0	0	0	0	0	0	0
Reserva Chelemhá, dpto. Alta Verapaz	15.3833°N 90.0667°W			See data for Montaña Yalijux																			
Reserva de Biosfera Montañas Mayas Chiquibul, dpto. Petén	16.4477°N 89.2910°W	AHT International (2000)	A	0	0	0	0	0	0	0	0	1	0	0	0	0	0	1	0	0	0	0	0
Reserva Heloderma, El Arenal, Cabañas, dpto. Zacapa	14.8625°N 89.7898°W	J. Berry, J. P. Cahill in Jones and Komar (2013); J. P. Cahill and K. V. Vusse, eBird (S21596821)	A	0	0	1	0	0	1	1	0	0	0	0	0	0	0	1	0	0	0	0	0
Reserva Los Andes, dpto. Suchitepéquez	14.5333°N 91.1833°W			See data for Volcán Atitlán																			
Reserva Los Tarrales, dpto. Suchitepéquez	14.5333°N 91.1667°W			See data for Volcán Atitlán																			
Reserva Pachuj and Cerro Iquitiú, dpto. Sololá	14.61188°N 91.1259°W	Valdez et al. (1999), Eisermann (2009a), Eisermann and Avendaño (2015)	H	0	0	0	1	0	0	1	0	1	0	0	0	0	0	1	0	0	0	0	0
Rincón Grande, dpto. Baja Verapaz	15.0352°N 90.3430°W	Eisermann and Avendaño (2015)	H	0	0	0	0	0	0	0	0	3	0	0	0	0	0	0	0	0	0	0	0

(continued)

Site	Geographic coordinates	References[a]	Region[b]	Relative abundance index for each species[c]																			
				TYT FUR	PSI FLA	MEG COO	MEG TRI	MEG BAR	MEG GUA	BUB VIR	PUL PER	STR SQU	CIC NIG	STR FUL	LOP CRI	GLA COB	GLA GRI	GLA RID	ATH CUN	AEG RID	ASI STY	ASI CLA	ASI FLA
Río Dulce (Fronteras), dpto. Izabal	15.6608°N 89.0070°W	Rotenberg (2007); E. Gobbato in Eisermann and Avendaño (2015), Eisermann and Avendaño (2015)	A	1	0	0	0	0	0	0	0	1	0	0	0	0	0	1	0	0	0	1	0
Sabana Grande, dpto. Chiquimula	14.7373°N 89.5566°W	Eisermann and Avendaño (2015)	A	0	0	0	0	0	0	1	0	0	0	0	0	0	0	1	0	0	0	0	0
Sacapulas, dpto. Quiché	15.2879°N 91.0902°W																						
San Benito, dpto. Petén	16.9237°N 89.9180°W																						
San Bernardo, dpto. Guatemala	14.8777°N 90.4387°W																						
San Carlos Sija, dpto. Quetzaltenango	14.9858°N 91.5426°W	Eisermann and Avendaño (2015)	H	0	0	0	0	0	0	0	0	0	0	0	0	0	0	0	0	1	0	0	0
San Gerónimo, dpto. Baja Verapaz	15.0605°N 90.2386°W																						
San Joaquín, dpto. Alta Verapaz	15.3349°N 90.4774°W	Eisermann and Avendaño (2015)	H	0	0	0	1	0	0	0	0	1	0	0	0	0	0	1	0	0	0	0	0
San Lucas Tolimán, dpto. Sololá	14.6335°N 91.1456°W																						
San Miguel Dueñas, dpto. Sacatepéquez	14.5238°N 90.7971°W	Eisermann and Avendaño (2015)	H	0	0	0	0	0	0	1	0	0	0	0	0	0	0	0	0	0	0	0	0
San Pedro Mártir, dpto. Escuintla	14.3442°N 90.7477°W																						

Santa Bárbara, dpto. Baja Verapaz	15.0715°N 90.1805°W																						
Santa María de Jesús, dpto. Quetzaltenango	14.7236°N 91.5254°W	J. Berry in Eisermann and Avendaño (2015)	H	0	0	0	0	0	0	0	0	4	0	0	0	0	0	0	0	0	0	0	0
Santa María Tzejá, dpto. Quiché	15.8662°N 90.8901°W	Eisermann (2011c)	A	1	0	0	0	0	2	0	0	2	1	0	0	0	2	0	0	0	0	0	0
Santa Rosa, Cuilco, dpto. Huehuetenango	15.4518°N 92.0089°W																						
Selempín, dpto. Izabal	15.3190°N 89.4071°W	Seglund and Conner (1997)	A	0	0	0	0	0	0	0	0	1	1	0	0	0	0	0	0	0	0	0	0
Semuc Champey, dpto. Alta Verapaz	15.5340°N 89.9587°W	A. Monroy Ojeda in Jones and Komar (2010b), Eisermann and Avendaño (2015)	A	0	0	0	0	0	1	0	0	1	0	0	0	0	0	1	0	0	1	0	0
Sierra Chinajá, dpto. Alta Verapaz	15.9896°N 90.2198°W	APROBA SANK (2006)	A	0	0	0	0	0	0	0	0	0	0	0	0	0	0	1	0	0	0	0	0
Sierra de las Minas (>1000 m), dptos. El Progreso and Zacapa	15.1426°N 89.8314°W	Land (1962b), Robbins and Dowell (1992), Eisermann (1999), Eisermann and Avendaño (2015)	H	0	0	0	0	1	0	0	0	1	0	1	0	1	0	0	0	1	0	0	0
Sierra del Merendón, incl. Sierra Caral and Sierra del Espíritu Santo, dpto. Izabal	15.4500°N 88.5667°W	Arrivillaga et al. (1992), Cerezo and Ramírez (2003), J. P. Cahill (eBird, S11305278)	A	0	0	0	0	0	1	0	0	1	1	0	1	0	2	1	0	0	0	0	0

(continued)

Site	Geographic coordinates	References[a]	Region[b]	Relative abundance index for each species[c]																			
				TYT FUR	PSI FLA	MEG COO	MEG TRI	MEG BAR	MEG GUA	BUB VIR	PUL PER	STR SQU	CIC NIG	STR FUL	LOP CRI	GLA COB	GLA GRI	GLA RID	ATH CUN	AEG RID	ASI STY	ASI CLA	ASI FLA
Sierra Santa Cruz, dpto. Izabal	15.6333°N 89.4167°W	Arrivillaga et al. (1992), Pérez (1998), Cerezo and Ramírez (2002)	A	0	0	0	0	0	3	0	0	3	0	0	1	0	3	1	0	0	0	0	0
Soloma, dpto. Huehuetenango	15.6575°N 91.4322°W																						
Sumpango, dpto. Chimaltenango	14.6471°N 90.7358°W	MUSHNAT	H	1	0	0	0	0	0	0	0	0	0	0	0	0	0	0	0	0	0	0	0
Takalik Abaj, dpto. Retalhuleu	14.6466°N 91.7362°W	J. Berry, C. Múnera, R. Schiele in Eisermann and Avendaño (2015), Eisermann and Avendaño (2015)	P	0	0	0	0	0	0	0	0	3	0	0	0	0	0	3	0	0	0	0	0
Tontem, dpto. Alta Verapaz	15.4153°N 90.4032°W	Eisermann and Avendaño (2015); KE, pers. obs.	H	0	0	0	0	0	1	0	0	1	0	0	0	1	0	0	0	0	0	0	0
Tulumaje, dpto. El Progreso	14.9276°N 90.0326°W																						
Unicornio Azul, Chiantla, dpto. Huehuetenango	15.3923°N 91.3809°W	Eisermann and Avendaño (2015), KE, pers. obs.	H	0	0	0	0	0	0	1	0	0	0	0	0	1	0	0	0	0	0	0	0
Unión Reforma, Sibinal, dpto. San Marcos	15.1633°N 92.0199°W	Eisermann and Avendaño (2015)	H	0	0	0	0	0	0	1	0	0	0	0	0	1	0	0	0	0	0	0	0
Uspantán, dpto. Quiché	15.3480°N 90.8698°W																						
Usumatlán, dpto. Zacapa	14.9469°N 89.7767°W																						
Polochic valley near Tucurú, dpto. Alta Verapaz	15.2744°N 90.0932°W	Sigüenza (1997)	A	0	0	0	0	0	0	0	0	1	0	0	0	0	0	1	0	0	0	0	0

Vega del Volcán, Sibinal, dpto. San Marcos	15.1523°N 92.0830°W	KE and CA, pers. obs.	H	0	0	0	0	0	0	0	0	0	0	1	0	2	0	0	0	0	0	0	0
Volcán Acatenango, dpto. Sacatepéquez	14.5068°N 90.8741°W	Tenez (2005a), Eisermann and Avendaño (2015)	H	1	0	0	0	0	0	1	0	1	0	0	0	1	0	0	0	1	0	0	0
Volcán de Agua (>1000 m), dpto. Sacatepéquez	14.4673°N 90.7407°W	Eisermann and Avendaño (2015); D. Aldana, eBird (S25925893, S17510117, S30623549)	H	0	0	0	1	0	0	0	0	2	1	1	0	1	0	2	0	1	0	0	0
Volcán Atitlán (<1000 m), dpto. Suchitepéquez	14.5821°N 91.1876°W	Eisermann (2008a, 2009a, 2010b), A.A. Anzueto, J. de León Lux in Eisermann and Avendaño (2015), Eisermann and Avendaño (2015); J. de León Lux, Personal Communication	P	1	0	0	0	0	0	0	1	4	2	0	0	0	1	3	0	0	0	0	0
Volcán Atitlán (>1000 m) dptos. Suchitepéquez and Sololá	14.5821°N 91.1876°W	Eisermann (2008a), Eisermann and Avendaño (2015), C. Múnera in Eisermann and Avendaño (2015)	H	1	0	0	0	0	0	1	0	4	2	3	0	2	0	2	0	1	1	0	0
Volcán Candelaria, dpto. Quetzaltenango	14.8000°N 91.5167°W	J. Berry in Eisermann and Avendaño (2015)	H	0	0	0	1	0	0	1	0	0	0	0	0	1	0	0	0	1	1	0	0

(continued)

Site	Geographic coordinates	References[a]	Region[b]	Relative abundance index for each species[c]																			
				TYT FUR	PSI FLA	MEG COO	MEG TRI	MEG BAR	MEG GUA	BUB VIR	PUL PER	STR SQU	CIC NIG	STR FUL	LOP CRI	GLA COB	GLA GRI	GLA RID	ATH CUN	AEG RID	ASI STY	ASI CLA	ASI FLA
Volcán Fuego, dptos. Sacatepéquez, Escuintla, and Chimaltenango	14.4750°N 90.8808°W	Tenez (2005a), C. Múnera in Eisermann and Avendaño (2015)	H	0	0	0	0	0	0	0	0	1	0	0	0	0	0	0	0	1	0	0	0
Volcán Lacandón, dpto. Quetzaltenango	14.8164°N 91.7170°W	Tenez (2005b)	H	0	0	0	0	0	0	0	0	4	0	0	0	0	0	3	0	0	0	0	0
Volcán San Pedro, dpto. Sololá	14.6594°N 91.2658°W	J. Rivas, J.S. Duerr, Carol Anderson in Eisermann and Avendaño (2015)	H	0	0	0	1	0	0	1	0	0	0	1	0	0	0	0	0	0	1	0	0
Volcán Santa María, dpto. Quetzaltenango	14.7569°N 91.5520°W																						
Volcán Santiaguito, dpto. Quetzaltenango	14.7426°N 91.5709°W																						
Volcán Santo Tomás-Zunil, dptos. Quetzaltenango and Suchitepéquez	14.7331°N 91.4694°W	Brooks and Gee (2006), J. Berry in Eisermann and Avendaño (2015), Eisermann and Avendaño (2015), M. Retter, eBird (S31493329); J. P. Cahill, eBird (S17051774); A. P. Oxom, eBird (S17399804)	H	0	0	0	2	0	0	0	0	3	1	3	0	1	0	1	0	2	0	0	0

Volcán Suchitán, dpto. Jutiapa	14.4000°N 89.7833°W	Valdez et al. (1999), Museum of Vertebrate Zoology-UC Berkeley- MVZ Bird Collection (2015)	H	0	0	0	1	0	0	0	0	1	0	0	0	1	0	0	0	1	0	0	0
Volcán Tolimán, dpto. Sololá	14.6230°N 91.1861°W																						

[a]Abbreviations used in references: *ARCAS* Asociación de Rescate y Conservación de Vida Silvestre, *CONAP* Consejo Nacional de Áreas Protegidas, *MUSHNAT* Museo de Historia Natural de la Universidad de San Carlos de Guatemala

[b]Ornithogeographic region: *A* Atlantic slope lowlands, *H* highlands, *P* pacific slope lowlands

[c]Index of relative abundance (based on data from 1989 to 2016): 0, not recorded; 1, rare (few records, low probability to detect the species in a week of observation, or single record for a site); 2, uncommon (high probability to detect the species in a week of observation); 3, fairly common (high probability to detect an individual/territory in a day of observation); 4, common (high probability to detect several individuals/territories in a day of observation). Species codes: TYTFUR, American barn owl; PSIFLA, flammulated owl; MEGCOO, Pacific screech owl; MEGTRI, whiskered screech owl; MEGBAR, bearded screech owl; MEGGUA, Guatemalan screech owl; BUBVIR, great horned owl; PULPER, spectacled owl; STRSQU, Mexican wood owl; CICNIG, black-and-white owl; STRFUL, fulvous owl; LOPCRI, crested owl; GLACOB, Guatemalan pygmy owl; GLAGRI, Central American pygmy owl; GLARID, Ridgway's pygmy owl; ATHCUN, burrowing owl; AEGRID, unspotted saw-whet owl; ASISTY, stygian owl; ASICLA, striped owl; ASIFLA, short-eared owl

Appendix 13.2

Common names for American barn owl (*Tyto furcata*) and owls (Strigidae) in the majority of languages in Guatemala

Language	Common name for American barn owl *Tyto furcata*	Common name for owl (*Strigidae*)[a]
Spanish	Lechuza	Tecolote, búho
Achí	Xooch'	Tukur
Kaqchikel	Xoch'	Tukre
K'iche'	Xooch'	Tukur
Poqomam	Kuxkux	Tuhkur
Poqomchii'	Xooch'	Tuhkur
Q'eqchi'	Hoob'aq	Warom
Sakapulteko	Xoch'	Tukur
Sipakapense	Sootz'	Tukr
Tz'utujiil	Xo'uuch'	Xken
Uspanteko	Wupup	Tukur
Akateko	Tonton	Tukulin
Chuj	Tzulpop	Tukul
Popti'	No'huh	No'ku'
Q'anjob'al	Chulpop	Tukur, tonton
Ch'orti'	Xo'ch	Tijkirin
Itzaj	Xooch	B'uj
Mopan	Ch'aaw	Buj
Ixhil	Xotx'	Tuuqul
Mam	Xitx'	Tukru
Awakateko	Tukuru	Tukuru
Garífuna	Duguyu	Duguyu

According to Cú Cab et al. (2003), B. Sánchez y T. Sánchez (Personal Communication)

[a]The common name *tecolote* derives from Nahuatl *tecolotl*, origin of the name of the town Teculután, dpto. Zacapa. In several Mayan languages, it translates to "tucur" (Cú Cab et al. 2003), an onomatopoeic word related to owl vocalizations. The name of the town Tucurú, dpto. Alta Verapaz, is based on this word (Sandoval 1942)

References

AHT International (2000) Diseño de un sistema de monitoreo y evaluación de indicadores biológicos para las áreas protegidas del sur de Petén. SEGEPLAN-PROSELVA, Guatemala (Unpublished report)

American Birding Association (2016) American birding association code of birding ethics. http://listing.aba.org/ethics/. Accessed 16 Oct 2016

American Museum of Natural History (2013a) AMNH bird collection. Record ID: SKIN-813288. Source: http://ipt.vertnet.org:8080/ipt/resource.do?r=amnh_birds (source published on 2013-10-30). Accessed 22 Nov 2016

American Museum of Natural History (2013b) AMNH bird collection. Record ID: SKIN-813289. Source: http://ipt.vertnet.org:8080/ipt/resource.do?r=amnh_birds (source published on 2013-10-30). Accessed 22 Nov 2016

AOU (1998) The American Ornithologists' Union check-list of North American birds, 7th edn. American Ornithologists' Union, Washington, DC

APROBA SANK (2006) Informe Final de la Consultoría: Caracterización Ecológica de la Sierra de Chinajá, Chisec, Alta Verapaz. Asociación Pro Bienestar en Acción SANK, Chisec (Unpublished report)

ARCAS (2002) Wildlife Rescue and Conservation Association 2002 annual report. Asociación de Rescate y Conservación de Vida Silvestre, Guatemala (Unpublished report)

ARCAS (2003) Wildlife Rescue and Conservation Association 2003 annual report. Asociación de Rescate y Conservación de Vida Silvestre, Guatemala (Unpublished report)

ARCAS (2004) Wildlife Rescue and Conservation Association 2004 annual report. Asociación de Rescate y Conservación de Vida Silvestre, Guatemala (Unpublished report)

ARCAS (2005) Wildlife Rescue and Conservation Association 2005 annual report. Asociación de Rescate y Conservación de Vida Silvestre, Guatemala (Unpublished report)

ARCAS (2006) Wildlife Rescue and Conservation Association 2006 annual report. Asociación de Rescate y Conservación de Vida Silvestre, Guatemala (Unpublished report)

ARCAS (2007) Wildlife Rescue and Conservation Association 2007 annual report. Asociación de Rescate y Conservación de Vida Silvestre, Guatemala (Unpublished report)

ARCAS (2008) Wildlife Rescue and Conservation Association 2008 annual report. Asociación de Rescate y Conservación de Vida Silvestre, Guatemala (Unpublished report)

ARCAS (2009) Wildlife Rescue and Conservation Association 20 year anniversary and 2009 annual report. Asociación de Rescate y Conservación de Vida Silvestre, Guatemala (Unpublished report)

ARCAS (2010) Wildlife Rescue and Conservation Association 2010 annual report. Asociación de Rescate y Conservación de Vida Silvestre, Guatemala (Unpublished report)

ARCAS (2011) Wildlife Rescue and Conservation Association 2011 annual report. Asociación de Rescate y Conservación de Vida Silvestre, Guatemala (Unpublished report)

ARCAS (2012) Wildlife Rescue and Conservation Association 2012 annual report. Asociación de Rescate y Conservación de Vida Silvestre, Guatemala (Unpublished report)

ARCAS (2013) Wildlife Rescue and Conservation Association 2013 annual report. Asociación de Rescate y Conservación de Vida Silvestre, Guatemala (Unpublished report)

ARCAS (2014) Wildlife Rescue and Conservation Association 2014 annual report. Asociación de Rescate y Conservación de Vida Silvestre, Guatemala (Unpublished report)

ARCAS (2015) Wildlife Rescue and Conservation Association 2015 annual report. Asociación de Rescate y Conservación de Vida Silvestre, Guatemala (Unpublished report)

Arrivillaga A, Billy S, Castro F et al (1992) Estudios técnicos, áreas de protección especial: Río Sarstún, Sierra de Santa Cruz, Bocas del Polochic, Espíritu Santo, Sierra Caral, Cumbre Alta. Proyecto Fase II, Izabal. CONAP, FUNDARY, AID, CI, TNC, Guatemala (Unpublished report)

Avendaño C (2001) Caracterización de la avifauna del Parque Nacional Laguna Lachuá y su zona de influencia, Cobán, Alta Verapaz. Tesis de Licenciatura, Universidad de San Carlos, Guatemala

Baepler DH (1962) The avifauna of the Soloma region in Huehuetenango, Guatemala. Condor 64:140–153

Balmori A (2005) Possible effects of electromagnetic fields from phone masts on a population of white stork (Ciconia ciconia). Electromagn Biol Med 24:109–119

Balmori A (2006) Efectos de las radiaciones electromagnéticas de la telefonía móvil sobre los insectos. Ecosistemas 15:87–95

Balmori A, Hallberg Ö (2007) The urban decline of the house sparrow (*Passer domesticus*): a possible link with electromagnetic radiation. Electromagn Biol Med 26:141–151

Bardolf J, Bauer GP (1992) Lista de aves del Biotopo del Quetzal. Ministerio de Comunicaciones, Guatemala

Baumgarten A (1998) Estudio de las aves del Parque Nacional Laguna del Tigre, Petén, durante la estación seca de 1998. In: Baumgarten A (ed) Informe final del Ejercicio Profesional Supervisado. Universidad de San Carlos, Guatemala
Beavers RA (1992) The birds of Tikal: an annotated checklist for Tikal National Park and Petén, Guatemala. Texas A&M University Press, College Station
Beavers RA, Delaney DJ, Leahy CW et al (1991) New and noteworthy bird records from Petén, Guatemala, including Tikal National Park. Bull Br Ornithol Club 111:77–90
Bibby CJ, Collar NJ, Crosby MJ et al (1992) Putting biodiversity on the map: priority areas for global conservation. International Council for Bird Preservation, Cambridge
Bland B (2007) Guatemala: the new birding Mecca in Central America. Birding World 20:210–215
Blus LJ (1996) Effects of pesticides on owls in North America. J Raptor Res 30:198–206
Bonta M, Anderson DL (2002) Birding Honduras: a checklist and guide. Ecoarte, Tegucigalpa
Boucard A (1878) Liste des oiseaux récoltés au Guatémala en 1877. Ann. Soc. Linnéenne, Lyon, pp 1–47
Brooks TM, Gee B (2006) Horned Guans and other birds at Fuentes Georginas, Guatemala. Pato-Poc 3:3–9
Bruce MD (1999) Family Tytonidae (barn-owls). In: del Hoyo J, Elliott A, Sargatal J (eds) Handbook of the birds of the world, Barn-owls to hummingbirds, vol 5. Lynx Edicions, Barcelona, pp 34–75
Budney GF, Iliff MJ, Iñigo-Elias EE et al (2008) Rapid bird surveys at the El Mirador and Tintal archaeological sites in the Maya Biosphere Reserve, Departamento del Petén, República de Guatemala. Cornell Laboratory of Ornithology, Ithaca. (Unpublished report)
Camacho-Varela P (2014) ¿Cómo identificar correctamente los búhos *Megascops* presentes en Costa Rica? Zeledonia 18(2):3–15
Carr DL (2004) Ladino and Q'eqchi' Maya land use and land clearing in the Sierra de Lacandón National Park, Petén, Guatemala. Agric Human Values 21:171–179
Carr DL (2005) Forest clearing among farm households in the Maya Biosphere Reserve. Prof Geogr 57:157–168
Carr DL (2008a) Farm households and land use in a core conservation zone of the Maya Biosphere Reserve, Guatemala. Hum Ecol 36:231–248
Carr DL (2008b) Migration to the Maya Biosphere Reserve, Guatemala: why place matters. Hum Organ 67:37–48
Carr DL, Barbieri A, Pan W et al (2006) Agricultural change and limits to deforestation in Central America. In: Brouwer F, McCarl BA (eds) Agriculture and climate beyond 2015: a new perspective on future land use patterns. Springer, Dordrecht, pp 91–107
Castillo Villeda ML (2001) Caracterización de la avifauna asociada al los sistemas acuáticos del Parque Nacional Laguna del Tigre, Petén, Guatemala. Tesis de Licenciatura, Universidad de San Carlos, Guatemala
CEPAL (Comisión Económica para América Latina y el Caribe) (2010) 2010 Anuario estadístico de América Latina y el Caribe/Statistical yearbook for Latin America and the Caribbean. Naciones Unidas, Santiago
CEPAL (Comisión Económica para América Latina y el Caribe) (2015) ECLAC logistics profile: Guatemala. Naciones Unidas, Santiago
Cerezo A (2001) Determinación y comparación de los ensambles de aves migratorias y residentes en cuatro hábitats (bosque, pastizal, cerco vivo y bosque ripario), en cinco fincas ganaderas, municipios de Puerto Barrios y Livingston, Departamento de Izabal. Tesis de Licenciatura, Universidad del Valle de Guatemala, Guatemala
Cerezo A, Ramírez M (2002) Evaluación de las comunidades de aves de dos zonas núcleo propuestas para el Area de Protección Especial Sierra Santa Cruz. Informe técnico. Fundación para el Ecodesarrollo y la Conservación, Guatemala (Unpublished report)
Cerezo A, Ramírez M (2003) Evaluación de la avifauna de dos zonas núcleo propuestas para el área de protección especial Sierra Caral. Fundación para el Ecodesarrollo y la Conservación, Guatemala (Unpublished report)

Cerezo A, Ramírez M, Enríquez H (2005) Aves de Cerro San Gil, listado de campo. Fundación para el Ecodesarrollo y la Conservación, Guatemala

Christensen JH, Hewitson B, Busuioc A et al (2007) Regional climate projections. In: Solomon S, Qin D, Manning M et al (eds) Climate change 2007: the physical science basis. Contribution of Working Group I to the Fourth Assessment Report of the Intergovernmental Panel on Climate Change. Cambridge University Press, Cambridge/New York, pp 847–940

Clark DA (2007) Detecting tropical forests' response to global climatic and atmospheric change: current challenges and a way forward. Biotropica 39:4–19

Cocker M (2008) Guatemala: a land of temples and toucans. BBC Wildlife 26(8):90–93

CONAP (2011) Lista de áreas protegidas inscritas en el SIGAP. Consejo Nacional de Áreas Protegidas (CONAP), Guatemala (database)

Constable H, Guralnick R, Wieczorek J et al (2010) VertNet: a new model for biodiversity data sharing. PLoS Biol 8(2):e1000309. doi:10.1371/journal.pbio.1000309

Cú Cab CH, Sacba Caal JC, Pérez Alonzo J et al (2003) Vocabulario comparativo. Okma, Cholsamaj, Guatemala

Dearborn N (1907) Catalogue of a collection of birds from Guatemala. Field Museum of Natural History, Chicago

Devenish C, Diaz Fernández DF, Clay RP et al (eds) (2009) Important bird areas Americas, priority sites for biodiversity conservation, BirdLife conservation series 16. Birdlife International, Quito

Dickerman RW (1975) Nine new specimen records for Guatemala. Wilson Bull 87:412–413

Dickerman RW (1987) Type localities of birds described from Guatemala. Proc West Found Vertebr Zool 3:51–107

Dickerman RW (2007) Birds of the southern Pacific lowlands of Guatemala, with a review of *Icterus gularis*. Spec Publ Mus Southwest Biol 7:1–45

Dickerman RW, Brash AR (1980) Barn owl thanatocenosis on Pacific lowlands of Guatemala. Biotropica 12:228–229

Dickey DR, van Rossem AJ (1938) The birds of El Salvador. Field Mus Nat Hist Zool Ser 23:1–609

Dickinson EC, Remsen Jr JV (2013) The Howard and Moore complete checklist of the birds of the world, 4th ed, vol. 1, non-passerines. Aves Press, Eastbourne

Dowell BA, Holroyd GL, Robbins CS (1994) Bird habitat survey of Cerro San Gil and Polochic delta, Guatemala, February 1994. National Biological Survey, Patuxent Wildlife Research Center, Laurel (Unpublished report)

eBird (2016) eBird: an online database of bird distribution and abundance (web application). eBird, Ithaca, New York. http://www.ebird.org. Accessed 15 Oct 2016

Eisermann K (1999) Avifaunistisch-ökologische Untersuchungen in einer Nebelwaldregion Guatemalas als Grundlage für die Entwicklung eines Biomonitoringprogamms. Diplomarbeit, Fachhochschule Eberswalde

Eisermann K (2001a) Caracterización de la avifauna del Área de Protección Especial Punta de Manabique, Izabal, Guatemala. Fundación Mario Dary, Guatemala (Unpublished report)

Eisermann K (2001b) Reporte de avance: investigaciones de la avifauna de las regiones: planicie del río Ik'bolay, Sierra Guaxac, Sierra Sacranix, Alta Verapaz. Proyecto Eco-Quetzal, National Fish and Wildlife Foundation, Cobán (Unpublished report)

Eisermann K (2006) Evaluación de la avifauna en las partes que formarán el área protegida trinacional de Montecristo en territorio guatemalteco y hondureño. Anexo 5. In: Komar O, Borjas G, Cruz GA et al (eds) Evaluación ecológica rápida del área protegida trinacional Montecristo en territorio guatemalteco y hondureño. SalvaNatura, San Salvador, El Salvador (Unpublished report)

Eisermann K (2007a) Important bird areas in Guatemala. Winging It 19(5):11

Eisermann K (2007b) Vogelreichtum in Guatemala: endemische Arten, IBAs und nachhaltige Bewirtschaftung. Falke 54:385–391

Eisermann K (2007c) Results, count data by count circle, Christmas Bird Count Tikal (GMTK) 2006. http://netapp.audubon.org/CBCObservation/Historical/ResultsByCount.aspx. Accessed 19 Sep 2016

Eisermann K (2008a) Results, count data by count circle, Christmas Bird Count Atitlan volcano (GMAV) 2007. http://netapp.audubon.org/CBCObservation/Historical/ResultsByCount.aspx. Accessed 19 Sep 2016

Eisermann K (2008b) Results, count data by count circle, Christmas Bird Count in Tikal (GMTK) 2007. http://netapp.audubon.org/CBCObservation/Historical/ResultsByCount.aspx. Accessed 19 Sep 2016

Eisermann K (2009a) Results, count data by count circle, Christmas Bird Count Atitlan volcano (GMAV) 2008. http://netapp.audubon.org/CBCObservation/Historical/ResultsByCount.aspx. Accessed 19 Sep 2016

Eisermann K (2009b) Results, count data by count circle, Christmas Bird Count in Tikal (GMTK) 2008. http://netapp.audubon.org/CBCObservation/Historical/ResultsByCount.aspx. Accessed 19 Sep 2016

Eisermann K (2010a) Results, count data by count circle, Christmas Bird Count in Antigua, Guatemala (GMAN) 2009. http://netapp.audubon.org/CBCObservation/Historical/ResultsByCount.aspx. Accessed 19 Sep 2016

Eisermann K (2010b) Results, count data by count circle, Christmas Bird Count Atitlan volcano (GMAV) 2009. http://netapp.audubon.org/CBCObservation/Historical/ResultsByCount.aspx. Accessed 19 Sep 2016

Eisermann K (2010c) Results, count data by count circle, Christmas Bird Count in Tikal (GMTK) 2009. http://netapp.audubon.org/CBCObservation/Historical/ResultsByCount.aspx. Accessed 19 Sep 2016

Eisermann K (2011a) Citizen science in the ornithology of a developing country: Christmas bird counts in Guatemala. Birding 43:38–47

Eisermann K (2011b) Azure-rumped Tanager *Tangara cabanisi*: from ornithological enigma to conservation flagship. Neotrop Birding 9:44–50

Eisermann K (2011c) Evaluación de la avifauna de la reserva forestal de Santa María Tzejá y sus alrededores, Ixcán, Quiché. PROEVAL RAXMU Bird Monitoring Program, Cobán, Alta Verapaz, Guatemala (Unpublished report)

Eisermann K (2013) Vocal field marks of unspotted saw-whet owl and Guatemalan pygmy owl. Neotrop Birding 13:8–13

Eisermann K (2016) Results, count data by count circle, Christmas Bird Count Atitlan volcano (GMAV) 2015. http://netapp.audubon.org/CBCObservation/Historical/ResultsByCount.aspx. Accessed 19 Sep 2016

Eisermann K, Avendaño C (2006) Diversidad de aves en Guatemala, con una lista bibliográfica. In: Cano E (ed) Biodiversidad de Guatemala, vol 1. Universidad del Valle de Guatemala, Guatemala, pp 525–623

Eisermann K, Avendaño C (2007) Lista comentada de las aves de Guatemala – annotated checklist of the birds of Guatemala. Lynx Edicions, Barcelona

Eisermann K, Avendaño C (2009a) Conservation priority-setting in Guatemala through the identification of important bird areas. In: Rich TD, Arizmendi C, Demarest D et al (eds) Tundra to tropics: connecting birds, habitats and people. Proceedings of the 4th International Partners in Flight Conference, 13–16 February 2008. Partners in Flight, McAllen, pp 315–327

Eisermann K, Avendaño C (2009b) Guatemala. In: Devenish C, Diaz Fernández DF, Clay RP et al (eds) Important bird areas Americas, priority sites for biodiversity conservation, BirdLife conservation series 16. Birdlife International, Quito, pp 235–242

Eisermann K, Avendaño C (2015) Los búhos de Guatemala. In: Enríquez PL (ed) Los búhos neotropicales: diversidad y conservación. ECOSUR, San Cristóbal de las Casas, pp 381–434

Eisermann K, Avendaño C, Tanimoto P (2013) Birds of the Cerro El Amay important bird area, Quiché, Guatemala. Cotinga 35:81–93

Eisermann K, Howell SNG (2011) Vocalizations of the Guatemalan pygmy-owl (*Glaucidium cobanense*). J Raptor Res 45:304–314

Eisermann K, Matías E, Avendaño C (2017) Primer registro de anidación de *Psiloscops flammeolus* en Guatemala. Cotinga 39 (in press)

Eisermann K, Schulz U (2005) Birds of a high-altitude cloud forest in Alta Verapaz, Guatemala. Rev Biol Trop 53:577–594

Enríquez PL (ed) (2015) Los búhos neotropicales: diversidad y conservación. ECOSUR, San Cristóbal de las Casas

Enríquez PL, Cheng KM (2008) Natural history of the threatened bearded screech-owl (*Megascops barbarus*) in Chiapas, Mexico. J Raptor Res 42:180–187

Enríquez PL, Cheng KM, Elliott JE (2010) The 'near threatened' bearded screech-owl *Megascops barbarus*: diet pattern and trophic assessment using δ13C and δ 15N stable-isotopes. Bird Conserv Int 20:25–33

Enríquez PL, Eisermann K, Mikkola H (2012) Los búhos de México y Centroamérica: necesidades en investigación y conservación. Ornitol Neotrop 23(Suppl):247–260

Enríquez PL, Eisermann K, Motta-Junior JC et al (2015) Una revisión de la taxonomía y sistemática de los búhos neotropicales. In: Enríquez PL (ed) Los búhos neotropicales: diversidad y conservación. ECOSUR, San Cristóbal de las Casas, pp 29–38

Enríquez PL, Johnson DH, Rangel-Salzar JL (2006) Taxonomy, distribution and conservation of owls in the Neotropics: a review. In: Rodríguez-Estrella R (ed) Current raptor studies in Mexico. Comision Nacional para el Conocimiento y Uso de la Biodiversidad (CONABIO), México, pp 254–307

Enríquez PL, Mikkola H (1997) Comparative study of general public owl knowledge in Costa Rica, Central America and Malawi, Africa. In: Duncan JR, Johnson DH, Nicholls TH (eds) Biology and conservation of owls of the northern hemisphere. General Technical Report NC-190. U.S. Dept. of Agriculture, Forest Service, North Central Forest Experiment Station, St. Paul, pp 160–166

Enríquez-Rocha P, Rangel-Salazar JL, Holt DW (1993) Presence and distribution of Mexican owls: a review. J Rapor Res 27:154–160

FAO (2011) State of the world's forests 2011. Food and Agriculture Organization of the United Nations, Rome

Fischlin A, Midgley GF, Price JT et al (2007) Ecosystems, their properties, goods, and services. In: Parry ML, Canziani OF, Palutikof JP et al (eds) Climate change 2007: impacts, adaptation and vulnerability. Contribution of Working Group II to the Fourth Assessment Report of the Intergovernmental Panel on Climate Change. Cambridge University Press, Cambridge, pp 211–272

Fitzner RE (1975) Owl mortality on fences and utility lines. J Raptor Res 9:55–57

Gallardo RJ (2014) Guide to the birds of Honduras. R. J. Gallardo, Tegucigalpa

García Barrientos A, Gaitán González LB (2003) Evaluación rápida de la avifauna en tres sitios identificados por PROCUCH como prioritarios para conservación de: Todos Santos Cuchumatán, Pepajau-Magdalena y Cerro Cruz Maltín, Sierra de Los Cuchumatanes, Huehuetenango, Guatemala. Fundación Interamericana de Investigación Tropical (FIIT), Guatemala (Unpublished report)

Gehlbach FR, Gehlbach NY (2000) Whiskered screech-owl (*Otus trichopsis*). In: Poole A, Gill F (eds) The birds of North America no. 507. The Birds of North America, Inc., Philadelphia

Gerhardt RP (1991) Response of Mottled Owls to broadcast of conspecific call. J Field Ornithol 62:239–244

Gerhardt RP, Gerhardt DM (1997) Size, dimorphism and related characteristics of *Ciccaba* owls from Guatemala. In: Duncan JR, Johnson DH, Nicholls TH (eds) Biology and conservation of owls of the Northern Hemisphere: 2nd international symposium: February 5–9, 1997, Winnipeg MBGen, Tech. Rep. NC-190. USDA Forest Service, North Central Research Station, St. Paul, pp 190–196

Gerhardt RP, Gerhardt DM (2012) Mexican wood owl. In: Whitacre DF (ed) Neotropical birds of prey: biology and ecology of a forest raptor community. Comstock Cornell, Ithaca, pp 313–319

Gerhardt RP, Bonilla N, Gerhardt DM et al (1994a) Breeding biology and home range of two *Ciccaba* owls. Wilson Bull 106:629–639

Gerhardt RP, Gerhardt DM, Flatten CJ et al (1994b) The food habits of sympatric *Ciccaba* owls in northern Guatemala. J Field Ornithol 65:258–264

Gerhardt RP, Gerhardt DM, Bonilla N et al (2012) Black-and-white owl. In: Whitacre DF (ed) Neotropical birds of prey: biology and ecology of a forest raptor community. Comstock Cornell, Ithaca, pp 320–327
Glutz von Blotzheim UN, Bauer KM (eds) (1994) Handbuch der Vögel Mitteleuropas, vol 9: Columbiformes – Piciformes, 2nd edn. Aula Verlag, Wiesbaden
Gómez de Silva H (2010) Spring migration, March through May 2010: Mexico. N Am Birds 64:503–507
Gómez de Silva H (2012) The winter season, December 2011 through February 2012: Mexico. N Am Birds 66:350–353
Gove B, Langston RHW, McCluskie A et al (2013) Wind farms and birds: an updated analysis of the effects of wind farms on birds, and best prectice guidance on integrated planning and impact assessment. Report written by BirdLife International on behalf of the Bern Convention. RSBP/BirdLife in the UK, Sandy
Griscom L (1930) Studies from the Dwight collection of Guatemala birds. III. Am Mus Novit 438:1–18
Griscom L (1932) Distribution of bird-life in Guatemala: a contribution to a study of the origin of Central American bird-life. Bull Am Mus Nat His 64
Griscom L (1935) Critical notes on Central American birds in the British museum. Ibis 77:541–554
Griscom L (1937) New name for *Otus flammeolus guatemalae* preoccupied. Auk 54:391
Haug EA, Millsap BA, Martell MS (1993) Burrowing owl (*Athene cunicularia*). In: Poole A, Gill F (eds) The birds of North America No. 61. The Birds of North America, Inc., Philadelphia
Herrera N (2005) Estudio ambiental: estado de conservación del complejo lago de Güija. Ministerio del Ambiente y Recursos Naturales, Centro Nacional de Registros, San Salvador, El Salvador (Unpublished report)
Herrera N, Ibarra Portillo R, Rivera R (1998) Estudio de la fauna vertebrada en la reserva La Fraternidad (El Salvador, Guatemala, Honduras). Informe de consultoría. Fundación para la Conservación de los Ecosistemas Mayas, Guatemala (Unpublished report)
Holt DW, Leasure SM (1993) Short-eared owl (*Asio flammeus*). In: Poole A, Gill F (eds) The birds of North America, no. 61. The Birds of North America, Inc., Philadelphia
Holt DW, Yuxó JL, Hiro S et al (2014) Nest record of the stygian owl (*Asio stygius*) in Guatemala. Spizaetus 17:14–17
Horch P, Keller V (2005) Windkraftanlagen und Vögel – ein Konflikt? Schweizerische Vogelwarte Sempach, Sempach
Houston CS, Smith DG, Rohner C (1998) Great horned owl (*Bubo virginianus*). In: Poole A, Gill F (eds) The birds of North America No. 372. The Birds of North America, Inc., Philadelphia
Howell SNG, Eisermann K (2011) Guatemalan pygmy-owl *Glaucidium cobanense* is a good species. Neotrop Birding 9:74–76
Howell SNG, Webb S (1995) A guide to the birds of Mexico and northern Central America. Oxford University Press, New York
Howell SNG, Dowell BA, James DA et al (1992) New and noteworthy bird records from Belize. Bull Brit Ornithol Club 112:235–244
INE (2002) Censos nacionales XI de población y VI de habitación 2002. Población y locales de habitación particulares censados según departamento y municipio (cifras definitivas). Instituto Nacional de Estadística, República de Guatemala (database)
IUCN (2003) Guidelines for Application of IUCN Red List Criteria at Regional Levels. Version 3.0. IUCN Species Survival Commission. IUCN, Gland/Cambridge
IUCN (2012a) IUCN red list categories and criteria. Versión 3.1. 2nd edn. IUCN, Gland/Cambridge
IUCN (2012b) Guidelines for application of IUCN red list criteria at regional and national levels. Version 4.0. IUCN, Gland
Johnsgard PA (2002) North American owls: biology and natural history, 2nd edn. Smithsonian Institution Press, Washington, DC
Jones HL, Komar O (2009) The nesting season, June through July 2008: Central America. N Am Birds 62:626–628

Jones HL, Komar O (2010a) Fall migration, August through November 2009: Central America. N Am Birds 64:164–168
Jones HL, Komar O (2010b) Winter season, December 2009 through February 2010: Central America. N Am Birds 64:333–336
Jones HL, Komar O (2013) Winter season, December 2012 through February 2013: Central America. N Am Birds 67:349–356
Jones HL, Komar O (2015) The winter season, December 2013 through February 2014: Central America. N Am Birds 68:284–289
Jones LE, Sutter J (1992) Results and comparisons of two years of census efforts at three units of the Maya Biosphere Reserve/Calakmul Biosphere Reserve complex. In: Whitacre DF, Thorstrom RK (eds) Maya project: use of raptors and other fauna as environmental indicators for design, management, and monitoring of protected areas and for building local capacity for conservation in Latin America. Progress report V. Peregrine Fund, Boise, pp 63–79 (Unpublished report)
Karl TR, Trenberth KE (2003) Modern global climate change. Science 302:1719–1723
Kaup JJ (1859) Monograph of the Strigidae. Trans Zool Soc Lond 4:201–260
Komar O (1998) Avian diversity in El Salvador. Wilson Bull 110:511–533
Komar O (2000) Lista de aves del Parque Nacional Montecristo, El Salvador. SalvaNatura, San Salvador
Komar O (2003) Lista de aves para el Parque Nacional El Imposible. Bird list for El Imposible National Park. In: Alvarez JM, Komar O (eds) El Parque Nacional El Imposible y su vida silvestre - El Imposible National Park and its wildlife. Biodiversity Series No. 2, SalvaNatura, San Salvador, pp 215–227
Komar O, Domínguez JP (2001) Lista de aves de El Salvador. Serie de Biodiversidad No. 1. SalvaNatura, San Salvador, El Salvador
König C, Weick F, Becking J-H (2008) Owls of the world, 2nd edn. Christopher Helm, London
La Rouche GP (2003) Birding in the United States: a demographic and economic analysis. USFWS Division of Federal Aid, Washington, DC
Land HC (1962a) A collection of birds from arid interior of eastern Guatemala. Auk 79:1–11
Land HC (1962b) A collection of birds from the Sierra de las Minas, Guatemala. Wilson Bull 74:267–283
Land HC (1963) A collection of birds from the Caribbean lowlands of Guatemala. Condor 65:49–65
Land HC (1970) Birds of Guatemala. Livingston, Wynnewood
Land HC, Wolf LL (1961) Additions to the Guatemalan bird list. Auk 78:94–95
Linkhart BD, McCallum DA (2013) Flammulated owl (*Psiloscops flammeolus*). In: Rodewald PG (ed) The birds of North America online. Cornell Lab of Ornithology, Ithaca. http://bna.birds.cornell.edu.bnaproxy.birds.cornell.edu/bna/species/093. Accessed 15 Oct 2016
Loening LJ, Markussen M (2003) Pobreza, deforestación y sus eventuales implicaciones para la biodiversidad en Guatemala. Econ Soc Territ 4:279–315
Loss SR, Will T, Marra PM (2014) Refining estimates of bird collision and electrocution mortality at power lines in the United States. PLoS One 9(7):e101565. doi:10.1371/journal.pone.0101565
Marks JS, Cannings RJ, Mikkola H (1999) Family Strigidae: typical owls. In: del Hoyo J, Elliott A, Sargatal J (eds) Handbook of the birds of the world. Vol. 5. Barn-owls to hummingbirds. Lynx Edicions, Barcelona, pp 76–242
Marques AT, Batalha H, Rodrigues S et al (2014) Understanding bird collisions at wind farms: an updated review on the causes and possible mitigation strategies. Biol Conserv 179:40–52
Marshall JT (1967) Parallel variation in North and Middle American screech-owls. Monogr West Found Vertebr Zool 1:1–72
Marshall JT (1978) Systematics of smaller Asian night birds based on voice. Ornithol Monogr 25:1–58
Marti CD (1992) Barn owl (*Tyto alba*). In: Poole A, Stettenheim P, Gill F (eds) The birds of North America, no. 1. The Birds of North America, Inc., Philadelphia
Mikkola H (1983) Owls of Europe. T. & A. D. Poyser, London

Mikkola H (2014) Owls of the world: a photographic guide, 2nd edn. Firefly Books, Buffalo/Richmond Hill
Ministerio de Agricultura, Ganadería y Alimentación (2006) Mapa de cobertura vegetal y uso de la tierra a escala 1:50,000 de la República de Guatemala, Año 2003 (Incluye cinco cultivos perennes actualizados al año 2005). Memoria técnica y descripción de resultados. Ministerio de Agricultura Ganadería y Alimentación –MAGA– / Unidad de Planificación Geográfica y Gestión de Riesgo –UPGGR–, Guatemala Ciudad, Guatemala
Ministerio de Educación (2016) Anuario estadístico de la edcuación, Guatemala, Centro América, 2015. Ministerio de Educación, Guatemala. http://estadistica.mineduc.gob.gt/anuario/2015/main.html. Accessed 16 Oct 2016
Ministerio de Energía y Minas (2011a) Catastro minero. Ministerio de Energía y Minas de Guatemala, Dirección General de Minería. http://www.mem.gob.gt. Accessed 15 Oct 2011
Ministerio de Energía y Minas (2011b) Área de petróleo: exploración y explotación. Ministerio de Energía y Minas de Guatemala, Dirección General de Hidrocarburos. http://www.mem.gob.gt. Accessed 15 Oct 2011
Molina WO (1998) Caracterización de la avifauna y estimación de las densidades poblacionales relativas de 54 especies de aves en Carmelita, San Andrés, Petén, Guatemala. Tesis de Licenciatura, Universidad de San Carlos, Guatemala
Morales R (2001) Monitoreo de aves en diferentes microhábitats del Parque Nacional Sierra del Lacandón, La Libertad, Petén, 2000–2001. Fundación Defensores de la Naturaleza, Consejo Nacional de Áreas Protegidas. USAID, TNC, Guatemala (Unpublished report)
Morales Cajas CL (1991) Actualización y ampliación de lista de ornitofauna de la finca El Faro, sur-occidente de Guatemala. In: Morales CL (ed) Informe final de Ejercicio Profesional Supervisado realizado en Fundación Interamericana de Investigación Tropical. Universidad de San Carlos, Guatemala
Museum of Vertebrate Zoology Berkeley (2015) Record ID: http://arctos.database.museum/guid/MVZ:Bird:188302?seid=3131696. Source: http://ipt.vertnet.org:8080/ipt/resource.do?r=mvz_bird (source published on 2015-10-27). Accessed 22 Nov 2016.
Naidoo R, Adamowicz WL (2005) Economic benefits of biodiversity exceed costs of conservation at an African rainforest reserve. PNAS 102:16712–16716
Nájera A (2010) Estudio preliminar de aves en la reserva biológica Refugio del Quetzal de la Universidad del Valle de Guatemala. Rev Univ del Valle de Guatemala 22:60–65
Natural History Museum of Los Angeles County (2016) LACM vertebrate collection. Record ID: 0e08ff4f-96c7-4e34-89f8-b3d12b8298fa. Source: http://ipt.vertnet.org:8080/ipt/resource.do?r=lacm_verts. Accessed 22 Nov 2016
Oberholser HC (1904) A revision of the American great horned owls. Proc US Natl Mus 27:177–192
Ordóñez N (1998) Estudio comparado de la avifauna del Parque Nacional Laguna del Tigre Petén durante las estaciones seca y lluviosa de 1998. In: Ordóñez N (ed) Informe final del Ejercicio Profesional Supervisado. Universidad de San Carlos, Guatemala
ParksWatch (2005) Perfil de Área Protegida – Guatemala: Parque Nacional Laguna del Tigre y Biotopo Protegido Laguna del Tigre-Río Escondido. ParksWatch, Guatemala. http://www.parkswatch.org. Accessed 12 Oct 2010
Parry ML, Canziani OF, Palutikof JP et al (eds) (2007) Climate change 2007: impacts, adaptation and vulnerability. Contribution of Working Group II to the Fourth Assessment Report of the Intergovernmental Panel on climate change. Cambridge University Press, Cambridge
Pérez ES (1998) Informe final del estudio técnico del área de protección especial Sierra Santa Cruz. Grupo aves. Fundación Defensores de la Naturaleza. Guatemala (Unpublished report)
Pérez ES (2006) Propuesta de areas importantes para la conservación de aves en Huehuetenango. Fundación para el Ecodesarrollo y la Conservación FUNDAECO, Guatemala (Unpublished report)
Pérez ES, Castillo Villeda ML (2000) A rapid assessment of avifaunal diversity in aquatic habitats of Laguna del Tigre National Park, Petén, Guatemala. In: Bestelmeyer BT, Alonso LA (eds) A biological assessment of Laguna del Tigre National Park, Petén, Guatemala. RAP Bull of Biological Assessment 16, 56–60 and 208–214

Pérez Consuegra SG, Kihn PH, Morales JE et al (2001) Caracterización ecológica de los Biotopos Chocón Machacas, Izabal y Cerro Cahuí, Petén. Informe final. Centro de Estudios Conservacionistas. Programa Universitario de Investigación en Recursos Naturales y Ambiente, Dirección General de Investigación. Universidad de San Carlos, Guatemala (Unpublished report)

Pérez León R, Vega I, Herrera N (2015) Los búhos de El Salvador. In: Enríquez PL (ed) Los búhos neotropicales: diversidad y conservación. ECOSUR, San Cristóbal de las Casas, pp 357–377

Peterson AT, Sánchez-Cordero V, Soberón J et al (2001) Effects of global climate change on geographic distributions of Mexican Cracidae. Ecol Model 144:21–30

Phillips AR (1942) Notes on the migrations of the elf and flammulated screech owls. Wilson Bull 54:132–137

Planqué B, Vellinga W-P (2005) xeno-canto. http://www.xeno-canto.org. Accessed 15 Oct 2016

Proudfoot GA, Johnson RR (2000) Ferruginous pygmy-owl (*Glaucidium brasilianum*). In: Poole A, Gill F (eds) The birds of North America no. 498. The Birds of North America, Inc., Philadelphia

Puebla-Olivares F, Rodriguez-Ayala E, Hsu YC et al (2002) Status and conservation of the avifauna of the Yaxchilán Natural Monument, Chiapas, México. Ornitol Neotrop 13:365–379

Radachowsky J, Aguirre O, Córdova M et al (2004) Birds. In: García R, Radachowsky J (eds) Ecological evaluation of the Mirador-Río Azul National Park, Petén, Guatemala. Wildlife Conservation Society, Flores, pp 61–78 (Unpublished report)

Ramírez-Julián R, González-García F, Reyes-Macedo G (2011) Registro del búho leonado *Strix fulvescens* en el estado de Oaxaca, México. Rev Mex Biodivers 82:727–730

Ramos, VH, Burgués I, Fleck LC et al (2007) Análisis económico y ambiental de carreteras propuestas dentro de la Reserva de la Biosfera Maya. Conservation Strategy Fund – Conservación Estratégica. Serie Técnica No. 8

Regalado O, Montenegro R, Urizandi R et al (2012) Mapa de cobertura forestal de Guatemala 2010 y dinámica de la cobertura forestal 2006-2010. Instituto Nacional de Bosques (INAB), Consejo Nacional de Áreas Protegidas (CONAP), Universidad del Valle de Guatemala, Universidad Rafael Landívar, Guatemala Ciudad

Renner SC, Waltert M, Mühlenberg M (2006) Comparison of bird communities in primary vs. young secondary tropical montane cloud forest in Guatemala. Biodivers Conserv 15:1545–1575

Ribeiro Gallo WL (2007) Perspectivas para el biodiesel en Centroamérica: Costa Rica, El Salvador, Guatemala y Honduras. Comisión Económica para America Latina y el Caribe, Naciones Unidas

Ridgway R (1914) The birds of North and Middle America. Part 6. Bull US Natl Mus 50:1–882

Robbins CS (1996) Birds observed at study sites on Cerro San Gil, Guatemala, February 4 – March 23, 1996. National Biological Service, Laurel (Unpublished report)

Robbins CS, Dowell BA (1992) Report on bird survey at Sierra de las Minas, March 1992. Patuxent Wildlife Research Center. U. S. Fish and Wildlife Service, Laurel (Unpublished report)

Robbins CS, Dowell BA (1993) Un informe sobre un reconocimiento de habitats para las aves en Cerro San Gil, Guatemala, Feb. 1 – Mar. 10, 1993. U. S. Fish and Wildlife Service, Patuxent Wildlife Research Center, Laurel (Unpublished report)

Robbins CS, Dowell BA (1995) Informe sobre el reconocimiento del habitat por las aves del Cerro San Gil, Guatemala, 9 Febrero – 17 Marzo de 1995. National Biological Service, Patuxent Environmental Science Center, Laurel (Unpublished report)

Rotenberg JA (2007) Ecological role of a tree (*Gmelina arborea*) plantation in Guatemala: an assessment of an alternative land use for tropical avian conservation. Auk 124:316–330

Salvin O (1866) A further contribution to the ornithology of Guatemala. Ibis 8:188–206

Salvin O (1874) Editorial notes of Mr. Salvin. Ibis 16:99–100

Salvin O, Godman FD (1897–1904) Biologia Centrali-Americana. Aves, vol 3. The authors, London

Salvin O, Sclater PL (1860) Contributions to the ornithology of Guatemala, III. Ibis 2:396–402

Sandoval L (1942) Semántica guatemalense o diccionario de guatemaltequismos. Tomo II, L-Z. Tipografía Nacional, Guatemala

Sandoval KJ (2000) Biodiversidad de Huehuetenango. Centro de Datos para la Conservación, Centro de Estudios Conservacionistas, Universidad de San Carlos, Guatemala
Sclater PL, Salvin O (1859) On the ornithology of Central America. Part III. Ibis 1:213–234
Sclater PL, Salvin O (1868) Descriptions of new species of birds of the families Dendrocolaptidae, Strigidae, und Columbidae. Proc Zool Soc Lond:53–60
Seavy NE, Whitacre DF, Córdova M (1995) Yaxhá/Nakum area of the Maya Biosphere Reserve, Guatemala: baseline ecological assessment, establishment of a framework for ecological monitoring, and training of local personnel. Report to TNC. The Peregrine Fund, Inc., Boise, ID (Unpublished report)
Seglund AE, Conner SA (1997) Bird survey of the Bocas del Polochic Wildlife Refuge and surrounding areas, Izabal, Guatemala. Cuerpo de Paz, Defensores de la Naturaleza, Guatemala (Unpublished report)
Sekercioglu CH (2002) Impacts of birdwatching on human and avian communities. Environ Conserv 29:282–289
Sergio F, Marchesi L, Pedrini P et al (2004) Electrocution alters the distribution and density of a top predator, the eagle owl (*Bubo bubo*). J Appl Ecol 41:836–845
Sharer RJ (1994) The ancient Maya, 5th edn. Stanford University Press, Stanford
Sharpe RB (1875a) Contributions to a history of the accipitres. The genus *Glaucidium*. Ibis 17:35–59
Sharpe RB (1875b) Contributions to a history of the accipitres. Notes on birds of prey in the museum at the Jardin des Plantes and in the collection of Mons. A. Bouvier. Ibis 17:253–261
Sharpe RB (1875c) Catalogue of birds in the British Museum, vol. 2 striges. Trustees of the British Museum (Natural History), London
Sigüenza R (1997) El impacto de las poblaciones de aves en plantaciones de café de la zona de influencia de la Reserva de la Biosfera Sierra de las Minas. Defensores de la Naturaleza, Guatemala (Unpublished report)
Smallwood KS, Thelander CG, Morrison ML et al (2007) Burrowing owl mortality in the Altamont Pass wind resource area. J Wildl Manag 71:1513–1524
Smithe FB, Paynter RA Jr (1963) Birds of Tikal, Guatemala. Bull Mus Comp Zool 128:245–324
Solomon S, Qin D, Manning M et al (eds) (2007) Climate change 2007: the physical science basis. Contribution of Working Group I to the Fourth Assessment Report of the Intergovernmental Panel on Climate Change. Cambridge University Press, Cambridge/New York
Stattersfield AJ, Crosby MJ, Long AJ et al (1998) Endemic bird areas of the world: priorities for biodiversity conservation, BirdLife conservation series no. 7. BirdLife International, Cambridge
Sullivan BL, Wood CL, Iliff MJ et al (2009) eBird: a citizen-based bird observation network in the biological sciences. Biol Conserv 142:2282–2292
Tashian RE (1953) The birds of southeastern Guatemala. Condor 55:198–210
Tejeda-Cruz C, Sutherland W (2005) Cloud forest bird responses to unusual severe storm damage. Biotropica 37:88–95
Tenez D (2005a) Aves del volcán de Acatenango y Fuego. Consejo Nacional de Áreas Protegidas, Guatemala (Unpublished report)
Tenez D (2005b) Avifauna del volcán Lacandón, Quetzaltenango. Consejo Nacional de Áreas Protegidas, Guatemala (Unpublished report)
Tenez D (2007) Avifauna del Parque Nacional Sierra del Lacandón, Reserva de la Biosfera Maya, Guatemala. Tesis de Licenciatura, Universidad de San Carlos, Guatemala
Terborgh J, Winter B (1983) A method for siting parks and reserves with special reference to Colombia and Ecuador. Biol Conserv 27:45–58
Thomas CD, Cameron A, Green RE et al (2004) Extinction risk from climate change. Nature 427:145–148
Trenberth K (2005) Uncertainty in hurricanes and global warming. Science 308:1753–1754
Tozzer AM, Allen GM (1910) Animal figures in the Maya codices. Pap Peabody Mus Amer Archaeol 4:283–372

Valdez OI, Marroquín DE, Orellana RE et al (1999) Fauna en peligro de extinción de Guatemala: inventarios rápidos para la conservación. Informe final (noviembre). Centro de Estudios Conservacionistas, Universidad de San Carlos, Guatemala (Unpublished report)

van Tyne J (1935) The birds of northern Petén, Guatemala, Misc Publ Mus Zool Univ Mich 27. University of Michigan Press, Ann Arbor

Vannini JP (1989) Preliminary checklist to the birds of Finca el Faro, Quetzaltenango, Guatemala. Fundación Interamericana de Investigación Tropical, Guatemala. Publicación Ocasional No. 2

Vannini JP, Morales Cajas C (1989) Captive reproduction and successful reintroduction of spectacled owls (*Pulsatrix perspicillata*) in Guatemala. In: Burnham WA, Jenny JP, Turley CW (eds) Maya project: use of raptors as environmental indices for design and management of protected areas and for building local capacity for conservation in Latin America. Progress report II. Peregrine Fund, Boise, pp 105–109 (Unpublished report)

VertNet (2016) VertNet Version 2016-09-29. http://vertnet.org. Accessed 22 Nov 2016

Vuilleumier F (2004) A critique of Neotropical ornithology: is research on Neotropical birds scientific? Ornitol Neotrop 15(Suppl):41–60

Wege DC, Long AJ (1995) Key areas for threatened birds in the Neotropics, BirdLife conservation series no. 5. BirdLife International, Cambridge

Weick F (2006) Owls (Strigiformes): annotated and illustrated checklist. Springer, Berlin/Heidelberg

Weidensaul S (2015) Owls of North America and the Caribbean. Houghton Mifflin Harcourt, Boston/New York

Wetmore A (1941) Notes on the birds of the Guatemala highlands. Proc US Natl Mus 89:523–581

Whitacre DF, Baker AJ, Jones LE et al (1991) Results of census efforts in three units of the Maya Biosphere Reserve/Calakmul Biosphere Reserve complex. In: Whitacre DF, Burnham WA, Jenny JP (eds) Maya project: use of raptors and other fauna as environmental indicators for design and management of protected areas and for building local capacity for conservation in Latin America. Progress report IV. Peregrine Fund, Boise, pp 43–58 (Unpublished report)

Whitacre DF, Jones LE, Sutter J (1992) Censusing raptors and other birds in tropical forest: further refinement of methodology. In: Whitacre DF, Thorstrom RK (eds) Maya project: use of raptors and other fauna as environmental indicators for design, management, and monitoring of protected areas and for building local capacity for conservation in Latin America. Progress report V. Peregrine Fund, Boise, pp 39–51 (Unpublished report)

Wigley TML, Jones PD, Raper SCB (1997) The observed global warming record: what does it tell us? PNAS 94:8314–8320

Wink M, Sauer-Gürth H, Fuchs M (2004) Phylogenetic relationships in owls based on nucleotide sequences of mitochondrial and nuclear marker genes. In: Chancellor RD, Meyburg B-U (eds) Raptors worldwide. World Working Group on Birds of Prey and Owls, Berlin, pp 517–526

Wink M, Heidrich P, Sauer-Gürth H et al (2008) Molecular phylogeny and systematics of owls (Strigiformes). In: König C, Weick F, Becking J-H (eds) Owls of the world, 2nd edn. Christopher Helm, London, pp 42–63

Winker K (1998) Recent geographic trends in Neotropical avian research. Condor 100:764–768

Chapter 14
The Owls of Hispaniola and Puerto Rico

Russell Thorstrom and Julio C. Gallardo

Abstract Hispaniola is the second largest island in the Caribbean that includes the countries of Dominican Republic and Haiti and has five resident owl species inhabiting open landscapes, rainforests, human-modified areas, and cities. The five species are the barn owl (*Tyto alba*), the endemic ashy-faced owl (*T. glaucops*), burrowing owl (*Athene cunicularia*), stygian owl (*Asio stygius*), and short-eared owl (*A. flammeus*). The stygian owl is the most threatened and least known of the five species. Puerto Rico represents the smallest and easternmost of the Greater Antilles, and it has a total of three species of owls inhabiting forest, open landscapes, and human-modified areas. The status of the barn owl on the island is not well known as the first breeding pair was documented in 2015, the short-eared owl is mostly restricted to lowlands and low montane open areas, and the Puerto Rican screech owl (*Megascops nudipes*) is the most common and widespread owl on the island. The Virgin Island race of the Puerto Rican screech owl (*M. nudipes newtoni*) that formerly inhabited a forested area of the Virgin Islands and Puerto Rico's satellite island of Vieques is apparently extinct. These owls are suffering from changing landscapes and loss of natural habitat, especially in Haiti where natural resources and forests are nearly gone. Natural habitat remaining on Hispaniola is mainly confined to protected areas and the majority lack enforcement and protection. On the other hand, in Puerto Rico, the abandonment of forested areas such as shade coffee plantations, mainly in the highlands, has resulted in forest regeneration benefitting the Puerto Rican screech owl, and the abandonment of sugar cane plantations in the lowlands has favored the local subspecies of short-eared owl and most likely the colonization by barn owls.

Keywords Dominican Republic • Puerto Rico • West Indies • Caribbean

---

R. Thorstrom (✉)
The Peregrine Fund, 5668 West Flying Hawk Lane, Boise, ID 83709, USA
e-mail: rthorstrom@peregrinefund.org

J.C. Gallardo
Department of Wildlife, Fisheries and Aquaculture, Mississippi State University, Box 9690, Mississippi State, MS 39762, USA

P.L. Enríquez (ed.), *Neotropical Owls*, DOI 10.1007/978-3-319-57108-9_14

Ashy-faced Owl (*Tyto glaucops*)

14.1 Introduction

Hispaniola includes the countries of Dominican Republic in the eastern two-thirds of the island and the Republic of Haiti in the western third (Fig. 14.1). It is the second largest island in the Caribbean region with a landmass of 77,842 km^2 (Latta et al. 2006). The major physiographic areas are as follows: the Central Mountains (Cordillera Central) runs north to south and includes the highest elevation on the island at Pico Duarte (3098 m), and it is the highest peak of the West Indies Islands; Sierra de Nieba runs west to east and rises to 2279 m; Northern Mountains (Cordillera Septentrional) runs north to south and reaches 1250 m; and Sierra de Bahoruco runs west to east and reaches 2367 m (Latta et al. 2006). In the Los Haitises area in northeastern Dominican Republic is an extensive moist karst region. The lowest elevation is 44 m below sea level at the saline Lake Enriquillo in western Dominican Republic (Latta et al. 2006). There are several small islands, notably, Saona on the southeastern coast, Beata on the southern coast, and Catalina on the southeastern coast. There are many rivers and the longest is the Yaque del Norte which extends 296 km originating from the Central Mountains.

The climate is humid and tropical and the island has several distinct ecoregions: the Hispaniolan moist forest composed of evergreen and montane broadleaf forests from the lowlands up to 2300 m with annual rainfall of 2000 mm or more, the

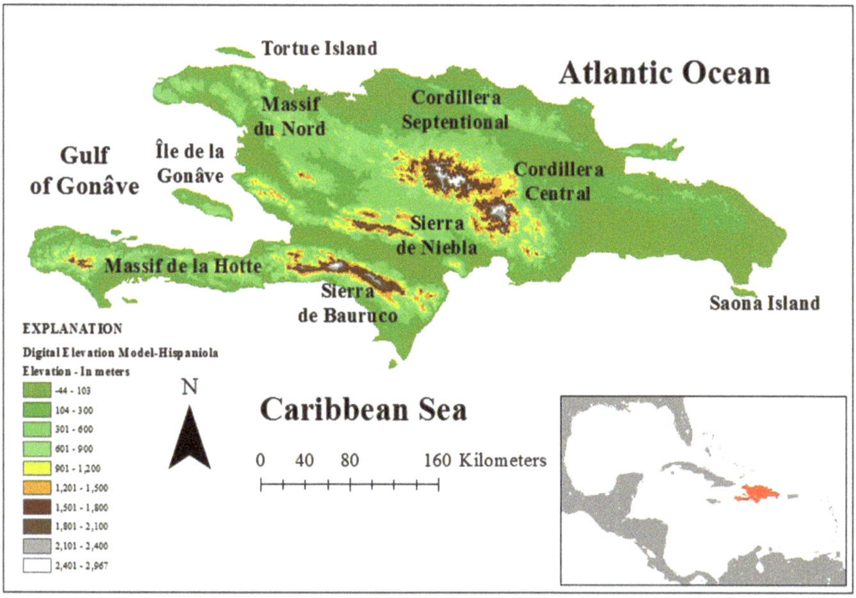

Fig. 14.1 Hispaniola map

Hispaniolan dry forest from 400 to 900 m elevation on foothills of mountains with annual precipitation 1000–1800 mm, Hispaniolan pine forest above 850 m elevation with annual precipitation from 1000 to 2000 mm, and the grasslands, savannas, and marshlands mostly in the lowlands with some seasonally flooded (Keith et al. 2003, Wikipedia contributors 2017c). Most of the native vegetation habitat types within these ecoregions have been human modified or degraded into agricultural lands, covering about 50% of the island; shrublands, formed from the loss of forested habitat; and dry scrubs, secondary growth of existing forested habitat (Keith et al. 2003).

There are 306 bird species on the island with 31 endemic species (Latta et al. 2006). Owls comprise five species including one endemic or 1.6% of the avian diversity in Hispaniola. As in nearly all tropical regions of the world, there are three major issues affecting biodiversity and avian conservation in Hispaniola: the loss and degradation of native habitats, the lack of enforcement and of environmental laws, and the general absence of an established environmental public education program, especially in schools and areas bordering protected areas (Latta et al. 2006). There is also a lack of a national environmental ethic, especially for raptors including owls. Raptors are persecuted for their predatory and ill-perceived nature. Recently, Hispaniola's importance to global biodiversity has been recognized as one of the highest priorities in the world for bird conservation based on the number of endemic, resident, and migratory species, and the island itself is considered an Endemic Bird Area (Stattersfield et al. 1998; BirdLife International 2003). The conservation issues in Dominican Republic and Haiti are nearly identical, but in Haiti the problems are much more severe due to the economical, social, and

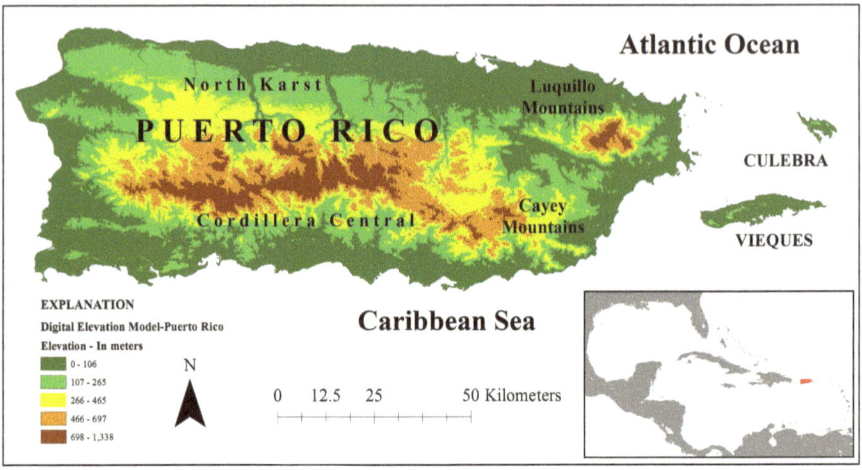

Fig. 14.2 Puerto Rico map

political situations with biodiversity conservation still in its infancy stage (Raffaele et al. 1998; Keith et al. 2003). The human population density ranges from nearly 200 inhabitants/km^2 in Dominican Republic to 325 in Haiti (Wikipedia contributors 2017a, Wikipedia contributors 2017b).

Haiti is almost entirely deforested. It has three national protected areas making up 87.5 km^2 (0.3% of the land area of the country) of forested habitat mainly above 1000 m elevation, but the future of these protected areas is uncertain (Raffaele et al. 1998). In Dominican Republic conservation actions and activities are more advanced with the government and nongovernmental organizations being more proactive in establishing 70 protected areas encompassing 13,000 km^2 (Latta et al. 2006).

The main island of Puerto Rico is the smallest and easternmost of the Greater Antilles; it comprises an area of 8740 km^2, with a length of 178 km and a width of 58 km (Daly et al. 2003). The territory of Puerto Rico includes several cays and satellite islands, where Culebra and Vieques are located in the east, and the islands of Desecheo, Mona, and Monito to the west. The landscape is predominately dominated by mountains and hills, comprising 75% of Puerto Rico's territory (Gould et al. 2007). The dominant physiographic formations of Puerto Rico include the northern moist karst region extending from Guajataca in the west to Bayamon in the east, the Central Mountain Range (Cordillera Central) that runs east-west from Mayaguez in the west to Orocovis in the east, the Cayey Mountains in the southeast region, and the Luquillo Mountains in the northeast (Fig. 14.2).

Puerto Rico typically shows two peaks of rainfall, one occurs between April and May and another from October to December. Mean annual rainfall ranges from below 900 mm in the subtropical dry forest zone to over 4000 mm in the wet rainforest (e.g., Luquillo Mountains). The island exhibits seasonality in rainfall and is most pronounced in the subtropical dry forest in the south (Daly et al. 2003; Gould et al. 2007). Six bioclimatic life zones have been described for Puerto Rico: the subtropical dry forest with a mean annual rainfall of 860 mm, which is mostly restricted to the southwestern regions; the subtropical lowland moist forest (0–400 m) covers more than half of the main island and has an average annual precipitation of 1000–2200 mm; the subtropical wet forest occurs at middle elevations (400–700 m) with an average annual rainfall of 2000–4000 mm; the subtropical lower montane wet forest occurs between 700 and 1000 m and annual rainfall averages 2250 mm; the subtropical lower rainforest is restricted to a small portion of the Luquillo Mountains and is similar to the subtropical lower montane wet forest with a mean annual rainfall up to 4000 mm; and the subtropical rainforest of Luquillo Mountains is the wettest region in the island averaging 4000 mm of annual rainfall (Ewel and Whitmore 1973; Murphy and Lugo 1986).

After a gradual deforestation after the Spanish colonization, Puerto Rico lost 94% of its forest by the 1940s, where the remaining forest fragments were localized at higher elevation, mainly associated with shade coffee plantations (Koenig 1953; Birdsey and Weaver 1987; Thomlinson et al. 1996; Franco et al. 1997). During the second half of the twentieth century, Puerto Rico's economy shifted from agriculture to manufacturing, leading to the abandonment of agricultural lands and resulting in the expansion of secondary forests that currently cover half of the island (Dietz 1986; López et al. 2001; Marcano-Vega et al. 2002). The island hosts 276 bird species of which 142 are resident breeders, including 16 endemic (Raffaele 1989). Perhaps in the deforestation history of Puerto Rico, only one species of birds has become extinct on recent historic times, an endemic parakeet (*Aratinga maugei*), and one extirpated, the white-necked crow (*Corvus leucognaphalus*) (Snyder et al. 1987; Raffaele 1989; William and Steadman 2001).

14.2 Taxonomic Diversity

Barn owl (*Tyto alba*) breeding resident and nonbreeding visitor.

Local names lechuza común, lechuza blanca (Dominican Republic); Frize, Fresaie (Haiti): lechuza (Puerto Rico).

Taxonomy represented by subspecies, *T. a. pratincola* (Bonaparte 1838).

Status and conservation this owl is not threatened and is least concern (CITES II – IUCN 2009). A common owl found in open areas throughout Hispaniola from sea level to the interior mountains. As with all owls in Hispaniola, most likely it is persecuted.

Distribution neotropical. This is breeding resident with some nonbreeding migrants arriving from North America. It may be a relatively recent colonist of both islands, possibly arriving from Bahamas or North America (Latta et al. 2006; König et al. 2008; Fig. 14.3). This owl was not recorded by the early ornithologists in Hispaniola, and it has been speculated that it became established in the mid-1900s with the spread of agriculture and deforestation increased the amount of suitable habitat for

Fig. 14.3 Barn owl (*Tyto alba*) distribution map in Hispaniola

Fig. 14.4 Barn owls nesting locality in Puerto Rico in 2015

this species, and the introduction of exotic rodents has increased its potential food supply (Latta et al. 2006). There are four old records of barn owls in Puerto Rico in the second part of the twentieth century (Raffaele 1989), but the first nesting record was made in August of 2015 in an abandoned sugar factory by a local birder and biologists in the town of Aguada, in the northwest of the island (J. Salgado and S. Colón 2016, pers. comm., Figs. 14.4 and 14.5). Specimen records indicate that a few owls from North America visit during the nonbreeding period (Keith et al. 2003).

Habitat occurs throughout Haiti and Dominican Republic in open areas that include rice fields, dry scrub, open woodlands, and villages. It has been reported from Navassa Island west of Haiti and Saona Island, southeastern Dominican Republic. The only nest recorded in Puerto Rico was located in a mosaic of active and abandoned pastures near the town of Aguada in the northwest Puerto Rico. It is a nocturnal species often found perching on fence posts or utility poles, especially in rice fields. Seen in large numbers at oil-palm plantations, this owl feeds on rats and mice that are attracted by the palm fruit.

Vocalization harsh long hissing screech or drawn-out scream clicking sounds.

Biology and ecology does not build a nest but lays white eggs, 2–8 per clutch, in a tree cavity, cave ledge, cliff pothole, or similar site. This owl may breed all year

Fig. 14.5 First documented nesting in August of Barn owls at Aguada, Puerto Rico (Photograph Julio Salgado)

round, but mainly from August to April (Raffaele et al. 1998). Of 4495 prey identified from regurgitated pellets, small mammals (77% frequency and 91% biomass), mainly introduced rodents, predominated in the diet (Wiley 2010).

Ashy-faced owl (*Tyto glaucops*) endemic breeding resident.

Local names lechuza cara ceniza (gris) (Dominican Republic); Effraie d'Haiti, Frize Figi Gri (Haiti).

Taxonomy *Strix glaucops* Kaup 1853, Hispaniola. Was considered a conspecific with *T. alba*, but after the probable colonization of Hispaniola by *T. a. pratincola*, it was found not to interbreed with nominate *glaucops* in the mid-1970s, and separation of the two taxa was first accepted in 1983 (AOU 1983; Bruce 1999; König et al. 2008).

Status and conservation this is an endemic owl and is not threatened and is least concern (CITES II – IUCN 2009). This is a restricted-range species fairly widespread and locally common on Hispaniola, and due to native habitat loss and direct persecution, it may have declined (Weidensaul 2015). Research and studies on this endemic species are urgently needed to better understand its population status and for conservation (Bruce 1999; König et al. 2008).

Distribution Hispaniola and Tortue Island, 8 km off the northwest coast of Haiti. In Dominican Republic common around Santo Domingo, Los Haitises region, the Samaná Province, the southeastern region, and throughout the Barahona Peninsula, especially in coastal forests south of the town of Barahona near limestone cliffs (Latta et al. 2006). In Haiti, its distribution is poorly documented and basically unknown (Bruce 1999; Latta et al. 2006, Fig. 14.6).

Habitat open woodlands, scrublands and caves, dry and humid open forests, agricultural areas and pastures, abandoned buildings, villages, and towns, from lowlands to over 2000 m (Raffaele et al. 1998; Bruce 1999).

Vocalization a hissing cry, and a series of high-pitched, ratchety clicks.

Fig. 14.6 Ashy-faced owl (*Tyto glaucops*) distribution map in Hispaniola

Biology and ecology nocturnal. This owl is found in habitat more heavily forested than those frequented by barn owls (Latta et al. 2006). It feeds on small mammals, especially rodents and bats, and occasionally on anole lizards (Iguanidae), tree frogs (Hylidae), and birds (from hummingbirds to domestic fowl). Of 3287 prey identified from regurgitated pellets, small mammals (52% frequency and 74% biomass), mainly introduced rodents, predominated in the diet which included 125 vertebrate species, and they had a more diverse prey base than barn owls (Wiley 2010). Nesting season is January to June, and nests are in natural cavities in trees and cliffs, and on ledges in caves, sinkholes, and cliffs and in artificial sites (Latta et al. 2006). Some nests are formed on an accumulation of prey remains on a ledge, in a sinkhole, or cave (Raffaele et al. 1998). Roost in palm trees and pines at higher elevation. This owl is less likely to be found around and in towns and cities than barn owls. It lays two to seven white unmarked white eggs (Bruce 1999; Latta et al. 2006).

Burrowing owl (*Athene cunicularia*) breeding resident.

Local names cucú (Dominican Republic), Koukou, Chevêche des terriers (Haiti).

Taxonomy represented by subspecies *A. c. troglodytes* (Wetmore and Swales 1931).

Status and conservation this owl is not threatened and is least concern (CITES II – IUCN 2009). This owl is a breeding resident, and it is locally common in the western half of Hispaniola, despite its persecution in Dominican Republic and Haiti. *A. c. troglodytes* is endemic to Hispaniola (Bruce 1999).

Distribution found throughout much of Haiti, and in western half of Dominican Republic from the 70° W longitude, but with a few recent observations in eastern Dominican Republic near Parque del Este, Punta Cana, and on Samana Peninsula (pers. comm. J. Brocca 2009). These recent observations suggest that burrowing owls may be found throughout the island in suitable habitat and not limited to areas west of the 70° W longitude as once thought and are occupying habitat that has become suitable nowadays. It occurs from below sea level to as high as 2200 m at an old sawmill site in Valle Nuevo, La Vega Province, Dominican Republic. It is most numerous in and along the sides of the Neiba Valley, Dominican Republic, and

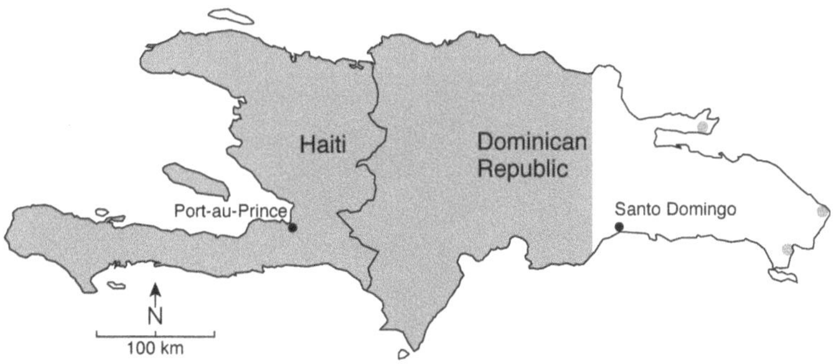

Fig. 14.7 Burrowing owl (*Athene cunicularia*) distribution map in Hispaniola

extending into Haiti, the Cul-de-Sac Plain, and in parts of Barahona and Pedernales provinces, DR (Fig. 14.7). Also found on the Île de la Gonâve (Haiti) and Isla Beata (Dominican Republic) (Bruce 1999; Latta et al. 2006).

Habitat resident of semi-open dry habitats, scrubby areas, sandy pine savannas, pastures, and limestone ravines (Latta et al. 2006).

Vocalization soft, high-pitched, 2-note *coo-cooo*. Alarm or distress calls consist of clucking chatter or buzzing scream.

Biology and ecology active day and night. It is an opportunistic predator capturing beetles, locusts, and large spiders and occasionally frogs, and lizards, small birds, and mammals. Insects are taken during the day and vertebrates at night. Birds mainly taken are todies (*Todus* spp.), bananaquits, hummingbirds, and ground-tanagers. In southwestern Dominican Republic at 20 nests diet data was collected in 1976, 1982 and 1996, and prey items recorded were made up of invertebrates (53%) and vertebrates (47%), but vertebrates (69%), manily birds and reptiles, comprised twice the total biomass as invertebrates (31%) (Wiley 1998). Nests single or in dispersed small colonies. Nest is a burrow from 1 to 2.5 m deep in sandy or loose soil with a mound at the entrance for a vantage point (Raffaele et al. 1998). Also utilizes banks of road cuts for nests. It lays four to six glossy white eggs. The nesting season takes place from March to July in northern Sierra de Bahoruco and on the south side of Lake Enriquillo (Wiley 1998; Bruce 1999; Latta et al. 2006).

Stygian owl (*Asio stygius*) critically endangered breeding resident.

Local names lechuza orejita, ciguapa, cu-chi, hu-hu (Dominican Republic); Mèt Bwa, Chouette (Haiti).

Taxonomy represented by subspecies *A. s. noctipetens* (Riley 1916).

Status and conservation This owl is considered least concern (CITES II – IUCN 2009), but the subspecies on Hispaniola *A. s. noctipetens* appears to be threatened. This owl is extremely rare and is a poorly known breeding resident with a sporadic distribution from the lowlands to the mountains. This species is mostly rare due to

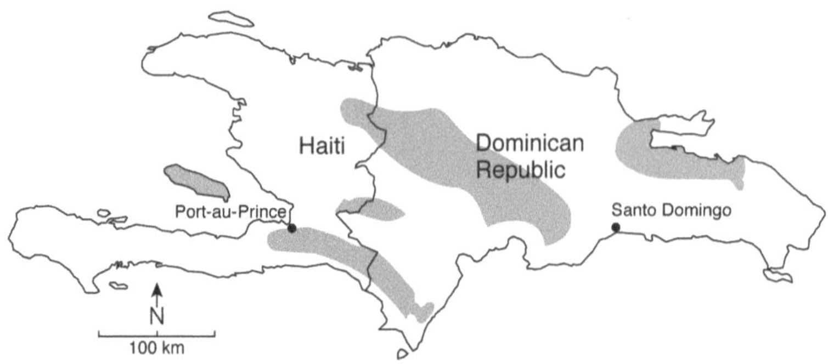

Fig. 14.8 Stygian owl (*Asio stygius*) distribution map in Hispaniola

deforestation and persecution. Local people persecute it because they believe it has supernatural powers and can transform itself into a witch and its call is considered a bad omen (Stockton Dod 1983; Raffaele et al. 1998). It has also been suggested that introduced predators have resulted in the disappearance of native small mammals on which this species depends (Raffaele et al. 1998). This subspecies is endemic to Hispaniola, including Île de la Gonâve, Haiti (Bruce 1999; Latta et al. 2006).

Distribution since the mid-1980s, this owl has been found infrequently in the pine forests of the Cordillera Central in Armando Bermúdez National Park, on the Samaná Peninsula, in Los Haitises National Park, in the Sierra de Bahoruco, and with many records from the eastern end of Dominican Republic (Stockton Dod 1983; Raffaele et al. 1998; Keith et al. 2003). In Haiti, it is basically unknown, but was known to occur there from studies that were conducted in 1807, and has been reported on Île de la Gonâve, and the last good sighting was in 1953 in the mountains of the southwest (Keith et al. 2003; Latta et al. 2006, Fig. 14.8).

Habitat in Dominican Republic, it is a rare resident in dense deciduous and pine forests in remote areas, from semiarid to humid and from sea level to mountains, and in recent years all sightings are from remote old forests, sometimes near caves or in wooded ravines, but never near human dwellings or in secondary forests (Stockton Dod 1983; Keith et al. 2003; Latta et al. 2006).

Voice generally silent; sometimes gives single deep *hu* or *whu*. During the breeding season, the male calls with a short, low-pitched *hoos* repeated regularly, and the female responds with a higher-pitched whistles and screams – *niek* or *quick*.

Biology and ecology lays possibly two white eggs. It is thought to breed April to June, but few data are available (Keith et al. 2003). A nest believed to be of this species at Hoyoncito, Dominican Republic, was 4.5–6 m aboveground in a cana palm and largely composed of grasses placed loosely together (Stockton Dod 1983). It may use old nests of other birds. In the neighboring island of Cuba, it is known to nest high in trees, including palms, and in cavities, building a platform nest of sticks, and may occasionally nest on the ground (Bruce 1999; Latta et al. 2006).

Fig. 14.9 Short-eared owl (*Asio flammeus*) near Salinas, Puerto Rico, 2014 (Photograph Julio Salgado)

Short-eared owl (*Asio flammeus*) breeding resident and nonbreeding visitor.

Local names Lechuza de Sabana (Dominican Republic); Chat-huant, Chwèt Savann (Haiti); múcaro real, múcaro sabanero (Puerto Rico).

Taxonomy represented by subspecies *A. f. domingensis* (Müller 1776) in the Hispaniola and the subspecies *A. f. portoricensis* (Ridgway 1882) in Puerto Rico (Holt and Leasure 1993; Weidensaul 2015, Fig. 14.9). The taxonomy of this polytypic species is still unresolved. Some authors lump both Antillean populations into a single subspecies either *A. f. domingensis* or *A. f. portoricensis* (Holt and Leasure 1993; König et al. 2008); other authors granted full species recognition as *A. domingensis*, with three subspecies *A. d. domingensis*, *A. d. cubensis*, and *A. d. portoricensis* (Garrido 2007). West Indian birds differ from North American continental forms in plumage and vocalizations, and this has led to suggest considering the West Indian form a separate species (Garrido 2007).

Status and conservation This owl is an endemic subspecies and is of least concern (CITES II – IUCN 2009). It is locally common on Hispaniola, where the population has dramatically increased since the 1930s, and resident birds are likely joined by North American birds, but this has not yet been confirmed (Latta et al. 2006). In Puerto Rico, it is considered an uncommon resident but is apparently increasing its range due to forest clearing and the abandonment of sugar plantations since the 1960s, especially in the lowlands (Wiley 1986; Raffaele 1989). There is some evidence based on plumage and vocalizations suggesting this endemic subspecies may be a separate species (Latta et al. 2006; Garrido 2007).

Distribution in Hispaniola it is more numerous in the eastern part of the Dominican Republic at Laguna Redondo, Laguna Limón, Sabana de la Mar, San Pedro de

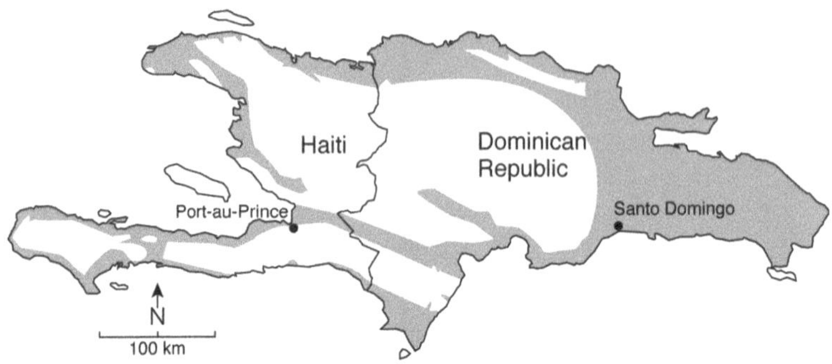

Fig. 14.10 Short-eared owl (*Asio flammeus*) distribution map in Hispaniola

Fig 14.11 Short-eared owl distribution map in Puerto Rico

Macorís, and Santo Domingo, and it has been registered farther west in Moca, Jarabacoa, San Juan de la Maguana, and Cabral (Latta et al. 2006; Fig. 14.10). There are some reports suggesting that it may be more abundant in the non-montane uplands up to 490 m than in the lowlands (Keith et al. 2003). The present status of this species in Haiti is unknown, as the only documented record for Haiti is a specimen collected in 1928 in the Plateau Central near Saint-Michel de L'Atalaye (Keith et al. 2003). In Puerto Rico it is found in open and semi-open habitats in lowlands and lower montane open areas and pastures, apparently more abundant in the southern lowlands from Cabo Rojo in the west to the grasslands of Salinas and Guayama in the east (Raffaele 1989; Fig. 14.11). This species has also been reported on Puerto Rico's satellite islands of Mona, Vieques, and Culebra, and on the island of Saint Thomas of the US Virgin Islands (Raffaele 1989; Fig. 14.11).

Habitat local resident on Hispaniola from sea level to 490 m elevation. It inhabits open country of the lowlands, including pastures, short-grass marshlands, savannas, rice fields, and citrus plantations (Latta et al. 2006). In Puerto Rico, it has been reported from sea level to 800 m of elevation, inhabiting grasslands/pastures, savannas, marshlands, rice fields, mangrove forest, openings in wet montane forest, and forest edges of second-grow subtropical rainforest (pers. comm. J. Salgado and J. Illanas 2016, pers. obs. 2013).

Voice short, emphatic barking doglike call, *bow-wow* or *uh-uh*. It gives a distinct wing-clap during courtship flights and nest defense.

Biology and ecology This owl is most active during the crepuscular period flying low over open areas and low vegetation in search of prey and sometimes observed perching on fence posts, low bushes, or ground, where it takes cover during the day (Bruce 1999; Latta et al. 2006). Its food habits are poorly known in the Antilles. Unpublished observations made by the authors found that this species fed on rodents (*Mus musculus*, *Rattus* spp.), American kestrel (*Falco sparverius*) and other birds, anole lizards (*Anolis* spp.), potentially green iguana (*Iguana iguana*), and common Puerto Rican ameiva (*Ameiva exsul*). Nests are found on the ground under bushes, in a thick clump of grass, and under *Agave* spp. Owls lay three to four white eggs (Keith et al. 2003). Its nesting period is from April to June but may breed throughout the year too, as one nesting attempt was reported in early December (Raffaele et al. 1998; Keith et al. 2003). A family group composed of two adults and a juvenile were sighted on 16 May 2014, and presumably the same territorial pair was sighted with two juveniles on 20 July 2015 in Cabo Rojo National Wildlife Refuge (M. Morel, pers. comm.).

Puerto Rican screech owl (*Megascops nudipes*) breeding resident (Puerto Rico).

Local names Mucarito, múcaro común.

Taxonomy represented by *M. n. nudipes* (Dandin 1800) in the main island of Puerto Rico and *M. n. newtoni* (Lawrence 1860) in Vieques Island (Puerto Rico), US Virgin Islands, and British Virgin Islands (König et al. 2008; Weidensaul 2015). The original description of *M. n. newtoni* was based on differences in coloration some authors suggest a revaluation of a possible synonymy of both taxa (Lawrence 1878; Moreno 1998).

Status and conservation Currently, *M. n. nudipes* is considered stable and of least concern. The current status of the Puerto Rican screech owl outside of the main island of Puerto Rico is still unknown (IUCN 2009). Between the 1950s and late 1970s, Newton's screech owl (*M. n. newtoni*) was recorded as rare from Puerto Rico's satellite islands of Vieques (see star symbol in Fig. 14.12), Culebra; the US Virgin Islands of St. Thomas, St. Johns; and the British Virgin Islands of Virgin Gorda, Tortola, and Guana (Robertson 1962; Wiley 1986; Moreno 1998; Nellis 1979; Raffaele 1998; König et al. 2008). In 1993, owl pellets were found on Guana Island, and this is the only sign that suggests that this subspecies still survives, though there have been no confirmed sightings on the island (Moreno 1998). Lack of reliable records suggests that if this subspecies survives, it is extremely rare or perhaps it is in fact extinct (Wiley 1986; Moreno 1998; König et al. 2008; Hume and Walters 2012).

Distribution Restricted to Puerto Rico and its satellite islands of Vieques and Culebra, the US Virgin Islands, and British Virgin Island (Wiley 1986; Moreno 1998; König et al. 2008; Weidensaul 2015). Currently, this species is found only on the main island of Puerto Rico, where it is common and widespread throughout all

Fig. 14.12 Puerto Rican screech owl (*Megascops nudipes*) map and distribution with location of last record of possible subspecies *Megascops nudipes newtoni* on Vieques Island

Fig. 14.13 Puerto Rican screech owl (*Megascops nudipes*) at Sierra Bermeja, Cabo Rojo 11 February 2013 and at Cueva del Convento, Guayanilla 9 May 2013 (Photograph Mike Morel)

wet and moist forest, forest edges, and urban parks, probably less abundant in the grasslands of Cabo Rojo in the southwest (Rivera-Milán 1995, Fig. 14.13). *Megascops nudipes newtoni* was recorded in forested areas and forest edges through the islands of Culebra, Guana, St. Johns, St. Thomas, Tortola, Vieques, and Virgin Gorda (Robertson 1962; Wiley 1986; Moreno 1998; Raffaele 1998; König et al. 2008). In the early 1900s, *M. n. newtoni* was reported by a local in the east of Vieques Island; this is questioned or attributed to the nominal subspecies (Wetmore 1916; Bond 1945; Gimmell 2015).

Habitat The Puerto Rican screech owl is considered the third most abundant and widespread raptor in the moist and wet zones of Puerto Rico after the red-tailed hawk (*Buteo jamaicensis jamaicensis*) and the American kestrel (*Falco sparverius caribaearum*). It is fairly common in dry woodlands and scarce or not recorded in dry grasslands (Rivera-Milán 1995; Arendt et al. 2015). This owl has increased throught the island where there are native and exotic trees and in edge habitats in suburban and agricultural areas with trees of suitable size for forming nesting habitat (cavities) (Wiley 1986; Rivera-Milán 1995). This species is tolerant to fragmentation; during a playback survey, the frequency of owl response was similar in fragmented habitat (36% of forest cover) and in large unbroken tract habitat (95% of forest cover) in Luquillo Mountains (Gannon et al. 1993). It is also fairly common in agricultural zones with scattered trees, shade coffee plantations, and urban and suburban areas with some wooded areas (Pardieck et al. 1996; Raffaele et al. 1998; Weidensaul 2015). As a secondary cavity nester, its abundance may be limited by the availability of suitable nesting cavities.

Voice low-pitched tremulous trill and chattered laugh, often common in duets. Calls during the evening and early in the morning.

Biology and ecology Strictly nocturnal, hides during the day in dense foliage, cavities, and caves. Its food habits are poorly known but presumably feeds on invertebrates and small vertebrates such as anole lizards (*Anolis* spp.), coqui frogs (*Eleutherodactylus* spp.), and birds (König et al. 2008; Weidensaul 2015). Breeding season apparently occurs from April to June; clutches can be between one and four eggs, but typically two white eggs (Raffaele 1998; König et al. 2008; Weidensaul 2015). Limited data on movements suggest that this species is highly sedentary with a small home range, with one adult using an area of 4.5 ha and a juvenile an area of 2.3 ha (Gannon et al. 1993). No abundance data but a playback survey estimated a mean owl detection of 1.1 per survey point in forested areas (95% of forest cover) and 0.96 in fragmented areas (36% of forest cover) in Luquillo Mountains (Gannon et al. 1993).

14.3 Conclusions

On Hispaniola there are five resident owl species inhabiting landscapes varying from arid to humid forests and from human-modified habitat to cities. They are poorly known and suffering from a changing environment and loss of natural habitat, especially in Haiti where natural resources and forests are nearly gone. Natural habitat left on Hispaniola is mainly confined to protected areas, which are practically nonexistent in Haiti, and the majority lack enforcement and protection. In Dominican Republic, there are some proactive initiatives by the government and conservation-interested organizations in creating a variety of protected areas to conserve habitat for biodiversity.

The stygian owl is the only species considered possibly threatened of the Hispaniolan owls. It is extremely scarce and poorly known, and its distribution is unknown throughout most of the island, and especially in Haiti where it may be extinct. It depends on mature and older forests and rare habitat types in Dominican Republic and is practically nonexistent in Haiti. The ashy-faced owl is the only endemic owl on Hispaniola but is easily confused with the common barn owl and likely to suffer from human persecution as with the other owl species in Haiti and Dominican Republic. The conservation status of Newton's screech owl is uncertain but presumably extinct, and it may have been a true subspecies. Confirmation of existing species or populations should be a conservation priority. There are no direct conservation activities or strategies for owls on any of the Caribbean Islands, but hopefully with the global interest in biodiversity conservation in the region, there will be an umbrella effect for owls. The first priorities for owl conservation in

Hispaniola and Puerto Rico will be to learn about their ecological needs, distributions, and population status throughout the islands.

References

American Ornithologists' Union (1983) Check-list of North American birds, 6th edn. American Ornithologists' Union, Washington, DC

Arendt WJ, Faaborg J, Canals M, Bauer J (2015) Bird checklist Guánica Biosphere Reserve Puerto Rico. U.S.D.A. Forest Service Research Note SRS-23. Southern Research Station, Ashville, NC

BirdLife International (2003) BirdLife's online World Bird Database: the site for bird conservation. Version 2.0. BirdLife International, Cambridge, UK

Birdsey RA, Weaver PL (1987) Forest area trends in Puerto Rico. Research Note SO-331. USDA Forest Service, Southern Forest Experiment Station, Ashville, NC

Bonaparte CL (1838) A geographical and comparative list of the birds of Europe and North America, p.7; based on Audubon, 1835, Ornithological Biography, 2, p. 403, pl. 171

Bond J (1945) Check-list of the birds of the West Indies. The Academy of Natural Sciences of Philadelphia, Philadelphia

Bruce MD (1999) Family Tytonidae (Barn-owls). In: del Hoyo J, Elliot A, Sargatal J (eds) Barn-owls to hummingbirds, Handbook of the birds of the world, vol 5. Lynx Edicions, Barcelona, pp 34–75

Daly C, Helmer EH, Quiñones M (2003) Mapping the climate of Puerto Rico, Vieques and Culebra. Int J Climatol 23:1359–1381

Dandin FM (1800) Traité élémentaire et complet d'Onithologie ou histoire naturelle des Oiseaux. Traité d'Ornitholigie, 2. p. 199

Dietz JL (1986) Economic history of Puerto Rico: institutional change and capitalist development. Princeton University Press, Princeton

Ewel JJ, Whitmore JL (1973) The ecological life zones of Puerto Rico and U.S. Virgin Islands. U.S.D.A. Forest Service Research Paper no. ITF-18, Río Piedras

Franco PA, Weaver PL, Eggen-McIntosh S (1997) Forest resources of Puerto Rico, 1990. Resource Bulletin SRS 11, USDA Forest Service, Southern Research Station, Ashville, NC

Gannon MR, Pardieck K, Willing MR (1993) Movement and home range of the Puerto Rican Screech-Owl (*Otus nudipes*) in Luquillo experimental forest. Caribb J Sci 29:174–178

Garrido OH (2007) Subespecie nueva de *Asio domingensis* para Cuba, con comentarios sobre *Asio flammeus* (Aves: Strigidae). Seledon 6:70–78

Gimmell D (2015) Birds of Vieques Island, Puerto Rico: status, abundance, and conservation. Special issue of The Journal of Caribbean Ornithology, Charlottesville

Gould W, Alarcón C, Fevold B et al (2007) Puerto Rico GAP analysis project – final report. USDA Forest Service, International Institute of Tropical Forestry, Rio Piedras

Holt DW, Leasure SM (1993) Short-eared owl (*Asio flammeus*). In: Poole A, Gill F (eds). The birds of North America, No. 62. Washington, DC, Philadelphia: The Academy of Natural Sciences and the American Ornithologists' Union

Hume JP, Walters M (2012) Extinct birds. T & AD Poyser, London

IUCN (2009) IUCN red list of threatened species. Version 2009.1. www.iucnredlist.org. Accessed 12 May 2009

Koenig N (1953) A comprehensive agricultural program for Puerto Rico. U.S. Department of Agriculture and Commonwealth of Puerto Rico, Washington, DC

König C, Weick F, Becking JH (2008) Owls of the world, 2nd edn. Yale University Press, New Haven/London

Keith AR, Wiley JW, Latta SC et al (2003) The birds of Hispaniola. Haiti and the Dominican Republic. An annotated checklist, BOU Checklist No 21. British Ornithologists' Union, Herts
Lawrence GN (1860) Annals of the Lyceum of Natural History, New Your Academy of Sciences, 7, p. 259
López TM, Aide TM, Thomlinson JR (2001) Urban expansion and the loss of prime agricultural lands in Puerto Rico. Ambio 30:49–54
Marcano-Vega H, Aide TM, Báez D (2002) Forest regeneration in abandoned coffee plantations and pastures in the Cordillera Central of Puerto Rico. Plant Ecol 161:75–87
Moreno JA (1998) Status of the Virgin Islands Screech-Owl. J Field Ornithol 69:557–562
Müller PLS (1776) Des Ritters Carl von Linné, Volständige Natursystem nack dem zwölften Latinischen Aufgabe. Suppl. p. 70
Murphy PG, Lugo AE (1986) Structure and biomass of a subtropical dry forest in Puerto Rico. Biotropica 18:89–96
Latta S, Rimmer C, Keith A et al (2006) Birds of the Dominican Republic and Haiti. Princeton University Press, Princeton/Oxford
Lawrence GN (1878) On the members of the genus *Gymnoglaux*. Ibis 20:184–187
Nellis DW (1979) Record of Puerto Rican Screech Owl, Turkey Vulture and Osprey from St. Croix, U.S. Virgin Islands. Wilson Bull 91:148–149
Pardieck KL, Meyers JM, Pagán M (1996) Surveys of Puerto Rican Screech-Owl populations in large-tract and fragmented forest habitats. Wilson Bull 108:776–782
Raffaele H (1989) A guide to the birds of Puerto Rico and the Virgin Islands. Princeton University Press, Princeton
Raffaele H, Wiley J, Garrido O et al (1998) A guide to the birds of the West Indies. Princeton University Press, Princeton, NJ, USA
Ridgway R (1882) Description of a new owl from Porto Rico. Proceedings of the U.S. National Museum, 4, pp. 366–369
Riley, JH (1916) Three remarkable new species of birds from Santo Domingo. Smithsonian Institution, publication 2435, Smithsonian Misceallaneous Collections, v. 66, no.5
Rivera-Milán FF (1995) Distribution and abundance of raptors in Puerto Rico. Wilson Bull 107:452–462
Robertson WB Jr (1962) Observations on the birds of St. John, Virgin Islands. Auk 79:44–76
Snyder NFR, Wiley JW, Kepler CB (1987) The parrots of Luquillo: natural history and conservation of the Puerto Rican Parrot. Western Foundation of Vertebrate Zoology, Los Angeles
Stattersfield AJ, Crosby MJ, Long AJ et al (1998) Endemic bird areas of the world: priorities for bird conservation. BirdLife International, Cambridge
Stockton Dod A (1983) The Stygian Owl (*Asio stygius noctipetens*) in the Dominican Republic. Am Birds 37:266–267
Thomlinson JR, Serrano MI, López T del M et al (1996) Land use dynamics in a post-agricultural Puerto Rican landscape (1936–1988). Biotropica 28:525–536
Weidensaul S (2015) Owls of North America and the Caribbean, Peterson reference guide series. Houghton Mifflin Harcourt Publishing Company, New York
Wetmore A (1916) Birds of Vieques Island, Porto Rico. Auk 33:403–419
Wetmore A, Swales BH (1931) Birds of Haiti and the Dominican Republic, vol 155. Bulletin U.S. National Museum, Washington, DC, pp 1–483
Wiley JW (1986) Habitat change and its effects on Puerto Rican raptors. Birds Prey Bull 3:51–56
Wiley J (1998) Breeding-season food habits of Burrowing Owls (Athene cunicularia) in southwestern Dominican Republic. J Raptor Res 32:241–245
Wiley J (2010) Food habits of the endemic Ashy-Faced Owl (*Tyto glaucops*) and recently arrived Barn Owl (*T. alba*) in Hispaniola. J Raptor Res 44:87–100
William MI, Steadman DW (2001) The historic and prehistoric distribution of parrots (Psittacidae) in the West Indies. In: Woods CA, Sergile FE (eds) Biogeography of the West Indies: patterns and perspectives, 2nd edn. CRC Press LLC, New York, pp 175–182

Wikipedia contributors (2017a) "Dominican Republic." Wikipedia, The Free Encyclopedia. Wikipedia, The Free Encyclopedia, 5 Apr. 2017. Web. 13 Apr. 2017
Wikipedia contributors (2017b) "Haiti." Wikipedia, The Free Encyclopedia. Wikipedia, The Free Encyclopedia, 13 Apr. 2017. Web. 13 Apr. 2017
Wikipedia contributors (2017c) "Hispaniola." Wikipedia, The Free Encyclopedia. Wikipedia, The Free Encyclopedia, 11 Apr. 2017. Web. 13 Apr. 2017

Chapter 15
The Owls of Mexico

Paula L. Enríquez and José Raúl Vázquez-Pérez

Abstract Mexico has 34 owl species, but little is known about their ecology particularly for tropical species. Of the 12 genera of owl species in the country, the genera *Megascops* and *Glaucidium* are the most diverse with eight species each. *Strix* has five species and *Asio* four. Four species are extensively distributed in the country, while others have distributions restricted to only one state in the country: *Megascops lambi* (endemic to the Pacific slope in the state of Oaxaca), *M. barbarus* (endemic to the highlands of Chiapas and Guatemala), and *G. hoskinsii* (endemic to the highlands of South Baja California). Another distributional pattern is altitudinal, where species with a wide latitudinal distribution also have a wide altitudinal distribution (e.g., *Tyto furcata, Bubo virginianus*). Thirteen species are distributed below 1500 m above sea level, and only seven species have a distribution over an altitude greater than 1500 m. The majority of owl species are forest species, and then the loss of these environments strongly affects their survival. Currently, secondary forests and ecotones have been considered important areas for owls. All of the owl species are included in Appendix II of CITES, and three species are listed on BirdLife International as near threatened (*Megascops barbarus*, *M. seductus*, and *S. occidentalis*). The Mexican Official Norm (NOM-059) currently considers 18 owls in a risk category, the majority is endangered, and three are at risk of extinction. Although these are national categories, there is little empirical information about the population trends or status of these species. The different impacts on and threats to owl populations are local, but also regional. The principal threats are habitat loss, degradation, and fragmentation; introduction of exotic species; pesticide contamination; illegal trafficking; and superstitious beliefs of bad luck. Empirical knowledge about this group has been increasing in recent years, primarily for species with species distributed in temperate zones. However, more effort in research should be considered necessary to improve our understanding from descriptive approaches but also functional and evolutionary ones.

Keywords Owl distribution • Endemic species • Threatened owls • Conservation strategies

P.L. Enríquez (✉) • J.R. Vázquez-Pérez
Departamento Conservación de la Biodiversidad, El Colegio de la Frontera Sur, Carretera Panamericana y Periférico Sur s/n. Barrio Ma. Auxiliadora, San Cristóbal de Las Casas, Chiapas, Mexico
e-mail: penrique@ecosur.mx

P.L. Enríquez (ed.), *Neotropical Owls*, DOI 10.1007/978-3-319-57108-9_15

Stygian Owl (*Asio stygius*)

15.1 Introduction

A few more than 1000 species of birds have been recorded in Mexico; this number varies by source: 1026 (Escalante-Pliego et al. 1998), 1050 (Howell and Webb 1995), and 1076 (Ceballos and Márquez-Valdelamar 2000). Given this high diversity, Mexico contains the 12th most bird species in the world, making it an important geographic location where two regions (Nearctic and Neotropical) converge, which allows for a center of diversification and evolution of different species (Navarro-Sigüenza and Sánchez-González 2003).

Although birds are one of the best-known terrestrial vertebrate groups, some bird groups have received very little attention, as is the case with nocturnal birds that include the Caprimulgiformes and Strigiformes. The lack of knowledge of these species is primarily due to the characteristics of the group. The majority of them are largely nocturnal; although some species are crepuscular and a few are diurnal, most of them live in forests or jungles, most of them are rare or uncommon species, and their behavior is very vigilant and secretive. It is a challenge to study them because of all of these characteristics. The objective of this work is to present the knowledge about the distribution and natural history of the owls in Mexico, as well as to analyze their threats and conservation strategies. The taxonomic nomenclature we use was König et al. (2008).

15.2 Study Area

The Mexican Republic (Mexico) is found in the northern hemisphere of the American continent and is bounded by its outermost coordinates: 32°43′06″ to the north with the border with the United States of America and 14°32′27″ to the south with the border with Guatemala. The area extent of the country is 1,964,375 km^2 which is politically divided into 31 states and one Mexico City.

Mexico's topography is very hilly with various mountains, plains, valleys, and plateaus that resulted from tectonic activity during the Cenozoic era, so that approximately 65% of the national territory is above 1000 m above sea level (de Alba and Reyes 1998). The highest altitude found are volcanoes such as Pico de Orizaba in Veracruz (5636 masl). Mexico contains two peninsulas (Baja California and the Yucatan); the Mexican Plateau, which is composed of two main mountain chains, the western Sierra Madre and the eastern Sierra Madre; and a Neovolcanic Transversal, the southern Sierra Madre, which derives the Sierra Madre of Chiapas, which extends through Central America (Fig. 15.1).

The Sierra Madre Occidental covers all of western Mexico (paralleling the Pacific coast) and has a length of 1500 km that runs through Arizona, part of Sonora, Chihuahua, Sinaloa, Durango, Zacatecas, Nayarit, and part of Jalisco where it joints the Neovolcanic Transversal. The Sierra Madre has a length of 1350 km that extends from the south of Rio Bravo and parallels the Gulf of Mexico until it joins with the Neovolcanic Transversal. The transverse volcanic axis is a chain of volcanoes that are a part of the Pacific ring of fire, which is characterized by its large volcanic activity and where some of the tallest mountains in the country are located, like Pico de Orizaba or Citlaltepetl, Popocatepetl, and Iztaccihuatl.

Between these two mountain chains and the transverse volcanic axis is the Mexican Plateau, which reaches an altitude of 1200 masl. The Chihuahua and Bolson de Mapimí deserts are located on this plateau and contain small mountains known as the transverse volcanic mountains. The Balsas Depression is located to the south of the Neovolcanic Transversal which is the lowest region in the entire country. The southern Sierra Madre mountain chain, which ends in the Isthmus of Tehuantepec, is located between the Balsas Depression and the Pacific Ocean. The Sierra Madre de Oaxaca (Sierra de Juárez) is located in the east which runs from the north of Oaxaca and ends in Veracruz, where it joins the Sierra Madre de Chiapas and the mountains of Soconusco, which form the central Plateau of Chiapas.

There are 2800 islands in Mexico, including rocks, keys, reefs, islets, and islands, of which most of them are in the Caribbean (667). Only 5% of these islands are inhabited.

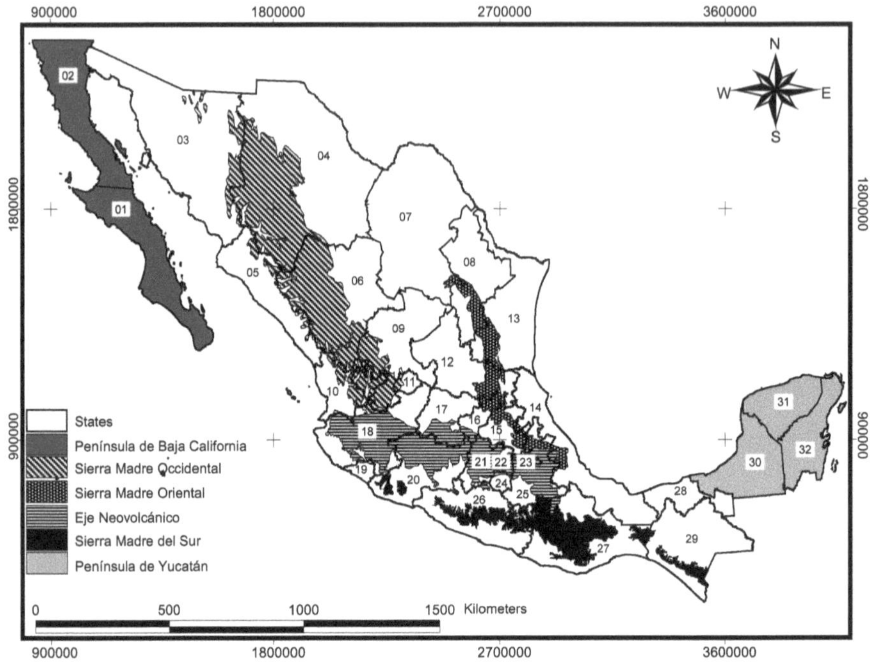

Fig. 15.1 Topography of Mexico and its states (*1* Baja California Sur, *2* Baja California, *3* Sonora, *4* Chihuahua, *5* Sinaloa, *6* Durango, *7* Coahuila, *8* Nuevo León, *9* Zacatecas, *10* Nayarit, *11* Aguascalientes, *12* San Luis Potosí, *13* Tamaulipas, *14* Veracruz, *15* Hidalgo, *16* Querétaro, *17* Guanajuato, *18* Jalisco, *19* Colima, *20* Michoacán, *21* México, *22* Mexico City, *23* Tlaxcala, *24* Morelos, *25* Puebla, *26* Guerrero, *27* Oaxaca, *28* Tabasco, *29* Chiapas, *30* Campeche, *31* Yucatán, *32* Quintana Roo)

15.3 Climate

Due to the particular characteristics of local geography, topographic complexity, ocean temperature and currents, trajectories of summer storms, and polar fronts in winter, Mexico has a great diversity of weather and environments, which includes practically all possible climatic groups and subgroups, from arid and semiarid climates to humid and subhumid climates (del Alba and Reyes 1998).

Approximately 56% of the country's territory contains very arid and semiarid zones located in the north and center of Mexico. Some 37% of the subhumid climate zones are located in the coastal plains (Gulf and Pacific) and the northeast of the Yucatan peninsula. The remaining 7% of the territory is humid, located in the foothills of the mountains (UNAM 1990). The precipitation in the country is highly variable. The annual rainfall in the north averages 100 mm, but 2000–4000 mm falls annually on average in the southeast and southern Pacific coast.

15.4 Environmental Diversity (Types of Vegetation)

Mexico has almost all different ecosystems, from deserts to tropical vegetation and from wild mountains to pasturelands of different types (Rzedowski 2006). All of this high diversity of vegetation and environmental types is because of the physiographic, geologic, and climatic conditions in the country, as well as the combination and influence of tropical South American environments with the boreal North American environments. The vegetation types vary by source. For example, Rzedowski (2006) considers ten vegetation types: forest or tropical evergreen forest, semi-deciduous, deciduous, thorn, xerophytic, oak, conifer, montane mesophytic, and aquatic and semiaquatic vegetation. Other vegetation types considered are mangroves, popales, tulares, palms, petenes vegetation, and chaparral.

15.5 Taxonomic Diversity and Distribution

The regional diversity and patterns of species richness have been discussed extensively (Wittaker et al. 2001) and many variables and mechanisms have been proposed to explain these patterns. Besides historic factors, recent factors like climate and topography also determine environmental heterogeneity. This determines the great variety of environments and conditions that permit diversity. The diversity of owl species in Mexico is 34 species, which represents 42.5% of the species distributed in the neotropics (König et al. 2008). While they have been considered as 32 species, a recent taxonomic revision by König et al. (2008) has proposed two more species in the country (i.e., *Glaucidium californicum* for the north of Sonora and *G. cobanense*, a species distributed in Chiapas).

This species richness of owls in Mexico is composed of 12 genera, of which the genera *Megascops* and *Glaucidium* are the most represented, with eight species in each. Following these two genera are *Strix* with five species, *Asio* with four, and *Aegolius* with two. Three genera are monospecific (*Psiloscops* (*Otus*) *flammeolus*, *Lophostrix cristata*, and *Micrathene whitneyi*). The rest of the genera (four) contain only one species in the country (Appendix 15.1). Of the most represented genera, *Megascops* is distributed in the Yucatan Peninsula only with one species, and two species are found in the Baja Peninsula. Only one species of *Glaucidium* is represented in each of the Peninsulas, *G. ridgwayi* in the Yucatan, and *G. hoskinsii* in Baja California.

The four most widely distributed species in the country are barn owl (*Tyto furcata*, previously *Tyto alba*), great horned owl (*Bubo virginianus*), burrowing owl (*Athene cunicularia*), and short-eared owl (*Asio flammeus*; Fig. 15.2). These species present a wide continental distribution. Meanwhile, there are six species with a distribution restricted to only one state in the country, some of which are endemic to Mexico, like Oaxaca screech owl (*Megascops lambi*, endemic in the Pacific slope in the state of Oaxaca), bearded screech owl (*M. barbarus*, endemic to the highlands

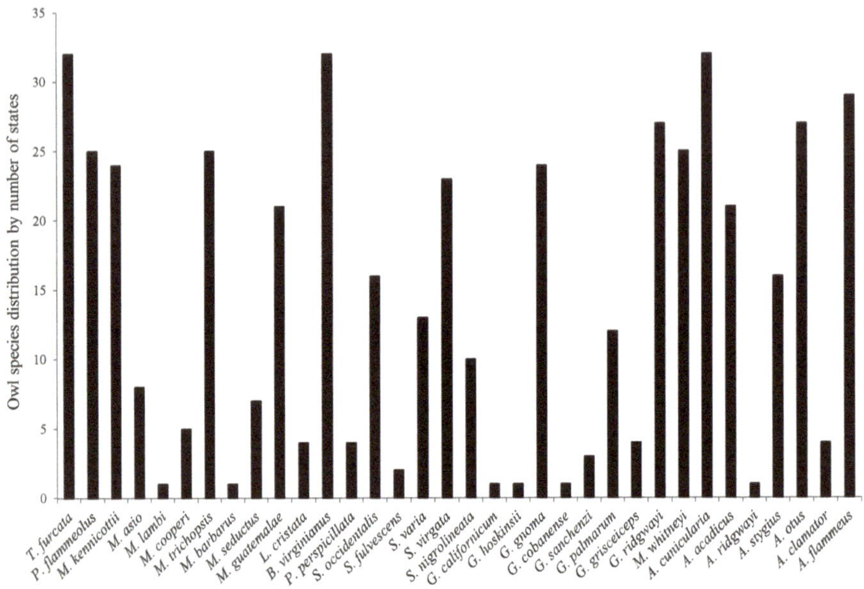

Fig. 15.2 Distribution of owl species by number of states of Mexico

Fig. 15.3 Bearded screech owl (*Megascops barbarus*) pair with two phases of color (gray and rufous). El Huitepec Ecological Reserve, San Cristóbal de Las Casas, Chiapas, 29 April 2004. Photograph José Luis Rangel-Salazar

of the state of Chiapas, Fig. 15.3), and cape pygmy owl (*G. hoskinsii*, endemic to the mountains of Baja California Sur), and the three remaining species are northern pygmy owl (*Glaucidium californicum*, distributed to the south of its range in the state of Sonora), Guatemalan pygmy owl (*G. cobanense*), and unspotted saw-whet owl (*Aegolius ridgwayi*; Fig. 15.4) (both species distributed to the northern of their range in Chiapas state; Fig. 15.2).

Of all the owl species distributed in Mexico, seven are endemic, of which three are *Megascops* species (*M. lambi*, *M. seductus*, and *M. barbarus*), three are *Glaucidium* (*G. hoskinsii*, *G. sanchezi*, *G. palmarum*), and *Micrathene* (Appendix 15.1). Two of these species are considered quasi-endemic (i.e., *M. barbarus* with a distribution in Chiapas, but shares its distribution with Guatemala, and *Micrathene*

Fig. 15.4 Unspotted saw-whet owl (*Aegolius ridgwayi*) in a cloud forest. El Huitepec Ecological Reserve, San Cristóbal de Las Casas, Chiapas, 13 March 2004. Photograph José Luis Rangel-Salazar

whitneyi, which although it presents the widest distribution in the country, it is only distributed in Mexico and the United States; Appendix 15.1).

Various species (21) present subspecies, of which only three species have subspecies with populations restricted to islands or islets:

- *Bubo virginianus mayensis*, which is distributed along the coasts of the Yucatan Peninsula.
- *Micrathene whitneyi graysoni* with a distribution on Socorro Island which is a volcanic island located in the Revillagigedo archipelago in the Pacific Ocean. The state of Colima controls this archipelago. However, it has not been observed since 1931 and is therefore considered extinct (BirdLife International 2012).
- *Athene cunicularia rostrata* with a distribution on Clarion Island also located in the Revillagigedo archipelago (Appendix 15.1).

Although the political/geographic borders of the states are more a political characteristic than a biological one, the species distribution does not follow these limits, and we can expect that the number of species increases according to the area of the state. The average owl species richness in each state is 15 ± 4.05. The state with the fewest reported species is Campeche with seven and Yucatan, Baja California, and

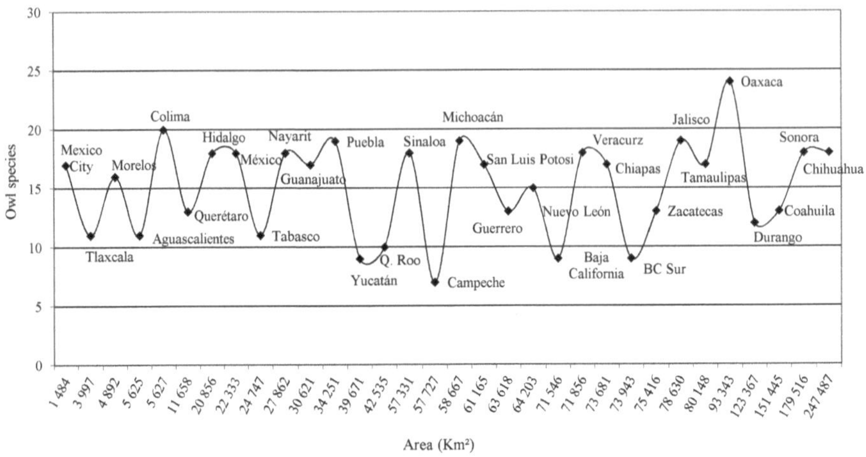

Fig. 15.5 Number of owl species registered by state. The order of the states is by area size, from smallest to largest area

Baja California Sur with nine species each. The se states are in the Peninsulas. On the other hand, the states with the highest species richness are Oaxaca with 24 species, Colima with 20, and Jalisco, Michoacán, and Puebla with 19 species in each state. There is no relation between the owl species richness in a state and its area (Fig. 15.5). Many of the records of owl species by state are accidental or occasional; therefore the species richness of a state will increase if the research with this group increases. Furthermore, the distributions of species are dynamic and in constant flux.

Five owl species distributed in Mexico migrate latitudinally, *Psiloscops* (*Otus*) *flammeolus*, *Micrathene whitneyi*, *Athene cunicularia*, *Asio otus*, and *A. flammeus*, but there are resident populations in the country (Appendix 15.1). Another pattern of distribution is altitudinal. Some species with a wide latitudinal distribution have a wide altitudinal distribution (i.e., *Tyto furcata*, *Bubo virginianus*; Fig. 15.6). Others, like *Strix occidentalis* and *Megascops kennicottii*, also have a wide altitudinal distribution from sea level until 2500 masl or greater. Thirteen species are distributed below 1500 m, and only seven species are distributed around 1500 masl; 25 species reach a distribution limited (maximum or minimum) at 1500 masl (Fig. 15.6).

The owl communities that are in temperate highland zones or tropical lowland zones generally have species of genera *Megascops*, *Glaucidium*, *Strix*, and *Asio*. In some communities there can be species congeneric that coexist, for example, *Megascops barbarus* and *M. trichopsis* are both found in temperate zones in Chiapas, but differ in habitat selection; the first uses environments that are more conserved and humid, while the latter uses environments that are more disturbed or forest edge habitat (pers. obs., Enríquez and Cheng 2008). For example, mottled owl (*Strix virgata*; Figs. 15.7 and 15.8) and black and white owl (*S. nigrolineata*) in tropical zones live in the same habitat coexisting in the same environments, but in distinct sites (Enríquez and Rangel-Salazar 2001, 2007). Depending on the altitude, other species can integrate into the community; in tropical regions lower than 1500

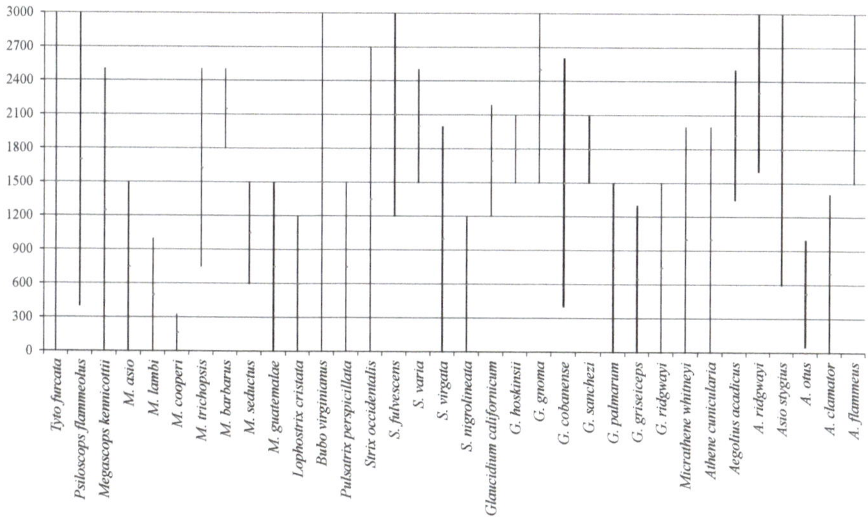

Fig. 15.6 Altitudinal distribution (masl) of 34 species that are distributed in Mexico

Fig. 15.7 Mottled owl (*Strix virgata*) in El Canelar Reserve, Acala, Chiapas, Mexico, 19 November 2008 (Photograph José Raúl Vázquez Pérez)

masl, crested owl (*Lophostrix cristata*) or spectacled (*Pulsatrix perspicillata*) can be found, while *Aegolius* species can be found in temperate zones higher than 1500 masl (Fig. 15.4). In the case of *Micrathene whitneyi* and *Athene cunicularia* that are distributed altitudinally up to 2000 masl, they use environments very specific to their ecological needs. *Micrathene whitneyi* is distributed in arid or semiarid environments with bushes, scrub, and chaparral. Primarily with saguaros, while *A. cunicularia* also inhabits arid and semiarid environments, it is found in pastures,

Fig. 15.8 Juvenile of mottled owl (*Strix virgata*) in El Silencio Reserve, Suchiate, Chiapas, 25 April 2014 (Photograph Noé Jiménez Lang)

agricultural areas, deserts, and savannas, but also in airports, cemeteries, and residential or industrial areas, to differentiate the rest of the owl species, it is an owl terrestrial and diurnal (Valdez-Gómez and Enríquez 2005).

15.6 Habitat Association

The majorities of owl species inhabit forests or jungles and are almost always associated to humid habitats and water bodies. The environments are not static and generally are heterogeneous in where they provide clearings or open areas promoting secondary growth and ecotones that are important occupation environments for raptor species, including owls (Enríquez and Rangel-Salazar 2007). While species inhabit old or mature forest environments and are important elements for their reproduction, secondary forests contain other elements for their survival, such as food or resting sites (Enríquez and Cheng 2008). Currently, secondary forests dominate the countryside landscape due to an increase of deforestation and fragmentation of mature forests; therefore they are considered important environments for the conservation of biodiversity in general and nocturnal birds in particular (Sekercioglu 2010; Dent 2010; Feely 2010; Chazdon 2014). A proposal is to develop a management in environments with secondary vegetation or acahuales with different levels of maturity, with an agroforestry system or model in which parcels are enriched with planted trees (Soto et al. 2011). These agroforestry systems are important for various animal species and particularly raptors which find most of their food in these environments (e.g., *Megascops guatemalae*, *Lophostrix cristata*, and *Strix virgata* use abandoned cacao agroforests; Enríquez and Rangel Salazar 2007).

The process of habitat fragmentation has caused the change in the distribution and abundance of species. Some nocturnal raptor species have been adapting to the new changes, and we can find them in environments modified by humans, in areas that are partially urban, depending on factors like vegetation, food availability, and nesting sites. However, other species have been extirpated locally and are joining international and national lists of at risk species (Enríquez et al. 2006).

15.7 State of Conservation at a National Level

One of the main threats for raptor species and particularly for owl species is the loss or fragmentation of forests for multiple reasons (Thiollay 1985a; Marcot 1995; Enríquez et al. 2006) which already have negatively affected the function and structure of these ecosystems. This has caused that various species are at risk. Raptors in Mexico are one of the most threatened groups of birds (NOM-059, SEMARNAT 2010). And, particularly for species of nocturnal raptors, all are found on the international CITES Appendix II, which means that they are species that are not necessarily at risk, but could become at risk if they continue to be sold without being regulated. BirdLife International (2016) considers various species of least concern, but two species are considered near threatened, *Megascops seductus* and *Strix occidentalis*, and one vulnerable, *Megascops barbarus* (Appendix 15.1).

On the other hand, the US Fish and Wildlife Service considers five species of owls in some risk category, which also have populations in Mexico. These species are *Strix occidentalis* which is considered at risk of extinction, *Glaucidium ridgwayi* which is considered threatened, *Tyto furcata* and *Athene cunicularia* which are considered threatened in some states, and *Asio flammeus* which is considered of special concern.

The Official Mexican Law NOM-059 (SEMARNAT 2010) currently considers 18 species of owls in some category of risk, which represents 53% of the species in the country and four subspecies. Four species have special protection, 11 species are threatened, and three are in danger of extinction (Appendix 15.1). Of the subspecies, *Bubo virginianus mayensis* is threatened, *Athene cunicularia hypugaea* has special protection, *A.c. rostrata* is threatened, and *Micrathene whitneyi graysoni* is considered apparently extinct (SEMARNAT 2010; BirdLife International 2016). These categories are generally for the country, and few studies have been completed to determine the populations of these species. The impacts or threats to the populations are local and perhaps at times regional, but little information exists about the population trends for these at risk species.

15.8 Threats

The main threats to the conservation of owls in Mexico, as in the rest of the world, are the loss, degradation, and fragmentation of habitat. It has been estimated that Mexico has the second highest deforestation rates in the world (FAO 2007), which is already occurring at a rate of 350,000 ha/year (FAO 2010), while these rates vary by vegetation type, the region, and period (Ochoa-Gaona 2001). The levels of threats to owl populations and communities vary in frequency and intensity depending on the species, state, region, and locality and are strongly determined by the human cultures, natural resource uses, and the politics of management in each region (Enríquez et al. 2006).

The main factors that cause this degradation and loss of habitat are extraction of natural resources, expansion of pastures and farms, as well as urban expansion. The natural events like hurricanes, tropical storms, and droughts also are factors that modify and fragment environments. During the last 50 years, Mexico has seen drastic changes in soil use due to rapid urbanization and industrialization, which has been poorly planned. This has eroded the natural environments and has increased the degradation and loss of biological diversity in the country. However, information about the direct impacts of these threats on the population loss and changes in distribution and abundance on species in long term is nonexistent.

Another important threat is the introduction of exotic species, some of which are domestic and have become feral populations with invasive behavior which displace native species and then drastically affect the entire ecosystem. Mexico established a National Strategy for Invasive Species in 2010 that aims to prevent, control, and eradicate this grave threat and plans to strengthen in 10 years (Comité Asesor Nacional sobre Especies Invasoras 2010). The invasive species are considered important threats and their consequences established, but the environmental impacts have not been widely studied. These invasive species have reached islands and eradicated native species, for example, in islands of Mexico, 12% of the endemic birds and 20% of the endemic mammals have disappeared due to introduced species (Aguirre-Muñoz et al. 2009). We know little about owl populations on islands, but two subspecies are distributed in the Revillagigedo archipelago where various exotic species have been reported, like sheep (*Ovis* sp.), pigs (*Sus scrofa domestica*), rabbits (*Sylvilagus* sp.) which destroy habitat, cats (*Felis domesticus*), and rats (*Mus* sp.) which prey on birds and eggs in nests. This site has been considered of extreme importance because of the endemic species or subspecies it supports; in the case of the birds on the island Socorro, one encounters a high level of endemism among birds (SEMARNAT 2016).

Other threats that have been little studied in Mexico are contaminants like pesticides (insecticides and rodenticides) and their effect on forest fauna. The wide use of pesticides as organochlorides is their persistence, because it remains active for a long time and is slowly degraded; those are very harmful. The second cause of their use is that they are economic. Mexico manufactured the majority of the organochlorides that were consumed for more than 30 years in the country; Mexico also was the principal exporter of DDT (Calva and Torres 1998). Currently, they continue to

Fig. 15.9 Ridgway's pygmy owl (*Glaucidium ridgwayi*) in El Silencio Reserve, Suchiate Chiapas, 25 April 2014 (Photograph Noé Jiménez Lang)

be used and of the 90 pesticides that are prohibited or restricted in the United States, 30 continue to be used in Mexico (INEGI 1992). Twelve of those are prohibited at an international scale. The use of pesticides in the country is a common practice, and the quantities and concentrations that are applied to crops are unknown (Ortiz et al. 2014). And while Mexico and international conventions regulate the use of these substances and limit their effects on populations and ecosystems, a better coordination between all institutions needs to be achieved (Ortiz et al. 2014).

Studies on the effects of organochloride pesticides on raptors in Mexico have been few. However, they have identified and discovered concentrations of organochlorides in chickens of Osprey (*Pandion haliaetus*; dieldrin 0.969 ± 0.724 pg/ul, DDE 0.922 ± 0.895 pg/ul), which represents a risk to the health of these organisms (e.g., Rivera-Rodriguez and Rodriguez-Estrella 2011). A study in the Delta of the Colorado River in Sonora found that such concentrations of organochlorides (biphenyl polychlorides #8 PCB 126) in eggs of burrowing owls (*Athene cunicularia)* are highly toxic and can affect the hatching of the eggs (García-Hernández et al. 2006). Recently, concentrations of organochlorides have been found in Ridgway's pygmy owl (*Glaucidium brasilianum*; now *G. ridgwayi*) in Chiapas (Arrona-Rivera 2015; Arrona-Rivera et al. 2016, Fig. 15.9).

The illegal market of wildlife species is one of the principal factors that contribute to the population declines of species. Mexico is considered one of the ten main countries that have this type of illegal market (Rangel-Salazar et al. 2013). In this case, birds compose 50% of the animal species that are illegally trafficked. The important groups are parrots and parakeets, but also raptors. The important raptor species are the Harris's hawk (*Parabuteo unicinctus*) and the prairie falcon (*Falco mexicanus*). In the case of owl species, the illegal market sells *Bubo virginianus*, *Strix virgata*, *Athene cunicularia*, *Glaucidium gnoma*, and *Micrathene whitneyi* (Sosa-Escalante 2011). Also small owls of the genus *Megascops* and *Glaucidium* have been reported (PLE pers. obs.).

The origin of the specimens is often unknown, but in the market of Sonora in Mexico City, specimens of *Strix virgata* are sold from Las Choapas, Veracruz, at a cost of 2000 pesos, around 130 USD (2013, PLE pers. obs.). Generally, the illegal sale is for pets, but also for homeopathic remedies. In indigenous communities, ritual cures use feathers and bones; therefore there are dissected organisms or animal parts of *Bubo virginianus* or some species like *Strix virgata* and *S. nigrolineata* and owls of the genus *Megascops* in esotericism markets (PLE pers. obs.).

The illegal trafficking of specimens, parts, and derivatives of forest plants and animals is a crime established in article 420 parts IV and V of the Federal Penal Code, which threatens a sentence of 1–9 years in prison or for an equivalent of 300–3000 days of minimum wage. Several illegal trafficking routes have been identified, and El Charco Cerrado in San Luis Potosi is one of the most important, but currently installed posts have been destroyed, and those responsible for this activity have been subject to legal proceedings (Sosa-Escalante 2011). In Mexico City, the Federal Attorney for Environmental Protection (PROFEPA) has established a permanent zero-tolerance policy for illegal trafficking of forest animals and plants and in 2013 has rescued more than 2569 specimens of plants and animals (Méndez 2013). Despite these strategies, birds are not exhibited publicly in markets but can be obtained illegally by request. Another problem is after the rescue of these specimens, it is difficult to release them in a forest environment because their origin is unknown.

Another important threat for those species that are little known is that we have created attitudes of fear or mystery due to their appearance, coloration, and nocturnal habits; there are diverse beliefs, legends, or myths relating them to death or bad luck. A common story in Mexico says that "when an owl sings, an Indian dies" and is associated with this happening. Other countries in Latin America also associate them with messengers of death or witches (Enríquez and Rangel-Salazar 2006; Restrepo Cardona and Enríquez 2014). Due to these beliefs, myths, and superstitions, owls are sacrificed. Also owls are hunted because they are associated with domestic damage (i.e., sometimes hunt poultry). However, these damages have not been quantified. Another motive for killing them includes target practice with rifles or slingshots (PLE pers. obs.).

Other evident, but unquantified, threats are deaths due to collisions with automobiles, high-voltage wires, or on barbed wire fences which trap and kill them. But also stochastic events like tropical storms, hurricanes, or forest fires (natural or human-caused) that modify structurally vegetation also influence in those owl population's decreases. However, they are global threats but not evaluated. Every species and population is exposed to different intensities of threats. For example, *Athene cunicularia* is threatened by habitat loss due to pasture expansion, which destroys their burrows, or poisoning due to agricultural chemicals in Chihuahua and Sonora (Chávez-Ramírez 1990; Rodríguez-Estrella and Granados 2006; García-Hernández et al. 2006).

15.9 Conservation Strategies

Mexico is a country with a large territory, resulting in high levels of biodiversity and a great variety of heterogeneous environments. These characteristics make conservation challenges difficult and complex. Thus, the establishment of conservation areas is not sufficient to protect owl species in particular and biological diversity in general. It is necessary to include social and economic aspects to conservation strategies. However, there are diverse concepts and methodologies regarding conservation among academics, managers, and administrators to establish the understanding of conservation of biological diversity (Rangel-Salazar et al. 2005).

Protected natural areas are considered important elements and provide knowledge of the function and conservation of ecosystems (Arcese and Sinclair 1997). The establishment of an important number of these areas is necessary. The most consolidated instrument of biodiversity conservation in Mexico is the National System of Protected Areas administered by the National Commission of Protected Natural Areas (CONANP) which includes 174 federal natural areas which represents more than 254,552.5 km^2, but only protects 13% of the national territory (CONANP 2014b). These areas are classified into six categories. Biosphere reserves cover the main part of the protected area, with a total of 41 reservations, and only cover 6.4% of the national area (Table 15.1). This National System of Protected Natural Areas is found in all of the states of the country, but the representation of each of the six categories varies in the states. For example, Chiapas is the only state with six categories; the majority of states have more than one category, including biosphere reserves, national parks, monuments, and protected natural resource areas, and four states (Tamaulipas, Zacatecas, Guanajuato, and Tabasco) only have one or two protected areas.

Categories and definitions are:

1. Flora and Fauna Protected Area: These areas allow exploitation of natural resources in accordance to the management program. This type covers 67,868.9 km^2 of the national territory.
2. Natural Resource Protected Area: These are areas for soil, watershed, and forest natural resource conservation and protection. Currently 45,359 km^2 is in this management category.
3. Natural Monument: These are areas that contain one or many important national natural elements. These areas are established for their beautiful scenery: educational, scientific, recreational, or historic value. Generally, these are small areas because they cannot be included in other management categories. They cover 162.7 km^2 of the national territory.
4. National Park: Natural areas of interest for conservation, but also considered potential sites for tourist development. These areas are for public use and, where it is permitted, the exploitation of natural resources in accordance to the management program. Currently 14,101.6 km^2 of the national territory is in this category of national protected area.
5. Biosphere Reserve: These areas are representatives of one or more environments undisturbed by humans that need to be conserved. The area of these reserves must be more than 100 km^2 and contains two zones (nucleus and buffer). At a national level, Biosphere Reserves protect 127,032 km^2 of the national territory.
6. Sanctuary: These are areas established in sites characterized by their rich plant and animal communities or because of the presence of species with restricted ranges. Some examples are glens, relics, caves, caverns, cenotes, cove, or other geographic formations that need to be conserved. These areas protect an area of 27.4 km^2.

Table 15.1 The categories and territorial extension (Km^2) of Natural Protected Areas (ANPs) in Mexico. Information from shapefile (CONANP 2014b)

ANP categories	States	Number of ANPs	Extension (km^2)
Flora and fauna protected area	Quintana Roo, Yucatán, Campeche, Baja California Sur, Baja California, Oaxaca, Chihuahua, Sonora, Chiapas, Tabasco, Estado de México, México City, Morelos Coahuila, Colima, Sinaloa, Jalisco, Tamaulipas, Michoacán, San Luis Potosí, Zacatecas, and Veracruz	38	67,868.9
Natural resource protected area	Aguascalientes, Zacatecas, Coahuila, Nuevo León, Jalisco, Durango, Nayarit, Colima, Chiapas, Hidalgo, Puebla, México State, and Michoacán	8	45,359.9
Natural monument	Chiapas, Nuevo León, Chihuahua, Oaxaca, and Coahuila	5	162.7
National park	Baja California, Yucatán, Quintana Roo, Baja California Sur, Michoacán, Oaxaca, Veracruz, Puebla, Chiapas, Chihuahua, México City, Querétaro, Nuevo León, Coahuila, Estado de México, Yucatán, Hidalgo, San Luis Potosí, Morelos, Guerrero, Nayarit, Tlaxcala, Zacatecas, Jalisco, and Colima	65	14,101.6
Biosphere reserve	Sonora, Colima, Quintana Roo, Baja California, Hidalgo, Campeche, Jalisco, Chiapas, Baja California Sur, Nayarit, Chihuahua, Durango, Veracruz, Coahuila, Nayarit, Tabasco, Campeche, Yucatán, Morelos, Puebla, Guerrero, San Luis Potosí, Tamaulipas, Guanajuato, Querétaro, Oaxaca, Michoacán, Hidalgo, and México State	40	127,032.0
Sanctuary	Jalisco, Sinaloa, Oaxaca, Yucatán, Quintana Roo, Michoacán, Chiapas, Tamaulipas, and Guerrero	16	27.4
Total (Km^2) = 254,552.5			

Despite the fact that Mexico has formed a National Commission of Protected Natural Areas, the number of protected areas is insufficient, but at the same time the number of personnel assigned to the management of these areas is limited as is the budget. Furthermore, the management and conservation effort is not acceptable for the objectives for each category. While some reserves like Biosphere Reserves have received much attention, others, like national parks or flora and fauna protected areas, could be in an abandoned state.

Another important tool for conservation of biological diversity is the Official Mexican Law (NOM-059 SEMARNAT 2010). The Official Mexican Law is elaborated by the National Consulting Committee of Normalization of Regulation and Promotion of Sanitation which establishes rules, attributes, and directives applicable to a product, process, system, or activity. The Official Mexican Law NOM-059-SEMARNAT-2010, which refers environmental protection of at risk forest plant and animal species native to Mexico, lists the species that have a conservation problem.

Furthermore, representatives and investigators from various universities, government agencies, associations, and nongovernmental organizations participate. They consider for categories of risk (E, probably extinct in wildlife; P, in peril of extinction; A, threatened or at risk; and Pr, special protection, SEMARNAT 2010). This law currently includes 22 species and subspecies of owls of which 5 have special protection, 13 are threatened, three are in peril of extinction, and one is probably extinct (Appendix 15.1). Of these, five are subspecies (SEMARNAT 2010, Appendix 15.1).

Another species and ecosystem conservation strategy or tool which is used in Mexico is the identification of priority sites with different criteria. Starting from 1995, they developed a national map of priority areas (terrestrial and marine) for the conservation of biodiversity; this map was developed by La Comisión Nacional para el Conocimiento y Uso de la Biodiversidad (CONABIO); 152 terrestrial priority regions currently exist that cover an area of 515,558 km^2, which corresponds to more than a quarter of the national territory (Arriaga et al. 2000). In recent years, Mexico has acquired various compromises with international community related with conservation of natural ecosystems through international conventions, agreements, and compromises. Particularly for Mexican bird conservation, 263 important areas for conservation of birds have been identified (AICAS; Arizmedi and Márquez-Valdelamar 2000). These areas are having a set of criteria of species richness, abundance, and seasonality. The proposal of this initiative, among others, is that it is a tool that can help prioritize resources for conservation.

All of these efforts are at an ecosystem or community level. Plans for recovery only exist for certain species where US initiatives include Mexico when the species are distributed in Canada, the United States, and Mexico or have neotropical migratory populations. For example, the Recovery Plan for the Spotted Owl (*Strix occidentalis lucida*) considers a recovery strategy of habitat management and species monitoring. The recovery strategy has five components: (1) protect current populations, (2) manage habitat for the future, (3) manage threats, (4) monitor populations and habitat, and (5) establish collaborations to facilitate the reestablishment of this species (US Fish and Wildlife Service 2012).

Besides this initiative, a lack of biological and ecological information still exists for populations in Mexico, and management strategies need to be developed at a large scale for the survival of this species.

15.10 State of Biological and Ecological Knowledge

The biological and ecological knowledge of different raptor species in tropical areas is very limited. In general, the distribution and basic aspects about their abundance are known. However, population trends and other information about their life history are little known. Since the article by Thiollay about the community composition of tropical forest raptors, which is 30 years old (1985a, 1994) and does not include owls because of a lack of information, it was mentioned that perhaps many of the raptor species will disappear before we learn about them, because there are very little information about their natural history.

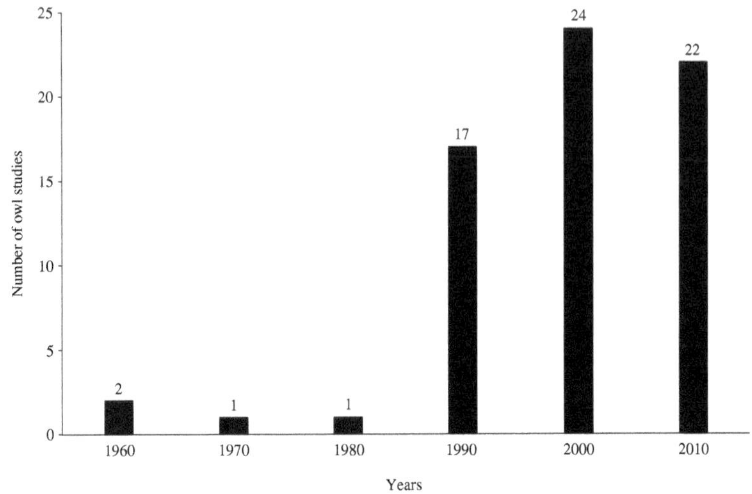

Fig. 15.10 Number of owl studies in Mexico for decades

Since this time, the situation has not changed significantly. Although there is more information about the distribution and ecology of species in certain areas, biological and ecological information is still limited in many areas (Enríquez et al. 2006, 2012). A factor that influences which birds are studied less is that they are difficult to study given their behavior, which is generally nocturnal and also cryptic, secretive, and stealthy, and many are rare. The information that exists about owls in Mexico are found on general bird lists or new registers of distribution or sightings, and only recently have studies begun that are aimed specifically at this group (Enríquez et al. 2006, Fig. 15.10). However, some studies are theses and have been not published and several are unavailable, and others are reports. Only few have been published.

Of the 34 owl species reported in Mexico, only 25 have been studied (Table 15.2). The Nearctic species show more information like *Strix occidentalis*, *Bubo virginianus*, *Tyto furcata*, *Psiloscops flammeolus*, and *Aegolius acadicus*. For the rest of the species, information are isolated or are only in lists (e.g., protected natural areas). After a revision of owl species publications, the states that have more studies are Hidalgo, Durango, Chiapas, and Tlaxcala (Table 15.2). On the other hand, there is a lack of information for nine species, which include *Megascops lambi*, *Glaucidium hoskinsii*, and *Asio clamator*, among others (Table 15.2, Fig. 15.11). In general, the studies about Mexican owls are grouped into the following topics: distribution, abundance, habitat use, reproduction, diet, and vegetation association.

The study of abundance and species distribution relationship is a basic topic of ecology (Krebs 2001). The descriptive approximation is the basis of ecology that still is limited for this group. The studies that have estimated the abundance of some species are Young et al. (1997), Garza (1999), Palacios et al. (2000), Enríquez (2002), Márquez (2005), González-Rojas et al. (2006), Flesch (2008), Alba-Zuñiga et al. (2009), Vázquez-Pérez et al. (2011), Rivera-Rivera et al. (2012), Fernández (2013), Ramírez (2014) and Ortiz-Pulido and Lara (2014). In total, these studies

Table 15.2 List of owl species with biological and ecological information produced in Mexican states

Species	Distribution	Abundance	Vegetation/landscape	Habitat use	Breeding	Diet	States
Tyto furcata (alba)	X	X	X		X	X	Tlaxcala, Oaxaca, Chiapas, Hidalgo, Chihuahua, Durango, Baja California, Baja California Sur, Jalisco, Michoacán, State of México
Psiloscops (Otus) flammeolus	X	X	X			X	Tlaxcala, Hidalgo, Durango, Nuevo León
Megascops kennicottii	X	X	X			X	Tlaxcala, Hidalgo, Baja California
Megascops asio	X						Hidalgo
Megascops lambi							
Megascops cooperi							
Megascops trichopsis	X	X	X				Chiapas, Hidalgo, Durango, Tlaxcala, Puebla
Megascops barbarus	X	X	X		X	X	Chiapas
Megascops seductus		X		X			Morelos
Megascops guatemalae	X	X	X	X			Chiapas, Hidalgo
Lophostrix cristata		X	X				Chiapas
Bubo virginianus	X	X	X		X	X	Tlaxcala, Hidalgo, Veracruz, Quintana Roo, Durango, Baja California, Oaxaca

Table 15.2 (continued)

Species	Distribution	Abundance	Vegetation/landscape	Habitat use	Breeding	Diet	States
Pulsatrix perspicillata		X	X			X	Chiapas, Oaxaca
Strix occidentalis	X	X	X			X	Zacatecas, Aguascalientes, Chihuahua, Durango, Sonora
Strix fulvescens	X	X					Oaxaca, Chiapas
Strix varia							
Strix (Ciccaba) virgata	X	X	X	X			Chiapas, Hidalgo, Oaxaca
Strix (Ciccaba) nigrolineata	X	X	X			X	Chiapas, Hidalgo
Glaucidium californicum							
Glaucidium hoskinsii							
Glaucidium gnoma	X	X					Hidalgo, Durango, Tlaxcala
Glaucidium cobanense							
Glaucidium sanchezi	X						Hidalgo
Glaucidium palmarum							
Glaucidium griseiceps							

Glaucidium ridgwayi	X	X	X	X	X		Chiapas, Campeche, Hidalgo, Sonora, Oaxaca
Micrathene whitneyi	X	X					Hidalgo, Oaxaca, México, Puebla, Guanajuato, Michoacán, Guerrero, Morelos
Athene cunicularia	X	X	X		X	X	Hidalgo, Durango, Baja California, Baja California Sur, Nuevo León, Jalisco, Chihuahua, Sonora, Tlaxcala, Puebla
Aegolius acadicus	X	X	X			X	Tlaxcala, Hidalgo, Oaxaca, Durango, Puebla
Aegolius ridgwayi	X	X	X				Chiapas
Asio stygius	X						Tamaulipas, Hidalgo
Asio otus	X						Hidalgo, Chihuahua
Asio clamator							
Asio flammeus	X					X	Hidalgo, Yucatán, Jalisco

Fig. 15.11 Striped owl (*Asio clamator*) in Villaflores, Chiapas, 23 October 2014 (Photograph José Raúl Vázquez Pérez)

have estimated the abundance of 20 owl species (e.g., *Tyto furcata*, *Megascops kennicottii*, *M. barbarus*, *Lophostrix cristata*, *Bubo virginianus*, *Pulsatrix perspicillata*, *Aegolius ridgwayi*, Table 15.2). These were realized in Tlaxcala, Hidalgo, Chihuahua, Durango, Chiapas, and Morelos, among others. The study areas have been very local and are generally done in protected natural areas like biosphere reserves. Recently the distribution and abundance of *Micrathene whitneyi* in Oaxaca has been studied (Flores-Dimas 2016).

The studies about the distribution of owl species have been done at different scales. Some studies have focused on only a level spatial (e.g., country, state, landscape, ecosystem) or temporal (e.g., monthly, seasonal) scale. The species that have been studied at different scales of distribution are *Megascops kennicottii*, *M. seductus*, *M. trichopsis*, *M. guatemalae*, *Bubo virginianus*, *Glaucidium gnoma*, *G. brasilianum*, *Micrathene whitneyi*, *Athene cunicularia*, *Strix occidentalis*, *S. virgata*, and *A. flammeus*, among others (Cirett-Galan and Díaz 1993; Arámbula 1994; Enríquez 1997; Peláez 1998; Garza 1999; Palacios et al. 2000; Valdez-Gómez and Holroyd 2000; Enríquez 2002; Rodríguez-Estrella and Careaga 2003; Martínez-Ortega 2009; Enríquez et al. 2010; Vázquez-Pérez et al. 2011; Rivera-Rivera et al. 2012; Valencia-Herveth et al. 2012; Fernández 2013; Ortíz-Pulido and Lara 2014, Fig. 15.12). Short communications have been published about new observations or changes in the distribution of *Psiloscops flammeolus*, *Aegolius acadicus*, *Asio flam-*

Fig. 15.12 Middle American screech owl (*Megascops guatemalae*) individuals in Biosphere Reserve Selva El Ocote, Ocozocoautla, Chiapas, 28 March 2014 (Photograph José Raúl Vázquez Pérez)

Fig. 15.13 Fulvous owl (*Strix fulvescens*) in El Triunfo Biosphere Reserve, Ángel Albino Corzo, Chiapas, 20 May 2013 (Photograph Pedro Ramírez Santos)

meus, *A. stygius*, *A. otus*, *Bubo virginianus*, *Strix fulvescens*, *Megascops trichopsis*, and *Athene cunicularia* (Contreras-Balderas 1991; McAndrews et al. 2006; Rodríguez-Ruiz and Herrera-Herrera 2009; Lavariega et al. 2011; Ramírez-Julián et al. 2011; Rueda-Hernández et al. 2012; Ruvalcaba-Ortega et al. 2014; Estay-Stange et al. 2015, Fig. 15.13).

Studies about the diet or feeding habits are of the topics that have been much studied for some species. Generally the species form pellets that they deposit below their roosts or nests; pellets are collected and analyzed. In temperate environments, pellets can remain for more time before they disintegrate. Therefore, most of the species whose diets have been studied have temperate distribution and include *Tyto furcata*, *Psiloscops flammeolus*, *Megascops kennicottii*, *Bubo virginianus*, *Strix occidentalis*, *Athene cunicularia*, *Aegolius acadicus*, and *Asio flammeus* (Anderson and Nelson 1960; Anderson and Long 1961; López-Forment and Urbano-V 1977; Babb-Stanley et al. 1991; Llinas-Gutierrez et al. 1991; Mejia-Zavala et al. 1991; Ibañez et al. 1992; Arámbula 1994; Morales 1997; Rodríguez-Estrella 1997; Young et al. 1997; Peláez 1998; Roman 1999; Gaona et al. 2000; Aragón et al. 2002;

Márquez 2002; Rodríguez-Vázquez 2002; Valdez-Gómez 2003; Velásquez 2003; Álvarez-Castañeda 2004; Bravo-Vinaja et al. 2005; González-Rojas et al. 2006; Santos-Moreno and Alfaro 2009; Medina-Romero et al. 2008; Valencia-Herveth et al. 2008; Valdez-Gómez et al. 2008; Enríquez et al. 2010). There are very few studies for tropical owl species (*Pulsatrix perspicillata*; Gomez de Silva et al. 1997, and *S. nigrolineata*; Ibañez et al. 1992). The diet of *Megascops barbarus* was determined by studying its feces, not its pellets (Enríquez and Cheng 2008). These studies about diet were done in 14 states (Table 15.2). The species most studied for this topic are *Tyto furcata* and *Strix occidentalis*.

Studies about habitat use have only been done for *Megascops seductus* (Alba-Zuñiga et al. 2009), *M. guatemalae*, *Strix virgata*, and *Glaucidium ridgwayi* (Vázquez-Pérez et al. 2011). These studies were done in Morelos and Chiapas, respectively. The study areas were in the regions of the Sierra de Huautla Biosphere and Selva El Ocote Biosphere Reserve. On the other hand, there is information about vegetation association for 16 owl species, most of which are temperate species like *Strix occidentalis*, *Bubo virginianus*, *Psiloscops flammeolus*, *Aegolius acadicus*, and *Tyto furcata* (Tarango 1994; Tarango et al. 1997; Young et al. 1998; Garza 1999; Tarango et al. 2001; Márquez-Olivas et al. 2002; González-Rojas et al. 2006; Ruiz-Ayma 2010; Fernández 2013). Other species which have been analyzed for such associations are *Megascops kennicottii*, *M. trichopsis*, *M. barbarus*, *M. guatemalae*, *Lophostrix cristata*, *Pulsatrix perspicillata*, *Strix virgata*, *S. nigrolineata*, *Glaucidium ridgwayi*, and *Aegolius ridgwayi* (Rodríguez-Estrella and Careaga 2003; Balan et al. 2008; Enríquez and Cheng 2008; Martínez-Ortega 2009; Vázquez-Pérez 2011; Rivera-Rivera et al. 2012; Fernández 2013). The majorities of these studies were done in protected natural areas and generally were short-term studies completed in less than a year.

Another theme of study in Mexico about owls is reproductive aspects, but there are few and in general for species with temperate distributions. Studies for *Tyto furcata*, *Megascops barbarus*, *Bubo virginianus*, *Glaucidium ridgwayi*, and *Athene cunicularia* have been realized (Rodríguez-Estrella and Hiraldo 1985; Rodríguez-Estrella and Ortega-Rubio 1993; Enríquez and Rangel-Salazar 1996; Enríquez and Cheng 2008; Hernández and Bonilla 2008; Ruiz-Ayma and González-Rojas 2008). These studies were done in the states of Chiapas, Oaxaca, Quintana Roo, Tlaxcala, Nuevo Leon, Jalisco, and Durango, among others (Table 15.2). Other studies on the relationships of environment factors (i.e., illumination, moon phase) associated to the owl vocalization have been considered (Vázquez-Pérez and Enríquez 2017). While we conducted an intensive search for studies about owls completed in Mexico, more unpublished information that is not available could exist.

15.11 Conclusions

Mexico is a country with a high number of owl species, but also with many threats that affect their survival. However, we know little about their ecology and how these threats influence population trends. The majority of the studies in Mexico have been done with Nearctic species, and few have been done with tropical species. Although studies about this group have increased recently, there still exists a lack of information. Following the Official Mexican Law (NOM-059), more than 50% of the species are found at some category of risk. Although protected natural areas are important for species conservation, these areas are insufficient and only cover 13% of the country's area. The increase in open areas and secondary growth has occurred in important areas for raptor conservation. In the case of owl species, many do use these environments. We suggest an increase in the population-level ecological studies, as well as the community level for these raptors in long term to understand ecological patterns, their function, and evolution.

Acknowledgments We thank to Esteban Méndez for your assistance with the different phases of this project and database management, Hermilo Cruz and Mario Zuñiga of the ECOSUR Library for your assistance with the literature search, Emmanuel Valencia of the Geographic Information Laboratory (LAIGE) at ECOSUR for your support with GIS systems, and Pedro Ramírez for your support with the revision of this manuscript. We thank Tomasz B. Falkowski for the translation to English of this chapter. We also thank all the national and international museums which offered information of their Mexican owls collection. We are especially thankful for all the national and international students that have been interested in the study of owls and have contributed to our understanding of this incredibly fascinating group that is little known. Many thanks to everyone.

Ridgway´s Pygmy Owl
(*Glaucidium ridgwayi*)

Appendix 15.1

Owl Species and Subspecies of Mexico

No.	Species	Subspecies	English name	Spanish name	SD	E	ST	NOM-059 (2010)	BirdLife 2012	USFWS	CITES	Distribution
1	*Tyto furcata (alba)*	*pratincola*	Barn owl	Lechuza de campanario	L				LC	E″	Appendix II	H
2	*Psiloscops (Otus) flammeolus*	*flammeolus*	Flammulated screech owl	Tecolote ojioscuro, Tlalquipatli	H		MLD*		LC		Appendix II	M
3	*Megascops kennicottii*	*aikeni, yumanensis, cardonensis, xanthusi, vinaceus*	Western screech owl	Tecolote occidental, Tecolotito chillón	L				LC		Appendix II	H
4	*Megascops asio*	*mccalli*	Eastern screech owl	Tecolote oriental	L			Pr	LC		Appendix II	R
5	*Megascops lambi*		Oaxaca screech owl	Tecolote de Oaxaca		E					Appendix II	R
6	*Megascops cooperi*		Pacific screech owl	Tecolote de Cooper, Tecolotito manglero	M			Pr	LC		Appendix II	R
7	*Megascops trichopsis*	*trichopsis, aspersus, mesoamericanus*	Whiskered screech owl	Tecolote rítmico, Tecolotito manchado, Tecolotito chillón, Zumaya, Ah-coo-akab, Kulte	H				LC		Appendix II	H
8	*Megascops barbarus*		Bearded screech owl	Tecolote barbudo, Tecolotito ocotero, Tecolotito bigotón	M	E*		P	VU		Appendix II	R

9	*Megascops seductus*		Balsas screech owl	Tecolote del Balsas	M	E		A	NT		Appendix II	R
10	*Megascops guatemalae*	*guatemalae, hastatus, cassini*	Middle American screech owl	Tecolote vermiculado, Tecolotito guatemalteco, Tecolotito maullador, Guía de León, Kulte	M				LC		Appendix II	M
11	*Lophostrix cristata*	*stricklandi*	Crested owl	Búho cuerno blanco, Tecolote cuerniblanco, Tecolote crestado	H			A	LC		Appendix II	M
12	*Bubo virginianus*	*pacificus, pallescens, elachistus, mayensis*	Great horned owl	Búho cornudo, Gran duque, Buho grande, Buho real, Ikim, Tunculuchu	L			A*	LC		Appendix II	H
13	*Pulsatrix perspicillata*	*saturata*	Spectacled owl	Búho de anteojos, Lechuza de anteojos	M			A	LC		Appendix II	M
14	*Strix occidentalis*	*lucida*	Spotted owl	Búho manchado	M			A	NT	T	Appendix II	M
15	*Strix fulvescens*		Fulvous owl	Búho lenonado, Lechuzón, Tecolote listado guatemalteco	H			P	LC		Appendix II	R
16	*Strix varia*	*sartori*	Barred owl	Búho listado	M			Pr	LC		Appendix II	M

(continued)

No.	Species	Subspecies	English name	Spanish name	SD	E	ST	NOM-059 (2010)	BirdLife 2012	USFWS	CITES	Distribution
17	*Strix (Ciccaba) squamulata (virgata)*	*squamulata, tamaulipensis, centralis*	Mottled owl	Búho café, Mochuelo rayado, Mochuelo llanero, Lechuza café	M				LC		Appendix II	M
18	*Strix (Ciccaba) nigrolineata*		Black and white owl	Búho blanquinegro, Mochuelo zarado, Lechuza listada	M			A	LC		Appendix II	M
19	*Glaucidium californicum*		Northern pygmy owl	Mochuelo norteamericano, Mochuelo californianio							Appendix II	
20	*Glaucidium hoskinsii*		Cape pygmy owl	Tecolote del Cabo		E		A			Appendix II	R
21	*Glaucidium gnoma*	*gnoma*	Mountain (northern) pygmy owl	Tecolote serrano, Picametate, Tecolotillo duende, Tlalquipatli, Toj-caj-xnuk	M				LC		Appendix II	H
22	*Glaucidium cobanense*		Guatemalan pygmy owl	Tecolote guatemalteco, Mochuelo guatemalteco								R
23	*Glaucidium sanchezi*		Tamaulipas pygmy owl	Tecolote tamaulipeco		E		P	LC		Appendix II	R
24	*Glaucidium palmarum*		Colima pygmy owl	Tecolote colimense		E		A	LC		Appendix II	M

25	*Glaucidium griseiceps*		Central American pygmy owl	Tecolote mesoamericano				A	LC		Appendix II	M
26	*Glaucidium ridgwayi*	*ridgway, cactorum*	Ridgway's pygmy owl	Tecolote bajeño, Tecolotillo rayado, Tecolotillo cuatrojos, Tecolotito rayado, Vieja, Maclovio, Aurorita, Toj-caj-xnuk					LC	E	Appendix II	H
27	*Micrathene whitneyi*	*whitneyi, idonea, sanfordi, graysoni*	Elf owl	Tecolote enano	M	E*	MLD*	E*	LC		Appendix II	H
28	*Athene cunicularia*	*hypugaea, rostrata*	Burrowing owl	Tecolote llanero, Lechucilla llanera, Lechuza de ojo, chicuate, chicuatotol, zacatecolotl	M		MLD*	Pr*, A*	LC	E″	Appendix II	H
29	*Aegolius acadicus*	*acadicus, brodkorbi*	Northern saw-whet owl	Tecolote afilador, Tecolotito cabezón, Lechuza cabezona	M				LC		Appendix II	M
30	*Aegolius ridgwayi*		Unspotted saw-whet owl	Tecolote canelo	M			A	LC		Appendix II	R
31	*Asio stygius*	*robustus*	Stygian owl	Búho cara oscura, Tecolote fusco, Lechuza estigia	M			A	LC		Appendix II	M
32	*Asio otus*	*wilsonianus*	Long-eared owl	Búho cara café, Lechuza barraquera	M		MLD*		LC		Appendix II	M

(continued)

No.	Species	Subspecies	English name	Spanish name	SD	E	ST	NOM-059 (2010)	BirdLife 2012	USFWS	CITES	Distribution
33	*Asio clamator*	*forbesi*	Striped owl	Búho cara clara, Tecolote gritón, Búho cornudo	L			A	LC		Appendix II	M
34	*Asio flammeus*	*flammeus*	Short-eared owl	Búho cuerno corto, Tecolote orejas cortas, Mochuelo	L		MLD	Pr	LC	SC	Appendix II	M

List in taxonomic order follows the American Ornithologists' Union (AOU 1998) y supplements and König et al. (2008). Spanish names follow Birkenstein and Tomlinson (1981) and Escalante-Pliego et al. (1996)

ST = seasonality refers to migratory species where *MLD* = latitudinal migratory and MLD* = with resident populations

SD = sensitivity to disturbed for human activities where *H* = high, M = media, and *L* = low follow Stotz et al. (1996). Blank species are not considered by Stotz et al. (1996)

Endemic species in the country (E) or quasi-endemics (E*) with distribution in Chiapas and Guatemala or Northern Mexico and the United States. Subspecies underlined are endemic to islands

NOM, Mexican Official Norm (NOM-059 ECOL 2010) where *E* = species possible extinct, *P* = peril of extinction, *A* = endangered, *Pr* = special protection and with *, subspecies considered in that category (SEMARNAT 2010)

BirdLife International: *LC* least concern, *NT* near threatened, *VU* vulnerable (BirdLife 2016)

USFWS: *T* threatened, *E* endangered, *E* endangered in some states, *SC* special concern (USFWS 2007)

CITES (the Convention on International Trade in Endangered Species of Wild Fauna and Flora): All the owl species are listed in the Appendix II, which are species not threatened with extinction, but may become so unless trade in specimens is subject to strict regulation in order to avoid utilization incompatible with the survival of the species in the wild

Distribution in Mexico: *R* (restricted to two or three states), *M* (medium, distributed less of 50% of the country), *H* (high, wide distribution in the country)

References

Aguirre-Muñoz A, Mendoza-Alfaro RE, Ponce-Bernal HA et al (2009) Especies exóticas invasoras: impactos sobre las poblaciones de flora y fauna, los procesos ecológicos y la economía. In: Dirzo R, González R, March IJ (eds) Capital natural de México. Vol. II: Estado de conservación y tendencias de cambio. Comisión Nacional para el Conocimiento y Uso de la Biodiversidad, México, DF, pp 277–318

Alba-Zuñiga A, Enríquez PL, Rangel-Salazar JL (2009) Population density and habitat use of the threatened Balsas screech owl (*Megascops seductus*) in the biosphere reserve Sierra de Huautla, México. Endanger Species Res 9:61–66

Álvarez-Castañeda ST, Cárdenas N, Méndez L (2004) Analysis of mammal remains from owl pellets (*Tyto alba*) in a suburban area in Baja California. J Arid Environ 59:59–69

American Ornithologists´ Union (1998) Check-list of north American birds, 7th edn. American Ornithologists' Union, Washington, DC

Anderson S, Long CA (1961) Small mammals in pellets of barn owls from Minaca, Chihuahua (México). Am Mus Novit 2052:3

Anderson S, Nelson CE (1960) Birds and mammals in barn owl pellets from near Laguna, Chihuahua, México. Southwest Nat 5:99–101

Aragón EE, Castillo B, Garza A (2002) Roedores en la dieta de dos aves rapaces nocturnas (*Bubo virginianus* y *Tyto alba*) en el noreste de Durango, México. Acta Zool Mex 86:29–50

Arambula L (1994) Mexican spotted owl distribution and habitat characterizations in southwestern Chihuahua, México. MS Thesis, New México State University, Las Cruces, NM

Arcese P, Sinclair A (1997) The role of protected areas as ecological baseline. J Wildl Manag 61:587–602

Arriaga L, Espinoza JM, Aguilar C et al (eds) (2000) Regiones terrestres prioritarias de México. Comisión Nacional para el Conocimiento y uso de la Biodiversidad, México

Arrona-Rivera AE (2015) Detección de plaguicidas organoclorados en el tecolote bajeño (*Glaucidium brasilianum*) en el Cerro Sonsonate, Chiapas. Tesis de Maestría en Recursos Naturales y Desarrollo Rural. El Colegio de la Frontera Sur, San Cristóbal de Las Casas, Chiapas, México

Arrona-Rivera AE, Enríquez PL, García-Feria LM et al (2016) Organochlorine pesticides in the ferruginous pygmy owl (*Glaucidium brasilianum*) in Chiapas, Mexico. Bull Environ Contam Toxicol. doi:10.1007/s00128-016-1861-0

Arizmendi MC, Márquez-Valdelamar L (eds) (2000) Áreas de Importancia para la Conservación de las Aves en México. CONABIO, México, DF

Babb-Stanley K, González-Alpizar R, Monterrubio-Mendoza JA (1991) Selección de presas y aporte energético de roedores a la dieta de *Tyto alba*. In: Resumen del XI Congreso Nacional de Zoología, Mérida, Yucatán, México

Balán GI, Escalona G, Cabrera P, del Carmen J (2008) Anidación de aves rapaces (Falconiformes y Strigiformes) en el sur del estado de Campeche, México. In: Resumen del CECAM (Congreso para el Estudio y Conservación de las Aves en México) de la Sociedad para el Estudio y Conservación de las Aves en México. Durango, Durango, México

BirdLife International (2012) *Micrathene whitneyi*. The IUCN Red List of Threatened Species. Version 2014.3. http://www.iucnredlist.org. Accessed 23 Feb 2015

BirdLife International (2016) IUCN Red List for birds. http://www.birdlife.org. Accessed 25 Sep 2016

Birkenstein R, Tomlinson E (1981) Native names of Mexican birds. Cross-referenced English/Spanish/Scientific. Resource Publication 139. United States Department of the Interior. Fish and Wildlife Service, Washington, DC

Bravo-Vinaja MG, Tarango-Arámbula LA, Clemente-Sánchez F et al (2005) Composición y variación de la dieta del tecolote moteado mexicano (*Strix occidentalis lucida*) en Valparaíso, Zacatecas, México. Agrociencia 39:509–515

Calva LG, Torres M (1998) Plaguicidas organoclorados. ContactoS 30:35–46

Chávez-Ramírez F (1990) Distribution of the burrowing owls in northern Mexico. Technical Report. Canadian Wildlife Service, Canada

Chazdon RL (2014) Second growth: the promise of tropical forest regeneration in an age of deforestation. The University of Chicago Press, Chicago
Ceballos G, Márquez-Valdelamar L (eds) (2000) Las aves de México en peligro de extinción. Universidad Nacional Autónoma de México. CONABIO y Fondo de Cultura Económica, México, DF
Cirett-Galan JM, Díaz ER (1993) Estatus y distribución del búho manchado mexicano (*Strix occidentalis lucida*) en Sonora, México. Centro Ecológico de Sonora, México
Comité Asesor Nacional sobre Especies Invasoras (2010) Estrategia Nacional sobre especies invasoras en México: Prevención, control y erradicación. Comisión Nacional para el Conocimiento y Uso de la Biodiversidad, Comisión Nacional de Áreas Protegidas, Secretaría de Medio Ambiente y Recursos Naturales. México. http://www.biodiversidad.gob.mx/especies/Invasoras/estrategia.html. Accessed 15 March 2015
Comisión Nacional de Áreas Naturales Protegidas (CONANP) (2014a) Sistema de información geográfica. http://sig.conanp.gob.mx/website/pagsig/informacion/info.htm. Accessed 12 Feb 2014
Comisión Nacional de Áreas Naturales Protegidas (CONANP) (2014b) Áreas Naturales Protegidas. http://www.conanp.gob.mx/. Accessed 12 Feb 2014
Contreras-Balderas A (1991) Second record of the Flammulated owl in Nuevo Leon, México. Wilson Bull 104:375
de Alba E, Reyes ME (1998) Contexto Físico. In: CONABIO (ed) La diversidad biológica de México: Estudio de país. Comisión Nacional para el Conocimiento y Uso de la Biodiversidad, México, DF, pp 4–22
Dent DH (2010) Defining the conservation value of secondary tropical forests. Anim Conserv 13:14–15
Enríquez PL (1997) Seasonal records of the Burrowing Owl in Mexico. J Raptor Res 9:49–51
Enríquez PL (2002) Distribution, abundance and genetic variability of the threatened Bearded Screech Owl (*Otus barbarus*). Lincoln Park Zoo Field Conservation Funds. University of British Columbia and El Colegio de la Frontera Sur. Final Report
Enríquez PL, Cheng KM (2008) Natural history of the threatened Bearded Screech-Owl (*Megascops barbarus*) in Chiapas, México. J Raptor Res 42:180–187
Enríquez PL, Cheng KM, Elliott JE (2010) The near threatened Bearded Screech-owl Megascops barbarus: diet pattern and trophic assessment using δ ^{13}C and δ ^{15}N stable-isotopes. Bird Conserv Int 20:25–33
Enríquez PL, Eisermann K, Mikkola H (2012) Los búhos de México y Centroamérica: necesidades en investigación y conservación. Ornitol Neotrop 23(Suppl):251–264
Enríquez PL, Johnson DH, Rangel-Salazar JL (2006) Taxonomy, distribution and conservation of owls in the neotropics: a review. In: Rodríguez-Estrella R (ed) Current raptor studies in México. Centro de Investigaciones Biológicas del Noroeste y CONABIO, México, pp 254–307
Enríquez PL, Rangel-Salazar JL (1996) Nest site records of the Yucatan Great Horned Owl in Sian Ka'an, México. Ornitol Neotrop 7:149–151
Enríquez PL, Rangel-Salazar JL (2001) Owl occurrence and calling behavior in a tropical rain forest. J Raptor Res 35:107–114
Enríquez PL, Rangel-Salazar JL (2006) Conocimiento popular sobre los búhos en los alrededores de un bosque húmedo tropical protegido en Costa Rica. Etnobiología 4:41–53
Enríquez, PL, Rangel-Salazar JL (2007) The intensity of habitat use by an owl assemblage of a Neotropical rain forest. In: Bildstein KL, Barber DR, Zimmerman A (eds) Proceeding of the II Neotropical Raptor Conference. Hawk Mountain Sanctuary. Raptor Conservation Science Series No. 1. Orwigsburg, PA, pp 88–98
Escalante-Pliego P, Sada AM, Robles Gil J (1996) Listado de nombres comunes de las aves de México. Agrupación Sierra Madre, SC
Escalante-Pliego P, Navarro-Sigüenza AG, Peterson AT (1998) Un análisis geográfico ecológico e histórico de la diversidad de aves terrestres de México. In: Ramamoorthy TP, Bye R, Lot A, Fa J (eds) Diversidad biológica de México: orígenes y distribución. Instituto de Biología, Universidad Nacional Autónoma de México, D. F., México, pp 279–304
Estay-Stange AE, Rodríguez-Estrella R, Bautista Ortega A (2015) Nuevos registros de aves en el Parque Nacional Iztaccíhuatl-Popocatépetl, Puebla, México. Acta Zool Mex 31(3):498–501

Feeley KJ (2010) The conservation value of secondary forests for tropical nocturnal bird species. Anim Conserv 13:16–18

Fernández CE (2013) Abundancia, ocupación y dieta de los búhos en el Área Natural Protegida del Bicentenario Piedra Canteada en San Felipe Hidalgo, Tlaxcala, México. Tesis de licenciatura en Biología. Facultad de estudios superiores Zaragoza. México, DF

Flesch AD (2008) Population and demographic trends of ferruginous pygmy owls in northern Sonora Mexico 2000–2008. Informe Técnico. Arizona Zoological Society Center for Biological Diversity – Global Owl Project Sierra Club – Grand Canyon Chapter

Flores-Dimas CA (2016) Distribución y abundancia de *Micrathene whitneyi* (Aves: Strigidae) en el rancho El Jabalín, Oaxaca, México. Bachelor Thesis. Universidad Autónoma del Estado de México, México

Food and Agriculture Organization (FAO) (2007) Situación de los bosques del mundo 2007. ONU para la Agricultura y la alimentación, Roma

Food and Agriculture Organization (FAO) (2010) Global Forest Resources Assessment 2010. FAO, Roma

Gaona S, Soto-Aquino H, Vega-Domínguez AD (2000) Restos de mamíferos recuperados en regurgitaciones de lechuza *Tyto alba* procedentes de la Hacienda Estipac, Jalisco, México. In: Resumen del V Congreso Nacional de Mastozoología. Mérida, Yucatán, México

García-Hernández J, Sapozhnikova YV, Schlenk D et al (2006) Concentration of contaminants in breeding bird eggs from the Colorado River Delta, Mexico. Environ Toxicol Chem 25:1640–1647

Garza A (1999) Situación actual del búho manchado mexicano *Strix occidentalis lucida* y de los Strigiformes de la Reserva de la Biosfera La Michilía. Informe Técnico. CONABIO, México

Gómez de Silva H, Pérez-Villafana M, Santos-Moreno A (1997) Diet of the spectacled owl *Pulsatrix perspicillata* during the rainy season in northern Oaxaca, México. J Raptor Res 31:385–386

González-Rojas JI, Ruiz-Ayma G, Ruvalcaba-Ortega I et al (2006) Estructura del hábitat, dieta y densidad de la población invernal de la lechuza llanera (*Athene cunicularia hypugaea*) en Galena, Nuevo León, México. In: Resumen del IV Congreso Norteamericano de Ornitología. Veracruz, México

Hernández R, Bonilla CR (2008) Hábitos reproductivos de la lechuza común (*Tyto alba*) en una localidad de valles centrales Oaxaca. In: Resumen del CECAM (Congreso para el Estudio y Conservación de las Aves en México) de la Sociedad para el Estudio y Conservación de las Aves en México. Durango, Durango, México

Howell SNG, Webb S (1995) A guide to the birds of México and northern Central America. Oxford University Press, New York

Ibañez C, Ramo C, Busto B (1992) Notes on food habits of the black and white owl. Conector 94:529–531

INEGI (1992) Estadística del medio ambiente. Instituto Nacional de Estadística, Geografía e Informática, México

Krebs CJ (2001) Ecology: the experimental analysis of distribution and abundance, 5th edn. Benajmin Cumming, San Francisco, CA

König C, Weick F, Becking J-H (2008) Owls of the world, 2ª edn. Christopher Helm, London

Lavariega MC, Martín N, Rodríguez CR et al (2011) Registro del tecolote afilador (*Aegolius acadicus*) en las montañas y valles del occidente de Oaxaca, México. Huitzil 12:48–52

Llinas-Gutiérrez J, Arnaud G, Acevedo M (1991) Food habits of the great horned owl (*Bubo virginianus*) in the cape region of lower California, México. J Raptor Res 25:140–141

López-Forment W, Urbano-V G (1977) The remains of small mammals recovered from owl pellets of *Tyto alba* in México. An Inst Biol Mex Serie Zoología 48:231–242

Márquez Gómez EI (2005) Comparación de dos técnicas para estimar la abundancia del tecolote ojioscuro del Balsas (*Megascops seductus*) en la Reserva de la Biosfera Sierra de Huautla, Morelos, México. Tesis Licenciatura. Facultad de Estudios Superiores Iztacala, UNAM, México, DF

Marcot B (1995) Owls of old forests of the World. General Technical Report PNW-343. U.S. Department of Agriculture, Forest Service, Pacific Northwest Research Station. Portland, OR

Márquez M (2002) Determinación de la dieta del tecolote moteado mexicano (*Strix occidentalis lucida*) en Sierra Fría, Aguascalientes. An Inst Biol Mex 73:205–211

Márquez-Olivas M, Tarango-Arámbula LA, Mendoza-Martínez GD (2002) Caracterización de hábitat del tecolote moteado mexicano (*Strix occidentalis lucida* (x) Nelson, 1903) en Sierra Fría, Aguascalientes. Agressologie 36:541–546
Martínez-Ortega JA (2009) Distribución, abundancia y asociaciones ambientales de un ensamble de búhos en un bosque de montaña en los Altos de Chiapas, México. Tesis de Licenciatura. Benemérita Universidad Autónoma de Puebla, Puebla
Medina-Romero M, García-Becerra A, Ramírez-Avilés Y, et al. (2008) Dieta de la lechuza común (*Tyto alba*) en el Xihuingo, Hidalgo. Resúmenes del CECAM (Congreso para el Estudio y Conservación de las Aves en México). Sociedad para el Estudio y Conservación de las Aves en México, AC (CIPAMEX). Durango, Durango
Mejía-Zavala M, Guzmán-Villa U, Aguilar-Balboa PA (1991) Preferencias alimenticias de la lechuza de campanario *Tyto alba* en la Ciénega de Chapala, Michoacán. Sociedad Mexicana de Zoología, A. C. In: Resumen del XI Congreso Nacional de Zoología, Mérida, Yucatán, México
Méndez E (2013) Asegura la Profepa más de dos mil animales en peligro de extinción. Excélsior. http://www.excelsior.com.mx/nacional/2013/07/01/906747
McAndrews A, Montejo-Diaz J, Tabasco-Contreras M (2006) First confirmed record of the Short-eared owl (*Asio flammeus*) for the state of Yucatan, México. Huitzil 7:32–34
Morales HS (1997) Hábitos alimenticios de la lechuza *Tyto alba* (Orden Strigiformes: Fam. Tytonidae) en la población de San Pedro Chichicasco, Estado de México, México
Navarro-Sigüenza AG, Sánchez-González LA (2003) La diversidad de las aves. In: Gómez de Silva H, Oliveras de Ita A (eds) Conservación de Aves: Experiencias en México. National Fish and Wildlife Fountation, CIPAMEX and CONABIO, México, DF, pp 24–56
Ochoa-Gaona S (2001) Traditional land-use systems and patterns of forest fragmentation in the highlands of Chiapas. México J Environ Manage 27:571–586
Ortiz I, Ávila-Chávez MA, Torres LG (2014) Plaguicidas en México: usos, riesgos y marco regulatorio. RELBAA 4(1):26–46
Ortiz-Pulido R, Lara C (2014) Owls in oak and pine forests in La Malinche National Park, México. Ornitol Neotrop 25:345–353
Palacios E, Anderson DW, Mellink E et al (2000) Distribution and abundance of burrowing owls on the peninsula and islands of Baja California. West Birds 31:89–99
Peláez CA (1998) Patrones de distribución y dieta de *Megascops kennicottii* en áreas de vegetación natural y áreas de influencia humana: ¿Es *Megascops* afectado o beneficiado? Tesis de Licenciatura. UNAM-Iztacala, México
Ramírez P (2014) Distribución y actividad vocal de *Strix fulvescens* en la Reserva de la Biosfera El Triunfo, Chiapas. Tesis de licenciatura en Biología. Instituto Tecnológico del Valle de Oaxaca, Oaxaca
Ramírez-Julián R, González-García F, Reyes-Macedo G (2011) Registro del búho leonado (*Strix fulvescens*) en el estado de Oaxaca, México. Rev Mex Biodivers 82:727–730
Rangel-Salazar JL, Enríquez PL, Altamirano MA et al (2013) Amenazas a la avifauna. In: CONABIO (ed) Amenazas a la Biodiversidad. Volumen I. La Biodiversidad en Chiapas. Estudio de Estado. Comisión Nacional para el Uso y Conocimiento de la Biodiversidad (CONABIO)/ Gobierno del Estado de Chiapas. México, pp 365–369
Rangel-Salazar JL, Enríquez PL, Will T (2005) Diversidad de aves en Chiapas: prioridades de investigación para su conservación. In: González-Espinosa M, Ramírez-Marcial N, Ruiz-Montoya L (eds) La Diversidad Biológica en Chiapas. Plaza y Valdés. México, DF, pp 265–323
Restrepo Cardona JS, Enríquez PL (2014) Conocimiento popular sobre los búhos en poblaciones rurales del suroccidente de Manizales, Caldas, Colombia. Etnobiología 12(3):41–48
Rivera-Rivera E, Enríquez PL, Flamenco-Sandoval A et al (2012) Ocupación y abundancia de aves rapaces nocturnas (Strigidae) en la Reserva de la Biosfera Selva El Ocote, Chiapas, México. Rev Mex Biodivers 83:742–752
Rivera-Rodríguez L, Rodríguez-Estrella R (2011) Incidence of organochlorine pesticides and the health condition of nestlings ospreys (*Pandion haliaetus*) at Laguna San Ignacio, a pristine area of Baja California Sur, México. Ecotoxicology 20:29–38
Rodríguez-Estrella R (1997) Nesting sites and feeding habits of the Burrowing Owl in the Biosphere Reserve of Mapimi, México. In: Lincer JL, Steenhof K (eds) The burrowing owl,

its biology and management including the proceedings of the first international burrowing owl symposium. J Raptor Res Report 9:99–106

Rodríguez-Estrella R, Careaga AP (2003) The western screech-owl and habitat alteration in Baja California: a gradient from urban and rural landscapes to natural habitat. Can J Zool 81:916–922

Rodríguez-Estrella R, Granados JB (2006) Resident and wintering populations of the burrowing owl (*Athene cunicularia*) in México. In: Rodríguez-Estrella R (ed) Current raptor studies in Mexico. Centro de Investigaciones Biológicas del Noroeste y CONABIO, México, pp 181–201

Rodríguez-Estrella R, Hiraldo F (1985) Reproductive biology of the Great Horned Owl (*Bubo virginianus*) in the desert of Mapimi, Durango, México. Raptor Research Found. Symposium on the Management of Birds of Prey, Raptor Research Foundation, Sacramento, CA, 9–10 Nov 1985, pp 4

Rodríguez-Estrella R, Ortega-Rubio A (1993) Nest site characteristics and reproductive success of Burrowing Owls (Strigiformes: Strigidae) in Durango, México. Rev Biol Trop 41:143–148

Rodríguez-Ruíz ER, Herrera-Herrera JR (2009) A un siglo del registro del búho cara oscura (*Asio stygius*) en el centro de Tamaulipas y notas sobre su distribución en México y los Estados Unidos de América. Huitzil 10:56–60

Rodríguez Vázquez FP (2002) Hábitos alimenticios de la lechuza de campanario *Tyto alba* en Xochitla, Tepotzotlán, Estado de México. Tesis Licenciatura. Facultad de Estudios Superiores Iztacala. UNAM, México

Román FI (1999) Variación estacional en la dieta de *Tyto alba* (Lechuza común) en el desierto Vizcaino, Baja California Sur, México. Tesis de Licenciatura. Facultad de Ciencias. Universidad Nacional Autónoma de México, México

Rueda-Hernández R, Ruiz-Sánchez A, Herrera-Alsina L (2012) Primer registro del búho cornudo (*Bubo virginianus*) para la ciudad de Xalapa, Veracruz. Huitzil 13:169–172

Ruiz-Aymá G (2010) Caracterización de madrigueras de anidación del tecolote llanero (*Athene cunicularia*) en pastizales del sur del Estado de Nuevo León, México. In: Resumen del X Congreso para el Estudio y Conservación de las Aves en México (CECAM). Sociedad para el Estudio y Conservación de las Aves en México, AC (CIPAMEX). Jalapa, Veracruz

Ruiz-Aymá G, González-Rojas JI (2008) Éxito reproductivo del tecolote llanero (*Athene cunicularia hypugea*), en pastizales de Galeana, Nuevo León, México. In: Resumen del CECAM (Congreso para el Estudio y Conservación de las Aves en México) de la Sociedad para el Estudio y Conservación de las Aves en México. Sociedad para el Estudio y Conservación de las Aves en México, AC (CIPAMEX). Durango

Ruvalcaba-Ortega I, Salinas-Rodríguez MM, Cruz-Nieto J et al (2014) First nesting record of the long-eared owl (*Asio otus*) for Chihuahua, Mexico. Southwest Nat 59(1):135–138

Rzedowski J (2006) Vegetación en México. Comisión Nacional para el Conocimiento y Uso de la Biodiversidad. México. 1ra. Edición digital. México DF México

Santos-Moreno A, Alfaro AM (2009) Mammalian prey of Barn Owl (*Tyto alba*) in southeastern Oaxaca, México. Acta Zool Mex 25:143–149

Sosa-Escalante JE (2011) Aplicación de la ley para el combate del tráfico ilegal de vida silvestre en México: El caso de Charco Cercado. Therya 2(3):245–262

Sekercioglu CH (2010) The mystery of nocturnal birds in tropical secondary forests. Anim Conserv 13:12–13

SEMARNAT (2010) Norma Oficial Mexicana 059- SEMARNAT 2010. Protección ambiental especies nativas de México de flora y fauna silvestres. Categorías de riesgo y especificaciones para su inclusión, exclusión, o cambio. Lista de especies en riesgo. Diario Oficial de la Federación. 30 diciembre 2010. México DF

SEMARNAT (2016) Reserva de la Biósfera Archipiélago de Revillagigedo. http://www.gob.mx/semarnat/articulos/reserva-de-la-biosfera-archipielago-de-revillagigedo. Accessed 05 Dec 2016

Soto L, Anzueto Martínez M, Quechulpa S (2011) El acahual mejorado un prototipo agroforestal. Ecosur, Redisa. San Cristóbal de Las Casas, Chiapas. México

Stotz DF, Fitzpatric JW, Parker TA III et al (1996) Neotropical birds: ecology and conservation. Conservation International and the Fiel Museum of Natural History. The University of Chicago Press, Chicago and London

Tarango LA (1994) Mexican spotted owl distribution and habitat characterization in southwestern Chihuahua, Mexico. MS thesis, New Mexico State University Las Cruces New Mexico USA

Tarango LA, Valdez R, Clemente F et al (2001) Roost-site characteristics of Mexican Spotted Owls in Sierra Fria, Aguascalientes, México. J Raptor Res 35:165–168

Tarango LA, Valdez R, Zwank PJ et al (1997) Mexican spotted owl habitat characteristics in southwestern Chihuahua, Mexico. Southwest Nat 42:132–136

Thiollay JM (1985a) Composition of falconiformes communities along successional gradients from primary rainforest to secondary habitats. In: Newton I, Chancellor RD (eds) Conservation Studies on raptors. International Council for Bird Preservation ICBP Technical Publication 5, Cambridge, pp 181–197

Thiollay JM (1985b) Falconiforms of tropical rainforest: a review. In: Newton I, Chancellor RD (eds) Conservation Studies on raptors. International Council for Bird Preservation ICBP Technical Publication 5, Cambridge, pp 155–165

Thiollay JM (1994) A world review of tropical forest raptors current trends, research objectives and conservation. In: Meyburg BU, Chancellor RD (eds) Raptor Conservation Today. WWG BP/The Pica Press, Mountfield, UK, pp 231–239

UNAM (1990) Atlas Nacional de México. Instituto de Geografía, UNAM. México, DF

US Fish and Wildlife Service (2007) Updates list of candidates for endangered species act listing. Department of the Interior. http://www.fws.gov/endangered/news/archives/2007.html. Accessed 15 March 2014

US Fish and Wildlife Service (2012) Final recovery plan for the Mexican Spotted Owl (*Strix occidentalis lucida*). U.S. Fish and Wildlife Service, Albuquerque, New Mexico

Valdéz-Gómez HE (2003) Dieta del tecolote llanero occidental, *Athene cunicularа hypugaea* (Bonaparte, 1825), durante su estancia invernal en el Bajío Mexicano. Tesis de Licenciatura. Universidad de Guadalajara, Centro Universitario de Ciencias Biológicas y Agropecuarias División de Ciencias Biológicas y Ambientales. Zapopan, Jalisco

Valdéz-Gómez H, Enríquez PL (2005) Ficha técnica de *Athene cunicularia*. In: Escalante-Pliego P (ed) Fichas sobre las especies de Aves incluidas en el Proyecto de Norma Oficial Mexicana PROY-NOM-ECOL-2000. Parte 2. Instituto de Biología, Universidad Nacional Autónoma de México. Bases de datos SNIB-CONABIO. Proyecto No. W042. México DF

Valdéz Gómez HE, Holroyd GL (2000) El tecolote llanero, sus hábitos y distribución en el centro oeste de México. Boletín de la Sociedad de Ciencias Naturales de Jalisco 1:57–63

Váldez-Gómez HE, Holroyd G, Trefry HE, et al. (2008) Ecología invernal del búho cuerno corto (*Asio flammeus*) en el municipio de Zapopan, Jalisco. In: Resumen del CECAM (Congreso para el Estudio y Conservación de las Aves en México) de la Sociedad para el Estudio y Conservación de las Aves en México. Durango, Durango, México

Valencia-Herverth J, Martínez S, Ignacio J et al (2008) Composición de la dieta del búho cornudo (*Bubo virginianus*) en la laguna de Santiaguillo, Nuevo Ideal, Durango, México. In: Resumen del CECAM (Congreso para el Estudio y Conservación de las Aves en México). Sociedad para el Estudio y Conservación de las Aves en México, AC (CIPAMEX). Durango, México

Valencia-Herverth J, Ortiz-Pulido R, Enríquez PL (2012) Riqueza y distribución espacial de rapaces nocturnas en Hidalgo, México. Huitzil 13:116–129

Vázquez-Pérez JR (2011) Densidad y uso de hábitat de búhos en la Selva El Ocote, Chiapas. Tesis de Maestría en Recursos Naturales y Desarrollo Rural. El Colegio de la Frontera Sur, San Cristóbal de las Casas, Chiapas, México

Vázquez-Pérez JR, Enríquez PL, Rangel-Salazar JL et al (2011) Densidad y uso de hábitat de búhos en la Reserva de la Biosfera Selva el Ocote, Chiapas, Sur de México. Ornitol Neotrop 22:577–587

Vázquez-Pérez JR, Enríquez PL (2017) Factores temporales y ambientales asociados a los llamados de búhos en la Selva el Ocote, Chiapas, México. El Hornero (in press)

Velásquez JS (2003) Dieta de *Tyto alba* Scopoli (Aves: Tytonidae) en la Cueva San Francisco, Municipio de la Trinitaria, Chiapas. Universidad de Ciencias y Artes de Chiapas. Escuela de Biología. Tuxtla Gutiérrez, Chiapas, México

Whittaker RJ, Willis KJ, Fiels R (2001) Scale and species richness: towards a general, hierarchical theory of species diversity. J Biogeogr 28:453–470

Young KE, Valdez R, Zwank PJ et al (1998) Density and roost site characteristics of spotted owls in the Sierra Madre Occidental, Chihuahua, Mexico. Conector 100:732–736

Young KE, Zwank PJ, Valdez R et al (1997) Diet of Mexican Spotted Owls in Chihuahua and Aguascalientes, México. J Raptor Res 31:376–380

Chapter 16
The Owls of Nicaragua

Ana Trejo and Martín Lezama-López

Abstract Despite being the largest country in Central America and having approximately 700 species of birds, Nicaragua's avifauna is probably the least known. Consensually, 14 owl species have been registered for the country, with representatives of both Nearctic and Neotropical avifauna. In addition to the incomplete knowledge of their distribution, nothing is known about the biology and ecology of the species in the country. Most owls in Nicaragua are typical of the forested areas that cover more than 40% of the total land area. This is the most threatened habitat due to the expansion of the agricultural frontier and pressure from human displacement into forested areas. In order to conserve biodiversity in general, Nicaragua has a system of protected areas (SINAP), including three important biosphere reserves. No species of owl found in Nicaragua is endemic or classified as globally at risk. However, we feel that an assessment of the conservation status of each species at a local level is needed. This requires a more detailed study of their distribution, but also an estimation of their abundance and the status of their populations, and habitat requirements and other aspects of their biology.

Keywords Status of knowledge • Owl conservation • Tradition • Beliefs

A. Trejo (✉)
Centro Regional Bariloche, Universidad Nacional del Comahue, 8400 Bariloche, Argentina
e-mail: ana.r.trejo@gmail.com

M. Lezama-López
Ecología y Manejo de Vida Silvestre, Casa Real, III etapa, casa 2G-26, Managua, Nicaragua
e-mail: nicapinol2002@yahoo.com

P.L. Enríquez (ed.), *Neotropical Owls*, DOI 10.1007/978-3-319-57108-9_16

Mottled Owl (*Ciccaba virgata*)

16.1 Introduction

Birdlife in Nicaragua consists of approximately 700 species (even though this number is constantly changing due to new research), and this listing of species is facing a period of inventory needed as a result of political, historical, and social changes (Martínez Sánchez 1990; Guillespie 2001; Wiedenfeld et al. 2001). Therefore, there is little information about general aspects of basic biology and ecology. Also, some regions (like tropical rainforests and cloud forests especially) have not been very well explored regarding their birdlife (Martínez Sánchez 1990). Almost half of Nicaraguan territory is covered by forest (Rueda Pereira 2007). In this bird diversity, owls belong to an unknown group due to their being nocturnal birds and usually inconspicuous. Most of owls' species in Nicaragua are distributed in forested habitats, and these species are, as in other parts of the Neotropics, the least known (Trejo and Bó 2015).

The knowledge of bird species in Nicaragua is still incomplete. However, it is reasonable that the efforts of many bird studies be conducted to try to establish conservation priorities and to study species with a great value of commercial and economic interest. If bird species' richness is insufficiently known, different aspects such as distribution, abundance, population ecology, diet, and reproductive behavior represent a significant lack of information about owl populations. Besides the lack of information that is caused by logistics procedures, it must be added that Strigiformes are of no economic interest because they are a threat to neither agriculture nor cattle raising. Furthermore, they have no commercial value as pets or interest as a source of animal proteins. Countries like Nicaragua that are characterized by their small-scale economy and medium levels of Human Development Index (HDI) for decades (United Nations Development Programme; UNDP 2010) view an investment in resources to study species outside of the priority lines as being a wasteful expense.

16.2 Taxonomic Diversity

Birds in Nicaragua are not very well known in spite of the ornithological importance of having species found in both Nearctic and Neotropical regions. In this review the list of birds of Nicaragua was taken as a base (Martínez Sánchez 2007; American Ornithologist Union AOU 2011). This list reports 14 Strigiformes species for Nicaragua (Table 16.1). None of these species are endemic to Nicaragua.

Two species are considered uncertain species: short-eared owl (*Asio flammeus*) and burrowing owl (*Athene cunicularia*) (Will 2004). Even though there are not resident populations of these two species, it is possible that both of them are occasionally present in Nicaragua. There are some records where *A. flammeus* has been observed in some other countries of Central America (Jones et al. 2000; Obando et al. 2009), and it may be that some are migrant individuals from North America. It is known that northern populations of these species are sudden migrants (Newton 2006) as a result of fluctuations of the food supply in their breeding territory. As for *A. cunicularia*, it is known that there are some resident populations in Honduras (Mejía Ordoñez and House 2009). There are records of this population migrating to the south (König et al. 1999). With regard to Nicaragua, it would not be strange to find records of individuals as well which has been found in Costa Rica (Obando et al. 2009).

16.3 Distribution and Habitat Association

Nicaragua is located in Central America, bordering Honduras to the north, Costa Rica to the south, the Atlantic Ocean to the west, and the Pacific Ocean to the east. Nicaragua covers a total area of 140,000 km^2 which makes it the largest country in Central America. Geographically, Nicaragua is divided into three regions with different environmental and demographic features (Taylor 1963; Incer 2002): Pacific, Central, and Atlantic region (Fig. 16.1).

Firstly, the Pacific region is flat; it is the driest and most deforested area between the three regions which originally was covered with tropical xeric forest, requiring very

Table 16.1 Strigiformes species registered for Nicaragua, according to Martínez Sánchez (2007). Common names follow Martínez Sánchez (2007). Names in English follow the American Ornithologists' Union (2011)

Species	Common name	Name in English	Habitat	Region
Tyto alba	Lechuza común	Barn owl	U, AZ, FB, DF[1], RF[2], CF[3]	A, C, P
Megascops cooperi	Tecolotito sabanero	Pacific screech owl	AZ, FB, DF, CF	C, P
Megascops trichopsis	Tecolotito manchado	Whiskered screech owl	FB, PF	C, P
Megascops guatemalae	Tecolotito vermiculado	Vermiculated screech owl	FB, CF	C
Lophostrix cristata	Búho penachudo	Crested owl	DF, CF	C
Pulsatrix perspicillata	Búho de anteojos	Spectacled owl	FB, DF, RF, W[4]	A, P
Bubo virginianus	Búho grande	Great horned owl	FB, CF, PS	A, C, P
Glaucidium gnoma	Mochuelo serrano	Northern pygmy owl	FB, CF	C
Glaucidium griseiceps	Mochuelo centroamericano	Central American pygmy owl	FB, RF	A
Glaucidium brasilianum	Mochuelo herrumbroso	Ferruginous pygmy owl	AZ, FB, RF, CF, DF, W, PS[5]	A, C, P
Ciccaba virgata	Cárabo café	Mottled owl	FB, CF, DF	A, C, P
Ciccaba nigrolineata	Cárabo negriblanco	Black-and-white owl	RF, DF	A, P
Asio stygius	Búho oscuro	Stygian owl	FB, CF	C
Pseudoscops clamator	Búho listado	Striped owl	FB, RF, DF, AZ[6]	A, P

The different kinds of habitats are explained for each species according to Martínez Sánchez (2007): *AZ* agricultural zones, open areas in general, scrublands, and dispersed trees that usually suffer the consequences of fire during the dry season. *FB* forests' borders and intervened areas by secondary growth. *RF* tropical rainforest from lowlands, broadleaf forests, and rainforest. *CF* cloud forest or tropical premontane forest. *DF* dry forest or deciduous forest. *W* wetland in general such interns as coastal zone. *PF* pure pinewood forest and mixed formations of pine-oak in mountain zone. *PS* pinewood savannah forest of low lands and grass formations. *U* urban centers, hamlet. In case of observation in habitat not mentioned by this author, the corresponding bibliographic references are indicated. Habitat types are detailed in the text. *A* Atlantic, *C* Central, *P* Pacific
[1]Castañeda et al. (2004) record *T. alba* in the xerophyte forest of Chocoyero-El Brujo Natural Reserve. [2]Howell (1957) cited *T. alba* in a palm tree plantation, and other species for commercial exploitation were located in the Atlantic region rainforest. [3]Chavarría and Durieux (2007) report *T. alba* in El Jaguar Natural Reserve cloud forest. [4]Flores Rocha (2000) registers *P. perspicillata* in riparian forest around the Atlantic Coast. Also, a specific species of *Glaucidium* (*G. minutissimum*) around swamps in the same area. M. Lezama registered twice *P. perspicillata* in freshwater wetlands, close to Cocibolca Lake, coincidently with Nutting (1884) in Sucuyá, east coast from the lake. [5]Howell (1971, 1972) recorded *G. brasilianum* presence in pine savannah in the Atlantic region, though closer to a tropical forest area. [6]Pérez et al. (2005) cited *P. clamator* in silvopasture systems in Matagalpa

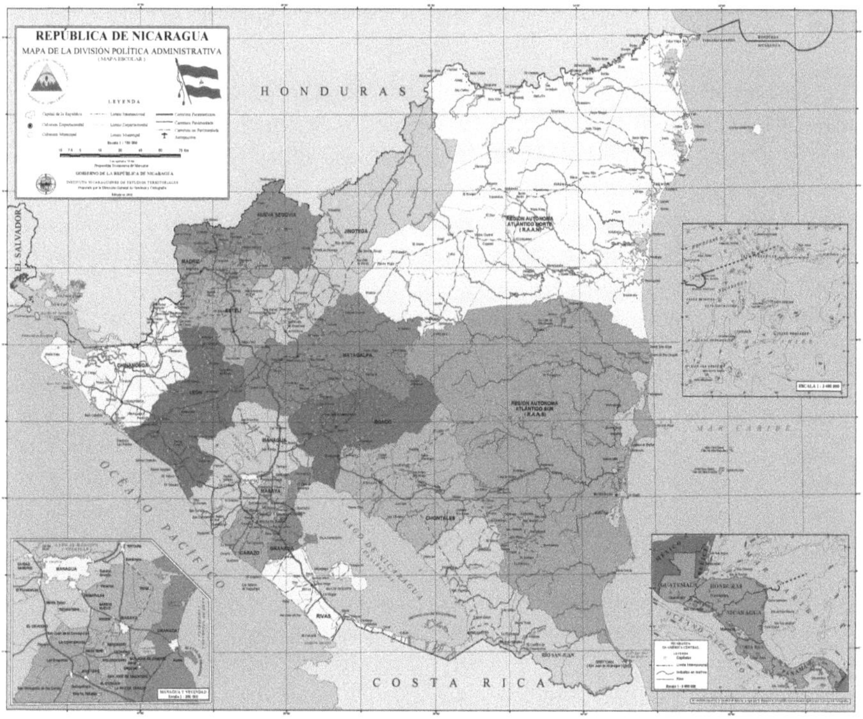

Fig. 16.1 Map of Nicaragua (Source: Nicaraguan Territorial Studies Institute (INETER), www.ineter.gob.ni)

little moisture, but currently contains many cultivated fields and pasture land. Secondly, the Central Region is separated from the Pacific region by the lake of Nicaragua (Cocibolca) and Managua (Xolotlán). This region includes the principal mountain ranges to the north (having a height above 2000 m). The sector NW has pine-oak forests, while the higher mountains have cloud forests in fragmented distribution. The southern land becomes flatter and is dominated by pasture land. Thirdly, the Atlantic region is more extensive; it is covered mostly by tropic humid forests and coniferous forests. Sparsely populated, the total distance of the East-West border of this region is also the border of the farming activity. The coastal zone has different wetland types.

Owls are currently distributed in all three regions in Nicaragua and in all the various habitats (Table 16.1). Barn owl (*Tyto alba*) is the species with the highest range of habitat coverage, including urban habitat. According to the consulted source (see footnote, Table 16.1), there are habitats with no owl records, such as cloud forest and pinewood of savanna, wetlands, and mountain pinewood, although the absence could be changed with intensive studies. On the other hand, ferruginous pygmy owl (*Glaucidium brasilianum*) is another species distributed widely in different habitats.

16.4 National Conservation and Conservation Strategies

The majority of owls' species in Nicaragua are found in forest areas that cover more than 40% of the land surface (Rueda Pereira 2007). Most of the existing forests consist of tropical rainforest and pine forest. Tropical dry forest (which covered a great part of the Pacific region in the past) has almost totally disappeared (FRA 2000). From the total extension of forest ecosystem, approximately 30% are under protection of the National System of Protected Areas (SINAP). The greatest threat to owl populations is the depletion and fragmentation of our forests.

Deforestation in Nicaragua began in the pre-Columbian era for the Pacific tropical dry forest, and it has expanded to the east, mainly due to the push to the east of the farming and livestock production. Other causes are wood dependency for energy consumption, inefficient forest industry, and productive land concentration in just a few people, provoking farmer migration, weak forestal legalization, and poor inter-institutional coordination (FRA 2000). Also Nicaragua's geographic position is susceptible to hurricanes and tropical storms which have devastating effects in the forest regions (Yih et al. 1991).

Nicaragua relies on a protected area system (SINAP) which is administrated by MARENA (Ministry of Environment and Natural Resources), protecting approximately 20% of the national territory. SINAP is organized in 72 areas of different management categories (MARENA 2011). Besides this, SINAP also manages the private wildlife reserves with legalities and administrative requirements for its development. Also Nicaragua has three biosphere reserves (UNESCO): Bosawás, Río San Juan (Nicaragua), and Ometepe.

The cloud forest of the Central Region has been reduced in the recent years due to the great increase in the farming industry, especially by the coffee plantations. However, these are special problems, because at the same time, the human communities, the agricultural industry, and the natural reserves all have special concerns which must be taken into consideration (Cooper 2007). The noncultivated lands are specially (sometimes literally) forest reserves. "La Reserva Silvestre Privada el Jaguar" is an example of such a situation. The barn owl (*T. alba*) has been registered in wide open areas, and the crested owl (*Lophostrix cristata*), mottled owl (*Ciccaba virgata*), and pygmy owl (*Glaucidium sp.*) have been recorded in forest borders (Chavarría y Duriaux 2007). Other species registered in coffee plantations are the vermiculated screech owl (*Megascops guatemalae*), stygian owl (*Asio stygius*), northern pygmy owl (*G. gnoma*), and ferruginous pygmy owl (*G. brasilianum*; Muñoz 2004; Cooper 2007).

All owls' species found in Nicaragua are categorized as "least concern" in the Red List from IUCN, except the Central American pygmy owl (*G. griseiceps*), the mottled owl (*C. virgata*), and the black-and-white owl (*C. nigrolineata*) whose populations have not been evaluated (IUCN 2012). However, we consider the conservation status of every species to be important, in order to determine the minimum ecological requirements for their existence. This will require studies in every environment of the country. Studies would be important to determine the abundance as well as the natural history of each species.

Guatemalan Screech Owl (*Megascops guatemalae*)

References

American Ornithologist Union (AOU) (2011) Check-list of North American birds. www.aou.org/checklis/north. Accessed 02 May 2015

Castañeda E, Medina A, Gámez JC (2004) El uso de la avifauna como herramienta para la conservación de áreas naturales de la Reserva Natural Chocoyero-El Brujo. Encuentro 36:7–24

Chavarría L, Durieux G (2007) Lista de las aves de la Reserva Silvestre Privada El Jaguar. ALAS (Alianza para las Áreas Silvestres), Managua

Cooper DS (2007) Ecological assessment of five coffee farms in north-central Nicaragua. Cooper Ecological Monitoring Inc., Pasadena

Flores Rocha O (2000) Evaluación preliminar de las diversas aves y sus hábitats en la zona de influencia de PROCODEFOR. Bluefields, R.A.A.S. Tesis de Licenciatura, Universidad Centroamericana, Managua

FRA (Programa de Evaluación de Recursos Forestales) (2000) Bibliografía comentada. Cambios en la cobertura Forestal, Nicaragua. Organización de las Naciones Unidas para la Agricultura y la Alimentación (FAO), Programa para la evaluación de Recursos Naturales (FRA). Doc.34. Roma, Italy

Guillespie TW (2001) Application of extinction and conservation theories for forest birds in Nicaragua. Conserv Biol 15:699–709

Howell TR (1957) Birds of a second growth rain forest area of Nicaragua. Condor 59:73–111

Howell TR (1971) An ecological study of the birds of the lowland pine savanna and adjacent rain forest in northeastern Nicaragua. The Living Birds, tenth annual: 185–242

Howell TR (1972). Birds of the lowland pine savanna of northeastern Nicaragua. Condor 74: 316–340

Incer J (2002) Geografía Básica de Nicaragua. HIspamer, Managua

IUCN (International Union for the Conservation of Nature) (2012) IUCN Red List of Threatened Species. http://iucn.redlist.org. Accessed 05 May 2015

Jones HL, McRae E, Meadows M et al (2000) Status updates for selected birds species in Belize, including several species previously undocumented from the country. Cotinga 13:17–31

König C, Weick F, Becking JH (1999) Owls. A guide to the owls of the world. Yale University Press, New Haven/London
MARENA (Ministerio del Ambiente y los Recursos Naturales) (2011) http://marena.gob.ni. Accessed 29 Jun 2014
Martínez Sánchez JC (1990) Biodiversidad en Nicaragua: estado actual del conocimiento sobre la fauna vertebrada. World Wildlife Fund, Washington, DC
Martínez Sánchez JC (2007) Lista patrón de las aves de Nicaragua. Con información de nuevos registros, distribución y localidades donde observar aves. Alianza para las aves Silvestres (ALAS), Managua
Mejía Ordóñez TM, House PR (eds) (2009) Especies de preocupación especial en Honduras. Secretaría de Recursos Naturales y Ambiente, Tegucigalpa
Muñoz FJ (2004) Investigación y sistematización de flora y fauna en zonas cafetaleras. Informe de consultoría para CAFENICA, Estelí, Nicaragua
Newton I (2006) Advances in the study of irruptive migration. Ardea 94:435–460
Nutting CC (1884) On a collection of birds from Nicaragua. Proc US Nat Mus 6:372–410
Obando Calderón G, Chaves Campos J, Garrigues R et al (2009) Lista Oficial de las aves de Costa Rica, Bol. Asoc. Ornitol. Costa Rica. San José, Costa Rica
Pérez AM, Sotelo M, Ramírez F et al (2005) Composición y diversidad en las comunidades de aves, moluscos y plantas asociadas con sistemas silvopastoriles de Matiguás y Río Blanco, Depto. de Matagalpa, Nicaragua. Encuentro (UCA) 71:1–36
PNUD (Programa de Naciones Unidas para el Desarrollo) (2010) Informe sobre el desarrollo humano 2010. La verdadera riqueza de las naciones: caminos al desarrollo humano. PNUD, New York
Rueda Pereira R (2007) Recopilación de la información sobre la biodiversidad de Nicaragua. Universidad Autónoma de Nicaragua-León, León
Taylor BW (1963) An outline of vegetation of Nicaragua. J Ecol 51:27–54
Trejo A, Bó MS (2015) Los búhos de Argentina. In: Enríquez PL (ed) Los Búhos Neotropicales: Diversidad y Conservación. El Colegio de la Frontera Sur, San Cristóbal de Las Casas, Chiapas México, pp 43–58
Wiedenfeld DA, Morales J, Lezama M (2001) Sight records of new species for Nicaragua and noteworthy records on range and occurrence. Continga 15:53–57
Will T (2004) Nica Checklist worksheet. U.S. Fish and Wildlife Service. Report. USA
Yih K, Boucher DH, Vandermeer JH et al (1991) Recovery of the rain forest of southeastern Nicaragua after destruction by hurricane Joan. Biotropica 23:106–113

Chapter 17
The Owls of Panama

Belkys Jiménez-Ruíz, Karla Aparicio-Ubillúz, Francisco Delgado-Botello, and Isis Tejada

Abstract Using existing information from museum collections, field studies, and bibliography regarding barn owls and typical owls of Panama, we have succeeded in documenting the status and ecology of at least 13 of the 15 species of the order Strigiformes across the country. In Panama, the order Strigiformes is represented by 15 species (1 barn owl and 14 typical owls), two of which are considered uncertain (*Bubo virginianus* and *Athene cunicularia*) because of scarce evidence. The first species has been reported in at least two protected areas and the second in none, thence the importance of further research studies in order to confirm the existence of these species in those areas. In terms of conservation, one species (*Bubo virginianus*) is considered endangered and the rest are considered vulnerable under the Panamanian laws of conservation. *Megascops clarkii*, *Glaucidium costaricanum*, and *Aegolius ridgwayi* are species restricted to the cloud forests of Panama, which are protected habitats in the eastern and western regions of Panama (reported in seven, six, and two areas, respectively). In contrast, species like *Megascops choliba* and *M. guatemalae, Lophostrix cristata, Pulsatrix perspicillata*, *Ciccaba virgata*, and *C. nigrolineata* are widely distributed in lowlands and foothills and are reported in at least 25 protected areas. Others, such as *Pseudoscops clamator, Tyto alba, Glaucidium griseiceps*, and *G. brasilianum*, are also widely distributed in lowlands and foothills of both Pacific and Atlantic slopes; these species are reported in at least 17, 14, 20, and 7 protected areas, respectively. None of the management plans reviewed presented conservation measures for the Strigiformes, thus making evident the need to implement

B. Jiménez-Ruíz (✉) • K. Aparicio-Ubillúz
Fundación Naturaleza y Ciencia 507, Urb. Altos del Rio, Calle C, No.87, Rio Abajo, Panamá
e-mail: b0311ys@yahoo.com

F. Delgado-Botello
Centro Regional Universitario de Veraguas, Departamento de Ciencias Naturales Exactas y Tecnología, Calle 10, Vía San Frsncisco, Santiago, Provincia de Veraguas, República de Panamá
e-mail: direccioncruv@hotmail.com; francisco.delgado2410@gmail.com

I. Tejada
La Chorrera, Bda. El Campesino #1380[a], Panamá, República de Panamá

P.L. Enríquez (ed.), *Neotropical Owls*, DOI 10.1007/978-3-319-57108-9_17

research studies with objectives that underpin programs for conservation and reasonable use of the national and private forest areas throughout the country, consequentially protecting the associated biodiversity, particularly avifauna.

Keywords Conservation • Diversity • Owl distribution • Knowledge status

Spectacled Owl (*Pulsatrix perspicillata*)

17.1 Introduction

Panama is located in the middle of the Western Hemisphere, between the coordinates 7°12′07″ and 9°38′46″ north latitude and 77°09′24″ and 83°03′07″ west longitude. It borders the Caribbean Sea in the north, the Republic of Colombia in the east, the Pacific Ocean in the south, and the Republic of Costa Rica in the west. It constitutes a link between Central and South America, forming an isthmus 80

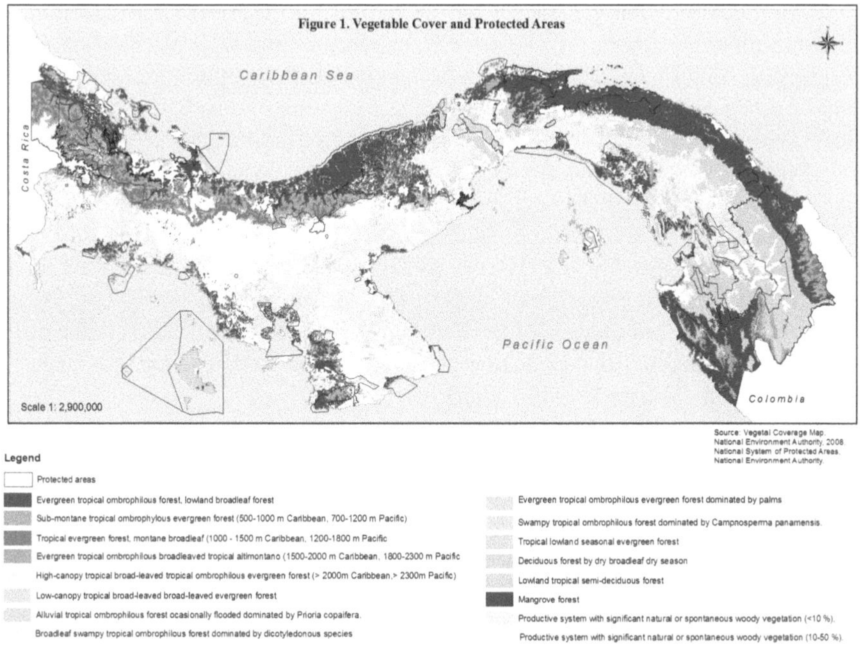

Fig. 17.1 Protected areas and forest cover of the Republic of Panama

kilometers wide in its narrowest section. Panama's total area is 75,517 km^2 (29,208 sq. mi.) and displays a mountainous topography toward the Caribbean coast, with smooth hills and large savannas toward the Pacific Ocean. Geographically, the country is divided into nine provinces and three indigenous territories called "comarcas" (Kuna Yala, Emberá, and Ngobe-Buglé) (Panamanian Institute of Tourism 2009) (Fig. 17.1). Birds in Panama have been studied for over a century, but in spite of being one of the best-known countries in terms of ornithological information, there is still very much to discover, some regions barely explored, and some little-known bird groups. For instance, no studies have been made specifically targeted at the order Strigiformes, even though there is distribution and ecology information within the collections made throughout the country dated back to the middle of the nineteenth century, which can be found mostly in North American and British museums. According to Ridgely and Gwynne (1993), in the mid-nineteenth century, James McLeannan, Chief of the Lion Hill Railroad Station in Panama (now lying under the waters of the Panama Canal), was the first to make a bird collection. Other collectors worked in different areas of Panama, like Enrique Arce in "Veragua," a region located in what is called western Panama (Bocas del Toro, Chiriquí, and Veraguas), and all that information was included in the volumes of Aves de la Biología Centrali-Americana written by O. Salvin and F DuC. Godman (1879–1904). But it was really the interest of some North Americans, spurred by the construction of the Canal, what prompted the intensive zoological exploration, giving way to the popular interest.

Collections like those made by W. W. Brown, T. Barbour, E. A. Goldman, L. Griscom, R. R. Benson, H. D. Bedel, H. E. Anthony and W. B. Richardson, F. H. Kennard, J. Aldrich, L. L. Jewel, and others were the starting point for the demarcation of the distribution of birds of Panama, a collection of data that was published later by them as well as by well-known ornithologists like O. Bangs, F. M. Chapman, J. L. Peters, and C. Hellmayr. An important study, published in several volumes, was that written by Robert Ridgway, *The Birds of North and Middle America* (1901–1950), which set the descriptive and taxonomic base for subsequent studies in Central America (Ridgely and Gwynne 1993). Other important studies were those made by W. Stone (the first list of birds of the old Canal Zone, 1918), L. Griscom (the first complete list of birds of the Republic, 1935), R. C. Murphy (marine birds of the Gulf of Panama), and E. R. Blake (1958), who worked in collections made by T. B. Mönniche (Volcan Massif in Chiriquí).

Ecological and behavioral studies were promoted upon the establishment of the biological research station in Barro Colorado Island, in the old Canal Zone (1923) under the supervision of J. Zetek. Some outstanding works are those of B. B. Sturgis, the *Field Book of Birds of the Panama Canal Zone* (1928); those of F. M. Chapman, *My Tropical Air Castle* (1929) and *Life in an Air Castle* (1938); and those of A. Skutch (1954–1960–1967), *Life Histories of Central American Birds*.

Later on, around 1940, collections made by A. Wetmore throughout the country contributed to the compilation of information that ended up with the publication of several volumes of *The Birds of the Republic of Panama* (1965, 1968, 1973); E. Eisenmann (1955) also made significant contributions regarding the ecology, distribution, and behavior of birds as well as E. Mendez (1969) did with "Una breve introducción a las aves de Panamá" and Edwards and Loftin (1971) with "Finding birds in Panama." This was then followed by Eisenmann and Loftin with the *Field checklist of birds of Panama Canal Zone Area* and the *Field checklist of birds of Western Chiriquí Highlands, Panama* (1972a, b). Other significant works were those of R. S. Ridgely along with J. A. Gwynne, based on the unpublished work of Eisenmann "Diagnostic list of Panama Birds"; Schauensee (1964), "Birds of Colombia"; and in their own research, which led to the publication of *The Birds of Panama* (1976). And last but not least, the work and information provided by J. E. Ambrose, F. O. Chapelle, E. S. Morton, R. Ryan, D. R. Sheets, E. O. Willis, P. Alden, D. O. Hill, J. R. Karr, N. G. Smith, F. G. Stiles, and G. Tudor, among others. The most recent publications are *Directory of Important Areas for Birds of Panama* (2003) and *Annotated checklist of the birds of Panama* (2006) by G. Angehr, which compile and update the information about the Panamanian bird life.

There are 15 species of Strigiformes distributed in Panama, which represent 7% of all Neotropical species. Little research has been conducted on this order in the country. The purpose in this chapter is to carry out a review of the existing information from museum collections, field work, and bibliography that shows information about typical owls and barn owls in Panama. The main goal is to be able to establish the status of this important family from the Neotropics.

17.2 Methodology

For the taxonomic diversity, the American Ornithologists' Union (2008) classification was used. Information regarding the historic and current distribution of the Panamanian species was obtained mainly from the publications of Wetmore (1968), Ridgely and Gwynne (1993), and Angehr (2006). The museum review included the Vertebrates Museum of the Universidad de Panama (MVUP), and the collections of Dr. Eustorgio Mendez; the Gorgas Memorial Laboratory (CDEM-LCG), at the local level; and internationally, the National Museum of Natural History of the Smithsonian Institution (NMNH-SI) in Washington D.C. in the United States and the Museum of Natural History in the United Kingdom. Just for that species with restricted distribution, Statterfield et al. (1998) and Angehr (2003, 2006) were used.

For the conservation status of species, we based mainly on the information published by the National Environmental Authority (ANAM) in the list of threatened and/or endangered species, Resolution No. AG-0051-2008 of January 22, 2008, which was published in the Official Gazette of April 7, 2008, with data provided by Solis et al. (1999). These data have been collected through the effort of government technicians and/or officials in charge of the elaboration of the official lists of threatened wildlife, with support from scientists that provide them with up-to-date information. For the analysis of the present species in the National System of Protected Areas of Panama, several land areas with surfaces larger than 4500 hectares were selected, thus leaving the marine surface of some coastal-marine-protected areas. The information regarding the amount of hectares and the altitude range of each protected area was used according to information from Angehr (2003) and ANAM (2006).

In order to know about the locations of the species in the protected areas, a bibliographic review was made of reports, publications, management plans, and gray literature obtained from the libraries of the National Environmental Authority (ANAM), the National Environment Conservation Association (ANCON), and the Smithsonian Tropical Research Institute (STRI). Additionally, the digital database of the STRI was reviewed (STRI 2008), and some expert communications and observations were gathered.

The management plans for the following protected areas were consulted: La Amistad International Park (ANCON and CEPSA 2004g), the San San Pond Sak Wetland (ANCON and CEPSA 2004f), Volcán Baru National Park (ANCON and CEPSA 2004h), Soberania National Park (Alvarado 2006), Bastimentos Island Marine National Park (PROARCA 2001), the Protective Forest of Palo Seco (ANCON and CEPSA 2004e), Portobelo National Park (Management Plan and Development of Portobelo National Park 1994), Altos de Campana National Park (Tovar 1999), Chagres National Park (Tovar et al. 2005), Darién National Park (OTSCorp 2003), Playa La Barqueta Wildlife Refuge (CEPSA 2005), and the San Lorenzo Protected Area (CEPSA 2002).

The list of owls species contained in the works of Ridgely and Gwynne (1993) and Angehr et al. (2008) were consulted as well, for the following protected areas: Protective Forest of Palo Seco (ANCON and CEPSA 2004e, Libsch 2002), San San

Pond Sak Wetland (ANCON and CEPSA 2004f), La Amistad International Park (ANCON and CEPSA 2004g), Volcan Baru National Park (ANCON and CEPSA 2004h), Fortuna Forest Reserve (Robbins et al. 1995), El Montuoso Forest Reserve (Arauz 2004), Major General Omar Torrijos Herrera National Park (CEPSA 2000), Santa Fe National Park (Santamaria 2000b), San Lorenzo National Park (Weaver and Bauer 2004), Kuna Yala Reserve (Marcus and Roldan 1984), and protected areas of the Canal watershed which include Camino de Cruces, Soberania and Chagres National Parks, as well as Barro Colorado Nature Monument (Gale et al. 1978, Karr 1990, Tejera 1995, Aparicio 1999a, b, c, Jimenez 1999a, b, Robinson 1999, Angehr et al. 1999), Jimenez and Aparicio 2005, the Darién area (Robbins et al. 1985, ANCON 1995 and Santamaria 2000a, b; ANAM 2008a, b), and Robinson (2001).

For the biological knowledge and status, Mendez (1987), Ridgely and Gwynne (1993), and Wetmore (1968) were consulted. For information regarding the local, native, or vernacular names, the following references were used: Ridgely and Gwynne (1993), Delgado (pers. comm.), and J. Kantule (pers. comm.). With regard to the English names, Del Hoyo et al. (1999) were consulted. For the activities related to the preparation and elaboration of the distribution maps of the different species of barn owls and typical owls, the following stages were developed.

17.2.1 Collection of Secondary Information Obtained from

Physical map of Panama (scale: 1:250,000) from the Tommy Guardia National Geographic Institute (Tommy Guardia National Geographic Institute 2007).
Digital map of Vegetation and Land Use of Panama (ANAM 2000).
National Atlas of the Republic of Panama (Tommy Guardia National Geographic Institute 2007).

For the elaboration of the base map, the physical map (scale: 1:250,000) from the Tommy Guardia National Geographic Institute was used, in which all the physical elements such as rivers, level curves, administrative borders, and elevation levels were defined and subsequently entered in the geographic information system (SIG) with the ArcGis 9.2 mapping software.

17.2.2 Drafting of the Distribution Polygons

According to the existing information about distribution and habitat preferences contained in the works of Ridgely and Gwynne (1993) for the barn owl and the 14 typical owl species registered in Panama, the areas were set apart according to elements such as elevation (level curves), land use and natural vegetation categories, sighting reports, and their respective toponymy. In addition, a geoprocessing

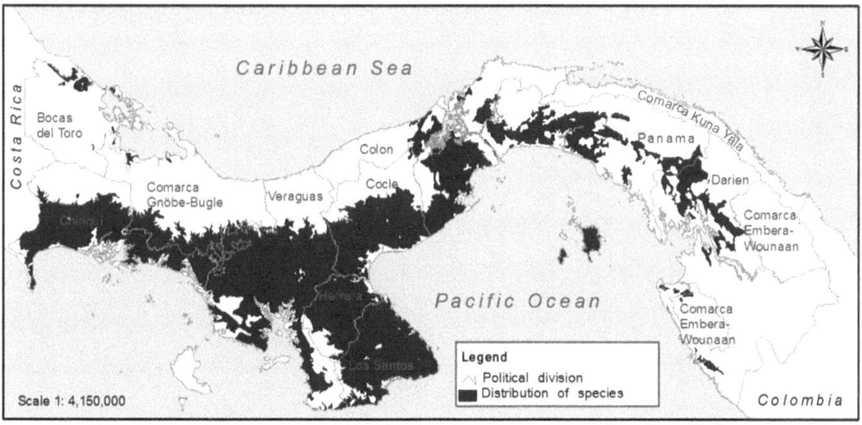

Fig. 17.2 Spatial distribution of *Tyto alba*

analysis was made through the overlaying of different polygons in such a way that those areas with no coincidence of elements could be segregated giving way to the formation of polygons of potential distribution for each species. Once the distribution analyses were made, a distribution map was elaborated for each species. These maps were created using the ArcGis 9.2 mapping software and exported in image format.

17.3 Taxonomy Diversity

The order Strigiformes (Appendix 17.1) is represented in Panama by two families: Tytonidae (1 species) and Strigidae (14 species), commonly called barn owl and typical owls, respectively (Wetmore 1968, Ridgely and Gwynne 1993, Angehr 2006, AOU 2008). The family Tytonidae is represented by *Tyto alba* and referred to as the common barn owl. It inhabits the productive system with natural woody vegetation or significantly spontaneous vegetation (0–50%) (Fig. 17.2, Table 17.1, Appendix 17.1).

On the other hand, the family Strigidae is represented by 14 species. Three species of the genus *Megascops* (formerly called *Otus*, AOU 1983, 2003): *M. choliba* lives in lowland ombrophilous and semi-deciduous forests and submontane ombrophilous forests up to 900 m amsl (Fig. 17.3, Table 17.1, Appendix 17.1); *M. guatemalae* has been reported in lowland forests, submontane forests up to 900 m amsl, and various types of floodable vegetation areas (Fig. 17.4, Table 17.1, Appendix 17.1), and *M. clarkii* is found in the submontane, montane, highland, and cloud forests between 1080 and 2100 m amsl (Fig. 17.5, Table 17.1, Appendix 17.1).

Three species of the genus *Glaucidium*: *G. costaricanum* (formerly *jardinii*, AOU 1983, 2000), in Panama, is a species whose distribution is restricted to submontane,

Table 17.1 Distribution of barn owl and typical owl species according to vegetation types

Types of vegetation (ANAM 2000)	*T. alba*	*M. choliba*	*M. guatemalae*	*M. clarkii*	*L. cristata*	*P. perspicillata*	*B. virginianus*	*G. costaricanum*	*G. griseiceps*	*G. brasilianum*	*A. cunicularia*	*C. virgata*	*C. nigrolineata*	*P. clamator*	*A. ridgwayi*
Lowland broadleaf tropical ombrophilous forest		X	X		X	X						X			
Submontane broadleaf tropical ombrophilous forest (500–1000 m Caribbean, 700–1200 m Pacific)			>900	X	>900	>1200		X				>900			
Montane broadleaf tropical ombrophilous forest (1000–1500 m Caribbean, 1200–1800 m Pacific)				X		>1200		X				>2100 Chiriquí			

Highland broadleaf tropical ombrophilous forest (1500–2000 m Caribbean, 1800–2300 m Pacific)				X				X				>2100 Chiriquí			X
Cloud broadleaf tropical ombrophilous forest (2000– 3000 m Caribbean, 2300–3000 m Pacific)				X								>2100 Chiriquí			X
Cloud broadleaf tropical ombrophilous forest (2000– 3000 m Caribbean, 2300–3000 m Pacific)				X								>2100 Chiriquí			X

(continued)

Table 17.1 (continued)

Types of vegetation (ANAM 2000)	*T. alba*	*M. choliba*	*M. guatemalae*	*M. clarkii*	*L. cristata*	*P. perspicillata*	*B. virginianus*	*G. costaricanum*	*G. griseiceps*	*G. brasilianum*	*A. cunicularia*	*C. virgata*	*C. nigrolineata*	*P. clamator*	*A. ridgwayi*
Alluvial tropical ombrophilous forest occasionally flooded dominated by *Prioria copaifera*			X		X	X									
Broadleaf swampy tropical ombrophilous forest dominated by dicotyledonous species			X		X	X									
Swampy tropical ombrophilous forest dominated by *Campnosperma panamensis*			X		X	X									
Lowland broadleaf tropical ombrophilous forest		X	X		X	X						X			

Lowland tropical semi-deciduous forest		X													
Mangrove forest			X		X	X									
Productive system with significant natural or spontaneous woody vegetation (10–50%)	X														
Productive system with significant natural or spontaneous woody vegetation (<10%)	X														

Note: some species have specific altitudes and locations

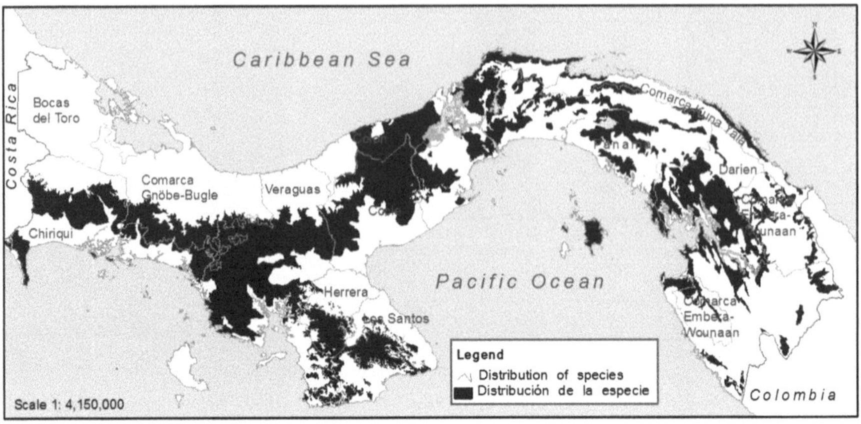

Fig. 17.3 Spatial distribution of *Megascops choliba*

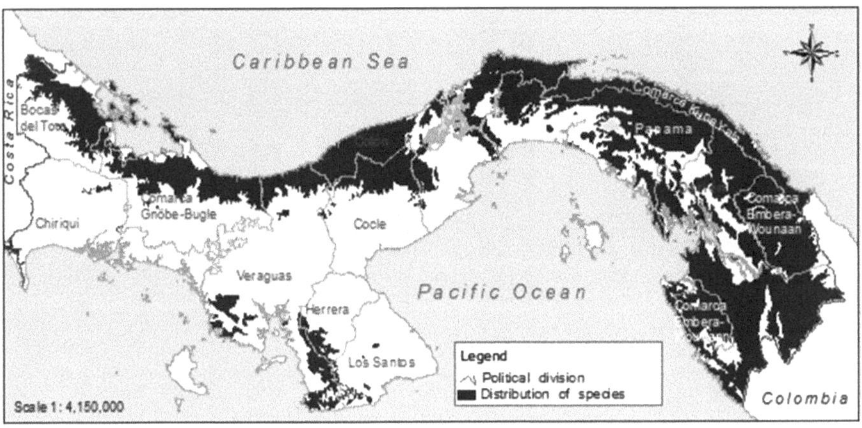

Fig. 17.4 Spatial distribution of *Megascops guatemalae*

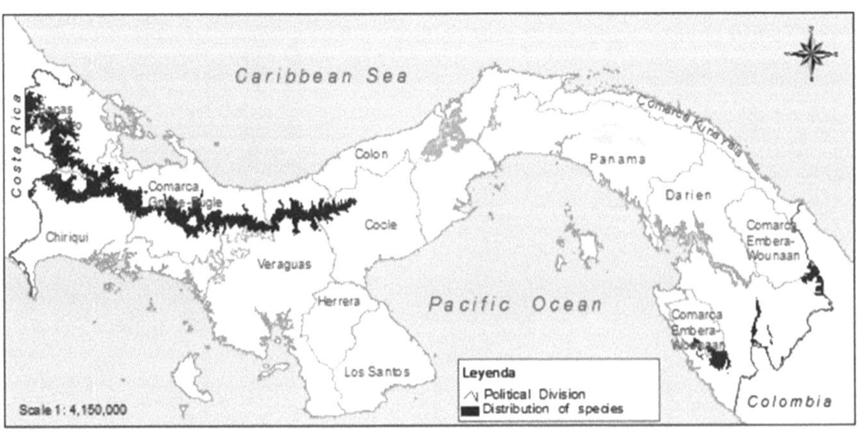

Fig. 17.5 Spatial distribution of *Megascops clarkii*

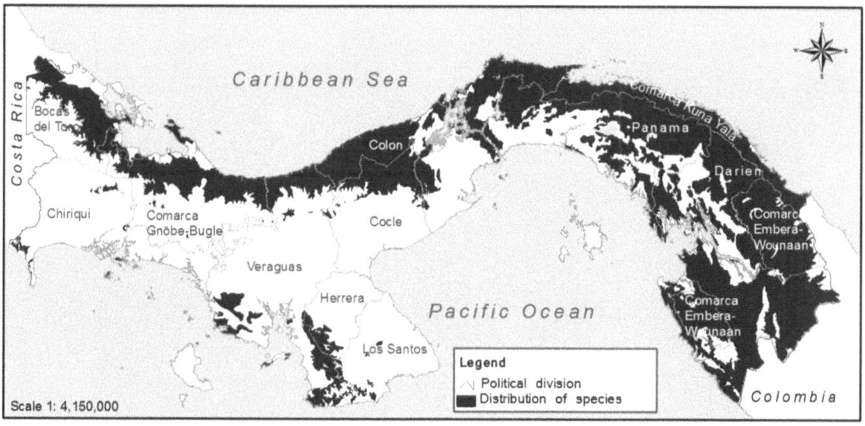

Fig. 17.6 Spatial distribution of *Lophostrix cristata*

montane, and highland broadleaf forest (Fig. 17.9, Table 17.1, Appendix 17.1); *G. griseiceps* (formerly *minutissimum*, AOU 1983, 1997) is found in lowland broadleaf and semi-deciduous forests and probably in flooded forests as well (Fig. 17.10, Table 17.1, Appendix 17.1), and *G. brasilianum* has restricted distribution in the central and Pacific areas of Panama, including lowland ombrophilous and semi-deciduous forests and productive systems (10–50%) (Fig. 17.11, Table 17.1, Appendix 17.1).

Two species of the genus *Ciccaba*: *C. virgata* is found in lowland ombrophilous forests, at the foothills of submontane forests, in highland and cloud forests, and probably in flooded forests as well (Fig. 17.12, Table 17.1, Appendix 17.1), and *C. nigrolineata* has a preference for lowland ombrophilous forests, foothills of submontane forests, montane forests, and highland and cloud forests and probably in flooded forests as well (Fig. 17.13, Table 17.1, Appendix 17.1).

Six genera, each one of them represented by one species: *Lophostrix crristata*'s habitat would correspond to the lowland broadleaf forest, submontane forest, and probably flooded forests of the country (Fig. 17.6, Table 17.1, Appendix 17.1); *Pulsatrix perspicillata* is found in lowland broadleaf forests, submontane and montane forests, and flooded forests (Fig. 17.7, Table 17.1, Appendix 17.1); *Pseudoscops clamator* (formerly called *Asio*, AOU 1983, 1997) and its habitat correspond to land uses known as productive systems with woody vegetation or with spontaneous vegetation (0–50%), and *Aegolius ridgwayi* is a species restricted to the highland and cloud forests in the Chiriquí province (Fig. 17.15, Table 17.1, Appendix 17.1). Finally, two species *Bubo virginianus* and *Athene cunicularia*, with rather poor records, have lack of information about the habitat use, showing only punctual records in some locations in the provinces of Veraguas and Chiriquí (Wetmore 1968, Ridgely and Gwynne 1993, Fig. 17.8, Appendix 17.1).

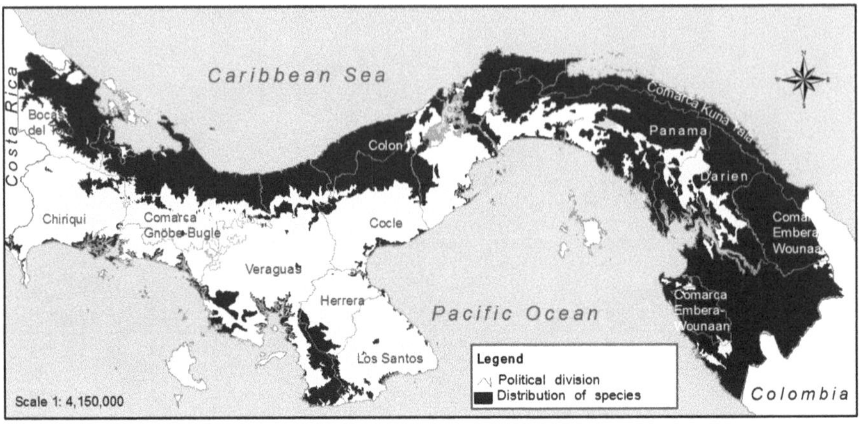

Fig. 17.7 Spatial distribution of *Pulsatrix perspicillata*

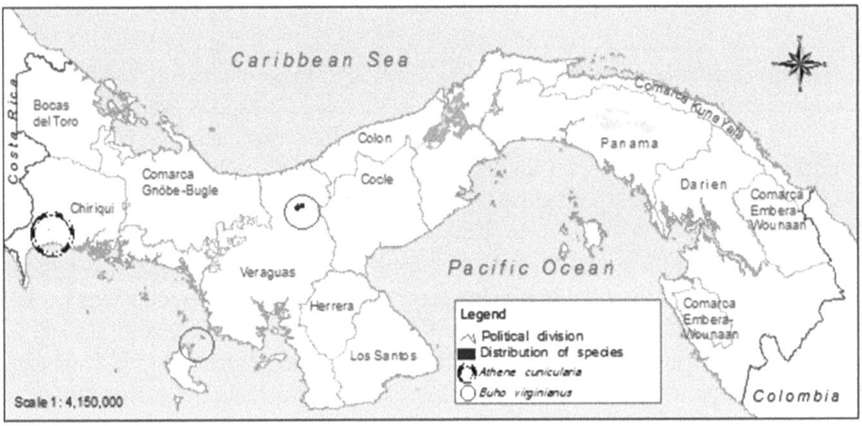

Fig. 17.8 Spatial distribution of *Bubo virginianus* and *Athene cunicularia*

17.4 Distribution and Habitat Associations

Tyto alba It prefers open areas (including pastures and agricultural areas like rice fields) in lowland areas (0–600 m amsl) in both the Pacific and Atlantic slopes, but mainly in the Pacific (Ridgely and Gwynne 1993; Fig. 2, Table 17.1). In the Pacific coast, it goes from the lowlands in the province of Chiriquí to as far as the highlands (one record in the Boquete mountains at 1550 m amsl) in the province of Coclé (one individual collected specimen, according to MVUP, in El Rocío), and another individual collected in Herrera (one individual collected according to CDEM-LCG); there are records of 16 specimens toward the province of Panama in four museums (Appendix 17.2). Besides, there is a record in the Las Perlas Archipelago (San Jose Island) and one record in the Chucunaque River in the province of Darién (Fig. 17.2).

On the Caribbean slope, it is only found in the province of Bocas del Toro and in the province of Colon, specifically in the Canal area (Fig. 17.2). This is one species that we also find in old churches of towns and cities since it likes to live in the bell towers or at the top of high palm trees near them.

Megascops choliba It prefers lowland mature secondary forests and the lower part of the foothills (0–900 m amsl) in both the Pacific and Atlantic slopes (Ridgely and Gwynne 1993; Fig. 17.3, Table 17.1). It has been reported mainly in the Pacific slope, from Chiriquí (lower area of the highlands of Boquete), central provinces, Canal area including Panama City, and as far as the province of Darién (Cerro Malí, 600 m asml) (Fig. 17.3). Some of these locations have been confirmed by the 36 specimens kept in the four museums, coming specifically from the provinces of Chiriquí, Los Santos, Herrera, Cocle, Darién, and Panama (Appendix 17.2), where it is frequently observed in some areas of the city such as the Natural Metropolitan Park and Cerro Ancon, or toward the east, around Campo Chagres in Chilibre. It has also been reported in the larger islands of the Las Perlas Archipelago (Del Rey Island and Pedro Gonzalez Island) (Fig. 17.3). In the Caribbean slope, it has not been found toward the west of Panama, but in the Canal area of the province of Colon as far as Armila, in the eastern part of the Kuna Yala region (Fig. 17.3).

Megascops guatemalae This species has been recorded in the lowland mature rainforests and in the lowland secondary forests and foothills (0–900 m amsl), in both the Pacific and Atlantic slopes, but mainly in the Caribbean (Ridgely and Gwynne 1993; Fig. 17.4, Table 17.1). In the Pacific slope, it is scarcely found in the provinces of Chiriquí, Veraguas, Canal area, Panama, and Darién (Cerro Malí), with 21 specimens in the museums from these locations (Fig. 17.3, Appendix 17.2). In the Caribbean slope, it has been recorded in Bocas del Toro, the Canal area, and the eastern part of Kuna Yala (Cerro Brewster, 899 m amsl) (Fig. 17.4).

Megascops clarkii This species is restricted to the rainforest at the highlands of Panama, between 1080 and 2100 m amsl (Fig. 17.5, Table 17.1); in the provinces of Chiriquí (Fortuna), Veraguas (Calobre), and Panama; and the northeastern part of Darién (Cerro Pirre), as shown by the nine specimens found in the museums from the last two locations listed (Birdlife International 2008, Ridgely and Gwynne 1993; Appendix 17.2). It is considered a regional endemic species being its distribution restricted to Costa Rica, Panama, and Colombia, recorded in endemic bird areas in Panama (020 Highlands of Costa Rica and Panama and 024 Highlands of Darién) (Ridgely and Gwynne 1993, Statterfield et al. 1998, Angehr 2003).

Lophostrix cristata This species has been recorded in mature and secondary rainforests in both the Pacific and Atlantic slopes, but mostly in the Caribbean, from sea level as far as 900 m amsl (Ridgely and Gwynne 1993; Fig. 17.6, Table 17.1). Reviews made through the museums showed 16 specimens collected in the following locations: Soná in the province of Veraguas and several sites in the provinces of Chiriquí, Panama, and Darién (Appendix 17.2).

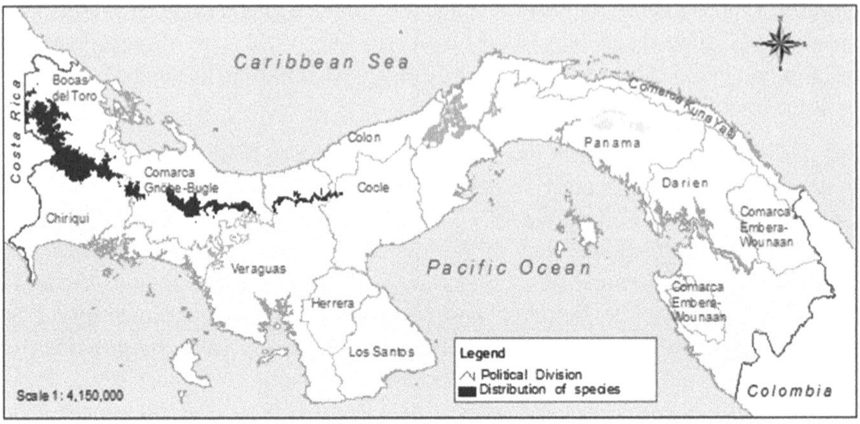

Fig. 17.9 Spatial distribution of *Glaucidium costaricanum*

Pulsatrix perspicillata This species is widely distributed, but has preference for mature and secondary rainforests in the Pacific and Atlantic slopes, humid lowlands, and foothills up to 1200 m amsl (Ridgely and Gwynne 1993; Fig. 17.7). Records from the museums show 30 specimens from the provinces of Bocas del Toro (Popa Island), Veraguas, Panama, and Colón (Appendix 17.2).

Bubo virginianus This species has three historic records, one in the surroundings of Chitra, in the province of Veraguas (specimen recorded in the MHN-UK), and two in Ranchería Island (Ridgely and Gwynne 1993; Fig. 17.8). No distribution map was made for this species due to the few records available.

Glaucidium costaricanum This species has only been reported in the highlands between 1500 and 3475 m amsl in the west region of Chiriquí (Cerro Punta, Boquete Trail) and in Veraguas (Ridgely and Gwynne 1993; Fig. 17.9, Table 17.1). The museum review reported three species found in these provinces (Appendix 17.2).

Glaucidium griseiceps This is a species of widespread distribution, with preference for lowland rainforests and the lower areas of the foothills (up to 900 m amsl) in both the Pacific and Atlantic slopes, in the Caribbean mainly, and toward the east of Panama and the province of Darién in the Pacific (Ridgely and Gwynne 1993; Fig. 17.10). Records from the Pacific slope come from the Canal area (Pipeline Road) and Darién (La Laguna, Cerro Pirre, low hillsides). Four specimens from both of these locations are found in the museums as well as one from the province of Veraguas (Appendix 17.2); in the Caribbean slope, this species has been recorded in the province of Bocas del Toro (lower part of the Changuinola River) and Colón (Peluca Hydrological Station), eastern area of the province of Panama, Majé-Bayano, and eastern area of Kuna Yala (Perné and Puerto Obaldía). The museum review confirms one specimen from the province of Colón (Appendix 17.2, Fig. 17.10).

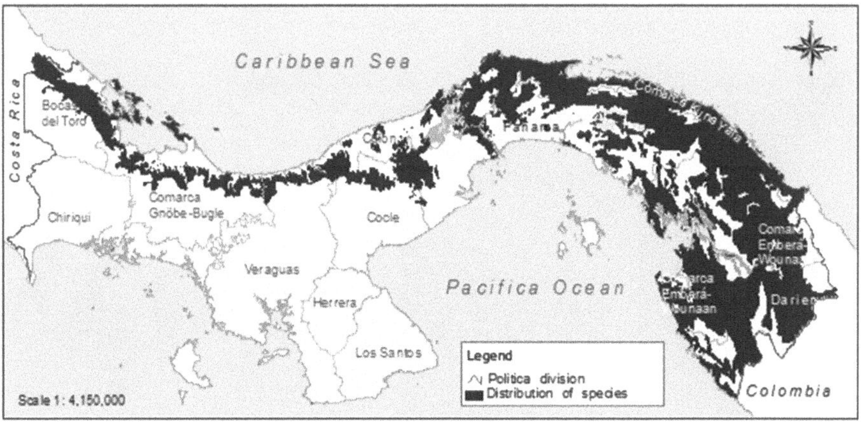

Fig. 17.10 Spatial distribution of *Glaucidium griseiceps*

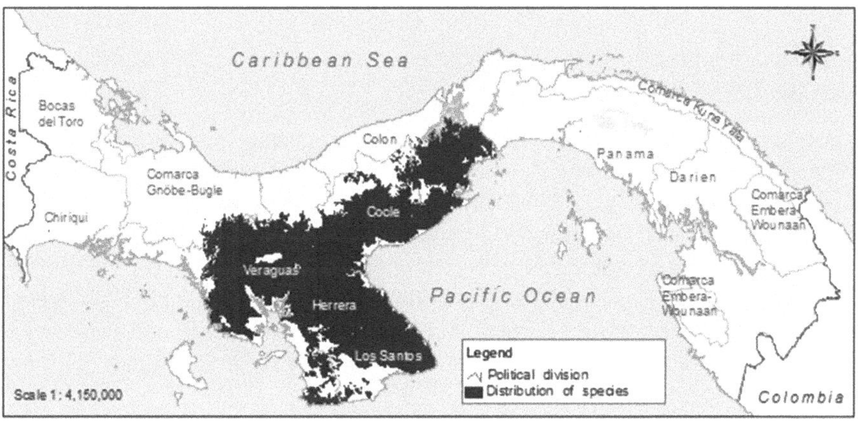

Fig. 17.11 Spatial distribution of *Glaucidium brasilianum*

Glaucidium brasilianum This species has only been reported in lowland meadows and secondary forests of the Pacific slope (lower than 600 m amsl) (Ridgely and Gwynne 1993; Fig. 17.11, Table 17.1). It is found in the province of Veraguas and the eastern part of the Azuero Peninsula (one specimen collected for the MVUP in La Evea, Guararé, province of Los Santos) and spreads as far as the west of the province of Panama. Two museum specimens come from the first location, another specimen comes from the province of Herrera, and from the last location, the MNHN-SI has 15 specimens (Appendix 17.2). There are no records of this species in the Caribbean slope (Fig. 17.11).

Athene cunicularia This species has only one historic record (December 13, 1900, Wetmore 1968), in Divalá, province of Chiriquí. There are no records in the muse-

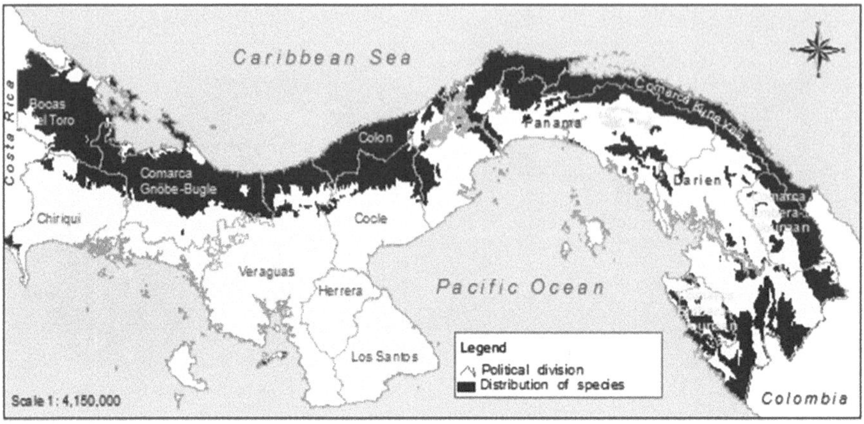

Fig. 17.12 Spatial distribution of *Ciccaba virgata*

ums nor a distribution map, due to the poor information available (Ridgely and Gwynne 1993; Fig. 17.8).

Ciccaba virgata This species prefers secondary and mature rainforests in both the Pacific and Atlantic slopes in lowlands and foothills, especially in the Caribbean slope (Ridgely and Gwynne 1993; Fig. 17.12, Table 17.1). Its widespread distribution includes lowland forests in the province of Chiriquí and extends up to 21,000 m amsl. Some locations where this species has been collected include the provinces of Chiriquí (three specimens), Bocas del Toro (Bastimentos Island, Popa Island, Chiriquí Grande) (seven specimens), Veraguas (two specimens), and Panama (29 specimens). It is almost absent in the Pacific slope, except for Darién, with one record (Cerro Malí) (Fig. 17.12, Appendix 17.2).

Ciccaba nigrolineata This species has been recorded in the rainforests and in the borders of the lowland forests in both the Pacific and Atlantic slopes. It is widely distributed throughout the lowlands of the province of Chiriquí and extends as far as 2100 m amsl (Ridgely and Gwynne 1993; Fig. 17.13, Table 17.1). There are records of this species from the provinces of Bocas del Toro (with eight specimens collected by the MNHN-SI in Bastimentos Island, Cayo de Agua, and Popa Island), Veraguas, and Herrera, the Canal area in the eastern part of Panama (12 specimens found in the museum review), Darién, and the Caribbean side of the Canal and Colón (Fig. 17.13, Appendix 17.2).

Pseudoscops clamator This species prefers open and wooded areas and clearings with shrubs in the lowlands in both the Pacific and Atlantic slopes, but mainly in the Pacific side (Ridgely and Gwynne 1993; Fig. 17.14, Table 17.1). In the Pacific slope, distribution goes from the province of Chiriquí (where it spreads up to 1050 m

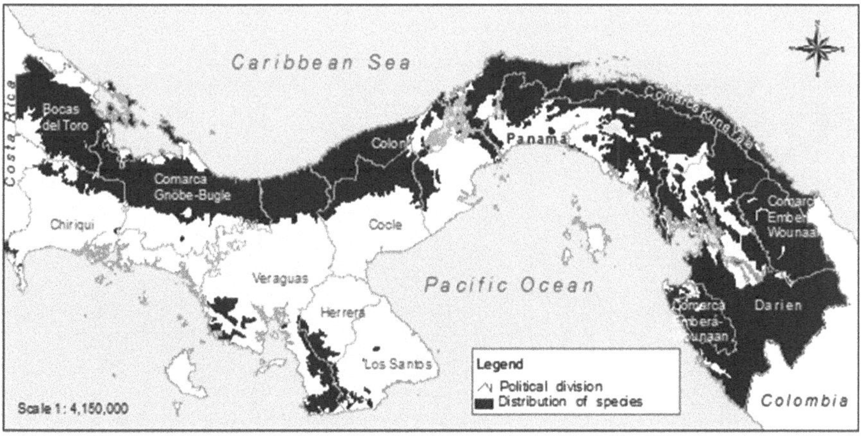

Fig. 17.13 Spatial distribution of *Ciccaba nigrolineata*

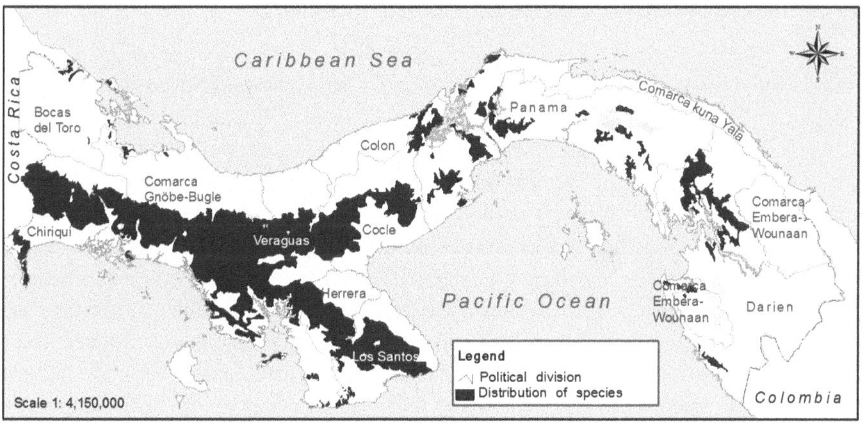

Fig. 17.14 Spatial distribution of *Pseudoscops clamator*

amsl) and central provinces (one specimen collected by the MNHN-SI in Herrera, in the surroundings of Santa María, and another one in Coclé), toward the east of Panama (13 specimens from the museums), and the Caribbean slope; it has only been recorded in the east of the province of Colón, near the Claro River, and in the east of Portobelo (Fig. 17.14, Appendix 17.2).

Aegolius ridgwayi This is a species with restricted distribution in Panama, since it has only been recorded in the upland and cloud forests in the highlands of the province of Chiriquí (western flank of Volcan Baru, 2280 m amsl), between 2100 and 3475 m amsl (Ridgely and Gwynne 1993, Statterfield et al. 1998; Fig. 17.15, Table 17.1). The only specimen reported in the MNHN-SI comes from that location (Appendix 17.2).

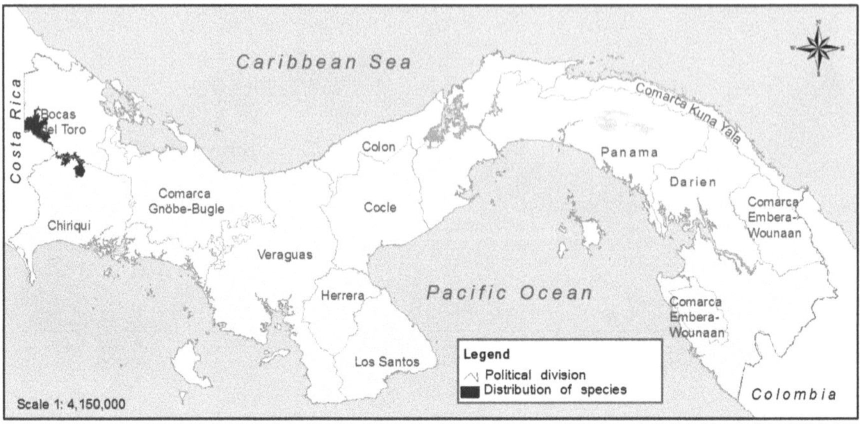

Fig. 17.15 Spatial distribution of *Aegolius ridgwayi*

17.5 Conservation Status at a National Level

At a national level and according to the National Environmental Authority (ANAM), Resolution 0051 of January 22, 2008, published in the Official Gazette No. 26013 of April 7, 2008, includes the 15 species of Strigiformes under the vulnerable category (VU), with the exception of *B. virginianus*, which has been classified as endangered (EN) due to the few existing records (Appendix 17.2). On the other hand, at the global level, according to the Convention on International Trade in Endangered Species (CITES), all owl species are listed in CITES II. Actually, the species are not in danger, but could reach that point if the trade of these species does not become subject to a strict regulation thus preventing the enforcing of actions to support their survival. In the red list of the International Union for Conservation of Nature (IUCN), these species are classified under the low risk category (CITES 1998, Solis et al. 1999, UICN 2009, UNEP-WCMC 2009; Appendix 17.2).

17.6 Threats

A large part of the country has been modified for agricultural and livestock development; thus new and extensive areas have been transformed to pastures (Heckadon 1983), mainly in the Pacific slope. This way, the elimination of the forest cover and the destruction of habitats modify the distribution of the owl species, since this affects the availability of natural food sources; consequently, this group of birds is restricted mainly to the edges of the forests or to riparian forests, becoming more flexible in their adaptation to the new habitats, as with common species like *T. alba*, *M. choliba*, and *G. brasilianum*.

In addition, this new scenario affects these and other owl species due to the development of activities associated to grazing or to agricultural practices. For instance, the ferruginous pygmy owl (*G. brasilianum*) eats diurnal insects; if fumigation is carried out in an area with extensive farming of rice, corn, or other products, this species is the first one to become affected due to the diet that usually eats. The situation turns even more critical when aerial fumigation is used – like it is the case with the large banana farms in Chiriquí and Bocas del Toro. The situation may become worse with aerial spraying in the highlands of the province of Chiriquí or Panama, since this involves the use of selective agrochemicals in the production of legumes, vegetables, and flowers in areas where these threatened species are present as well as those other species mentioned above.

Extensive rice farming promotes the proliferation of harmful herbaceous plants. The traditional choice in Panama has been the use of conventional pesticides such as herbicides and fungicides, due to the type of anthropogenic wetland required. On the other hand, this type of farming favors the proliferation of wild mice in the cultivated area and its surroundings, thus the arrival of predators like the barn owl and some small snakes. The use of rodenticides is necessary to control the mice, which consequently affects the owl populations due to the accumulation of pesticide in their bodies. Besides, it is worth noting that in rice farming areas near roads with continuous vehicle traffic, the impact of owls hit by cars when flying low during evening hours is high, as happens in areas like Los Llanos de Santa María in Herrera, San Lorenzo, and Gualaca in Chiriquí, especially during the mating season. The accidental deaths of these species again favors the proliferation of more mice in the area, giving way to the outbreak of diseases like hantavirus, as has already occurred in Aguadulce, Parita, Guararé, and Tonosí in the Azuero Peninsula.

A third negative element that affects owls is the use of rodenticides for the control of wild mice and rats in nontraditional farming, since the owls are large consumers of these agricultural pests and they become seriously affected by the consumption of poisoned rodents. This is the case for the control of the excavator mouse (*Macrogeomys excavator*) found in the highlands of the country, which has strictly nocturnal habits and is well known for causing damages in certain areas of Cerro Punta and Volcán and in the surroundings of Boquete in the province of Chiriquí.

17.7 Conservation Strategies

17.7.1 Protected Areas

Some efforts of Panama's environmental authority (National Environmental Authority – ANAM) geared at the protection, conservation, and recovery of the environment include the creation of the National System of Protected Areas (SINAP) in 1992, with the purpose of strengthening the protected area. In 1999, the National

Strategy for the Environment was developed, which contains the public policies that further set the framework for national, sectorial, and regional policies. Complementing this, the National Strategy for Biodiversity is being prepared, which in turn is part of the decisions of the Second Conference of Parts of the agreement over Biological Diversion (ANAM 2006).

In 2004 Panama had 50 protected areas, and between 1998 and 2003, the protected area was increased by 2% (ANAM 2004). According to a report from the ANAM (2006), the SINAP had 65 protected areas, representing 34.43% of the national territory, which represented 2,600,018 hectares. According to the National Environmental Strategy, national parks had over 80% of forest areas (except for Sarigua and Altos de Campana National Parks, which are not suitable for the development of extensive forest areas due to their natural conditions with salt marshes and volcanic plains, respectively; therefore they are not appropriate for the development of large forest areas).

Besides, the SINAP has a document called Technical Guides for the Elaboration of Management Plans, and 26 (40%) of those 65 protected areas currently have management plans; of these 26 areas, 19 of them (29%) have been approved (ANAM 2006). For this work, at least 12 of those management plans were reviewed, and in the section about the research programs included in those management plans, there are no measures or specific actions taken for owl species, but they make efforts to promote research on threatened species and communities.

Since the SINAP has become a relevant instrument for the protection and conservation of owls in situ, this analysis shows that these species are potentially distributed in at least 29 protected areas, which constitute 2,094,296.50 hectares, Darién being the province with the largest extension of protected areas (54.8%). In this case, a total of ten species may potentially be found; however, there are important information gaps in the provinces of Los Santos, Coclé, and Herrera (Table 17.2).

In the case of the owl species that inhabit the cloud forests, like *M. clarkii*, *G. costaricanum*, and *A. ridgwayi*, they are very well represented in the protected areas that extend as far as the highlands, like La Amistad International Park, the Protective Forest of Palo Seco, Volcan Barú National Park, and Darién National Park, among others. The same thing happens with species of wide altitudinal distribution which are typical of lowlands and foothills, like *P. perspicillata*, *C. nigrolineata*, and *M. guatemalae*, among others that are present in more than 25 protected areas, and consequently, their distribution polygons are the most extensive (Table 17.2). On the other hand, Coiba National Park and Santa Fe National Park protect the habitat of *B. virginianus*, where the first specimen was collected in 1868 (E. Arce), and later on, it was recorded in Ranchería in 1956 (Wetmore 1968; Table 17.2).

As a result, owl species are protected by the different management categories of the ANAM. In addition, a significant fact has been the creation of the new political-administrative category that constitutes the indigenous reservations of the Ngöbe, Bugle, Kuna, and Emberá groups, and the future reservations of the Naso-Teribes and the Bribrís groups in Bocas del Toro. The positive attitude of the indigenous groups toward nature or mother earth constitutes a guarantee for the subsistence of this group of birds of cryptic behavior, even when they still keep some traditions that instill fear toward them (Tables 17.3 and 17.4).

Table 17.2 Distribution of owl species in Panama by protected area

Provinces and Indigenous comarcas	Protected Areas	Altitude (amsl)	Land Area (hectares)	*T. alba*	*M. choliba*	*M. guatemalae*	*M. clarkii*	*L. cristata*	*P. perspicillata*	*B. virginianus*	*G. costaricanum*	*G. griseiceps*	*G. brasilianum*	*A. cunicularia*	*C. virgata*	*C. nigrolineata*	*P. clamator*	*A. ridgwayi*	Total number of species
		Altitude Ranges (amsl)		0-600	0-900	0-900	1080-2100	0-900	0-1200	0-600	1500-3475	0-900	0-600	0-600	0-900	0-2100	0-600	2100-3475	
Bocas del Toro	Protective Forest of Palo Seco	40-2300	125000			x	x	x	x			x			x	x	x		10
	San San Pond Sak Wetland of International Importance	0-20	16125	x		x			x						x				5
Bocas del Toro and Chiriquí	La Amistad International Park	90-3325	207000			x	x	x				x			x	x			9
Chiriquí	Volcán Barú National Park	1400-3475	14000		x		x				x				x	x		x	6
	Fortuna Forest Reserve	700-2213	19500		x		x				x				x	x			5
Veraguas	Santa Fe National Park	80-1964	72636		x	x	x	x	x	x			x		x				12
	Coiba National Park	0-416	53528							x									1
	Gulf of Montijo Wetland of International Importance	0-194	38435																8
Herrera	El Montuoso Forest Reserve	100-987	10375		x										x				9
Los Santos	Cerro Hoya National Park	0-1559	27900																8
	La Tronosa Forest Reserve	s/i	20579																8
Coclé	General de Division Omar Torrijos Herrera National Park	80-1710	25275		x		x	x					x		x	x			12
Colón	Protective Forest and Protected Landscape of San Lorenzo	0-189	12000	x	x	x		x	x						x	x			8
	Portobelo National Park	0-979	24000																8

(continued)

Table 17.2 (continued)

Colón and Panama	Chagres National Park	90-1007	129585	x	x	x		x	x						x				9
Panama	Altos de Campana National Park	220-1007	4925		x										x				10
	Camino de Cruces National Park	15-269	4590	x	x														9
	Soberanía National Park	30-332	19541		x	x		x	x			x			x	x	x		8
	Barro Colorado National Monument	30-171	4900		x	x		x	x						x	x	x		7
	Majé Hydrological Reserve	67-1074	16732		x	x		x	x			x			x	x			8
Comarca of Kuna Yala	Corregimiento Nargana 1 Wildlife Area	0-899	100000		x	x			x			x			x	x			7
Darién	Darién National Park	0-1875	579000	x	x	x	x	x	x			x			x	x	x		10
	Punta Patiño Wetland of International Importance	0-336	4800		x			x	x										7
	Serranía de Bagre Biological Corridor	n/i	31275					x							x	x			7
	Protective Forest of Alto Darién	n/i	211958																7
	Canglón Forest Reserve	n/i	31650																8
	Chepigana Forest Reserve	n/i	257219																7
	Serranía de Darién Hydrological Reserve	n/i	7046																7
	Serranía Filo del Tallo Hydric Reserve	n/i	24722																9
	Total Protected Area			14	25	26	7	25	26	2	6	20	7	0	25	27	17	2	

n/i = no information, X = presence recorded in literature and direct observations, blank spaces = no record for that species, gray = potential presence

Table 17.3 Extension and altitudinal range of the protected areas by province and number of recorded and potential species

Provinces	Protected hectares	%	Altitudinal range (m amsl)	# rec sp/# pot sp
Darien[a]	1,147,670.20	54.8	0–1875	10/10
Bocas del Toro[b]	348,125.00	16.6	0–3325	9/11
Panamá[c]	180,273.30	8.6	15–1074	9/10
Veraguas	164,599.00	7.9	0–1964	8/13
Comarca Kuna Yala	100,000.00	4.8	0–899	5/7
Los Santos	48,479.00	2.3	0–1559	0/8
Colón	36,000.00	1.7	0–979	7/9
Chiriquí	33,500.00	1.6	700–3475	6/6
Coclé	25,275.00	1.2	80–1710	3/12
Herrera	10,375.00	0.5	100–987	2/9
Total	2,094,296.50	100		

[a]Includes part of the two Emberá-Wounaan "comarcas"
[b]Includes the Chiriquí's section of La Amistad National Park
[c]Includes the Colón's section of the Chagres National Park

Table 17.4 Number of protected areas and extension of the distribution polygon of the different species of typical owls and the barn owl, with their altitudinal ranges

Species	Altitudinal range (m amsl)	Number of protected areas (29)	Extension of distribution area (ha)
Tyto alba	0–600	14	2,688,303.96
Megascops choliba	0–900	25	2,679,985.96
Megascops guatemalae	0–900	26	3,308,083.65
Megascops clarkii	1080–2100	7	497,034.77
Lophostrix cristata	0–900	25	3,298,702.72
Pulsatrix perspicillata	0–1200	26	4,007,435.52
Bubo virginianus	0–600	2	
Glaucidium costaricanum	1500–3475	6	264,241.61
Glaucidium griseiceps	0–900	20	2,512,412.32
Glaucidium brasilianum	0–600	7	1,764,014.67
Athene cunicularia	0–600	0	
Ciccaba virgata	0–900	25	2,703,694.67
Ciccaba nigrolineata	0–2100	27	3,966,340.38
Pseudoscops clamator	0–600	17	1,744,788.38
Aegolius ridgwayi	2100–3475	2	44,039.33

17.8 Future Conservation Strategy

The general strategy for the conservation of owls in protected areas must be the safeguarding of their habitat and the integrity of their communities. Immediate actions must include the control of subsistence or commercial invasions, logging activities, regulations for expansion of cattle farming, supervision in the opening of access roads, and supervision of visitors to the protected areas with regard to the scientific collections of threatened species or species with restricted distribution. A second component must include an environmental education program that allows people to learn about the natural history of this group of birds in the buffer zone of the protected areas and in the cities as well. The conservation strategy in the agricultural ecosystems of Panama must be aimed to ensure the subsistence of the local bird populations. Immediate and permanent actions must include a public awareness program through the different social communication media, emphasizing the important role of owls (and snakes) in agricultural areas; precise recommendations about the use of biodegradable pesticides, as well as the control of pests through environmentally friendly practices; and surveillance from the authorities involved in the protection of the environment and the agricultural production to ensure the compliance with the existing agricultural and phytosanitary laws for these cases. The program must also include the posting of sign warning about the owls flying at low altitude in rice fields next to high-traffic roads, where they frequently are hit by vehicles, as well as recommendations to NGOs and to the civil society to contribute in a practical and effective way with these actions in favor of the group of birds.

Finally, it is important to point out the role that bird watchers may play in their periodical records about the natural history of owls of Panama, especially in the recording of new sightings, nests, chicks, and predators. Likewise, scientists and institutions must become engaged in participating in research about the accumulation of pesticides in owls, with special reference to the annual eggs and litter of their preys. Only with the integral help of the groups involved, we will be able to ensure the survival of the Panamanian and American Strigidae populations.

17.9 Status of the Biological and Ecological Knowledge

T. alba Nocturnal, active at dawn and dusk. It feeds mainly on small mammals (bats, mice, and shrews), small birds, and large insects and occasionally reptiles and amphibians. It usually nests in cavities (in trees, underground holes, wells), abandoned buildings, church bell towers, or palm trees (in agricultural zones or grasslands), and it lays three to seven eggs. To hunt, it remains silent on a tree branch, where it observes and listens to any movement. When it hears or sees an animal, it pounces silently near enough to catch it with its claws (Wetmore 1968, Mendez 1987, Ridgely and Gwynne 1993, Ventocilla 2004). It frequently makes a strong, harsh chirp; occasionally during the day, it can be seen in a tree branch, and if disturbed, it tends to bend down and extend its wings, making hissing and clicking sounds with its beak (Ridgely and Gwynne 1993).

M. choliba Nocturnal. It feeds primarily on large insects (ants, leafcutter ants, cockroaches, butterflies, grasshoppers, beetles, cicadas, praying mantis, scorpions, spiders) and small mammals (mice, bats). It nests in cavities in trees and lays two to four eggs. To hunt, it poses in naked branches at low altitude or close to lamps (Wetmore 1968, Ridgely and Gwynne 1993). Its characteristic call, a sort of purr like *prrrrrrr* or *hurrrrrrr*, which usually ends up in an *uuk* or *¿uuk? ¿uuk?* in an abrupt and whiny manner, made just right after nightfall and very early before dawn (Ridgely and Gwynne 1993).

M. guatemalae Nocturnal. It feeds on large insects (Tettigoniidae beetles). There is no information about the nesting habits of this species (Wetmore 1968, Ridgely and Gwynne 1993). Two different types of calls have been recorded: one is a short and guttural *k-r-r-r-r-o*, similar to the one of a toad, repeated at very long intervals, and the other one is a very fast and tremulous monotone sound (Ridgely and Gwynne 1993). There is no information about the nesting habits of this species (Wetmore 1968).

M. clarkii Nocturnal. It feeds on large insects (beetles and Orthoptera) and small mammals (rodents and shrews). There are no records about their nesting habits or egg laying in Panama (Wetmore 1968, Ridgely and Gwynne 1993). Wetmore (1968) recorded a call made in two parts; the first one is a simple note with a quick pause, followed by a short and quick repetition up to three times, *coo, coo-coo-coo*, high and somehow musical. Another call is a series of deep notes like *uuu*, or *cuu*, repeated in an irregular manner, without a specific pattern, yet occasionally it is repeated in a regular pattern, like an *uuu, uuu-uuu-uuu; uuu-uuu-uuu…*; or *cuu, cuu-cu, cuu-cu…* (Ridgely and Gwynne 1993). Field observations record bird couples singing in a duo (Robbins et al. 1985 and B. Jiménez pers. obs.).

L. cristata Nocturnal. It feeds on beetles, Orthoptera, cockroaches, and caterpillars. There is no information about local nesting or egg laying (Wetmore 1968, Ridgely and Gwynne 1993). It usually rests during the day along streams in the forest, at heights between 3 and 9 m above the ground; its call is a deep and penetrating *gruurrr* or *buoorrr*, which is made while posing in the canopy of big trees (Ridgely and Gwynne 1993).

P. perspicillata Nocturnal, with occasional activity during cloudy sunsets. It feeds on large insects (Orthoptera), small and medium mammals (mice: *Oryzomys bicolor* and *Tylomis* sp.), lizards, and birds. There is no information about nesting habits for this species (Wetmore 1968, Ridgely and Gwynne 1993). Its call is a rapid series of 6–8 deep hoots, *bubububububububu*, which can be heard from afar and sounds like a machine gun; juveniles also make a strong and harsh hiss, *juiiiu* (Ridgely and Gwynne 1993).

B. virginianus There is no information about this species in Panama.

G. costaricanum It is active during the day and at night, very often at dusk or at dawn (Ridgely and Gwynne 1993). There is little information about this species in the country.

G. griseiceps Active during the day and at night, in the mid-level or at the canopy level of the forest. It makes between three and five notes, sometimes between eight and nine, *puup*, when feeling disturbed; it can be easily attracted by imitating its call (Ridgely and Gwynne 1993).

G. brasilianum Active during the day and at night. It usually poses in fence posts or telephone cables. It feeds on large insects (cicadas) and small lizards and birds. It is frequently hounded by small birds (Wetmore 1968, Ridgely and Gwynne 1993). Its call, which is mainly made during the night and seldom during the day, is a long series of *puup* and *tuut*, which is repeated in a rapid manner for several minutes. It also responds to imitated calls or recordings (Ridgely and Gwynne 1993).

A. cunicularia Active during the day and at night. It usually poses at low altitude (on the ground or in branches), and when it is nervous, it balances up and down (Wetmore 1968, Ridgely and Gwynne 1993).

C. virgata Nocturnal (it rests during the day within the dense forest). It feeds on small insects, Orthoptera and large beetles, small mammals (rodents), snakes, and small amphibians (Wetmore 1968, Ridgely and Gwynne 1993). Among its calls, there is one with a deep and dull *juu-au, juu-au,* or *juuuu, juuuu*, duplicated or triplicated (sometimes repeated longer); another call sounds like a *kiyáuu* or *kiiauiíyu*, similar to the chirping of a cat, and an abrupt *bru bru*, or *bu. bu. bu.*, also repeated two or three times with one or more second intervals between each sound (Wetmore 1968, Ridgely and Gwynne 1993).

C. nigrolineata Nocturnal. It feeds on large insects (beetles and Orthoptera), small rodents, and bats (*Myotis nigricans*) (Wetmore 1968, Ridgely and Gwynne 1993). It has a variety of calls, the most frequent being a nasal and whiny *kii-yáu* or *kiyáu, a buj-buj-buj-buj-buj-buój-buo*, with an accent in the seventh note (occasionally, the last note is not heard or is not made at all); another call that has been recorded is a deep and resounding *juuf, juuf, juuf* (Ridgely and Gwynne 1993).

P. clamator Nocturnal (it rests during the day in the trees and in low altitude wires or dense forests, looking straightly downward). It nests in the ground and lay two eggs (Wetmore 1968, Ridgely and Gwynne 1993). Its call is strong, penetrating, and semi-hiss that sounds like *a juiiyu*; there is also some sort of barking sound *au,au,au,au* (Ridgely and Gwynne 1993).

A. ridgwayi There is no information about this species in Panama.

17.10 Conclusions

1. There are 15 species distributed in Panama (one barn owl and 14 typical owls), of which two of them are considered uncertain (*B. virginianus* and *A. cunicularia*) due to the poor existing records. The first species is considered endangered, and the second one is considered vulnerable, along with the rest of the species by the Panamanian conservation laws.
2. Information about the current status and ecology of at least 13 of the 15 species of Strigiformes was obtained at a national level.
3. The management plans of at least 12 protected areas were reviewed, in which no information was found regarding specific conservation actions for owls. Thus, there are no conservation strategies for this group of birds.
4. Of the 65 protected areas of the SINAP that were identified in 2006, this analysis records typical owl and barn owl species in at least 20 of the 29 potential protected areas, with land extensions larger than 4500 hectares.
5. Of the nine provinces and the Kuna Yala "comarca," where the protected areas are located, the province of Darién has more than 50% of the potential habitat for these species. All the 15 species that should potentially be found were recorded in this province. In provinces like Bocas del Toro, Chiriquí, and Panama, the richness of the barn owl and the typical owls is well known. A different situation happens in the provinces of Herrera, Los Santos, and Coclé, where there are significant information gaps.
6. The owl species that are restricted to the cloud forests of Panama (*M. clarkii*, *G. costaricanum*, and *A. ridgwayi*) are well represented in the protected areas at the east and west regions of Panama, being *A. ridgwayi* with the most restricted distribution.
7. The species that are widely distributed of lowland forests and foothills are also largely found in at least 25 protected areas (*M. guatemalae*, *P. perspicillata*, *L. cristata*, *C. virgata*, and *C. nigrolineata*).
8. Predictive distribution maps were established for the 15 species of owl recorded in Panama. Of all these species, ten of them present distribution polygons of more than one million hectares, which coincides with the fact that they are species with wide altitudinal ranges.
9. Coiba National Park and Santa Fe National Park protect the habitat of *B. virginianus*, a species whose existence in the country needs to be confirmed. The same situation applies to *A. cunicularia*, which has been registered only once, in Divala, province of Chiriquí (December 13, 1900); hence there is the need for confirmation of this species. Due to the few existing records, these species do not have a distribution polygon in this study.
10. The main threats for the owls are the destruction of their habitats, the decrease of the natural feeding areas, the use of agricultural chemicals and/or pesticides that affect them, either directly or indirectly (by accumulation of these substances in their bodies or by the ingestion of poisoned animals), and the impact against vehicles in roads next to rice fields.

11. Throughout the present work, it was possible to get a glance of the status of the owl species according to the existing information from museum collections, field work, and bibliography published up to now, but it is precisely the human behavior what have been causing the drastic changes in the natural habitat of this group of birds; therefore, it is imperative that short- and long-term research be made to allow us to determine the current status of these species, especially those two species that are considered uncertain (*B. virginianus* and *A. cunicularia*), aiming at the implementation of conservation programs.

Striped Owl (*Pseudoscops clamator*)

Spectacled Owl (*Pulsatrix perspicillata*)

Tropical Scheech Owl (*Megascops choliba*)

Tropical Scheech Owl (*Megascops choliba*)

Acknowledgments We are grateful to Yolani Holmes and Katherine Arauz, for helping in the search for information in different documentation centers. We are also thankful to Jacobo Arauz, Alvin Zapata, Arturo Morris, and the National Association for Nature Conservancy (ANCON), for providing the lists of birds in specific areas. Our special gratitude goes to Prof. Eustorgio Mendez, one of the main experts in biology of Panama, who gave us the drawing of striped owl (*Pseudoscops clamator*) for the illustration of the present article. In this sense, we give special thanks to Christian Ziegler, Gary† and Daisy Vaucher, Andrés Hernandez, and Nessim Bassam, who very gently provided us with representative photographs of some of the species. Likewise, we want to thank Dr. Rafael Samudio, Research Director of the Gorgas Memorial Laboratory for Health Studies, who allowed us to visit and review the zoological collection of Dr. Eustorgio Méndez, and to Mr. Miguel Morales and Mr. Erick Quintero, both of them in charge of the care and maintenance of the said collection. Our thanks go also to Dr. Victor Tejera, Director of the Vertebrate Museum of Panama, and to Ricardo Pérez, for allowing us to review the museum's collection. Our eternal gratitude goes to Mr. Jesus Kantule as well, who gave us the local names of this fascinating group of birds in Kuna dialect; also to Dr. Mark Adams of the Natural History Museum of the United Kingdom and to Mr. Hill Pakenham, who gently gave us the list of specimens collected in Panama that are kept in their records; and to Nefertaris Daguerre, who helped us in the elaboration of the photograph archives and to Maria Eugenia De Diego.

Appendix 17.1

Taxonomic diversity of the species of owls recorded in Panama

				Local common names				Actual status of the wild population	
									Globally
Scientific name[a]	Subspecies[f]	Common Spanish name[b]	Common English name[a, b, c]	Central provinces[d]	Kuna dialect [e]	Condition [b, f]	Status[b, e, f]	National[g]	CITES [h]/ UICN [i]
Order Strigiformes		Búhos y lechuzas	Owls and barn owls						
Family Tytonidae		Lechuza de campanario	Barn owls						
Tyto alba	*T. a. guatemalae*	Lechuza campanario	Barn owl	*leana* (Coclé)		R	Pc a Fr	VU	CITES II/ LR
FAMILY STRIGIDAE		Búhos típicos	Typical owls						
Megascops choliba	*M. c. luctisonus*; *M. c. crucigerus*	Autillo tropical	Tropical screech owl	*cocorito* (Veraguas) *turrututú* (Las Tablas)	*paluqua*	R	Fr	VU	CITES II/ LR
Megascops guatemalae	*M. g. vermiculatus*	Autillo vermiculado	Vermiculated screech owl			R	Pc/L	VU	CITES II/ LR
Megascops clarkii		Autillo serranero	Bare-shanked screech owl			R	R	VU	CITES II/ LR
Lophostrix cristata	*L. c. stricklandi*; *L. c. wedeli*	Búho penachudo	Crested owl			R	Pc/L	VU	CITES II/ LR

(continued)

Scientific name[a]	Subspecies[f]	Common Spanish name[b]	Common English name[a, b, c]	Local common names		Condition[b, f]	Status[b, e, f]	Actual status of the wild population	
				Central provinces[d]	Kuna dialect[e]			National[g]	Globally CITES[h]/ UICN[i]
Pulsatrix perspicillata	*P. p. saturata; P. p. chapmani*	Búho de anteojos	Spectacled owl	*leona* (Coclé)	*irmon*	R	Pc a Fr	VU	CITES II/ LR
Bubo virginianus	*B. v. mesembrinus*	Búho cornado americano	Great horned owl			V	In	EN	CITES II/ LR
Glaucidium costaricanum		Mochuelo montañero	Costa Rican pygmy owl			R	R	VU	CITES II/ LR
Glaucidium griseiceps	*G. g. rarum*	Mochuelo enano	Central American pygmy owl			R	R a Pc	VU	CITES II/ LR
Glaucidium brasilianum	*G. b. ridgwayi*	Mochuelo ferruginoso	Ferruginous pygmy owl	*cocorito* (Veraguas) *lebruna* (Herrera) *silampera* (Las Tablas) *picapiedra* (Coclé)		R	Pc a Fr	VU	CITES II/ LR
Athene cunicularia	*A. c. hypugaea*	Búho terrestre	Burrowing owl			V	Ac	VU	CITES II/ LR
Ciccaba virgata	*C. v. centralis; C. v. virgata*	Búho moteado	Mottled owl		*ueko*	R	Fr	VU	CITES II/ LR

Ciccaba nigrolineata		Búho blanquinegro	Black-and-white owl		*ilaqua*	R	Pc	VU	CITES II/ LR
Pseudoscops clamator	*P. c. forbesi*	Búho listado	Striped owl		*amma noo kukualet*	R	Pc/L	VU	CITES II/ LR
Aegolius ridgwayi	*A. ridgwayi*	Buhito pardo	Unspotted Saw-whet owl			R	R	VU	CITES II/ LR

Codes – condition: *R* resident, *V* vagrant (species recorded less than 10 times). Status: *Fr* frequent, *Pc* uncommon, *R* rare, *L* local; In, uncertain; *Ac* accidental; actual status of the wild population. National: *VU* vulnerable species, *EN* endangered species. Global: CITES, CITES II Appendix II; IUCN, *LR* low risk

[a]AOU (1993, 1997, 2000, 2002, 2003, 2004, 2005)

[b]Ridgely and Gwynne (1993)

[c]Del Hoyo et al. (1999)

[d]Kantule (pers. comm.)

[e]Angehr (2006)

[f]Wetmore (1968)

[g]ANAM (2008b)

[h]UNEP – WCMC (2009)

[i]UICN (2009)

Appendix 17.2

Distribution of the specimens recorded in Panama in two local and in two international museums

	West			Middle region					East		
	Bocas del Toro	Chiriquí	Veraguas	Los Santos	Herrera	Coclé	Panamá	Colón	Darién	San Blas	Total
T. alba						1	10				11
M. choliba		1		1		1	16				19
M. guatemalae		1			1		9		2		13
M. clarkii							7		1		8
L. cristata			1				8				9
P. perspicillata	1						19				20
B. virginianus	No records available										0
G. costaricanum		1					2				3
G. griseiceps							2	1	1		4
G. brasilianum				1			12				13
A. cunicularia	No records available										0
C. virgata	7						21	1			29
C. nigrolineata	8						4				12
P. clamator					1	1	10				12
A. ridgwayi							1				1
Total	16	3	1	2	2	3	121	2	4	0	154

References

Alvarado R (2006) Actualización del Plan de Manejo del Parque Nacional Soberanía. Technical Report. ANAM and USAID, p 211.

ANAM (Autoridad Nacional del Ambiente) (2000) Mapa de Vegetación de Panamá – Corredor Biológico Mesoamericano del Caribe Panameño. Scale 1: 250,000. Panama. Louis Berger Internacional

ANAM (2004) Informe del Estado del Ambiente. GEO Panamá, p 175

ANAM (2006) Indicadores Ambientales de La República de Panamá, p 145

ANAM (2008a) Plan de Manejo del Corredor Biológico de la Serranía de Bagre Integrado a la Reserva Forestal de Chepigana. Technical Report, p 204

ANAM (2008b) Resolución AG-0051-2008. Por la cual se reglamenta lo relativo a las especies de fauna y flora amenazadas y en peligro de extinción, y se dictan otras disposiciones. In: Gaceta Oficial N° 26,013

ANCON (Asociación Nacional para la Conservación de la Naturaleza) (1993) Evaluación Ecológica Rápida de Darién. ANCON. Technical Report. Panama, p 312

ANCON (1995) Monitoreo Biológico y Forestal de la Reserva Natural Punta Patiño. ANCON-OIMT. Technical Report. Panama, p 74

ANCON, CEPSA (Consultores Ecológicos Panameños, S. A.) (2004a) Diagnóstico Biológico y Sociocultural del Bosque Protector Palo Seco. Elaboración del Plan de Manejo del Bosque Protector Palo Seco. Final Report. ANAM/CBMAP. Panama, p 337

ANCON, CEPSA (2004a) Diagnóstico Biológico y Sociocultural del Humedal San San Pond Sak. Elaboración del Plan de Manejo del Humedal San San Pond Sak. Final Report. ANAM/ CBMAP. Panama, p 266

ANCON, CEPSA (2004b) Diagnóstico Biológico y Sociocultural del Parque Internacional La Amistad, *En:* Elaboración del Plan de Manejo del Parque Internacional La Amistad. Final Report. ANAM/CBMAP. Chiriqui, Panama, p 325

ANCON, CEPSA (2004c) Diagnóstico Biológico y Sociocultural del Parque Nacional Volcán Barú, *In:* Elaboración del Plan de Manejo del Parque Nacional Volcán Barú. Final Report. ANAM/CBMAP. Chiriqui, Panama, p 288

ANCON, CEPSA (2004d) Plan de Manejo del Bosque Protector de Palo Seco, provincia de Bocas del Toro. ANAM-CBMAP. Bocas del Toro, Panama, p 140

ANCON, CEPSA (2004e) Plan de Manejo: Humedal San San Pond Sak, provincia de Bocas del Toro. Technical Report, ANAM/CBMAP. Panama, p 155

ANCON, CEPSA (2004f) Plan de Manejo: Parque Internacional La Amistad, provincias de Bocas del Toro y Chiriquí. Technical Report. ANAM/CBMAP. Panama, p 158

ANCON, CEPSA (2004g) Plan de Manejo: Parque Nacional Volcán Barú, provincia de Chiriquí. Technical Report. ANAM/CBMAP. Panama. p 176

ANCON, CEPSA (2004h) Plan de Manejo: Parque Nacional Volcan Baru, provincia de Chiriquí. Technical Report. ANAM/CBMAP. Panama, p. 176

Angehr G (2003) Directorio de Áreas Importantes para Aves en Panamá. Sociedad Audubon de Panamá y Bird Life/Vogelbescherming Nederland, Panama

Angehr G (2006) Annotated checklist of the Birds of Panama. Panama Audubon Society, Panamá

Angehr G, Domínguez I, Gallardo MM et al (1999) Proyecto de Monitoreo de la Cuenca del Canal de Panamá. Componente de Poblaciones de Vertebrados. Aves. Technical Report. Vol. 5. Appendix 14. Panama, p 26

Angehr GR, Engleman D, Engleman L (2008) A bird-finding guide to Panama. Comstock Pub. Associates/Cornell University Press, Ithaca

AOU (American Ornithologists' Union) (1983) Checklist of north American Birds, 6th edn. Allen Press, Lawrence, p 877

AOU (1993) Thirty-ninth Supplement to the American Ornithologists´ Union Check-list of North American Birds. AUK 110:675–682

AOU (1997) Forty-first supplement to the American ornithologists' Union *checklist* of north American birds. Auk 114(3):542–552
AOU (2000) Forty-second supplement to the American ornithologists' Union *checklist* of north American birds. Auk 117:847–858
AOU (2002) Forty-third Supplement to the American Ornithologists´s Union Check-list of North American Birds. AUK 119(3):897–906
AOU (2003) Forty-fourth supplement to the American ornithologists' Union *checklist* of north American birds. Auk 120:923–931
AOU (2004) Forty-fifth Supplement to the American Ornithologists´ Union Check-list of North American Birds. AUK 121(3):985–995
AOU (2005) Forty-sixth supplement to the American ornithologists' Union checklist of north American birds. Auk 122:1026–1031
AOU (2008) Check-list of North American birds 7th edition. http://www.aou.org/checklist/index.php3. Accessed 9 Jan 2009
Aparicio K (1999a) Aves. Informe Final del inventario de vertebrados terrestres realizado en los ríos Pequení y San Miguel, Parque Nacional Chagres In: Proyecto Monitoreo de la Cuenca del Canal. Final Report for the monitoring Project of the Canal Watershed, Appendix 10. Vol 4, USAID-ANAM-STRI, Panama, pp 13–34
Aparicio K (1999b) Aves. Informe Final del Inventario de Vertebrados Terrestres Realizado en la quebrada Las Pavas, río Chagres, Parque Nacional Chagres, In: Proyecto Monitoreo de la Cuenca del Canal. Final Report for the monitoring Project of the Canal Watershed, Appendix 13. Vol 5, USAID-ANAM-STRI, Panama, p 14–28
Aparicio K (1999c) Aves. Informe Final del Inventario de Vertebrados Terrestres Realizado en Río Trinidad, Parque Nacional Campana, In: Proyecto Monitoreo de la Cuenca del Canal. Final Report for the monitoring Project of the Canal Watershed, Appendix 9. Vol 4, USAID-ANAM-STRI. Panama, p 12–25
Araúz GJ (2004) "Inventario de aves en los bosques fragmentados de la Reserva Forestal El Montuoso, Herrera-Panamá". In: Garibaldi C (ed) Diversidad biológica y servicios ambientales de los fragmentos de bosque en la Reserva Forestal El Montuoso, Panamá. Agencia de Cooperación Internacional del Japón (JICA), pp 139–159
BirdLifeInternational (2008) Species factsheet: *Megascops clarkii*. http://www.birdlife.org. Accessed 2 Dec 2008
Blake ER (1958) Birds of Volcán de Chiriquí, Panama. Fieldiana Zool 36:499–577
CITES (Convention on International Trade in Endangered Species) (1998) Checklist of CITES species: a reference to the appendices to the convention on international trade in endangered species of wild fauna and flora. CITES Secretariat/World Conservation Monitoring Centre. Unwin Brothers, Martins Printing Group, Old Working, Surrey, Genève
CEPSA (Consultores Ecológicos Panameños, S.A.) (2000) Evaluación Ecológica Rápida del Parque Nacional General de División Omar Torrijos Herrera. Technical Report. ANAM. Panama, p 169
CEPSA (Consultores Ecológicos Panameños, S.A.) (2002) Plan de Manejo del Área Protegida San Lorenzo. Technical Report. ANAM-ARI-INAC-IPAT. Colon, p 180
CEPSA (Consultores Ecológicos Panameños, S.A.) (2005) Plan de Manejo: Refugio de Vida Silvestre La Playa de La Barqueta Agrícola, provincia de Chiriquí. Technical Report. ANAM. Chiriquí, Rep. de Panamá, p 181
Chapman FM (1929) My tropical air castle. D. Appleton & Co., New York
Chapman FM (1938) Life in an air castle. Appleton-Century Co., New York
Del Hoyo J, Elliott A, Sargatal J (1999) Handbook of the birds of the world. Barn owls to hummingbirds. Lynx Editions, Barcelona
Edwards E, Loftin H (1971) Finding birds in Panama, 2nd edn. Ernest P. Edwards, Sweet Briar
Eisenmann E (1955) The species of middle American birds. Trans Linnaean Soc 7:1–128
Eisenmann E, Loftin H (1972a) Field checklist of birds of Panama canal zone area, 2nd edn. Russ Mason's Flying Carpet Tours, Inc., Kissimmee

Eisenmann E, Loftin H (1972b) Field checklist of the western Chiriquí highlands, Panama, 2nd edn. Russ Mason's Flying Carpet Tours, Inc., Kissimmee
Gale N, Karr J, Morton E, Ridgely R, Smith N, Smythe N (1978) Birds of the pipeline road Are*a. Panama* Audubon Society – Smithsonian Tropical Research Institute, pp 4
Griscom L (1935) The ornithology of the Republic of Panama. Bull Mus Comp Zool 3:261–382
Heckadon MS (1983) Cuando se acaban los montes los campesinos santeños y la colonización de Tonosí. Editorial Universitaria Panamá. Smithsonian Tropical Research Institute, Panama, p 172
Instituto Geográfico Nacional Tommy Guardia (2007) Atlas Nacional de la República de Panamá, 4th edn. Ministery of Public Services, Panama, p 290
Instituto Panameño de Turismo (2009) Información general de Panamá. http://www.atp.gob.pa/index.php?option=com_content&view=article&id=51&Itemid=68. Accessed Nov 2009
Jiménez B (1999a) Aves. Informe Final del Inventario de Vertebrados Terrestres realizado en Cerro Bruja, Parque Nacional Chagres. In: Proyecto Monitoreo de la Cuenca del Canal. Final Report for the monitoring Project of the Canal Watershed. USAID-ANAM-STRI, Panama, vol 4 Appendix 11, pp 16–29
Jiménez B (1999b) Aves. Informe Final del Inventario de Vertebrados Terrestres realizado en Cerro Brewster, Parque Nacional Chagres. In: Proyecto Monitoreo de la Cuenca del Canal. Final Report for the monitoring Project of the Canal Watershed. USAID-ANAM-STRI, Panama, vol 4 Appendix 12, pp 16–30
Jiménez B, Aparicio K (2005) Aves del Área de Bahía Honda (Veraguas, Panamá). In: Castroviejo S, Ibáñez A (eds) XI. Estudios sobre la Biodiversidad de la región de Bahía Honda (Veraguas, Panamá). Superior Council for Scientific Research, Spain Institute, Real Academia de Ciencias Exactas, Físicas y Naturales, Madrid, pp 627–742
Karr J (1990) The Avifauna of Barro Colorado Island and the Pipeline Road, Panama. In: Gentry AH (ed) Four neotropical rainforests. Yale University Press, New Haven, pp 183–198
Libsch M (2002) Lista Preliminar de las Aves de la Reserva Forestal Palo Seco. University of Illinois. Smithsonian Tropical Research Institute. Technical Report. Panama, p 4
Marcus MJ, Roldán J (1984) Informe sobre la fauna de la Reserva Kuna Yala. STRI-University of Panama. Panama, p 55
Méndez E (1969) Una breve introducción a las Aves de Panamá. Private Edition. Panama, p 57
Méndez E (1987) Elementos de la Fauna Panameña. Panama, p 216
OTSCorp (2003) Plan de Manejo: Parque Nacional Darién. Technical Report. ANAM y Sustainable Development Program for Darien, Panama. Panama, p 77
Plan de Manejo y Desarrollo del Parque Nacional de Portobelo (1994) IPAT-FAO (Protected Area Management Program for Environmental Conservation)-MIDA. Panama, Panama
PROARCA (2001) Plan de Manejo: Parque Nacional Marino Isla Bastimentos. Technical Report. ANAM. Panama. p 128
Ridgely RS (1976) A guide to the Birds of Panama. Princeton University Press, Princeton, p 394
Ridgely RS, Gwynne Jr JA (1993) Guía de las Aves de Panamá, incluyendo Costa Rica, Nicaragua y Honduras. 1st Spanish Ed. ANCON. Panama, p 614
Ridgway R (1901–1950) The birds of north and middle America, parts 1–11. U.S. National Museum Bulletin 50, Washington, DC
Robbins MT, Parker TA III, Allen SE (1985) The Avifauna of Cerro Pirre, Darién, Eastern Panama. In: Buckley PA (ed) Neotropical ornithology, vol 36. American Ornithologist´ Union. Ornithol Monogr, Washington, DC, pp 198–232
Robbins M, Braun MJ, Graves GR, et al (1995) Lista preliminar de las aves de la Reserva Fortuna, Chiriquí, Panamá. University of Kansas. Natural History Museum. Technical Report, p 10
Robinson WD (1999) Long-term changes in the avifauna of a tropical forest isolate, Barro Colorado Island. Conserv Biol 13:85–97
Robinson WD (2001) Changes in abundance of birds in a Neotropical forest fragment over 25 years: a review. Animl Biodivers Conserv 24:51–65

Salvin O, Godman FD (1879–1904) Biologia Centrali-Americana. Birds, vol. 1, 2 and 3. Taylor and Francis, London

Santamaría DE (2000a) Conservación y consolidación de la diversidad biológica y cultural del Darién. Series 1: Biological Diversity. Asociación Nacional para la Conservación de la Naturaleza (ANCON). Technical Report, p 350

Santamaría DE (2000b) Propuesta para la Creación de un Área Protegida en Santa Fe de Veraguas, República de Panamá. ANCON, PROARCA/CAPAS, p 389

Schauensee RM (1964) The Birds of Colombia. Livingston Publ. Co., Wynnewood

Skutch A (1954-1960-1967) Life histories of central American Birds, vol 1–3. Pacific coast avifauna series 31, 34 and 35. Berkeley, Cooper Ornithological Society

Solís R, Jiménez EA, Brenes O, Vilnitzy SL (eds) (1999) Lista de Fauna de Importancia en conservación en Centroamérica y México. UICN-HORMA, WWF Central America, p 224

Statterfield AJ, Crosby MJ, Long AJ et al (1998) Endemic bird areas of the world, priorities for biodiversity conservation, Series 7. Birdlife International, Cambridge, p 846

STRI (2008) Base de datos de Especies. Instituto Smithsonian de Investigaciones Tropicales (STRI) http://biogeodb.stri.si.edu/bocas_database/search/class/29/. Accessed 3 Dec 2008

Sturgis BB (1928) Field book of Birds of the Panama Canal zone. G.P. Putnam's Sons, New York

Tejera V (1995) Inventario Biológico del Canal de Panamá. Estudio Ornitológico SCIENTIA (Panamá) 2:5–106

Tovar D (1999) Plan de Manejo: Parque Nacional Altos de Campana. Technical Report. ANAM, p 170

Tovar D, Samudio R, Dam Lau S, et al (2005) Plan de Manejo: Parque Nacional Chagres. Technical Report: Parks in Danger. ANAM-TNC-USAID. Panama, p 131

UICN (Union Internacional para la Conservación de la Naturaleza) (2009) UICN Lista Roja de Especies Amenazadas, Versión 2009.1. (http:// www.iucnredlist.org). Accessed 9 Mar 2009

UNEP-WCMC (2009) UNEP-WCMC species database: CITES-listed species. http://old.unep-wcmc.org/checklist-of-cites-species_499.html. Accessed 9 Jan 2009

Ventocilla, J (2004) ¿Qué vuela ahí?. Smithsonian Tropical Research Institute y Sociedad Audubon de Panamá. Panama

Weaver PL, Bauer GP (2004) El Parque Nacional San Lorenzo: Resumen de Recursos Culturales y Naturales. CEASPA. *Panama*, p 113

Wetmore A (1965) The birds of the Republic of Panama. Part 1: Tinamidae (Tinamous) to Rhynchopidae (Skimmers). Smith Misc Coll 150:1–483

Wetmore A (1968) The birds of the Republic of Panama. Part 2: Columbidae (Pigeons) to Picidae (Woodpeckers). Smith Misc Coll 150:1–605

Wetmore A (1973) The Birds of the Republic of Panama. Part 3: Passeriformes: Dendrocolaptidae (Woodcreepers) to Oxyruncidae (Sharpbills). Smith Misc Coll 150:1–631

Chapter 18
The Owls of Paraguay

Alejandro Bodrati, Paul Smith, Hugo del Castillo, and Ana Trejo

Abstract Paraguay is located at the center of South America at the confluence of several biomes: Atlantic Forest, Cerrado, Mesopotamian Grasslands, Chaco, and Pantanal. Some 719 bird species are documented for the country, of which 16 are owls. No owls are endemic to Paraguay, but one species is endemic to the Chaco biome and three species and one subspecies are endemic to the Atlantic Forest. In Paraguay, ornithological research has experienced productive periods and other long periods without advances; currently there are no studies underway that focus on the ecology or conservation of owls, although some studies on diet and behavior have been published. Only one species, Rusty-barred Owl (*Strix hylophila*), is considered at risk internationally, but seven species are considered at risk at the national level. The principal threats to the owls of Paraguay are habitat destruction, lack of protected areas or poor implementation of existing areas, popular beliefs that nocturnal birds bring bad luck, forest management incompatible with owl conservation, and lack of information about the basic biology of these birds.

Keywords Atlantic Forest • ChacoCerrado • Local Attitudes • Strigidae

A. Bodrati (✉)
Proyecto Selva de Pino Paraná, Vélez Sarsfield y San Jurjo s/n (3352),
San Pedro, Misiones, Argentina
e-mail: alebodrati@yahoo.com.ar

P. Smith
FAUNA Paraguay, www.faunaparaguay.com, Encarnación, Paraguay

Para La Tierra, Centro IDEAL, Mariscal Estigarribia 321 c/ Tte. Capurro, Pilar, Paraguay
e-mail: faunaparaguay@gmail.com

H. del Castillo
Guyra Paraguay, Parque Ecológico Capital Verde, Av. Cnel. Carlos Bóveda, Viñas Cué,
Asunción, Paraguay
e-mail: hugofdelcastillo@gmail.com

A. Trejo
Centro Regional Bariloche, Universidad Nacional del Comahue, 8400 Bariloche, Argentina
e-mail: ana.r.trejo@gmail.com

P.L. Enríquez (ed.), *Neotropical Owls*, DOI 10.1007/978-3-319-57108-9_18

Rusty-barred Owl (*Strix hylophilla*)

18.1 Introduction

Paraguay is a landlocked country located in the heart of South America between 54°19′ and 62° 38′ west and 18°18′ and 27° 30′ south. It is divided into two regions by the Paraguay River: the Occidental or Chaco region (with 61% of the land surface and less than 3% of the population) and the Oriental region (with 39% of the territory and 97% of the populace). Despite being one of the five smallest countries in South America with a total area of 406,752 km^2, Paraguay is home to a notable diversity of bird species as a result of its location at the interface of the Atlantic Forest, Cerrado, Mesopotamian Grasslands, Chaco (Dry and Humid Chaco), and Pantanal (Hayes 1995, Fig.18.1). An important network of rivers also traverses the country, formed principally by the Paraguay, Paraná, and Pilcomayo and their tributaries. The gallery forests that line the banks of these rivers act as subhumid, mesic corridors which allow the exchange of forest species among biomes (Cardoso da Silva 1996).

To date, 719 species have been reported from Paraguay, including 16 species of owls (Guyra Paraguay 2004, 2005; Table 18.1). Five species are widespread throughout the country: Barn Owl (*Tyto alba*), Tropical Screech Owl (*Megascops choliba*), Ferruginous Pygmy Owl (*Glaucidium brasilianum*), Striped Owl (*Pseudoscops clamator*), and Burrowing Owl (*Athene cunicularia*). Burrowing Owl is a species of open habitats (grasslands, savannas, and human-modified areas), and

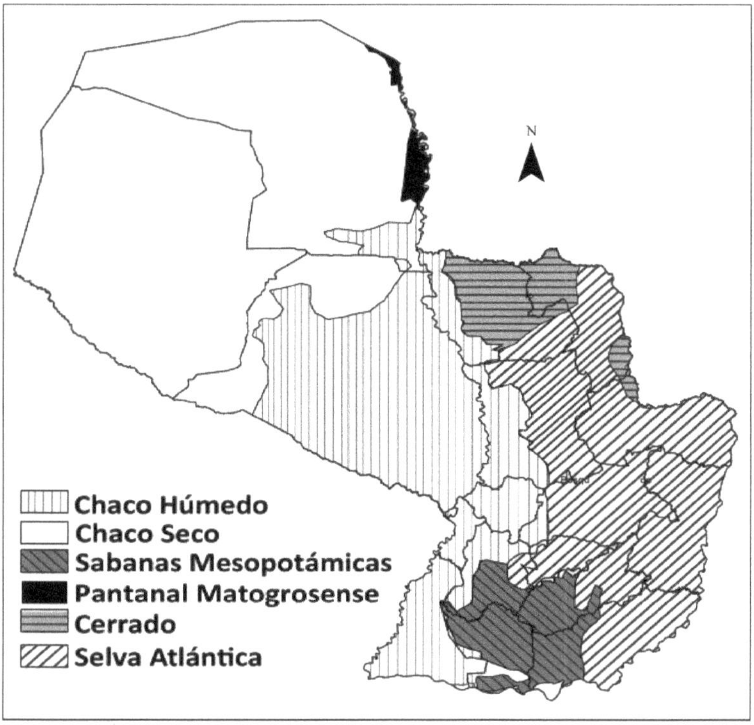

Fig. 18.1 Ecoregions of Paraguay (Guayra Paraguay 2005)

its distribution in Paraguay has increased dramatically in modern times as a result of the loss of much of the original coverage of the Atlantic Forest, a biome that it is only able to marginally penetrate.

Although no species of owl is endemic to Paraguay, five species are considered to be biome endemics. Chaco Owl (*Strix chacoensis*) is endemic to the Chaco biome with a distribution that also extends to Bolivia and Argentina. It is a relatively common species in much of the Chaco region, occurring in natural habitats as well as degraded forests and even close to human habitation (Guyra Paraguay 2004).

Black-capped Screech Owl (*Megascops atricapilla*), Rusty-barred Owl (*Strix hylophila*), Tawny-browed Owl (*Pulsatrix koeniswaldiana*), and a subspecies of the Black-banded Owl (*Ciccaba huhula albomarginata*) are endemic to the Atlantic Forest and share their distribution with Brazil and Argentina (Parker et al. 1996; Brooks et al. 1999; Marks et al. 1999; Bodrati and Cockle 2013). Black-capped Screech Owl (*M. atricapilla*) is locally common in Atlantic Forest, being more numerous in the remnants of forest in the northern and eastern Oriental region and less common in those farther south (Lowen et al.1997; Guyra Paraguay 2004; Cockle et al. 2005). It has also been recorded in transitional areas with Cerrado (Robbins et al. 1999). Rusty-barred Owl (*S. hylophila*) occurs in larger remnants of mature and degraded Atlantic Forest and is one of the more frequently encountered large owls in forests of the southern Oriental region (Bodrati and Cockle 2006).

Table 18.1 Species of owl known to occur in Paraguay (Guyra-Paraguay 2004, 2005)

English	Scientific name	Guaraní	Spanish name	Biomes
Barn Owl	*Tyto alba*	Suindá	Lechuza de campanario	AR
Tropical Screech Owl	*Megascops choliba*	Kavuré	Lechucita común	AR
Black-capped Screech Owl	*Megascops atricapilla*	Kavuré	Lechucita grande	AF, CE
Spectacled Owl	*Pulsatrix perspicillata*	Urukuré'a guasú	Lechuzón mocho grande	AF, PA, HCH, CE
Tawny-browed Owl	*Pulsatrix koeniswaldiana*	Urukure'a mini	Lechuzón mocho chico	AF
Great Horned Owl	*Bubo virginianus*	Ñakurutû guasú	Ñacurutú	DCH, HCH, CE, PA, MG
Rusty-barred Owl	*Strix hylophila*	Suindá ka'aguy o guasú	Lechuza listada	AF
Chaco Owl	*Strix chacoensis*	Suindá chaco	Lechuza chaqueña	DCH, PA
Mottled Owl	*Ciccaba virgata*	Kavure guasú, Suindá kaagui	Lechuza estriada	AF, CE
Black-banded Owl	*Ciccaba huhula*	Suindá hû	Lechuza negra	AF
Ferruginous Pygmy Owl	*Glaucidium brasilianum*	Kavure'i	Caburé	AR
Burrowing Owl	*Athene cunicularia*	Urukurea chichi	Lechucita vizcachera	AR
Buff-fronted Owl	*Aegolius harrisii*	Kavure'i pytâ	Lechucita canela	HCH, DCH, PA, AF
Striped Owl	*Pseudoscops clamator*	Ñakurutû'i	Lechuzón orejudo	AR
Stygian Owl	*Asio stygius*	Ñakurutû hû	Lechuzón negruzco	AF, HCH
Short-eared Owl	*Asio flammeus*	Suindá ñu	Lechuzón de campo	HCH, DCH, AF, CE, MG

Taxonomy follows Remsen et al. (2012). Common names in Guaraní and Spanish (the official languages of Paraguay) follow Guyra Paraguay (2004, 2005). The following abbreviations are used for biomes: *HCH* Humid Chaco, *DCH* Dry Chaco, *AF* Atlantic Forest (includes Alto Paraná forest and Paraguay Central forest), *CE* Cerrado, *MG* Mesopotamian Grasslands, *PA* Pantanal, *AR* species found in all regions of the country. Distribution data is compiled from Bertoni (1901, 1939), Short (1976), Storer (1989), Hayes (1995), Brooks et al. (1995), Cardoso da Silva (1996), Ericson and Amarilla (1997), Lowen et al. (1997), Robbins et al. (1999), Capper et al. (2001), Zyskowski et al. (2003), Cockle et al. (2005), Guyra Paraguay (2004, 2005), Tierno de Figueroa and Padial (2005), Bodrati and Cockle (2006), Esquivel et al. (2007), Ramírez Llorens and Bellocq (2007), and Bodrati et al. (2012)

Tawny-browed Owl (*P. koeniswaldiana)* also appears to be more common in the south of the country and is able to inhabit even small patches of well-preserved forest (Guyra Paraguay 2004). The presence of Black-banded Owl (*C. huhula*) was first confirmed in the 1990s. It remains extremely scarce and seems to have a restricted range in a few well-preserved blocks of the Atlantic Forest (Brooks et al.1995; Lowen et al.1997; Ericson and Amarilla 1997; Cockle et al. 2005; Velázquez and Bodrati in Guyra Paraguay 2004).

The remaining six species of owls can be found in various biomes. Known records of Spectacled Owl (*Pulsatrix perspicillata*) suggest a distribution associated with gallery forests in the Paraguay and Paraná River watersheds (Ramírez Llorens and Bellocq 2007). It occurs in Cerrado in the northern Oriental region (Concepción Department) (Robbins et al. 1999), Atlantic Forest at the Reserva de Bosque Mbaracayú (Canindeyú Department), islands of forest in the Mesopotamian Grasslands (Itapúa department), and in the northern Dry Chaco (Capper et al. 2001; Zyskowski et al. 2003).

Great Horned Owl (*Bubo virginianus*) is a common owl in the forests of the Dry and Humid Chaco but occurs only marginally east of the Paraguay River in Cerrado, where it is much less numerous (Short 1976; Robbins et al.1999; Zyskowski et al. 2003; Guyra Paraguay 2004; Tierno de Figueroa and Padial 2005). Mottled Owl (*Ciccaba virgata*) occurs predominantly in remnants of the Atlantic Forest in eastern Paraguay, being more frequent in the north and east of that region and less common in the south (Guyra Paraguay 2004; Cockle et al. 2005). There are recent reports from gallery forest in the Cerrado region (Concepción Department along the Apa River), and marginal presence in the Chaco is suspected (Bodrati et al. 2012). Buff-fronted Owl (*Aegolius harrisii*) is a rarely recorded species in Paraguay, being uncommon in the Atlantic Forest and having been recently recorded in the Pantanal and Dry and Humid Chaco (Guyra Paraguay 2004; Bodrati and Cockle 2006b). Striped Owl (*Pseudoscops clamator*) is widespread and apparently fairly common, with records from across the country in all the major biomes and even urban areas. However, the species is likely under-recorded because of its secretive habits. Stygian Owl (*Asio stygius*) is known from few records mainly in the east and south of the Oriental region, and there have been few modern reports. The species would seem to be rare (Guyra Paraguay 2004, 2005), though again it is probably overlooked.

Four other species of owls have been reported for Paraguay but lack documentation. König et al. (1999) include southeastern Paraguay in the range of Long-tufted Screech Owl (*Megascops sanctaecatarinae*), but there are no reports of the species from the country and this appears to be an error or overextrapolation of distribution. Short (1976), König et al. (1999), and Mikkola (2012) treat Chaco Pygmy Owl (*Glaucidium tucumanum*) as a distinct species, but the majority of authors consider this taxon to be a subspecies of Ferruginous Pygmy Owl (*G. brasilianum*) (Guyra Paraguay 2004; Remsen et al. 2012) and that arrangement is followed here. Howell and Robbins (1995), König et al. (1999), and Marks et al. (1999) report the Least Pygmy Owl (*G. minutissimum*) for Paraguay, but Guyra Paraguay (2004) affirms that no reliable evidence exists that the species occurs in Paraguay. The reference to Sick's Pygmy Owl (*G. sicki*) in eastern Paraguay by Mikkola (2012) refers to the same erroneous reports. Short-browed Owl (*Pulsatrix pulsatrix*) was listed for Paraguay by Ihering (1904), Kelso (1934), and Bertoni (1914, 1939), but this is a misapplication of the name in reference to Spectacled Owl (*P. perspicillata*). Perhaps as a result of this, the form has also been listed more recently for Paraguay in error by Marks et al. (1999) who treated it as a subspecies of Spectacled Owl (*P. perspicillata*).

18.2 A Short History of Paraguayan Ornithology

The Paraguayan avifauna has been studied since colonial times, due principally to the contribution of various explorers, Jesuits, and European, North American, and Argentinian naturalists (Hayes 1995). The pioneering work of Félix de Azara (1805), considered the "father of ornithology" in southern South America, stands out for its meticulous attention to detail at a time when the discipline was still in its infancy. He listed eight species of owls: the Ñacurutú (*Bubo virginianus*), Ñacurutú mocho (*Pulsatrix perspicillata*), Ñacurutú chorreado (*Pseudoscops clamator*), Suindá (*Asio flammeus*), Lechuza (*Tyto alba*), Urucureá (*Athene cunicularia*), Choliba (*Megascops choliba*), and Caburé (*Glaucidium brasilianum*).

Almost a century had passed before the next great "Paraguayan" naturalist emerged Arnaldo de Winkelried Bertoni, who plied his trade from his base at Puerto Bertoni in Alto Paraná Department, collecting birds and other local fauna. Being the most prolific Paraguayan ornithologist of the early twentieth century, his publications were frequent (Hayes 1995), and his contribution to regional ornithology was significant, including the discovery and description of Tawny-browed Owl (*Pulsatrix koeniswaldiana*) (Bertoni 1901), as well as two subspecies still considered valid today, of Mottled Owl (*Ciccaba virgata borelliana*) (Bertoni 1901) and Stygian Owl (*Asio stygius barberoi*) (Bertoni 1930). Other substantial contributions to the Paraguayan avifauna around this time were published by Laubmann (1939–1940a, b), Podtiaguin (1941–1945), and Schade and Masi Pallarés (1967, 1970a, b, c, 1971).

More recently, Hayes (1995) published the first modern revision of the Paraguayan avifauna, providing a list with distribution and abundance data comprising 645 species (Guyra Paraguay 2004). Beyond this point there was a notable increase in field work aimed at improving our knowledge of the Paraguayan avifauna. Various projects performed inventories of different areas of the country, and their results were published in ornithological journals (Brooks et al. 1993, 1995; Lowen et al. 1996). The Asociación Guyra Paraguay was created in 1997 and brought together numerous naturalists and ornithologists, both Paraguayan and foreign, to carry out hundreds of field expeditions covering all of the national territory. These campaigns contributed up-to-date information about distribution, abundance, and status, raising the number of species documented as occurring in the country to almost 700 (Guyra Paraguay 2004, 2005). The latter of these two works included for the first time a review of the threat categories of all Paraguayan species, and this was later followed by a book on the Important Bird Areas (IBAs) of Paraguay (Cartes and Clay 2009). In the last few years, there have been occasional publications on the diet of some Paraguayan owls (*A. cunicularia*, Andrade et al. 2004; *T. alba*, Pardiñas et al. 2005; Teta and Contreras 2003; *A. flammeus*, Torres et al. 2014), one on reproduction (del Castillo 2014) and one on behavior (*A. cunicularia*, Austin et al. 2016).

However, Paraguay continues to be one of the least ornithologically studied of the Neotropical countries (Esquivel Mattos 2010). As with other countries in the region, owls are among the most poorly known species of Paraguayan birds due to their inconspicuous behavior, nocturnal habits, and lack of commercial importance.

18.3 Conservation

All Paraguayan Strigiformes are included on CITES (2012) Appendix II, but only one species (*S. hylophila*) is classified as Near Threatened at international level as a result of presumed declining populations (BirdLife International 2012). Seven species are considered to be under some degree of threat at the national level. Tawny-browed Owl (*P. koeniswaldiana*) is "Endangered." Rusty-barred (*S. hylophila*) and Black-banded Owls (*C. huhula*) are treated as "Vulnerable" (Guyra Paraguay 2005). The principle threat to these species is continued loss of habitat coupled with their natural low density, and they do not adapt well to fragmented habitats. Spectacled Owl (*P. perspicillata*), Black-capped Screech Owl (*M. atricapilla*), and Mottled Owl (*C. virgata*) are all classified as "Near Threatened" for essentially the same reasons. Stygian Owl (*A. stygius*) is an extremely rare species, and little concrete data about its status is available. As a result it is considered "Data Deficient" nationally, though it may be suspected that the species is under some degree of threat (Guyra Paraguay 2004, 2005).

18.4 Local Attitudes to Owls

The nocturnal habits of owls have attached to them an air of mystery, sometimes associated with malign forces. The calls of *Megascops choliba* and *Tyto alba*, for example, have been considered predictors of illness or harbingers of death (Laprovitta 2016). Consequently in contemporary times owls have often been seen as a bad omen, and some are killed as a result (Bodrati and Cockle 2012).

However the perceived strength of owls also wins them admiration. *Athene cunicularia* was considered a celestial protector whose role was to prevent the uncontrolled reproduction of bats which will signal the end times. The possession of amulets made from owls can thus be understood to pass the gifts of strength on to the wearer (Laprovitta 2016). As an extension of this idea, in some areas, the possession of feathers or wings of *Glaucidium* and *Megascops* is thought to bring good luck, especially in business dealings, and may even assist in finding a life partner (Laprovitta 2016). Such beliefs are still strongly held in Paraguay and neighboring countries and may be an important source of mortality for owls in some rural areas. Small owls such as *Glaucidium*, *Megascops*, and *Athene cunicu-*

laria are occasionally offered for sale as pets in markets in Asunción, with the presumed "luckiness" of these birds perhaps also contributing to the willingness to own one.

The Aché indigenous group of Canindeyú Department hunts and consumes several species of owls including *Glaucidium brasilianum*, *Megascops atricapilla*, *Strix virgata*, and *Pulsatrix perspicillata*. The feathers of the latter species are also used to make arrows (Chachugi 2013). In Guaraní cosmology, owls were seen as celestial "observers," sent by the God Tamandú to judge those that are worthy of a place in the heavens (Micó 2001).

18.5 Threats

The principal threat to Paraguayan owls is the massive scale deforestation and conversion of land to monoculture and cattle ranches. Eleven of the sixteen species of Paraguayan owls depend on forested habitats, and four of these are found exclusively in Atlantic Forest which, since 1945, has seen its coverage reduced from 88,000 km^2 to 12,000 km^2 (Cartes 2006). Today, little of Paraguay's remaining Atlantic Forest is protected by law, and many of the protected areas that do exist are poorly administrated or virtually unpoliced. The degradation of existing native forests through selective logging is a serious threat to many species of Atlantic Forest owls, as they require, for both nesting and roosting, holes or platforms in large trees – the same trees that are targeted by selective loggers (Cockle et al. 2010, 2011, 2012; Bodrati and Cockle 2013).

The habit of many owls of hunting near to roads means that many owls fall victim to roadkill. This is a significant source of mortality in some species such as *Athene cunicularia*, *Bubo virginianus*, *Megascops choliba*, *Strix chacoensis*, and *Tyto alba*.

Unfortunately, the conservation of owls is made more complex by the near total lack of knowledge of their reproductive biology, habitat requirements, and diet, making it difficult to propose effective conservation strategies to combat the challenges posed to them by the modern world.

Acknowledgments To Paula Enríquez, for her enthusiasm and good grace which made possible the publication of this chapter. Thanks to Kristina Cockle for critically reviewing the manuscript and Gonzalo Ignazi for assistance in preparing the map. The authors would like to thank Myriam Velázquez and Rob Clay who generously shared publications and information about Paraguay. Part of the data presented here was gathered during fieldwork commissioned by the Asociación Guyra Paraguay and during FAUNA Paraguay field expeditions. Paul Smith and Hugo del Castillo are supported by the PRONII program of CONACYT.

Black-banded Owl (*Ciccaba huhula*)

Black-capped Screech Owl (*Megascops atricapilla*)

Chaco Owl (*Strix chacoensis*)

Rusty-barred Owl (*Strix hylophilla*)

References

Andrade A, Teta P, Contreras JR (2004) Dieta de la lechucita vizcachera (*Speotyto cunicularia*) en el Parque Nacional Médanos del Chaco (Paraguay). Ornitol Neotrop 15:87–92

Austin V, Sarvary J, Smith P (2016) Burrowing owls *Athene cunicularia* (Strigidae) respond with increased vigilance to calls of the curl-crested jay *Cyanocorax cristatellus* (Corvidae) in the Paraguayan Cerrado. Braz J Ornithol 24:1–8

Azara F (1805) Apuntamientos para la historia natural de los páxaros del Paraguay y Río de la Plata. Imprenta de la Viuda de Ibarra, Madrid

Bertoni A de W (1901) Aves nuevas del Paraguay. Catálogo de las aves del Paraguay. An Científicos Paraguayos 1:1–126

Bertoni A de W (1914) Fauna Paraguaya. Catálogos sistemáticos de los vertebrados del Paraguay. M. Brossa, Asunción

Bertoni A de W (1930) Sobre ornitología del Chaco Paraguayo. Aves colectadas por Félix Posner en la Colonia "Monte Sociedad", hoy Benjamin Aceval (Villa Hayes). Rev Soc Científica Paraguay 2(6):241–257

Bertoni A de W (1939) Catálogos sistemáticos de los vertebrados del Paraguay. Rev Soc Científica Paraguay 4:1–59

BirdLife International (2012) Species factsheet: *Strix hylophila*. http://www.Birdlife.org. Accessed 1 Apr 2012

Bodrati A, Cockle KL (2006) New records of rare and threatened birds from the Atlantic Forest of Misiones, Argentina. Cotinga 26:20–24

Bodrati A, Cockle KL (2012) El Atajacaminos Coludo *Macropsalis forcipata* en Argentina: ¿una especie amenazada o en expansión? Cotinga 34:46–54

Bodrati A, Cockle KL (2013) Distribution, nesting and vocalization of the black-banded owl (*Ciccaba huhula albomarginata*) in Argentina. Ornitol Neotrop 24:169–182

Bodrati A, Areta JI, Cowper Coles P et al (2012) La suinda kaaguí (*Strix virgata*) habita la región chaqueña. Nuestras Aves 57:29–31

Brooks TM, Barnes R, Bartrina L et al (1993) Bird survey and conservation in the Paraguayan Atlantic forest. Project Canopy 92′: final report. BirdLife International Study Report No. 57. BirdLife International, Cambridge

Brooks TM, Clay RP, Lowen JC et al (1995) New information on nine birds from Paraguay. Ornitol Neotrop 6:129–134

Brooks TM, Tobias JA, Balmford A (1999) Deforestation and bird extinctions in the Atlantic Forest. Anim Conserv 2:211–111

Cardoso da Silva JM (1996) Distribution of Amazonian and Atlantic birds in gallery forests of the Cerrado región, South America. Ornitol Neotrop 7:1–18

Capper DR, Clay RP, Madroño A et al (2001) New information on the distribution of twenty-two bird species in Paraguay. Ararajuba 9:57–59

Cartes JL (2006) Breve historia de la conservación en el bosque Atlántico. In: Cartes JL (ed) El Bosque Atlántico en Paraguay, Biodiversidad, Amenazas y Perspectivas. State of Hotspots Series. Conservation International – Center for Applied Biodiversity Science, Guyra Paraguay, Asunción, pp 37–57

Cartes JL, Clay RP (2009) Paraguay. In: Devenish C, Díaz Fernández DF, Clay RP, Davidson I, Yépez Zabala I (eds) Important Bird areas of the Americas. Priority sites for biodiversity conservation, BirdLife Conservation series, vol 16. BirdLife International, Quito, pp 297–306

Chachugi R (2013) Las aves y el conocimiento tradicional Ache. Fundación Moises Bertoni, Asunción

CITES (Convention on International Trade in Endangered Species of Wild fauna and Flora) (2012). http://www.cites.org/. Accessed 15 Mar 2012

Cockle KL, Leonard ML, Bodrati AA (2005) Presence and abundance of birds in an Atlantic forest reserve and adjacent plantation of shade-grown yerba mate, in Paraguay. Biodivers Conserv 14:3265–3288

Cockle KL, Martin K, Drever MC (2010) Supply of tree-holes limits nest density of cavity-nesting birds in primary and logged subtropical Atlantic forest. Biol Conserv 143:2851–2857
Cockle KL, Martin K, Wiebe K (2011) Selection of nest trees by cavity-nesting birds in the Neotropical Atlantic forest. Biotropica 43:228–236
Cockle KL, Martin K, Robledo G (2012) Linking fungi, trees, and hole-nesting birds in a Neotropical tree-cavity network: pathways of cavity production and implications for conservation. Forest Ecol Manag 264:210–219
Del Castillo F (2014) Nido de Lechuza de campanario (*Tyto alba*) (Aves: Tytonidae) en cavidad de palma (*Acrocomia aculeata*). Paraquaria Natural 3(1):31–32
Ericson PGP, Amarilla LA (1997) First observations and new distributional data for birds in Paraguay. Bull Br Ornithol Club 117:60–67
Esquivel Mattos A (2010) Comunidades de aves del bosque atlántico del Paraguay. Tesis Doctoral, Universidad de Salamanca, Spain
Esquivel Mattos A, Velázquez MC, Bodrati A et al (2007) Status of the avifauna of San Rafael National Park, one of the last large fragments of Atlantic Forest in Paraguay. Bird Conserv Int 17:301–317
Paraguay G (2004) Lista comentada de las aves de Paraguay. Annotated checklist of birds of Paraguay. Guyra Paraguay, Asunción
Paraguay G (2005) Atlas de las aves de Paraguay. Guyra Paraguay, Asunción
Hayes FE (1995) Status, distribution and biogeography of the birds of Paraguay. American Birding Association, Colorado Springs
Howell SNG, Robbins MB (1995) Species limits of the least pygmy-owl (*Glaucidium minutissimum*) complex. Wilson Bull 107:7–25
Ihering H v (1904) As aves do Paraguai em comparação as de. São Paulo. Rev Mus Paulista 6:310–344
Kelso L (1934) A key to the owls of the genus *Pulsatrix* Kaup. Auk 34:234–236
König C, Weick F, Becking JH (1999) Owls. A guide to the owls of the world. Yale University Press, New Haven/London
Laprovitta F (2016) Guyra: Reminiscencias míticas desde la fauna guaraní. El Argos, Corrientes
Laubmann A (1939) Die Vögel von Paraguay. I. Strecker & Schröder, Stuttgart
Laubmann A (1940) Die Vögel von Paraguay. II. Strecker & Schröder, Stuttgart
Lowen JC, Bartrina L, Brooks TM et al (1996) Project YACUTINGA '95: birds surveys and conservation priorities in eastern Paraguay. Cotinga 5:14–19
Lowen J, Mazar Barnett J, Pearman M et al (1997) New distributional information for 25 species in eastern Paraguay. Ararajuba 5:234–237
Marks JS, Canning RJ, Mikkola H (1999) Family Strigidae (typical owls). In: del Hoyo J, Elliot A, Sargatal J (eds) Handbook of the birds of the world, vol 5. Lynx Edicions, Barcelona, pp 76–242
Micó TL (2001) Aves en la mitología Guaraní. Servilibro, Asunción
Mikkola H (2012) Owls of the world. A photographic guide. Christopher Helm, London
Pardiñas UFJ, Teta P, Heinonen Fortabat S (2005) Vertebrate prey of the barn owl (*Tyto alba*) in subtropical wetlands of northeastern Argentina and eastern Paraguay. J Raptor Res 39:65–69
Parker TA, Stotz DF, Fitzpatrick JW (1996) Ecological and distributional databases. In: Stotz DF, Fitzpatrick J, Parker TA, Moskovits DK (eds) Neotropical birds:ecology and conservation. University of Chicago Press, Chicago, pp 132–291
Podtiaguin B (1941) Catálogo sistemático de las aves del Paraguay. Aumentado por las contribuciones al conocimiento de la ornitología paraguaya. Rev Soc Científica Paraguay 5:1–109
Podtiaguin B (1945) Catálogo sistemático de las aves del Paraguay. Aumentado por las contribuciones al conocimiento de la ornitología paraguaya. Rev Soc Científica Paraguay 6:63–80
Ramírez Llorens P, Bellocq MI (2007) New records clarify the southern distribution of the spectacled owl (*Pulsatrix perspicillata*). J Raptor Res 41:268–276
Remsen JV, Cadena CD, Jaramillo A et al (2012) A classification of the bird species of South America. American Ornithologists'Union http://www.museum.lsu.edu/~Remsen/SACCBaseline.html. Accessed Apr 2012

Robbins MB, Faucett RC, Rice NH (1999) Avifauna of a Paraguayan Cerrado locality: Parque Nacional Serranía San Luis, Depto. Concepción. Wilson Bull 111:216–228
Schade F, Masi Pallarés R (1967) Las Aves del Paraguay 1a Parte. Rev Paraguaya Microbiol 2:72–85
Schade F, Masi Pallarés R (1970a) Las Aves del Paraguay 2a Parte. Rev Paraguaya Microbiol 3:86–105
Schade F, Masi Pallarés R (1970b) Las Aves del Paraguay 3a Parte. Rev Paraguaya Microbiol 4:77–96
Schade F, Masi Pallarés R (1970c) Las Aves del Paraguay y un índice 4a Parte. Rev Paraguaya Microbiol 5:35–58
Schade F, Masi Pallarés R (1971) Las Aves del Paraguay y un índice 5a Parte. Rev Paraguaya Microbiol 6:103–128
Short LL (1976) Notes on a collection of birds from the Paraguayan Chaco. Am Mus Novit 2597:1–16
Storer RW (1989) Notes on Paraguayan birds. Occ Pap Mus Zool Univ Michigan 719:1–21
Teta P, Contreras JR (2003) Primeros antecedentes de la dieta de la lechuza de campanario (*Tyto alba*) en el Departamento Ñeembucú (Paraguay). Hornero 18:57–59
Tierno de Figueroa JM, Padial JM (2005) Avifauna invernal en ambientes de Chaco y Pantanal en Bahía Negra, norte de Paraguay. Hornero 20:153–162
Torres JC, Teta P, de la Sancha NU (2014) Presas del Búho Campestre (*Asio flammeus*) en un agroecosistema subtropical de Paraguay. Nuestras Aves 59:24–27
Zyskowski K, Robbins MB, Peterson AT et al (2003) Avifauna of the northern Paraguayan Chaco. Ornitol Neotrop 14:247–262

Chapter 19
The Owls of Suriname

Serano Ramcharan and Otte Ottema

Abstract Suriname is one of the smallest countries in South America. This country has a bird diversity of approximately 720 species, 15 of which are owls. Little ecological and biological information exist for the owl species. There is only information on their distribution and abundance. There is no information on the conservation status of owls in the country. Suriname has several biosphere reserves and national parks where owls are distributed. More information is needed to determine the conservation status of those species in the country.

Keywords Owl conservation • Ecological knowledge • Taxonomy diversity • South America

---

S. Ramcharan (✉)
STINASU Foundation of Nature Conservation in Suriname. Research Department, Cornelis Jongbawstraat 14, Postal Box 12252, Paramaribo, Suriname
e-mail: research@stinasu.sr

O. Ottema
STINASU Foundation of Nature Conservation in Suriname. Research Department, Cornelis Jongbawstraat 14, Postal Box 12252, Paramaribo, Suriname

Komeetstraat 2, Ephraimszegen, Paramaribo, Suriname
e-mail: otteottema@gmail.com

P.L. Enríquez (ed.), *Neotropical Owls*, DOI 10.1007/978-3-319-57108-9_19

Amazonian Pygmy-Owl (*Glaucidium hardy*)

19.1 Introduction

The Republic of Suriname is situated on the Atlantic coast of northeastern South America, roughly between 2° and 6° north latitude and between 54° and 58° west longitude, and has a terrestrial area of approximately 170,000 km^2 (Fig. 19.1). It is bordered in the west by Guyana, in the east by French Guiana, and in the south by Brazil. The human population is small (approximately 500,000 inhabitants) and largely concentrated in the coastal plain (mainly in the capital of Paramaribo and its surroundings). Suriname is politically divided in ten districts; most of them are in the northern part of the country. Sipaliwini is the most extensive district; it is situated in the southern part of the country and occupies 79.7% of the total area. As a result, most of the land in Suriname is still covered with pristine tropical rainforest. Approximately 80% of Suriname is covered by dense tropical rainforest.

The ornithological history of Suriname goes back to the seventeenth century when the first information on the birds of the country was published (Haverschmidt and Mees 1994). Publications dealing with a complete survey of birds known to occur in Suriname, however, did not appear before the beginning twentieth century when the brothers F.P. Penard and A.P. Penard (1908–1910) published their

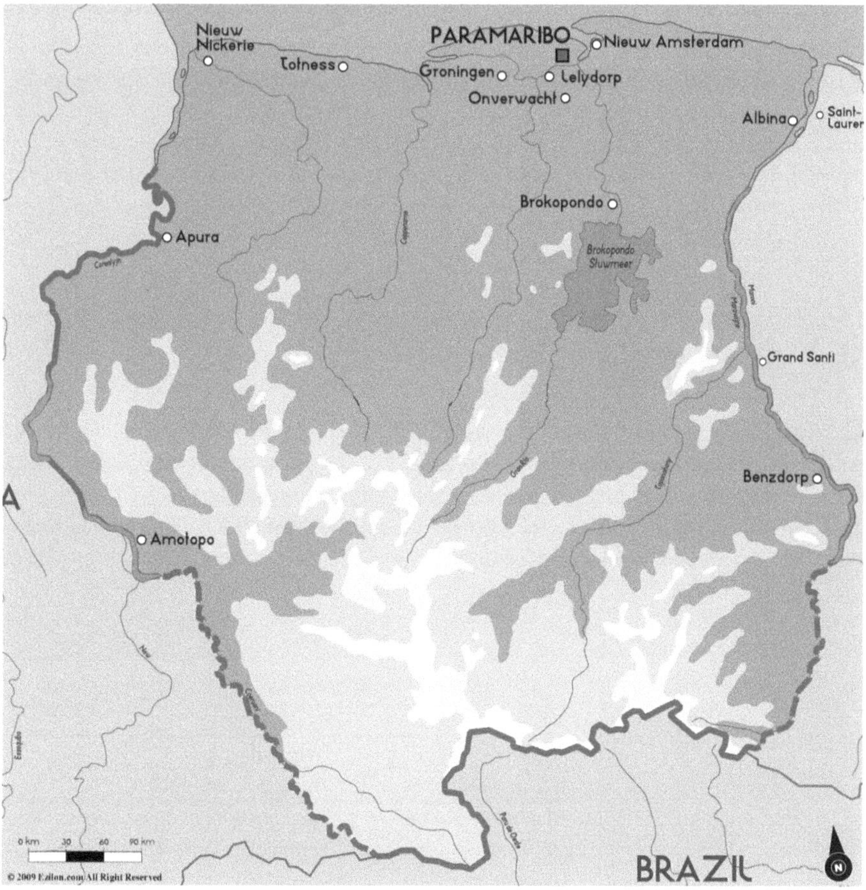

Fig. 19.1 Suriname map (http://catmargar.blogspot.mx/2013_06_01_archive.html)

book in two volumes about “De Vogels van Guyana” (“The Birds of Guyana”), dealing with the birds of Suriname and, as the subtitle indicates, “Cayenne” and “Demerara.” Important bird information on Suriname became known with the arrival of Francois Haverschmidt (1906–1987) in 1946. During his stay in Suriname, he put together a large collection of bird skins, most of them prepared by his wife Mrs. W. Haverschmidt-Liong A San. He wrote his famous book “Birds of Suriname” (Haverschmidt 1968), richly illustrated with splendid color plates drawn by P. Barruel. A further step forward was made when various ornithologists, such as H.A. Beatty, R. Freund, and G.F. Mees and his wife Mrs. V.J. Mees-Balchin, visited the country and made important bird collections in areas of the interior that F. Haverschmidt had never been able to visit. In the 1990s, a second edition of Haverschmidt’s “bible” was published, a largely rewritten edition with G.F. Mees as coauthor.

Since the 1970s, an increasing number of expert birders have been visiting Suriname, in particular from the United States of America, such as T.A. Davis, P.K. Donahue, S.L. Hilty, B.J. O'Shea, D. Stejskal, B.M. Whitney, and K. Zyskowski, almost all of whom often brought and guided bird tours. By video recording and audio recording calls and songs of birds, they were able to establish the presence of several new species for Suriname. Another positive development is that from the 1990s onward; a small but growing number of Surinamese people are studying birds in the field.

19.2 Taxonomy Diversity

Currently there are 15 owl species known for Suriname, of which one is a member of the Tytonidae family and the others are from the Strigidae family (Table 19.1). Representatives' species include *Megascops*, with three species, two *Ciccaba* species, two *Glaucidium* species, and two *Asio* species (Table 19.1). The local name for barn owl is poespoesi owroekoekoe and for owl species in general it is owroekoekoe.

Table 19.1 Suriname owl species (scientific and English names)

Scientific name	English name
Tyto alba	Barn owl
Megascops choliba	Tropical screech owl
Megascops watsonii	Northern tawny-bellied screech owl
Megascops guatemalae	Vermiculated screech owl
Lophostrix cristata	Crested owl
Pulsatrix perspicillata	Spectacled owl
Bubo virginianus	Great horned owl
Ciccaba virgata	Mottled owl
Ciccaba huhula	Black-banded owl
Glaucidium hardyi	Amazonian pygmy owl
Glaucidium brasilianum	Ferruginous pygmy owl
Athene cunicularia	Burrowing owl
Pseudoscops clamator	Striped owl
Asio stygius	Stygian owl
Asio flammeus	Short-eared owl

19.3 Distribution and Habitat Association

Four life zones are recognized in Suriname (coastal plain, savanna, highland, and lowland forest). However, Ottema et al. (2009) have distinguished six life zones, described as follows:

19.3.1 Estuarine Zone

This zone comprises the marine waters, including the ocean waters of the 200 nautical miles exclusive economic zone, and the adjacent estuarine area. The latter consists of a mosaic of soft, tidal mudflats; hard, firm clay banks eroded from older deposits; sandy beaches; mangrove forests (mainly consisting of black mangrove forests *Avicennia germinans*); and coastal lagoons and swamps with either salt, brackish, or fresh water, which are interrupted by low, sandy ridges or cheniers running parallel to the coastline. The estuarine zone includes the western part of the coastal area north of the lower Nickerie River and the East-West Connection (EWC) up to the Coppename River, the area north of the lower Saramacca River, and the EWC up to the Suriname River. In the eastern part of the coastal area it pertains to the area north of the lower Commewijne River and the Cottica River up to the Marowijne River. Cultivated areas such as rice fields, pastures, and vegetable fields are excluded, as are the developed areas found on the southern border in this zone. This zone covers about 16,200 km^2.

19.3.2 Rest of Coastal Plain

This zone consists mainly of freshwater swamps and swamp forests, which are interrupted by forested sand ridges. In the westernmost (Nickerie District), central (Paramaribo and surrounding area), and eastern (Mungo and surrounding area) parts, large areas have been brought under cultivation. This zone covers about 4300 km^2.

19.3.3 Northern Savanna Area

This includes the white sand savanna belt situated south of the Old Coastal Plain. This vegetation has mainly open sand savannas, xerophytic and dry land forests, and swamp forests in the wet areas in-between.

19.3.4 Southern Savanna Area

This life zone includes the open grassy savannas in the south of Suriname (10 to 100 msl), interspersed with some xerophytic forests and swamp forests. The main part of this zone consists of the Sipaliwini Savanna, which is the northern extension of the Brazilian Paru Savanna into Suriname. The vegetation is rain forest, arid forests, and open savannas. This zone covers about 3378 square miles (8750 km^2).

19.3.5 Lowland Forest

This zone comprises the lower forested parts of the interior (with hills up to an elevation of 400 m). It is almost completely covered with high dry land forest, dotted here and there with small rocky outcrops and small open savannas.

19.3.6 Highland Forest

The higher hills and mountains (400–1230 msl) in the interior are covered with mountain forest, with a tepui landscape at the Tafelberg, Central Suriname Nature Reserve. In some localities, e.g., at the Bakhuisgebergte, however, birds normally associated with mountain forest are already seen at an elevation of 200–300 m. It is mainly covered with undisturbed rain forests. This is the largest zone, covering an area of 51,738 square miles (134,000 km^2).

19.4 Vegetation Types

There are 16 different vegetation types recorded in Suriname following Ottema et al. (2009), which can be found within the six life zones:

Coastal brackish lagoons and swamps (CO), dwarf forests on white sand savannas and sand stone savannas (DW) (Tafelberg), forest creeks and overgrown trenches of former coffee plantations (FC), wooded sand ridges, savanna forests, and forests of the interior (includes clearings and forest edges; FO), freshwater swamps, lakes, ponds, canals, trenches, rain pools, and other temporarily wet places (FW), human-altered landscape (towns, villages, gardens, agricultural land, and industrial areas; HU), marine waters and estuaries (MA), mangroves (MN), soft tidal mudflats (MU), palm trees and forests (PA), riverine habitats, including streams, rocks, islands, banks, waterfalls, and riparian forests (RI), rocky outcrops (RO), scrub and brush habitats, such as liana forests and low secondary growth (SC), bare and overgrown sandy beaches, sand spits, and firm, hard clay banks from eroded, older deposits (SN), open savanna with short or long grasses, thickets or scattered bushes, and/ or trees (also includes airfields and airstrips constructed in these savannas (SV), and swampy, wet, and temporarily flooded forests (SW).

19.5 Owl Species Distribution

Barn owl (*Tyto alba*): This owl species has a wide distribution and can be found in human-altered landscapes. It breeds in Suriname but is seldom found in the estuarine and northern savanna area. It commonly occurs in the rest of the coastal plain. The barn owl is also very common in Paramaribo (Appendix 19.1).

Tropical screech owl (*Megascops choliba*) This owl is known to occur in human-altered landscapes. This species is uncommon in the estuarine zone and common in the rest of the coastal plain. It is rarely seen in the northern savanna area, lowland forest, highland forest, and the southern savanna area. This species is known to breed in Suriname (Appendix 19.1).

Tawny-bellied screech owl (*Megascops watsonii*) In Suriname this species is occasionally found in the rest of the coastal plain, northern savanna area, lowland forest, and in highland forest. The breeding of this species in Suriname is unconfirmed (Appendix 19.1).

Vermiculated screech owl (*Megascops guatemalae*) This species is rarely encountered in lowland and highland forests. Breeding is unconfirmed in Suriname (Appendix 19.1).

Crested owl (*Lophostrix cristata*) This species is uncommon in lowland forests and where might occasionally be encountered in the coastal plain, in northern savanna area, and in highland forests. Breeding is unconfirmed in Suriname (Appendix 19.1).

Spectacled owl (*Pulsatrix perspicillata*) This species is known to be present in human-altered landscapes and in mangrove forests. Breeding is known to occur in Suriname. Among the life zones it is rarely encountered in the estuarine, highland forest, and northern and southern savanna areas. It is uncommon in the rest of the coastal area. This species is common in the lowland forest. It breeds in Paramaribo (Appendix 19.1).

Great horned owl (*Bubo virginianus*) In Suriname, this is an owl of mangrove forests and coastal brackish lagoons and swamps. Breeding is known to occur in Suriname. It is uncommon in the estuarine zone (Appendix 19.1).

Mottled owl (*Ciccaba virgata*) This owl is known to occur in human-altered landscapes; it can occasionally be encountered in the rest of the coastal area and lowland or highland forests. The breeding status of this owl for Suriname is unknown (Appendix 19.1).

Black-banded owl (*Ciccaba huhula*) This owl species is known to occur in human-altered landscapes. It is rarely found in the rest of the coastal area, northern savanna area, and lowland and the highland forests. Breeding status for this owl is unknown in Suriname (Appendix 19.1).

Amazonian pygmy owl (*Glaucidium hardyi*) It is uncommon in lowland forest and is rarely found in the southern savanna area and highland forests. Breeding status of this species is not known for Suriname (Appendix 19.1).

Ferruginous pygmy owl (*Glaucidium brasilianum*) This species occurs in human-altered landscape; it is seldom found in lowland forest and is uncommon in the southern savanna area. Breeding status is unknown in Suriname (Appendix 19.1).

Burrowing owl (*Athene cunicularia*) This species has been reported in Suriname, but its breeding status is not clear in the country. It can occasionally be encountered in the northern savanna area. On March 7, 11, and 17, 2007, one bird was sighted and photographed at the J.A. Pengel International Airport, Para District (F. Chin Joe, K.D.B. Dijkstra, O.H. Ottema). On February 11, 2009, this owl species was seen at Zanderij at the J.A. Pengel International Airport (O.H. Ottema and J. Timmer) (Appendix 19.1).

Striped owl (*Pseudoscops clamator*) This species is known to occur in human-altered landscapes, although rarely; it can be encountered in the estuarine zone, the northern savanna area, and lowland forest. It is uncommon in the rest of the coastal area. This owl species breeds in Suriname (Appendix 19.1).

Stygian owl (*Asio stygius*) Breeding status is not clear in Suriname, and it can rarely be found in lowland forests. There are two important records: at the Bakhuisgebergte near the very upper Nickerie River, October 20, 2005, by Brian O'Shea and Otte Ottema, and near the Voltzberg on November 8, 2008, by Brian O'Shea (Appendix 19.1).

Short-eared owl (*Asio flammeus*) This owl species is known to occur in human-altered landscapes. It is rarely found within the coastal plain. Short-eared owl seems to be a wanderer. There is definite record: rice fields south of Nieuw Nickerie, Nickerie District, February 23, 2005. By J. Verkerk (Appendix 19.1).

19.6 Owl Conservation

All owl species occurring in Suriname are protected by law. Since there are no known threats to the occurring owl species, locally, regionally, or at the national level, no strategies of owl conservation are being implemented in protected areas, national parks, or biosphere reserves. There are three different kinds of protected areas in Suriname: nature reserves, multiple-use management areas, and parks. The Suriname Forest Service (LBB) is in charge of the management of the protected areas, while the Nature Conservation Division (NB) is responsible for the day to day management. Suriname Forest Service has appointed STINASU to develop and conduct educational and touristic aspects in the protected areas.

The protected areas in Suriname are Hertenrits, Coppename Monding Nature Reserve, Wia Wia Nature Reserve, Galibi, Peruvia, Boven Coesewijne Nature Reserve, Copi Nature Reserve, Wane Kreek Nature Reserve, Brinck Heuvel, Sipaliwini Savanna, and Central Suriname Nature Reserve. The Brownsberg Nature Park and four protected areas have been proposed: Nanni and Kaburi Nature Reserve, Mac Clemen and Snake Creek Protected Forest. Furthermore there are four multiple-use management areas (Bigi Pan, Noord Coronie, Noord Saramacca, and Noord Commewijne/Marowijne MUMA, including Matapica beach). The total protected areas represent 16% of the Suriname terrestrial area. The smallest protected area has 100 ha and the largest, Central Suriname Nature Reserve, is 1.2 million ha (Westerman et al. 2002).

Information on bird species and owl species, including identification, occurrence, breeding behavior, food, and their range distribution in Suriname, can be found in Birds of Suriname (Haverschmidth and Mees 1994). In addition, field guides that are useful for identifying the birds of Suriname are Birds of Venezuela (Hilty 2003) and Birds of Northern South America (Restall et al. 2006).

19.7 Conclusions

There are 15 owl species distributed in Suriname; however, little information exists on their ecology and biology, and there is only information on distribution and abundance. All owl species are protected by law, but no conservation strategies are being implemented in protected areas in the country. However, protected areas can help to conserve owl species populations in Suriname.

Appendix 19.1

Suriname owl species, habitat occurrence and abundance, reliability, and breeding status

				Abundance in life zone					
Species	Habitat	Reliability of species in Suriname	Breeding status of species	Estuarine zone	Rest of coastal plain	Northern savanna area	Lowland forest	Highland forest	Southern savanna area
Tyto alba Barn owl	HU	1	B	R	C	R			
Megascops choliba Tropical screech owl	HU, MN, SC, SV	1	B	U	C	R	R	R	R
Megascops watsonii Tawny-bellied screech owl	FO	1	(B)		R	R	R	R	
Megascops guatemalae Vermiculated screech owl	FO	1	(B)				R	R	
Lophostrix cristata Crested owl	FO	1	(B)		R	R	U	R	
Pulsatrix perspicillata Spectacled owl	FO, HU, MN	1	B	R	U	R	C	R	R
Bubo virginianus Great horned owl	MN, CO	1	B	U					
Ciccaba virgata Mottled owl	FO, HU	1	(B)		R		R	R	
Ciccaba huhula Black-banded owl	FO, HU	1	(B)		R	R	R	R	
Glaucidium hardyi Amazonian pygmy owl	FO	1	(B)			R	U	R	

Glaucidium brasilianum Ferruginous pygmy owl	SV, HU	1	(B)				R		U
Athene cunicularia Burrowing owl	SV	1	U			R			
Pseudoscops clamator Striped owl	HU , SV, FO	1	B	R	U	R	R		
Asio stygius Stygian owl	FO	1	U				R		
Asio flammeus Short-eared owl	HU	2	W		R				

References

Haverschmidt F, Mees GF (1994) The birds of Suriname. VACO NV, Paramaribo
Hilty SL (2003) Birds of Venezuela. Princeton University Press, Princeton
Ottema OH, Ribot JM, Spaans AL (2009) Annotated checklist of the birds of Suriname. WWF Guiana, Paramaribo
Restall R, Rodner C, Lentino M (2006) Birds of northern South America. Christopher Helm, London
Westermann JH, Schulz JP, Spaans A et al (2002) Natuurbescherming en Natuurreservaten in Suriname. BHPBilliton, Paramaribo

Chapter 20
The Owls of Uruguay

Heimo Mikkola

Abstract Only ten owl species and two more subspecies are known to live in Uruguay. Knowledge of actual populations of all species is extremely limited as only food of some species has been studied. It is important to carry out more studies on owls and their habitats, in order to understand how they tolerate numerous changes in the environment. The National Institute for Agricultural Research (INIA) has started in 2014 an interesting experiment by putting up nest boxes for the owls. Preliminary results indicate that at least American barn owl (*Tyto furcata*) and tropical screech owl (*Megascops choliba uruguaiensis*) have started to accept the boxes for breeding and roosting. It is awaited that INIA will soon publish some photos and more detailed results on this experiment.

Keywords Uruguay • Owl distribution • Cultural beliefs • Conservation

Tropical Screech Owl (*Megascops choliba*)

H. Mikkola (✉)
Department of Biology, Eastern Finland University, FIN-70211 Kuopio, Finland
e-mail: heimomikkola@yahoo.co.uk

P.L. Enríquez (ed.), *Neotropical Owls*, DOI 10.1007/978-3-319-57108-9_20

20.1 Introduction

Uruguay is the third smallest country in South America, where it is situated 34°53′0″ S and 56° 10′0″ W and has territory of 176,220 km^2 including 660 km of Atlantic coastline (Fig. 20.1). The country is divided into 19 departments. Charles Darwin concluded in 1832 that Uruguay is one of the most uninteresting landscapes he had ever seen. "Not a tree or a house or traces of cultivation give cheerfulness to scene. An undulating green plain and large herd of cattle has not even the charm of novelty." However, Darwin admitted that some of the smallest birds are most brilliantly colored, much more so than those in Brazil (Keynes 1988). And that is why we now call Uruguay "country of painted birds."

Fig. 20.1 Map of Uruguay (MVOTMA 2015)

20.2 Diversity of Birds and Biotopes

The number of bird species this far registered in Uruguay is 477, out of which 18 are globally threatened and five human-introduced species (Wikipedia 2016). Within this diversity of birds, 12 subspecies out of ten owl species are known to live in the country (Table 20.1).

Azpiroz (2003) and Rocha (2006a) have established basic classification of 11 biotopes and main vegetation types where the distribution of the owls is described:

20.2.1 Park and Spinal Forests

Park and spinal forests are formed by a group of relatively low and spiny trees. This type of vegetation forms a parallel strip with River Uruguay from the departments of Colonia to Artigas. This is the biotype for most of the forest owls including the rare Santa Catarina screech owl (*Megascops sanctaecatarinae*).

20.2.2 Riparian Forest

Riparian forest is another type of forest, which is also called **gallery forest**. It occurs at river- and streamsides all over the country. This biotype is the home for striped owl (*Asio clamator*) and all *Megascops* and *Glaucidium* species. In the northern part of the country, also rare buff-fronted owl (*Aegolius harrisii*) lives in this habitat.

20.2.3 Meadows

Meadows present the most widely distributed biotope in Uruguay in which grass plants and low grass form the main vegetation. Large part of meadows is used as pasture land for sheep and cattle herds. Typical owls for these open fields are short-eared owl (*Asio flammeus*) and burrowing owl (*Athene cunicularia*).

*20.2.4 Jelly Palm (***Butia capitata***)*

Jelly palm (*Butia capitata*) forms dense palm forests, especially in Rocha and some sectors of Treinta y Tres department. This biotope is typical for great horned owl (*Bubo virginianus*) and would be excellent for American barn owl (*Tyto furcata*), if artificial nest boxes would be provided (Project FPTA 314 INIA-COPAGRAN-

Table 20.1 Distribution of the owl species and subspecies occurring in Uruguay

Scientific names (subspecies)	Common and local names	Frequency	Distribution
Tyto alba tuidara (J.E.Gray, 1829)	Lechuza de campanario, lechuzón de campanario, lechuza común, barn owl	IV	T
Megascops (Otus) choliba uruguaiensis (Hekstra, 1982)	alicuco común, alilicucu común, autillo choliba, tamborcito común, tropical screech owl	IV	T
Megascops (Otus) atricapillus (Temminck, 1822)	Alicuco tropical, alilicucu grande, autillo capirotado, tamborcito grande, black-capped screech owl	I	N
Megascops (Otus) sanctaecatarinae (Salvin, 1897)	Alicuco grande, autillo de Santa Catarina, tamborcito grande, long-tufted screech owl	I	O&E[a]
Bubo virginianus nacurutu (Vieillot, 1817)	Búho americano, ñacurutú, great horned owl	III	T
Glaucidium brasilianum brasilianum (Gmelin, 1788)	Caburé, caburé chico, capuré común, mochuelo caburé, ferruginous pygmy owl	II	N
Glaucidium brasilianum stranecki (König & Wink, 1995)	Caburé, caburé chico, capuré común, mochuelo caburé, ferruginous pygmy owl	II	S
Athene cunicularia cunicularia (Molina, 1782)	Búho llanero, lechuza común, lechucita de campo, lechucita vizcachera, mochuelo de madriguera, vizachera, burrowing owl	IV	T[b]
Aegolius harrisii iheringi (Sharpe, 1899)	Lechucita acanelada, lechucita canela, mochuelo canela, buff-fronted owl	II	N[a]
Asio clamator midas (Vieillot, 1817)	Búho, búho cornudo cariblanco, búho gritón, lechuzón orejudo, striped owl	III	T
Asio flammeus suinda (Vieillot, 1817)	Búho campestre, lechuza campestre, lechuzón de campo, lechuzón de las pajas, short-eared owl	III	T

I = rare or difficult to find, II = very few or rarely seen, III = low density or only few opportunities to observe, and IV = frequent or easy to find

T in entire Uruguay, *N* only in the North, *S* only in the South, *E* only in the East and *O* only in the West

[a]No breeding information on *Aegolius harrisii* and *Megascops sanctaecatarinae* in Uruguay (Rocha 2006b)

[b]According to Azpiroz (2003), there are two *Athene* owl subspecies in Uruguay, but König et al. (2008) include *Athene cunicularia partridgei* as a synonym of *Athene cunicularia cunicularia*

DGSSAA). Many different woodpecker species (*Picidae* family) are making suitable nest holes for the small owls, like ferruginous pygmy owl (*Glaucidium brasilianum*).

20.2.5 Hill Forests

Hill forests generally occur on low-altitude hills and hillsides. Buff-fronted owl would be a typical owl for this biotope, but obviously it is extremely rare, if not even endangered in Uruguay.

20.2.6 Humid Ravine Forests

Humid ravine forests have a very special microclimate in north and northeast of the country. These forests are excellent biotopes for tropical screech owl (*Megascops choliba*) and also for black-capped screech owl (*Megascops atricapilla*) although it is very rare in the country.

20.2.7 Seashore and Estuaries

Seashore and estuaries are widely available environments in Uruguay. Most remote coastal sand dunes can attract burrowing and short-eared owls.

20.2.8 Sea and Islands

Sea and islands cover almost 45% of the total surface area of Uruguay, but these serve only for a few owls if any. Short-eared owl is not afraid of crossing the water, and it is known to prey on shorebird colonies (e.g., Mikkola 1983; Holt 1994).

20.2.9 Marshland

Marshland is a wide term but in Uruguay it includes inundated lowlands, mainly with freshwater, generally near river, stream, or lagoon. Only a few owl species like barn owl and short-eared owl benefit from this type of biotope, mainly as a hunting ground.

There are many **lagoons** in Uruguay, but these are not so important for the breeding owls, but also serve as hunting grounds for some owl species, including great horned owl.

20.2.10 Man-Made Environments

Human influence is very clear in practically all biotopes in Uruguay. In many cases this influence is negative for most owl species as human population has destroyed many natural habitats. Loss of habitats has made many species very rare if not extinct. However, some owls have adapted to live in the man-made environment and structures. Especially deserted or rarely visited farm sheds are used by owls like tropical screech owl and American barn owl. Reforestation in the form of pine and eucalyptus (all exotic species) has been done in a large scale in Uruguay by the plywood and paper companies. These huge forest plantations are not ideal for any birds but give new hiding and breeding places for some owls. This is the case with great horned owl and some smaller owls, like *Megascops* species.

During 1990–2005 Uruguay added with the tree plantations its forest-covered areas by 5.1% (Mongabay 2010). The World Rainforest Movement (2010) has, however, paid attention that these plantations have caused the loss of bird diversity and disappearance of many forest owls. Introduced tree species are accused to cause pollution, groundwater-level sinking, and other serious ecosystem affecting changes. In one interview undertaken in Tranqueras, Uruguay, people mentioned that one *Scarabaeidae* beetle, which owls used to eat, is becoming a nuisance and pest as the owls have disappeared from the area.

20.3 Owl Distribution

Only very general and mainly diurnal bird distribution studies exist in Uruguay, so distribution of the largely nocturnal owls is badly known. However, two owls are thought to be fairly common in the entire country, the barn owl (*Tyto alba*) and the tropical screech owl (*Megascops choliba*), when three more species can be found everywhere, but are not frequent anywhere, short-eared owl (*Asio flammeus*), striped owl (*Asio clamator*), and great horned owl (*Bubo virginianus*).

During my time in Uruguay (2004–2008), burrowing owl was fairly common but only in the northern parts of the country. My first contact with the burrowing owls was when I saw two pairs on meadows with communal burrow systems of subterranean rodents like collared tuco-tuco (*Ctenomys torquatus*) between Salto and the Brazilian border in April 2005. In October 2006 I counted 20 burrowing owls/50 km (max. 4 pairs/km) next to the sandy roads in Artigas not far from the Brazilian border. Same time between Artigas and Salto, I counted only four pairs/200 km at the tarmac road. The conclusion for this is that the owls depend on the burrows of the subterranean rodents which obviously fall more easily as the victims of the traffic at tarmac roads (as do the owls as well), explaining why sandy roads had ten times higher owl densities. Always owls were near the other animals, cows, sheep, and greater rheas (*Rhea americana intermedia*), and not too far from the water.

After my time in the country, burrowing owl has been listed almost endangered in Uruguay together with short-eared owl (Azpiroz et al. 2012a). Rests of the species (four) are very little known and difficult to find: black-capped screech owl (*Megascops atricapillus*), Santa Catarina screech owl (*M. sanctaecatarinae*), ferruginous pygmy owl (*Glaucidium brasilianum*), and buff-fronted owl (*Aegolius harrisii*). There is no information available on the breeding of the black-capped screech owl and buff-fronted owl in Uruguay, and therefore they are now listed as endangered (Azpiroz et al. 2012a).

Local authority in the Uruguay birds Gabriel Rocha (2006a) did not list black-capped screech owl at all, but Argentineans Tito Narosky and Dario Yzurieta (2003) listed it for the northern part of Uruguay as not common and difficult to see. König et al. (2008) state that black-capped screech owl is an endemic owl of SE Brazil, Uruguay, and NE Argentina (Misiones) and that the southernmost limit seems to be north of the river Urugua-i (not to be confused with the River Uruguay!). However, the distribution map in the König et al. (2008) does not show this species into any areas in Uruguay, although the distribution comes almost until the Uruguay–Brazilian border near the Laguna Merín and Rio Yaguarón. According to Lepage (2009) both species black-capped screech owl and Santa Catarina screech owl occur in Uruguay.

Narosky and Yzurieta (2003) don't have Santa Catarina screech owl at all in their book *Aves de Argentina y Uruguay*. However, this species has been found in Uruguay according to König et al. (2008), who marked the species for western and eastern parts of the country, but without giving any further details for the Uruguay observations. More recently Wikipedia (2010) lists black-capped screech owl for Uruguay as one of the ten owl species. Interestingly, Rocha (2006b) only states that there are no breeding records of this owl in Uruguay. It is clear that more detailed studies are urgently needed on distribution and biology of both owl species in Uruguay.

20.4 National-Level Conservation Status

A small interview study was undertaken in 2007 on general public owl knowledge and beliefs in the coastal department of Maldonado, Uruguay. The respondents were asked to classify owls according to their personal knowledge and beliefs (Fig. 20.2). This study demonstrated very well that unfortunately many common people in Uruguay still hold very negative views on owls.

Very negative classification came almost from every third person, which resembles the situation in Africa where deep taboos about owls are still powerful and living traditions (Mikkola 1997a, b). In Uruguay 11% of the interviewed people ($N = 99$) knew some people who had killed an owl, and similarly 11% knew someone who used owl feathers as an amulet/talisman or owl bones for magic power purposes or as traditional medicine for different diseases. But luckily the old Indian habit to eat the owls as delicacy has now obviously vanished in the modern Uruguay. According to Wetmore (1926), the flesh of the burrowing owl was served as a delicacy to those convalescing from illness in the belief that it produces appetite for other food.

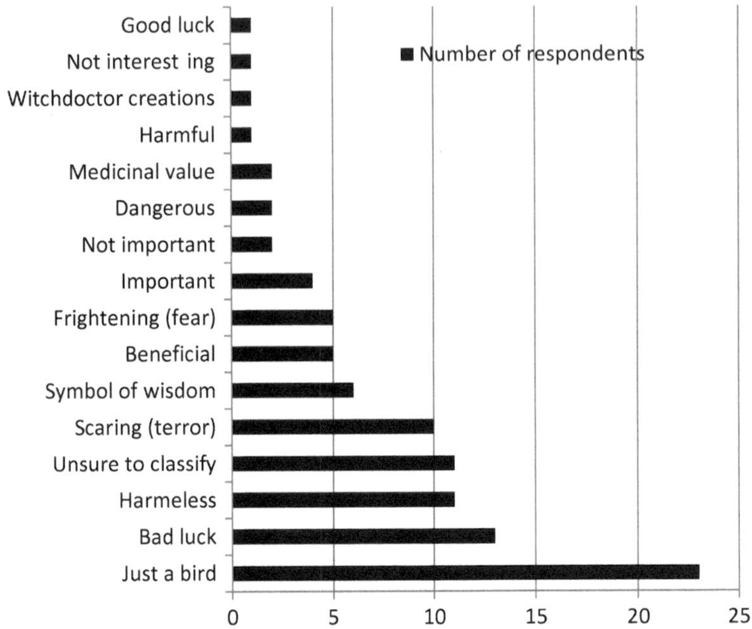

Fig. 20.2 Classification of owls in Maldonado, Uruguay. Total of 99 interviews were undertaken in 2007. People were asked: how would you classify the owls? And the answers were as follows

Study shows how important it is to know peoples' attitudes toward owl as it affects their action toward conservation of species and ecosystems. Understanding both environmental problems and the influence of human behavior is indispensable to achieve success in the conservation of owl populations in situ. There is no hope to get any area to be protected for owls or wildlife in general without proper awareness building and education efforts. Only by educating local people, for instance, through schools and television, of the roles owls have in nature will superstitious beliefs in them be overcome (Enriquez and Mikkola 1997).

20.5 Important Threats

It is clear that livestock- and agriculture-related clearings as well as harvesting of natural and planted forests have been the major risk in the survival of various bird species in Uruguay (Rocha 2006b). Since 1990 some 24% of the original primary forests have been harvested, and between 1990 and 2000 Uruguay lost 57,000 ha of its natural forest cover (Mongabay 2010). Total losses of natural forested habitats and change in the habitat structure have both alone and together affected negatively the populations of some owl species in the country. Fragmented habitats obviously affect the general bird diversity less, but can be fatal to some sensible owl species (Mikkola 2013).

Earlier so common, burrowing owl will suffer when the new roads will be surfaced and the traffic gets faster, as this will affect the owls directly but maybe even more through killing the subterranean rodents building communal burrow systems which the owl is using as breeding holes.

Largest impact will be due to the intensive agriculture production to feed the planned biodiesel plants. Until today very little attention has been given to devastating problems what large-scale energy plantations like the soya been (a legume native to East Asia) have caused in countries like Uruguay. Only the future will show final impacts on biodiversity, but already now the excessive use of agrochemicals and rodenticides has killed most of the rodents. After the rodents as food are lost, the owls will disappear. Luckily Uruguay is now testing the idea to use barn owl and other owls to control rodent and bird pests instead of conventional pesticides (Charter 2016).

20.6 Owl Conservation Strategies

Although the Ministry of Livestock, Agriculture, and Fisheries controls through its wildlife department the illegal trade and trafficking of wild birds and other animals, Uruguay has no real national system to protect the birds in the country (Rocha 2006b). However, there are some isolated efforts to protect small areas in the departments of Treinta y Tres and Rocha. Uruguay has 68 protected areas covering some 778,602 hectares. Unfortunately these areas have no real conservation status although they are called protected and some are even misleadingly named as "national parks."

It is important to note that Uruguay has protected only 0.6% of its territory. This percentage is the lowest in the new continent. In comparison with the other countries, a neighboring Argentina has legally protected 1.6% of its territory, Chile 14%, and Ecuador 38%, respectively (Chiappe 2005).

Luckily Uruguay now has a very strong environmental movement (PNUMA, 2008), including NGOs and the government departments which will surely improve the situation with the owl and nature conservation in the country. Lately this movement has been supported through ONE UN (United Nations) funds.

20.6.1 Ministry of Environment and Natural Resources (MVOTMA)

This state authority is in charge of protection of the natural resources in the country through its National Directorate for the Environment.

20.6.2 National Directorate for the Environment

National Directorate for the Environment is formulating, executing, supervising, and evaluating national environment protection plans. It implements national policies for natural resources protection in the country.

20.6.3 Vida Silvestre Uruguay (VSU)

Vida Silvestre Uruguay (VSU): nongovernmental organization which started in 1995 species and ecosystem protection and natural resources conservation in Uruguay. It participates in the development of new protected areas, and it is organizing the training for the technical personnel and producing necessary environment protection-related materials. Already in 1999 VSU organized first training course for the national park ranchers together with Park Rancher Association after they got funds from the International Monetary Fund.

20.6.4 Ministry of Livestock, Agriculture, and Fisheries (MGAP)

Ministry of Livestock, Agriculture, and Fisheries (MGAP) is in charge of natural resources department which develops nature conservation policies in the country. Some 2 years ago the Department of Wildlife was moved to MVOTMA, so that Ministry now oversees the policies.

Since 1980s the Vertebrate Pest Unit from the Ministry develops, adapts, and tests different management strategies to decrease bird and rodent damage on grain crops. This unit invited Dr. Motti Charter (University of Tel-Aviv and currently the University of Haifa) to visit Uruguay in 2010. Dr. Charter provided the professional knowledge of how to use the raptors as biological pest control agents and taught the Uruguayans how to build and use the nest boxes. INIA/MGAP provided the funds to invite him and materials to build the nest boxes. The idea is to use barn owl and other owls to control rodents and bird pests instead of conventional pesticides.

While in Uruguay Dr. Charter evaluated the feasibility of development of an artificial nest box program in one of the National Institute for Agricultural Research (INIA) experimental station (La Estanzuela) at the Department of Colonia. Estanzuela station hosts many grain crop experiment plots every year and has a long history of bird and rodent damage studies. Dr. Charter; Ethel Rodriguez, PhD, and Lourdes Olivera, MSc (from the Vertebrate Pest Unit); Lic. Guillermo Tellechea; and Ing. MSci Sergio Ceretta from INIA surveyed the issue on the spot by visiting different field sites and developed final nest box plan in August 2010.

On September 2014 the Uruguayan collaborators built 11 nest boxes for the screech owls and eight for barn owls in the INIA experimental station. The nest boxes were hanged to trees following the experience gained in Israel. After just 6 months, the fruits of the collaboration were seen as the first barn owl pair bred in one of the nest boxes and successfully raised four fledglings. Also two nest boxes were found with tropical screech owls (*Megascops choliba uruguaiensis*). After 2014 more nest boxes have been built, and INIA aims soon to publish interesting photos and all breeding results in this experiment.

Unlike in many continents of the world that have used nest boxes for owls for projects that dealt mainly in conservation, there is very little experience of using nest boxes for birds in South America. The main goal of this initial project is to determine which nest box breeding species are suitable to be used as biological pest control agents of rodents. Since the current project has already had one successful breeding of barn owls, Dr. Charter wants to expand the project first in Uruguay and hopefully afterwards introduce it to other countries in South America.

20.6.5 Directorate of Renewable Natural Resources (RENARE)

It is responsible to promote rational use and management of the renewable natural resources. It has a department for the national parks and protected areas which develops conservation policies for 14 conservation units falling under the jurisdiction of MGAP.

20.6.6 Program for the Biodiversity Conservation and Sustainable Development of the Wetlands in East Uruguay (PROBIDES 2008)

Its aim is to protect the biodiversity of East Uruguay wetlands through sustainable development of the region. It participates also in the regional conservation efforts through environment education and capacitation. Environment workers have been trained since 2008 when first 20 park ranchers participated in the regional PROBIDES office to learn how to manage the different protected areas in the country.

20.6.7 Aves Uruguay (GUPECA)

It is the official BirdLife International representative in Uruguay for the research and conservation of birds. It is a nongovernmental group (NGO) which studies and protects wild birds in the country as well as the habitats birdlife requires, respecting the rules of sustainable development for the natural resources.

It is very important to note that the Federación Ornitológica Uruguay has nothing to do with the bird conservation or ornithology as they claim. They only sell parrots, canaries, and other exotic birds. It is rare to see young owls to be offered for sale as excellent night hunters and bird "watchdogs."

20.6.8 Averaves

This nonprofit-making association was formed in 2001 in order to add public awareness and knowledge on Uruguay birdlife. Averaves has undertaken many research projects and environment education having its headquartering in the Faculty of Science of the National University in Montevideo. It has received funding assistance from the American Ornithologists' Union.

20.7 Status of Biological and Ecological Knowledge on Owl Species

There are countries in the world where the owl populations are well studied and where owls are understood as indicator species reflecting the conditions of the environment (Bart and Forsman 1992; Mikkola 2013). So owls associated with old forest have been seen by modern natural resource management agencies as real prognosticators of the health and fate of such environments. These owls have been called management indicator species by agencies such as the USDA Forest Service, who has identified the northern spotted owl (*Strix occidentalis caurina*) as an indicator of old-growth conifer forests of northwestern United States (Starkey 1994). Throughout the world, some 82 owl species are closely associated with old forests (Marcot 1995), most awaiting recognition as useful indicators of old-forest conditions.

In Uruguay there are no real surveys of the natural forest resources and no detailed studies on the role of owls in the forests. So there is a need for more research before making any final. Cortés et al. (2013) revealed only some 15 specific owl studies from Uruguay (Mones et al. 1973; Cuello 1980; Gonzáles et al. 1995; Gonzáles and Saralegui 1996; Claramunt and Gonzáles 1999; Gonzáles 1998; Gonzáles and Altuna 1999; Altuna and Gonzáles 1999; Vázquez 2003; Garcia-Olaso 2005; Altuna and Gonzáles 2006; Montenegro Tourón and Caballero Sadi 2009,Cruces et al. 2010; Rodríguez-Cajarville and Bessonart 2011), and none of them covered the biology or distribution of the owls in Uruguay. Only Azpiroz et al. (2012b) have given some distribution data for conclusions. Research knowledge is minimal even when compared with the other countries in the region, partly because of the rarity of many owl species and owl investigators in the country.

It is extremely urgent to start owl distribution and population studies taking into account how the negative environmental changes affect the owl species in the country. Common people have fairly limited knowledge on owls in Uruguay. In my interview study (2007), 14% of people knew no owl species at all, and 17% were able to identify one owl species, and 39% recognized two or three species. One person said that she knows seven out of ten possible species. Clearly the best known species was burrowing owl (75% of the respondents knew this species) and the next came short-eared owl (61%), barn owl (53%), and great horned owl (42%). Six of the ten species were familiar only to a few persons, and nobody mentioned Santa Catarina screech owl, and only one mentioned fairly common ferruginous pygmy owl (Mikkola 2015).

20.8 Conclusions

Destruction and fragmentation of natural habitats are equally serious in Uruguay as in Europe and the United States, covering loss of the natural forests, building of the coastal areas, and using meadows for livestock and agriculture. In many cases this human influence is negative for most owl species, and clearly loss of habitats has made many species very rare if not extinct. Unfortunately our knowledge of the real situation is extremely limited to draw any final conclusions. Therefore, it is extremely important to create owl population studies taking into account how the environmental changes and problems have affected and will affect the owls in the country.

It is very encouraging the National Institute for Agricultural Research, Uruguay, has started an owl nest box project, preliminary results of which are indicated earlier in this chapter. It would be highly important also to involve at least a few young Uruguay birders/investigators to become members of the global owl team and World Working Group on Birds of Prey and Owls.

Burrowing Owl (*Athene cunicularia*)

References

Altuna C, Gonzáles EM (1999) Roedores fossoriales (Rodentia, Octodontidae) depredados por lechuzas (Aves, *Strigiformes*) en Uruguay. In: Resumen XIV Jornadas Argentinas de Mastozoologia. Salta, pp 46–47

Altuna C, Gonzáles EM (2006) Owl predation on the subterranean rodent *Ctenomys* (Rodentia, *Octodontidae*) in Uruguay, with an assessment of relative age classes. Comm Zool Mus Hist Nat Montevideo 203:1–5

Azpiroz AB (2003) Aves del Uruguay. Lista e introducción a su biología y conservación. Aves Uruguay-GUPECA, Montevideo

Azpiroz AB, Alfaro M, Jiménez S (2012a) Lista roja de las aves del Uruguay. DINAMA, Montevideo

Azpiroz AB, Menéndez JL, Jaramillo A et al (2012b) New information on the distribution and status of birds in Uruguay. Bull British Ornithol Club 132:46–54

Bart J, Forsman ED (1992) Dependence of northern spotted owls *Strix occidentalis caurina* on old-growth forests in the western USA. Biol Conserv 62:95–100

Charter M (2016) Israel barn owl project flies abroad to Uruguay and scored a goal! Barn owl pairs and screech owls breed. Unpublished MS shared by e-mail with HM. University of Haifa University, 4p

Chiappe L (2005) Es fundamental crear "áreas naturales protegidas". Portada: Ecología social. http://www.rebelion.org/noticia.php?id=19000. Accessed 19 Nov 2008

Claramunt SJ, Gonzáles EM (1999) Mamíferos y aves depredados por *Tyto alba* (Aves: *Strigiformes*) en zonas urbanas y rurales de Montevideo. Bol Soc Zool Uruguay 11:14

Cortés GD, Rodríguez-Cajarville MJ et al (2013) Estado del conocimiento sobre las aves rapaces de Uruguay. Ornitol Neotrop 24:243–256

Cruces S, Larrobla G, Rojas S et al (2010). Micromamíferos en la dieta estival de *Tyto alba* (Aves, *Strigiformes*) en el Parque Lecocq (Montevideo). In: Resumen I Congr. Urug. Zool., Montevideo, p 86

Cuello J (1980) Los Búhos y Lechuzas. Almanaque del Banco de Seguros del Estado, Montevideo, p 192–196

Enríquez PL, Mikkola H (1997) Comparative study of general public owl knowledge in Costa Rica, Central America and Malawi, Africa. In: Duncan JR, Johnson DH, Nicholls TH (eds) Biology and conservation of owls of the northern hemisphere: Winnipeg. Gen. Tech. Rep. NC-190. US Department of Agriculture and Forest Service. North Central Research Station, St. Paul, pp 160–166

Garcia-Olaso F (2005) Vertebrados depredados por la lechuza de campanario *Tyto alba* durante una estación reproductiva en Balneario Solís, Maldonado. Tesis de licenciatura, Facultad de Ciencias, Univ. de la República, Montevideo.

Gmelin JF (1788) (*Strix brasiliana*) in Systematica Natural 1, p 289

Gonzáles EM (1998) Do burrowing owls eat toads? With brief comments about hunting selection theory. Alauda 66:316–317

Gonzáles EM, Altuna CA (1999) Mamíferos hallados en la dieta de tres especies de *Strigiformes* en el Uruguay. Bol Soc Zool Uruguay 11:20

Gonzáles EM, Saralegui A (1996) Análisis de componentes masto- zoológicos en regurgitados de *Athene cunicularia* (Aves, *Strigiformes*) del Parque Santa Teresa, Rocha, Uruguay. Contrib Biol Centro Invest. Prom. Franciscano Ecol. 16:4

Gonzáles EM, Gonzáles J, Fregueiro G et al (1995) Mamíferos encontrados en regurgitados de lechuzas en el Noreste de Uruguay (Mammalia: *Rodentia, Marsupialia* y *Quiroptera*). Com Zool Mus Hist Nat Montevideo 12(181):1–4

Gonzáles EM, Rodríquez-Cajarville MJ, Bessonart J (2011) Mamíferos en la dieta de aves rapaces en Uruguay. Rev Mus La Plata, Zool 18(172):42

Gray JE (1829) (*Strix tuidara*) in Griffiths Animal Kingdom of Cuvier. JSocBibtphy nat Hist 6:75. reprinted by the University Press, Edinburg

Hekstra CE (1982) Description of twenty new subspecies of American *Otus* (Aves, *Strigidae*). Bull Zool Mus Amsterdam 9:49–63
Holt DW (1994) Effects of short-eared owls on common tern colony desertion, reproduction, and mortality. Colon Waterbirds 17:1–6
Keynes RD (ed) (1988) Charles Darwin's beagle diary. Cambridge University Press, Cambridge
König C, Wink M (1995) Eine neue Unterart des Brasil-Sperlingskauzes aus Zentralargentinien: *Glaucidium brasilianum stranecki n. ssp.* J Ornith 136:461–465
König C, Weick F, Becking J-H (2008) Owls of the world, 2nd edn. Christopher Helm, London
Lepage D (2009) Avibase-bird checklists of the world. Uruguay. http://avibase.bsc-eoc.org/checklist.jsp?region=uy&list=clements&lang=EN&synlang=ES. Accessed 5 Feb 2009
Marcot BG (1995) Owls of old forests of the world. USDA Forest Service General Technical Report PNW-GTR-343. Portland
Mikkola H (1983) Owls of Europe. A. & T. Poyser, Calton
Mikkola H (1997a) General public owl knowledge in Malawi. J Malawi Hist Sci Soc 50(1):13–35
Mikkola H (1997b) Comparative study on general public owl knowledge in Malawi and in eastern and southern Africa. Nyala 20:25–35
Mikkola H (2013) Owls of the world. A photographic guide, 2nd rev. edn. Bloomsbury/Christopher Helm, London
Mikkola H (2015) Los Búhos de Uruguay. In: Enríquez PL (ed) Los Búhos Neotropicales: Diversidad y Conservación. El Colegio de la Frontera Sur, México, pp 589–601
Molina GI (1782) (*Strix cunicularia*) Sagg Stor Nat Chili 263
Mones A, Ximénez A, Cuello J (1973) Análisis del contenido de bolos regurgitación de *Tyto alba tuidara* (J.E.Gray) con el hallazgo de un nuevo mamífero para el Uruguay. In V Congreso Latinoamericano de Zoología. Montevideo, p 166–167
Mongabay (2010) Uruguay deforestation rates and related forestry figures, 1–8. http://rainforests.mongabay.com/deforestation/archive/Uruguay.htm
Montenegro Tourón FH, Caballero Sadi D (2009) Nuevas contribuciones al conocimiento del las aves del Uruguay: primer registro de *Athene cunicularia* (Molina 1782) (Aves: *Strigiformes*) melanina. Bol Soc Zool Uruguay 18:39–42
MVOTMA (2015) Sistema Nacional de Áreas Protegidas. Montevideo. 14p
Narosky T, Yzurieta D (2003) Guía para la Identificación de las Aves de Argentina y Uruguay. Vazquez Mazzini Editores, Buenos Aires
PNUMA (2008) Organizaciones Relacionadas con el Manejo de Áreas Protegidas en América Latina. 8 p. http://www.ots.ar.cr/index.php?option=com_content&task=view&id=295&Itemid=317. Accessed 19 Nov 2008
PROBIDES (2008) Probides-Uruguay 6 p. http://www.probides.org.uy/inicio.htm. Accessed 19 Nov 2008
Project FPTA 314 INIA-COPAGRAN-DGSSAA. Aumentando la protección de los cultivos de secano y feedlots al daño de aves mediante nuevas técnicas de repelencia. Rodriguez E, Olivera, L and Tellechea G. Uruguay
Rocha G (2006a) Aves del Uruguay 1: El país de los Pájaros Pintados. Banda Oriental, Montevideo
Rocha G (2006b) Aves del Uruguay2: El país de los Pájaros Pintados. Banda Oriental, Montevideo
Rodríguez-Cajarville MJ, Bessonart J (2011) Composición de la dieta de *Tyto alba* en tres localidades de Uruguay. In Resumen IX Congr. Ornitol. Neotrop., Cusco, Perú, p 343–344
Salvin O (1897) (*Scops santa-catarinae*) Bulletin of the British Ornithologists' Club 6:37
Schlegel H (1862) (*Otus midas*) Museum Pays-Pas 2, Oti: 2
Sharpe RB (1899) On a new species of owl (*Gisella iheringi*) from Sao Paulo, Brazil. Bull British Ornithol Club 8:39–40
Starkey EE (1994) Forest planning and Northern spotted owls. Wingspan, p 11
Vázquez A (2003) Posible predación diferencial sobre individuos agoutis y melánicos de *Ctenomys rionegrensis,* reflejada en bolos de *Athene cunicularia.* Tesis de licenciatura, Univ. de la República, Montevideo

Vieillot LJP (1817) (*Strix suinda*) and (*Strix nacurutu*) Nouveau Dictionnaire Histoire Naturelle 7: 34 and 44
Wetmore A (1926) Observations on the birds of Argentina, Paraguay, Uruguay and Chile. Bull US Nat Mus 133:1–448
Wikipedia (2010) Long-tufted screech-owl, 1–4 p. http://en.wikipedia.org/wiki/Long-tufted_Screech_owl. Accessed 24 Nov 2010
Wikipedia (2016) List of birds of Uruguay. https://en.wikipedia.org/wiki/List_of_birds_of_Uruguay. Accessed 15 Aug 2016
World Rainforest Movement (2010) Uruguay: a place where the lies about plantations are all too obvious, 1–4 p. http://www.wrm.org.uy/bulletin/150/Uruguay.html. Accessed 24 Nov 2010

Index

P.L. Enríquez (ed.), *Neotropical Owls*, DOI 10.1007/978-3-319-57108-9

G

H

FSC
www.fsc.org
MIX
Papier | Fördert
gute Waldnutzung
FSC® C083411

Kolibri 360 GmbH
Ferdinand-Jühlke-Straße 7
99095 Erfurt, Deutschland
produktsicherheit@kolibri360.de